JN446726

제3판

FUNDAMENTALS OF
LINEAR ALGEBRA

핵심 선형대수학

곽진호 지음

KM 경문사

To my grandchildren
Minseo, Ian, Jeehan, and Logan

서론

행렬과 행렬식의 개념은 기원전 200-100년경에 쓰인 중국의 수학고서 "구장산술(九章算術)"에서 선형연립방정식 문제를 다루면서 처음으로 소개되었다고 알려져 있다. 여기서 소개된 행렬과 행렬식의 개념은 가우스-조르당 소거법과 같은 많은 동·서양 수학자들의 역사적 연구과정을 거치면서 오늘날 선형대수학으로 불리는 학문으로 발전해왔다. 하지만 선형대수학은 1950년대가 되어서야 비로소 미국의 학부 과정에서 교과목으로 도입되기 시작하였다.

이제 선형대수학은 수학의 거의 모든 분야, 특히 기하학과 해석학에서 핵심적인 역할을 하고 있다. 또한 선형대수학은 대부분의 과학과 공학에서도 중요한 도구로 이용되고 있다. 이는 주로 선형 모델링의 연구 기법이 광범위하게 응용되고 있기 때문이다. 특히 최근에는 통계학, 양자역학, 심지어 데이터 마이닝(Data Mining), 머신 러닝(machine learning) 등에서도 응용되고 있다.

이 책은 1997년과 2004년에 Birkhäuser에서 출간된 '선형대수학(S. Hong 공저)'에 뿌리를 두고 있다. 새로운 출판을 위하여 미흡했던 내용들을 보완하여 추가하고 기존 내용들을 보다 쉽게 재정리 하였다. 예를 들어, 왼쪽이나 오른쪽의 편측 역행렬과 양측 역행렬, 기저와 좌표계의 관계, QR 분해와 특이값 분해(SVD) 등의 다양한 행렬의 분해와 그 응용, 선형계차방정식(線型階差方程式), 선형미분방정식(線型微分方程式)의 해법을 효율적으로 소개하기 위해 최소 다항식과 그 응용을 보완하였다. 이차선형형식(bilinear form)을 다루는 마지막 장에서는 Hermite의 1.5차선형형식(Hermitian sesquilinear)과 불변질(不變質) 이차선형형식(nondegenerate bilinear form) 등을 추가하여 복소 이차선형형식을 보완하였다. 또한 거의 모든 예제와 연습문제를 독자들의 이해를 넓히고 흥미를 유도할 수 있도록 새롭게 정리하고, 집합 혹은 함수와 같은 수학의 기본 개념과 기초적인 수학 표기법은 독자들의 기억을 되살리기 위해 각주에 간단히 기술하였다. 선형대수학 발전에 기여한 수학자들의 역사적인 업적과 약력도 독자들이 과학과 수학의 역사적 발전 추세를 읽을 수 있도록 각주에 추가하였다.

특히 독자들이 책 내의 수학적 개념에 쉽게 익숙해지고, 계산 능력을 향

상시킬 수 있도록 많은 예제와 연습문제를 포함하였다. 반복적인 훈련을 유도하고 독자들의 본문 이해 여부를 검증하기 위해 본문 중간중간에 많은 예제와 더불어 신중하게 선택한 연습문제를 풀이 또는 힌트와 함께 배치하였다. 대부분의 기초개념과 각 장의 서두에 학습 동기부여는 2차원 평면 혹은 3차원 공간에서 유클리드 기하학 문제를 다루는 주제들로 시작하였다. 이러한 주제는 보다 다양한 관점에서 검토되고 점차 공리적·추상적 개념으로 진화하여 수학에서 일반적 정리나 원리를 찾아가는 길잡이가 되어주고 있다.

이 책은 한 학기 또는 두 학기 학부과정의 선형대수학 교재로 채택될 수 있도록 구성되었다. 이론을 지향하는 한 학기 교과과정으로는 1장 1.2-1.9절, 2장 2.3-2.6절, 3장 3.1-3.8절, 4장 4.1-4.7절, 5장 5.1-5.9절, 6장 6.1-6.3절과 6.5절, 7장 7.1-7.7절을 다룰 수 있을 것이며, 여유가 있다면 추가로 8장 8.1-8.6절 또는 9장 9.1-9.6절도 가능할 것이다.

이론과 응용 모두를 지향할 수 있는 두 학기 교과과정이라면 위의 과정을 포함해서 수강생들의 전공이나 흥미로운 분야를 참고하여, 강의 교수의 재량으로 교재의 나머지 부분을 적절히 포함할 수 있을 것이다. 또한 학부 수준의 선형대수학 과정 복습이 필요한 학부 상급생이나 대학원생들에게도 도움이 될 수 있기를 바란다.

이 책을 집필하는 동안, 그리고 저자가 오랜 기간 선형대수학을 강의하는 동안 많은 분들의 도움과 조언이 있었다. 특히 마지막 원고에서 문법적·구술적 검증을 위하여 노스캐롤라이나대학교(UNC at Charlotte)의 Bruno Wichnoski교수의 도움이 있었다. 또한 교재내용 편집을 위하여 아헨공과대학교(RWTH)의 Yubao Guo교수의 조언도 있었다. 저자는 도움을 준 모든 분들에게 신세를 졌으며, 이에 감사를 드린다. 마지막으로 이 책을[1] 한국에서 출판하기 위해 힘써주신 경문사 편집부에도 감사 인사를 전한다.

곽진호
E-mail:jinkwak@postech.ac.kr

[1]이 책을 참고도서로 인용 할 때는 Jin Ho Kwak, Fundamentals of Linear Algebra, Third Edition, ISBN 979-11-6073-624-3, 2023 KyungMoon Publishing, Seoul로 표기 가능합니다.

Acronyms and Key notations

LRR	Linear recurrence relation
LdE	Linear difference equation
LDE	A system of Linear differential equations
EVD	Eigenvalue decomposition
SVD	Singular value decomposition
$\mathbb{R}$	The set of real numbers
$\mathbb{C}$	The set of complex numbers
$[a_{ij}]$	The matrix having entries a_{ij}
$M_{m\times n}(\mathbb{R})$	The set of all $m \times n$ matrices with real number entries
$M_{m\times n}(\mathbb{C})$	The set of all $m \times n$ matrices with complex number entries
$\det A$	The determinant of a matrix A
adj A	The adjugate (or adjunct) of A
$P_n(\mathbb{R})$	The set of all real polynomials in x of degree $\leq n$
$\dim V$	The dimension of a vector space V
$\mathcal{R}(A)$	The row space of a matrix A
$\mathcal{C}(A)$	The column space of a matrix A
$\mathcal{N}(A)$	The null space of a matrix A
Ker(T)	The kernel of a linear transformtion T
Im(T)	The image of a linear transformtion T
$\langle \mathbf{x}, \mathbf{y} \rangle$	The inner product of vectors $\mathbf{x}$ and $\mathbf{y}$
$\|x\|$	The absolute value of a number x
$\|\|\mathbf{x}\|\|$	The length (or magnitude) of a vector $\mathbf{x}$
$\mathrm{Proj}_U(\mathbf{x})$	The orthogonal projection of a vector $\mathbf{x}$ onto a subspace U
e^A or $\exp A$	The exponential of a matrix A
In(A)	The inertia of a matrix A
$T : V \to W$	A linear transformation from V to W
$\langle S \rangle$ or **Span** S	The subspace spanned by a set S.
$x \approx a$	x is approximately equal to a.
$V \cong W$	The vector spaces V and W are isomorphic.

Contents

Chapter 1

Systems of linear equations with Matrices

1.1 From *Nine Chapters on the Mathematical Art*

The beginnings of matrices and determinants goes back to the second century BC. The Chinese mathematics textbook *Nine Chapters on the Mathematical Art* [1] written in 200 BC-100 BC is the first known example of the use of array (or matrix) methods to solve a system of linear equations.

A problem is set up in the textbook as follows.

There are three types of corn (like barley, rice, wheat).

Three bundles of the first type of corn, two bundles of the second type of corn, and one bundle of the third type of corn make 39 measures. [2]

Two bundles of the first type, three bundles of the second type, and one bundle of the third type make 34 measures.

And one bundle of the first type, two bundles of the second type, and three bundles of the third type make 26 measures.

How many measures of corn are contained in one bundle of each type?

Now the author does something quite remarkable. He sets up the coefficients of the system of three linear equations in three unknowns as an array as follows. [3]

---

[1] 九章算術 in Chinese

[2] Here, one measure means about 20 liters.

[3] The book *Nine Chapters on the Mathematical Art* was written vertically in columns going from top to bottom and ordered from right to left in Chinese, while English is written horizontally in rows going from left to right.

$$\begin{array}{ccc} 1 & 2 & 3 \\ 2 & 3 & 2 \\ 3 & 1 & 1 \\ 26 & 34 & 39 \end{array}$$

Our late 21st Century methods would have us write the linear equations as the rows of the matrix rather than the columns as seen above but of course the method is identical.

Most remarkably the author, writing in 200 BC, instructs the reader to *multiply the middle column by* 3 and *subtract the right column as many times as possible, the same is then done subtracting the right column as many times as possible from 3 times the first column. This gives*

$$\begin{array}{ccc} 0 & 0 & 3 \\ 4 & 5 & 2 \\ 8 & 1 & 1 \\ 39 & 24 & 39. \end{array}$$

Next the left most column is multiplied by 5 *and then the middle column is subtracted as many times as possible. This gives*

$$\begin{array}{ccc} 0 & 0 & 3 \\ 0 & 5 & 2 \\ 36 & 1 & 1 \\ 99 & 24 & 39 \end{array}$$

from which the solution can be found for the third type of corn, then for the second, then the first by backward substitution.

The solution is $37/4, 17/4$, and $11/4$, respectively. (See Problem 3 on page 10.)

This method, now known as Gaussian elimination, is well known in high-school algebra nowadays, but would not become popularized until the early 19th Century.

1.2 Three types of solution sets of systems of linear equations

A linear equation in two unknowns x and y is

$$ax + by = c,$$

in which at least one of a and b is assumed to be nonzero. Geometrically this equation represents a straight line in the xy-plane. Therefore, a point $P = (x, y)$ (in the rectangular coordinates x and y) is a solution of the equation if and only if the point P lies on the line. Thus there are infinitely many solutions, which are all the points on the line.

If the coefficients a, b are all zero in the equation $ax+by = c$, then the equation becomes $0 = c$. Thus it has no solutions if $c \neq 0$, or has infinitely many solutions if $c = 0$. In the latter case, any two numbers x and y can be a solution. In either case, the equation is vacuously trivial.

Throughout this book, whenever we speak of a linear equation $a_1x_1 + a_2x_2 + \cdots + a_nx_n = b$ in unknowns $x_1, x_2, \ldots, x_n$, we always assume that not all the coefficients $a_1, a_2, \ldots, a_n$ are zero unless otherwise specified.

Example 1 (*High-school algebra*) Solve the system of two equations in two unknowns x and y:

$$\left\{\begin{array}{rcrcl} a_1x & + & b_1y & = & c_1 \\ a_2x & + & b_2y & = & c_2. \end{array}\right. \tag{1.1}$$

Solution: (I) *Geometric method.* Two linear equations represent two straight lines in the xy-plane and a solution of the system is a point (x, y) lying on both straight lines simultaneously.

Hence, only three different types of solution sets [4] are possible.

Case (1) The straight line itself if two lines coincide (infinitely many solutions), as in the case of

$$\left\{\begin{array}{rcl} x + y & = & 1 \\ 2x + 2y & = & 2 \end{array}\right.$$

Case (2) The empty set if two lines are parallel but distinct (no solutions), as in the case of

$$\left\{\begin{array}{rcl} x + y & = & 1 \\ x + y & = & 2 \end{array}\right.$$

[4] Set notation: The set of the first five natural numbers is commonly written as $\{1, 2, 3, 4, 5\}$. We also use the notation $\{n : n \text{ is a natural number}, 1 \leq n \leq 5\}$

Case (3) Only one point if two lines cross at a point (exactly one solution), as in the case of

$$\begin{cases} x+y &= 1 \\ x-y &= 2\,. \end{cases}$$

(II) *Algebraic method.*

Case (1) (Two lines coincide): Let the two equations represent the same straight line. Equivalently, one equation is a nonzero constant multiple of the other. This condition is also equivalent to

$$a_2 = \lambda a_1,\ b_2 = \lambda b_1,\ c_2 = \lambda c_1 \ \text{ for some } \ \lambda \neq 0.$$

In this case, if a point (s,t) satisfies one equation, then it automatically satisfies the other too. Thus, there are infinitely many solutions which are all the points on the line.

Case (2) (Two lines are parallel but distinct): In this case,

$$a_2 = \lambda a_1,\ b_2 = \lambda b_1,\ \text{ but } \ c_2 \neq \lambda c_1 \ \text{ for some } \lambda \neq 0.$$

(Here, the first two equalities are equivalent to $a_1b_2 - a_2b_1 = 0$). Then no points (s,t) can satisfy both equations simultaneously, so that there are no solutions.

Case (3) (Two lines cross at a point): Let the two lines have distinct slopes, which means $a_1b_2 - a_2b_1 \neq 0$. In this case, they cross at a point (the unique solution), which can be found by the elementary method of *elimination* and *substitution.*

The following computation shows how to do *elimination* and *substitution.*

Without any loss of generality, one may assume $a_1 \neq 0$ by interchanging the two equations if necessary. (If both a_1 and a_2 are zero, the system reduces to a system of one variable y.)

(1) *Elimination*: The variable x can be eliminated from the second equation in Eq. (1.1) by adding $-\frac{a_2}{a_1}$ times the first equation to the second one, to get

$$\begin{cases} a_1x \ + & b_1y &= c_1, \\ & (b_2 - \dfrac{a_2}{a_1}b_1)y &= c_2 - \dfrac{a_2}{a_1}c_1. \end{cases}$$

(2) *Multiply or divide both sides of an equation by a nonzero constant*: Since $a_1b_2 - a_2b_1 \neq 0$, y can be found by multiplying the second equation by the nonzero number $\dfrac{a_1}{a_1b_2 - a_2b_1}$ to get

$$\begin{cases} a_1x + b_1y = c_1, \\ \qquad\qquad y = \dfrac{a_1c_2 - a_2c_1}{a_1b_2 - a_2b_1}. \end{cases}$$

(3) *Substitution*: Now, x can be solved by substituting the value of y into the first equation, and then the solutions are

$$\begin{cases} x = \dfrac{c_1b_2 - c_2b_1}{a_1b_2 - a_2b_1}, \\ y = \dfrac{a_1c_2 - a_2c_1}{a_1b_2 - a_2b_1}. \end{cases} \tag{1.2}$$

It is shown that the condition $a_1b_2 - a_2b_1 \neq 0$ is necessary for the system to have only one solution. □

As in Example 1, it will be shown later that any system of linear equations has either *no solution, exactly one solution, or infinitely many solutions.* (See Theorem 1.16 on page 24.)

A linear equation $ax + by + cz = d$, $(a, b, c) \neq (0, 0, 0)$, in three unknowns represents a plane in the 3-dimensional space $\mathbb{R}^3$. [5] The solution set includes three lines

$$\begin{array}{ll} \{(x, y, 0) \mid ax + by = d\} & \text{in the } xy\text{-plane}, \\ \{(x, 0, z) \mid ax + cz = d\} & \text{in the } xz\text{-plane}, \\ \{(0, y, z) \mid by + cz = d\} & \text{in the } yz\text{-plane}. \end{array}$$

---

[5] The notations $\mathbb{R}$ and $\mathbb{C}$ stand for the sets of real and complex numbers, respectively. The real numbers include all of the rational numbers, such as the integers $3, -4$ and the fraction $2/3$, and all of the irrational numbers, such as $\sqrt{2} = 1.41421356\cdots$. Within the irrationals, there are the transcendental numbers, such as the (Euler's) number $e = \sum_{n=0}^{\infty} \frac{1}{n!} = 2.71828\cdots$ or $\pi = 3.1415926535\cdots$.

A complex number is a number of the form $a + bi$, where a and b are real numbers, and i is the imaginary unit which satisfies the equation $i^2 = -1$. The complex number $a + bi$ can be identified with the point (a, b) in the (complex) plane. A complex number whose real part a is zero is said to be purely imaginary, whereas a complex number whose imaginary part b is zero is a real number.

One can also examine the various possible geometric types of the solution set of a system of three equations in three unknowns.

In general, a system of m linear equations in n unknowns is written as

$$\begin{cases} a_{11}x_1 & + & a_{12}x_2 & + & \cdots & + & a_{1n}x_n & = & b_1 \\ a_{21}x_1 & + & a_{22}x_2 & + & \cdots & + & a_{2n}x_n & = & b_2 \\ & & & & & & & \vdots & \\ a_{m1}x_1 & + & a_{m2}x_2 & + & \cdots & + & a_{mn}x_n & = & b_m, \end{cases}$$

where $x_1, x_2, \ldots, x_n$ are the unknowns, and a_{ij}'s and b_i's denote constant (real or complex) numbers.

A sequence of numbers $(s_1, s_2, \ldots, s_n)$ is called a **solution** of the system if $x_1 = s_1,\ x_2 = s_2,\ \ldots,\ x_n = s_n$ satisfy each equation in the system simultaneously. When $b_1 = b_2 = \cdots = b_m = 0$, the system is called **homogeneous**.

We are interested in examining whether or not a given system has a solution, and in finding all solutions if it has at least one.

For instance, every homogeneous system has at least one solution $x_1 = x_2 = \cdots = x_n = 0$, called the **trivial solution**. Naturally, one may ask whether such a homogeneous system has an extra nontrivial solution or not,. If so, we would like to find a systematic method of getting all of the solutions.

A system of linear equations is said to be **consistent** if it has a solution, and **inconsistent** if it has no solutions.

In the next section, we show how a system of linear equations can be related to a matrix theory.

Problem 1 For a system of three linear equations in three unknowns

$$\begin{cases} a_{11}x & + & a_{12}y & + & a_{13}z & = & b_1 \\ a_{21}x & + & a_{22}y & + & a_{23}z & = & b_2 \\ a_{31}x & + & a_{32}y & + & a_{33}z & = & b_3\,, \end{cases}$$

show the three different possibilities of the solution set in the 3-dimensional space $\mathbb{R}^3$; (i) no solution, (ii) exactly one solution, or (iii) infinitely many solutions.

Hint: Each linear equation represents a plane in the 3-dimensional space $\mathbb{R}^3$.

1.3 Elementary operations; Eliminations and substitutions

In Example 1, the basic procedures used in solving a system of linear equations are essentially based on the following three operations, called **elementary operations**:

(1) multiply an equation by a nonzero constant,
(2) interchange two equations,
(3) add a constant multiple of an equation to another equation.

Certainly, none of these operations alters the solution set. That is, if x_i's satisfy the given equations, then they also satisfy those equations altered by any of the three operations, and vice-versa.

Moreover, each of the three elementary operations has its *inverse* operation which is also an elementary operation:

(1′) multiply the equation by the reciprocal of the same nonzero constant,
(2′) interchange two equations again,
(3′) add the negative of the same constant multiple of the equation to the other.

By applying a finite sequence of the elementary operations to a given system, one can obtain a new (reduced) system, and by applying the corresponding inverse operations in reverse order to the new system, one can recover the original system. Since none of the three elementary operations alters the solutions, the original and the new systems have the same set of solutions.

In performing any of the three elementary operations, notice that only the coefficients of the unknowns are involved in the operations, while the unknowns $x_1, x_2, \ldots, x_n$ and the signs "+" and "=" are simply repeated.

Keeping the places of the unknowns and the signs in mind, we just pick off the coefficients from the system of linear equations (as in the historical example in Section 1.1) and make a rectangular array of numbers, called a **matrix** (plural matrices) [6], as follows:

---

[6]The first to use the term 'matrix' was the English mathematician James Joseph Sylvester (September 3, 1814– March 15, 1897) in 1848. Sylvester defined a matrix to be a rectangular arrangement of entries. After leaving America and returning to England in 1851, Sylvester became a lawyer and met Cayley, a fellow lawyer who shared his interest in mathematics. Cayley quickly saw the significance of the matrix concept and in 1853 he published a note giving, for the first time, the inverse of a matrix.

$$\begin{bmatrix} a_{11} & a_{12} & \cdots & a_{1n} & b_1 \\ a_{21} & a_{22} & \cdots & a_{2n} & b_2 \\ \vdots & \vdots & & \vdots & \vdots \\ a_{m1} & a_{m2} & \cdots & a_{mn} & b_m \end{bmatrix}.$$

This matrix is called the **augmented matrix** of the system. The numbers a_{ij} and b_j in this array are called the *entries* of the matrix.

In the following sections, we shall introduce matrices in a general sense. For the moment, we restrict our attention to the augmented matrix of a system.

Within an augmented matrix, the horizontal and vertical subarrays

$$[a_{i1}\ a_{i2}\ \cdots\ a_{in}\ b_i] \quad \text{and} \quad \begin{bmatrix} a_{1j} \\ a_{2j} \\ \vdots \\ a_{mj} \end{bmatrix}$$

are called the i-th **row**, which represents the i-th equation, and the j-th **column**, which are the coefficients of j-th unknown x_j, of the augmented matrix, respectively.

The matrix consisting of the first n columns of the augmented matrix

$$\begin{bmatrix} a_{11} & a_{12} & \cdots & a_{1n} \\ a_{21} & a_{22} & \cdots & a_{2n} \\ \vdots & \vdots & & \vdots \\ a_{m1} & a_{m2} & \cdots & a_{mn} \end{bmatrix}$$

is called the **coefficient matrix** of the system (or of the augmented matrix), usually denoted by A. Following it, the augmented matrix is commonly written by $[A\ \mathbf{b}]$.

It is easy to see that there is a one-to-one correspondence between the columns of the coefficient matrix and unknown variables of the system. The last column with entries $b_1, b_2, \ldots, b_m$ of the augmented matrix $[A\ \mathbf{b}]$ represents the right-hand side of the system and so no unknown variable corresponds to it.

In the notation $[A\ \mathbf{b}]$, A in roman stands for the coefficient matrix and $\mathbf{b}$ in bold stands for the last column in the augmented matrix.

Since each row of the augmented matrix contains all the information of the corresponding equation of the system, one may deal with this augmented matrix instead of handling the whole system of linear equations.

The elementary operations may be applied to an augmented matrix just like they are applied to a system of linear equations.

In matrix terminology, the elementary operations are called the **elementary row operations** for the augmented matrix:

(Type I) multiply a row by a nonzero constant,
(Type II) interchange two rows,
(Type III) add a constant multiple of a row to another row.

The *inverse* row operations are

(Type I) multiply the row by the reciprocal of the same constant,
(Type II) interchange the two rows again,
(Type III) add the negative of the same constant multiple of the row to the other,

which are also elementary row operations.

Definition 1.1 Two augmented matrices (or systems of linear equations) are said to be **row-equivalent** if one can be transformed to the other by a finite sequence of elementary row operations.

If an augmented matrix $[A\ \mathbf{b}]$ is transformed to the other $[B\ \mathbf{c}]$ by a finite sequence of elementary row operations, then $[A\ \mathbf{b}]$ can obviously be recovered from $[B\ \mathbf{c}]$ by applying the corresponding inverse elementary row operations in reverse order.

Therefore, the two systems have the same set of solutions.

Theorem 1.2 *If two systems of linear equations are row-equivalent, then they have the same set of solutions.*

Problem 2 Suppose that the augmented matrix $[A\ \mathbf{b}]$ of a system of linear equations has been transformed to the following matrices by elementary row operations. Find the set of solutions of each system. (Just use your brain without a pencil).

$$(1)\begin{bmatrix}1&0&0&5\\0&1&0&-2\\0&0&2&4\end{bmatrix},\quad (2)\begin{bmatrix}1&0&0&5\\0&1&0&-2\\0&0&0&4\end{bmatrix},\quad (3)\begin{bmatrix}1&0&0&1\\0&0&1&2\\0&1&0&3\end{bmatrix}.$$

Problem 3 'The problem of three types of corn' introduced in the book *Nine Chapters on the Mathematical Art* can be stated in recent terminologies as follows. The problem is to solve a system of linear equations

$$\begin{cases} 3x + 2y + z = 39 \\ 2x + 3y + z = 34 \\ x + 2y + 3z = 26\,. \end{cases}$$

Its augmented matrix is

$$[A\,\mathbf{b}] = \begin{bmatrix} 3 & 2 & 1 & 39 \\ 2 & 3 & 1 & 34 \\ 1 & 2 & 3 & 26 \end{bmatrix},$$

which is transformed by elementary row operations to

$$\begin{bmatrix} 3 & 2 & 1 & 39 \\ 0 & 5 & 1 & 24 \\ 0 & 0 & 36 & 99 \end{bmatrix}.$$

Now, one can get the solution $x = \frac{37}{4}$, $y = \frac{17}{4}$, and $z = \frac{11}{4}$ by backward substitution.

1.4 Gauss-Jordan elimination

A general procedure for finding the solutions of a system of linear equations will be illustrated by the following example. This is a part of most high-school algebra courses, but readers are supposed to be familiar with the corresponding matrix expressions.

Example 2 (*High-school algebra, but in matrix notation*) Solve the system of linear equations

$$\begin{cases} \quad\;\; 2y + 3z = 8 \\ x + 2y + 2z = 3 \\ 3x + 4y + 7z = 1\,. \end{cases}$$

Solution: Most readers may already be familiar with a procedure to solve the system. However, we like to give a demonstration of a solving procedure in matrix notation.

To compare the preceding operations on the system of linear equations with those on the augmented matrix, we work on the system and on the augmented matrix in parallel. The augmented matrix of the system is

$$[A\,\mathbf{b}] = \begin{bmatrix} 0 & 2 & 3 & 8 \\ 1 & 2 & 2 & 3 \\ 3 & 4 & 7 & 1 \end{bmatrix}.$$

(1) To have an x-term in the first equation, *interchange the first two equations (or rows in the augmented matrix)*:

$$\begin{cases} x + 2y + 2z = 3 \\ \quad\; 2y + 3z = 8 \\ 3x + 4y + 7z = 1 \end{cases} \qquad \begin{bmatrix} 1 & 2 & 2 & 3 \\ 0 & 2 & 3 & 8 \\ 3 & 4 & 7 & 1 \end{bmatrix}.$$

(2) To *eliminate the x-term* in the third equation (or row), *add -3 times the first equation (or row) to the third equation (or row)*:

$$\begin{cases} x + 2y + 2z = 3 \\ \quad\; 2y + 3z = 8 \\ \; - 2y + z = -8 \end{cases} \qquad \begin{bmatrix} 1 & 2 & 2 & 3 \\ 0 & 2 & 3 & 8 \\ 0 & -2 & 1 & -8 \end{bmatrix}.$$

Thus, the first variable x is eliminated from the second and the third equations (or rows). In this procedure, the coefficient 1 of the first unknown x in the first equation (or row) is called the first **pivot**.

The second and the third equations (or rows) have only the two unknowns y and z. Leave the first equation (or row) alone, and the same elimination procedure can be applied to the second and the third equations (or rows): The pivot to eliminate y from the third equation is the coefficient 2 of y in the second equation.

(3) To *eliminate the y-term* in the third equation (or row), *add 1 times the second equation (or row) to the third equation (or row)*:

$$\begin{cases} x + 2y + 2z = 3 \\ \quad\; 2y + 3z = 8 \\ \qquad\quad 4z = 0 \end{cases} \qquad \begin{bmatrix} 1 & 2 & 2 & 3 \\ 0 & 2 & 3 & 8 \\ 0 & 0 & 4 & 0 \end{bmatrix}.$$

The elimination procedure (*i.e.,* (1) row interchanging, (2) elimination of x from the last two equations (or rows), and then (3) elimination of y from the last equation (or row)) done so far is called **forward elimination**.

After the forward elimination, the first nonzero coefficients in the nonzero rows are the **pivots**. Thus the pivots of the second and third rows are 2 and 4, respectively.

(4) *Normalize nonzero rows* by dividing them by their pivots. Then the pivots are replaced by the number 1:

$$\begin{cases} x + 2y + 2z = 3 \\ \qquad y + \frac{3}{2}z = 4 \\ \qquad\quad\;\; z = 0 \end{cases} \qquad \begin{bmatrix} 1 & 2 & 2 & 3 \\ 0 & 1 & \frac{3}{2} & 4 \\ 0 & 0 & 1 & 0 \end{bmatrix}.$$

The resulting matrix on the right-hand side is called a **row-echelon form** of the augmented matrix $[A\ \mathbf{b}]$, and the (first nonzero) number 1's at the pivotal positions are called the **leading 1's**. The procedure so far is called **Gaussian elimination.** [7]

The last equation is $z = 0$. Substituting $z = 0$ into the second equation gives $y = 4$. Now, putting these two values into the first equation, we get $x = -5$. This procedure is called **backward substitution**.

(5) *Backward substitution* of $z = 0$ into the first two equations (or rows) and then $y = 4$ into the first:

$$\begin{cases} x & & & = & -5 \\ & y & & = & 4 \\ & & z & = & 0 \end{cases} \qquad \begin{bmatrix} 1 & 0 & 0 & -5 \\ 0 & 1 & 0 & 4 \\ 0 & 0 & 1 & 0 \end{bmatrix}.$$

The solution is unique and it is in matrix notation

$$\begin{bmatrix} x \\ y \\ z \end{bmatrix} = \begin{bmatrix} -5 \\ 4 \\ 0 \end{bmatrix}.$$

The resulting augmented matrix in the last step (5) is called the **reduced row-echelon form** of the augmented matrix $[A\ \mathbf{b}]$. It is row-equivalent to the augmented matrix $[A\ \mathbf{b}]$, and gives the solution to the system. The whole procedure to obtain the reduced row-echelon form is called **Gauss-Jordan elimination.** [8] □

By applying a finite sequence of elementary row operations, the augmented matrix $[A\ \mathbf{b}]$ can be transformed into its reduced row-echelon form, which is row-equivalent to the matrix $[A\ \mathbf{b}]$. Hence the two corresponding systems have the same solution set. From the reduced row-echelon form, one can easily decide whether the system has a solution or not, and find the solution set of the given system if it is consistent.

[7] Gaussian elimination is named after Carl Friedrich Gauss (April 30, 1777– February 23, 1855). He was a German mathematician. He developed Gaussian elimination around 1800 and used it to solve least squares problems. Gauss is famous for (1) proving the fundamental theorem of algebra which states that *every non-constant single-variable polynomial with complex coefficients has at least one complex root,* (2) the Gaussian curvature in differential geometry, and (3) the Carl Friedrich Gauss Prize.

[8] Wilhelm Jordan (March 1, 1842– April 17, 1899) was a German geodesist who conducted surveys in Germany and Africa. Wilhelm Jordan is not to be confused with the mathematician Camille Jordan (Jordan curve theorem, Jordan canonical matrix), nor with the German physicist Pascual Jordan (Jordan algebras). Gaussian elimination was popularized by Wilhelm Jordan, who attached his name to the process by which row reduction is used to compute matrix inverses.

Definition 1.3 An augmented matrix $[A\ \mathbf{b}]$ is in **row-echelon form** if it satisfies the following conditions.

(1) The zero rows, if there exist, come last in the order of rows.
(2) The first nonzero entry in each nonzero row is 1, called the **leading** 1.
(3) Below each leading 1 in the coefficient part A is a column of zeros. Thus, in any two consecutive nonzero rows, the leading 1 in the lower row appears farther to the right than the leading 1 in the upper row.

An augmented matrix $[A\ \mathbf{b}]$ is in **reduced row-echelon form** if it is in row-echelon form and satisfies the following additional condition.

(4) Each column of A containing a leading 1 has zeros everywhere else.

Example 3 (*Reduced row-echelon form or not?*) Consider four augmented matrices $[A\ \mathbf{b}]$;

$$\begin{bmatrix} 1 & 0 & 0 \\ 0 & 1 & 0 \\ 0 & 0 & 0 \end{bmatrix}, \begin{bmatrix} 1 & 2 & 0 & a & b \\ 0 & 0 & 1 & 1 & 1 \\ 0 & 0 & 0 & 0 & 0 \end{bmatrix}, \begin{bmatrix} 1 & 3 & 0 & 4 \\ 0 & 1 & k & 1 \\ 0 & 0 & 1 & 2 \end{bmatrix}, \begin{bmatrix} 1 & 0 & 3 & 0 \\ 0 & 2 & 2 & 2 \\ 0 & 0 & 0 & 1 \end{bmatrix}.$$

The first two augmented matrices are in reduced row-echelon form with two leading 1's, and the third one is just in row-echelon form. The last one is not in row-echelon form because the first nonzero entry in the second row is not 1, but by dividing the second row by 2 it can be in reduced row-echelon form. In this case, the third row has the leading 1 but it does not belong to the coefficient matrix A, and hence above this leading 1 needs not be zeros.

In an augmented matrix $[A\ \mathbf{b}]$, the last column $\mathbf{b}$ does not correspond to any variable. Thus, if the reduced row-echelon form of an augmented matrix $[A\ \mathbf{b}]$ has a row of the form $[0\ \cdots\ 0\ b]$ with $b \neq 0$, then the associated equation is $0x_1 + 0x_2 + \cdots + 0x_n = b$ with $b \neq 0$, which means the system is inconsistent.

If $b = 0$, it is a zero row, and it can be neglected and deleted since all values of the variables would give a solution to this equation. In this example, the fourth matrix shows the former case, and the first two matrices show the latter case. □

In the following example, we use Gauss-Jordan elimination again to solve a system which has more unknowns than equations.

Example 4 (*Describe infinitely many solutions in matrix notation*) Use Gauss-Jordan elimination to solve

$$\begin{cases} x_1 + 2x_2 + x_4 = 5 \\ - x_3 + 2x_4 = -4 \\ x_2 + x_4 = 2. \end{cases}$$

Solution: The augmented matrix of the system is

$$[A\ \mathbf{b}] = \begin{bmatrix} 1 & 2 & 0 & 1 & 5 \\ 0 & 0 & -1 & 2 & -4 \\ 0 & 1 & 0 & 1 & 2 \end{bmatrix}.$$

The Gaussian elimination begins with:

(1) The coefficient of x_2 in the second equation is zero and that in the third equation is not. Thus, interchanging the second and the third rows produces

$$\begin{bmatrix} 1 & 2 & 0 & 1 & 5 \\ 0 & 1 & 0 & 1 & 2 \\ 0 & 0 & -1 & 2 & -4 \end{bmatrix}.$$

(2) Dividing the third row by the pivot -1 produces a row-echelon form

$$\begin{bmatrix} 1 & 2 & 0 & 1 & 5 \\ 0 & 1 & 0 & 1 & 2 \\ 0 & 0 & 1 & -2 & 4 \end{bmatrix}.$$

We now continue the backward substitution.

(3) Adding -2 times the second row to the first produces the reduced row-echelon form

$$\begin{bmatrix} 1 & 0 & 0 & -1 & 1 \\ 0 & 1 & 0 & 1 & 2 \\ 0 & 0 & 1 & -2 & 4 \end{bmatrix}.$$

The corresponding system of equations is

$$\begin{cases} x_1 - x_4 = 1 \\ x_2 + x_4 = 2 \\ x_3 - 2x_4 = 4. \end{cases}$$

This system can be rewritten as

$$\begin{cases} x_1 = 1 + x_4 \\ x_2 = 2 - x_4 \\ x_3 = 4 + 2x_4. \end{cases}$$

Since there are no other conditions on x_4, one can see that all the other variables x_1, x_2, and x_3 will be uniquely determined if an arbitrary real value $t \in \mathbb{R}$ is assigned to x_4. For this reason, the variables x_1, x_2, and x_3 will be called the *basic variables*, while x_4 is called a *free variable*.

Thus the solutions can be written as

$$(x_1, x_2, x_3, x_4) = (1 + t, 2 - t,\ 4 + 2t, t)$$

for any real $t \in \mathbb{R}$. In matrix notation, these are

$$\begin{bmatrix} x_1 \\ x_2 \\ x_3 \\ x_4 \end{bmatrix} = \begin{bmatrix} 1 \\ 2 \\ 4 \\ 0 \end{bmatrix} + t \begin{bmatrix} 1 \\ -1 \\ 2 \\ 1 \end{bmatrix},$$

where t is any real number. Certainly, this gives infinitely many solutions. □

In the reduced row-echelon form in Example 4, one can see that the (basic) variables x_1, x_2, and x_3 correspond to the columns containing leading 1's, while the column corresponding to the (free) variable x_4 contains no leading 1.

Definition 1.4 Among the variables in a system, the ones corresponding to the columns containing leading 1's in its reduced row-echelon form are called the **basic variables**, and the ones corresponding to the columns without leading 1's, if there are any, are called the **free variables**.

Clearly the sum of the number of basic variables and the number of free variables is equal to the total number of unknowns, which is the number of columns of the coefficient matrix.

Example 5 (*Example 3 continued*) In Example 3, the first and the third augmented matrices have only basic variables without free variables, while the second one has two basic variables x_1 and x_3, and two free variables x_2 and x_4. The fourth one has two basic variables x_1 and x_2, and one free variable x_3. □

As in Example 4, a consistent system has infinitely many solutions if it has a free variable, and has a unique solution if it has no free variables.

As a special case, one can have the following theorem.

Theorem 1.5 *Every homogeneous system having more unknowns than equations has always infinitely many solutions.*

Problem 4 Suppose that the augmented matrix $[A\ \mathbf{b}]$ of a systems of linear equations has been reduced to one of the following matrices by elementary row operations. Find the set of solutions of each system. Write it in matrix notation.

$$(1)\begin{bmatrix}1&0&0&4&10\\0&1&0&5&20\\0&0&1&6&30\end{bmatrix},\qquad (2)\begin{bmatrix}1&0&2&0&10\\0&1&1&0&30\\0&0&0&1&50\end{bmatrix}.$$

Problem 5 Solve the following systems of linear equations by Gaussian elimination. What are the pivots? Write the solutions in matrix notation.

$$(1)\left\{\begin{array}{rcrcrcr} & & 2y & - & z & = & 10\\ 2x & - & 5y & + & 3z & = & 40\\ & & 2y & & & = & 80.\end{array}\right.\qquad (2)\left\{\begin{array}{rcrcrcr} -x & + & y & + & 2z & = & 0\\ 3x & + & 4y & + & z & = & 0\\ 2x & + & 5y & + & 3z & = & 0.\end{array}\right.$$

Problem 6 (*Trivial, but in matrix notation*) Solve the following systems of linear equations. Write the solutions in matrix notation. How many basic variables and free variables are there in each case?

$$(1)\left\{\begin{array}{rcrcrcr} x & & & + & z & = & 2\\ & & y & + & 2z & = & 0.\end{array}\right.\qquad (2)\left\{\begin{array}{rcrcrcrcr} w & + & x & + & y & & & = & 2\\ & & x & + & 2y & + & z & = & 0.\end{array}\right.$$

1.5 Matrices; Their sums and scalar multiplications

Definition 1.6 An m by n (written $m \times n$) **matrix** is a rectangular array of numbers arranged into m (horizontal) rows and n (vertical) columns. The **size** of a matrix is specified by the number m of the rows and the number n of the columns.

In general, a matrix is written in the following form:

$$A = \begin{bmatrix} a_{11} & a_{12} & \cdots & a_{1n}\\ a_{21} & a_{22} & \cdots & a_{2n}\\ \vdots & \vdots & & \vdots\\ a_{m1} & a_{m2} & \cdots & a_{mn}\end{bmatrix} = [a_{ij}]_{m\times n},$$

or just $A = [a_{ij}]$ if the size $m \times n$ of the matrix is clear from the context. The number a_{ij} is called the (i, j)**-entry** of the matrix A, and is also written as $a_{ij} = [A]_{ij}$ sometimes.

An $m\times 1$ matrix is called a **column (matrix)** or a **column vector**, and a $1\times n$ matrix is called a **row (matrix)**, or a **row vector**.

As before, we commonly use italic capital letters like A, B, C for matrices and boldface small letters like $\mathbf{x}, \mathbf{y}, \mathbf{z}$ for (column or row) vectors, while the italic small letters x, y, z are used for variables or numbers (sometimes called scalars) as usual.

Definition 1.7 A matrix $A = [a_{ij}]$ is called a **square matrix of order** n if the number of rows and the number of columns are both equal to n. In contrast to a square matrix, an $m \times n$ (square or not,) matrix A is called a **rectangular matrix**. An $m \times n$ matrix A is said to be **wide** if $m \leq n$, and is **long** if $m \geq n$.

For example, every row matrix is wide, while every column matrix is long.

Definition 1.8 Let $A = [a_{ij}]$ be a square matrix of order n.

(1) The entries $a_{11}, a_{22}, \ldots, a_{nn}$ are called the **diagonal entries** of A.

(2) A is called a **diagonal** matrix if the entries except for the diagonal entries are all zero.

(3) A is called an **upper (lower) triangular** matrix if the entries below (above, respectively) the diagonal are all zero.

The following matrices U and L are the general forms of the upper triangular and lower triangular matrices, respectively:

$$U = \begin{bmatrix} a_{11} & * & \cdots & * \\ 0 & a_{22} & \cdots & * \\ \vdots & & \ddots & \vdots \\ 0 & 0 & \cdots & a_{nn} \end{bmatrix}, \quad L = \begin{bmatrix} a_{11} & 0 & \cdots & 0 \\ * & a_{22} & \cdots & 0 \\ \vdots & & \ddots & \vdots \\ * & * & \cdots & a_{nn} \end{bmatrix}.$$

Both matrices cannot be diagonal if $a_{ij} \neq 0$ for some $i \neq j$. A matrix which is both upper and lower triangular should be diagonal.

Example 6 (*Two special diagonal matrices*)

(1) A matrix whose entries are all zeros is called the **zero matrix**, denoted by the symbol **0** as a bold letter (or $\mathbf{0}_{m \times n}$ when the size is emphasized and $\mathbf{0}_n$ when $m = n$).

(2) The **identity matrix** of order n, denoted by I_n (or I if the order is clear from the context), is a diagonal matrix whose diagonal entries are all 1, *i.e.,*

$$I_n = \begin{bmatrix} 1 & 0 & \cdots & 0 \\ 0 & 1 & & \vdots \\ \vdots & & \ddots & 0 \\ 0 & \cdots & 0 & 1 \end{bmatrix}.$$

All matrices of $\mathbf{0}_n$ and I_n are diagonal matrices, which are both upper and lower triangular. □

Definition 1.9 Let A be an $m \times n$ matrix. The **transpose** of A is the $n \times m$ matrix, denoted by A^T, whose j-th column is taken from the j-th row of A: *i.e.*, $[A^T]_{ij} = [A]_{ji}$.

Example 7 **(1)** If $A = \begin{bmatrix} 1 & 2 & 3 \\ x & y & z \end{bmatrix}$, then $A^T = \begin{bmatrix} 1 & x \\ 2 & y \\ 3 & z \end{bmatrix}$.

(2) The transpose of a column vector is a row vector, and vice-versa:

$$\mathbf{x} = \begin{bmatrix} x_1 \\ x_2 \\ \vdots \\ x_n \end{bmatrix} \quad \Longleftrightarrow \quad \mathbf{x}^T = [x_1 \; x_2 \; \cdots \; x_n].$$

(3) The transpose of an upper (lower, respectively) triangular matrix is lower (upper, respectively) triangular. □

Definition 1.10 Two matrices A and B are said to be **equal**, written $A = B$, if their sizes are the same and their corresponding entries are equal: *i.e.*, $[A]_{ij} = [B]_{ij}$ for all i and j.

This definition allows us to write a matrix equation. A simple example is $(A^T)^T = A$ by definition. Or $\begin{bmatrix} 1+x & 2+y \\ 3+z & 4+w \end{bmatrix} = \begin{bmatrix} x+1 & 3+y-1 \\ z+3 & 2+w+2 \end{bmatrix}$.

Let $M_{m \times n}(\mathbb{R})$ denote the set of all $m \times n$ matrices with real number entries. [9] Among the elements of $M_{m \times n}(\mathbb{R})$, one can define two operations, called the scalar multiplication and the sum of matrices, as follows.

Definition 1.11 **(1)** (*Scalar multiplication*) For an $m \times n$ matrix $A = [a_{ij}] \in M_{m \times n}(\mathbb{R})$ and a *scalar* $k \in \mathbb{R}$ (which is simply a real number), the **scalar multiplication** of k and A is defined to be the matrix kA such that $[kA]_{ij} = k[A]_{ij}$ for

[9] In the same way, one can define $M_{m \times n}(\mathbb{C})$ with entries being complex numbers.

all i and j. In an expanded form,

$$k\begin{bmatrix} a_{11} & \cdots & a_{1n} \\ \vdots & \ddots & \vdots \\ a_{m1} & \cdots & a_{mn} \end{bmatrix} = \begin{bmatrix} ka_{11} & \cdots & ka_{1n} \\ \vdots & \ddots & \vdots \\ ka_{m1} & \cdots & ka_{mn} \end{bmatrix}.$$

(2) (*Sum of matrices*) For two matrices $A = [a_{ij}]$ and $B = [b_{ij}]$ in $M_{m\times n}(\mathbb{R})$, the **sum** of A and B is defined to be the matrix $A + B$ such that $[A + B]_{ij} = [A]_{ij} + [B]_{ij}$ for all i and j. In an expanded form,

$$\begin{bmatrix} a_{11} & \cdots & a_{1n} \\ \vdots & \ddots & \vdots \\ a_{m1} & \cdots & a_{mn} \end{bmatrix} + \begin{bmatrix} b_{11} & \cdots & b_{1n} \\ \vdots & \ddots & \vdots \\ b_{m1} & \cdots & b_{mn} \end{bmatrix} = \begin{bmatrix} a_{11}+b_{11} & \cdots & a_{1n}+b_{1n} \\ \vdots & \ddots & \vdots \\ a_{m1}+b_{m1} & \cdots & a_{mn}+b_{mn} \end{bmatrix}.$$

The resulting matrices kA and $A + B$ from these two operations also belong to $M_{m\times n}(\mathbb{R})$. In this sense, we say $M_{m\times n}(\mathbb{R})$ is *closed* under the two operations.

Matrices of different sizes cannot be added. For example, the sum $\begin{bmatrix} 1 & 2 \end{bmatrix} + \begin{bmatrix} 3 & 4 & 5 \end{bmatrix}$ is not defined.

For any matrix B, its *negative* $-B$ is by definition the multiplication $(-1)B$. Moreover, if A and B are two matrices of the same size, then the subtraction $A - B$ is by definition the sum $A + (-1)B$.

Clearly, the sum of matrices has the same properties as the sum of real numbers. The real numbers in the context here are called **scalars** even though 'number' is a perfectly good name and 'scalar' sounds more technical.

The following theorem lists the basic arithmetic properties of the sum and scalar multiplication of matrices.

Theorem 1.12 *Suppose that the sizes of A, B, and C are the same, and let k, ℓ be arbitrary scalars. Then the following arithmetic rules of matrices are valid:*

(1) $A + B = B + A$ (commutativity in matrix sum),

(2) $(A + B) + C = A + (B + C)$, (*written as* $A + B + C$) (associativity in matrix sum),

(3) $A + \mathbf{0} = \mathbf{0} + A = A$,

(4) $A + (-A) = (-A) + A = \mathbf{0}$, *the zero matrix,*

(5) $k\,(A + B) = kA + kB$ (*distributivity with respect to matrix sum*),

(6) $(k+\ell)A = kA + \ell A$ (*distributivity with respect to scalar sum*),
(7) $(k\ell)A = k(\ell A)$ (*associativity with respect to scalar multiplication*),
(8) $1A = A$.

Here, the first four rules are the properties of the sum of matrices, while the last four rules are the properties of the scalar multiplication of matrices.

Proof: We prove only the equality (5) and the remaining ones are left for exercises. By comparing their (i,j)-entries, one can have

$$\begin{aligned} [k(A+B)]_{ij} &= k[A+B]_{ij} = k([A]_{ij} + [B]_{ij}) = k[A]_{ij} + k[B]_{ij} \\ &= [kA]_{ij} + [kB]_{ij} = [kA + kB]_{ij}. \end{aligned}$$

Hence, $k(A+B) = kA + kB$. □

In particular, $A+A = 2A$, $A+(A+A) = 3A = (A+A)+A$, and inductively $nA = (n-1)A + A$ for any positive integer n.

Definition 1.13 A square matrix A is said to be **symmetric** if $A^T = A$, or **skew-symmetric** if $A^T = -A$.

For example, the matrices

$$A = \begin{bmatrix} a & x & y \\ x & b & z \\ y & z & c \end{bmatrix}, \quad B = \begin{bmatrix} 0 & x & y \\ -x & 0 & z \\ -y & -z & 0 \end{bmatrix}$$

are symmetric and skew-symmetric, respectively. Notice that all of the diagonal entries of a skew-symmetric matrix should be zero, since $a_{ii} = -a_{ii}$.

The following properties of transpose of matrices are trivial.

Theorem 1.14 *Let A and B be $m \times n$ matrices. Then*

$$(kA)^T = kA^T \quad \textit{and} \quad (A+B)^T = A^T + B^T.$$

Problem 7 Prove the remaining parts of Theorem 1.12.

Problem 8 Find a matrix B such that $A + B^T = (A-B)^T$, where $A = \begin{bmatrix} 1 & 2 \\ 3 & 4 \end{bmatrix}$.

1.6 Products of vectors and products of matrices

We now introduce the product of matrices. Unlike the sum of two matrices, the product of matrices is a little bit complicated in the sense that it can also be defined for two matrices of different sizes. The product of matrices is defined in three steps.

Step (1) Product of a row vector and a column vector: For a $1 \times n$ row vector $\mathbf{a} = [a_1\ a_2\ \cdots\ a_n]$ and an $n \times 1$ column vector $\mathbf{x} = [x_1\ x_2\ \cdots\ x_n]^T$, the **product ax** is a 1×1 matrix [10] defined by the rule [11]

$$\mathbf{ax} = [a_1\ a_2\ \cdots\ a_n] \begin{bmatrix} x_1 \\ x_2 \\ \vdots \\ x_n \end{bmatrix} = [a_1x_1 + a_2x_2 + \cdots + a_nx_n] = \left[\sum_{i=1}^{n} a_i x_i \right].$$

Here, the number of entries of the row vector **a** should be equal to the number of entries of the column vector **x**, so that an entrywise multiplication is possible.

Step (2) Product of a matrix and a column vector: For an $m \times n$ matrix

$$A = \begin{bmatrix} a_{11} & a_{12} & \cdots & a_{1n} \\ a_{21} & a_{22} & \cdots & a_{2n} \\ \vdots & \vdots & & \vdots \\ a_{m1} & a_{m2} & \cdots & a_{mn} \end{bmatrix} = \begin{bmatrix} \mathbf{a}_1 \\ \mathbf{a}_2 \\ \vdots \\ \mathbf{a}_m \end{bmatrix}$$

with the row vectors $\mathbf{a}_i$'s and for an $n \times 1$ column vector $\mathbf{x} = [x_1\ x_2\ \cdots\ x_n]^T$, the **product** $A\mathbf{x}$ is by definition an $m \times 1$ matrix, whose m rows are computed according to Step (1):

$$A\mathbf{x} = \begin{bmatrix} a_{11} & a_{12} & \cdots & a_{1n} \\ a_{21} & a_{22} & \cdots & a_{2n} \\ \vdots & \vdots & & \vdots \\ a_{m1} & a_{m2} & \cdots & a_{mn} \end{bmatrix} \begin{bmatrix} x_1 \\ x_2 \\ \vdots \\ x_n \end{bmatrix} = \begin{bmatrix} \mathbf{a}_1\mathbf{x} \\ \mathbf{a}_2\mathbf{x} \\ \vdots \\ \mathbf{a}_m\mathbf{x} \end{bmatrix} = \begin{bmatrix} \sum_{i=1}^{n} a_{1i}x_i \\ \sum_{i=1}^{n} a_{2i}x_i \\ \vdots \\ \sum_{i=1}^{n} a_{mi}x_i \end{bmatrix}.$$

Therefore, for a system of m linear equations in n unknowns, by writing the n unknowns as an $n \times 1$ column vector **x** and the coefficients as an $m \times n$ matrix A, the system can be expressed as a matrix equation $A\mathbf{x} = \mathbf{b}$ with unknown vector **x**.

---

[10] A 1×1 matrix $[x]$ is sometimes identified with the entry number x.

[11] The sigma notation $\sum$ is used for a summation. For example, the summation of the consecutive integers from 1 to 10 can be written as $1 + 2 + 3 + 4 + \cdots + 9 + 10 = \sum_{i=1}^{10} i$.

Step (3) Product of two matrices: Let A be an $m\times n$ matrix and B an $n \times \ell$ matrix with columns $\mathbf{b}_1, \mathbf{b}_2, \ldots, \mathbf{b}_\ell$, written as $B = [\mathbf{b}_1\ \mathbf{b}_2\ \cdots\ \mathbf{b}_\ell]$. The **(matrix) product** AB is defined to be an $m \times \ell$ matrix whose j-th column is the product $A\mathbf{b}_j$, as computed in Step (2), for $j = 1, 2, \ldots, \ell$. That is,

$$AB = \begin{bmatrix} A\mathbf{b}_1 & A\mathbf{b}_2 & \cdots & A\mathbf{b}_\ell \end{bmatrix} = \begin{bmatrix} \mathbf{a}_1\mathbf{b}_1 & \mathbf{a}_1\mathbf{b}_2 & \cdots & \mathbf{a}_1\mathbf{b}_\ell \\ \mathbf{a}_2\mathbf{b}_1 & \mathbf{a}_2\mathbf{b}_2 & \cdots & \mathbf{a}_2\mathbf{b}_\ell \\ \vdots & & \ddots & \vdots \\ \mathbf{a}_m\mathbf{b}_1 & \mathbf{a}_m\mathbf{b}_2 & \cdots & \mathbf{a}_m\mathbf{b}_\ell \end{bmatrix},$$

which is an $m \times \ell$ matrix. Therefore, the (i, j)-entry $[AB]_{ij}$ of AB is: [12]

$$[AB]_{ij} = \mathbf{a}_i\mathbf{b}_j = a_{i1}b_{1j} + a_{i2}b_{2j} + \cdots + a_{in}b_{nj} = \sum_{k=1}^{n} a_{ik}b_{kj}.$$

This (i, j)-entry can be recognized as the sum of entrywise multiplications of the i-th row and j-th column vectors, as shadowed in Figure 1.1.

$$AB = \begin{bmatrix} a_{11} & a_{12} & \cdots & a_{1n} \\ \vdots & \vdots & & \vdots \\ a_{i1} & a_{i2} & \cdots & a_{in} \\ \vdots & \vdots & & \vdots \\ a_{m1} & a_{m2} & \cdots & a_{mn} \end{bmatrix} \begin{bmatrix} b_{11} & \cdots & b_{1j} & \cdots & b_{1r} \\ b_{21} & \cdots & b_{2j} & \cdots & b_{2r} \\ \vdots & & \vdots & & \vdots \\ b_{n1} & \cdots & b_{nj} & \cdots & b_{nr} \end{bmatrix}$$

Figure 1.1: The entry $[AB]_{ij}$

Example 8 Consider the matrices

$$A = \begin{bmatrix} 1 & 2 \\ 3 & 4 \end{bmatrix}, \qquad B = \begin{bmatrix} x & 2 & 0 \\ y & 1 & 0 \end{bmatrix}.$$

[12]One may ask why not define the product AB by the entrywise multiplications: For matrices A, B of the same size, $[AB]_{ij} = [A]_{ij}[B]_{ij}$ for all i, j? In fact, such product is also possible, called the *Hadamard product*. However, Hadamard product is not as useful as the matrix product defined in Step (3). For this reason, we only deal with the matrix product in this book, but not the Hadamard product.

The columns of AB are the products of A and each column of B:

$$\begin{bmatrix}1 & 2\\3 & 4\end{bmatrix}\begin{bmatrix}x\\y\end{bmatrix} = \begin{bmatrix}1\cdot x+2\cdot y\\3\cdot x+4\cdot y\end{bmatrix} = \begin{bmatrix}x+2y\\3x+4y\end{bmatrix},$$

$$\begin{bmatrix}1 & 2\\3 & 4\end{bmatrix}\begin{bmatrix}2\\1\end{bmatrix} = \begin{bmatrix}1\cdot 2+2\cdot 1\\3\cdot 2+4\cdot 1\end{bmatrix} = \begin{bmatrix}4\\10\end{bmatrix},$$

$$\begin{bmatrix}1 & 2\\3 & 4\end{bmatrix}\begin{bmatrix}0\\0\end{bmatrix} = \begin{bmatrix}1\cdot 0+2\cdot 0\\3\cdot 0+4\cdot 0\end{bmatrix} = \begin{bmatrix}0\\0\end{bmatrix}.$$

Therefore, AB is

$$\begin{bmatrix}1 & 2\\3 & 4\end{bmatrix}\begin{bmatrix}x & 2 & 0\\y & 1 & 0\end{bmatrix} = \begin{bmatrix}x+2y & 4 & 0\\3x+4y & 10 & 0\end{bmatrix}.$$

Since A is a 2×2 matrix and B is a 2×3 matrix, the product AB is a 2×3 matrix. If one needs, for example, the $(2,1)$-entry of AB, just single out the second row from A and the first column from B as in Figure 1.1, and then multiply corresponding entries together and add them up, which gives $3\cdot x+4\cdot y = 3x+4y$. □

The product AB is not defined if the number of columns of A and the number of rows of B are not equal. For example, the product $\begin{bmatrix}1 & 2\end{bmatrix}\begin{bmatrix}3 & 4 & 5\end{bmatrix}$ is not defined.

By a computation, one can see that $AI_n = A$ is true for any $m\times n$ matrix A and the identity matrix I_n. Also, $I_nB = B$ for any $n\times \ell$ matrix B.

The operations of scalar multiplication, sum, and product of matrices satisfy many, but not all, of the same arithmetic rules that real or complex numbers have. The zero matrix $\mathbf{0}_{m\times n}$ plays the role of the number 0, and the identity matrix I_n plays that of the number 1 in the set of usual numbers.

The rule that does not hold for matrices in general is commutativity $AB = BA$ of the product of matrices, while commutativity $A+B = B+A$ of the sum of matrices holds always.

The next example illustrates non-commutativity of the product of matrices.

Example 9 (*A matrix multiplication is not commutative*) Let $A = \begin{bmatrix}1 & 0\\0 & 0\end{bmatrix}$ and $B = \begin{bmatrix}0 & 1\\1 & 0\end{bmatrix}$. Then, $AB = \begin{bmatrix}0 & 1\\0 & 0\end{bmatrix}$ and $BA = \begin{bmatrix}0 & 0\\1 & 0\end{bmatrix}$ implying $AB \neq BA$. □

The following theorem lists some arithmetic rules for the matrix product.

Theorem 1.15 *Let A, B, C be arbitrary (square or rectangular) matrices for which the matrix operations below can be defined, and let k be an arbitrary scalar. Then*

(1) $A(BC) = (AB)C$, *(written as ABC)* (associativity in matrix product),
(2) $I_m A = A$ *and* $A = AI_n$, *where A is an $m \times n$ matrix,*
(3) $A(B + C) = AB + AC$ *and* $(A + B)C = AC + BC$
(distributivity with respect to matrix product),
(4) $k(BC) = (kB)C = B(kC)$,
(5) $(AB)^T = B^T A^T$.

Proof: Each equality can be shown by direct comparison of the (i, j)-entries of both sides of the equality.

We prove only (1), and leave the others as an exercise.

Assume that $A = [a_{ij}]$ is an $m{\times}n$ matrix, $B = [b_{k\ell}]$ is an $n{\times}p$ matrix, and $C = [c_{st}]$ is a $p{\times}r$ matrix. We now compute the (i, j)-entry of each side of the equality. Note that BC is an $n \times r$ matrix whose (i, j)-entry is $[BC]_{ij} = \sum_{\lambda=1}^{p} b_{i\lambda} c_{\lambda j}$. Thus

$$[A(BC)]_{ij} = \sum_{\mu=1}^{n} a_{i\mu}[BC]_{\mu j} = \sum_{\mu=1}^{n} a_{i\mu} \sum_{\lambda=1}^{p} b_{\mu\lambda} c_{\lambda j} = \sum_{\mu=1}^{n} \sum_{\lambda=1}^{p} a_{i\mu} b_{\mu\lambda} c_{\lambda j}.$$

Similarly, AB is an $m{\times}p$ matrix with the (i, j)-entry $[AB]_{ij} = \sum_{\mu=1}^{n} a_{i\mu} b_{\mu j}$, and

$$[(AB)C]_{ij} = \sum_{\lambda=1}^{p} [AB]_{i\lambda} c_{\lambda j} = \sum_{\lambda=1}^{p} \sum_{\mu=1}^{n} a_{i\mu} b_{\mu\lambda} c_{\lambda j} = \sum_{\mu=1}^{n} \sum_{\lambda=1}^{p} a_{i\mu} b_{\mu\lambda} c_{\lambda j}.$$

This shows that $[A(BC)]_{ij} = [(AB)C]_{ij}$ for all i, j, and consequently $A(BC) = (AB)C$ as desired. □

Now, one can show that there are three different types of solution sets of a system $A\mathbf{x} = \mathbf{b}$.

Theorem 1.16 *A system $A\mathbf{x} = \mathbf{b}$ of linear equations has either no solution, exactly one solution, or infinitely many solutions.*

Proof: A given system $A\mathbf{x} = \mathbf{b}$ may have either no solutions or at least one solution. If it has no solutions, then there is nothing to prove.

Suppose that the system has at least two solutions and let $\mathbf{x}_1$ and $\mathbf{x}_2$ be two distinct solutions, that is, $A\mathbf{x}_1 = \mathbf{b}$ and $A\mathbf{x}_2 = \mathbf{b}$, but $\mathbf{x}_1 \neq \mathbf{x}_2$. Let $\mathbf{x}_0 = \mathbf{x}_1 - \mathbf{x}_2$. Then $\mathbf{x}_0 \neq \mathbf{0}$ and $A\mathbf{x}_0 = A(\mathbf{x}_1 - \mathbf{x}_2) = A\mathbf{x}_1 - A\mathbf{x}_2 = \mathbf{b} - \mathbf{b} = \mathbf{0}$. Thus

$$A(\mathbf{x}_1 + k\mathbf{x}_0) = A\mathbf{x}_1 + kA\mathbf{x}_0 = \mathbf{b} + k\mathbf{0} = \mathbf{b}.$$

This means that $\mathbf{x}_1 + k\mathbf{x}_0$ is also a solution of $A\mathbf{x} = \mathbf{b}$ for any k. Since there are infinitely many choices of k, $A\mathbf{x} = \mathbf{b}$ has infinitely many solutions. □

Problem 9 Prove or disprove: If A is not a zero matrix and $AB = AC$, then $B = C$. Similarly, is it true or not, that $AB = \mathbf{0}$ implies $A = \mathbf{0}$ or $B = \mathbf{0}$?

Problem 10 For a square matrix A, show that

(1) AA^T and $A + A^T$ are symmetric,
(2) $A - A^T$ is skew-symmetric, and
(3) A can be expressed as the sum of the *symmetric part* $B = \frac{1}{2}(A+A^T)$ and the *skew-symmetric part* $C = \frac{1}{2}(A - A^T)$, so that $A = B + C$. Moreover, this expression is unique, which means that if $B_1 + C_1 = A = B_2 + C_2$ with symmetric B_i and skew-symmetric C_i, then $B_1 = B_2$ and $C_1 = C_2$.

1.7 Inverse matrices; One-sided and two-sided inverses

A matrix form $A\mathbf{x} = \mathbf{b}$ of a system of linear equations resembles a simple linear equation $ax = b$ whose solution is $x = a^{-1}b$ when $a \neq 0$. Thus it is tempting to write the solution to the system as $\mathbf{x} = A^{-1}\mathbf{b}$. To do this, we first have to assign a meaning to A^{-1}.

To discuss this, we begin with the following definition.

Definition 1.17 For a given $m \times n$ matrix A, an $n \times m$ matrix B is called a **left inverse** of A if $BA = I_n$, and an $n \times m$ matrix C is called a **right inverse** of A if $AC = I_m$. The matrices B and C are called **one-sided inverses.**

The following example shows that a rectangular matrix can have infinitely many one-sided inverses.

Example 10 (*Some non-square matrices have infinitely many one-sided inverses*) For two (rectangular) matrices

$$A = \begin{bmatrix} 1 & 0 & x \\ 0 & 1 & y \end{bmatrix} \quad \text{and } B = \begin{bmatrix} 1 & 0 \\ 0 & 1 \\ 0 & 0 \end{bmatrix},$$

a direct computation gives $AB = \begin{bmatrix} 1 & 0 \\ 0 & 1 \end{bmatrix} = I_2$ and $BA = \begin{bmatrix} 1 & 0 & x \\ 0 & 1 & y \\ 0 & 0 & 0 \end{bmatrix} \neq I_3$.

Thus, the matrix B is a right inverse but not a left inverse of A, while A is a left inverse but not a right inverse of B.

By taking any $x, y \in \mathbb{R}$, the matrix B has infinitely many different left inverses, but cannot have any right inverse (because B has a zero row). See Problem 13. □

Later, it will be shown in Theorems 3.27 and 3.28 that if a matrix A is not square, A cannot have both a right inverse and a left inverse. However, this is not true for a square matrix.

If a square matrix A has a left inverse, then it has a right inverse too, and vice-versa (see Theorem 1.22). Moreover, the following lemma shows that this should be unique.

Lemma 1.18 *If an $n \times n$ square matrix A has a left inverse B and a right inverse C, then B and C are equal, i.e., $B = C$.*

Proof: Easy to show $B = BI_n = B(AC) = (BA)C = I_nC = C$. □

By Lemma 1.18, one can say that if a *square* matrix A has both left and right inverses, then any two left inverses should be both equal to a right inverse, and hence any two left inverses should be the same. For the same reason, any two right inverses should be the same too. So there exists *only one* left and right inverse.

Definition 1.19 An $n \times n$ square matrix A is said to be **invertible** (or **nonsingular**) if there exists a square matrix B of the same size such that

$$AB = I_n = BA.$$

Such a matrix B is called the (two-sided) **inverse** of A, and is denoted by A^{-1}. A matrix A is said to be **singular** if it is not invertible.

The inverse of a square matrix is unique, and hence call it 'the' inverse of A.

For instance, consider a 2×2 matrix $A = \begin{bmatrix} a & b \\ c & d \end{bmatrix}$. If $ad - bc \neq 0$,

$$A^{-1} = \frac{1}{ad-bc}\begin{bmatrix} d & -b \\ -c & a \end{bmatrix} = \begin{bmatrix} \dfrac{d}{ad-bc} & \dfrac{-b}{ad-bc} \\ \dfrac{-c}{ad-bc} & \dfrac{a}{ad-bc} \end{bmatrix},$$

since $AA^{-1} = I_2 = A^{-1}A$.

Certainly, not all square matrices have an inverse, like the zero matrix. However, for a system of linear equations $A\mathbf{x} = \mathbf{b}$ with a square matrix A, if A is invertible, then

$$\mathbf{x} = I\mathbf{x} = (A^{-1}A)\mathbf{x} = A^{-1}(A\mathbf{x}) = A^{-1}\mathbf{b}$$

is the unique solution.

Theorem 1.20 *The product of invertible matrices is also invertible. Its inverse is the product of the individual inverses in reversed order:*

$$(AB)^{-1} = B^{-1}A^{-1}.$$

Proof: Suppose that A and B are invertible matrices of the same size. Then

$$(AB)(B^{-1}A^{-1}) = A(BB^{-1})A^{-1} = AIA^{-1} = AA^{-1} = I,$$

and similarly $(B^{-1}A^{-1})(AB) = I$. Thus, AB has the inverse $B^{-1}A^{-1}$. □

The inverse of A is written as 'A to the power -1', so one can give the meaning of A^k for any integer k. Let A be a square matrix. Define $A^0 = I$, the identity matrix. For any positive integer k, the matrix power A^k of A is inductively defined as

$$A^k = A(A^{k-1}) \quad \text{for } k > 0.$$

If A is invertible, then the negative integer power is also defined as

$$A^{-k} = (A^{-1})^k \quad \text{for } k > 0.$$

It is easy to check that $A^{k+\ell} = A^k A^\ell$ whenever the right-hand side is defined. (If A is not invertible, $A^{3+(-1)}$ is defined but A^3A^{-1} is not.)

Problem 11 Let A be an invertible matrix and c any nonzero scalar. Show that

(1) A^{-1} is invertible and $(A^{-1})^{-1} = A$,
(2) the matrix cA is invertible and $(cA)^{-1} = \frac{1}{c}A^{-1}$, when $c \neq 0$,
(3) A^T is invertible and $(A^T)^{-1} = (A^{-1})^T$.

Problem 12 (1) For any rectangular (square or not,) matrix A, show that A has a right inverse if and only if A^T has a left inverse.
(2) For any square and invertible matrix A, is it true that $(A^k)^T = (A^T)^k$ for any (positive or negative) integer k? Justify your answer.

Problem 13 Show that

(1) if A has a zero row, so does AB,
(2) if B has a zero column, so does AB,
(3) every matrix having a zero row or a zero column can not be invertible.

Hint: Use Figure 1.1.

1.8 Elementary matrices for Gauss-Jordan elimination

In this section we review again the elementary row operations in terms of matrix notation.

Forward elimination is a procedure in which the augmented matrix is transformed into its row-echelon form by a finite sequence of elementary row operations.

In the following, one can see that each elementary row operation can be expressed as a nonsingular matrix, called an *elementary matrix*, so that the procedure of forward elimination is the same as multiplying a finite sequence of corresponding elementary matrices on the left-hand side of the augmented matrix.

Definition 1.21 An **elementary matrix** is a matrix obtained from an identity matrix I_n by applying *only one time* operation of any type of elementary row operation.

For example, the following matrices are three elementary matrices corresponding to the three types of elementary row operations.

$$\text{(Type I): } \begin{bmatrix} 1 & 0 \\ 0 & 3 \end{bmatrix} \text{ or } \begin{bmatrix} 1 & 0 & 0 \\ 0 & 3 & 0 \\ 0 & 0 & 1 \end{bmatrix}, \text{ the second row of } I \text{ is multiplied by } 3,$$

$$\text{(Type II): } \begin{bmatrix} 0 & 0 & 1 \\ 0 & 1 & 0 \\ 1 & 0 & 0 \end{bmatrix}, \text{ the first and the third rows of } I_3 \text{ are interchanged,}$$

$$\text{(Type III): } \begin{bmatrix} 1 & 0 & 0 \\ -2 & 1 & 0 \\ 0 & 0 & 1 \end{bmatrix}, \text{ add } (-2) \times \text{ the first row to the second row of } I_3.$$

There are infinitely many elementary matrices of type I and also of type III, but there are only $\binom{3}{2}$ different 3×3 elementary matrices of type II. [13, 14] In a similar way, one can get many elementary matrices from the identity matrix I_4 or I_5, in fact from I_n of any order n.

It is a quite interesting and important fact that, if E is an elementary matrix obtained by applying a certain elementary row operation to the identity matrix I_m, then for any $m \times n$ matrix A,

the product EA is exactly the matrix that is obtained from A when the same elementary row operation to get E is applied to A.

Hence, for any $m \times n$ matrix A,

Elementary row operations in A	1-to-1 ←---→	Matrix multiplications EA by an elementary matrix E of size m

The following example illustrates this argument. (The reversed product AE is not what we want. For this, see Problem 15.)

Example 11 (*Elementary operation by an elementary matrix*) Choose any matrix B having three rows, say B is

$$B_1 = \begin{bmatrix} b_1 \\ b_2 \\ b_3 \end{bmatrix} \quad \text{or} \quad B_2 = \begin{bmatrix} b_{11} & b_{12} \\ b_{21} & b_{22} \\ b_{31} & b_{32} \end{bmatrix}.$$

[13]By definition, $\binom{n}{0} = 1$, $\binom{n}{1} = n$, $\binom{n}{2} = \frac{n(n-1)}{2}$, and $\binom{n}{k} = \frac{n(n-1)\cdots(n-k+1)}{k!}$ for $k \geq 3$, known as the *binomial coefficients*. By definition, $0! = 1$, $n! = n(n-1)\cdots 2 \cdot 1$ for a positive integer n, called *n factorial*.

[14]A *k-combination* of a set X is a subset of k distinct elements of X. If the set X has n elements, the number of k-combinations of X is equal to the binomial coefficient $\binom{n}{k}$ (often read as "n choose k"). This is also sometimes denoted by $C(n,k)$, C^n_k, ${}_nC_k$, $C_{n,k}$, or even C^k_n.

Suppose that it is needed to apply the type III elementary operation

'adding $(-2) \times$ the first row to the second row'

to the matrix B. To do this, first apply this operation to the identity matrix I_3 to get an elementary matrix E;

$$E = \begin{bmatrix} 1 & 0 & 0 \\ -2 & 1 & 0 \\ 0 & 0 & 1 \end{bmatrix},$$

and product this elementary matrix E on the left side of B. Then it produces the desired result. For example,

$$EB_1 = \begin{bmatrix} 1 & 0 & 0 \\ -2 & 1 & 0 \\ 0 & 0 & 1 \end{bmatrix} \begin{bmatrix} b_1 \\ b_2 \\ b_3 \end{bmatrix} = \begin{bmatrix} b_1 \\ b_2 - 2b_1 \\ b_3 \end{bmatrix},$$

or

$$EB_2 = \begin{bmatrix} 1 & 0 & 0 \\ -2 & 1 & 0 \\ 0 & 0 & 1 \end{bmatrix} \begin{bmatrix} b_{11} & b_{12} \\ b_{21} & b_{22} \\ b_{31} & b_{32} \end{bmatrix} = \begin{bmatrix} b_{11} & b_{12} \\ b_{21} - 2b_{11} & b_{22} - 2b_{12} \\ b_{31} & b_{32} \end{bmatrix}.$$

Similarly, the type II elementary operation

'interchanging the first and the third rows'

applied to the matrix B can be achieved by multiplying an elementary matrix P obtained from I_3 by interchanging the same two rows on the left side of B:

$$PB_1 = \begin{bmatrix} 0 & 0 & 1 \\ 0 & 1 & 0 \\ 1 & 0 & 0 \end{bmatrix} \begin{bmatrix} b_1 \\ b_2 \\ b_3 \end{bmatrix} = \begin{bmatrix} b_3 \\ b_2 \\ b_1 \end{bmatrix},$$

or

$$PB_2 = \begin{bmatrix} 0 & 0 & 1 \\ 0 & 1 & 0 \\ 1 & 0 & 0 \end{bmatrix} \begin{bmatrix} b_{11} & b_{12} \\ b_{21} & b_{22} \\ b_{31} & b_{32} \end{bmatrix} = \begin{bmatrix} b_{31} & b_{32} \\ b_{21} & b_{22} \\ b_{11} & b_{12} \end{bmatrix}.$$

□

Every elementary row operation has an inverse operation, which is also an elementary operation, that brings an elementary matrix back to the original identity matrix.

In other words, for any elementary matrix E, let E' denote the elementary matrix corresponding to the 'inverse' elementary row operation of E. Then $E'E = I$, because

(1) if E multiplies a row by $c \neq 0$, then E' multiplies the same row by $1/c$,
(2) if E interchanges two rows, then E' interchanges them again,
(3) if E adds a multiple of one row to another, then E' subtracts that multiple from the same row.

Furthermore, one can say that $E'E = I = EE'$. In a word, *every elementary matrix is invertible and the inverse matrix $E^{-1} = E'$ is also an elementary matrix.*

Example 12 (*Every elementary matrix is invertible and the inverse is also elementary*) If

$$E_1 = \begin{bmatrix} 1 & 0 & 0 \\ 0 & c & 0 \\ 0 & 0 & 1 \end{bmatrix} \text{ with } c \neq 0, \quad E_2 = \begin{bmatrix} 0 & 0 & 1 \\ 0 & 1 & 0 \\ 1 & 0 & 0 \end{bmatrix}, \quad E_3 = \begin{bmatrix} 1 & 0 & 0 \\ -2 & 1 & 0 \\ 0 & 0 & 1 \end{bmatrix},$$

then

$$E_1^{-1} = \begin{bmatrix} 1 & 0 & 0 \\ 0 & 1/c & 0 \\ 0 & 0 & 1 \end{bmatrix}, \quad E_2^{-1} = \begin{bmatrix} 0 & 0 & 1 \\ 0 & 1 & 0 \\ 1 & 0 & 0 \end{bmatrix}, \quad E_3^{-1} = \begin{bmatrix} 1 & 0 & 0 \\ 2 & 1 & 0 \\ 0 & 0 & 1 \end{bmatrix}.$$

All of these three inverse matrices are also elementary matrices. □

The following example illustrates a matrix form of a forward elimination.

Example 13 (*Matrix form of a forward elimination*) (Visit Example 2 again) Consider the system of linear equations given in Example 2 again

$$A\mathbf{x} = \begin{bmatrix} 0 & 2 & 3 \\ 1 & 2 & 2 \\ 3 & 4 & 7 \end{bmatrix} \begin{bmatrix} x_1 \\ x_2 \\ x_3 \end{bmatrix} = \begin{bmatrix} 8 \\ 3 \\ 1 \end{bmatrix} = \mathbf{b} \text{ or } \left\{ \begin{array}{rcrcrcl} & & 2y & + & 3z & = & 8 \\ x & + & 2y & + & 2z & = & 3 \\ 3x & + & 4y & + & 7z & = & 1. \end{array} \right.$$

In solving the system $A\mathbf{x} = \mathbf{b}$ in Example 2, it begins with a row-interchanging.

(i) To have an x-term in the first row, interchange the first two rows:

$$\left\{ \begin{array}{rcrcrcl} x & + & 2y & + & 2z & = & 3 \\ & & 2y & + & 3z & = & 8 \\ 3x & + & 4y & + & 7z & = & 1 \end{array} \right. ; \ PA = \begin{bmatrix} 1 & 2 & 2 \\ 0 & 2 & 3 \\ 3 & 4 & 7 \end{bmatrix} \text{ with } P = \begin{bmatrix} 0 & 1 & 0 \\ 1 & 0 & 0 \\ 0 & 0 & 1 \end{bmatrix}.$$

After that, two more forward elimination procedures are needed:

(ii) $(-3) \times \text{row } 1 + \text{row } 3$, and
(iii) $1 \times \text{row } 2 + \text{row } 3$.

The elementary matrices for these two forward eliminations are

$$E_1 = \begin{bmatrix} 1 & 0 & 0 \\ 0 & 1 & 0 \\ -3 & 0 & 1 \end{bmatrix} \quad \text{and} \quad E_2 = \begin{bmatrix} 1 & 0 & 0 \\ 0 & 1 & 0 \\ 0 & 1 & 1 \end{bmatrix},$$

so that

$$E_2 E_1 PA = \begin{bmatrix} 1 & 2 & 2 \\ 0 & 2 & 3 \\ 0 & 0 & 4 \end{bmatrix} = U,$$

which is an upper triangular matrix. Thus, if we set

$$L = (E_2 E_1)^{-1} = E_1^{-1} E_2^{-1} = \begin{bmatrix} 1 & 0 & 0 \\ 0 & 1 & 0 \\ 3 & -1 & 1 \end{bmatrix},$$

then L is a lower triangular matrix with 1's on the diagonal and

$$PA = LU = \begin{bmatrix} 1 & 0 & 0 \\ 0 & 1 & 0 \\ 3 & -1 & 1 \end{bmatrix} \begin{bmatrix} 1 & 2 & 2 \\ 0 & 2 & 3 \\ 0 & 0 & 4 \end{bmatrix}.$$

Hence, PA is written as a product of a lower triangular matrix L and an upper triangular matrix U. □

In solving an arbitrarily given system of linear equations $A\mathbf{x} = \mathbf{b}$, if no interchangings of rows (type II row operations) are necessary through the entire procedure of forward elimination, then the forward elimination is just multiplications of the augmented matrix $[A\ \mathbf{b}]$ by finitely many elementary matrices $E_1, \ldots, E_k$ on the left;

$$[E_k \cdots E_1 A \quad E_k \cdots E_1 \mathbf{b}] = [U\ \mathbf{y}],$$

where each E_i represents a forward elimination step, all of which reduce A to a row-echelon form U.

Hence, each E_i is a lower triangular elementary matrix whose diagonal entries are all 1's and $[U\ \mathbf{y}]$ is a row-echelon form of the augmented matrix $[A\ \mathbf{b}]$ without divisions of the rows by the pivots. If A is a square matrix, U is also square and so should be an upper triangular matrix.

Let $L = (E_k \cdots E_1)^{-1} = E_1^{-1} \cdots E_k^{-1}$. Then it follows $A = LU$, where *L is a lower triangular matrix whose diagonal entries are still all 1's*. (In fact, each

E_i^{-1} is a lower triangular matrix, and their product L is also lower triangular, see Problem 14).

Such factorization $A = LU$ is called an LU **factorization** or an LU **decomposition** of A.

As in Example 13, if some interchangings of rows (type II row operations) are needed through the procedure of forward elimination, let P be the product of their type II elementary matrices, called a *permutation matrix*. Then the matrix PA has an LU decomposition. [15]

An application of an LU decomposition of a matrix A (or of PA) will be shown later in Subsection 1.10.3.

Problem 14 Let A and B be two lower triangular matrices. Show that

(1) their product AB is also a lower triangular matrix,
(2) if A is invertible, then its inverse is also a lower triangular matrix,
(3) if the diagonal entries of A and B are all 1's, then the same holds for their product AB and their inverses.

Show that the same holds for upper triangular matrices, and for the product of more than two matrices.

Problem 15 Define the **elementary column operations** for a matrix by just replacing 'row' by 'column' in the definition of the elementary row operations. Show that if A is an $m \times n$ matrix and if E is a matrix obtained by applying an elementary column operation to I_n, then AE is exactly the matrix that is obtained from A when the same column operation is applied to A.

Consequently, if D is an $n \times n$ diagonal matrix with diagonal entries $d_1, d_2, \ldots, d_n$, then AD is obtained by multiplying the n columns of A by $d_1, d_2, \ldots, d_n$, respectively.

1.9 When and how can one find the inverse of a matrix?

We again return to a system of linear equations $A\mathbf{x} = \mathbf{b}$.

If (any rectangular matrix) A has a right inverse B so that $AB = I_m$, then $\mathbf{x} = B\mathbf{b}$ is a solution to the system since

$$A\mathbf{x} = A(B\mathbf{b}) = (AB)\mathbf{b} = \mathbf{b}.$$

[15] The LU decomposition was introduced by a Polish mathematician Tadeusz Banachiewicz (February 13, 1882– November 17, 1954) in 1938.

(Compare this with Problem 19). In particular, if A is an invertible square matrix, then it has only one inverse A^{-1} and $\mathbf{x} = A^{-1}\mathbf{b}$ is the only solution to the system.

The next theorem establishes some fundamental relations between $n \times n$ square matrices and systems of n linear equations in n unknowns.

Theorem 1.22 *Let A be an $n \times n$ matrix. The following are equivalent.*

(1) *A has a left inverse,*
(2) *$A\mathbf{x} = \mathbf{0}$ has only the trivial solution $\mathbf{x} = \mathbf{0}$,*
(3) *A is row-equivalent to I_n,*
(4) *A is a product of elementary matrices,*
(5) *A is invertible,*
(6) *A has a right inverse.*

In particular, if a square matrix A has a one-sided inverse, then it becomes a two-sided inverse, and either of its one-sided inverses is the unique two-sided inverse (by Lemma 1.18).

Proof: (1) $\Rightarrow$ (2): Let $\mathbf{x}$ be a solution of the homogeneous system $A\mathbf{x} = \mathbf{0}$, and let B be a left inverse of A. Then

$$\mathbf{x} = I_n\mathbf{x} = (BA)\mathbf{x} = B(A\mathbf{x}) = B\mathbf{0} = \mathbf{0}.$$

(2) $\Rightarrow$ (3): Suppose that the homogeneous system $A\mathbf{x} = \mathbf{0}$ has only the trivial solution $\mathbf{x} = \mathbf{0}$:

$$\begin{cases} x_1 & & & = & 0 \\ & x_2 & & = & 0 \\ & & \ddots & & \\ & & x_n & = & 0. \end{cases}$$

This means that the augmented matrix $[A\ \mathbf{0}]$ of the system $A\mathbf{x} = \mathbf{0}$ is reduced to the augmented matrix $[I_n\ \mathbf{0}]$ by Gauss-Jordan elimination. Hence, A is row-equivalent to I_n.

(3) $\Rightarrow$ (4): Assume A is row-equivalent to I_n, so that A can be reduced to I_n by a finite sequence of elementary row operations. Thus, one can find elementary matrices $E_1, E_2, \ldots, E_k$ such that

$$E_k \cdots E_2 E_1 A = I_n.$$

By multiplying successively both sides of this equation by $E_k^{-1}, \ldots, E_2^{-1}, E_1^{-1}$ on the left, one can get

$$A = E_1^{-1}E_2^{-1} \cdots E_k^{-1} I_n = E_1^{-1}E_2^{-1} \cdots E_k^{-1},$$

which expresses A as a product of elementary matrices.

(4) $\Rightarrow$ (5) is trivial, because any elementary matrix is invertible. In fact, $A^{-1} = E_k \cdots E_2 E_1$.

(5) $\Rightarrow$ (1) and (5) $\Rightarrow$ (6) are trivial.

(6) $\Rightarrow$ (5): If B is a right inverse of A, then A is a left inverse of B and one can apply (1) $\Rightarrow$ (2) $\Rightarrow$ (3) $\Rightarrow$ (4) $\Rightarrow$ (5) to the matrix B and conclude that B is invertible, with A as its unique inverse, by Lemma 1.18. That is, B is the inverse of A and so A is invertible. $\square$

If a triangular matrix A has a zero diagonal entry, then the reduced row-echelon form of A has at least one column which does not have leading 1's. Hence, the system $A\mathbf{x} = \mathbf{0}$ has at least one free variable, so that it has infinitely many solutions. However, if A has no zero diagonal entries, $A\mathbf{x} = \mathbf{0}$ has only the trivial solution.

It gives the following corollary.

Corollary 1.22.1 *A triangular matrix is invertible if and only if it has no zero diagonal entries.*

Problem 16 For a product of three elementary matrices

$$\begin{bmatrix} 1 & 0 & 0 \\ 0 & 1 & 0 \\ 0 & -c & 1 \end{bmatrix} \begin{bmatrix} 1 & 0 & 0 \\ 0 & 1 & 0 \\ -b & 0 & 1 \end{bmatrix} \begin{bmatrix} 1 & 0 & 0 \\ -a & 1 & 0 \\ 0 & 0 & 1 \end{bmatrix},$$

what is the inverse of each factor matrix? (Try it without a pencil.) What is the inverse of the product matrix? Express the inverse as a product of three elementary matrices.

Problem 17 When is a diagonal matrix $D = \begin{bmatrix} d_1 & & \mathbf{0} \\ & \ddots & \\ \mathbf{0} & & d_n \end{bmatrix}$ nonsingular, and what is D^{-1} in the nonsingular case?

As an application of Theorem 1.22, one can find an easy method for finding the inverse A^{-1} of an invertible $n \times n$ matrix A.

If A is invertible, then A is row-equivalent to I_n and so there are elementary matrices $E_1, E_2, \ldots, E_k$ such that $E_k \cdots E_2 E_1 A = I_n$. Hence,

$$A^{-1} = E_k \cdots E_2 E_1 = E_k \cdots E_2 E_1 I_n.$$

This means that A^{-1} *can be obtained by performing on I_n the same sequence of the elementary row operations that reduces A to I_n.*

Practically, one first constructs an $n \times 2n$ augmented matrix $[A \mid I_n]$ and then performs a Gaussian–Jordan elimination that reduces A to I_n on $[A \mid I_n]$, which gives $\left[I_n \mid A^{-1}\right]$. That is,

$$\begin{aligned} [A \mid I_n] &\rightarrow [E_\ell \cdots E_1 A \mid E_\ell \cdots E_1 I_n] = [U \mid K] \\ &\rightarrow [F_k \cdots F_1 U \mid F_k \cdots F_1 K] = [I_n \mid A^{-1}], \end{aligned} \tag{1.3}$$

where the product $E_\ell \cdots E_1$ represents a Gaussian elimination that reduces A to a row-echelon form U of A and the product $F_k \cdots F_1$ represents the backward substitution to reduce U to I_n.

The following example illustrates the computation of the inverse matrix.

Example 14 (*Computing A^{-1} by Gauss-Jordan elimination*) Find the inverse of

$$A = \begin{bmatrix} 1 & 1 & 1 \\ 1 & 2 & 1 \\ 1 & 3 & 0 \end{bmatrix}.$$

Solution: Apply Gauss-Jordan elimination to

$$\begin{aligned}
[A \mid I_3] &= \left[\begin{array}{ccc|ccc} 1 & 1 & 1 & 1 & 0 & 0 \\ 1 & 2 & 1 & 0 & 1 & 0 \\ 1 & 3 & 0 & 0 & 0 & 1 \end{array}\right] \begin{array}{l} \\ (-1) \times \text{row } 1 + \text{row } 2 \\ (-1) \times \text{row } 1 + \text{row } 3 \end{array} \\
&\rightarrow \left[\begin{array}{ccc|ccc} 1 & 1 & 1 & 1 & 0 & 0 \\ 0 & 1 & 0 & -1 & 1 & 0 \\ 0 & 2 & -1 & -1 & 0 & 1 \end{array}\right] \quad (-2) \times \text{row } 2 + \text{row } 3 \\
&\rightarrow \left[\begin{array}{ccc|ccc} 1 & 1 & 1 & 1 & 0 & 0 \\ 0 & 1 & 0 & -1 & 1 & 0 \\ 0 & 0 & -1 & 1 & -2 & 1 \end{array}\right] \quad (-1) \times \text{row } 3 \\
&\rightarrow \left[\begin{array}{ccc|ccc} 1 & 1 & 1 & 1 & 0 & 0 \\ 0 & 1 & 0 & -1 & 1 & 0 \\ 0 & 0 & 1 & -1 & 2 & -1 \end{array}\right].
\end{aligned}$$

This is $[U \mid K]$ obtained by Gaussian elimination. Now continue the backward substitution to reduce $[U \mid K]$ to $\left[I_3 \mid A^{-1}\right]$.

$$[U \mid K] = \left[\begin{array}{ccc|ccc} 1 & 1 & 1 & 1 & 0 & 0 \\ 0 & 1 & 0 & -1 & 1 & 0 \\ 0 & 0 & 1 & -1 & 2 & -1 \end{array}\right] \qquad (-1) \times \text{row } 3 + \text{row } 1$$

$$\rightarrow \left[\begin{array}{ccc|ccc} 1 & 1 & 0 & 2 & -2 & 1 \\ 0 & 1 & 0 & -1 & 1 & 0 \\ 0 & 0 & 1 & -1 & 2 & -1 \end{array}\right] \qquad (-1) \times \text{row } 2 + \text{row } 1$$

$$\rightarrow \left[\begin{array}{ccc|ccc} 1 & 0 & 0 & 3 & -3 & 1 \\ 0 & 1 & 0 & -1 & 1 & 0 \\ 0 & 0 & 1 & -1 & 2 & -1 \end{array}\right] = \left[I_3 \mid A^{-1}\right].$$

Thus,

$$A^{-1} = \begin{bmatrix} 3 & -3 & 1 \\ -1 & 1 & 0 \\ -1 & 2 & -1 \end{bmatrix}.$$

(The reader may try to verify $AA^{-1} = I_3 = A^{-1}A$ by a direct multiplication.) □

The Gaussian elimination process in Example 14 shows that if A is a lower triangular matrix with nonzero diagonal entries, then A^{-1} is also lower triangular.

If A is not invertible, then, at some step in Gaussian elimination, a zero row will show up on the left-hand side in $[U \mid K]$. For example, the matrix $A = \begin{bmatrix} 1 & 2 & 3 \\ 4 & 5 & 6 \\ 7 & 8 & 9 \end{bmatrix}$ is row-equivalent to $\begin{bmatrix} 1 & 2 & 3 \\ 0 & 3 & 6 \\ 0 & 0 & 0 \end{bmatrix}$, which has a zero row and is not invertible. See Problem 13.

By Theorem 1.22, a square matrix A is invertible if and only if $A\mathbf{x} = \mathbf{0}$ has only the trivial solution. That is, a square matrix A is not invertible if and only if $A\mathbf{x} = \mathbf{0}$ has a nontrivial solution, say $\mathbf{x}_0$. In this case, for any column vector $\mathbf{b} = [b_1 \ \cdots \ b_n]^T$, if $\mathbf{x}_1$ is a solution of $A\mathbf{x} = \mathbf{b}$, so is $k\mathbf{x}_0 + \mathbf{x}_1$ for any k, since

$$A(k\mathbf{x}_0 + \mathbf{x}_1) = k(A\mathbf{x}_0) + A\mathbf{x}_1 = k\mathbf{0} + \mathbf{b} = \mathbf{b}.$$

This argument strengthens Theorem 1.16 as follows when A is a square matrix.

Theorem 1.23 *If A is an invertible $n \times n$ matrix, then for any column vector $\mathbf{b} = [b_1 \ \cdots \ b_n]^T$, the system $A\mathbf{x} = \mathbf{b}$ has exactly one solution $\mathbf{x} = A^{-1}\mathbf{b}$. If A is not invertible, then the system has either no solution or infinitely many solutions depending on what* $\mathbf{b}$ *is.* □

A computing of a one-sided (right or left) inverse of a rectangular matrix will be discussed separately in Subsection 7.8.3.

Problem 18 Express A^{-1} as a product of elementary matrices for A given in Example 14.

Problem 19 True or false: If a matrix A has a left inverse C so that $CA = I_n$, then $\mathbf{x} = C\mathbf{b}$ is a solution to a system $A\mathbf{x} = \mathbf{b}$. Justify your answer.

1.10 Applications

1.10.1 Permutation matrices

A *permutation* of a set is an arrangement of its members into a sequence of linear order. For example, there are $3! = 3 \cdot 2 \cdot 1 = 6$ permutations of the set $\{1, 2, 3\}$, namely

$$(1,2,3), (1,3,2), (2,1,3), (2,3,1), (3,1,2), \text{ and } (3,2,1),$$

written as tuples [16]. One can see that each of them can be converted to $(1, 2, 3)$ by taking a finite number of interchangings of two elements in the set $\{1, 2, 3\}$.

A formal definition of a permutation will be given again as Definition 2.10 on page 72 in Chapter 2.

Definition 1.24 A **permutation matrix** is a square matrix obtained from the identity matrix by a rearrangement of the rows, that is, by a permutation of the rows.

Clearly, for any natural number n, there are exactly $n!$ permutation matrices of order n. For $n = 3$, following the six permutations of the set $\{1, 2, 3\}$, they are

$$\begin{bmatrix} 1 & 0 & 0 \\ 0 & 1 & 0 \\ 0 & 0 & 1 \end{bmatrix}, \quad \begin{bmatrix} 1 & 0 & 0 \\ 0 & 0 & 1 \\ 0 & 1 & 0 \end{bmatrix}, \quad \begin{bmatrix} 0 & 1 & 0 \\ 1 & 0 & 0 \\ 0 & 0 & 1 \end{bmatrix},$$

[16] Here, a tuple $(3, 2, 1)$ means a permutation of $\{1, 2, 3\}$ of linear order $3, 2, 1$, not a coordinate of a point in the 3-space $\mathbb{R}^3$.

$$\begin{bmatrix} 0 & 1 & 0 \\ 0 & 0 & 1 \\ 1 & 0 & 0 \end{bmatrix}, \quad \begin{bmatrix} 0 & 0 & 1 \\ 1 & 0 & 0 \\ 0 & 1 & 0 \end{bmatrix}, \quad \begin{bmatrix} 0 & 0 & 1 \\ 0 & 1 & 0 \\ 1 & 0 & 0 \end{bmatrix}.$$

Note that each of these six permutation matrices has a single 1 in each row and a single 1 in each column, all other entries being 0. In fact, this is also true for all $n!$ permutation matrices of any order n.

Lemma 1.25 **(1)** *A permutation matrix is a product of a finite sequence of type II elementary matrices each of them corresponds to a 'row-interchanging' elementary row operation.*

(2) *Let E be any type II elementary matrix. Then E is symmetric and $EE = I$, that is, $E^{-1} = E$.*

(3) *Every permutation matrix P is invertible and $P^{-1} = P^T$. (Such a matrix is called an* orthogonal matrix.*)*

(4) *The product of any two permutation matrices is also a permutation matrix.*

(5) *For any permutation matrix P, P^{-1} and P^T are also permutation matrices.*

Proof: (1) is left for exercise. To get a hint, see Problem 20.

(2) Let E be a type II elementary matrix, say interchanging i-th and j-th rows. Then the (i, j)- and (j, i)-entries of E are 1, and all other off-diagonal entries are 0, showing that $E = E^T$. The identity $EE = I$ is clear.

(3) comes from (1) and (2), while (4) comes from (1).

(5) comes from (1), (2), and (3). □

Problem 20 For each of $3! = 6$ permutation matrices of order 3, express it as a product of type II elementary matrices.

Hint: $\begin{bmatrix} 0 & 0 & 1 \\ 0 & 1 & 0 \\ 1 & 0 & 0 \end{bmatrix} = \begin{bmatrix} 0 & 1 & 0 \\ 1 & 0 & 0 \\ 0 & 0 & 1 \end{bmatrix} \begin{bmatrix} 1 & 0 & 0 \\ 0 & 0 & 1 \\ 0 & 1 & 0 \end{bmatrix} \begin{bmatrix} 0 & 1 & 0 \\ 1 & 0 & 0 \\ 0 & 0 & 1 \end{bmatrix}$, which comes from $(3, 2, 1)$ $\mapsto (2, 3, 1) \mapsto (2, 1, 3) \mapsto (1, 2, 3)$.

Problem 21 Show that any square matrix obtained from the identity matrix by a permutation of the columns is also a permutation matrix.

Hint: P is a permutation matrix if and only if P^T is a permutation matrix.

1.10.2 Sums and products of block matrices

A **submatrix** of a matrix A is a matrix obtained from A by deleting certain rows and/or columns of A. Using some horizontal and/or vertical lines, one can partition a matrix A into submatrices, called **blocks** of A.

For example, a matrix

$$A = \left[\begin{array}{ccc|c} a_{11} & a_{12} & a_{13} & a_{14} \\ a_{21} & a_{22} & a_{23} & a_{24} \\ \hline a_{31} & a_{32} & a_{33} & a_{34} \end{array}\right]$$

is partitioned into four blocks by the dashed lines shown. Now, if we write

$$A_{11} = \begin{bmatrix} a_{11} & a_{12} & a_{13} \\ a_{21} & a_{22} & a_{23} \end{bmatrix}, \quad A_{12} = \begin{bmatrix} a_{14} \\ a_{24} \end{bmatrix},$$

$$A_{21} = \begin{bmatrix} a_{31} & a_{32} & a_{33} \end{bmatrix}, \quad A_{22} = \begin{bmatrix} a_{34} \end{bmatrix},$$

then A can be written as

$$A = \begin{bmatrix} A_{11} & A_{12} \\ A_{21} & A_{22} \end{bmatrix},$$

called a **block matrix**, whose entries are blocks.

Clearly, the sum of block matrices $A = [A_{ij}]$ and $B = [B_{ij}]$ is the block matrix

$$[A_{ij}] + [B_{ij}] = [A_{ij} + B_{ij}]$$

if the blocks A_{ij} and B_{ij} have the same size for all i, j. Moreover,

the product of block matrices $A = [A_{ij}]$ and $B = [B_{jk}]$ also follows the matrix product formula, as if the blocks A_{ij} and B_{jk} were numbers.

If

$$A = \begin{bmatrix} A_{11} & A_{12} \\ A_{21} & A_{22} \end{bmatrix} \quad \text{and} \quad B = \begin{bmatrix} B_{11} & B_{12} \\ B_{21} & B_{22} \end{bmatrix}$$

are block matrices and the number of columns in A_{ij} is equal to the number of rows in B_{jk} for all i, j, k, then

$$AB = \begin{bmatrix} A_{11}B_{11} + A_{12}B_{21} & A_{11}B_{12} + A_{12}B_{22} \\ A_{21}B_{11} + A_{22}B_{21} & A_{21}B_{12} + A_{22}B_{22} \end{bmatrix}.$$

This will be true only if the columns of A are partitioned in the same way as the rows of B.

It is not hard to see that the matrix product by blocks is correct. As a simple example, suppose that a 3×3 matrix A is partitioned as

$$A = \left[\begin{array}{cc|c} a_{11} & a_{12} & a_{13} \\ a_{21} & a_{22} & a_{23} \\ \hline a_{31} & a_{32} & a_{33} \end{array}\right] = \begin{bmatrix} A_{11} & A_{12} \\ A_{21} & A_{22} \end{bmatrix},$$

and a 3×2 matrix B is partitioned as

$$B = \left[\begin{array}{cc} b_{11} & b_{12} \\ b_{21} & b_{22} \\ \hline b_{31} & b_{32} \end{array}\right] = \begin{bmatrix} B_{11} \\ B_{21} \end{bmatrix}.$$

Then the (i, j)-entry of AB is

$$[AB]_{ij} = (a_{i1}b_{1j} + a_{i2}b_{2j}) + a_{i3}b_{3j}\,.$$

The quantity $a_{i1}b_{1j} + a_{i2}b_{2j}$ is simply the (i, j)-entry of $A_{11}B_{11}$ if $i \le 2$, and is the (i, j)-entry of $A_{21}B_{11}$ if $i = 3$. Similarly, $a_{i3}b_{3j}$ is the (i, j)-entry of $A_{12}B_{21}$ if $i \le 2$, and of $A_{22}B_{21}$ if $i = 3$. Thus AB can be written as

$$AB = \begin{bmatrix} A_{11} & A_{12} \\ A_{21} & A_{22} \end{bmatrix} \begin{bmatrix} B_{11} \\ B_{21} \end{bmatrix} = \begin{bmatrix} A_{11}B_{11} + A_{12}B_{21} \\ A_{21}B_{11} + A_{22}B_{21} \end{bmatrix}.$$

Example 15 If the blocks A and B are both square matrices, then

$$\begin{bmatrix} A & \mathbf{0} \\ \mathbf{0} & B \end{bmatrix}^k = \begin{bmatrix} A^k & \mathbf{0} \\ \mathbf{0} & B^k \end{bmatrix}$$

for any natural number $k = 1, 2, \ldots$. As the process of mathematical induction, [17] this is clear for $k = 1$. By using the induction hypothesis, one can see that

$$\begin{aligned} \begin{bmatrix} A & \mathbf{0} \\ \mathbf{0} & B \end{bmatrix}^k \begin{bmatrix} A & \mathbf{0} \\ \mathbf{0} & B \end{bmatrix} &= \begin{bmatrix} A^k & \mathbf{0} \\ \mathbf{0} & B^k \end{bmatrix} \begin{bmatrix} A & \mathbf{0} \\ \mathbf{0} & B \end{bmatrix} \\ &= \begin{bmatrix} A^kA + \mathbf{0}\mathbf{0} & A^k\mathbf{0} + \mathbf{0}B \\ \mathbf{0}A + B^k\mathbf{0} & \mathbf{0}\mathbf{0} + B^kB \end{bmatrix} = \begin{bmatrix} A^{k+1} & \mathbf{0} \\ \mathbf{0} & B^{k+1} \end{bmatrix}. \end{aligned}$$

[17] *Deduction* uses general rules to draw conclusion about specifics, while *induction* uses particular instances to reveal generalities. *Mathematical induction* is a technique to prove that *a statement involving a natural number n holds for all values of n*. The proof consists of two steps. The basis case: prove that the statement holds for the first natural number n. Usually, $n = 0$ or $n = 1$. The inductive step: prove that, if the statement holds for some natural number k, then the statement also holds for $k + 1$.

Example 16 If A and B are both square and invertible matrices, then

$$\begin{bmatrix} A & \mathbf{0} \\ \mathbf{0} & B \end{bmatrix}^{-1} = \begin{bmatrix} A^{-1} & \mathbf{0} \\ \mathbf{0} & B^{-1} \end{bmatrix},$$

because

$$\begin{bmatrix} A & \mathbf{0} \\ \mathbf{0} & B \end{bmatrix} \begin{bmatrix} A^{-1} & \mathbf{0} \\ \mathbf{0} & B^{-1} \end{bmatrix} = \begin{bmatrix} AA^{-1} + \mathbf{0}\mathbf{0} & A\mathbf{0} + \mathbf{0}B^{-1} \\ \mathbf{0}A^{-1} + B\mathbf{0} & \mathbf{0}\mathbf{0} + BB^{-1} \end{bmatrix} = \begin{bmatrix} I & \mathbf{0} \\ \mathbf{0} & I \end{bmatrix}$$

is the identity matrix.

Example 17 ($A\mathbf{x} = \mathbf{b}$ *is the same as the vector equation* $x_1\mathbf{c}_1 + \cdots + x_n\mathbf{c}_n = \mathbf{b}$) If an $m \times n$ matrix A is partitioned into blocks of column vectors: *i.e.,* $A = [\mathbf{c}_1\ \mathbf{c}_2\ \cdots\ \mathbf{c}_n]$, where each block $\mathbf{c}_j$ is the j-th column of A, then the product $A\mathbf{x}$ with $\mathbf{x} = [x_1\ \cdots\ x_n]^T$ is the sum of the block matrices (or column vectors) with coefficients x_j's:

$$A\mathbf{x} = [\mathbf{c}_1\ \mathbf{c}_2\ \cdots\ \mathbf{c}_n] \begin{bmatrix} x_1 \\ x_2 \\ \vdots \\ x_n \end{bmatrix} = x_1\mathbf{c}_1 + x_2\mathbf{c}_2 + \cdots + x_n\mathbf{c}_n, \tag{1.4}$$

where $x_j\mathbf{c}_j = x_j[a_{1j}\ a_{2j}\ \cdots\ a_{nj}]^T$. Hence, a matrix equation $A\mathbf{x} = \mathbf{b}$ is just the vector equation $x_1\mathbf{c}_1 + x_2\mathbf{c}_2 + \cdots + x_n\mathbf{c}_n = \mathbf{b}$. □

Example 18 Let A be an $m \times n$ matrix partitioned into the row vectors $\mathbf{a}_1, \mathbf{a}_2, \ldots,$ $\mathbf{a}_n$ as its blocks, and let B be an $n \times \ell$ matrix so that their product AB is well defined. By considering the matrix B as a single block, the product AB can be written as

$$AB = \begin{bmatrix} \mathbf{a}_1 \\ \mathbf{a}_2 \\ \vdots \\ \mathbf{a}_m \end{bmatrix} B = \begin{bmatrix} \mathbf{a}_1 B \\ \mathbf{a}_2 B \\ \vdots \\ \mathbf{a}_m B \end{bmatrix} = \begin{bmatrix} \mathbf{a}_1\mathbf{b}_1 & \mathbf{a}_1\mathbf{b}_2 & \cdots & \mathbf{a}_1\mathbf{b}_\ell \\ \mathbf{a}_2\mathbf{b}_1 & \mathbf{a}_2\mathbf{b}_2 & \cdots & \mathbf{a}_2\mathbf{b}_\ell \\ \vdots & & \ddots & \vdots \\ \mathbf{a}_m\mathbf{b}_1 & \mathbf{a}_m\mathbf{b}_2 & \cdots & \mathbf{a}_m\mathbf{b}_\ell \end{bmatrix},$$

where $\mathbf{b}_1, \mathbf{b}_2, \ldots, \mathbf{b}_\ell$ denote the columns of B for the last equality. Hence, the (i, j)-entry of AB is $\mathbf{a}_i\mathbf{b}_j$, which is the (block) product of the row (block) $\mathbf{a}_i$ of A and the column (block) $\mathbf{b}_j$ of B, as in Figure 1.1. □

Problem 22 Compute AB and BA as a product of block matrices, where

$$A = \left[\begin{array}{rr|rr} 1 & 2 & 1 & 0 \\ -3 & 4 & 0 & 1 \\ \hline 0 & 0 & 2 & -1 \end{array}\right], \qquad B = \left[\begin{array}{rr|r} 1 & 0 & 2 \\ 0 & 1 & 3 \\ \hline 2 & 3 & 4 \\ 3 & -2 & 1 \end{array}\right].$$

Problem 23 Let A be an $m \times n$ matrix and let B be an $n \times \ell$ matrix with the columns $\mathbf{b}_1, \mathbf{b}_2, \ldots, \mathbf{b}_\ell$, so that their product AB is well defined. Show that the product AB can be written as

$$AB = A\,[\mathbf{b}_1\ \mathbf{b}_2\ \ldots\ \mathbf{b}_\ell] = [A\mathbf{b}_1\ A\mathbf{b}_2\ \ldots\ A\mathbf{b}_\ell]$$

as a product of block matrices.

Hint: Consider the matrix A as a single block.

Problem 24 Let A be an $m \times n$ matrix partitioned into the columns $\mathbf{c}_1, \mathbf{c}_2, \ldots, \mathbf{c}_n$ as its blocks and let B be any $n \times n$ diagonal matrix, say

$$B = \begin{bmatrix} \lambda_1 & & \mathbf{0} \\ & \ddots & \\ \mathbf{0} & & \lambda_n \end{bmatrix}.$$

Show that the product AB is a block matrix with columns $\lambda_j \mathbf{c}_j$'s: *i.e.*,

$$AB = [\lambda_1 \mathbf{c}_1\ \lambda_2 \mathbf{c}_2\ \cdots\ \lambda_n \mathbf{c}_n],$$

where $\lambda_j \mathbf{c}_j = \lambda_j [a_{1j}\ a_{2j}\ \cdots\ a_{mj}]^T$.

Hint: First use Problem 23 for the product AB and use Example 17 for each block $A\mathbf{b}_i$.

1.10.3 How to use LU decompositions

As a continuation of LU decomposition of a matrix introduced in Section 1.8, let $A = LU$ be an LU decomposition. Then the system $A\mathbf{x} = \mathbf{b}$ can be written as $LU\mathbf{x} = \mathbf{b}$. Set $U\mathbf{x} = \mathbf{y}$. Now, the system

$$A\mathbf{x} = LU\mathbf{x} = \mathbf{b}$$

can be solved by the following two (substitution) steps:

Step 1 Solve $L\mathbf{y} = \mathbf{b}$ (for $\mathbf{y}$) by forward substitution.
Step 2 Solve $U\mathbf{x} = \mathbf{y}$ (for $\mathbf{x}$) by backward substitution.

The next example illustrates how to factor a matrix A into LU and then how to use its LU decomposition for solving a system $A\mathbf{x} = \mathbf{b}$.

Example 19 (*Solving $A\mathbf{x} = \mathbf{b}$ by an LU decomposition*) Solve the system of linear equations

$$A\mathbf{x} = \begin{bmatrix} 2 & 1 & 1 \\ 4 & 1 & 0 \\ -2 & 2 & 1 \end{bmatrix} \begin{bmatrix} x_1 \\ x_2 \\ x_3 \end{bmatrix} = \begin{bmatrix} 1 \\ 2 \\ 7 \end{bmatrix} = \mathbf{b} \text{ or } \left\{ \begin{array}{rcrcrcr} 2x_1 & + & x_2 & + & x_3 & = & 1 \\ 4x_1 & + & x_2 & & & = & 2 \\ -2x_1 & + & 2x_2 & + & x_3 & = & 7 \end{array} \right.$$

by using an LU decomposition of a (square) matrix A.

Solution: To solve the system $A\mathbf{x} = \mathbf{b}$, one needs three steps of forward elimination procedures; (i) $(-2) \times$ row $1 +$ row 2, (ii) $1 \times$ row $1 +$ row 3, and (iii) $3 \times$ row $2 +$ row 3.

The elementary matrices for these three forward eliminations on the augmented matrix $[A\ \mathbf{b}]$ are

$$E_1 = \begin{bmatrix} 1 & 0 & 0 \\ -2 & 1 & 0 \\ 0 & 0 & 1 \end{bmatrix}, \quad E_2 = \begin{bmatrix} 1 & 0 & 0 \\ 0 & 1 & 0 \\ 1 & 0 & 1 \end{bmatrix}, \quad \text{and} \quad E_3 = \begin{bmatrix} 1 & 0 & 0 \\ 0 & 1 & 0 \\ 0 & 3 & 1 \end{bmatrix},$$

so that

$$E_3E_2E_1A = \begin{bmatrix} 2 & 1 & 1 \\ 0 & -1 & -2 \\ 0 & 0 & -4 \end{bmatrix} = U,$$

which is an upper triangular matrix. Thus, if we set

$$L = (E_3E_2E_1)^{-1} = E_1^{-1}\,E_2^{-1}\,E_3^{-1} = \begin{bmatrix} 1 & 0 & 0 \\ 2 & 1 & 0 \\ -1 & -3 & 1 \end{bmatrix},$$

which is also a lower triangular matrix with 1's on the diagonal, then

$$A = LU = \begin{bmatrix} 1 & 0 & 0 \\ 2 & 1 & 0 \\ -1 & -3 & 1 \end{bmatrix} \begin{bmatrix} 2 & 1 & 1 \\ 0 & -1 & -2 \\ 0 & 0 & -4 \end{bmatrix}.$$

Now, one can solve the system $A\mathbf{x} = LU\mathbf{x} = \mathbf{b}$ by using the LU decomposition. The system

$$L\mathbf{y} = \mathbf{b} : \quad \left\{ \begin{array}{rcrcrcr} y_1 & & & & & = & 1 \\ 2y_1 & + & y_2 & & & = & 2 \\ -y_1 & - & 3y_2 & + & y_3 & = & 7 \end{array} \right.$$

can be solved immediately by forward substitution to get $\mathbf{y} = (1, 0, 8)$, and then the system

$$U\mathbf{x} = \mathbf{y}: \quad \left\{ \begin{array}{rcrcrcr} 2x_1 & + & x_2 & + & x_3 & = & 1 \\ & - & x_2 & - & 2x_3 & = & 0 \\ & & & - & 4x_3 & = & 8 \end{array} \right.$$

can also be immediately solved by backward substitution. It gives $x_1 = -1/2$, $x_2 = 4$, $x_3 = -2$, as a solution. □

Example 20 (*Solving $A\mathbf{x} = \mathbf{b}$ by an LU decomposition*) Solve

$$A\mathbf{x} = \begin{bmatrix} 1 & 0 & 1 & 0 \\ 0 & 1 & 0 & 1 \\ -1 & 1 & 1 & -1 \end{bmatrix} \begin{bmatrix} x_1 \\ x_2 \\ x_3 \\ x_4 \end{bmatrix} = \begin{bmatrix} 1 \\ 2 \\ 7 \end{bmatrix} = \mathbf{b}$$

by using an LU decomposition of a (rectangular) matrix A.

Solution: In solving the system $A\mathbf{x} = \mathbf{b}$, the elementary matrices for forward elimination are

$$E_1 = \begin{bmatrix} 1 & 0 & 0 \\ 0 & 1 & 0 \\ 1 & 0 & 1 \end{bmatrix} \quad \text{and} \quad E_2 = \begin{bmatrix} 1 & 0 & 0 \\ 0 & 1 & 0 \\ 0 & -1 & 1 \end{bmatrix},$$

so that

$$E_2 E_1 A = \begin{bmatrix} 1 & 0 & 1 & 0 \\ 0 & 1 & 0 & 1 \\ 0 & 0 & 2 & -2 \end{bmatrix} = U,$$

where U is not a square matrix and so not upper triangular, but is a row-echelon form of A. If we set

$$L = (E_2 E_1)^{-1} = E_1^{-1} E_2^{-1} = \begin{bmatrix} 1 & 0 & 0 \\ 0 & 1 & 0 \\ -1 & 1 & 1 \end{bmatrix},$$

which is a lower triangular matrix with 1's on the diagonal, then

$$A = LU = \begin{bmatrix} 1 & 0 & 0 \\ 0 & 1 & 0 \\ -1 & 1 & 1 \end{bmatrix} \begin{bmatrix} 1 & 0 & 1 & 0 \\ 0 & 1 & 0 & 1 \\ 0 & 0 & 2 & -2 \end{bmatrix}.$$

Now, the system

$$L\mathbf{y}=\mathbf{b}:\quad \begin{cases} y_1 & & & = 1 \\ & y_2 & & = 2 \\ -y_1 & + \; y_2 & + \; y_3 & = 7 \end{cases}$$

can be solved by forward substitution to get $\mathbf{y}=(1,2,6)$, and then the system

$$U\mathbf{x}=\mathbf{y}:\quad \begin{cases} x_1 & & + \; x_3 & & = 1 \\ & x_2 & & + \; x_4 & = 2 \\ & & 2x_3 & - \; 2x_4 & = 6 \end{cases}$$

can be solved by backward substitution, which gives

$$\mathbf{x}=\begin{bmatrix} x_1 \\ x_2 \\ x_3 \\ x_4 \end{bmatrix}=\begin{bmatrix} -2 - t \\ 2 - t \\ 3 + t \\ t \end{bmatrix}=\begin{bmatrix} -2 \\ 2 \\ 3 \\ 0 \end{bmatrix}+t\begin{bmatrix} -1 \\ -1 \\ 1 \\ 1 \end{bmatrix},$$

for $t \in \mathbb{R}$. This is the solution for the given system. □

As shown in Examples 19 and 20, it is quite straightforward to solve the systems $L\mathbf{y}=\mathbf{b}$ and $U\mathbf{x}=\mathbf{y}$, because L is lower triangular and U is a row-echelon form of A after forward elimination.

The matrix U in the decomposition $A = LU$ can further be factored as the product $U = D\bar{U}$, where D is a diagonal matrix whose diagonal entries are the pivots of U or zeros and $\bar{U}$ is a row-echelon form of A with leading 1's, so that $A = LD\bar{U}$. For example,

$$\begin{aligned} A &= \begin{bmatrix} 1 & 0 & 0 & 0 \\ * & 1 & 0 & 0 \\ * & * & 1 & 0 \\ * & * & * & 1 \end{bmatrix}\begin{bmatrix} d_1 & * & * & * & * & * \\ 0 & 0 & d_2 & * & * & * \\ 0 & 0 & 0 & 0 & d_3 & * \\ 0 & 0 & 0 & 0 & 0 & 0 \end{bmatrix} = LU \\ &= \begin{bmatrix} 1 & 0 & 0 & 0 \\ * & 1 & 0 & 0 \\ * & * & 1 & 0 \\ * & * & * & 1 \end{bmatrix}\begin{bmatrix} d_1 & 0 & 0 & 0 \\ 0 & d_2 & 0 & 0 \\ 0 & 0 & d_3 & 0 \\ 0 & 0 & 0 & 0 \end{bmatrix}\begin{bmatrix} 1 & */d_1 & */d_1 & */d_1 & */d_1 & */d_1 \\ 0 & 0 & 1 & */d_2 & */d_2 & */d_2 \\ 0 & 0 & 0 & 0 & 1 & */d_3 \\ 0 & 0 & 0 & 0 & 0 & 0 \end{bmatrix} \\ &= LD\bar{U}. \end{aligned}$$

For notational convenience, we replace $\bar{U}$ again by U and write $A = LDU$. This decomposition is called an **LDU factorization** or an **LDU decomposition** of A.

For example, the matrix A in Example 20 is factored as

$$A = LU = \begin{bmatrix} 1 & 0 & 0 \\ 0 & 1 & 0 \\ -1 & 1 & 1 \end{bmatrix} \begin{bmatrix} 1 & 0 & 1 & 0 \\ 0 & 1 & 0 & 1 \\ 0 & 0 & 2 & -2 \end{bmatrix}.$$

It can be further factored as $A = LDU$ by taking

$$\begin{bmatrix} 1 & 0 & 1 & 0 \\ 0 & 1 & 0 & 1 \\ 0 & 0 & 2 & -2 \end{bmatrix} = \begin{bmatrix} 1 & 0 & 0 \\ 0 & 1 & 0 \\ 0 & 0 & 2 \end{bmatrix} \begin{bmatrix} 1 & 0 & 1 & 0 \\ 0 & 1 & 0 & 1 \\ 0 & 0 & 1 & -1 \end{bmatrix} = DU.$$

The LDU decomposition of a matrix A is always possible when no row interchanges are needed in the forward elimination procedure. In general, if a permutation matrix is necessary in the forward elimination, then the matrix A may not have any LDU decomposition.

Example 21 (*Some matrices may not have any LU decomposition*) Consider $A = \begin{bmatrix} 0 & 1 \\ 1 & 0 \end{bmatrix}$. For forward elimination, it is necessary to interchange the first row with the second row. Without this interchanging, A has no LU decompositions. In fact, the matrix A cannot be expressed as a product of any lower triangular matrix L and any upper triangular matrix U. □

Suppose now that a row interchange is necessary for forward elimination in solving the system $A\mathbf{x} = \mathbf{b}$. In this case, one can first do all of the row interchanges before any forward elimination, since the interchange of rows can be done at any time, before or after the other elementary operations. Those row-interchanging elementary matrices altogether form a permutation matrix P so that no more row interchanges are necessary during the forward elimination for PA. Now, the matrix PA has an LU or LDU decomposition.

Following Example 13, we give another example of a decomposition of $PA = LDU$.

Example 22 ($PA = LDU$ *decomposition*) Consider a matrix

$$A = \begin{bmatrix} 0 & 1 & 2 \\ 0 & 1 & 0 \\ 1 & 0 & 2 \end{bmatrix}.$$

For forward elimination, it is necessary to interchange the first row with the third row, that is, we need to multiply A on the left by the permutation matrix $P = \begin{bmatrix} 0 & 0 & 1 \\ 0 & 1 & 0 \\ 1 & 0 & 0 \end{bmatrix}$ so that

$$PA = \begin{bmatrix} 1 & 0 & 2 \\ 0 & 1 & 0 \\ 0 & 1 & 2 \end{bmatrix} = \begin{bmatrix} 1 & 0 & 0 \\ 0 & 1 & 0 \\ 0 & 1 & 1 \end{bmatrix} \begin{bmatrix} 1 & 0 & 0 \\ 0 & 1 & 0 \\ 0 & 0 & 2 \end{bmatrix} \begin{bmatrix} 1 & 0 & 2 \\ 0 & 1 & 0 \\ 0 & 0 & 1 \end{bmatrix} = LDU.$$

Here, U is a row-echelon form of the matrix PA. □

Of course, if we choose a different permutation P', then the LDU decomposition of $P'A$ may be different from that of PA, even if there is a permutation matrix P'' that changes $P'A$ to PA. Moreover, even if a permutation matrix is not necessary in the Gaussian elimination, the LDU decomposition of A needs not be unique.

Example 23 (*Some matrices may have infinitely many different LDU decompositions*) A (singular) matrix

$$A = \begin{bmatrix} 1 & 1 & 0 \\ 1 & 2 & 0 \\ 0 & 0 & 0 \end{bmatrix}$$

has an LDU decomposition

$$A = \begin{bmatrix} 1 & 0 & 0 \\ 1 & 1 & 0 \\ 0 & 0 & 1 \end{bmatrix} \begin{bmatrix} 1 & 0 & 0 \\ 0 & 1 & 0 \\ 0 & 0 & x \end{bmatrix} \begin{bmatrix} 1 & 1 & 0 \\ 0 & 1 & 0 \\ 0 & 0 & 0 \end{bmatrix} = LDU$$

for any value x. This shows that the (singular) matrix A has infinitely many different LDU decompositions. □

However, if the matrix A is invertible and if the permutation matrix P is fixed when it is needed, then the matrix PA has the unique LDU decomposition.

Theorem 1.26 *Let A be an invertible matrix. Then for a fixed suitable permutation matrix P, the matrix PA has the unique LDU decomposition.*

Proof: Suppose that

$$PA = L_1D_1U_1 = L_2D_2U_2,$$

where the L's are lower triangular, the U's are upper triangular whose diagonals are all 1's, and the D's are diagonal matrices with no zeros on the diagonal (because A is invertible). In this case, the U's are all square matrices because the matrix A is square.

One needs to show $L_1 = L_2$, $D_1 = D_2$, and $U_1 = U_2$ for the uniqueness.

It is shown that the inverses of lower triangular L_i's, upper triangular U_i's, and diagonal matrices D_i's are lower triangular, upper triangular, and diagonal, respectively. Therefore, by multiplying $(L_1D_1)^{-1} = D_1^{-1}L_1^{-1}$ on the left and U_2^{-1} on the right, the equation $L_1D_1U_1 = L_2D_2U_2$ becomes

$$U_1U_2^{-1} = D_1^{-1}L_1^{-1}L_2D_2\,.$$

The left-hand side is upper triangular, while the right-hand side is lower triangular. Hence, both sides should be diagonal. However, since the diagonal entries of the upper triangular matrix $U_1U_2^{-1}$ are all 1's, it should be the identity matrix I (see Problem 14). Thus

$$U_1U_2^{-1} = I, \quad \textit{i.e.,}\ U_1 = U_2.$$

Now, the condition $L_1D_1U_1 = L_2D_2U_2$ gives $L_1D_1 = L_2D_2$, equivalently $L_1^{-1}L_2 = D_1D_2^{-1}$, which is diagonal with diagonal entries all 1's. Hence, $L_1 = L_2$ and $D_1 = D_2$. □

In particular, if an invertible matrix A is symmetric (*i.e.,* $A = A^T$), and if it can be factored into $A = LDU$ without row interchange, then we have

$$LDU = A = A^T = (LDU)^T = U^TD^TL^T = U^TDL^T.$$

By the uniqueness of decompositions, we have $U = L^T$ and $A = LDL^T$.

Remark: (1) The efficiency of LU decomposition is improved when one needs to solve several systems of linear equations $A\mathbf{x} = \mathbf{b}_i$ with the same coefficient matrix A. Let $A\mathbf{x} = \mathbf{b}_i$ for $i = 1, 2, \ldots, \ell$ be given. Instead of performing the forward elimination repeatedly, one can use an LU decomposition of A: Solve first $L\mathbf{y} = \mathbf{b}_i$ for each i to get solutions $\mathbf{y}_i$ and then solve $U\mathbf{x} = \mathbf{y}_i$ for each i. From an algorithmic point of view, the method based on the LU decomposition of A is much more efficient than doing the forward elimination repeatedly, in particular, when ℓ is large. Actually, in many engineering problems solving $A\mathbf{x} = \mathbf{b}$, the matrix A remains unchanged, while the right hand side vector $\mathbf{b}$ keeps changing.

(2) Like LU or LDU decompositions, one sometimes needs to factor a matrix into a product of two or three matrices, called a **matrix decomposition** or **matrix factorization**.

Throughout this book, the readers will meet several other types of matrix decompositions like QR decompositions, eigenvalue decompositions, singular value decompositions, or Jordan decompositions, etc.

Problem 25 For the matrix $A = \begin{bmatrix} 1 & 1 & 1 \\ 1 & 2 & 1 \\ 1 & 3 & 0 \end{bmatrix}$ given in Example 14, determine its LU decomposition. Use it to solve $A\mathbf{x} = \mathbf{b}$ for $\mathbf{b} = [1\ 1\ 1]^T$.

1.11 Exercises

1.1. *Every system of linear equations can be solved by Gauss-Jordan elimination.* Solve each of the following systems of linear equations by Gauss-Jordan elimination on its augmented matrix $[A\ \mathbf{b}]$.

$$(1) \left\{ \begin{array}{rcrcl} x & + & y & = & 5 \\ 2x & - & y & = & 7, \end{array} \right.$$

$$(2) \left\{ \begin{array}{rcrcrcl} 2x & - & 3y & & & = & 8 \\ 4x & - & 5y & + & z & = & 15 \\ 2x & & & + & 4z & = & 1, \end{array} \right.$$

$$(3) \left\{ \begin{array}{rcrcrcrcl} x_1 & + & x_2 & + & x_3 & - & x_4 & = & -2 \\ 2x_1 & - & x_2 & + & x_3 & + & x_4 & = & 0 \\ 3x_1 & + & 2x_2 & - & x_3 & - & x_4 & = & 1 \\ x_1 & + & x_2 & + & 3x_3 & - & 3x_4 & = & -8. \end{array} \right.$$

1.2. *Every system of linear equations can be solved by Gauss-Jordan elimination.* Solve each of the following linear equations. Write the solutions in matrix notation. Here, each system has only one linear equation, having more unknowns than equations.

(1) $x + y = 5,$
(2) $x_1 + x_2 + x_3 - x_4 = -2.$

1.3. Which of the following homogeneous systems has a nontrivial solution, with a justification on the augmented matrix?

$$(1) \left\{ \begin{array}{rcrcrcl} & & 2y & + & 2z & = & 0 \\ x & + & 2y & + & 3z & = & 0, \end{array} \right. \qquad (2) \left\{ \begin{array}{rcrcrcl} 2x & + & y & - & z & = & 0 \\ x & - & 2y & - & 3z & = & 0 \\ 3x & + & y & - & 2z & = & 0. \end{array} \right.$$

1.4. Which of the following augmented matrices $[A\ \mathbf{b}]$ are in row-echelon form? Which are in reduced row-echelon form?

$$(1)\ [A\ \mathbf{b}] = \begin{bmatrix} 1 & 0 & 0 & 0 & 10 \\ 0 & 1 & 0 & 0 & 20 \\ 0 & 0 & 0 & 1 & 30 \end{bmatrix}, \quad (2)\ [A\ \mathbf{b}] = \begin{bmatrix} 1 & 0 & 0 & 0 & 20 \\ 0 & 0 & 1 & 0 & 10 \\ 0 & 0 & 0 & 1 & 30 \\ 0 & 0 & 0 & 0 & 0 \end{bmatrix},$$

$$(3)\ [A\ \mathbf{b}] = \begin{bmatrix} 0 & 0 & 0 & 0 & 0 \\ 0 & 0 & 1 & 2 & 6 \\ 0 & 0 & 0 & 1 & 0 \\ 0 & 0 & 0 & 0 & 0 \end{bmatrix}, \quad (4)\ [A\ \mathbf{b}] = \begin{bmatrix} 0 & 1 & 0 & 0 & -5 \\ 0 & 0 & 1 & 1 & -4 \\ 0 & 0 & 0 & 1 & 30 \end{bmatrix},$$

$$(5)\ [A\ \mathbf{b}] = \begin{bmatrix} 0 & 1 & 0 & 0 & 10 \\ 0 & 0 & 1 & 0 & 30 \\ 0 & 1 & 0 & 2 & 50 \end{bmatrix}, \quad (6)\ [A\ \mathbf{b}] = \begin{bmatrix} 1 & 0 & 0 & 0 & 10 \\ 0 & 1 & 0 & 0 & 20 \\ 0 & 0 & 0 & 1 & -7 \\ 0 & 0 & 0 & 0 & 30 \end{bmatrix}.$$

For which of $[A\ \mathbf{b}]$, the system $A\mathbf{x} = \mathbf{b}$ has infinitely many solutions? For which of $[A\ \mathbf{b}]$, the system $A\mathbf{x} = \mathbf{b}$ has no solutions?

1.5. Find a row-echelon form and the reduced row-echelon form of

$$(1)\ [A\ \mathbf{b}] = \begin{bmatrix} 3 & 1 & 2 & 3 \\ 2 & 1 & 1 & 9 \\ 3 & 3 & 2 & 0 \end{bmatrix}, \qquad (2)\ [A\ \mathbf{b}] = \begin{bmatrix} 1 & 2 & 3 & 4 & 10 \\ 2 & 3 & 4 & 1 & 10 \\ 3 & 4 & 1 & 2 & 10 \\ 4 & 1 & 2 & 3 & 10 \end{bmatrix}.$$

Solve $A\mathbf{x} = \mathbf{b}$.

1.6. Construct five different matrices which are row-equivalent to

$$[A\ \mathbf{b}] = \begin{bmatrix} 2 & 1 & -3 & 4 \\ 0 & 1 & 2 & -1 \\ 2 & 2 & -3 & 2 \end{bmatrix}.$$

Hint: Its (reduced) row-echelon form is one of them.

1.7. Let A and B be matrices of the same size. Show that

(1) if $A\mathbf{x} = \mathbf{0}$ for all $\mathbf{x}$, then A is the zero matrix,
(2) if $A\mathbf{x} = B\mathbf{x}$ for all $\mathbf{x}$, then $A = B$.

Hint for (1): Take $\mathbf{x}$ as transpose of each row vector of A.

1.8. Compute ABC, BCA, and CAB for

$$A = \begin{bmatrix} 1 & 1 & 0 \\ 1 & 0 & 1 \end{bmatrix}, \qquad B = \begin{bmatrix} 2 \\ 1 \\ -1 \end{bmatrix}, \qquad C = \begin{bmatrix} 1 & -1 \end{bmatrix}.$$

1.9. Compute $(A - B)C^T$ and B^TB for

$$A = \begin{bmatrix} 1 & 0 & 1 \\ 0 & 1 & 0 \\ 0 & 0 & 1 \end{bmatrix}, \quad B = \begin{bmatrix} 1 & 0 & 0 \\ -2 & 1 & 0 \\ 0 & 0 & 1 \end{bmatrix}, \quad C = \begin{bmatrix} 0 & 1 & 0 \\ 0 & 0 & 1 \\ 1 & 0 & 0 \end{bmatrix}.$$

1.10. Let $f(x) = a_nx^n + a_{n-1}x^{n-1} + \cdots + a_1x + a_0$ be a polynomial. For any square matrix A, a *matrix polynomial* $f(A)$ is defined as

$$f(A) = a_nA^n + a_{n-1}A^{n-1} + \cdots + a_1A + a_0I.$$

For $f(x) = 3x^3 + x^2 - 2x + 1$, find $f(A)$ for

$$(1)\ A = \begin{bmatrix} 1 & 0 & 0 \\ 0 & 2 & 0 \\ 0 & 0 & -1 \end{bmatrix}, \qquad (2)\ A = \begin{bmatrix} 1 & -1 & 2 \\ 0 & 2 & -1 \\ 0 & 0 & 3 \end{bmatrix}.$$

1.11. Show that if A is a 3×3 matrix such that $AB = BA$ for every 3×3 matrix B, then $A = cI_3$ for some constant c.

1.12. Let D be a 4×4 diagonal matrix with distinct nonzero diagonal entries and let A be any 4×4 matrix. Show that $DA = AD$ if and only if A is diagonal.

Hint: First, assume that D is the diagonal matrix with diagonal entries $1, 2, 3, 4$, and compare the (i, j)-entries on both sides of $DA = AD$.

1.13. For $A = \begin{bmatrix} 1 & 2 & 0 \\ 0 & 1 & 3 \\ 0 & 0 & 1 \end{bmatrix}$, compute A^k for $k = 1, 2, 3, 4, \ldots$.

1.14. Show that *the sum and difference of any two symmetric matrices are also symmetric, but this is not true in general for the product.* Given two symmetric matrices A and B, show that AB is symmetric if and only if $AB = BA$.

Hint: As a product of symmetric matrices, $\begin{bmatrix} 0 & 1 \\ 1 & 0 \end{bmatrix} \begin{bmatrix} 2 & 0 \\ 0 & 3 \end{bmatrix} = \begin{bmatrix} 0 & 3 \\ 2 & 0 \end{bmatrix}$ is not symmetric.

1.15. Find the symmetric part and the skew-symmetric part of each of

$$A = \begin{bmatrix} 2 & 3 & 1 \\ 2 & 5 & 9 \\ 1 & 3 & 2 \end{bmatrix}, \qquad B = \begin{bmatrix} 1 & 3 & 4 \\ 0 & 2 & 1 \\ 4 & 0 & 3 \end{bmatrix}.$$

Hint: Use Problem 10(3). The symmetric part of A is $\frac{1}{2}(A + A^T)$.

1.16. Compute AA^T and A^TA for the matrix $A = \begin{bmatrix} 0 & 0 & -5 & 7 \\ 0 & 9 & 1 & 8 \\ 1 & 3 & 9 & 2 \end{bmatrix}$. Are they symmetric?

1.17. Show that any (lower or upper) triangular matrix A satisfying $AA^T = A^TA$ is a diagonal matrix.

Hint: Compare the (i,j)-entries of AA^T and A^TA.

1.18. Let $A^{-1} = \begin{bmatrix} 1 & 1 & 0 \\ 0 & 1 & 3 \\ 0 & 0 & 1 \end{bmatrix}$.

(1) Find a matrix B such that $AB = \begin{bmatrix} 1 & 0 \\ 0 & 1 \\ 1 & 1 \end{bmatrix}$.

(2) Find a matrix C such that $AC = A^2 - AA^T$.

1.19. Find all possible values of a, b, c so that $A = \begin{bmatrix} a & b \\ c & 0 \end{bmatrix}$ is invertible and $A^{-1} = A$.

1.20. Find the inverse of each of the following (permutation) matrices.

$$A = \begin{bmatrix} 0 & 0 & 0 & 1 \\ 0 & 0 & 1 & 0 \\ 0 & 1 & 0 & 0 \\ 1 & 0 & 0 & 0 \end{bmatrix}, \qquad B = \begin{bmatrix} 0 & 1 & 0 & 0 \\ 1 & 0 & 0 & 0 \\ 0 & 0 & 0 & 1 \\ 0 & 0 & 1 & 0 \end{bmatrix}.$$

1.21. Decide whether or not, each of the following matrices is invertible. Find the inverses for invertible ones.

$$A = \begin{bmatrix} 1 & 2 & 1 \\ 3 & 2 & 0 \\ 4 & 4 & 1 \end{bmatrix}, \qquad B = \begin{bmatrix} 1 & 1 & 1 \\ 0 & 2 & 0 \\ 3 & 0 & 1 \end{bmatrix}, \qquad C = \begin{bmatrix} 1 & 2 & 3 & 4 \\ 0 & 2 & 3 & 4 \\ 0 & 0 & 3 & 4 \\ 0 & 0 & 0 & 4 \end{bmatrix}.$$

Hint: Use Theorem 1.22.

1.22. Find the inverse of each of the following matrices.

$$A = \begin{bmatrix} 1 & 1 & 1 \\ 2 & 4 & 8 \\ 3 & 9 & 27 \end{bmatrix}, \quad B = \begin{bmatrix} 1 & 0 & 0 & 0 \\ 1 & 2 & 0 & 0 \\ 1 & 2 & 4 & 0 \\ 1 & 2 & 4 & 8 \end{bmatrix}, \quad C = \begin{bmatrix} k & 1 & 0 & 0 \\ 0 & k & 1 & 0 \\ 0 & 0 & k & 1 \\ 0 & 0 & 0 & k \end{bmatrix}, \ (k \neq 0).$$

1.23. Let A be any 2×1 matrix and let B be any 1×2 matrix. Show that the product AB cannot be invertible.

1.24. Write the following systems of linear equations in matrix forms $A\mathbf{x} = \mathbf{b}$ and solve them by computing $A^{-1}\mathbf{b}$.

$$(1) \left\{ \begin{array}{rcrcrcl} 2x_1 & - & x_2 & + & 3x_3 & = & 3 \\ & & x_2 & - & 4x_3 & = & 6 \\ 2x_1 & + & x_2 & - & 2x_3 & = & 6, \end{array} \right. \qquad (2) \left\{ \begin{array}{rcrcrcl} x_1 & - & x_2 & + & x_3 & = & 1 \\ x_1 & + & x_2 & - & x_3 & = & 2 \\ x_1 & - & 3x_2 & + & 2x_3 & = & 1. \end{array} \right.$$

1.25. Find an LU decomposition of each of the following matrices.

$$(1)\ \begin{bmatrix} 1 & 2 & 3 \\ 3 & 3 & 4 \\ 2 & 4 & 0 \end{bmatrix}, \qquad (2)\ \begin{bmatrix} 1 & 2 & 3 & 0 \\ 3 & 3 & 2 & 4 \\ 1 & 2 & 1 & 2 \\ 2 & 4 & 0 & 2 \end{bmatrix}.$$

1.26. (Revisit Exercise **1.25**) Find the inverse of each of the matrices given in Exercise **1.25** by using their LU decompositions.

An LU or LDU decomposition of a square matrix is sometimes useful for finding its inverse matrix, and $(LU)^{-1} = U^{-1}L^{-1}$.

1.27. Find the LDU decomposition of each of the following symmetric matrices.

$$(1)\ \begin{bmatrix} a & b \\ b & d \end{bmatrix},\ a \neq 0, \qquad (21\ \begin{bmatrix} 1 & 2 & 3 \\ 2 & 1 & 3 \\ 3 & 3 & 0 \end{bmatrix}.$$

Can you find any relation between L and U in your LDU decompositions?

1.28. *An LU decomposition of A is useful for solving a system* $A\mathbf{x} = \mathbf{b}$. Solve $A\mathbf{x} = \mathbf{b}$ when $A = LU$ is given, where

$$L = \begin{bmatrix} 1 & 0 & 0 \\ -1 & 1 & 0 \\ 0 & -1 & 1 \end{bmatrix}, \quad U = \begin{bmatrix} 1 & -1 & 0 \\ 0 & 1 & -1 \\ 0 & 0 & 1 \end{bmatrix}, \quad \mathbf{b} = \begin{bmatrix} 2 \\ -3 \\ 4 \end{bmatrix}.$$

Use a forward substitution for $L\mathbf{y} = \mathbf{b}$, and a backward substitution for $U\mathbf{x} = \mathbf{y}$.

1.29. Let $A = \begin{bmatrix} 1 & 1 & 1 \\ 1 & 4 & 5 \\ 1 & 4 & 7 \end{bmatrix}$ and $\mathbf{b} = \begin{bmatrix} 6 \\ 60 \\ 600 \end{bmatrix}$.

(1) Solve $A\mathbf{x} = \mathbf{b}$ by Gauss-Jordan elimination.
(2) Find an LU factorization of A.
(3) Write A as a product of elementary matrices.
(4) Find the inverse of A.

An LU or LDU decomposition of a square matrix is sometimes useful for finding its inverse matrix, and $(LU)^{-1} = U^{-1}L^{-1}$.

1.30. Compute $A^k, k = 1, 2, 3, \ldots$ for matrices

$$A = \begin{bmatrix} 0 & 1 \\ 0 & 0 \end{bmatrix}, \quad A = \begin{bmatrix} 0 & 1 & 0 \\ 0 & 0 & 1 \\ 0 & 0 & 0 \end{bmatrix}, \quad A = \begin{bmatrix} 0 & 1 & 0 & 0 \\ 0 & 0 & 1 & 0 \\ 0 & 0 & 0 & 1 \\ 0 & 0 & 0 & 0 \end{bmatrix}, \quad A = \begin{bmatrix} 0 & x & 0 & 0 \\ 0 & 0 & y & 0 \\ 0 & 0 & 0 & z \\ 0 & 0 & 0 & 0 \end{bmatrix}.$$

1.31. A square matrix A is said to be *nilpotent* if $A^k = \mathbf{0}$ for a positive integer k. Show that

(1) every invertible matrix cannot be nilpotent,
(2) all matrices in Exercise **1.30** are nilpotent, and more generally, every triangular matrix whose diagonal entries are all zero is nilpotent,
(3) every upper triangular matrix U can be written as a sum of a diagonal matrix D and a nilpotent matrix N; $U = D + N$, while the same is true for every lower triangular matrix.
(4) if A is a nilpotent with $A^k = \mathbf{0}, k \geq 2$, then $I - A$ is invertible with its inverse $I + A + \cdots + A^{k-1}$.

1.32. A square matrix A is said to be *idempotent* if $A^2 = A$.

(1) Find an example of an idempotent matrix other than $\mathbf{0}$ or I.
(2) For any idempotent A, show that $I - A$ is also idempotent.
(3) Show that, if a matrix A is both idempotent and invertible, then $A = I$.

1.33. Determine whether the following statements are true or false, in general, and justify your answers.

(1) For any (square or not,) matrix A, a system $A\mathbf{x} = \mathbf{0}$ of linear equations has at least one solution.
(2) A system $A\mathbf{x} = \mathbf{b}$ of linear equations has a solution if and only if $\mathbf{b} = \mathbf{0}$.
(3) Let A and B be row-equivalent square matrices. Then A is invertible if and only if B is invertible.
(4) For any two (square or not,) matrices A and B, if AB is the identity, then BA is also the identity.
(5) Every permutation matrix is invertible.
(6) Let A be a square matrix such that $AA = A$. Then A is the identity.
(7) If A and B are invertible matrices such that $A^2 = I$ and $B^2 = I$, then $(AB)^{-1} = BA$.
(8) If A and B are invertible matrices, $A + B$ is also invertible.
(9) $\begin{bmatrix} a & b & 0 \\ c & 0 & b \\ 0 & c & a \end{bmatrix}$ is symmetric and $\begin{bmatrix} a & b & 0 \\ c & 0 & -b \\ 0 & -c & -a \end{bmatrix}$ is skew-symmetric.
(10) If A, B, and AB are all symmetric, then $AB = BA$.
(11) The difference $A - B$ of any two symmetric matrices A and B is also symmetric.
(12) The product of any two skew-symmetric matrices is either symmetric or skew-symmetric.
(13) If A is invertible and symmetric, then A^{-1} is also symmetric.
(14) Let $AB^T = I$. Then A is invertible if and only if B is invertible.
(15) If a square matrix A is not invertible, then neither is AB for any B.
(16) If E_1 and E_2 are elementary matrices, then $E_1E_2 = E_2E_1$.
(17) The product of any two elementary matrices is again an elementary matrix.
(18) The product of any number of elementary matrices is invertible.
(19) The inverse of an invertible upper triangular matrix is also upper triangular.

(20) Any invertible matrix A can be written as $A = LU$, where L is lower triangular and U is upper triangular.

Hint: Exactly eleven of them are true.

Chapter 2

Determinants; The signed areas and the signed volumes

The *determinant* of a square matrix A is a certain scalar value that can be computed from the entries of A. The determinant of a square matrix A *determines* whether a system $A\mathbf{x} = \mathbf{b}$ has a unique solution or not. This depends completely on whether the determinant of A is zero or not.

2.1 Leibniz's letter to de l'Hôpital

In 1693, the German philosopher Gottfried Wilhelm (von) Leibniz wrote to the French mathematician Guillaume de l'Hôpital. [1, 2] He explained that

the system of equations

$$\begin{array}{ccccccc} 10 & + & 11x & + & 12y & = & 0 \\ 20 & + & 21x & + & 22y & = & 0 \\ 30 & + & 31x & + & 32y & = & 0 \end{array}$$

has a solution because

$$10.21.32 + 11.22.30 + 12.20.31 = 10.22.31 + 11.20.32 + 12.21.30.$$

This is exactly the condition that the coefficient matrix has determinant 0. In the letter, Leibniz is not using numerical coefficients a_{ij} but only two characters ij

[1]Reference: D.E. Smith, *A Source Book in Mathematics,* Dover Publ. Inc. New York, 1984.

[2]Guillaume Francois Antoine, Marquis de l'Hôpital (1661– February 2, 1704) was a French mathematician. He is famous for l'Hôpital's rule for calculating limits.

in his letter, the first i marking in which equation it occurs, the second j marking which letter it belongs to. For example, Leibniz uses two characters 21 in his letter for what we might write as the coefficient a_{21} nowadays, not for the number 21.

The determinant of a matrix arises through the study of systems of linear equations and matrices, as is shown in a historical remark above.

In geometry, determinants express the 'signed volumes' of high dimensional parallelepipeds. In the following section, we show how to express the 'signed areas' of parallelograms spanned by the row vectors of 2×2 matrices using determinants.

2.2 Areas of parallelograms and determinants

There are several formulae from ancient mathematics for computing the areas of two-dimensional objects such as a triangle, a rectangular, or a parallelogram. These are developed in elementary geometry.

A point in the space $\mathbb{R}^3$ with rectangular coordinate (x, y, z) is commonly denoted by

$$\text{a row vector } \mathbf{x} = [x\ y\ z] \quad \text{or} \quad \text{a column vector} \quad \mathbf{x} = \begin{bmatrix} x \\ y \\ z \end{bmatrix}.$$

As is familiar, we temporarily use the row vector notation $\mathbf{x} = [x\ y\ z]$ or $\mathbf{x} = (x, y, z)$ for a point in the space $\mathbb{R}^3$.

Let $\mathbf{a}, \mathbf{b}$ be any two vectors in the space $\mathbb{R}^3$ and let $\mathcal{P}(\mathbf{a}, \mathbf{b})$ be the parallelogram with the sides $\mathbf{a}, \mathbf{b}$ in the space $\mathbb{R}^3$. Since the area of a parallelogram is invariant under any translation and any rotation in the space $\mathbb{R}^3$, one can assume that the parallelogram $\mathcal{P}(\mathbf{a}, \mathbf{b})$ is placed in the xy-plane with the base point $\mathbf{0} = (0, 0) \in \mathbb{R}^2$:

$$\mathcal{P}(\mathbf{a}, \mathbf{b}) = \{s\,\mathbf{a} + t\,\mathbf{b} \,:\, 0 \le s, t \le 1\}.$$

The area of the parallelogram $\mathcal{P}(\mathbf{a}, \mathbf{b})$ is

$$Area(\mathcal{P}(\mathbf{a}, \mathbf{b})) = \|\mathbf{a}\| h, \quad \text{where} \quad h = \|\mathbf{b}\| \sin\theta$$

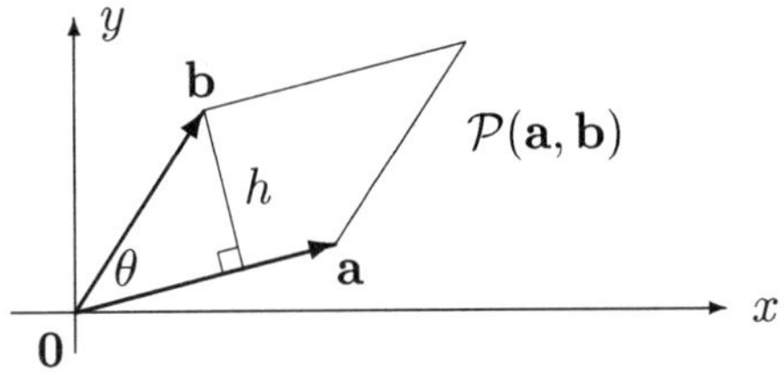

Figure 2.1: The parallelogram $\mathcal{P}(\mathbf{a}, \mathbf{b})$ with the sides $\mathbf{a}, \mathbf{b}$ in the xy-plane

and θ is the angle between $\mathbf{a}$ and $\mathbf{b}$. Here $\|\mathbf{v}\|$ represents the length of the vector $\mathbf{v}$. In fact, the height h is the (shortest) distance from the point $\mathbf{b}$ to the side line $\mathbf{a}$, see Figure 2.1.

The following lemma is immediate from an elementary geometry.

Lemma 2.1 **(1)** $Area(\mathcal{P}(\mathbf{a}, \mathbf{a})) = 0$ *for any* $\mathbf{a} \in \mathbb{R}^2$.

(2) $Area(\mathcal{P}(\mathbf{a}, \mathbf{b})) = 1$ *for the unit square* $\mathcal{P}(\mathbf{a}, \mathbf{b})$*, with* $\mathbf{a} = (1, 0), \mathbf{b} = (0, 1)$.

(3) $Area(\mathcal{P}(k\mathbf{a}, \mathbf{b})) = |k| Area(\mathcal{P}(\mathbf{a}, \mathbf{b}))$ *for any real number* k.

As shown in $Area(\mathcal{P}(k\mathbf{a}, \mathbf{b})) = |k| Area(\mathcal{P}(\mathbf{a}, \mathbf{b}))$, $Area(\mathcal{P}(\mathbf{a}, \mathbf{b}))$ is not *linear* in the first vector $\mathbf{a}$. This means that $Area(\mathcal{P}(k\mathbf{a}, \mathbf{b})) = k Area(\mathcal{P}(\mathbf{a}, \mathbf{b}))$ does *not* hold when $k < 0$ and

$$Area(\mathcal{P}(\mathbf{a}_1 + \mathbf{a}_2, \mathbf{b})) = Area(\mathcal{P}(\mathbf{a}_1, \mathbf{b})) + Area(\mathcal{P}(\mathbf{a}_2, \mathbf{b}))$$

does *not* hold for some $\mathbf{a}_1, \mathbf{a}_2$, (e.g., when $\mathbf{a}_2 = -\mathbf{a}_1$).

Let $\mathbf{a} = (a, b), \mathbf{b} = (c, d)$ be any two vectors in the plane $\mathbb{R}^2$. Then by using the *law of cosines*, one can show that

$$\|\mathbf{a}\| \|\mathbf{b}\| \cos\theta = \mathbf{a} \cdot \mathbf{b} = ac + bd,$$

where $\mathbf{a} \cdot \mathbf{b}$ is the 'dot product' of $\mathbf{a}$ and $\mathbf{b}$. It follows that

$$\begin{aligned} Area(\mathcal{P}(\mathbf{a}, \mathbf{b}))^2 &= (\|\mathbf{a}\| \|\mathbf{b}\| \sin\theta)^2 \\ &= \|\mathbf{a}\|^2 \|\mathbf{b}\|^2 (1 - \cos^2\theta) \\ &= \|\mathbf{a}\|^2 \|\mathbf{b}\|^2 - (\mathbf{a} \cdot \mathbf{b})^2 \\ &= (a^2 + b^2)(c^2 + d^2) - (ac + bd)^2 \\ &= (ad - bc)^2. \end{aligned}$$

Now for a 2×2 matrix A, its determinant is defined as follows.

Definition 2.2 For a 2×2 matrix $A = \begin{bmatrix} \mathbf{a} \\ \mathbf{b} \end{bmatrix} = \begin{bmatrix} a & b \\ c & d \end{bmatrix} \in M_{2\times 2}(\mathbb{R})$, the **determinant** of A is defined as

$$\det A = ad - bc,$$

where $\mathbf{a} = (a, b), \mathbf{b} = (c, d)$ in the plane $\mathbb{R}^2$.

From the definition of a determinant, one may expect to have $Area(\mathcal{P}(\mathbf{a}, \mathbf{b})) = \det \begin{bmatrix} \mathbf{a} \\ \mathbf{b} \end{bmatrix}$, but this is not true in general. For example,

- $Area(\mathcal{P}(\mathbf{a}, \mathbf{b})) = Area(\mathcal{P}(\mathbf{b}, \mathbf{a}))$, but $\det \begin{bmatrix} \mathbf{a} \\ \mathbf{b} \end{bmatrix} = -\det \begin{bmatrix} \mathbf{b} \\ \mathbf{a} \end{bmatrix}$.
- $Area(\mathcal{P}(\mathbf{a}, \mathbf{b}))$ is not *linear* in the first vector $\mathbf{a}$ (or in $\mathbf{b}$), but $\det \begin{bmatrix} \mathbf{a} \\ \mathbf{b} \end{bmatrix}$ is *linear* in the first vector $\mathbf{a}$ (also in the second vector $\mathbf{b}$). This means that by definition,

  $$\det \begin{bmatrix} k\mathbf{a} \\ \mathbf{b} \end{bmatrix} = k \det \begin{bmatrix} \mathbf{a} \\ \mathbf{b} \end{bmatrix}$$

  for every real number k, and

  $$\det \begin{bmatrix} \mathbf{a}_1 + \mathbf{a}_2 \\ \mathbf{b} \end{bmatrix} = \det \begin{bmatrix} \mathbf{a}_1 \\ \mathbf{b} \end{bmatrix} + \det \begin{bmatrix} \mathbf{a}_2 \\ \mathbf{b} \end{bmatrix}$$

  for any two vectors $\mathbf{a}_1, \mathbf{a}_2$.

However, it is a fact that

$$Area(\mathcal{P}(\mathbf{a}, \mathbf{b})) = \left| \det \begin{bmatrix} \mathbf{a} \\ \mathbf{b} \end{bmatrix} \right|,$$

which is always nonnegative. In this sense, we say that $\det \begin{bmatrix} \mathbf{a} \\ \mathbf{b} \end{bmatrix}$ is the *signed area* of $\mathcal{P}(\mathbf{a}, \mathbf{b})$. It means

$$Area(\mathcal{P}(\mathbf{a}, \mathbf{b})) = \begin{cases} \det \begin{bmatrix} \mathbf{a} \\ \mathbf{b} \end{bmatrix} & \text{if } \det \begin{bmatrix} \mathbf{a} \\ \mathbf{b} \end{bmatrix} \geq 0, \\ -\det \begin{bmatrix} \mathbf{a} \\ \mathbf{b} \end{bmatrix} & \text{if } \det \begin{bmatrix} \mathbf{a} \\ \mathbf{b} \end{bmatrix} \leq 0. \end{cases}$$

Moreover, as is shown on page 27 in Chapter 1,

$$\text{a } 2 \times 2 \text{ matrix } A = \begin{bmatrix} a & b \\ c & d \end{bmatrix} \text{ is invertible if and only if } \det A = ad - bc \neq 0.$$

Problem 26 In the plane $\mathbb{R}^2$, the three points $P = (1,2), Q = (-1,0), R = (0,-1)$ determine a unique triangle, say $\triangle PQR$. Find its area $Area(\triangle PQR)$.

In the space $\mathbb{R}^3$, the three points $P = (1,2,0), Q = (-1,0,1), R = (0,-1,1)$ determine a unique triangle in the space $\mathbb{R}^3$, say $\triangle PQR$. Try to find its area $Area(\triangle PQR)$.

Problem 27 Given a real valued (continuous) function [3] $f(x)$ defined on an interval $[a,b]$ of the real line, the definite integral $\int_a^b f(x)\,dx$ is defined informally as the 'signed area' of the region in the xy-plane that is bounded by the graph of f, the x-axis, and the vertical two lines $x = a$ and $x = b$. Here, the 'signed area' means that the area of the region equals

$$\begin{cases} \int_a^b f(x)\,dx & \text{if } f(x) \geq 0 \text{ on } [a,b], \\ -\int_a^b f(x)\,dx & \text{if } f(x) \leq 0 \text{ on } [a,b]. \end{cases}$$

Show that the integral operation $f \mapsto \int_a^b f(x)\,dx$ is *linear* on the set of continuous functions defined on a closed interval $[a,b]$. This means that

$$\int_a^b (kf + \ell g)(x)\,dx = k\int_a^b f(x)\,dx + \ell\int_a^b g(x)\,dx$$

for any scalars k, ℓ and for any two continuous functions $f(x)$ and $g(x)$.

2.3 Definition and properties of the determinant

In this section we introduce the notion of *determinant* as a real-valued function of higher order square matrices by certain axiomatic rules. It will be also shown that *a square matrix A is invertible if and only if its determinant* $\det A$ *is not zero.*

The formula $\det A = ad - bc$ in Definition 2.2 does not provide any clue of how to extend this formula of the determinant to higher order matrices. Hence, we first examine some further fundamental properties of the determinant function $\det A = ad - bc$.

[3] A *function* (*mapping* or *transformation*) f from a set X to a set Y is a rule that assigns to each element x in X exactly one element y in Y. We use the notation $f : X \to Y$ to mean that f is a function from X to Y, and write $f(x) = y$ or $f : x \mapsto y$ to indicate that f assigns to x in X an element y in Y. If $X = Y$, f is called a function on X.

By a direct computation, one can easily verify that the function det in Definition 2.2 satisfies the three properties in the following lemma.

Lemma 2.3 **(1)** $\det I_2 = \det \begin{bmatrix} 1 & 0 \\ 0 & 1 \end{bmatrix} = 1.$

(2) $\det \begin{bmatrix} c & d \\ a & b \end{bmatrix} = -\det \begin{bmatrix} a & b \\ c & d \end{bmatrix}.$

(3) $\det \begin{bmatrix} ka + \ell a' & kb + \ell b' \\ c & d \end{bmatrix} = k \det \begin{bmatrix} a & b \\ c & d \end{bmatrix} + \ell \det \begin{bmatrix} a' & b' \\ c & d \end{bmatrix}$ *for any real numbers k and ℓ.*

Proof: (1) and (2) are clear. For (3),

$$\begin{aligned} \det \begin{bmatrix} ka + \ell a' & kb + \ell b' \\ c & d \end{bmatrix} &= (ka + \ell a')d - (kb + \ell b')c \\ &= k(ad - bc) + \ell(a'd - b'c) \\ &= k \det \begin{bmatrix} a & b \\ c & d \end{bmatrix} + \ell \det \begin{bmatrix} a' & b' \\ c & d \end{bmatrix}. \end{aligned}$$ □

In Lemma 2.9 later, it will be shown that if a function $f : M_{2\times2}(\mathbb{R}) \to \mathbb{R}$ satisfies the properties (1)-(3) in Lemma 2.3, then it should be the function det defined in Definition 2.2, that is, $f(A) = ad - bc$.

The properties (1)-(3) in Lemma 2.3 of the determinant function on $M_{2\times2}(\mathbb{R})$ enable us to define the determinant function for square matrix of any order. [4]

Definition 2.4 A real-valued function $f : M_{n\times n}(\mathbb{R}) \to \mathbb{R}$ of all $n \times n$ square matrices is called a **determinant** if it satisfies the following three rules:

(R$_1$) The value of f of the identity matrix is 1, *i.e.*, $f(I_n) = 1$,
(R$_2$) the value of f changes sign if any two rows are interchanged, (alternating),

[4] An axiomatic definition of a determinant was used by Weierstrass in his lectures and, after his death, it was published in 1903 in the note "On determinant theory". In the same year Kronecker's lectures on determinants were also published, again after his death. With these two publications the modern theory of determinants was in place.

(R$_3$) f is *linear* in the first row, *i.e.*, for any real numbers k and ℓ, (linearity),

$$f\begin{bmatrix} k\mathbf{r}_1 + \ell\mathbf{r}'_1 \\ \mathbf{r}_2 \\ \vdots \\ \mathbf{r}_n \end{bmatrix} = kf\begin{bmatrix} \mathbf{r}_1 \\ \mathbf{r}_2 \\ \vdots \\ \mathbf{r}_n \end{bmatrix} + \ell f\begin{bmatrix} \mathbf{r}'_1 \\ \mathbf{r}_2 \\ \vdots \\ \mathbf{r}_n \end{bmatrix},$$

where $\mathbf{r}_i$'s denote the row vectors $[a_{i1} \ \cdots \ a_{in}]$ of a matrix.

As an example of the linearity rule **(R$_3$)**, one can say that

$$f\begin{bmatrix} a+x & b+y & c+z \\ a_{21} & a_{22} & a_{23} \\ a_{31} & a_{32} & a_{33} \end{bmatrix} = f\begin{bmatrix} a & b & c \\ a_{21} & a_{22} & a_{23} \\ a_{31} & a_{32} & a_{33} \end{bmatrix} + f\begin{bmatrix} x & y & z \\ a_{21} & a_{22} & a_{23} \\ a_{31} & a_{32} & a_{33} \end{bmatrix},$$

$$f\begin{bmatrix} 33x & 33y & 33z \\ a_{21} & a_{22} & a_{23} \\ a_{31} & a_{32} & a_{33} \end{bmatrix} = 33f\begin{bmatrix} x & y & z \\ a_{21} & a_{22} & a_{23} \\ a_{31} & a_{32} & a_{33} \end{bmatrix}.$$

It has been already shown that the det on 2×2 matrices satisfies the rules **(R$_1$)–(R$_3$)**. In the next section, it will be shown that for each positive integer n,

there always exists a function $f : M_{n\times n}(\mathbb{R}) \to \mathbb{R}$ *satisfying the rules* **(R$_1$)–(R$_3$)**

and such a function is unique (*existence and uniqueness*). Therefore, we say 'the' determinant and designate it as 'det' instead of the function f.

Let us first derive some direct consequences of the rules **(R$_1$)–(R$_3$)**.

Theorem 2.5 *The determinant satisfies the following properties.*

(1) *The determinant is linear in each row* (*multilinearity*).
(2) *If A has either a zero row or two identical rows, then* $\det A = 0$ (*alternating*).
(3) *The (type III) elementary row operation that adds a constant multiple of one row to another row leaves the determinant unchanged.*

Proof: (1) Any row can be placed in the first row by interchanging it with the first row, with a change of sign in the determinant by the rule **(R$_2$)**, and next using the linearity rule **(R$_3$)**, and finally **(R$_2$)** once more by interchanging the same two rows.

(2) If A has a zero row, then this row is zero times the zero row, $0\mathbf{0} = \mathbf{0}$, so that $\det A = 0 \det A = 0$ by (1). If A has two identical rows, then interchanging

those two identical rows does not change the matrix itself but $\det A = -\det A$ by the rule **(R$_2$)**, so that $\det A = 0$.

(3) By a direct computation using the linearity in (1), one can get

$$\det \begin{bmatrix} \vdots \\ \mathbf{r}_i + k\mathbf{r}_j \\ \vdots \\ \mathbf{r}_j \\ \vdots \end{bmatrix} = \det \begin{bmatrix} \vdots \\ \mathbf{r}_i \\ \vdots \\ \mathbf{r}_j \\ \vdots \end{bmatrix} + k \det \begin{bmatrix} \vdots \\ \mathbf{r}_j \\ \vdots \\ \mathbf{r}_j \\ \vdots \end{bmatrix},$$

in which the second term on the right-hand side is zero by (2). □

The rule **(R$_2$)** of the determinant function is said to be the **alternating** property, and the property (1) in Theorem 2.5 is said to be **multilinearity** or **n-linearity**.

Now, one can see how the elementary row operations affect the determinant.

The type I elementary row operation that 'multiplies a row by a nonzero k' changes the determinant to k times the determinant, by Theorem 2.5(1). The rule **(R$_2$)** says that if 'two rows are interchanged', then the sign of the determinant is changed. The type III elementary row operation that 'adds a constant multiple of a row to another' leaves the determinant unchanged, as shown in Theorem 2.5(3).

Consequently, one has the following lemma.

Lemma 2.6 *Let E be any elementary matrix. Then,*

(1) $\det E \neq 0$.
(2) $\det(EA) = \det E \ \det A$ *for any square matrix A if EA is well defined. Consequently, for any LU decomposition $A = LU$ of A,* $\det A = \det U$.
(3) $\det P = \pm 1$ *for any permutation matrix P by the rule* **(R$_2$)**.
(4) $\det(PA) = \det P \ \det A = \pm \det A$ *for any permutation matrix P and for any square matrix A if PA is well defined.*

As a consequence, if two matrices A and B are row-equivalent, then $\det A = k \det B$ *for some nonzero number k, which means either* $\det A = 0 = \det B$ *or* $\det A \neq 0 \neq \det B$.

Example 24 Consider a matrix

$$A = \begin{bmatrix} 1 & 1 & 1 \\ a & b & c \\ b+c & c+a & b+a \end{bmatrix}.$$

If one adds the second row to the third, then the third row becomes

$$[a+b+c \quad a+b+c \quad a+b+c],$$

which is a scalar multiple of the first row. Thus, $\det A = 0$. □

Example 25 ($\det E = \det E^T$ *for any elementary matrix* E) There are three different types of elementary matrices corresponding to the types of the elementary row operations.

(Type I): $\begin{bmatrix} 1 & 0 & 0 \\ 0 & k & 0 \\ 0 & 0 & 1 \end{bmatrix}$, the second row of I_3 is multiplied by $k \neq 0$.

For such an elementary matrix E, $\det E = k$ and $E^T = E$.

(Type II): $\begin{bmatrix} 1 & 0 & 0 & 0 \\ 0 & 0 & 0 & 1 \\ 0 & 0 & 1 & 0 \\ 0 & 1 & 0 & 0 \end{bmatrix}$, the second and the fourth rows of I_4 are interchanged.

For such an elementary matrix E, $\det E = -1$ and $E^T = E$.

(Type III): $\begin{bmatrix} 1 & 0 & 3 \\ 0 & 1 & 0 \\ 0 & 0 & 1 \end{bmatrix}$, 3 times the third row is added to the first row of I_3.

For such an elementary matrix E, $E^T \neq E$ but $\det E = 1 = \det E^T$.

Consequently, in any case $\det E = \det E^T$ and it is not zero. □

Problem 28 Show that, for an $n \times n$ matrix A and $k \in \mathbb{R}$, $\det(kA) = k^n \det A$.

Problem 29 Using only the properties of the determinant, explain why $\det A = 0$ for

$$(1)\, A = \begin{bmatrix} 1 & 4 & 7 \\ 2 & 5 & 8 \\ 3 & 6 & 9 \end{bmatrix}, \quad (2)\, A = \begin{bmatrix} a+1 & a+4 & a+7 \\ a+2 & a+5 & a+8 \\ a+3 & a+6 & a+9 \end{bmatrix}, \quad (3)\, A = \begin{bmatrix} a & a^4 & a^7 \\ a^2 & a^5 & a^8 \\ a^3 & a^6 & a^9 \end{bmatrix}.$$

Can you do it in your brain without a pencil?

Further properties of the determinant are stated in the following theorem.

Theorem 2.7 **(1)** *The determinant of a triangular matrix is the product of the diagonal entries.*

(2) *The matrix A is invertible if and only if* $\det A \neq 0$.

(3) *For any two $n \times n$ matrices A and B,* $\det(AB) = \det A \det B$.

(4) $\det A^T = \det A$.

Proof: (1) If A is a diagonal matrix, then clearly $\det A = a_{11} \cdots a_{nn}$ by the multilinearity in Theorem 2.5(1) and rule **($\mathbf{R}_1$)**.

Suppose that A is a lower triangular matrix. If A has a zero diagonal entry, then a forward elimination, which does not change the determinant, produces a zero row, so that $\det A = 0$. If A does not have any zero diagonal entry, forward elimination reduces the lower triangular matrix A to the diagonal matrix D but leaves the diagonal entries a_{ii}'s unchanged. Hence, $\det A = \det D = a_{11} \cdots a_{nn}$.

Similar arguments can be applied to an upper triangular matrix.

(2) From Corollary 1.22.1, it is known that a triangular matrix is invertible if and only if it has no zero diagonal entries. Hence, if A is a diagonal matrix, (2) is clear because of (1).

Now, let A be any square matrix and let U be the reduced row-echelon form of A. Then U is diagonal and hence (2) is true for the matrix U. Moreover, A and U are row-equivalent, which gives either $\det A = 0 = \det U$ or $\det A \neq 0 \neq \det U$, by Lemma 2.6. Also as known already, the matrix A is invertible if and only if the matrix U is invertible. Consequently, (2) holds for any square matrix A; saying that the matrix A is invertible if and only if $\det A \neq 0$.

(3) If A is not invertible, neither is AB. It gives $\det(AB) = 0 = \det A \det B$ by (2).

If A is invertible, it can be expressed as a product of elementary matrices by Theorem 1.22, say $A = E_1 E_2 \cdots E_k$. Then by using Lemma 2.6(2) repeatedly,

$$\begin{aligned} \det(AB) &= \det(E_1 E_2 \cdots E_k B) \\ &= \det E_1 \ \det E_2 \ \cdots \ \det E_k \ \det B \\ &= \det(E_1 E_2 \cdots E_k) \det B \\ &= \det A \det B. \end{aligned}$$

(4) If A is not invertible, neither is A^T. So, $\det A^T = 0 = \det A$ by (2).

If A is invertible, then write it again as a product of elementary matrices, say $A = E_1 E_2 \cdots E_k$. However, $\det E = \det E^T$ for any elementary matrix E as shown in Example 25. Hence, by (3),

$$\det A^T = \det E_k^T \cdots \det E_2^T \ \det E_1^T = \det E_k \cdots \det E_2 \ \det E_1 = \det A. \ \square$$

Remark: From the equality $\det A = \det A^T$, one could define the determinant in terms of columns instead of rows in Definition 2.4, and Theorem 2.5 is also true with 'columns' instead of 'rows'. That is,

(1) *the determinant is linear in each column.*

(2) *If A has either a zero column or two identical columns, then* $\det A = 0$.

(3) *The (type III) elementary column operation that adds a constant multiple of one column to another column leaves the determinant unchanged.*

Example 26 (*Computing* $\det A$ *by using a forward elimination*) Compute the determinant of

$$A = \begin{bmatrix} 1 & 0 & 1 & 1 \\ 2 & 4 & 2 & 1 \\ 3 & 0 & 6 & 0 \\ 4 & 8 & 4 & 0 \end{bmatrix}.$$

Solution: By using forward elimination, A can be transformed to an upper triangular matrix U. Since the forward elimination does not change the determinant,

$$\det A \;=\; \det U = \det \begin{bmatrix} 1 & 0 & 1 & 1 \\ 0 & 4 & 0 & -1 \\ 0 & 0 & 3 & -3 \\ 0 & 0 & 0 & -2 \end{bmatrix} = 1 \cdot 4 \cdot 3 \cdot (-2) = -24.$$

$\square$

Problem 30 Show that if A is invertible, then $\det A^{-1} = 1/\det A$.
Hint: Use Theorem 2.7(3).

Problem 31 Compute the determinant of each of the following matrices.

$$(1)\ A = \begin{bmatrix} 1 & 0 & 2 \\ 0 & 2 & 4 \\ 2 & 4 & x \end{bmatrix}, \quad (2)\ B = \begin{bmatrix} 0 & 4 & 2 \\ -4 & 0 & 1 \\ -2 & -1 & 0 \end{bmatrix}, \quad (3)\ C = \begin{bmatrix} 11 & 12 & 13 \\ 21 & 22 & 23 \\ 31 & 32 & 33 \end{bmatrix}.$$

Hint: A is symmetric and B is a skew-symmetric matrix.

2.4 The determinant function det is well defined

Now, we show that a determinant function $f : M_{n\times n}(\mathbb{R}) \to \mathbb{R}$ in Definition 2.4 is well defined by proving the existence and uniqueness of such a function f.

Theorem 2.8 *For any natural number n,*

(1) (**Existence**) *there exists a function $f : M_{n\times n}(\mathbb{R}) \to \mathbb{R}$ which satisfies the three rules $(\mathbf{R}_1)$–$(\mathbf{R}_3)$ in Definition 2.4.*

(2) (**Uniqueness**) *Such a function is unique.*

Existence and Uniqueness for 1×1 matrices: It is clear with $\det[a] = a$.

Existence and Uniqueness for 2×2 matrices: When $n = 2$, the existence comes from Lemma 2.3. The next lemma shows the uniqueness of the determinant function on $M_{2\times 2}(\mathbb{R})$.

Lemma 2.9 (Uniqueness) *If a function $f : M_{2\times 2}(\mathbb{R}) \to \mathbb{R}$ satisfies the rules $(\mathbf{R}_1)$–$(\mathbf{R}_3)$, then $f\begin{bmatrix} a & b \\ c & d \end{bmatrix} = ad - bc$. That is, $f(A) = \det A$.*

Proof: First, $f\begin{bmatrix} 0 & 1 \\ 1 & 0 \end{bmatrix} = -1$ by the rules $(\mathbf{R}_1)$ and $(\mathbf{R}_2)$.

$$
\begin{aligned}
f(A) &= f\begin{bmatrix} a & b \\ c & d \end{bmatrix} \\
&= f\begin{bmatrix} a+0 & 0+b \\ c & d \end{bmatrix} \\
&= f\begin{bmatrix} a & 0 \\ c & d \end{bmatrix} + f\begin{bmatrix} 0 & b \\ c & d \end{bmatrix} \\
&= f\begin{bmatrix} a & 0 \\ 0 & d \end{bmatrix} + f\begin{bmatrix} a & 0 \\ c & 0 \end{bmatrix} + f\begin{bmatrix} 0 & b \\ 0 & d \end{bmatrix} + f\begin{bmatrix} 0 & b \\ c & 0 \end{bmatrix} \\
&= ad + 0 + 0 - bc = ad - bc,
\end{aligned}
$$

where the third and fourth equalities come from the multilinearity in rows and the fifth equality comes from the statement on columns in the Remark on page 67. □

Existence and Uniqueness for 3×3 matrices: *There exists a function $f : M_{3\times 3}(\mathbb{R}) \to \mathbb{R}$ which satisfies the rules $(\mathbf{R}_1)$–$(\mathbf{R}_3)$, and such a function is unique.*

This can be proved by repeating a similar procedure as in the case of $n = 2$. In fact, the next process to drive an *explicit value* of $\det A$ for a 3×3 matrix A gives the existence and uniqueness, and then one can write $f(A) = \det A$ for any 3×3 matrix A.

For a 3×3 matrix $A = [a_{ij}]$,

$$
\begin{aligned}
& f\begin{bmatrix} a_{11} & a_{12} & a_{13} \\ a_{21} & a_{22} & a_{23} \\ a_{31} & a_{32} & a_{33} \end{bmatrix} \\
&= f\begin{bmatrix} a_{11} & 0 & 0 \\ 0 & a_{22} & 0 \\ 0 & 0 & a_{33} \end{bmatrix} + f\begin{bmatrix} 0 & a_{12} & 0 \\ 0 & 0 & a_{23} \\ a_{31} & 0 & 0 \end{bmatrix} + f\begin{bmatrix} 0 & 0 & a_{13} \\ a_{21} & 0 & 0 \\ 0 & a_{32} & 0 \end{bmatrix} \\
&\quad + f\begin{bmatrix} a_{11} & 0 & 0 \\ 0 & 0 & a_{23} \\ 0 & a_{32} & 0 \end{bmatrix} + f\begin{bmatrix} 0 & a_{12} & 0 \\ a_{21} & 0 & 0 \\ 0 & 0 & a_{33} \end{bmatrix} + f\begin{bmatrix} 0 & 0 & a_{13} \\ 0 & a_{22} & 0 \\ a_{31} & 0 & 0 \end{bmatrix} \\
&= a_{11}a_{22}a_{33} + a_{12}a_{23}a_{31} + a_{13}a_{21}a_{32} \\
&\qquad - a_{11}a_{23}a_{32} - a_{12}a_{21}a_{33} - a_{13}a_{22}a_{31}. \qquad (2.1)
\end{aligned}
$$

The first equality comes from the multilinearity in the three rows stated in Theorem 2.5(1). Indeed, by applying the linearity in the first row with

$$
\begin{bmatrix} a_{11} & a_{12} & a_{13} \end{bmatrix} = \begin{bmatrix} a_{11} & 0 & 0 \end{bmatrix} + \begin{bmatrix} 0 & a_{12} & 0 \end{bmatrix} + \begin{bmatrix} 0 & 0 & a_{13} \end{bmatrix},
$$

$\det A$ becomes the sum of determinants of three matrices. Observe that, in each of those three matrices, the first row has just one entry from a_{1j} and two others are zero. Subsequently, by applying the linearity in the second and third rows of each of the three matrices, one gets the sum of the determinants of $3^3 = 27$ matrices, each of which has exactly three entries from a_{ij}, one in each of three rows, and all the other entries are zero.

In each of those 27 matrices, if any two of the three entries of a_{ij} are in the same column, then the matrix contains a zero column so that its determinant is zero.

Consequently, the determinants of only six matrices are left to get the first equality.

The second equality is just the computation of the six determinants by using the alternating rule $(\mathbf{R}_2)$ for row or column interchange and Theorem 2.7(1). In fact, in each of those six matrices, no two entries from a_{ij} are in the same row or in the same column, so one can take suitable 'column interchanges' to convert it into a diagonal matrix. Thus the determinant of each of them is just the product of the three entries with $\pm$ sign which is determined by the number of column interchanges.

Sarrus' Method: The explicit formula Eq. (2.1) for the determinant of a 3×3 matrix can easily be memorized by the following scheme. Copy the first two columns and put them on the right of the matrix, and compute the determinant by multiplying entries on six diagonals with a $+$ sign or a $-$ sign as in Figure 2.2. This is known as Sarrus' method for 3×3 matrices. It has no analogue for matrices of higher order $n \geq 4$.

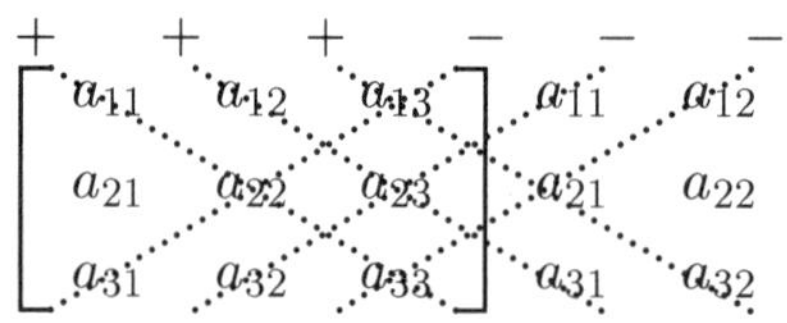

Figure 2.2: Sarrus' method

In summary, the explicit formula Eq. (2.1) for $f(A)$ of a 3×3 matrix A satisfies the three rules $(\mathbf{R}_1)$–$(\mathbf{R}_3)$ (see Problem 32), which proves the existence of $f(A)$ when $n = 3$. Moreover, the computation of the explicit formula Eq. (2.1) for $f(A)$ shows the uniqueness of $f(A)$.

It proves Theorem 2.8 and $f(A) = \det A$ when $n = 3$. □

Problem 32 Show that the explicit formula Eq. (2.1) of the determinant for 3×3 matrices satisfies the three rules $(\mathbf{R}_1)$–$(\mathbf{R}_3)$.

Problem 33 Use Sarrus' method to compute the determinants of

$$(1) \begin{bmatrix} 1 & 2 & 3 \\ 2 & 3 & 2 \\ 3 & 1 & 1 \end{bmatrix}, \quad (2) \begin{bmatrix} x & y & z \\ 2 & 1 & 1 \\ -2 & 2 & 3 \end{bmatrix}, \quad (3) \begin{bmatrix} x-4 & -2 & 2 \\ 1 & x-3 & -1 \\ -2 & 6 & x-4 \end{bmatrix}.$$

Existence and Uniqueness for matrices of higher order $n > 3$: By repeating a similar procedure for $\det A$ of any matrix $A = [a_{ij}]$ of order $n > 3$, one can get an explicit formula of $\det A$. This can be roughly sketched as follow.

(Step 1) Use the multilinearity in each row of A to get $\det A$ as the sum of the determinants of n^n matrices and neglect the determinants of matrices which have a zero column in the summation. Then, in remain terms, no two of the n entries from a_{ij} are in the same row or in the same column.

(Step 2) In each remain term, those n entries from a_{ij} are of the form

$$a_{1i}, a_{2j}, a_{3k}, \ldots, a_{n\ell}$$

for some column indices i, j, k, ..., ℓ. Since no two of these n entries are in the same column, the column indices $i, j, k, \ldots, \ell$ in a linear order is just a permutation of 1, 2, ..., n and so only $n!$ matrices remain for further consideration.

(Step 3) Each of the $n!$ remaining matrices can be converted into a diagonal matrix by 'suitable' column interchanges. Hence, its determinant will be of the form

$$\pm a_{1i}a_{2j}a_{3k}\cdots a_{n\ell}.$$

Here the sign '$\pm$' will be determined by the number of the column interchanges. To determine the sign, let us once again look back to the case of $n = 3$.

Example 27 (*Converting into a diagonal matrix by column interchanges*) Suppose that one of the six matrices is of the form

$$\begin{bmatrix} 0 & 0 & a_{13} \\ 0 & a_{22} & 0 \\ a_{31} & 0 & 0 \end{bmatrix}.$$

Then, one can convert it into a diagonal matrix by interchanging the first and the third columns. That is,

$$\det \begin{bmatrix} 0 & 0 & a_{13} \\ 0 & a_{22} & 0 \\ a_{31} & 0 & 0 \end{bmatrix} = -\det \begin{bmatrix} a_{13} & 0 & 0 \\ 0 & a_{22} & 0 \\ 0 & 0 & a_{31} \end{bmatrix} = -a_{13}a_{22}a_{31}.$$

A column interchange is the same as an interchange of the corresponding column indices. Moreover, in order to convert the given matrix having the entries a_{13}, a_{22}, a_{31} into a diagonal matrix, one has to convert the column indices $3, 2, 1$ to the standard order $1, 2, 3$.

Clearly, this can be done by only one time interchange of indices 3 and 1. However, there are many other different ways of converting the indices $3, 2, 1$ to $1, 2, 3$. For example, it can be done by using three interchanges of indices: 3 and 2, 3 and 1, and then 2 and 1. In either case, the parity is odd so that the "$-$" sign in the computation of the determinant comes from the equality $(-1)^1 = (-1)^3$, where the exponents mean the numbers of interchanges of the column indices. □

To drive an explicit formula for $\det A$ of any $n \times n$ matrix A, we redefine the terminology *permutation* as a function for an ordered rearrangement of n objects.

Definition 2.10 A **permutation** of n objects is a one-to-one function [5] from the set of n objects onto itself. [6]

In most cases, we use the set of integers $N_n = \{1, 2, \ldots, n\}$ for a set of n objects. A permutation σ of N_n assigns a number $\sigma(i)$ in N_n to each number i in N_n. We adopt a notation for a permutation σ denoted by

$$\sigma = \langle \sigma(1),\ \sigma(2), \ldots, \sigma(n) \rangle, \quad \text{or} \quad \sigma = \begin{pmatrix} 1 & 2 & \cdots & n \\ \sigma(1) & \sigma(2) & \cdots & \sigma(n) \end{pmatrix}.$$

Here, the first row is the usual lay-out of N_n as the domain set, and the second row is the image set showing a rearrangement of the numbers in N_n in a linear order without repetitions or omissions.

Let S_n denote the set of all permutations of N_n. As mentioned already, one can see that S_n has exactly $n!$ permutations. For example, S_2 has $2 = 2!$, S_3 has $6 = 3!$, and S_4 has $24 = 4!$ permutations.

Definition 2.11 A permutation $\sigma = \langle j_1, j_2, \ldots, j_n \rangle$ is said to have an **inversion** if $j_s > j_t$ for $s < t$, that is, a larger number precedes a smaller number.

For example, the permutation $\sigma = \langle 3, 1, 5, 4, 2 \rangle$ has five inversions, since 3 precedes 1 and 2; 5 precedes 4 and 2; and 4 precedes 2. Clearly, the identity $\langle 1, 2, \ldots, n \rangle$ is the only permutation which has no inversions.

Definition 2.12 A permutation is said to be **even** if it has an even number of inversions, and it is said to be **odd** if it has an odd number of inversions. For a permutation σ in S_n, the **sign** of σ is defined as

$$\operatorname{sign}(\sigma) = \left\{ \begin{array}{rl} 1 & \text{if } \sigma \text{ is an even permutation} \\ -1 & \text{if } \sigma \text{ is an odd permutation} \end{array} \right\} = (-1)^k,$$

where k is the number of inversions of σ.

[5] A function $f : X \to Y$ is *one-to-one* (or *injective*) if for any two elements x_1, x_2 in X, $f(x_1) = f(x_2)$ implies $x_1 = x_2$. Or equivalently, $x_1 \neq x_2$ in X implies $f(x_1) \neq f(x_2)$ in Y.

[6] The concept of a permutation admits a generalization that is actually not a permutation in most cases. For $0 \leq k \leq n$, a k-**permutation of** n is an ordered rearrangement of a k-element subset of n objects. The number of such k-permutations of n is denoted by $P(n, k)$ which is given by

$$P(n,k) = \frac{n!}{(n-k)!} = n \cdot (n-1) \cdot (n-2) \cdots (n-k+1).$$

For example, for $\sigma = \langle 3, 1, 5, 4, 2 \rangle$, $\mathrm{sign}(\sigma) = (-1)^5 = -1$ and σ is an odd permutation.

When $n = 3$, the permutations $\langle 1, 2, 3\rangle$, $\langle 2, 3, 1\rangle$, and $\langle 3, 1, 2\rangle$ are even, while the permutations $\langle 1, 3, 2\rangle$, $\langle 2, 1, 3\rangle$, and $\langle 3, 2, 1\rangle$ are odd. All of these six permutations form the set S_3.

In general, one can convert a permutation $\sigma = \langle \sigma(1), \sigma(2), \ldots, \sigma(n)\rangle$ in S_n into the identity permutation $\langle 1, 2, \ldots, n\rangle$ by transposing each inversion of σ. However, the number of necessary transpositions to convert the given permutation into the identity permutation is not unique as is shown in Example 27.

An interesting fact is that, *even though the number of necessary transpositions is not unique, the parity (even or odd) is always the same as that of the number of inversions.* (This may not be clear and readers are suggested to convince themselves with couples of examples. Also, one can refer to some abstract algebra books.)

We now go back to Step 3 to compute the determinants of those remaining $n!$ matrices. Each of them has n entries of the form $a_{1\sigma(1)}, a_{2\sigma(2)}, \ldots, a_{n\sigma(n)}$ for a permutation $\sigma \in S_n$. Moreover, this can be converted to a diagonal matrix by column interchanges corresponding to the inversions in the permutation $\sigma = \langle \sigma(1), \sigma(2), \ldots, \sigma(n)\rangle$. Hence, its determinant is equal to

$$\mathrm{sign}(\sigma) a_{1\sigma(1)} a_{2\sigma(2)} \cdots a_{n\sigma(n)}.$$

This is called a **signed elementary product** of A.

It gives It gives the following.

Theorem 2.13 (Leibniz formula) [7] *For an $n \times n$ matrix A,*

$$\det A = \sum_{\sigma \in S_n} \mathrm{sign}(\sigma) a_{1\sigma(1)} a_{2\sigma(2)} \cdots a_{n\sigma(n)}. \tag{2.2}$$

That is,

det A *is the sum of all signed elementary products of* A.

[7] **Gottfried Wilhelm (von) Leibniz** (July 1, 1646 – November 14, 1716) was a German polymath and philosopher who occupies a prominent place in the history of mathematics and the history of philosophy, having developed differential and integral calculus independently of Isaac Newton. Leibniz formula for π is $1 - \frac{1}{3} + \frac{1}{5} - \frac{1}{7} + \frac{1}{9} - \cdots = \frac{\pi}{4}$.

This shows that the determinant should be unique if there exists. On the other hand, one can show that the Leibniz formula for $\det A$ in Theorem 2.13 satisfies the three rules $(\mathbf{R}_1)$–$(\mathbf{R}_3)$.

It gives both the *existence* and the *uniqueness* for the determinant function of square matrices of any order $n \geq 1$, which proves Theorem 2.8.

It is not hard to see that the number of even permutations is equal to that of odd permutations in S_n, so it is $n!/2$. See Problem 36. In the case $n = 3$, one notices that there are three terms with $+$ sign and three terms with $-$ sign in the Leibniz formula for $\det A$.

Problem 34 By column interchanges, convert each matrix into a diagonal matrix and then compute its determinant.

$$A = \begin{bmatrix} 0 & a_{12} & 0 & 0 \\ 0 & 0 & 0 & a_{24} \\ a_{31} & 0 & 0 & 0 \\ 0 & 0 & a_{43} & 0 \end{bmatrix}, \quad B = \begin{bmatrix} 0 & 0 & x & 0 & 0 \\ 0 & y & 0 & 0 & 0 \\ z & 0 & 0 & 0 & 0 \\ 0 & 0 & 0 & 0 & k \\ 0 & 0 & 0 & \ell & 0 \end{bmatrix}.$$

Problem 35 Let $A = [\mathbf{c}_1 \ \cdots \ \mathbf{c}_n]$ be an $n \times n$ matrix with the column vectors $\mathbf{c}_j$'s. Show that $\det[\mathbf{c}_j \ \mathbf{c}_1 \ \cdots \ \mathbf{c}_{j-1} \ \mathbf{c}_{j+1} \ \cdots \ \mathbf{c}_n] = (-1)^{j-1} \det[\mathbf{c}_1 \ \cdots \ \mathbf{c}_j \ \cdots \ \mathbf{c}_n] = (-1)^{j-1} \det A$. Also, the same kind of equality holds when A is written in row vectors.

Hint: Count the number of inversions.

Problem 36 Show that the number of even permutations and the number of odd permutations in S_n are equal.

Hint: Fix a transposition σ in S_n, say $\sigma = \langle \mathbf{2}, \mathbf{1}, 3, 4, \ldots, n \rangle$. Then the composition of σ with an even (odd, respectively) permutation in S_n is an odd (even, respectively) permutation.

2.5 Cofactor expansions make it easy to compute the determinant

Even if one has the Leibniz formula explicitly for $\det A$, it is not quite useful in computing because one has to sum up $n!$ terms and $n!$ grows much faster than n when n grows large; even for a 5×5 matrix, one needs $5! = 120$ times of summations.

Fortunately, one can reformulate the formula by writing it in an inductive way, by which the summing time can be reduced in most cases.

First, look at the case of $n = 3$ with Eq. (2.1) for $\det A$:

$$\begin{aligned}
&\det \begin{bmatrix} a_{11} & a_{12} & a_{13} \\ a_{21} & a_{22} & a_{23} \\ a_{31} & a_{32} & a_{33} \end{bmatrix} \\
&= a_{11}a_{22}a_{33} + a_{12}a_{23}a_{31} + a_{13}a_{21}a_{32} - a_{11}a_{23}a_{32} - a_{12}a_{21}a_{33} - a_{13}a_{22}a_{31} \\
&= a_{11}(a_{22}a_{33} - a_{23}a_{32}) - a_{12}(a_{21}a_{33} - a_{23}a_{31}) + a_{13}(a_{21}a_{32} - a_{22}a_{31}) \\
&= a_{11} \det \begin{bmatrix} a_{22} & a_{23} \\ a_{32} & a_{33} \end{bmatrix} - a_{12} \det \begin{bmatrix} a_{21} & a_{23} \\ a_{31} & a_{33} \end{bmatrix} + a_{13} \det \begin{bmatrix} a_{21} & a_{22} \\ a_{31} & a_{32} \end{bmatrix}. \qquad (2.3)
\end{aligned}$$

This can be written as $a_{11}A_{11} + a_{12}A_{12} + a_{13}A_{13}$ with

$$A_{11} = \det \begin{bmatrix} a_{22} & a_{23} \\ a_{32} & a_{33} \end{bmatrix}, \quad A_{12} = -\det \begin{bmatrix} a_{21} & a_{23} \\ a_{31} & a_{33} \end{bmatrix}, \quad A_{13} = \det \begin{bmatrix} a_{21} & a_{22} \\ a_{31} & a_{32} \end{bmatrix}.$$

For a generalization to the case of an $n \times n$ matrix A, look at again the Leibniz formula for $\det A$.

$$\det A = \sum_{\sigma \in S_n} \operatorname{sign}(\sigma) a_{1\sigma(1)} a_{2\sigma(2)} \cdots a_{n\sigma(n)}.$$

Here, the first factor $a_{1\sigma(1)}$ in each of the $n!$ terms of the signed elementary products is one of $a_{11}, a_{12}, \ldots, a_{1n}$ in the first row of A. Hence, one can divide the $n!$ terms of the expansion of $\det A$ into n groups according to the value of $\sigma(1)$: Say,

$$\begin{aligned}
\det A &= \sum_{\sigma \in S_n} \operatorname{sign}(\sigma) a_{1\sigma(1)} a_{2\sigma(2)} \cdots a_{n\sigma(n)} \\
&= a_{11}A_{11} + a_{12}A_{12} + \cdots + a_{1n}A_{1n},
\end{aligned}$$

where, for $j = 1, 2, \ldots, n$, the number A_{1j} is

$$A_{1j} = \sum_{\sigma \in S_n, \sigma(1) = j} \operatorname{sign}(\sigma) a_{2\sigma(2)} \cdots a_{n\sigma(n)}.$$

The number A_{1j} can be shown to be $(-1)^{j-1}$ times the determinant of the submatrix of A obtained by deleting the first row and the j-th column. We denote this submatrix by M_{1j}. (See the case of $n = 3$, or use the Leibniz formula for $\det M_{1j}$).

As a result, one can get an expansion of $\det A$ with respect to the first row:

$$\begin{aligned}\det A &= a_{11}A_{11} + a_{12}A_{12} + \cdots + a_{1n}A_{1n}\\ &= a_{11}\det M_{11} + (-1)^1 a_{12}\det M_{12} + \cdots + (-1)^{n-1}a_{1n}\det M_{1n}.\end{aligned}$$

This is called the **cofactor expansion** of $\det A$ along the first row. [8]

There is a similar expansion with respect to any other row, say the i-th row. To show this, first construct a new matrix C from A by moving the i-th row of A up to the first row by interchanges with its preceding $i-1$ rows one by one. Then $\det A = (-1)^{i-1}\det C$ by the rule **(R$_2$)**.

Now, the cofactor expansion of $\det C$ along its first row $[a_{i1} \ \cdots \ a_{in}]$ can be written as

$$\det C = a_{i1}C_{11} + a_{i2}C_{12} + \cdots + a_{in}C_{1n}.$$

Here, C_{1j} is $(-1)^{j-1}$ times the determinant of the submatrix of C obtained from C by deleting the first row and the j-th column. But, the submatrix of C obtained from C by deleting the first row and the j-th column is just the submatrix M_{ij} of A obtained from A by deleting the i-th row and the j-th column, by the construction of the matrix C. That is, $C_{1j} = (-1)^{j-1}\det M_{ij}$.

Hence, we have

$$\begin{aligned}\det A &= (-1)^{i-1}\det C\\ &= (-1)^{i-1}(a_{i1}C_{11} + a_{i2}C_{12} + \cdots + a_{in}C_{1n})\\ &= (-1)^{i-1}\left(a_{i1}\det M_{i1} - a_{i2}\det M_{i2} + \cdots + (-1)^{n-1}a_{in}\det M_{in}\right)\\ &= a_{i1}A_{i1} + a_{i2}A_{i2} + \cdots + a_{in}A_{in},\end{aligned}$$

where

$$A_{ij} = (-1)^{i+j}\det M_{ij}.$$

The submatrix M_{ij} of A is called the **minor** of the entry a_{ij} and the number

$$A_{ij} = (-1)^{i+j}\det M_{ij}$$

is called the **cofactor** of the entry a_{ij}.

Also, one can do the same along the columns because $\det A^T = \det A$. This gives the following theorem.

[8] Similar looking notation: Use $[A]_{ij}$ for the (i, j)-entry of a matrix A; A_{ij} for a block of a block matrix A and also for the cofactor of the entry a_{ij}.

Theorem 2.14 *Let A be an $n \times n$ matrix and let A_{ij} be the cofactor of a_{ij}. Then,*

(1) *for each $1 \leq i \leq n$,*

$$\det A = a_{i1}A_{i1} + a_{i2}A_{i2} + \cdots + a_{in}A_{in}, \tag{2.4}$$

called the **cofactor expansion** *of* $\det A$ *along the i-th row.*

(2) *For each $1 \leq j \leq n$,*

$$\det A = a_{1j}A_{1j} + a_{2j}A_{2j} + \cdots + a_{nj}A_{nj}, \tag{2.5}$$

called the **cofactor expansion** *of* $\det A$ *along the j-th column.*

The cofactor expansion also gives an alternative way of defining the determinant inductively.

Remark: The sign $(-1)^{i+j}$ of the cofactor can be visualized as follows:

$$\begin{bmatrix} + & - & + & \cdots & (-1)^{1+n} \\ - & + & - & \cdots & (-1)^{2+n} \\ + & - & + & \cdots & (-1)^{3+n} \\ \vdots & & & \ddots & \vdots \\ (-1)^{n+1} & (-1)^{n+2} & (-1)^{n+3} & \cdots & (-1)^{n+n} \end{bmatrix}.$$

In summary, the determinant of an $n \times n$ matrix A is the sum of the products of the entries in any given row (or, any given column) with their cofactors.

The cofactor expansion formula for $\det A$ suggests that the computing of A_{ij} can be avoided whenever $a_{ij} = 0$, because the product $a_{ij}A_{ij}$ is zero regardless of the value of A_{ij}. It shows that the computation of the determinant will be easier by taking the cofactor expansion along a row or a column that contains as many zero entries as possible.

Moreover, by using the elementary row (or column) operations which do not alter the determinant, a matrix A can be reduced into another one having more zero entries in a row or in a column that makes easier to compute the determinant.

For example, forward elimination reduces a square matrix A to its row-echelon form U which is an upper triangular matrix. Hence, the determinant of A will be just the product of the diagonal entries of U up to the sign caused by possible row interchanges.

The following two examples illustrate that *the cofactor expansions with elementary row/column operations make it easy to compute the determinant.*

Example 28 (*Computing* $\det A$ *by using forward elimination and a cofactor expansion*) Compute the determinant of

$$A = \begin{bmatrix} 1 & -2 & -1 & 4 \\ -5 & 9 & 8 & -5 \\ -3 & 4 & 7 & -3 \\ 2 & -3 & -5 & 8 \end{bmatrix}.$$

Solution: Apply the elementary operations

$$5 \times \text{row } 1 + \text{row } 2, \qquad 3 \times \text{row } 1 + \text{row } 3, \quad \text{and} \quad (-2) \times \text{row } 1 + \text{row } 4$$

to A, then

$$\det A = \det \begin{bmatrix} 1 & -2 & -1 & 4 \\ 0 & -1 & 3 & 15 \\ 0 & -2 & 4 & 9 \\ 0 & 1 & -3 & 0 \end{bmatrix} = \det \begin{bmatrix} -1 & 3 & 15 \\ -2 & 4 & 9 \\ 1 & -3 & 0 \end{bmatrix}.$$

Now apply the operation $1 \times$ row $1 +$ row 3 to the matrix on the right-hand side, and take the cofactor expansion along the third row to get

$$\begin{aligned} \det \begin{bmatrix} -1 & 3 & 15 \\ -2 & 4 & 9 \\ 1 & -3 & 0 \end{bmatrix} &= \det \begin{bmatrix} -1 & 3 & 15 \\ -2 & 4 & 9 \\ 0 & 0 & 15 \end{bmatrix} \\ &= (-1)^{3+3} \cdot 15 \cdot \det \begin{bmatrix} -1 & 3 \\ -2 & 4 \end{bmatrix} = 30. \end{aligned}$$

□

Problem 37 Use forward elimination and a cofactor expansion to compute

$$\det \begin{bmatrix} 0 & -2 & 2 & 0 \\ 1 & 3 & 6 & 8 \\ 1 & 6 & 4 & 4 \\ 1 & 9 & 4 & 9 \end{bmatrix}.$$

Example 29 (Vandermonde matrix) [9] The determinants of **Vandermonde matrices** of order 2, 3, and 4 are

$$\det \begin{bmatrix} 1 & x \\ 1 & y \end{bmatrix} = y - x, \qquad \det \begin{bmatrix} 1 & x & x^2 \\ 1 & y & y^2 \\ 1 & z & z^2 \end{bmatrix} = (y-x)(z-x)(z-y),$$

[9] **Alexandre-Theophile Vandermonde** (February 28, 1735– January 1, 1796) was a French mathematician, musician, and chemist.

$$\det\begin{bmatrix}1 & x & x^2 & x^3\\ 1 & y & y^2 & y^3\\ 1 & z & z^2 & z^3\\ 1 & w & w^2 & w^3\end{bmatrix} = (y-x)(z-x)(z-y)(w-x)(w-y)(w-z).$$

Solution: Just show it for the matrix A of order 4. Use a forward elimination. To begin with, add $(-1)\times$ row 1 to rows 2, 3, and 4 of A:

$$\begin{aligned}
\det A &= \det\begin{bmatrix}1 & x & x^2 & x^3\\ 0 & y-x & y^2-x^2 & y^3-x^3\\ 0 & z-x & z^2-x^2 & z^3-x^3\\ 0 & w-x & w^2-x^2 & w^3-x^3\end{bmatrix}\\
&= \det\begin{bmatrix}y-x & y^2-x^2 & y^3-x^3\\ z-x & z^2-x^2 & z^3-x^3\\ w-x & w^2-x^2 & w^3-x^3\end{bmatrix}\\
&= (y-x)(z-x)(w-x)\det\begin{bmatrix}1 & y+x & y^2+xy+x^2\\ 1 & z+x & z^2+xz+x^2\\ 1 & w+x & w^2+xw+x^2\end{bmatrix}\\
&= (y-x)(z-x)(w-x)\det\begin{bmatrix}1 & y+x & y^2+xy+x^2\\ 0 & z-y & (z-y)(z+y+x)\\ 0 & w-y & (w-y)(w+y+x)\end{bmatrix}\\
&= (y-x)(z-x)(w-x)\det\begin{bmatrix}z-y & (z-y)(z+y+x)\\ w-y & (w-y)(w+y+x)\end{bmatrix}\\
&= (y-x)(z-x)(w-x)(z-y)(w-y)\det\begin{bmatrix}1 & z+y+x\\ 1 & w+y+x\end{bmatrix}\\
&= (y-x)(z-x)(w-x)(z-y)(w-y)(w-z),
\end{aligned}$$

as wanted. □

Problem 38 Use cofactor expansions to compute the determinants of

$$A=\begin{bmatrix}0 & 1 & 2 & 3\\ 1 & 0 & 2 & 3\\ 2 & 2 & 0 & 3\\ 3 & 3 & 3 & 0\end{bmatrix},\quad B=\begin{bmatrix}1 & 1 & 1 & 1\\ -1 & 2 & 4 & 8\\ 1 & 3 & 9 & 27\\ -1 & 4 & 16 & 64\end{bmatrix},\quad C=\begin{bmatrix}1 & 1 & 1 & 1\\ 1 & 2 & 4 & 8\\ 1 & 3 & 9 & 27\\ 1 & 4 & 16 & 64\end{bmatrix}.$$

2.6 Cramer's rule for solving a system of linear equations

In Chapter 1, we have studied two ways of solving a system of linear equations $A\mathbf{x} = \mathbf{b}$: (i) by Gauss-Jordan elimination (or by LU factorization) and (ii) by using A^{-1} if A is invertible.

In this section we introduce the third method for solving the system $A\mathbf{x} = \mathbf{b}$ for an invertible matrix A.

Fix a row of A arbitrarily, say j-th row, and let B be the matrix obtained from A by replacing the j-th row by a new row

$$\mathbf{x} = [x_1\ x_2\ \dots\ x_n].$$

Then, along the j-th row, the cofactors of the matrix B are the same as those of A: that is, $B_{jk} = A_{jk}$ for all $k = 1, 2, \dots, n$. Therefore, the cofactor expansion of $\det B$ along the j-th row gives

$$\begin{aligned} \det B &= x_1 B_{j1} + x_2 B_{j2} + \cdots + x_n B_{jn} \\ &= x_1 A_{j1} + x_2 A_{j2} + \cdots + x_n A_{jn}, \end{aligned}$$

see Eq. (2.4). Now, if the row vector $\mathbf{x}$ is chosen to be the j-th row of A, that is,

$$\mathbf{x} = [x_1\ x_2\ \dots\ x_n] = [a_{j1}\ a_{j2}\ \dots\ a_{jn}],$$

then clearly, $B = A$ and $\det B = \det A$.

For any i, $i \neq j$, if the vector $\mathbf{x}$ is chosen to be the i-th row of A, that is,

$$\mathbf{x} = [x_1\ x_2\ \dots\ x_n] = [a_{i1}\ a_{i2}\ \dots\ a_{in}],$$

then the i-th and j-th rows of B are the same and hence $\det B$ should be zero.

The computing of $\det B$ gives the following lemma.

Lemma 2.15

$$a_{i1}A_{j1} + a_{i2}A_{j2} + \cdots + a_{in}A_{jn} = \begin{cases} \det A & \textit{if } i = j \\ 0 & \textit{if } i \neq j. \end{cases}$$

Definition 2.16 Let A be an $n \times n$ matrix and let A_{ij} denote the cofactor of a_{ij}. Then the new matrix

$$\begin{bmatrix} A_{11} & A_{12} & \cdots & A_{1n} \\ A_{21} & A_{22} & \cdots & A_{2n} \\ \vdots & \vdots & \ddots & \vdots \\ A_{n1} & A_{n2} & \cdots & A_{nn} \end{bmatrix}$$

is called the **matrix of cofactors** of A. Its transpose is called the **adjugate** (or **adjunct**) of A and is denoted by adj A. [10]

It follows from Lemma 2.15 that

$$A \cdot \text{adj } A = \begin{bmatrix} \det A & 0 & \cdots & 0 \\ 0 & \det A & \cdots & 0 \\ \vdots & \vdots & \ddots & \vdots \\ 0 & 0 & \cdots & \det A \end{bmatrix} = (\det A)I. \tag{2.6}$$

If A is invertible, then $\det A \neq 0$ and hence one may write

$$A^{-1} = \frac{1}{\det A} \text{adj } A, \quad \text{and} \quad A = (\det A) \text{ adj } (A^{-1}) \tag{2.7}$$

by replacing A with A^{-1}. It also shows that

$$\text{adj } (A^{-1}) = (\text{adj } A)^{-1}.$$

Example 30 (*Computing A^{-1} with* adj A) For $A = \begin{bmatrix} a & b \\ c & d \end{bmatrix}$, adj $A = \begin{bmatrix} d & -b \\ -c & a \end{bmatrix}$. If $\det A = ad - bc \neq 0$, then

$$A^{-1} = \frac{1}{\det A} \text{adj } A = \frac{1}{ad - bc} \begin{bmatrix} d & -b \\ -c & a \end{bmatrix},$$

as shown on page 27 in Chapter 1.

Problem 39 Show that A is invertible if and only if adj A is invertible, and that if A is invertible, then

$$(\text{adj } A)^{-1} = \frac{A}{\det A} = \text{adj}(A^{-1}).$$

[10] The *adjugate* has sometimes been called the 'adjoint', but today the 'adjoint' of a (complex) matrix normally refers to its conjugate transpose.

The next theorem establishes the third method for solving the system $A\mathbf{x} = \mathbf{b}$ for an invertible matrix A.

Theorem 2.17 (Cramer's rule) [11] *Let* $A\mathbf{x} = \mathbf{b}$ *be a system of* n *linear equations in* n *unknowns and let* $\det A \neq 0$. *Then* $A\mathbf{x} = \mathbf{b}$ *has the unique solution given by*

$$x_j = \frac{\det C_j}{\det A}, \qquad j = 1, 2, \ldots, n,$$

where C_j *is the matrix obtained from* A *by replacing the* j*-th column with the column vector* $\mathbf{b} = [b_1\ b_2\ \cdots\ b_n]^T$.

Proof: If $\det A \neq 0$, then A is invertible and $\mathbf{x} = A^{-1}\mathbf{b}$ is the unique solution of $A\mathbf{x} = \mathbf{b}$. Since

$$\mathbf{x} = A^{-1}\mathbf{b} = \frac{1}{\det A}(\text{adj } A)\mathbf{b},$$

it follows that the j-th entry of $\mathbf{x}$ is

$$x_j = \frac{b_1 A_{1j} + b_2 A_{2j} + \cdots + b_n A_{nj}}{\det A} = \frac{\det C_j}{\det A}$$

by the cofactor expansion of $\det C_j$ along the j-th column. □

Example 31 Use Cramer's rule to solve

$$\begin{cases} 2x_1 + x_2 + x_3 = 2000 \\ x_1 + 3x_2 + x_3 = 1000 \\ 4x_1 + 2x_2 + x_3 = 3000. \end{cases}$$

Solution:

$$A = \begin{bmatrix} 2 & 1 & 1 \\ 1 & 3 & 1 \\ 4 & 2 & 1 \end{bmatrix},$$

$$C_1 = \begin{bmatrix} 2000 & 1 & 1 \\ 1000 & 3 & 1 \\ 3000 & 2 & 1 \end{bmatrix}, \quad C_2 = \begin{bmatrix} 2 & 2000 & 1 \\ 1 & 1000 & 1 \\ 4 & 3000 & 1 \end{bmatrix}, \quad C_3 = \begin{bmatrix} 2 & 1 & 2000 \\ 1 & 3 & 1000 \\ 4 & 2 & 3000 \end{bmatrix}.$$

[11] This is named after Gabriel Cramer (July 31, 1704 ?– January 4, 1752), a Swiss mathematician, born in Geneva. At 18 he received his doctorate and at 20 he was co-chair of mathematics at the University of Geneva. In 1750, he published Cramer's rule.

Therefore,

$$x_1 = \frac{\det C_1}{\det A} = \frac{-3000}{-5} = 600, \quad x_2 = \frac{\det C_2}{\det A} = -200, \quad x_3 = \frac{\det C_3}{\det A} = 1000,$$

as the quotients of determinants. □

Cramer's rule provides a convenient method for writing down the solution of a system of n linear equations in n unknowns in terms of determinants. To find the solution, one needs to compute $n+1$ determinants of order n. Computing even two of them generally involves more computations than solving the system by using Gauss-Jordan elimination. However, Cramer's rule can be used to study properties of the solution without solving the system.

Problem 40 Use Cramer's rule to solve the following systems.

$$(1)\begin{cases} 3x + 2y = 3z + 1 \\ 3x + 2z = 8 - 5y \\ 3z - 1 = x - 2y, \end{cases} \qquad (2)\begin{cases} 4y + 3z = -20 \\ 3x + 4y + 5z = 60 \\ 2x - 5y + 2z = -10. \end{cases}$$

2.7 Applications

2.7.1 Determinants of Vandermonde and circulant matrices

Example 32 (Vandermonde matrix) As in Example 29, one can compute the determinant of the **Vandermonde matrix** of order n by using cofactor expansions with proper elementary row operations.[12]

$$\det\begin{bmatrix} 1 & x_1 & x_1^2 & \cdots & x_1^{n-1} \\ 1 & x_2 & x_2^2 & \cdots & x_2^{n-1} \\ \vdots & \vdots & \vdots & & \vdots \\ 1 & x_n & x_n^2 & \cdots & x_n^{n-1} \end{bmatrix} = \begin{array}{l} (x_2-x_1)(x_3-x_1) \quad \cdots \quad \cdots (x_n-x_1) \\ (x_3-x_2)(x_4-x_2) \quad \cdots \quad (x_n-x_2) \\ \vdots \\ (x_n-x_{n-1}) \end{array}$$

$$= \prod_{1\le i<j\le n} (x_j - x_i).$$

This matrix will be extensively used in "Curve fitting theory". See Subsection 2.7.2.

[12]The Pi notation $\prod$ is used for a multiplication. For example, the multiplication of the consecutive integers from 1 to 10 is written as: $1 \times 2 \times 3 \times 4 \times \cdots \times 9 \times 10 = \prod_{i=1}^{10} i$.

Example 33 Let A_{ij} denote the cofactor of a_{ij} in a matrix $A = [a_{ij}]$. If $n > 1$, [13]

$$\begin{aligned} \det \begin{bmatrix} 0 & x_1 & x_2 & \cdots & x_n \\ x_1 & a_{11} & a_{12} & \cdots & a_{1n} \\ x_2 & a_{21} & a_{22} & \cdots & a_{2n} \\ \vdots & \vdots & \vdots & \ddots & \vdots \\ x_n & a_{n1} & a_{n2} & \cdots & a_{nn} \end{bmatrix} &= -\left(\sum_{j=1}^{n} A_{1j}x_1x_j + \sum_{j=1}^{n} A_{2j}x_2x_j + \cdots + \sum_{j=1}^{n} A_{nj}x_nx_j \right) \\ &= -\sum_{i=1}^{n}\sum_{j=1}^{n} A_{ij}x_ix_j. \end{aligned}$$

Solution: Take the cofactor expansion along the first row, and then take the cofactor expansion along the first column of each $n \times n$ submatrix. □

Example 34 (Circulant matrix) An $n \times n$ matrix A is called a **circulant matrix** if the i-th row of A is obtained from the first row of A by a cyclic shift of $i-1$ steps, *i.e.*, the general form of the circulant matrix of order n is

$$A = \begin{bmatrix} a_1 & a_2 & a_3 & \cdots & a_n \\ a_n & a_1 & a_2 & \cdots & a_{n-1} \\ a_{n-1} & a_n & a_1 & \cdots & a_{n-2} \\ \vdots & \vdots & \vdots & \ddots & \vdots \\ a_2 & a_3 & a_4 & \cdots & a_1 \end{bmatrix}.$$

(1) The determinant of **circulant matrix** of order 2 is

$$\det \begin{bmatrix} a_1 & a_2 \\ a_2 & a_1 \end{bmatrix} = a_1^2 - a_2^2 = (a_1 + a_2)(a_1 - a_2).$$

(2) The determinant of **circulant matrix** of order 3 is

$$\begin{aligned} \det \begin{bmatrix} a_1 & a_2 & a_3 \\ a_3 & a_1 & a_2 \\ a_2 & a_3 & a_1 \end{bmatrix} &= a_1^3 + a_2^3 + a_3^3 - 3a_1a_2a_3 \\ &= (a_1 + a_2 + a_3)(a_1 + a_2\omega + a_3\omega^2)(a_1 + a_2\omega^2 + a_3\omega), \end{aligned}$$

[13] Double sum notation: $\sum_{i=1}^{n}\sum_{j=1}^{n} A_{ij}x_ix_j = \sum_{j=1}^{n} A_{1j}x_1x_j + \sum_{j=1}^{n} A_{2j}x_2x_j + \cdots + \sum_{j=1}^{n} A_{nj}x_nx_j$.

where $\omega = e^{2\pi i/3} = -1/2 + (i\sqrt{3})/2$ is the primitive third root of unity. [14, 15]

(3) The determinant of **circulant matrix** of order 4 is

$$\begin{aligned}\det\begin{bmatrix} a_1 & a_2 & a_3 & a_4 \\ a_4 & a_1 & a_2 & a_3 \\ a_3 & a_4 & a_1 & a_2 \\ a_2 & a_3 & a_4 & a_1 \end{bmatrix} &= -a_4^4 + 4a_4^2a_1a_3 + 2a_4^2a_2^2 - 4a_4a_1^2a_2 - 4a_4a_2a_3^2 \\ &\quad + a_1^4 - 2a_1^2a_3^2 + 4a_1a_2^2a_3 - a_2^4 + a_3^4 \\ &= (a_1 + a_2 + a_3 + a_4)(a_1 - a_2 + a_3 - a_4) \\ &\quad \times ((a_4 - a_2)^2 + (a_1 - a_3)^2) \\ &= (a_1 + a_2 + a_3 + a_4)(a_1 - a_2 + a_3 - a_4) \\ &\quad \times (a_1 + a_2 i - a_3 - a_4 i)(a_1 - a_2 i - a_3 + a_4 i),\end{aligned}$$

which can be written as

$$\prod_{j=0}^{3}\left(a_1 + a_2\omega^j + a_3\omega^{2j} + a_4\omega^{3j}\right)$$

with $\omega = e^{2\pi i/4} = \cos 2\pi/4 + i \sin 2\pi/4 = i$, the primitive fourth root of unity. In this case, $\omega^2 = -1, \omega^3 = -i, \omega^4 = 1$, etc.

(4) In general, for the circulant matrix of order n,

$$\det A = \prod_{j=0}^{n-1}\left(a_1 + a_2\omega^j + a_3\omega^{2j} + \cdots + a_n\omega^{(n-1)j}\right),$$

---

[14] The (real) *exponential function* is defined as a real-valued function

$$e^x = \sum_{k=0}^{\infty}\frac{x^k}{k!} = 1 + x + \frac{x^2}{2!} + \frac{x^3}{3!} + \cdots = \lim_{N\to\infty}\sum_{k=0}^{N}\frac{x^k}{k!}.$$

Similarly, the *complex exponential function* is defined as a complex-valued function $e^z = \sum_{k=0}^{\infty}\frac{z^k}{k!}$.

As the fundamental relation between the real trigonometric functions and the complex exponential function, it is well known that $e^{ix} = \cos x + i \sin x$ for any real number x, called *Euler's formula.* As a special case, $e^{\pi i} + 1 = 0$, called *Euler's identity*,. This gives a beautiful relation of the most important five constants; the number 0, the real unity 1, the (Euler's) number e, the ratio of a circle's circumference to its diameter π, and the imaginary unity i.

As another complex constant, the primitive n-th root of unity is $e^{2\pi i/n} = \cos 2\pi/n + i \sin 2\pi/n$.

[15] **Leonhard Euler** (April 15, 1707– September 18, 1783) was a Swiss mathematician, physicist, astronomer, logician and engineer, who made important and influential discoveries in many branches of mathematics, such as calculus and graph theory. He also made pioneering contributions to topology and analytic number theory. He also introduced much of modern mathematical terminology and notation, such as $f(x)$, including modern notation for the trigonometric functions, the letter e, the Greek letter $\sum$ for summations and the letters i and π etc in mathematics.

where $\omega = e^{2\pi i/n} = \cos 2\pi/n + i \sin 2\pi/n$ is the primitive n-th root of unity. (See Exercise **8.23** for a proof.) □

Problem 41 In the definition of a circulant matrix given in Example 34, if we take a cyclic shift of the first row in the opposite direction, one can get different types of **circulant matrices**;

$$\begin{bmatrix} a_1 & a_2 \\ a_2 & a_1 \end{bmatrix}, \quad \begin{bmatrix} a_1 & a_2 & a_3 \\ a_2 & a_3 & a_1 \\ a_3 & a_1 & a_2 \end{bmatrix}, \quad \begin{bmatrix} a_1 & a_2 & a_3 & a_4 \\ a_2 & a_3 & a_4 & a_1 \\ a_3 & a_4 & a_1 & a_2 \\ a_4 & a_1 & a_2 & a_3 \end{bmatrix}, \ldots.$$

Compute their determinants.

Hint: In the second 3×3 matrix, by interchanging the second and the third rows, one can get

$$\begin{bmatrix} a_1 & a_2 & a_3 \\ a_3 & a_1 & a_2 \\ a_2 & a_3 & a_1 \end{bmatrix}.$$

2.7.2 Exact polynomial curve fitting to data

In a scientific experiment, let a scientist put finitely many input data x_i into an experimental device and obtain output data y_i.

Let a certain $n+1$ distinct experimental input data x_i be given and obtain $n+1$ output values y_i as follows.

input x_i	x_0	x_1	$\cdots$	x_n
output y_i	y_0	y_1	$\cdots$	y_n

If $n = 1$, there is a (unique) polynomial $y = ax + b$ of degree ≤ 1 that exact fits to two points (x_0, y_0) and (x_1, y_1).

If $n = 2$, there is a (unique) polynomial $y = ax^2 + bx + c$ of degree ≤ 2 that exact fits to three points (x_0, y_0), (x_1, y_1), and (x_2, y_2), etc.

Here, 'exact fits to' means 'passes exactly through'. If $y_0 = y_1 = y_2 = c$, a constant, then the polynomial $y = ax^2 + bx + c$ should be the constant function c, even the x_i's are assumed to be all distinct.

In general, if $n + 1$ experimental input/output data (x_i, y_i) are given, there is the (unique) polynomial $y = p(x)$ of degree $\leq n$ that exact fits to the given data.

To find such polynomial $y = p(x)$ of degree $\leq n$, let

$$p(x) = a_0 + a_1 x + \cdots + a_n x^n$$

with $n + 1$ unknown coefficients a_i's. Then,

$$p(x_i) = a_0 + a_1 x_i + \cdots + a_n x_i^n = y_i$$

for $i = 0, 1, \ldots, n$. In matrix notation, it gives

$$\begin{bmatrix} 1 & x_0 & \cdots & x_0^n \\ 1 & x_1 & \cdots & x_1^n \\ \vdots & \vdots & & \vdots \\ 1 & x_n & \cdots & x_n^n \end{bmatrix} \begin{bmatrix} a_0 \\ a_1 \\ \vdots \\ a_n \end{bmatrix} = \begin{bmatrix} y_0 \\ y_1 \\ \vdots \\ y_n \end{bmatrix}.$$

The coefficient matrix A is a Vandermonde matrix of order $n+1$, (see Example 32), whose determinant is

$$\det A = \prod_{0 \leq i < j \leq n} (x_j - x_i).$$

Since the x_i's are assumed to be all distinct, $\det A \neq 0$. Hence, the system $A\mathbf{x} = \mathbf{b}$ has a unique solution, which determines the unique polynomial $p(x)$ of degree $\leq n$ that passes through the given $n + 1$ data points $(x_0, y_0), (x_1, y_1), \ldots, (x_n, y_n)$ in the plane $\mathbb{R}^2$. This is the exact fitting polynomial to the given data.

Example 35 (*Exact polynomial curve fitting to data*) Let four experimental input/output data (x_i, y_i) be given as

$$(0, 3), \ (1, 0), \ (-1, 2), \ (3, 6) \ ,$$

and let $p(x)$ be the exact fitting polynomial to the given data. Then, $p(x)$ should be a polynomial of degree ≤ 3, say $p(x) = a_0 + a_1 x + a_2 x^2 + a_3 x^3$. It gives a system of linear equations

$$\begin{cases} a_0 & & & & & & & = & 3 \\ a_0 & + & a_1 & + & a_2 & + & a_3 & = & 0 \\ a_0 & - & a_1 & + & a_2 & - & a_3 & = & 2 \\ a_0 & + & 3a_1 & + & 9a_2 & + & 27a_3 & = & 6. \end{cases}$$

Solving the system, one can get $a_0 = 3$, $a_1 = -2$, $a_2 = -2$, $a_3 = 1$, and the (exact) fitting polynomial is $p(x) = 3 - 2x - 2x^2 + x^3$. □

Problem 42 Find the exact fitting polynomial $p(x)$ to the five input/output data $(-1, -14)$, $(0, -5)$, $(1, -4)$, $(2, 1)$, and $(3, 22)$.

Hint: $p(x)$ should be a polynomial of degree ≤ 4.

2.7.3 Cauchy-Binet formula to evaluate the determinant $\det(AB)$

The product of a row vector and a column vector has been defined as

$$[a_1\ a_2\ \cdots\ a_n]\begin{bmatrix}x_1\\x_2\\\vdots\\x_n\end{bmatrix} = [a_1x_1 + a_2x_2 + \cdots + a_nx_n] = \left[\sum_{i=1}^{n} a_ix_i\right].$$

This is a 1×1 matrix and its determinant is clearly $\sum_{i=1}^{n} a_ix_i$.

As its generalization, it is sometimes necessary to compute the determinant of a product of two rectangular matrices. The next theorem shows how one can do it.

Theorem 2.18 (Cauchy-Binet formula) [16] *Let A and B be matrices of size $n \times m$ and $m \times n$, respectively, and $n \leq m$ (so, A is wide and B is long). Then*

$$\begin{aligned}\det(AB) &= \sum_{1\leq k_1<k_2<\cdots<k_n\leq m} \det(A_{k_1,k_2,\ldots,k_n} B^{k_1,k_2,\ldots,k_n})\\ &= \sum_{1\leq k_1<k_2<\cdots<k_n\leq m} \det A_{k_1,k_2,\ldots,k_n} \det B^{k_1,k_2,\ldots,k_n},\end{aligned}$$

where $A_{k_1,k_2,\ldots,k_n}$ is the $n \times n$ submatrix of A consisting of the columns indexed by $k_1, k_2, \ldots, k_n$ and $B^{k_1,k_2,\ldots,k_n}$ is the $n \times n$ submatrix of B consisting of the rows indexed by $k_1, k_2, \ldots, k_n$.

In particular, if A and B are both square matrices of the same size, then Cauchy-Binet formula states $\det(AB) = \det A \det B$, which is already known.

Proof: Let $C = AB$, $c_{ij} = \sum_{k=1}^{m} a_{ik}b_{kj}$, it is an $n \times n$ matrix. Then by the

[16]This is named after Augustin-Louis Cauchy and Jacques Philippe Marie Binet. **Baron Augustin-Louis Cauchy** (August 21, 1789– May 23, 1857) was a French mathematician and physicist who made pioneering contributions to (complex) analysis. **Jacques Philippe Marie Binet** (February 2, 1786– May 12, 1856) was a French mathematician, physicist, and astronomer.

Leibniz formula for determinants,

$$\begin{aligned}
\det C &= \sum_{\sigma \in S_n} \operatorname{sign}(\sigma) c_{1\sigma(1)} c_{2\sigma(2)} \cdots c_{n\sigma(n)} \\
&= \sum_{\sigma \in S_n} \operatorname{sign}(\sigma) \left(\sum_{k_1=1}^{m} a_{1k_1} b_{k_1\sigma(1)} \sum_{k_2=1}^{m} a_{2k_2} b_{k_2\sigma(2)} \cdots \sum_{k_n=1}^{m} a_{nk_n} b_{k_n\sigma(n)} \right) \\
&= \sum_{j=1}^{n} \sum_{k_j=1}^{m} a_{1k_1} a_{2k_2} \cdots a_{nk_n} \sum_{\sigma \in S_n} \operatorname{sign}(\sigma) b_{k_1\sigma(1)} b_{k_2\sigma(2)} \cdots b_{k_n\sigma(n)} \\
&= \sum_{j=1}^{n} \sum_{k_j=1}^{m} a_{1k_1} a_{2k_2} \cdots a_{nk_n} \det B^{k_1,k_2,\ldots,k_n},
\end{aligned}$$

where $\det B^{k_1,k_2,\ldots,k_n}$ can be nonzero only when the indices $k_1, k_2, \ldots, k_n$ are all distinct. Thus, the summation can be performed over all distinct indices $k_1, k_2, \ldots,$ k_n, for which

$$\det B^{k_1,k_2,\ldots,k_n} = \operatorname{sign}(\tau) \det B^{\tau(k_1),\tau(k_2),\ldots,\tau(k_n)},$$

where τ is a permutation of the indices k_1, k_2, $\ldots$, k_n and $\tau(k_1) < \tau(k_2) < \cdots < \tau(k_n)$. Hence, we have

$$\begin{aligned}
&\sum_{j=1}^{n} \sum_{k_j=1}^{m} a_{1k_1} a_{2k_2} \cdots a_{nk_n} \det B^{k_1,k_2,\ldots,k_n} \\
&\quad = \sum_{\{k_j\}\,:\,\text{all distinct}} a_{1k_1} a_{2k_2} \cdots a_{nk_n} \operatorname{sign}(\tau) \det B^{\tau(k_1),\tau(k_2),\ldots,\tau(k_n)} \\
&\quad = \sum_{1 \le k_1 < k_2 < \cdots < k_n \le m} \sum_{\tau} \operatorname{sign}(\tau) a_{1\tau(1)} \cdots a_{n\tau(n)} \det B^{k_1,k_2,\ldots,k_n} \\
&\quad = \sum_{1 \le k_1 < k_2 < \cdots < k_n \le m} \det A_{k_1,k_2,\ldots,k_n} \det B^{k_1,k_2,\ldots,k_n}.
\end{aligned}$$

This completes the proof. □

Corollary 2.18.1 *For any square or wide rectangular matrix A,*

$$\begin{aligned}
\det(AA^T) &= \sum_{1 \le k_1 < k_2 < \cdots < k_n \le m} \det A_{k_1,k_2,\ldots,k_n} \det(A^T)^{k_1,k_2,\ldots,k_n} \\
&= \sum_{1 \le k_1 < k_2 < \cdots < k_n \le m} \left(\det A_{k_1,k_2,\ldots,k_n}\right)^2.
\end{aligned}$$

As a result, for any wide matrix A, $\det(AA^T)$ *is always nonnegative, while for any long matrix A,* $\det(A^TA)$ *is always nonnegative.*

For example, if $A = \begin{bmatrix} 1 & 1 & 1 & 1 \\ a & b & c & d \end{bmatrix}$ and $B = \begin{bmatrix} x & 100 \\ y & 200 \\ z & 300 \\ w & 400 \end{bmatrix}$, then

$$\begin{aligned} \det(AB) = \ & \det\begin{bmatrix} 1 & 1 \\ a & b \end{bmatrix} \det\begin{bmatrix} x & 100 \\ y & 200 \end{bmatrix} + \det\begin{bmatrix} 1 & 1 \\ a & c \end{bmatrix} \det\begin{bmatrix} x & 100 \\ z & 300 \end{bmatrix} \\ & + \det\begin{bmatrix} 1 & 1 \\ a & d \end{bmatrix} \det\begin{bmatrix} x & 100 \\ w & 400 \end{bmatrix} + \det\begin{bmatrix} 1 & 1 \\ b & c \end{bmatrix} \det\begin{bmatrix} y & 200 \\ z & 300 \end{bmatrix} \\ & + \det\begin{bmatrix} 1 & 1 \\ b & d \end{bmatrix} \det\begin{bmatrix} y & 200 \\ w & 400 \end{bmatrix} + \det\begin{bmatrix} 1 & 1 \\ c & d \end{bmatrix} \det\begin{bmatrix} z & 300 \\ w & 400 \end{bmatrix}, \end{aligned}$$

which can be computed without computing the product AB.

Remark: In the condition of the Cauchy-Binet formula in Theorem 2.18, if $n > m$ instead of $n \leq m$, what is $\det(AB)$? In this case, A is long and B is wide, but both are not square. Then $\det(AB) = 0$ and AB cannot be invertible. This can be shown later by using Corollaries 3.23.1 and 3.24.1.

Problem 43 For the matrices $A = \begin{bmatrix} 1 & -1 & 3 & 3 \\ 2 & 2 & -1 & 2 \end{bmatrix}$ and $B = \begin{bmatrix} 1 & 2 \\ 2 & -1 \\ -3 & 1 \\ 1 & 2 \end{bmatrix}$, show that $\det(AB) = -167$ by the Cauchy-Binet formula, while $\det(BA) = 0$ by a direct computation. Also, compute $\det(AA^T)$ and $\det(A^TA)$.

2.7.4 Areas and volumes by determinants

As a continuation of the area mentioned in Section 2.2, we give a geometrical interpretation of the determinant of an $n \times n$ matrix A as the *signed volume* (or *signed area* for $n = 2$) of the parallelepiped spanned by the row vectors of A.

For an $m \times n$ matrix A, its *row vectors* $\mathbf{r}_i = [a_{i1} \ \cdots \ a_{in}]$ can be considered as vectors in the n-dimensional space

$$\mathbb{R}^n = \{(a_1, a_2, \ldots, a_n) : a_i \in \mathbb{R}, i = 1, 2, \ldots, n\}.$$

The set

$$\mathcal{P}(A) = \left\{ \sum_{i=1}^{m} t_i \mathbf{r}_i \,:\, 0 \le t_i \le 1,\ i = 1, 2, \ldots, m \right\}$$

is called a **parallelogram** if $m = 2$ [17], and a **parallelepiped** of dimension m spanned by the row vectors $\mathbf{r}_1, \mathbf{r}_2, \ldots, \mathbf{r}_m$ of A if $m \ge 3$.

In this case, the row vectors of A form the edges of $\mathcal{P}(A)$ ending at the origin $\mathbf{0}$, and changing the order of row vectors does not alter the shape of $\mathcal{P}(A)$.

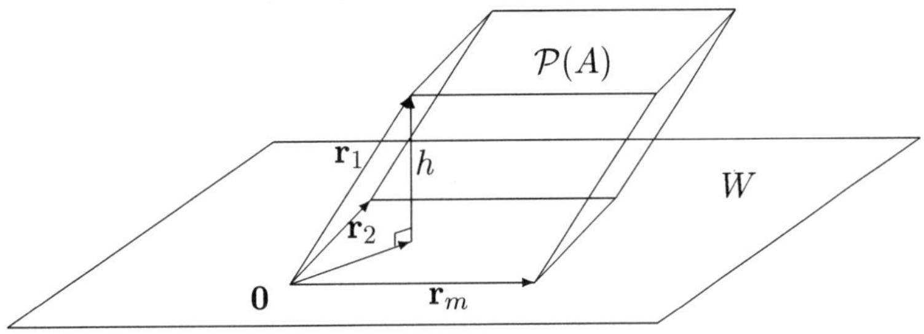

Figure 2.3: An n-dimensional parallelepiped $\mathcal{P}(A)$ and the height h

The *volume* of the m-dimensional parallelepiped $\mathcal{P}(A)$ spanned by the row vectors $\mathbf{r}_1, \mathbf{r}_2, \ldots, \mathbf{r}_m$ in $\mathbb{R}^n$ is defined inductively as the multiplication of the volume of the $(m-1)$-dimensional parallelepiped (base) spanned by $\mathbf{r}_2, \ldots, \mathbf{r}_m$ and the height h of $\mathbf{r}_1$ from the (sub)space W spanned by $\mathbf{r}_2, \ldots, \mathbf{r}_m$.

Intuitively, the height h is the (shortest) distance from the point $\mathbf{r}_1$ to the space W (see Figure 2.3). If the vector $\mathbf{r}_1$ is contained in the space W, then $h = 0$ and the parallelepiped is contained in a subspace of dimension less than m. In this case, the volume should be 0. (The subspace W and the computing of height h will be discussed formally in Theorem 3.9 and Theorem 5.16, respectively.)

As a trivial example, for the identity matrix $A = I_n$ of order n, the parallelepiped $\mathcal{P}(A)$ spanned by the row vectors of I_n is the standard n-dimensional **hypercube**, or simply the n-**cube**. From the definition, its volume is 1 inductively.

[17] If $m = 2$, $\mathcal{P}(A) = \{t_1\mathbf{r}_1 + t_2\mathbf{r}_2 \,:\, 0 \le t_1, t_2 \le 1\}$ is a parallelogram with edges $\mathbf{r}_1$ and $\mathbf{r}_2$.

Theorem 2.19 *For any square or non-square (but wide rectangular) matrix A, the volume of the parallelepiped $\mathcal{P}(A)$ spanned by the row vectors $\mathbf{r}_1, \mathbf{r}_2, \ldots, \mathbf{r}_m$ of A is*

$$\mathrm{vol}(\mathcal{P}(A)) = \sqrt{\det(AA^T)}\,. \tag{2.8}$$

In particular, if A is a square matrix, then

$$\mathrm{vol}(\mathcal{P}(A)) = \sqrt{\det(AA^T)} = \sqrt{\det A\ \det(A^T)} = |\det A|. \tag{2.9}$$

Its proof will be given later in Example 93 on page 216.

As an equivalent version, by taking a transpose of a long matrix, one can state the following corollary.

Corollary 2.19.1 *Assume that A is a long matrix with $m \geq n$. Then, the volume of the n-dimensional parallelepiped $\mathcal{P}(A)$ spanned by the columns $\mathbf{c}_j$'s in $\mathbb{R}^m$ is given by*

$$\mathrm{vol}(\mathcal{P}(A)) = \sqrt{\det(A^TA)}\,. \tag{2.10}$$

In particular, if A is a square matrix, then

$$\mathrm{vol}(\mathcal{P}(A)) = |\det(A^T)| = |\det A|.$$

Remark: **(1)** If we construct the parallelepiped $\mathcal{P}(A)$ using the column vectors of a square matrix A instead of the rows, then the shape of the parallelepiped is totally different from the one spanned by the row vectors. However, $\det A = \det A^T$ gives that their volumes are equal, which is a totally nontrivial fact.

(2) For any wide matrix A, Cauchy-Binet formula states that $\det(AA^T) = \mathrm{vol}(\mathcal{P}(A))^2$ is equal to the sum of the squares of the volumes that arise when the parallelepiped $\mathcal{P}(A)$ is orthogonally projected onto the m-dimensional coordinate subspaces $\mathbb{R}^m$, of which there are $\binom{n}{m}$ subspaces $\mathbb{R}^m$'s in $\mathbb{R}^n$.

Problem 44 Show that the area of a triangle PQR in the plane $\mathbb{R}^2$, where $P = (x_1, y_1)$, $Q = (x_2, y_2)$, $R = (x_3, y_3)$, is equal to *the absolute value of*

$$\frac{1}{2}\det\begin{bmatrix} x_1 & y_1 & 1 \\ x_2 & y_2 & 1 \\ x_3 & y_3 & 1 \end{bmatrix}.$$

Hint: Use

$$\det\begin{bmatrix} x_1 & y_1 & 1 \\ x_2 & y_2 & 1 \\ x_3 & y_3 & 1 \end{bmatrix} = \det\begin{bmatrix} x_2 - x_1 & y_2 - y_1 \\ x_3 - x_1 & y_3 - y_1 \end{bmatrix},$$

and its absolute value is equal to $Area(\mathcal{P}(Q - P, R - P))$.

Problem 45 Find the area of the triangle with vertices at $P = (1, 2, 3)$, $Q = (4, 4, 4)$, and $R = (5, 6, 7)$ in the space $\mathbb{R}^3$.

Hint: $Q - P = (3, 2, 1)$ and $R - P = (4, 4, 4)$. Take these two vectors as the rows, set $A = \begin{bmatrix} 3 & 2 & 1 \\ 4 & 4 & 4 \end{bmatrix}$. Then the area of the triangle PQR is $\frac{1}{2}\sqrt{\det(AA^T)}$ by Eq. (2.8).

Problem 46 Show that the volume of a tetrahedron in the space $\mathbb{R}^3$ whose edge vectors are (x_1, y_1, z_1), (x_2, y_2, z_2), (x_3, y_3, z_3) is equal to *the absolute value of*

$$\frac{1}{6}\det\begin{bmatrix} x_1 & y_1 & z_1 \\ x_2 & y_2 & z_2 \\ x_3 & y_3 & z_3 \end{bmatrix}.$$

Hint: Every three-dimensional parallelepiped $\mathcal{P}$ has six faces, all of which are parallelograms. See Figure 2.3. Fix one of them as the base. This base parallelogram can be partitioned into two congruent triangles. Following this partition, the parallelepiped $\mathcal{P}$ is partitioned into two congruent triangular prisms, and each of these two prisms can be further partitioned into three congruent tetrahedrons.

2.8 Exercises

2.1. *Every square matrix has a determinant.* Compute the determinants of

$$A = \begin{bmatrix} 3 & 0 \\ 0 & 2 \end{bmatrix}, \qquad B = \begin{bmatrix} 3x & 0 & 5 \\ 0 & 2 & 8 \\ 0 & 0 & 7 \end{bmatrix}.$$

Hint: Both are upper triangular.

2.2. *Every square matrix has a determinant.* Compute the determinants of

$$A = \begin{bmatrix} 3 & 2 \\ 4 & 2 \end{bmatrix}, \qquad B = \begin{bmatrix} 3x & 0 & 5 \\ -5 & 2 & 8 \\ -3 & 4 & 7 \end{bmatrix}.$$

2.3. *Every square matrix has a determinant.* Find all permutations of $\{1, 2\}$ and also of $\{1, 2, 3\}$. Use your answer to repeat Exercise **2.2** with the Leibniz formula

$$\det A = \sum_{\sigma \in S_2} \operatorname{sign}(\sigma) a_{1\sigma(1)} a_{2\sigma(2)} \text{ and } \det B = \sum_{\sigma \in S_3} \operatorname{sign}(\sigma) a_{1\sigma(1)} a_{2\sigma(2)} a_{3\sigma(3)}.$$

2.4. Determine the values of x for which $\det\begin{bmatrix} x-1 & -3 \\ -2 & x-2 \end{bmatrix} = 0$.

2.5. Compute $\det A$, $\det B$, $\det(A^2BA^{-1})$, and $\det(B^{-1}A^3)$ for the matrices

$$A = \begin{bmatrix} 1 & 2x & 3 \\ -2 & 10 & 1 \\ 0 & 1 & 0 \end{bmatrix}, \qquad B = \begin{bmatrix} 1 & 0 & 2 \\ 0 & -2 & 5 \\ 2 & 1 & 3 \end{bmatrix}.$$

Hint: It is not necessary to compute A^2 or A^3.

2.6. *Cofactor expansions make it easy to compute the determinant.* Compute the determinants of

$$A = \begin{bmatrix} 1 & 2 & 3 & 4 \\ -2 & 1 & 2 & 3 \\ -3 & -2 & 1 & 2 \\ -4 & -3 & -2 & 1 \end{bmatrix}, \qquad B = \begin{bmatrix} 0 & 0 & -5 & 7 \\ 0 & 9 & 1 & 8 \\ 1 & 3 & 9 & 2 \\ 1 & 9 & 4 & 8 \end{bmatrix}.$$

Use forward elimination and a cofactor expansion as you like.

2.7. Compute $\det A$ for an $n \times n$ matrix $A = [a_{ij}]$ where

$$(1)\ a_{ij} = \begin{cases} 1 & \text{if } i \neq j, \\ 0 & \text{if } i = j, \end{cases} \qquad (2)\ a_{ij} = i + 2j.$$

Hint: First try it when $n = 2, 3$.

2.8. Compute the determinant of

$$(1) \begin{bmatrix} x-a & b \\ -1 & x \end{bmatrix}, \ (2) \begin{bmatrix} x-a_{11} & -a_{12} & 0 \\ -a_{21} & x-a_{22} & -a_{23} \\ -a_{31} & -a_{32} & x-a_{33} \end{bmatrix}, \ (3) \begin{bmatrix} 0 & x & x & x \\ -x & 0 & x & x \\ -x & -x & 0 & x \\ -x & -x & -x & 0 \end{bmatrix},$$

as polynomials in x.

2.9. An **anti-diagonal** of a square matrix is the diagonal going from the lower left corner to the upper right corner. An **anti-diagonal matrix** is an $n \times n$ matrix where all entries are zero except for the anti-diagonal entries, that is the (i, j)-entry is zero if $i + j \neq n + 1$. For example,

$$A = \begin{bmatrix} 0 & a_{12} \\ a_{21} & 0 \end{bmatrix}, \qquad B = \begin{bmatrix} 0 & 0 & a_{13} \\ 0 & a_{22} & 0 \\ a_{31} & 0 & 0 \end{bmatrix}, \qquad C = \begin{bmatrix} 0 & 0 & 0 & a_{14} \\ 0 & 0 & a_{23} & 0 \\ 0 & a_{32} & 0 & 0 \\ a_{41} & 0 & 0 & 0 \end{bmatrix}$$

are anti-diagonal matrices. Compute their determinants.

2.10. *A square matrix is a square array of numbers.* Rotate the following matrices clockwise $\pi/2, \pi, 3\pi/2$, and compute their determinants. [18]

$$A = \begin{bmatrix} a_{11} & a_{12} & a_{13} \\ a_{21} & a_{22} & a_{23} \\ a_{31} & a_{32} & a_{33} \end{bmatrix}, \qquad B = \begin{bmatrix} 0 & 0 & x & a_{14} \\ 0 & y & a_{23} & 0 \\ z & a_{32} & 0 & 0 \\ a_{41} & 0 & 0 & 0 \end{bmatrix}.$$

Hint: For example,

$$A_{\pi/2} = \begin{bmatrix} a_{31} & a_{21} & a_{11} \\ a_{32} & a_{22} & a_{12} \\ a_{33} & a_{23} & a_{13} \end{bmatrix}, \qquad A_{\pi} = \begin{bmatrix} a_{33} & a_{32} & a_{31} \\ a_{23} & a_{22} & a_{21} \\ a_{13} & a_{12} & a_{11} \end{bmatrix}.$$

2.11. (Revisit Exercise **1.25**) Compute the determinant for each of the matrices given in Exercise **1.25** by using their LU decompositions. (Use your answer of Exercise **1.26**).

$$A = \begin{bmatrix} 1 & 2 & 3 \\ 3 & 3 & 4 \\ 2 & 4 & 0 \end{bmatrix}, \qquad B = \begin{bmatrix} 1 & 2 & 3 & 0 \\ 3 & 3 & 2 & 4 \\ 1 & 2 & 1 & 2 \\ 2 & 4 & 0 & 2 \end{bmatrix}.$$

An LU or LDU decomposition of a square matrix can be used for computing the determinant.

2.12. There are $24 = 4!$ permutations on the set $N_4 = \{1, 2, 3, 4\}$. Choose any six of them, say σ, and compute their signed elementary products

$$\text{sign}(\sigma) a_{1\sigma(1)} a_{2\sigma(2)} a_{3\sigma(3)} a_{4\sigma(4)}$$

for each matrix given in Exercise **2.7** with $n = 4$.

2.13. Find all solutions of the equation $\det(ABC) = 0$ for

$$A = \begin{bmatrix} x-1 & 2x \\ 2 & x-1 \end{bmatrix}, \qquad B = \begin{bmatrix} x & 0 \\ 4 & x+2 \end{bmatrix}, \qquad C = \begin{bmatrix} 4x & x-1 \\ 4 & x+2 \end{bmatrix}.$$

Hint: It is not necessary to compute the product ABC.

2.14. Show that if A is an $n \times n$ skew-symmetric matrix and n is odd, then $\det A = 0$. Give an example of 4×4 skew-symmetric matrix A with $\det A \neq 0$.

Hint: See Problem 31(2) and Exercise **2.8**(3).

---

[18] The ratio of the circumference of a circle to its diameter is called *Pi*, denoted by the Greek letter π. $\pi = 3.1415926\cdots$. The *radian* is the standard unit for measuring angles, by definition π rad $= 180$ degrees.

2.15. *The matrix $[A_{ij}]$ of cofactors is called the matrix of cofactors of A. Its transpose is called the adjugate of A and is denoted by* adj A. Calculate the cofactors A_{11}, A_{12}, A_{13}, and A_{33} for the matrix A:

$$(1)\ A = \begin{bmatrix} 1 & 2 & 1 \\ 0 & 1 & 3 \\ 2 & 3 & 4 \end{bmatrix}, \quad (2)\ A = \begin{bmatrix} 1 & 2 & 0 \\ 1 & 0 & 2 \\ 3 & 4 & 5 \end{bmatrix}, \quad (3)\ A = \begin{bmatrix} 2 & -1 & 3 \\ -1 & 2 & 2 \\ 3 & 2 & 1 \end{bmatrix}.$$

2.16. In an $n \times n$ matrix $A = [a_{ij}]$, we replace the entries $a_{11}, a_{12}, \ldots, a_{1n}$ in the first row as unknown variables $x_1, x_2, \ldots, x_n$ to obtain a new matrix B. Show that $\det B$ becomes a polynomial in the variables x_i, and the cofactor A_{1j} becomes the coefficient of the variable x_j in this polynomial $\det B$.

Hint: Use the cofactor expansion of $\det B$ along the first row. Or read Page 80.

2.17. Let A be the matrix obtained from the identity matrix I_n with i-th column replaced by the column vector $\mathbf{x} = [x_1 \ \cdots \ x_n]^T$. Show that $\det A = x_i$.

Hint: Use Cramer's rule to solve $I_n\mathbf{x} = \mathbf{x}$.

2.18. Find a constant k so that the system of linear equations

$$\begin{cases} kx - 2y - z = 0 \\ (k+1)y + 4z = 0 \\ (k-1)z = 0 \end{cases}$$

has more than one solution. Is it possible to apply Cramer's rule here?

2.19. Solve the following system of linear equations in two ways; (1) by using Cramer's rule and (2) by using Gaussian elimination.

$$\begin{bmatrix} 1 & 1 & 1 & 1 \\ 1 & 2 & 1 & 1 \\ 1 & 1 & 2 & 1 \\ 1 & 1 & 1 & 2 \end{bmatrix} \mathbf{x} = \begin{bmatrix} 1 \\ 2 \\ 3 \\ 4 \end{bmatrix}.$$

2.20. *The matrix $[A_{ij}]$ of cofactors is called the matrix of cofactors of A. Its transpose is called the adjugate of A and is denoted by* adj A. Compute adj A, $\det A$, $\det(\text{adj } A)$, and A^{-1} for

$$(1)\ A = \begin{bmatrix} 1 & 0 & 0 \\ 4 & 2 & 0 \\ 6 & 5 & 3 \end{bmatrix}, \quad (2)\ A = \begin{bmatrix} 2 & 1 & 3 \\ -1 & 2 & 0 \\ 3 & -2 & 1 \end{bmatrix}, \quad (3)\ A = \begin{bmatrix} 1 & 2 & 3 \\ 2 & 3 & 4 \\ 1 & 5 & 7 \end{bmatrix}.$$

Hint: If A is invertible, adj $A = (\det A)A^{-1}$.

2.21. Show that if A is symmetric, so is adj A. In addition, if A is invertible, then the inverse of A is also symmetric.

2.22. *The matrix* $[A_{ij}]$ *of cofactors is called the matrix of cofactors of* A. *Its transpose is called the adjugate of* A *and is denoted by* adj A. Let A be an $n \times n$ invertible matrix with $n > 1$. Show that

(1) $\det(\text{adj } A) = (\det A)^{n-1}$,
(2) $\text{adj}(\text{adj } A) = (\det A)^{n-2}A$.

Hint: (1) Take the determinant in both sides of $A \cdot \text{adj } A = (\det A)I$. (2) $A \cdot \text{adj } A = (\det A)I$ for any square matrix A. Hence, $\text{adj } A \cdot \text{adj}(\text{adj } A) = \det(\text{adj } A)\ I$.

2.23. For any two invertible matrices A and B, show that

(1) $\text{adj } AB = \text{adj } B \cdot \text{adj } A$,
(2) $\text{adj } QAQ^{-1} = Q(\text{adj } A)Q^{-1}$ for any invertible matrix Q,
(3) if $AB = BA$, then $(\text{adj } A)B = B(\text{adj } A)$.

Hint: If A is invertible, $\text{adj } A = (\det A)A^{-1}$.

In fact, these three properties are satisfied for any (invertible or not) two square matrices A and B. (See Exercise **6.28**.)

2.24. Let J be the $n \times n$ matrix whose entries are all 1. Show that $\det(J - nI_n) = 0$.

Hint: The sum of rows (or the columns) of $J - nI_n$ is zero, so its determinant is zero (by using type III elementary operation repeatedly).

2.25. Compute the determinant of

$$A = \begin{bmatrix} 1 & 2 & 2^2 & 2^3 \\ 1 & 0 & 0 & 0 \\ 1 & 1 & 1 & 1 \\ 1 & 3 & 3^2 & 3^3 \end{bmatrix}, \qquad B = \begin{bmatrix} 1 & 20 & 20^2 & 20^3 \\ 1 & 13 & 13^2 & 13^3 \\ 1 & 12 & 12^2 & 12^3 \\ 1 & 3 & 3^2 & 3^3 \end{bmatrix}.$$

Hint: Both are Vandermonde matrices of order 4.

2.26. Compute the determinants of

$$A = \begin{bmatrix} 1 & x^2 & x^3 & x^4 \\ 1 & y^2 & y^3 & y^4 \\ 1 & z^2 & z^3 & z^4 \\ 1 & w^2 & w^3 & w^4 \end{bmatrix}, \qquad B = \begin{bmatrix} 1 & 1 & 1 & 1 \\ x & y & z & w \\ x^2 & y^2 & z^2 & w^2 \\ x^4 & y^4 & z^4 & w^4 \end{bmatrix}.$$

Hint for (1): $\det A$ is the coefficient of $-v$ in the determinant of the Vandermonde matrix

$$\begin{bmatrix} 1 & x & x^2 & x^3 & x^4 \\ 1 & y & y^2 & y^3 & y^4 \\ 1 & z & z^2 & z^3 & z^4 \\ 1 & w & w^2 & w^3 & w^4 \\ 1 & v & v^2 & v^3 & v^4 \end{bmatrix}.$$

2.27. Compute the determinants of circulant matrices

$$A = \begin{bmatrix} 1 & x & x^2 & x^3 \\ x^3 & 1 & x & x^2 \\ x^2 & x^3 & 1 & x \\ x & x^2 & x^3 & 1 \end{bmatrix}, \qquad B = \begin{bmatrix} 1 & x & x^2 & x^3 \\ x & x^2 & x^3 & 1 \\ x^2 & x^3 & 1 & x \\ x^3 & 1 & x & x^2 \end{bmatrix}.$$

Hint: Use the formula for the determinant of circulant matrix of order 4 given in Page 85.

2.28. *Cofactor expansions make it easy to compute the determinant.* Compute the determinant of *tridiagonal matrices*

$$A = \begin{bmatrix} x & b & 0 & 0 \\ c & x & b & 0 \\ 0 & c & x & b \\ 0 & 0 & c & x \end{bmatrix}, \qquad B = \begin{bmatrix} x & 1 & 0 & 0 \\ -1 & y & 1 & 0 \\ 0 & -1 & z & 1 \\ 0 & 0 & -1 & w \end{bmatrix}.$$

Hint: Take the cofactor expansion of $\det(\cdot)$ along the last row.

2.29. *Cofactor expansions make it easy to compute the determinant.* In comparison with Exercise **2.28**, compute the determinants of

$$A = \begin{bmatrix} 0 & 0 & b & x \\ 0 & b & x & c \\ b & x & c & 0 \\ x & c & 0 & 0 \end{bmatrix}, \qquad B = \begin{bmatrix} 0 & 0 & 1 & x \\ 0 & 1 & y & -1 \\ 1 & z & -1 & 0 \\ w & -1 & 0 & 0 \end{bmatrix}.$$

Hint: Take the cofactor expansion of $\det(\cdot)$ along the last row.

2.30. Use the determinant function to find

(1) the area of the parallelogram spanned by $(4, 3)$ and $(7, 5)$,
(2) the volume of the parallelepiped spanned by the vectors $(1, 0, 4)$, $(0, -2, 2)$ and $(3, 1, -1)$.

Hint: Use Eq. (2.9) $\text{vol}(\mathcal{P}(A)) = |\det A|$ in Page 92 or see Theorem 2.19.

2.31. Let $A = \begin{bmatrix} 1 & 3 & 0 \\ 3 & 1 & 1 \end{bmatrix}$, a wide matrix. Find the area of the parallelogram spanned by two row vectors.

Hint: Use Eq. (2.8) $\text{vol}(\mathcal{P}(A)) = \sqrt{\det(AA^T)}$ in Page 92.

2.32. Find the area of the triangle with vertices at $(0, 0, 0)$, $(1, 1, 2)$, and $(2, 2, 1)$ in $\mathbb{R}^3$.

Hint: Set $A = \begin{bmatrix} 1 & 1 & 2 \\ 2 & 2 & 1 \end{bmatrix}$. Then the area is $\frac{1}{2}\sqrt{\det(AA^T)} = 3\sqrt{2}/2$ by Eq. (2.8).

2.33. Determine whether or not the following statements are true in general, and justify your answers.

(1) $\det(kI_n) = k\det(I_n) = k$ for any positive integers k and n.
(2) For any square matrices A and B of the same size, $\det(A + B) = \det A + \det B$.
(3) For any square matrices A and B of the same size, $\det(AB) = \det(BA)$.
(4) For any rectangular (square or not) matrices A and B, $\det(AB) = \det(BA)$ if both AB and BA are well defined.
(5) For an invertible matrix A, $\det A^n = (\det A)^n$ for any positive or negative integer n.
(6) For any square matrix A, if $A \neq \mathbf{0}$ then $\det(AA^T)$ is always positive.
(7) The multilinearity of det holds in any two rows at the same time:

$$\det\begin{bmatrix} a+u & b+v & c+w \\ d+x & e+y & f+z \\ \ell & m & n \end{bmatrix} = \det\begin{bmatrix} a & b & c \\ d & e & f \\ \ell & m & n \end{bmatrix} + \det\begin{bmatrix} u & v & w \\ x & y & z \\ \ell & m & n \end{bmatrix}.$$

(8) If A is an $n \times n$ square matrix, then for any scalar c, $\det(cI_n - A) = c^n - \det A$.
(9) If A is an $n \times n$ square matrix, then for any scalar c, $\det(cI_n - A^T) = \det(cI_n - A)$.
(10) If E is an elementary matrix, then $\det E = \pm 1$.
(11) There are no matrices A of order 3 such that $A^2 = -I_3$.
(12) Let A be a nilpotent matrix, *i.e.*, $A^k = \mathbf{0}$ for some natural number k. Then $\det A = 0$.
(13) Let A be an idempotent matrix, *i.e.*, $A^2 = A$. Then $\det A = 1$ or -1.
(14) A system $A\mathbf{x} = \mathbf{b}$ of linear equations has a solution if and only if $\det A \neq 0$.
(15) For any $n \times 1$, $n \geq 2$, column vectors $\mathbf{u}$ and $\mathbf{v}$, $\det(\mathbf{u}\mathbf{v}^T) = 0$.
(16) If the entries of A are all integers and $\det A = 1$ or -1, then the entries of A^{-1} are also integers.
(17) If the entries of A are 0's or 1's, then $\det A = 1, 0$, or -1.
(18) Every system of n linear equations in n unknowns can be solved by Cramer's rule.
(19) If A is a permutation matrix, then $\det A^T = -\det A$.
(20) If A is a skew-symmetric matrix, $A^T = -A$, then $\det A = \det A^T = -\det A$.

Hint: Exactly seven of them are true.

Chapter 3

Vector Spaces; Abstraction of the plane $\mathbb{R}^2$ and the space $\mathbb{R}^3$

Many physical quantities, such as length, area, mass, temperature are described by only real numbers as magnitudes. Other physical quantities like force or velocity have directions as well as magnitudes. Such quantities with direction are called **vectors**, while the numbers are called **scalars**. For instance in mathematics, a real or complex number is just a scalar.

A point $\mathbf{x}$ in the 3-dimensional space $\mathbb{R}^3$ (say simply 3-space) is a vector which is usually written as a triple of real numbers

$$\mathbf{x} = (x_1, x_2, x_3),$$

called the **Cartesian coordinate** of $\mathbf{x}$, and pictorially such a point in $\mathbb{R}^3$ is identified with an arrow from the origin to the point $\mathbf{x}$. In this way, a point in the 3-space $\mathbb{R}^3$ can be understood as a vector. The direction of the arrow specifies the direction of the vector, and the length of the arrow describes its magnitude.

In this chapter, we introduce the notion of a vector space, which is an abstraction of the usual algebraic structures of the 3-space $\mathbb{R}^3$. We then elaborate to build a coordinate system for an (abstract) vector space just like the Cartesian coordinate system given by the x, y, z-axes for $\mathbb{R}^3$. [1]

---

[1]**René Descartes** (Renatus Cartesius in Latin)(March 31, 1596– February 11, 1650) was a French philosopher, mathematician, and scientist. He is well remembered for his highly quotable "I THINK, THEREFORE I AM", but he is also responsible for the Cartesian coordinate system, introducing (x, y) as the coordinate of a point in the Euclidean plane. It enables us to know where we are once we have realized we exist.

3.1 From the 3-space $\mathbb{R}^3$ to vector spaces

In order to have a more general definition of vectors, we extract the most basic properties of those arrows in the 3-space $\mathbb{R}^3$. For all vectors (or points) in $\mathbb{R}^3$, there are two natural algebraic operations, the *sum of two vectors* and the *scalar multiplication* of a vector.

For any two vectors $\mathbf{x} = (x_1, x_2, x_3)$ and $\mathbf{y} = (y_1, y_2, y_3)$ in $\mathbb{R}^3$ and a scalar k, we define

$$\begin{aligned} \mathbf{x}+\mathbf{y} &= (x_1+y_1, x_2+y_2, x_3+y_3), \\ k\mathbf{x} &= (kx_1, kx_2, kx_3). \end{aligned}$$

Geometrically, they are illustrated as in Figure 3.1.

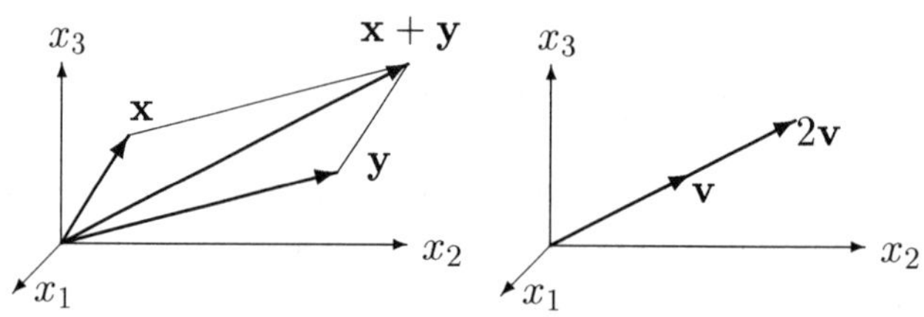

Figure 3.1: The sum of two vectors and the scalar multiplication of a vector

Even though our geometric visualization of vectors does not go beyond the 3-space $\mathbb{R}^3$, it is possible to extend these two algebraic operations of vectors in the 3-space $\mathbb{R}^3$ to the n-space $\mathbb{R}^n$ for any positive integer n.

The n-space $\mathbb{R}^n$ is defined to be the set of all ordered n-tuples $(a_1, a_2, \ldots, a_n)$ of real numbers, called *vectors, i.e.,*

$$\mathbb{R}^n = \{(a_1, a_2, \ldots, a_n) \,:\, a_i \in \mathbb{R}, i = 1, 2, \ldots, n\}.^2$$

For any two vectors $\mathbf{x} = (x_1, x_2, \ldots, x_n)$ and $\mathbf{y} = (y_1, y_2, \ldots, y_n)$ in $\mathbb{R}^n$, and a scalar k, the *sum* $\mathbf{x}+\mathbf{y}$ and the *scalar multiplication* $k\mathbf{x}$ are defined by

$$\begin{aligned} \mathbf{x}+\mathbf{y} &= (x_1+y_1, x_2+y_2, \ldots, x_n+y_n), \\ k\mathbf{x} &= (kx_1, kx_2, \ldots, kx_n). \end{aligned}$$

It is not difficult to verify the following arithmetical rules of these operations.

[2] An n-tuple $(a_1, a_2, \ldots, a_n)$ is the *Cartesian coordinate* of a vector in $\mathbb{R}^n$.

Theorem 3.1 *For any scalars k and ℓ, and vectors $\mathbf{x} = (x_1, x_2, \ldots, x_n)$, $\mathbf{y} = (y_1, y_2, \ldots, y_n)$, and $\mathbf{z} = (z_1, z_2, \ldots, z_n)$ in the n-space $\mathbb{R}^n$, the following rules hold:*

(1) $\mathbf{x} + \mathbf{y} = \mathbf{y} + \mathbf{x}$,
(2) $\mathbf{x} + (\mathbf{y} + \mathbf{z}) = (\mathbf{x} + \mathbf{y}) + \mathbf{z}$,
(3) $\mathbf{x} + \mathbf{0} = \mathbf{x} = \mathbf{0} + \mathbf{x}$,
(4) $\mathbf{x} + (-1)\mathbf{x} = \mathbf{0}$,
(5) $k(\mathbf{x} + \mathbf{y}) = k\mathbf{x} + k\mathbf{y}$,
(6) $(k + \ell)\mathbf{x} = k\mathbf{x} + \ell\mathbf{x}$,
(7) $k(\ell\mathbf{x}) = (k\ell)\mathbf{x}$,
(8) $1\mathbf{x} = \mathbf{x}$,

where $\mathbf{0} = (0, 0, \ldots, 0)$ is the **zero** *vector. The first four rules are of the sum of vectors and the last four rules are of the scalar multiplication of a vector.*

A vector $(a_1, a_2, \ldots, a_n)$ in the n-space $\mathbb{R}^n$ is usually identified with an $n \times 1$ column vector

$$(a_1, a_2, \ldots, a_n) = \begin{bmatrix} a_1 \\ a_2 \\ \vdots \\ a_n \end{bmatrix} = [a_1 \ a_2 \ \cdots \ a_n]^T.$$

Sometimes a vector in $\mathbb{R}^n$ is also identified with a $1 \times n$ row vector (see Sections 2.2 and 3.7). The two algebraic operations of the sum of matrices and the scalar multiplication of matrices coincide with those of vectors in $\mathbb{R}^n$, and Theorem 3.1 rephrases Theorem 1.12.

The eight arithmetic rules in these two theorems are sometimes called *the* '4+4' *rules* in this book. They are the most important four properties for the sum of matrices and another four for the scalar multiplication of matrices, and commonly needed to manipulate vectors in the n-space $\mathbb{R}^n$.

For this reason, an (abstract) vector space is defined with respect to the '$4 + 4$' rules of operations of vectors in the n-space $\mathbb{R}^n$ so that $\mathbb{R}^n$ itself becomes a vector space. [3]

---

[3]Peano was the first to give the modern axiomatic definition of vector spaces and linear maps in 1888. **Giuseppe Peano** (August 27, 1858– April 20, 1932) was an Italian mathematician. As a founder of mathematical logic and set theory, he was interested in the foundations of mathematics. The standard axiomatization of the natural numbers is named the Peano axioms in his honor.

Definition 3.2 A (real) **vector space** (also called a **linear space**) is a nonempty set V of elements, called **vectors**, with two algebraic operations that satisfy the following rules.

(A) There is an operation called the **vector sum** that associates to every pair $\mathbf{x}$ and $\mathbf{y}$ of vectors in V a unique new vector $\mathbf{x}+\mathbf{y}$ in V, called the **sum** of $\mathbf{x}$ and $\mathbf{y}$, [4] so that the following four rules hold for all vectors $\mathbf{x}, \mathbf{y}, \mathbf{z}$ in V:

(1) $\mathbf{x}+\mathbf{y}=\mathbf{y}+\mathbf{x}$ (commutativity in sum),
(2) $\mathbf{x}+(\mathbf{y}+\mathbf{z})=(\mathbf{x}+\mathbf{y})+\mathbf{z}$, (written by $\mathbf{x}+\mathbf{y}+\mathbf{z}$) (associativity in sum),
(3) there is a unique vector $\mathbf{0}$ in V such that $\mathbf{x}+\mathbf{0}=\mathbf{x}=\mathbf{0}+\mathbf{x}$ for all $\mathbf{x}\in V$, (called the **zero vector**),
(4) for any $\mathbf{x}\in V$, there is a vector $-\mathbf{x}\in V$, called the **negative** of $\mathbf{x}$, such that $\mathbf{x}+(-\mathbf{x})=(-\mathbf{x})+\mathbf{x}=\mathbf{0}$.

(B) There is another operation called the **scalar multiplication** that associates to each vector $\mathbf{x}$ in V and each scalar k a unique vector $k\mathbf{x}$ in V [5] so that the following another four rules hold for all vectors $\mathbf{x}, \mathbf{y}$ in V and all scalars k, ℓ:

(5) $k(\mathbf{x}+\mathbf{y})=k\mathbf{x}+k\mathbf{y}$ (distributivity with respect to vector sum),
(6) $(k+\ell)\mathbf{x}=k\mathbf{x}+\ell\mathbf{x}$ (distributivity with respect to scalar sum),
(7) $k(\ell\mathbf{x})=(k\ell)\mathbf{x}$ (associativity with respect to scalar multiplication),
(8) $1\mathbf{x}=\mathbf{x}$.

Clearly, the n-space $\mathbb{R}^n$ is a vector space by Theorem 3.1.

A **complex vector space** is defined if, instead of real numbers, we take complex numbers for scalars. For example, the set $\mathbb{C}^n$ of all ordered n-tuples of complex numbers is a complex vector space. In Chapter 7, the complex vector spaces will be discussed separately, and until then we restrict ourselves to the real vector spaces unless otherwise stated.

Example 36 (*Miscellaneous examples of vector spaces*)

(1) For any two positive integers m and n,

$$\text{the set } M_{m\times n}(\mathbb{R}) \text{ of all } m\times n \text{ matrices is a vector space}$$

under the sum of matrices and the scalar multiplication defined in Section 1.5. The zero vector in this space is the zero matrix $\mathbf{0}_{m\times n}$ and $-A$ is the negative of a matrix A. See Theorem 1.12 again.

[4]Here, the sum $+$ is not just the usual sum of numbers as happened in $3+5=8$.
[5]Here, the multiplication in $k\mathbf{x}$ is not just the usual multiplication of numbers like $3\cdot 5=15$.

(2) Let A be any $m \times n$ matrix. Then

the set of solutions to the homogeneous system $A\mathbf{x} = \mathbf{0}$ is a vector space

(under the sum and the scalar multiplication of matrices). It is left as an exercise to verify the '$4 + 4$' arithmetic rules in the definition of a vector space.

(3) Let $C(\mathbb{R})$ denote

the set of all real-valued continuous functions defined on the real line $\mathbb{R}$.

For two functions f and g, and a real number k, the sum $f + g$ and the scalar multiplication kf of them are defined as usual. That is,

$$\begin{aligned}(f+g)(x) &= f(x)+g(x),\\(kf)(x) &= kf(x).\end{aligned}$$

Then the set $C(\mathbb{R})$ is a vector space under these operations. The zero vector in this space is the constant function whose value at each point is zero.

By taking a closed interval $[a, b]$ instead of the whole real line $\mathbb{R}$, the set $C[a, b]$ of all continuous real-valued functions defined on the interval $[a, b]$ is also a vector space with the same operations as ones in the vector space $C(\mathbb{R})$.

(4) Let $S(\mathbb{R})$ denote

the set of real-valued functions defined on the set of integers.

A function $f \in S(\mathbb{R})$ can be expressed as a doubly infinite sequence of real numbers:

$$\ldots,\ x_{-2}, x_{-1},\ x_0,\ x_1,\ x_2,\ \ldots,$$

where $x_k = f(k)$ for each k. These kinds of sequences appear frequently in engineering, and is called a discrete signal or a digital signal. [6] One can define the sum of two functions and the scalar multiplication of a function just as in $C(\mathbb{R})$ in (3) so that $S(\mathbb{R})$ also becomes a vector space. □

Theorem 3.3 *Let V be any vector space and let $\mathbf{x}$, $\mathbf{y}$ be any vectors in V. Then*

(1) $\mathbf{x} + \mathbf{y} = \mathbf{y}$ *implies* $\mathbf{x} = \mathbf{0}$*,*

(2) $0\mathbf{x} = \mathbf{0}$*,*

[6] The word *digital* is everywhere today. However, digital just means 'using numbers' and more often than not just two numbers, 0 and 1.

(3) $k\mathbf{0} = \mathbf{0}$ *for any* $k \in \mathbb{R}$,
(4) *A negative* $-\mathbf{x}$ *of* $\mathbf{x}$ *is unique, and* $-\mathbf{x} = (-1)\mathbf{x}$,
(5) *if* $k\mathbf{x} = \mathbf{0}$, *then* $k = 0$ *or* $\mathbf{x} = \mathbf{0}$.

Here, the symbol $\mathbf{0}$ *as a bold letter is the zero vector* [7] *in* V *and* 0 *as a roman letter is the number zero as a scalar.*

Proof: (1) By adding $-\mathbf{y}$ to both sides of $\mathbf{x} + \mathbf{y} = \mathbf{y}$, one can see

$$\mathbf{x} = \mathbf{x} + \mathbf{0} = \mathbf{x} + \mathbf{y} + (-\mathbf{y}) = \mathbf{y} + (-\mathbf{y}) = \mathbf{0}.$$

(2) $0\,\mathbf{x} = (0+0)\mathbf{x} = 0\,\mathbf{x} + 0\,\mathbf{x}$ implies $0\,\mathbf{x} = \mathbf{0}$ by (1).
(3) Similarly, $k\mathbf{0} = k(\mathbf{0} + \mathbf{0}) = k\mathbf{0} + k\mathbf{0}$ implies $k\mathbf{0} = \mathbf{0}$ by (1).
(4) Clearly, $(-1)\mathbf{x}$ is a negative of $\mathbf{x}$ because

$$\mathbf{x} + (-1)\mathbf{x} = 1\mathbf{x} + (-1)\mathbf{x} = (1-1)\mathbf{x} = 0\,\mathbf{x} = \mathbf{0}.$$

To show the uniqueness of the negative of $\mathbf{x}$, let $\bar{\mathbf{x}}$ is another negative of $\mathbf{x}$. Then

$$(-1)\mathbf{x} = (-1)\mathbf{x} + \mathbf{0} = (-1)\mathbf{x} + (\mathbf{x} + \bar{\mathbf{x}}) = ((-1)\mathbf{x} + \mathbf{x}) + \bar{\mathbf{x}} = \mathbf{0} + \bar{\mathbf{x}} = \bar{\mathbf{x}},$$

showing that $(-1)\mathbf{x}$ is the unique negative of $\mathbf{x}$, written simply by $-\mathbf{x}$.
(5) Suppose $k\mathbf{x} = \mathbf{0}$ and $k \neq 0$. Then $\mathbf{x} = 1\mathbf{x} = \frac{1}{k}(k\mathbf{x}) = \frac{1}{k}\mathbf{0} = \mathbf{0}$. □

Problem 47 Let V be the set of all pairs (x, y) of real numbers. Suppose that a sum and a scalar multiplication are defined by

$$(x, y) + (u, v) = (x + 2u,\ y + 2v), \quad k(x, y) = (kx,\ ky).$$

Is the set V a vector space under these operations? Try to verify the '4 + 4' rules in the definition of a vector space.

3.2 Creating new vector spaces from old ones

In this section, we introduce several ways of constructing new vector spaces.

[7] The symbol $\mathbf{0}$ is also used for the zero matrix.

3.2.1 Which subsets can be vector spaces?

The 3-space $\mathbb{R}^3$ is a founding example of a vector space. One can see that the x-axis is a vector space as the real line $\mathbb{R}^1$, and the xy-plane is also a vector space as the plane $\mathbb{R}^2$, with the same algebraic operations: the sum of (column) matrices and the scalar multiplication of (column) matrices in $\mathbb{R}^3$.

This concept can be generalized as follows.

Definition 3.4 A subset W of a vector space V is called a **subspace** of V if W itself is a vector space under the sum and the scalar multiplication defined in the space V.

In order to show that a (nonempty) subset W is a vector space to be a subspace of V, it is needed to verify all of the '$4+4$' rules in the definition of a vector space for the vectors in W. However, it is not necessary to verify all of these rules, and it is just needed to check *whether the subset W is closed under the vector sum and scalar multiplication* defined in V.

This is due to the fact that certain rules satisfied in a larger space are automatically satisfied in every smaller space. For example, the commutativity rule $\mathbf{x}+\mathbf{y}=\mathbf{y}+\mathbf{x}$ in sum is automatically satisfied for all $\mathbf{x}, \mathbf{y}$ in W because they are also vectors in the vector space V.

Theorem 3.5 *A nonempty subset W of a vector space V is a subspace if and only if $\mathbf{x}+\mathbf{y}$ and $k\mathbf{x}$ are contained in W (or equivalently, $\mathbf{x}+k\mathbf{y} \in W$) for any vectors $\mathbf{x}$ and $\mathbf{y}$ in W and any scalar $k \in \mathbb{R}$.*

Proof: We need only to prove the sufficiency. Assume both conditions hold in W and let $\mathbf{x}$ be any vector in W. Since W is closed under scalar multiplication, $\mathbf{0} = 0\mathbf{x}$ and $-\mathbf{x} = (-1)\mathbf{x}$ are in W, so rules (3) and (4) for a vector space hold. All the others in the '$4+4$' rules for a vector space are clearly (and automatically) satisfied in the subset W. □

The next example shows that a plane W in the 3-space $\mathbb{R}^3$ is a subspace if and only if it contains the origin $\mathbf{0}$.

Example 37 (*Which planes in the 3-space $\mathbb{R}^3$ can be subspaces?*) Let

$$W_0 = \{(x, y, z) \in \mathbb{R}^3 \ : \ ax + by + cz = 0\},$$

where a, b, c are constants, not all zero. Then W_0 is a plane in $\mathbb{R}^3$ passing through the origin $\mathbf{0}$. If $\mathbf{x} = (x_1, x_2, x_3)$, $\mathbf{y} = (y_1, y_2, y_3)$ are points in W_0, then clearly $\mathbf{x} + \mathbf{y} = (x_1 + y_1, x_2 + y_2, x_3 + y_3)$ is also a point in W_0, because it satisfies the equation $ax + by + cz = 0$. Similarly, $k\mathbf{x}$ also lies in W_0 for any scalar k. Hence, W_0 is a subspace of $\mathbb{R}^3$.

However, one can show that the subset

$$W_1 = \{(x, y, z) \in \mathbb{R}^3 \,:\, ax + by + cz = 1\}$$

is not a subspace of $\mathbb{R}^3$, because for any two points $\mathbf{x} = (x_1, x_2, x_3)$, $\mathbf{y} = (y_1, y_2, y_3)$ in W_1, their sum $\mathbf{x} + \mathbf{y} = (x_1 + y_1, x_2 + y_2, x_3 + y_3)$ is not contained in W_1 since $a(x_1 + y_1) + b(x_2 + y_2) + c(x_3 + y_3) = 1 + 1 = 2$.

Consequently, a plane $ax + by + cz = d$ is a subspace of the 3-space $\mathbb{R}^3$ if and only if $d = 0$, because all subspaces must contain the zero vector $\mathbf{0}$. □

A vector space V itself and the zero vector $\{\mathbf{0}\}$ are trivially subspaces, called the **trivial subspaces** of V. Several nontrivial subspaces are given in the following examples.

Example 38 (*The space of polynomials of degree* $\le n$) For a nonnegative integer n, let

$P_n(\mathbb{R})$ denote the set of all real polynomials in x of degree $\le n$.

Then $P_n(\mathbb{R})$ is a subspace of the vector space $C(\mathbb{R})$ of all continuous functions defined on $\mathbb{R}$, because $P_n(\mathbb{R})$ is closed under the sum $f+g$ of two real polynomials and the scalar multiplication kf.

In other words, for any two f and g in $P_n(\mathbb{R})$ and any scalar $k \in \mathbb{R}$, $f + g$ and kf are contained in $P_n(\mathbb{R})$. Use Theorem 3.5. □

Example 39 (*The space of symmetric or skew-symmetric matrices*) Let W be *the set of all $n \times n$ real symmetric matrices.* Then W is a subspace of the vector space $M_{n\times n}(\mathbb{R})$ of all $n \times n$ matrices because the sum of two symmetric matrices is symmetric and a scalar multiple of a symmetric matrix is also symmetric.

Similarly, *the set of all $n \times n$ skew-symmetric matrices* is also a subspace of $M_{n\times n}(\mathbb{R})$. Also, use Theorem 3.5. □

Example 40 (*The solutions of* $A\mathbf{x} = \mathbf{0}$ *form a subspace*) Let A be an $m \times n$ matrix. Then, the set

$$W = \{\mathbf{x} \in \mathbb{R}^n \ : \ A\mathbf{x} = \mathbf{0}\}$$

of solutions to the homogeneous system $A\mathbf{x} = \mathbf{0}$ is a subspace of the n-space $\mathbb{R}^n$ because it is closed under the sum of (column) matrices and the scalar multiplication of (column) matrices. Telling again, it is not necessary to verify all of the '$4 + 4$' rules as mentioned in Example 36(2). □

Problem 48 Show that a straight line in the 3-space $\mathbb{R}^3$ is a subspace of $\mathbb{R}^3$ if and only if it passes through the origin $\mathbf{0}$.

Hint: In Example 40, if $A = \begin{bmatrix} a_{11} & a_{12} & a_{13} \\ a_{21} & a_{22} & a_{23} \end{bmatrix}$ as a 2×3 matrix, then the set of solutions to $A\mathbf{x} = \mathbf{0}$ is a (straight) line or a plane in $\mathbb{R}^3$ which passes through the origin $\mathbf{0}$.

Problem 49 Which of the following sets are subspaces of the 3-space $\mathbb{R}^3$? Justify your answer.

(1) $W = \{(x, y, z) \in \mathbb{R}^3 \ : \ xyz = 0\}$,
(2) $W = \{(x, y, z) \in \mathbb{R}^3 \ : \ x + y + z = 0\}$,
(3) $W = \{(2t, 3t, 4t) \in \mathbb{R}^3 \ : \ t \in \mathbb{R}\}$,
(4) $W = \{(x, y, z) \in \mathbb{R}^3 : x^2 + y^2 - z^2 = 0\}$,

Can you describe all subspaces of the 3-space $\mathbb{R}^3$?

Problem 50 Let $V = C(\mathbb{R})$ be the vector space of all continuous functions on $\mathbb{R}$. Which of the following sets W are subspaces of V? Justify your answer.

(1) W is the set of all bounded continuous functions defined on $\mathbb{R}$.
(2) W is the set of all continuous nonnegative-valued functions on $\mathbb{R}$, *i.e.*, $f(x) \geq 0$ for any $x \in \mathbb{R}$.
(3) W is the set of all continuous *odd* functions defined on $\mathbb{R}$, *i.e.*, $f(-x) = -f(x)$ for any $x \in \mathbb{R}$.
(4) W is the set of all polynomials with integer coefficients.

3.2.2 Sums and intersections of subspaces

Let U and W be two subspaces of a vector space V. Then, their intersection $U \cap W$ is clearly a subspace of V, by Theorem 3.5. [8]

[8] The *intersection* of two sets A and B is written as $A \cap B$. Formally: $A \cap B = \{x : x \in A \text{ and } x \in B\}$. The *union* of A and B is written as $A \cup B$: $A \cup B = \{x : x \in A \text{ or } x \in B\}$. Hence, the intersection $A \cap B$ is the set of all elements that are members of both A and B, and the union $A \cup B$ is the set of elements which are members of A, of B, or of both A and B.

As another way of constructing vector spaces, we introduce a sum of two subspaces.

Definition 3.6 Let U and W be two subspaces of a vector space V.

(1) The **sum** of U and W is defined by

$$U + W = \{\mathbf{u} + \mathbf{w} \ : \ \mathbf{u} \in U, \ \mathbf{w} \in W\},$$

as the set of vectors $\mathbf{u} + \mathbf{w}$ in V for any $\mathbf{u} \in U$ and $\mathbf{w} \in W$.

(2) When $U \cap W = \{\mathbf{0}\}$, the sum $U + W$ is called the **direct sum** of subspaces U and W, written as $U \oplus W$.

The sum $U + W$ and the direct sum $U \oplus W$ are both subspaces of V (by Theorem 3.5). If $V = \mathbb{R}^2$ (the xy-plane), $U = \{x\mathbf{i} : x \in \mathbb{R}\}$ (the x-axis), and $W = \{y\mathbf{j} : y \in \mathbb{R}\}$ (the y-axis) where $\mathbf{i} = (1, 0)$ and $\mathbf{j} = (0, 1)$, then it is easy to see that $\mathbb{R}^2 = U \oplus V = \mathbb{R} \oplus \mathbb{R}$, by considering the x-axis (and also y-axis) as $\mathbb{R}$.

Similarly, one can define the sum or direct sum of more than two subspaces of V. For example, $\mathbb{R}^3 = \mathbb{R}^2 \oplus \mathbb{R}^1 = \mathbb{R}^1 \oplus \mathbb{R}^1 \oplus \mathbb{R}^1$, as the sum of the x-axis, the y-axis, and the z-axis.

Theorem 3.7 *A vector space V is the direct sum of subspaces U and W, i.e., $V = U \oplus W$, if and only if for any $\mathbf{v} \in V$ there exist* unique *$\mathbf{u} \in U$ and $\mathbf{w} \in W$ such that $\mathbf{v} = \mathbf{u} + \mathbf{w}$.*

Proof: ($\Rightarrow$) Suppose that $V = U \oplus W$. Then, for any $\mathbf{v} \in V$, there exist vectors $\mathbf{u} \in U$ and $\mathbf{w} \in W$ such that $\mathbf{v} = \mathbf{u} + \mathbf{w}$, since $V = U + W$. To show the uniqueness, suppose that $\mathbf{v}$ is also expressed as a sum $\mathbf{u}' + \mathbf{w}'$ for $\mathbf{u}' \in U$ and $\mathbf{w}' \in W$. Then $\mathbf{u} + \mathbf{w} = \mathbf{u}' + \mathbf{w}'$ implies

$$\mathbf{u} - \mathbf{u}' = \mathbf{w}' - \mathbf{w} \in U \cap W = \{\mathbf{0}\}.$$

Hence, $\mathbf{u} = \mathbf{u}'$ and $\mathbf{w} = \mathbf{w}'$.

($\Leftarrow$) Clearly, $V = U + W$. Suppose that there exists a nonzero vector $\mathbf{v}$ in $U \cap W$. Then $\mathbf{v}$ can be expressed as a sum of vectors in U and W in many different ways: $\mathbf{v} = \mathbf{v} + \mathbf{0} = \mathbf{0} + \mathbf{v} = \frac{1}{2}\mathbf{v} + \frac{1}{2}\mathbf{v} = \frac{1}{3}\mathbf{v} + \frac{2}{3}\mathbf{v} \in U + W$. $\square$

Example 41 (*Sum, but not direct sum*) In the 3-space $\mathbb{R}^3$, consider the three vectors $\mathbf{e}_1 = (1,0,0)$, $\mathbf{e}_2 = (0,1,0)$, and $\mathbf{e}_3 = (0,0,1)$. These three vectors are also well known as **i**, **j**, and **k**, respectively. Let $U = \{a\mathbf{i} + c\mathbf{k} : a, c \in \mathbb{R}\}$ be the xz-plane, and let $W = \{b\mathbf{j} + c\mathbf{k} : b, c \in \mathbb{R}\}$ be the yz-plane, which are both subspaces of $\mathbb{R}^3$. Then a vector in $U + W$ is of the form

$$\begin{aligned}(a\mathbf{i} + c_1\mathbf{k}) + (b\mathbf{j} + c_2\mathbf{k}) &= a\mathbf{i} + b\mathbf{j} + (c_1 + c_2)\mathbf{k} \\ &= a\mathbf{i} + b\mathbf{j} + c\mathbf{k} = (a, b, c),\end{aligned}$$

where $c = c_1 + c_2$ and a, b, c can be arbitrary scalars. Thus $U + W = \mathbb{R}^3$.

However, $\mathbb{R}^3 \neq U \oplus W$ since $\mathbf{k} \in U \cap W \neq \{\mathbf{0}\}$. In fact, the vector $\mathbf{k} \in \mathbb{R}^3$ can be expressed as linear combinations of vectors in U and W in many different ways:

$$\mathbf{k} = \frac{1}{2}\mathbf{k} + \frac{1}{2}\mathbf{k} = \frac{1}{3}\mathbf{k} + \frac{2}{3}\mathbf{k} \in U + W.$$

If we had taken $W = \{y\mathbf{j} : y \in \mathbb{R}\}$ to be the y-axis, then $\mathbb{R}^3 = U \oplus W$. However, there are infinitely many different choices of W for a direct sum $\mathbb{R}^3 = U \oplus W$. In fact, W can be any straight line passing through the origin **0** which does not lie in the xz-plane U. □

Problem 51 Let U and W be subspaces of a vector space V.

(1) Show that both $U + W$ and $U \cap W$ are subspaces of V.
(2) Show that if Z is a subspace of V contained in both U and W, then Z is also contained in $U \cap W$ as a subspace.
(3) Show that if Z is a subspace of V containing both U and W, then Z also contains $U + W$ as a subspace.

Problem 52 Let U and W be the subspaces of the vector space $M_{n\times n}(\mathbb{R})$ consisting of all symmetric matrices and all skew-symmetric matrices, respectively. Show that $M_{n\times n}(\mathbb{R}) = U \oplus W$. Therefore, the decomposition of a square matrix A into the sum of symmetric part and skew-symmetric part given in Problem 10(3) is unique (as shown already in Problem 10).

3.2.3 Subspace spanned by a set

Any vector (x_1, x_2, x_3) in the 3-space $\mathbb{R}^3$ is of the form

$$(x_1, x_2, x_3) = x_1(1,0,0) + x_2(0,1,0) + x_3(0,0,1).$$

That is, any vector in $\mathbb{R}^3$ can be expressed as a sum of scalar multiples of three vectors $\mathbf{e}_1 = (1,0,0)$, $\mathbf{e}_2 = (0,1,0)$, and $\mathbf{e}_3 = (0,0,1)$.

The following definition gives a name to such an expression.

Definition 3.8 Let V be a vector space and let $\{\mathbf{x}_1, \mathbf{x}_2, \ldots, \mathbf{x}_m\}$ be a set of vectors in V. Then a vector $\mathbf{y}$ in V of the form

$$\mathbf{y} = a_1\mathbf{x}_1 + a_2\mathbf{x}_2 + \cdots + a_m\mathbf{x}_m$$

with scalars a_i is called a **linear combination** of the vectors $\mathbf{x}_1, \mathbf{x}_2, \ldots, \mathbf{x}_m$.

The next theorem shows that the set of all linear combinations of a given m vectors in a vector space forms a subspace. This is a new way of constructing subspaces or vector spaces.

Theorem 3.9 *Let $\mathbf{x}_1, \mathbf{x}_2, \ldots, \mathbf{x}_m$ be given in a vector space V. Then the set*

$$W = \{a_1\mathbf{x}_1 + a_2\mathbf{x}_2 + \cdots + a_m\mathbf{x}_m \; : \; a_i \in \mathbb{R}\}$$

of all linear combinations of $\mathbf{x}_1, \mathbf{x}_2, \ldots, \mathbf{x}_m$ is a subspace of V.

The subspace W is called the **subspace of V spanned** by $\mathbf{x}_1, \mathbf{x}_2, \ldots, \mathbf{x}_m$. Or, $\mathbf{x}_1, \mathbf{x}_2, \ldots, \mathbf{x}_m$ **span** the subspace W. The subspace W is sometimes written by $\langle \mathbf{x}_1, \mathbf{x}_2, \ldots, \mathbf{x}_m \rangle$.

Proof: It is enough to show that W is closed under the vector sum and the scalar multiplication. Let $\mathbf{u}$ and $\mathbf{w}$ be any two vectors in W. Then

$$\begin{array}{rcccccccc} \mathbf{u} & = & a_1\mathbf{x}_1 & + & a_2\mathbf{x}_2 & + & \cdots & + & a_m\mathbf{x}_m, \\ \mathbf{w} & = & b_1\mathbf{x}_1 & + & b_2\mathbf{x}_2 & + & \cdots & + & b_m\mathbf{x}_m \end{array}$$

for some scalars a_i's and b_j's. It gives

$$\mathbf{u} + \mathbf{w} = (a_1 + b_1)\mathbf{x}_1 + (a_2 + b_2)\mathbf{x}_2 + \cdots + (a_m + b_m)\mathbf{x}_m,$$

and for any scalar k,

$$k\mathbf{u} = (ka_1)\mathbf{x}_1 + (ka_2)\mathbf{x}_2 + \cdots + (ka_m)\mathbf{x}_m.$$

Thus, $\mathbf{u} + \mathbf{w}$ and $k\mathbf{u}$ are also linear combinations of $\mathbf{x}_1, \mathbf{x}_2, \ldots, \mathbf{x}_m$ and hence contained in W. This shows that W is a subspace of V, by Theorem 3.5 again. □

Example 42 (*For any $m \leq n$, $\mathbb{R}^m$ is a subspace of $\mathbb{R}^n$*) Let

$$\begin{array}{rcl} \mathbf{e}_1 & = & (1, 0, 0, \ldots, 0), \\ \mathbf{e}_2 & = & (0, 1, 0, \ldots, 0), \\ & \vdots & \\ \mathbf{e}_n & = & (0, 0, 0, \ldots, 1) \end{array}$$

be n vectors in the n-space $\mathbb{R}^n$ $(n \geq 3)$. Then a linear combination of $\mathbf{e}_1, \mathbf{e}_2, \mathbf{e}_3$ is of the form

$$a_1\mathbf{e}_1 + a_2\mathbf{e}_2 + a_3\mathbf{e}_3 = (a_1, a_2, a_3, 0, \ldots, 0).$$

Hence, the set

$$W = \{(a_1, a_2, a_3, 0, \ldots, 0) \in \mathbb{R}^n \ : \ a_1, a_2, a_3 \in \mathbb{R}\}$$

is a subspace of the n-space $\mathbb{R}^n$ spanned by the vectors $\mathbf{e}_1, \mathbf{e}_2, \mathbf{e}_3$. ($W$ is also closed under the sum and the scalar multiplication of column matrices.) Certainly, the subspace W *can be identified* with the 3-space $\mathbb{R}^3$ through the identification

$$(a_1, a_2, a_3, 0, \ldots, 0) \equiv (a_1, a_2, a_3)$$

with $a_i \in \mathbb{R}$. (Strictly speaking, $(a_1, a_2, a_3, 0, \ldots, 0) \neq (a_1, a_2, a_3)$.) In this sense, for any $m \leq n$, the m-space $\mathbb{R}^m$ *can be considered* as a subspace of $\mathbb{R}^n$. □

Example 43 (*All $A\mathbf{x}$'s form the column space*) Let $A = [\mathbf{c}_1 \ \mathbf{c}_2 \cdots \mathbf{c}_n]$ be an $m \times n$ matrix with column $\mathbf{c}_i$'s. These column vectors belong to $\mathbb{R}^m$, and the matrix product $A\mathbf{x}$ represents the linear combination of the column vectors $\mathbf{c}_i$ whose coefficients are the components of $\mathbf{x} \in \mathbb{R}^n$, *i.e.*,

$$A\mathbf{x} = x_1\mathbf{c}_1 + x_2\mathbf{c}_2 + \cdots + x_n\mathbf{c}_n,$$

(see Eq. (1.4) in Page 42). Therefore, the set

$$W = \{A\mathbf{x} \in \mathbb{R}^m \ : \mathbf{x} \in \mathbb{R}^n\}$$

of all linear combinations of the column vectors of A is a subspace of $\mathbb{R}^m$, called the **column space** of A.

Consequently, $A\mathbf{x} = \mathbf{b}$ *has a solution* $\mathbf{x} = (x_1, x_2, \ldots, x_n)$ *in* $\mathbb{R}^n$ *if and only if the vector* $\mathbf{b}$ *belongs to the column space of A.* □

Remark: One can take another point of view on the equation $A\mathbf{x} = \mathbf{b}$. For a vector $\mathbf{x} \in \mathbb{R}^n$, an $m \times n$ matrix A assigns a vector $\mathbf{b}$ in $\mathbb{R}^m$ which is given as $A\mathbf{x}$. That is, A can be considered as a function from $\mathbb{R}^n$ into $\mathbb{R}^m$; $\mathbf{x} \mapsto A\mathbf{x}$, and then the column space of A is just the image of this function $A : \mathbb{R}^n \to \mathbb{R}^m$.

Example 44 (*A vector space can be spanned in many different ways*)

(1) For a nonzero vector $\mathbf{v}$ in a vector space V, a linear combination of $\mathbf{v}$ is simply a scalar multiple of $\mathbf{v}$. Thus the subspace W of V spanned by $\mathbf{v}$ is

$W = \{k\mathbf{v} : k \in \mathbb{R}\}$. The subspace W can also be spanned by $-\mathbf{v}$, or by any of $2\mathbf{v}, 3\mathbf{v}$, etc. By notation, $\langle \mathbf{v} \rangle = \langle -\mathbf{v} \rangle = \langle 2\mathbf{v} \rangle$, etc.

(2) Consider three vectors $\mathbf{e}_1 = (1,0,0)$, $\mathbf{e}_2 = (0,1,0)$, and $\mathbf{d} = \mathbf{e}_1 + \mathbf{e}_2 = (1,1,0)$ in $\mathbb{R}^3$. The subspace spanned by $\mathbf{e}_1$ and $\mathbf{e}_2$ is

$$W_1 = \{a_1\mathbf{e}_1 + a_2\mathbf{e}_2 = (a_1, a_2, 0) : a_i \in \mathbb{R}\}.$$

The subspace spanned by $\mathbf{e}_1$, $\mathbf{e}_2$, and $\mathbf{d}$ is

$$W_2 = \{a_1\mathbf{e}_1 + a_2\mathbf{e}_2 + a_3\mathbf{d} = (a_1 + a_3, a_2 + a_3, 0) : a_i \in \mathbb{R}\}.$$

Clearly, $W_1 \subseteq W_2$, a subset. On the other hand, since $\mathbf{d} = \mathbf{e}_1 + \mathbf{e}_2 \in W_1$, $W_2 \subseteq W_1$. Thus $W_1 = W_2$ which is just the xy-plane in $\mathbb{R}^3$ and hence the vector $\mathbf{d} = \mathbf{e}_1 + \mathbf{e}_2$ is redundant in the spanning set.

In general, a subspace in a vector space has (infinitely) many different spanning sets. □

As shown in Theorem 3.9 and Example 44, any nonempty subset of a vector space V spans a subspace through the linear combinations of the vectors, and two different subsets may span the same subspace.

In Example 44(2), we saw that the xy-plane W is spanned by two vectors $\mathbf{e}_1 = (1,0,0)$ and $\mathbf{e}_2 = (0,1,0)$, in which every vector $(x, y, 0)$ in W is *uniquely* expressed as a linear combination of $\mathbf{e}_1$ and $\mathbf{e}_2$: $(x, y, 0) = x\mathbf{e}_1 + y\mathbf{e}_2$.

The space W is also spanned by $\mathbf{e}_1, \mathbf{e}_2$, and $\mathbf{d} = \mathbf{e}_1 + \mathbf{e}_2 = (1,1,0)$. In this case, every vector in W can be expressed as a linear combination of $\mathbf{e}_1$, $\mathbf{e}_2$, and $\mathbf{d}$ in infinitely many different ways. For example, $\mathbf{x} = (k, k, 0) = t\mathbf{e}_1 + t\mathbf{e}_2 + (k-t)\mathbf{d}$ for any real t. In fact, the redundant vector $\mathbf{d}$ gives the possibility of infinitely many different ways of constructing a linear combination for any vector in the xy-plane.

Problem 53 Does the vector $(3, -1, 3)$ belong to the subspace of $\mathbb{R}^3$ spanned by two vectors $(1, -1, 2)$, and $(1, 1, 1)$?

Hint: Are there scalars a, b satisfying $(3, -1, 3) = a(1, -1, 2) + b(1, 1, 1)$?

Problem 54 Let A be any $m \times n$ matrix and let

$$\begin{aligned} \mathbf{e}_1 &= (1,0,0,\ldots,0), \\ \mathbf{e}_2 &= (0,1,0,\ldots,0), \\ &\vdots \\ \mathbf{e}_n &= (0,0,0,\ldots,1) \end{aligned}$$

be n vectors in the n-space $\mathbb{R}^n$. Show that the matrix product $A\mathbf{e}_i$ is just the i-th column of the matrix A for $i = 1, 2, \ldots, n$. Thus, $A\mathbf{e}, A\mathbf{e}_2, \ldots, A\mathbf{e}_n$ span the column space of A.

Hint: Use Example 43.

3.3 Linear dependence and Linear independence

What kinds of subsets of a vector space V span a subspace in which every vector should be expressed *uniquely* as a linear combination of the vectors in the spanning set? We try to find an answer to this question.

Definition 3.10 A set of vectors $\{\mathbf{x}_1, \mathbf{x}_2, \ldots, \mathbf{x}_m\}$ in a vector space V is said to be **linearly independent** if the vector equation, called the **linear dependence** of $\mathbf{x}_i$'s,

$$c_1\mathbf{x}_1 + c_2\mathbf{x}_2 + \cdots + c_m\mathbf{x}_m = \mathbf{0}$$

with unknowns c_i has only the trivial solution $c_1 = c_2 = \cdots = c_m = 0$. In other words, the zero vector $\mathbf{0}$ is expressed *uniquely* as a linear combination (with zero coefficients) of the vectors $\mathbf{x}_1, \mathbf{x}_2, \ldots, \mathbf{x}_m$.

Otherwise, it is said to be **linearly dependent**.

By definition, a set of vectors $\{\mathbf{x}_1, \mathbf{x}_2, \ldots, \mathbf{x}_m\}$ is linearly dependent if and only if the linear dependence

$$c_1\mathbf{x}_1 + c_2\mathbf{x}_2 + \cdots + c_m\mathbf{x}_m = \mathbf{0}$$

has a nontrivial solution $(c_1, c_2, \ldots, c_m) \neq (0, 0, \ldots, 0)$. Equivalently, the zero vector $\mathbf{0}$ can be expressed as a linear combination of the vectors $\mathbf{x}_1, \mathbf{x}_2, \ldots, \mathbf{x}_m$ in *two or more different ways*.

In this case, by assuming $c_m \neq 0$, the equation can be rewritten as

$$\mathbf{x}_m = -\frac{c_1}{c_m}\mathbf{x}_1 - \frac{c_2}{c_m}\mathbf{x}_2 - \cdots - \frac{c_{m-1}}{c_m}\mathbf{x}_{m-1}.$$

That is,

a set of vectors is linearly dependent if and only if at least one of the vectors in the set can be expressed as a linear combination of all the others.

Example 45 (*Linear independence in the 3-space* $\mathbb{R}^3$)

(1) Let $\mathbf{x} = (1, 2, 3)$ and $\mathbf{y} = (3, 2, 1)$ in $\mathbb{R}^3$. Then clearly $\mathbf{y} \neq \lambda\mathbf{x}$ for any $\lambda \in \mathbb{R}$ (or $a\mathbf{x} + b\mathbf{y} = \mathbf{0}$ is possible only when $a = b = 0$). Hence, $\{\mathbf{x}, \mathbf{y}\}$ is linearly independent in $\mathbb{R}^3$.

(2) Let $\mathbf{w} = (3, 6, 9)$. Then $\{\mathbf{x}, \mathbf{w}\}$ is linearly dependent since $\mathbf{w} - 3\mathbf{x} = \mathbf{0}$.

(3) Let $\mathbf{x}, \mathbf{y}$ be any two non-collinear vectors in the 3-space $\mathbb{R}^3$. Then, the set of all linear combinations of $\mathbf{x}$ and $\mathbf{y}$ determines a plane W through the origin in $\mathbb{R}^3$, *i.e.*,

$$W = \{a\mathbf{x} + b\mathbf{y} : a, b \in \mathbb{R}\}.$$

Let $\mathbf{z}$ be another nonzero vector in the 3-space $\mathbb{R}^3$. If $\mathbf{z} \in W$, then there are some scalars $a, b \in \mathbb{R}$, not all of them are zero, such that $\mathbf{z} = a\mathbf{x}+b\mathbf{y}$, so the set $\{\mathbf{x}, \mathbf{y}, \mathbf{z}\}$ is linearly dependent. If $\mathbf{z} \notin W$, then $a\mathbf{x} + b\mathbf{y} + c\mathbf{z} = \mathbf{0}$ is possible only when $a = b = c = 0$ (prove it).

As a result, *the set* $\{\mathbf{x}, \mathbf{y}, \mathbf{z}\}$ *is linearly independent if and only if* $\mathbf{z} \notin \mathbf{W}$. □

By abuse of language, sometimes we say that 'the vectors $\mathbf{x}_1, \mathbf{x}_2, \ldots, \mathbf{x}_m$ are linearly independent,' although this is really a property of a set $\{\mathbf{x}_1, \mathbf{x}_2, \ldots, \mathbf{x}_m\}$.

Example 46 The columns of the matrix

$$A = \begin{bmatrix} 2 & -2 & 0 \\ 4 & 2 & 6 \\ 2 & 3 & 5 \end{bmatrix}$$

are linearly dependent in the 3-space $\mathbb{R}^3$, since the third column is the sum of the first and the second.

As shown in Example 46, the concept of linear dependence can be applied to the row or column vectors of any matrix.

Example 47 Consider an upper triangular matrix

$$A = \begin{bmatrix} 2 & 3 & 4 \\ 0 & 1 & 6 \\ 0 & 0 & 7 \end{bmatrix}.$$

The linear dependence of the column vectors of A can be written as

$$c_1 \begin{bmatrix} 2 \\ 0 \\ 0 \end{bmatrix} + c_2 \begin{bmatrix} 3 \\ 1 \\ 0 \end{bmatrix} + c_3 \begin{bmatrix} 4 \\ 6 \\ 7 \end{bmatrix} = \begin{bmatrix} 0 \\ 0 \\ 0 \end{bmatrix}.$$

In matrix notation, this can be expressed as a homogeneous system

$$\begin{bmatrix} 2 & 3 & 4 \\ 0 & 1 & 6 \\ 0 & 0 & 7 \end{bmatrix} \begin{bmatrix} c_1 \\ c_2 \\ c_3 \end{bmatrix} = \begin{bmatrix} 0 \\ 0 \\ 0 \end{bmatrix}.$$

From the third row, $c_3 = 0$, then from the second row $c_2 = 0$, and substitution of them into the first row forces $c_1 = 0$, *i.e.*, the homogeneous system $A\mathbf{x} = \mathbf{0}$ has only the trivial solution, so that the column vectors of A are linearly independent.

For a similar reason, the n vectors $\mathbf{e}_1, \mathbf{e}_2, \ldots, \mathbf{e}_n$ defined in Example 42 are linearly independent in the n-space $\mathbb{R}^n$. □

Let $A = [\mathbf{c}_1\ \mathbf{c}_2\ \cdots\ \mathbf{c}_n]$ be any $m \times n$ matrix with columns $\mathbf{c}_1, \mathbf{c}_2, \ldots, \mathbf{c}_n$. Then, their linear dependence $c_1\mathbf{c}_1 + c_2\mathbf{c}_2 + \cdots + c_n\mathbf{c}_n = \mathbf{0}$ is just the homogeneous system $A\mathbf{x} = \mathbf{0}$, where $\mathbf{x} = (c_1, c_2, \ldots, c_n)$.

It gives the following.

Lemma 3.11 *The column vectors $\mathbf{c}_i$'s of an $m \times n$ matrix A are linearly independent in $\mathbb{R}^m$ if and only if the homogeneous system $A\mathbf{x} = \mathbf{0}$ has only the trivial solution.*

Equivalently they are linearly dependent if and only if $A\mathbf{x} = \mathbf{0}$ has a nontrivial solution.

The following theorem can be proved by the same argument.

Theorem 3.12 **(1)** *The nonzero rows of any rectangular matrix in the (reduced) row-echelon form are linearly independent, and so are the columns that contain leading 1's.*

(2) *If U is the (reduced) row-echelon form of A, then the columns of U are linearly independent if and only if the columns of A are linearly independent.*

Corollary 3.12.1 *The rows of any triangular or diagonal matrix with nonzero diagonal entries are linearly independent, and so are the columns.*

If U is the reduced row-echelon form of A, then $A\mathbf{x} = \mathbf{0}$ and $U\mathbf{x} = \mathbf{0}$ have the same set of solutions. Moreover, a homogeneous system $U\mathbf{x} = \mathbf{0}$ with more unknowns than equations always has at least one free variable and hence has at least one nontrivial solution by Theorem 1.5.

It gives the following theorem.

Theorem 3.13 *If $n > m$, any set of n vectors in the m-space $\mathbb{R}^m$ is linearly dependent.*

For example, any four, five, or six vectors in the 3-space $\mathbb{R}^3$ are always linearly dependent. This theorem will be used frequently later.

Problem 55 Consider four vectors $\mathbf{e}_1 = (1,0,0)$, $\mathbf{e}_2 = (0,1,0)$, $\mathbf{e}_3 = (0,0,1)$, and $\mathbf{d} = \mathbf{e}_1 + \mathbf{e}_2 + \mathbf{e}_3 = (1,1,1)$ in the 3-space $\mathbb{R}^3$. Show that any one of these four vectors can be expressed as a linear combination of all the other vectors. It means that any three vectors of them span the space $\mathbb{R}^3$.

Problem 56 Let W be the subspace of a vector space V spanned by $\mathbf{x}_1, \mathbf{x}_2, \ldots, \mathbf{x}_m$. Show that every vector in W is expressed *uniquely* as a linear combination of the vectors $\mathbf{x}_1, \mathbf{x}_2, \ldots, \mathbf{x}_m$ if and only if the zero vector $\mathbf{0}$ is expressed *uniquely* as a linear combination of the same vectors $\mathbf{x}_1, \mathbf{x}_2, \ldots, \mathbf{x}_m$, that is, $\mathbf{x}_1, \mathbf{x}_2, \ldots, \mathbf{x}_m$ are linearly independent.

Problem 57 Let W be the subspace of a vector space V spanned by $\mathbf{x}_1, \mathbf{x}_2, \ldots, \mathbf{x}_m$. Show that W is the smallest subspace of V containing $\mathbf{x}_1, \mathbf{x}_2, \ldots, \mathbf{x}_m$. In other words, if U is a subspace of V containing $\mathbf{x}_1, \mathbf{x}_2, \ldots, \mathbf{x}_m$, then $W \subseteq U$, a subset.

3.4 Bases and coordinate systems for vector spaces

Let's get back to the founding example of a vector space $\mathbb{R}^3$ again. The vectors $\mathbf{e}_1 = (1,0,0)$, $\mathbf{e}_2 = (0,1,0)$, and $\mathbf{e}_3 = (0,0,1)$ span the 3-space $\mathbb{R}^3$, by the identity

$$(x_1, x_2, x_3) = x_1(1,0,0) + x_2(0,1,0) + x_3(0,0,1).$$

Moreover, it is clear from this equality that $x_1(1,0,0) + x_2(0,1,0) + x_3(0,0,1) = (0,0,0)$ if and only if $x_1 = x_2 = x_3 = 0$. It shows that the set of vectors $\{\mathbf{e}_1, \mathbf{e}_2, \mathbf{e}_3\}$ is linearly independent. (See Corollary 3.12.1 or Example 47).

Definition 3.14 Let V be a vector space. A **basis** for V is any set of linearly independent vectors that spans V.

Example 48 (*The standard basis for* $\mathbb{R}^n$ *induces the standard rectangular coordinate system for* $\mathbb{R}^n$) The vectors

$$\begin{aligned}
\mathbf{e}_1 &= (1,0,0,\ldots,0), \\
\mathbf{e}_2 &= (0,1,0,\ldots,0), \\
&\vdots \\
\mathbf{e}_n &= (0,0,0,\ldots,1)
\end{aligned}$$

in the n-space $\mathbb{R}^n$ are linearly independent by Corollary 3.12.1 or Example 47). Moreover, they span the n-space $\mathbb{R}^n$, by the identity

$$\mathbf{x} = (x_1, x_2, \ldots, x_n) = x_1\mathbf{e}_1 + x_2\mathbf{e}_2 + \cdots + x_n\mathbf{e}_n.$$

Hence, the set $\{\mathbf{e}_1, \mathbf{e}_2, \ldots, \mathbf{e}_n\}$ is a basis, called the **standard basis** for the n-space $\mathbb{R}^n$.

It gives the rectangular coordinate system for $\mathbb{R}^n$, like the xyz-coordinates in the Euclidean space $\mathbb{R}^3$. The readers are suggested to find a non-standard basis for $\mathbb{R}^n$. □

Example 49 (*Which one is a basis? Which one is not?*)

(1) The set of vectors $(1,1,0)$, $(0,-1,1)$, and $(1,0,1)$ is not a basis for the 3-space $\mathbb{R}^3$, since this set is linearly dependent (the third is the sum of the first two vectors), and cannot span $\mathbb{R}^3$. The vector $(1,0,0)$ cannot be expressed as a linear combination of them (prove it). This set does not have enough vectors to span $\mathbb{R}^3$.

(2) The set of vectors $(1,0,0)$, $(0,1,1)$, $(1,0,1)$, and $(0,1,0)$ is not a basis either, since they are not linearly independent by Theorem 3.13. In fact, the sum of the first two minus the third makes the fourth. However, they span $\mathbb{R}^3$. This set of vectors has a redundant vector to span $\mathbb{R}^3$.

(3) The set of vectors $(1,1,1)$, $(0,1,1)$, and $(0,0,1)$ is linearly independent by Corollary 3.12.1 and also spans $\mathbb{R}^3$. Hence, it is a basis for $\mathbb{R}^3$, different from the standard basis. This set has the proper number of vectors to span $\mathbb{R}^3$, since the set cannot be reduced to a smaller set to span $\mathbb{R}^3$. □

The following theorem shows that a basis for a vector space induces a coordinate system just like the rectangular coordinate system for $\mathbb{R}^3$ given by the x, y, z-axes.

Theorem 3.15 *Let $\alpha = \{\mathbf{v}_1, \mathbf{v}_2, \ldots, \mathbf{v}_n\}$ be a basis for a vector space V. Then each vector $\mathbf{x}$ in V can be uniquely written as a linear combination of $\mathbf{v}_1, \mathbf{v}_2, \ldots, \mathbf{v}_n$, i.e., there are unique scalars a_i's, $i = 1, 2, \ldots, n$, such that*

$$\mathbf{x} = a_1\mathbf{v}_1 + a_2\mathbf{v}_2 + \cdots + a_n\mathbf{v}_n.$$

In this case, the coefficient vector $(a_1, a_2, \ldots, a_n)$ is called the **coordinate vector** of a vector $\mathbf{x}$ with respect to the basis α.

Proof: If $\mathbf{x}$ can also be written as $\mathbf{x} = b_1\mathbf{v}_1 + b_2\mathbf{v}_2 + \cdots + b_n\mathbf{v}_n$, then we have $\mathbf{0} = (a_1 - b_1)\mathbf{v}_1 + (a_2 - b_2)\mathbf{v}_2 + \cdots + (a_n - b_n)\mathbf{v}_n$. Now, the linear independence of $\mathbf{v}_i$'s gives $a_i = b_i$ for all $i = 1, 2, \ldots, n$. □

Example 50 (*Thee are infinitely many different coordinate systems for* $\mathbb{R}^3$) Let $\alpha = \{\mathbf{e}_1, \mathbf{e}_2, \mathbf{e}_3\}$ be the standard basis for $\mathbb{R}^3$, and let $\beta = \{\mathbf{v}_1, \mathbf{v}_2, \mathbf{v}_3\}$ with $\mathbf{v}_1 = (1,1,1) = \mathbf{e}_1 + \mathbf{e}_2 + \mathbf{e}_3$, $\mathbf{v}_2 = (0,1,1) = \mathbf{e}_2 + \mathbf{e}_3$, $\mathbf{v}_3 = (0,0,1) = \mathbf{e}_3$. Then, β is also a basis for $\mathbb{R}^3$ (see Example 49(3)). For any $\mathbf{x} = (x_1, x_2, x_3) \in \mathbb{R}^3$, one can easily show that

$$\begin{aligned} \mathbf{x} &= x_1\mathbf{e}_1 + x_2\mathbf{e}_2 + x_3\mathbf{e}_3 \\ &= x_1\mathbf{v}_1 + (x_2 - x_1)\mathbf{v}_2 + (x_3 - x_2)\mathbf{v}_3. \end{aligned}$$

Hence, (x_1, x_2, x_3) is the standard rectangular coordinate vector of $\mathbf{x}$ with respect to the standard basis, while $(x_1, x_2 - x_1, x_3 - x_2)$ is the coordinate vector of $\mathbf{x}$ with respect to the basis $\beta = \{\mathbf{v}_1, \mathbf{v}_2, \mathbf{v}_3\}$. Hence, the coordinate vector of $\mathbf{x}$ depends on the choice of a basis.

In fact. for any natural number k, $k\alpha = \{k\mathbf{e}_1, k\mathbf{e}_2, k\mathbf{e}_3\}$ is also a basis for $\mathbb{R}^3$. It shows that there are infinitely many different rectangular coordinate systems for $\mathbb{R}^3$, but not standard if $k \neq 1$. □

The coordinates and the coordinate vectors will be examined again more carefully in Section 4.4.

Example 51 (*Basis for the column space of a triangular matrix*) (Example 47 again) Given the upper triangular matrix

$$A = \begin{bmatrix} 2 & 3 & 4 \\ 0 & 1 & 6 \\ 0 & 0 & 7 \end{bmatrix}$$

in Example 47, no diagonal entries are zero and hence their column vectors are linearly independent and form a basis for the column space of A.

In general, for any triangular or diagonal matrix with nonzero diagonal entries, their columns form a basis for the column space of A. See Corollary 3.12.1. □

Problem 58 If $\{\mathbf{v}_1, \mathbf{v}_2, \ldots, \mathbf{v}_n\}$ is a basis for a vector space V, so does $\{k\mathbf{v}_1, \ell\mathbf{v}_2, \ldots, m\mathbf{v}_n\}$ for any nonzero $k, \ell, \ldots, m$. In fact, there are infinitely many different bases for any (nontrivial) vector space, inducing infinitely many different coordinate systems.

Problem 59 Show that the set $\{1, x, x^2, \ldots, x^n\}$ is a basis for the vector space $P_n(\mathbb{R})$ of all polynomials $p(x) = a_n x^n + a_{n-1}x^{n-1} + \cdots + a_1 x + a_0$ of degree $\leq n$. Here, 1 denotes the constant function $p(x) = 1$, x denotes the identity function $p(x) = x$,[9] x^2 denotes the quadratic function $p(x) = x^2$, etc.

Hint: Use the fact that $p(x) = a_n x^n + a_{n-1}x^{n-1} + \cdots + a_1 x + a_0 = 0$ for all x if and only if all a_i are zero.

Problem 60 Let $\mathbf{x}_k$ be the vector in the n-space $\mathbb{R}^n$ whose first $k-1$ entries are zero and whose last $n-k+1$ entries are 1. Show that the set $\{\mathbf{x}_1, \mathbf{x}_2, \ldots, \mathbf{x}_n\}$ is a basis for $\mathbb{R}^n$. Hence, $\{k\mathbf{x}_1, \ell\mathbf{x}_2, \ldots, m\mathbf{x}_n\}$ is also a basis for any nonzero $k, \ell, \ldots, m$.

Hint: Use Theorem 3.12(1) or Corollary 3.12.1.

Problem 61 Let α be a basis for the space $\mathbb{R}^4$ consisting of $\mathbf{x}_1 = (1, 1, 1, 1), \mathbf{x}_2 = (0, 1, 1, 1), \mathbf{x}_3 = (0, 0, 1, 1), \mathbf{x}_4 = (0, 0, 0, 1)$. What are the coordinate vectors of $\mathbf{u} = (1, 5, 9, 6)$ and $\mathbf{v} = (1, 6, 5, 0)$ in $\mathbb{R}^4$ with respect to the basis α?

3.5 Every vector space has its dimension

We often say,

the line $\mathbb{R}^1$ is one-dimensional;
the plane $\mathbb{R}^2$ is two-dimensional;
the space $\mathbb{R}^3$ is three-dimensional.

This is mostly due to the fact that the freedom in choosing coordinates for each element in the space is $1, 2$, or 3, respectively. This means that the *dimension* is closely related to the bases.

For any vector space there are infinitely many ways in choosing a basis. But, there is something common to all bases for a vector space, and this is related to the notion of dimension.

From the following lemma, one can define the dimension of a vector space.

Lemma 3.16 *Let V be a vector space and let $\alpha = \{\mathbf{x}_1, \mathbf{x}_2, \ldots, \mathbf{x}_m\}$ be a set of m vectors in V.*

(1) *If α spans V, then every set of vectors with more than m vectors cannot be linearly independent.*

[9] The *identity function* f on the real line is a function $f : \mathbb{R} \to \mathbb{R}$ which is defined by $f(x) = x$ for all x in $\mathbb{R}$. In general, for any nonempty set X, the function $f : X \to X$ defined by $f(x) = x$ for all x is called the *identity function* on a set X, denoted by id_X.

(2) *If α is linearly independent, then any set of vectors with fewer than m vectors cannot span V.*

Proof: Since (2) follows from (1) directly, we prove only (1). Let $\alpha = \{\mathbf{x}_1, \mathbf{x}_2, \ldots, \mathbf{x}_m\}$ span V and let $\beta = \{\mathbf{y}_1, \mathbf{y}_2, \ldots, \mathbf{y}_n\}$ be any set of n vectors in V with $n > m$. To show that β is linearly dependent, consider its linear dependence

$$c_1\mathbf{y}_1 + c_2\mathbf{y}_2 + \cdots + c_n\mathbf{y}_n = \mathbf{0}. \tag{3.1}$$

By expressing each $\mathbf{y}_j$ as a linear combination of the spanning vectors $\mathbf{x}_i$'s, say,

$$\mathbf{y}_j = a_{1j}\mathbf{x}_1 + a_{2j}\mathbf{x}_2 + \cdots + a_{mj}\mathbf{x}_m,$$

Eq. (3.1), the linear dependence of $\mathbf{y}_j$'s, can be written as a linear combination of the vectors $\mathbf{x}_1, \mathbf{x}_2, \ldots, \mathbf{x}_m$:

$$\begin{aligned}
\mathbf{0} = c_1\mathbf{y}_1 + c_2\mathbf{y}_2 &+ \cdots + c_n\mathbf{y}_n \\
&= c_1(a_{11}\mathbf{x}_1 + a_{21}\mathbf{x}_2 + \cdots + a_{m1}\mathbf{x}_m) \\
&\quad + c_2(a_{12}\mathbf{x}_1 + a_{22}\mathbf{x}_2 + \cdots + a_{m2}\mathbf{x}_m) \\
&\quad\quad \vdots \\
&\quad + c_n(a_{1n}\mathbf{x}_1 + a_{2n}\mathbf{x}_2 + \cdots + a_{mn}\mathbf{x}_m) \\
&= (a_{11}c_1 + a_{12}c_2 + \cdots + a_{1n}c_n)\mathbf{x}_1 \\
&\quad + (a_{21}c_1 + a_{22}c_2 + \cdots + a_{2n}c_n)\mathbf{x}_2 \\
&\quad\quad \vdots \\
&\quad + (a_{m1}c_1 + a_{m2}c_2 + \cdots + a_{mn}c_n)\mathbf{x}_m.
\end{aligned} \tag{3.2}$$

Thus, β is linearly dependent if and only if the linear dependence $c_1\mathbf{y}_1 + c_2\mathbf{y}_2 + \cdots + c_n\mathbf{y}_n = \mathbf{0}$ has a nontrivial solution $(c_1, c_2, \ldots, c_n) \neq (0, 0, \ldots, 0)$. This can happen if all of the coefficients of $\mathbf{x}_i$'s in the last term of Eq. (3.2) are zero but not all of c_i's are zero. This means that the homogeneous system of linear equations in c_i's

$$\begin{bmatrix} a_{11} & a_{12} & \cdots & a_{1n} \\ a_{21} & a_{22} & \cdots & a_{2n} \\ \vdots & \vdots & & \vdots \\ a_{m1} & a_{m2} & \cdots & a_{mn} \end{bmatrix} \begin{bmatrix} c_1 \\ c_2 \\ \vdots \\ c_n \end{bmatrix} = \begin{bmatrix} 0 \\ 0 \\ \vdots \\ 0 \end{bmatrix}$$

have a nontrivial solution. But it is guaranteed by Theorem 3.13, since the coefficient matrix $[a_{ij}]$ is wide with $m < n$. (Or, this homogeneous system has at least one free variable.) □

By Lemma 3.16, if a set $\alpha = \{\mathbf{x}_1, \mathbf{x}_2, \ldots, \mathbf{x}_n\}$ of n vectors is a basis for a vector space V, then any other sets of k vectors cannot be a basis for V if $k \neq n$. However, $\{k\mathbf{x}_1, \ell\mathbf{x}_2, \ldots, m\mathbf{x}_n\}$ is another basis for any nonzero $k, \ell, \ldots, m$, as mentioned in Problem 58.

This means that all bases for a vector space V should have the same number of vectors, even though there are infinitely many different bases for a vector space.

Theorem 3.17 *If a basis for a vector space V consists of n vectors, so does every other basis.*

Definition 3.18 The **dimension** of a vector space V is the number, say n, of vectors in a basis for V, denoted by $\dim V = n$.

As a special case, the dimension of a trivial vector space $V = \{\mathbf{0}\}$ is defined to be zero.

Clearly, $\dim \mathbb{R}^1 = 1, \dim \mathbb{R}^2 = 2,$ and $\dim \mathbb{R}^3 = 3.$

Example 52 (*Can you compute the dimension of any vector space?*) **(1)** The standard basis $\{\mathbf{e}_1, \mathbf{e}_2, \ldots, \mathbf{e}_n\}$ for the n-space $\mathbb{R}^n$ gives $\dim \mathbb{R}^n = n$.

(2) For the vector space $P_n(\mathbb{R})$ of all polynomials of degree $\leq n$, $\dim P_n(\mathbb{R}) = n + 1$ since $\{1, x, x^2, \ldots, x^n\}$ is a basis for $P_n(\mathbb{R})$. (See Problem 59.)

(3) For the vector space $M_{m\times n}(\mathbb{R})$ of all $m \times n$ matrices, $\dim M_{m\times n}(\mathbb{R}) = mn$ since $\{E_{ij} : i = 1, \ldots, m, j = 1, 2, \ldots, n\}$ is a basis for $M_{m\times n}(\mathbb{R})$, where E_{ij} is the $m \times n$ matrix whose (i, j)-entry is 1 and all the others are zero. □

When a vector space V has a basis of a finite number of vectors, V is said to be **finite dimensional**. Otherwise, V is called **infinite dimensional**

If $V = C(\mathbb{R})$, the vector space of all real-valued continuous functions defined on the real line, then V is not finite dimensional because the set $\{1, x, x^2, \ldots, x^n\}$ is linearly independent for all n.

In this book, we are concerned only with finite dimensional vector spaces unless otherwise stated.

Problem 62 (Exercise 52(3), again) Let $M_{m\times n}(\mathbb{R})$ be the vector space of all $m \times n$ matrices and let E_{ij} be the $m \times n$ matrix in $M_{m\times n}(\mathbb{R})$ whose (i, j)-entry is 1 and all the others are zero. Show that

(1) the set $\{E_{ij} : i = 1, \ldots, m, j = 1, 2, \ldots, n\}$ is linearly independent, and
(2) the set $\{E_{ij} : i = 1, \ldots, m, j = 1, 2, \ldots, n\}$ spans the vector space $M_{m\times n}(\mathbb{R})$.

Consequently, $\dim M_{m\times n}(\mathbb{R}) = mn$.

Hint: First, try it when either $m = 1$ or $n = 1$. Next, try it when $m = n = 2$.

Problem 63 Show that the set of all $n \times n$ (real) diagonal matrices is a subspace of the vector space $M_{n\times n}(\mathbb{R})$. Find a bases and the dimension of this subspace.

3.6 Always one can construct a basis for a vector space

The following theorem shows that *every finite dimensional vector space has at least one basis* and the proof shows how one can practically construct a basis.

Theorem 3.19 (Existence of a basis) *Let V be a finite dimensional vector space.*

(1) *Any linearly independent set in V (even any one nonzero vector) can be extended to a basis by adding more vectors if necessary.*

(2) *Any set of vectors that spans V can be reduced to a basis by discarding vectors if necessary.*

Proof: We prove (1) and leave (2) as an exercise. Let $\alpha = \{\mathbf{x}_1, \mathbf{x}_2, \ldots, \mathbf{x}_k\}$ be any linearly independent set in V. (As a particular case, one can choose $\alpha = \{\mathbf{x}\}$ with any nonzero vector $\mathbf{x}$.) If α spans V, then α is a basis. If α does not span V, then there exists a vector, say $\mathbf{x}_{k+1}$, in V that is not contained in the subspace spanned by the vectors in α. Now $\{\mathbf{x}_1, \ldots, \mathbf{x}_k, \mathbf{x}_{k+1}\}$ is linearly independent (check why). If $\{\mathbf{x}_1, \ldots, \mathbf{x}_k, \mathbf{x}_{k+1}\}$ spans V, then this is a basis for V. If not span V, then the same procedure can be repeated to add one more vector, yielding finally a linearly independent set that spans V, *i.e.*, a basis for V.

This procedure must stop in a finite number of steps because of Lemma 3.16 for a finite dimensional vector space V. □

It follows from Theorem 3.19 that a basis for a vector space V is a set of vectors in V which is *maximally independent* and *minimally spanning* in a sense. In particular, if W is a subspace of V, then any basis for W is also linearly independent in V, and can be extended to a basis for V. Thus $\dim W \le \dim V$.

Corollary 3.19.1 *Let V be a vector space of dimension n. Then*

(1) *any set of n vectors that spans V is a basis for V, and*
(2) *any set of n linearly independent vectors is a basis for V.*

Proof: Again we prove only (1). If a spanning set of n vectors were not linearly independent, then one of them can be written as a linear combination of all the other $n-1$ vectors. It means these $n-1$ vectors span the space V. It is impossible. □

The definition of the direct sum of subspaces gives the following corollary.

Corollary 3.19.2 *For any subspace U of V, there is a subspace W of V such that $V = U \oplus W$.*

Proof: Choose a basis $\{\mathbf{u}_1, \ldots, \mathbf{u}_k\}$ for U, and extend it to a basis $\{\mathbf{u}_1, \ldots, \mathbf{u}_k, \mathbf{u}_{k+1}, \ldots, \mathbf{u}_n\}$ for V. Then the subspace W spanned by $\{\mathbf{u}_{k+1}, \ldots, \mathbf{u}_n\}$ satisfies the requirement. □

To compute the dimension of a vector space, it is enough to find any one basis for the vector space, as shown in the following examples.

Example 53 (*Constructing a basis by adding more vectors*) Let W be the subspace of $\mathbb{R}^4$ spanned by the vectors

$$\mathbf{x}_1 = (1, 4, 9, 2), \qquad \mathbf{x}_2 = (1, 5, 0, 2), \qquad \mathbf{x}_3 = (1, 7, 7, 6).$$

Find a basis for W and extend it to a basis for $\mathbb{R}^4$.

Solution: Clearly, $\dim W \leq 3$ since W is spanned by three vectors $\mathbf{x}_i$'s. To see their independence, let A be the 3×4 matrix whose rows are $\mathbf{x}_1, \mathbf{x}_2$, and $\mathbf{x}_3$:

$$A = \begin{bmatrix} 1 & 4 & 9 & 2 \\ 1 & 5 & 0 & 2 \\ 1 & 7 & 7 & 6 \end{bmatrix}.$$

Reduce A to a row-echelon form:

$$U = \begin{bmatrix} 1 & 4 & 9 & 2 \\ 0 & 1 & -9 & 0 \\ 0 & 0 & 25 & 4 \end{bmatrix}.$$

By a forward elimination procedure, each of three row vectors of U can be expressed as a linear combination of the three row vectors $\mathbf{x}_1$, $\mathbf{x}_2$, $\mathbf{x}_3$ of A, and vice-versa.

Hence, the subspace W is also spanned by the three nonzero row vectors of U. These three row vectors of U are clearly linearly independent. Hence, they form a basis for W. (This implies $\dim W = 3$ and hence the row vectors $\mathbf{x}_1, \mathbf{x}_2, \mathbf{x}_3$ of A is also a basis for W by Corollary 3.19.1. The linear independence of $\mathbf{x}_i$'s is a by-product of this fact).

To extend it to a basis for $\mathbb{R}^4$, just add any nonzero vector of the form $\mathbf{x}_4 = (0, 0, 0, t)$ to the rows of U (hence also of A). □

Example 54 (*Constructing bases for $V + W$ and $V \cap W$*) Let V be the subspace of $\mathbb{R}^5$ spanned by

$$\begin{aligned}\mathbf{v}_1 &= (1,\ 1,\ 9,\ 7,\ 2),\\ \mathbf{v}_2 &= (1,\ 4,\ 3,\ 4,\ 2),\\ \mathbf{v}_3 &= (2, 5, 12, 14, 5),\end{aligned}$$

and let W be the subspace spanned by

$$\begin{aligned}\mathbf{w}_1 &= (1, 2, 0, 2, 0),\\ \mathbf{w}_2 &= (1, 4, 4, 2, 2),\\ \mathbf{w}_3 &= (2, 7, 2, 6, 3).\end{aligned}$$

Find a basis for V, W, $V + W$, and for $V \cap W$.

Solution: The matrix A whose row vectors are $\mathbf{v}_i$'s can be reduced to a row-echelon form

$$U_A = \begin{bmatrix} 1 & 1 & 9 & 7 & 2 \\ 0 & 1 & -2 & -1 & 0 \\ 0 & 0 & 0 & 3 & 1 \end{bmatrix},$$

so that the three row vectors of U_A are not zero and form a basis for V and hence $\dim V = 3$.

Similarly, the matrix B whose row vectors are $\mathbf{w}_j$'s can be reduced to a row-echelon form

$$U_B = \begin{bmatrix} 1 & 2 & 0 & 2 & 0 \\ 0 & 1 & 2 & 0 & 1 \\ 0 & 0 & -2 & 1 & 0 \end{bmatrix},$$

so that the row vectors of U_B are not zero and form a basis for W, and $\dim W = 3$.

Now, let Q be the 6×5 matrix whose row vectors are $\mathbf{v}_i$'s and $\mathbf{w}_j$'s. By Gaussian elimination, one can reduce Q to a row-echelon form, or can show that Q is row-equivalent to a matrix excluding zero rows:

$$\begin{bmatrix} 1 & 1 & 9 & 7 & 2 \\ 0 & 1 & -2 & -1 & 0 \\ 0 & 0 & 1 & 4 & 1 \\ 0 & 0 & 0 & 3 & 1 \\ 0 & 0 & 0 & 0 & 1 \end{bmatrix}.$$

Thus, its five nonzero row vectors form a basis for $V+W$, so that $\dim(V+W) = 5$.

Finally, find a basis for $V \cap W$. A vector $\mathbf{x} \in \mathbb{R}^5$ is contained in $V \cap W$ if and only if $\mathbf{x}$ can be written as a linear combination of the row vectors of A and also of the row vectors of B.

Or, equivalently, $\mathbf{x}$ can be written as a linear combination of the row vectors of the row-echelon form U_A and also of the row vectors of the row-echelon form U_B.

Let

$$U_A = \begin{bmatrix} \mathbf{r}_1 \\ \mathbf{r}_2 \\ \mathbf{r}_3 \end{bmatrix} \quad \text{and} \quad U_B = \begin{bmatrix} \mathbf{t}_1 \\ \mathbf{t}_2 \\ \mathbf{t}_3 \end{bmatrix}$$

with row vectors $\mathbf{r}_i, \mathbf{t}_j$, respectively and let $\mathbf{x} \in V \cap W$ be written as

$$\ell \mathbf{r}_1 + m\mathbf{r}_2 + n\mathbf{r}_3 = \mathbf{x} = q\mathbf{t}_1 + r\mathbf{t}_2 + s\mathbf{t}_3$$

with scalars ℓ, m, n and q, r, s in $\mathbb{R}$. By comparing the entries on both sides, one can get

$$(\ell, m, n) = (0, t, t) \quad \text{and} \quad (q, r, s) = (0, t, 2t)$$

for any $t \in \mathbb{R}$. It gives

$$\mathbf{x} = \ell \mathbf{r}_1 + m\mathbf{r}_2 + n\mathbf{r}_3 = t(\mathbf{r}_2 + \mathbf{r}_3) = t(0, 1, -2, 2, 1).$$

Hence, the vector $\{(0, 1, -2, 2, 1)\}$ is a basis for $V \cap W$ and $\dim(V \cap W) = 1$. □

Problem 64 Let W be a subspace of V. Show that if $\dim W = \dim V$, then $W = V$.
Hint: Any basis for W should be a basis for V too, by Corollary 3.19.1.

Problem 65 Let W be the subspace of $\mathbb{R}^5$ spanned by the vectors

$$\mathbf{x}_1 = (1, 0, -2, 2, 0), \quad \mathbf{x}_2 = (1, 1, 6, 4, 4), \quad \mathbf{x}_3 = (1, 1, 9, 4, 9).$$

Find a basis for W and extend it to a basis for $\mathbb{R}^5$.

Problem 66 Let $U = \{(x, y, z) : 2x + 3y + z = 0\}$ and $V = \{(x, y, z) : x + 2y - z = 0\}$ be subspaces of the 3-space $\mathbb{R}^3$.

(1) Find a basis for $U \cap V$.
(2) Determine the dimension of $U + V$.
(3) Describe $U, V, U \cap V$, and $U + V$ geometrically.

Hint: First, find a basis for U and a basis for V.

3.7 To every matrix, there are Row, Column, and Null spaces

An $m \times n$ matrix A can be abbreviated by the row or column vectors as follows:

$$\begin{aligned} A &= \begin{bmatrix} a_{11} & a_{12} & \cdots & a_{1n} \\ a_{21} & a_{22} & \cdots & a_{2n} \\ \vdots & \vdots & & \vdots \\ a_{m1} & a_{m2} & \cdots & a_{mn} \end{bmatrix} = \begin{bmatrix} \mathbf{r}_1 \\ \mathbf{r}_2 \\ \vdots \\ \mathbf{r}_m \end{bmatrix} \\ &= \begin{bmatrix} \mathbf{c}_1 & \mathbf{c}_2 & \cdots & \mathbf{c}_n \end{bmatrix}, \end{aligned}$$

where $\mathbf{r}_i$ is the i-th row of A in $\mathbb{R}^n$ and $\mathbf{c}_j$ is the j-th column of A in $\mathbb{R}^m$.

Definition 3.20 Let A be an $m \times n$ matrix with row vectors $\{\mathbf{r}_1, \mathbf{r}_2, \ldots, \mathbf{r}_m\}$ and column vectors $\{\mathbf{c}_1, \mathbf{c}_2, \ldots, \mathbf{c}_n\}$.

(1) The **row space** of A is the subspace in $\mathbb{R}^n$ spanned by the row vectors $\{\mathbf{r}_1, \mathbf{r}_2, \ldots, \mathbf{r}_m\}$, denoted by $\mathcal{R}(A)$.

(2) The **column space** of A is the subspace in $\mathbb{R}^m$ spanned by the column vectors $\{\mathbf{c}_1, \mathbf{c}_2, \ldots, \mathbf{c}_n\}$, denoted by $\mathcal{C}(A)$.

(3) The solution set of the homogeneous system $A\mathbf{x} = \mathbf{0}$ is called the **null space** of A, denoted by $\mathcal{N}(A)$; that is, $\mathcal{N}(A) = \{\mathbf{x} \in \mathbb{R}^n : A\mathbf{x} = \mathbf{0}\}$.

The null space $\mathcal{N}(A)$ is a subspace of the n-space $\mathbb{R}^n$. Its dimension is called the **nullity** of A. Later, it will be shown in Theorem 3.22 the row space and the column space of A have the same dimension, and it is called the **rank** of A.

Since the row vectors of A are just the column vectors of its transpose A^T and the column vectors of A are the row vectors of A^T, it follows

$$\mathcal{R}(A) = \mathcal{C}(A^T) \quad \text{and} \quad \mathcal{C}(A) = \mathcal{R}(A^T).$$

Since $A\mathbf{x} = x_1\mathbf{c}_1 + x_2\mathbf{c}_2 + \cdots + x_n\mathbf{c}_n$ for any $\mathbf{x} = (x_1, x_2, \ldots, x_n) \in \mathbb{R}^n$, we get

$$\mathcal{C}(A) = \{A\mathbf{x} : \mathbf{x} \in \mathbb{R}^n\}.$$

Thus, for a vector $\mathbf{b} \in \mathbb{R}^m$, *the system* $A\mathbf{x} = \mathbf{b}$ *has a solution if and only if* $\mathbf{b} \in \mathcal{C}(A) \subseteq \mathbb{R}^m$. In other words, the column space $\mathcal{C}(A)$ is the set of vectors $\mathbf{b} \in \mathbb{R}^m$ for which $A\mathbf{x} = \mathbf{b}$ has a solution.

It is natural to ask what the dimensions of these subspaces are, and how one can find a basis for each of them.

This will help us to understand the structure of the solution set $\mathcal{N}(A)$ of the system $A\mathbf{x} = \mathbf{b}$. Since the set of the row vectors and the set of the column vectors of A are spanning the row space and the column space, respectively, one needs to find a minimal spanning subset of each of them to find a basis for each space.

This is not difficult if the matrix A is diagonal, triangular, or of a (reduced) row-echelon form.

Example 55 (*Bases for* $\mathcal{R}(U)$, $\mathcal{N}(U)$, *and* $\mathcal{C}(U)$) Let U be in a reduced row-echelon form given as

$$U = \begin{bmatrix} 1 & 0 & 0 & 0 & 3 \\ 0 & 1 & 2 & 0 & 4 \\ 0 & 0 & 0 & 1 & 5 \\ 0 & 0 & 0 & 0 & 0 \end{bmatrix}.$$

Clearly, the first three nonzero row vectors contain the leading 1's. Hence they are linearly independent and form a basis for the row space $\mathcal{R}(U)$, so that $\dim \mathcal{R}(U) = 3$.

Similarly, the first, the second, and the fourth columns containing leading 1's are also linearly independent (see Theorem 3.12(1)), and the other two columns can be expressed as linear combinations of the three linearly independent columns.

Hence, the first, the second, and the fourth columns form a basis for $\mathcal{C}(U)$ and $\dim \mathcal{C}(U) = 3$.

To find a basis for the null space $\mathcal{N}(U)$, we first solve the system $U\mathbf{x} = \mathbf{0}$ to get the solution

$$\begin{bmatrix} x_1 \\ x_2 \\ x_3 \\ x_4 \\ x_5 \end{bmatrix} = \begin{bmatrix} & - & 3t \\ -2s & - & 4t \\ s & & \\ & - & 5t \\ & & t \end{bmatrix} = s\begin{bmatrix} 0 \\ -2 \\ 1 \\ 0 \\ 0 \end{bmatrix} + t\begin{bmatrix} -3 \\ -4 \\ 0 \\ -5 \\ 1 \end{bmatrix} = s\mathbf{n}_s + t\mathbf{n}_t,$$

where $\mathbf{n}_s = (0, -2, 1, 0, 0, 0)$, $\mathbf{n}_t = (-3, -4, 0, -5,\ 1)$, and s, t are arbitrary values for the free variables x_3, x_5, respectively. This shows that these two vectors $\mathbf{n}_s$ and $\mathbf{n}_t$ span the null space $\mathcal{N}(U)$, and they are clearly linearly independent, since $(1, 0)$ and $(0, 1)$, as the (x_3, x_5)-coordinates of $\mathbf{n}_s$ and $\mathbf{n}_t$ respectively, are linearly independent.

Hence, the set $\{\mathbf{n}_s, \mathbf{n}_t\}$ forms a basis for the null space $\mathcal{N}(U)$. So $\dim \mathcal{N}(U) = 2$. □

For any matrix A, we first investigate the row space $\mathcal{R}(A)$ and the null space $\mathcal{N}(A)$ of A by comparing them with those of the reduced row-echelon form U of A. Since $A\mathbf{x} = \mathbf{0}$ and $U\mathbf{x} = \mathbf{0}$ have the same solution set by Theorem 1.2, we clearly have $\mathcal{N}(A) = \mathcal{N}(U)$.

Let $\{\mathbf{r}_1, \mathbf{r}_2, \ldots, \mathbf{r}_m\}$ be the row vectors of an $m \times n$ matrix A. The three elementary row operations transform A into the matrices A_i of the following three types:

$$A_1 = \begin{bmatrix} \mathbf{r}_1 \\ \vdots \\ k\mathbf{r}_i \\ \vdots \\ \mathbf{r}_m \end{bmatrix} \text{ for } k \neq 0; \quad A_2 = \begin{bmatrix} \vdots \\ \mathbf{r}_j \\ \vdots \\ \mathbf{r}_i \\ \vdots \end{bmatrix} \text{ for } i < j; \quad A_3 = \begin{bmatrix} \mathbf{r}_1 \\ \vdots \\ \mathbf{r}_i + k\mathbf{r}_j \\ \vdots \\ \mathbf{r}_m \end{bmatrix}.$$

Clearly, the row vectors of each of the three matrices A_1, A_2, and A_3 are linear combinations of the row vectors of A. On the other hand, by the inverse elementary row operations, these matrices can be reversed back to A.

Thus, $\mathcal{R}(A) = \mathcal{R}(A_i)$ for $i = 1, 2, 3$.

This means that if matrices A and B are row-equivalent, then their row spaces should be the same. In particular, $\mathcal{R}(A) = \mathcal{R}(U)$.

Now the nonzero row vectors in the reduced row-echelon matrix U are always linearly independent and span the row space of U (see Theorem 3.12(1) or Example 55). Thus they form a basis for the row space $\mathcal{R}(A)$ $(= \mathcal{R}(U))$.

This gives the following theorem.

Theorem 3.21 *Let U be a (reduced) row-echelon form of a matrix A. Then*

$$\mathcal{R}(A) = \mathcal{R}(U) \quad \text{and} \quad \mathcal{N}(A) = \mathcal{N}(U).$$

Moreover, if U has r nonzero row vectors (containing leading 1's), then they form a basis for the row space $\mathcal{R}(A)$, so that the dimension of $\mathcal{R}(A)$ is r.

The next two examples show how to find bases for the row space and the null space, and separately how to find a basis for the column space of any rectangular matrix A.

Example 56 (*Bases for $\mathcal{R}(A)$ and $\mathcal{N}(A)$*) Let

$$A = \begin{bmatrix} 1 & 1 & 2 & 0 & 7 \\ 0 & 1 & 2 & 0 & 4 \\ 1 & 0 & 0 & 1 & 8 \\ 1 & 1 & 2 & -1 & 2 \end{bmatrix} = \begin{bmatrix} \mathbf{r}_1 \\ \mathbf{r}_2 \\ \mathbf{r}_3 \\ \mathbf{r}_4 \end{bmatrix}$$

with row vectors $\mathbf{r}_i$. Find bases for the row space $\mathcal{R}(A)$ and the null space $\mathcal{N}(A)$.

Solution: (1) *Find a basis for $\mathcal{R}(A)$*: By Gauss-Jordan elimination on A, one can get its reduced row-echelon form U:

$$U = \begin{bmatrix} 1 & 0 & 0 & 0 & 3 \\ 0 & 1 & 2 & 0 & 4 \\ 0 & 0 & 0 & 1 & 5 \\ 0 & 0 & 0 & 0 & 0 \end{bmatrix}.$$

This is the matrix given in Example 55. As before, the three nonzero row vectors

$$\begin{aligned} \mathbf{v}_1 &= (1,0,0,0,3), \\ \mathbf{v}_2 &= (0,1,2,0,4), \\ \mathbf{v}_3 &= (0,0,0,1,5) \end{aligned}$$

of U are linearly independent, they form a basis for the row space $\mathcal{R}(U) = \mathcal{R}(A)$, so $\dim \mathcal{R}(A) = 3$.

As a result, $\dim \mathcal{R}(A) = 3 =$ *the number of basic variables in* $U\mathbf{x} = \mathbf{0}$.

(2) *Find a basis for* $\mathcal{N}(A)$: It is needed to solve the homogeneous system $U\mathbf{x} = \mathbf{0}$, since $\mathcal{N}(A) = \mathcal{N}(U)$. However, in Example 55, two vectors $\mathbf{n}_s = (0, -2, 1, 0, 0)$ and $\mathbf{n}_t = (-3, -4, 0, -5,\ 1)$ are given to form a basis for the null space $\mathcal{N}(U) = \mathcal{N}(A)$.

As a result, $\dim \mathcal{N}(A) = 2 =$ *the number of free variables in* $U\mathbf{x} = \mathbf{0}$. □

As shown in Example 51, for any triangular or diagonal matrix A with nonzero diagonal entries, the column vectors form a basis for the column space of A.

The following example shows how to find a basis for the column space $\mathcal{C}(A)$ of any rectangular matrix A.

Example 57 (*Basis for* $\mathcal{C}(A)$) Given the matrix

$$A = \begin{bmatrix} 1 & 1 & 2 & 0 & 7 \\ 0 & 1 & 2 & 0 & 4 \\ 1 & 0 & 0 & 1 & 8 \\ 1 & 1 & 2 & -1 & 2 \end{bmatrix} = [\mathbf{c}_1\ \mathbf{c}_2\ \mathbf{c}_3\ \mathbf{c}_4\ \mathbf{c}_5]$$

in Example 56 with columns $\mathbf{c}_i$, find a basis for the column space $\mathcal{C}(A)$.

Solution: The reduced row-echelon form of A is already given as

$$U = \begin{bmatrix} 1 & 0 & 0 & 0 & 3 \\ 0 & 1 & 2 & 0 & 4 \\ 0 & 0 & 0 & 1 & 5 \\ 0 & 0 & 0 & 0 & 0 \end{bmatrix}$$

in Example 56. Since the column vectors of A span $\mathcal{C}(A)$, we only need to discard the columns which can be expressed as linear combinations of the other column vectors.

But, the linear dependence

$$x_1\mathbf{c}_1 + x_2\mathbf{c}_2 + x_3\mathbf{c}_3 + x_4\mathbf{c}_4 + x_5\mathbf{c}_5 = \mathbf{0}, \quad i.e.,\ A\mathbf{x} = \mathbf{0},$$

holds if and only if $\mathbf{x} = (x_1, x_2, \ldots, x_5) \in \mathcal{N}(A)$. By taking $\mathbf{x} = \mathbf{n}_s = (0, -2, 1, 0, 0)$ or $\mathbf{x} = \mathbf{n}_t = (-3, -4, 0, -5,\ 1)$, the basis vectors for $\mathcal{N}(A)$ given in Example 56(2), one can get two nontrivial linear dependencies of $\mathbf{c}_i$'s:

$$\begin{array}{rrrrrl} & -\ 2\mathbf{c}_2 & +\ \mathbf{c}_3 & & & = \mathbf{0}, \\ -\ 3\mathbf{c}_1 & -\ 4\mathbf{c}_2 & & -\ 5\mathbf{c}_4 & +\ \mathbf{c}_5 & = \mathbf{0}, \end{array}$$

respectively.

This shows that the column vectors $\mathbf{c}_3$ and $\mathbf{c}_5$ corresponding to the free variables in $A\mathbf{x} = \mathbf{0}$ can be written as linear combinations of the column vectors $\mathbf{c}_1, \mathbf{c}_2, \mathbf{c}_4$, which correspond to the basic variables in $A\mathbf{x} = \mathbf{0}$. Hence, $\{\mathbf{c}_1, \mathbf{c}_2, \mathbf{c}_4\}$ spans the column space $\mathcal{C}(A)$.

Now, need to show that $\{\mathbf{c}_1, \mathbf{c}_2, \mathbf{c}_4\}$ is linearly independent. Let $\tilde{A} = [\mathbf{c}_1\ \mathbf{c}_2\ \mathbf{c}_4]$ and $\tilde{U} = [\mathbf{u}_1\ \mathbf{u}_2\ \mathbf{u}_4]$ be submatrices of A and U, respectively, where $\mathbf{u}_j$ is the j-th column vector of the reduced row-echelon form U of A:

$$\tilde{U} = \begin{bmatrix} 1 & 0 & 0 \\ 0 & 1 & 0 \\ 0 & 0 & 1 \\ 0 & 0 & 0 \end{bmatrix} \quad \text{and} \quad \tilde{A} = \begin{bmatrix} 1 & 1 & 0 \\ 0 & 1 & 0 \\ 1 & 0 & 1 \\ 1 & 1 & -1 \end{bmatrix}.$$

Then clearly $\tilde{U}$ is the reduced row-echelon form of $\tilde{A}$ so that $\mathcal{N}(\tilde{A}) = \mathcal{N}(\tilde{U})$. Since the vectors $\mathbf{u}_1, \mathbf{u}_2, \mathbf{u}_4$ are just the columns of U containing leading 1's, they are linearly independent by Theorem 3.12(1).

Hence, equivalently the columns $\{\mathbf{c}_1, \mathbf{c}_2, \mathbf{c}_4\}$ of A is linearly independent by Theorem 3.12(2), (or by $\tilde{A}\mathbf{x} = \mathbf{0} \Leftrightarrow \tilde{U}\mathbf{x} = \mathbf{0}$). Therefore, this is a basis for the column space $\mathcal{C}(A)$ and $\dim \mathcal{C}(A) = 3 =$ *the number of basic variables.*

That is, *the column vectors of A corresponding to the basic variables in $U\mathbf{x} = \mathbf{0}$ form a basis for the column space $\mathcal{C}(A)$.* □

In summary, given a matrix A, first find the (reduced) row-echelon form U of A by Gauss-Jordan elimination. Then

- the set of nonzero row vectors of U is a basis for $\mathcal{R}(A) = \mathcal{R}(U)$,
- a basis for $\mathcal{N}(A) = \mathcal{N}(U)$ can be found by solving $U\mathbf{x} = \mathbf{0}$,
- for a basis for the column space $\mathcal{C}(A)$, one notices that $\mathcal{C}(U) \neq \mathcal{C}(A)$ in general, since the column space of A is not preserved by Gauss-Jordan elimination. (See Problem 68 below.) However, we have $\dim \mathcal{C}(A) = \dim \mathcal{C}(U)$, and a basis for $\mathcal{C}(A)$ can be formed by selecting the columns in A, *not in* U, which correspond to the basic variables (or the leading 1's in U).

Alternatively, a basis for the column space $\mathcal{C}(A)$ can be found with the elementary column operations, which is the same as finding a basis for the row space $\mathcal{R}(A^T)$ of A^T.

Problem 67 Find a basis for $\mathcal{R}(A)$ of the matrix

$$A = \begin{bmatrix} 1 & 2 & 0 & 0 & 1 \\ 1 & 0 & 3 & 2 & 1 \\ 2 & 2 & 3 & 2 & 0 \end{bmatrix}.$$

Also find a basis for $\mathcal{C}(A)$ in two ways: (i) by finding a basis for $\mathcal{R}(A^T)$, (ii) by using the reduced row-echelon form U of A.

Problem 68 Find an example of a matrix A and its row-echelon form U such that $\mathcal{C}(A) \neq \mathcal{C}(U)$. What is wrong in $\mathcal{C}(A) = \mathcal{R}(A^T) = \mathcal{R}(U^T) = \mathcal{C}(U)$?

Hint: For $A = \begin{bmatrix} 1 \\ 1 \end{bmatrix}$, $\mathcal{C}(A)$ is the diagonal line in $\mathbb{R}^2$, but $\mathcal{C}(U)$ is the x-axis in $\mathbb{R}^2$.

3.8 Rank-Nullity theorem; Relation between rank and nullity

Given every $m \times n$ matrix A, there are three vector spaces; a row space $\mathcal{R}(A)$, the null space $\mathcal{N}(A)$, and the column space $\mathcal{C}(A)$. Although they are quite different spaces, there are some relations among their dimensions.

The first relation comes from the argument in Examples 56 and 57, which can be used to prove the next theorem: $\dim \mathcal{R}(A) = \dim \mathcal{C}(A)$. The proof given here is just a repetition of the argument in Examples 56 and 57 in a general case, and so it may be skipped.

Theorem 3.22 *For any $m \times n$ matrix A, the row space and the column space of A have the same dimension; that is,* $\dim \mathcal{R}(A) = \dim \mathcal{C}(A)$.

$\dim \mathcal{R}(A) = \dim \mathcal{C}(A)$ is called the **rank** of a matrix A, denoted by rank A.

Proof: Let U be the reduced row-echelon form of A. Then $\dim \mathcal{R}(A) = r$ is just the number of the nonzero row (or column) vectors of U containing leading 1's. This is also equal to the number of basic variables in $U\mathbf{x} = \mathbf{0}$ or $A\mathbf{x} = \mathbf{0}$.

We aim to prove that the r columns of A corresponding to the basic variables (or leading 1's) form a basis for $\mathcal{C}(A)$, so that $\dim \mathcal{C}(A) = r = \dim \mathcal{R}(A)$.

(1) They are linearly independent: Let $\tilde{A}$ denote the submatrix of A whose columns are those of A corresponding to the r basic variables (or leading 1's) in U, and let $\tilde{U}$ denote the submatrix of U consisting of the r columns containing leading 1's. Then, it is clear that $\tilde{U}$ is the reduced row-echelon form of $\tilde{A}$, so that

$\tilde{A}\mathbf{x} = \mathbf{0}$ if and only if $\tilde{U}\mathbf{x} = \mathbf{0}$. However, $\tilde{U}\mathbf{x} = \mathbf{0}$ has only the trivial solution since the columns of U containing the leading 1's are linearly independent by Theorem 3.12(1). Therefore, $\tilde{A}\mathbf{x} = \mathbf{0}$ also has only the trivial solution, so the columns of $\tilde{A}$ are linearly independent.

(2) They span $\mathcal{C}(A)$: The columns of A corresponding to the free variables are not contained in $\tilde{A}$, and each of these column vectors of A can be written as a linear combination of the column vectors of $\tilde{A}$ (see Example 57). To show this, let $\{\mathbf{c}_{i_1}, \mathbf{c}_{i_2}, \ldots, \mathbf{c}_{i_k}\}$ be the columns of A (not contained in $\tilde{A}$) corresponding to the free variables $\{x_{i_1}, x_{i_2}, \ldots, x_{i_k}\}$, and let x_{i_j} be any of these free variables. Then, by assigning the value 1 to x_{i_j} and 0 to all the other free variables, one can get a nontrivial solution of

$$A\mathbf{x} = x_1\mathbf{c}_1 + x_2\mathbf{c}_2 + \cdots + x_n\mathbf{c}_n = \mathbf{0}.$$

When such a solution is substituted into this equation, one can see that the column $\mathbf{c}_{i_j}$ of A corresponding to a free variable $x_{i_j} = 1$ is written as a linear combination of the columns of $\tilde{A}$. This can be done for each free variable x_{i_j}, $j = 1, 2, \ldots, k$, so the columns of A corresponding to those free variables are redundant in the spanning set of $\mathcal{C}(A)$. □

Clearly, $\operatorname{rank} I_n = n$ and $\operatorname{rank} A = \operatorname{rank} A^T$. For any $m \times n$ matrix A, $\dim \mathcal{R}(A) \le m$ and $\dim \mathcal{C}(A) \le n$.

Hence, one can have the following corollary.

Corollary 3.22.1 *For any $m \times n$ matrix A,* $\operatorname{rank} A \le \min\{m, n\}$.

An $m \times n$ matrix A is said to have **full rank** if $\operatorname{rank} A = \min\{m, n\}$.

Remark: In the proof of Theorem 3.22, we have shown that the r columns in $\tilde{A}$ are linearly independent as the step (1), which implies $\dim \mathcal{C}(A) \ge r = \dim \mathcal{R}(A)$. With this inequality, one may replace the step (2) by the following argument: Since the step (1) holds for an arbitrary matrix, applying this to A^T to get $\dim \mathcal{C}(A^T) \ge \dim \mathcal{R}(A^T)$. But, we know $\mathcal{C}(A^T) = \mathcal{R}(A)$ and $\mathcal{R}(A^T) = \mathcal{C}(A)$, which implies $\dim \mathcal{C}(A) \le \dim \mathcal{R}(A)$. Consequently, $\dim \mathcal{C}(A) = \dim \mathcal{R}(A)$.

In summary, the following equalities are now clear from Theorems 3.21 and 3.22: If U is the reduced row-echelon form of a matrix A,

$$\begin{aligned}\dim \mathcal{N}(A) &= \dim \mathcal{N}(U)\\ &= \text{the number of free variables in } U\mathbf{x} = \mathbf{0}.\end{aligned}$$

$$\begin{aligned}\dim \mathcal{R}(A) &= \dim \mathcal{R}(U)\\ &= \text{the number of nonzero row vectors of } U\\ &= \text{the number of leading 1's in } U\\ &= \text{the number of basic variables in } U\mathbf{x} = \mathbf{0}\\ &= \dim \mathcal{C}(A).\end{aligned}$$

Since $\dim \mathcal{R}(A) = \dim \mathcal{C}(A) = \text{rank}\, A$ is the number of basic variables in $A\mathbf{x} = \mathbf{0}$ and $\dim \mathcal{N}(A) = \text{nullity of}\, A$ is the number of free variables $A\mathbf{x} = \mathbf{0}$, one can get the following theorem, as the second relation among the dimensions of three spaces $\mathcal{R}(A), \mathcal{N}(A)$, and $\mathcal{C}(A)$.

Theorem 3.23 (Rank-Nullity Theorem) *For any $m \times n$ matrix A,*

$$\begin{aligned}\dim \mathcal{R}(A) + \dim \mathcal{N}(A) &= \text{rank}\, A + \text{nullity of}\, A &= n,\\ \dim \mathcal{C}(A) + \dim \mathcal{N}(A^T) &= \text{rank}\, A + \text{nullity of}\, A^T &= m.\end{aligned}$$

As a special case, if A is a square matrix of order n, then $\dim \mathcal{N}(A) = 0$ (or $\mathcal{N}(A) = \{\mathbf{0}\}$) if and only if $\dim \mathcal{R}(A) = n$ by the Rank-Nullity Theorem 3.23.

By Theorem 1.22, we get the following corollary.

Corollary 3.23.1 *Let A be an $n \times n$ square matrix. Then A is invertible if and only if rank $A = n$. In other words, any square matrix is invertible if and only if it has full rank.*

The next example shows that the rank and the nullity of a matrix can be found by using Gaussian elimination.

Example 58 (*Find the rank and the nullity*) For the 4×5 matrix

$$A = \begin{bmatrix} 1 & 1 & 2 & 0 & 7\\ 0 & 1 & 2 & 0 & 4\\ 1 & 0 & 0 & 1 & 8\\ 1 & 1 & 2 & -1 & 2 \end{bmatrix}$$

again given in Example 56, find the rank and the nullity of A.

Solution: The reduced row-echelon form of A is given as

$$U = \begin{bmatrix} 1 & 0 & 0 & 0 & 3 \\ 0 & 1 & 2 & 0 & 4 \\ 0 & 0 & 0 & 1 & 5 \\ 0 & 0 & 0 & 0 & 0 \end{bmatrix}$$

in Example 56. The first three nonzero rows containing leading 1's in U form a basis for $\mathcal{R}(U) = \mathcal{R}(A)$. Therefore, rank $A = \dim \mathcal{R}(A) = \dim \mathcal{C}(A) = 3$, and the nullity of $A = \dim \mathcal{N}(A) = 5 - \dim \mathcal{R}(A) = 2$. Two basis vectors for $\mathcal{N}(A)$ are given in Example 56, □

Problem 69 Find the nullity and the rank for each of the following matrices.

$$(1)\ A = \begin{bmatrix} 1 & 4 & 9 & 2 \\ 1 & 5 & 0 & 2 \\ 1 & 7 & 7 & 6 \end{bmatrix}, \qquad (2)\ A = \begin{bmatrix} 1 & 2 & 0 & 2 & 1 \\ -1 & -2 & 1 & 1 & 0 \\ 1 & 2 & -3 & -7 & 2 \\ 1 & 2 & -2 & -4 & 3 \end{bmatrix}.$$

Theorem 3.24 *For any two matrices A and B for which AB is well defined,*

(1) $\mathcal{N}(AB) \supseteq \mathcal{N}(B)$,

(2) $\mathcal{N}((AB)^T) \supseteq \mathcal{N}(A^T)$,

(3) $\mathcal{C}(AB) \subseteq \mathcal{C}(A)$,

(4) $\mathcal{R}(AB) \subseteq \mathcal{R}(B)$.

Proof: (1) and (2) are clear, since $B\mathbf{x} = \mathbf{0}$ implies $(AB)\mathbf{x} = A(B\mathbf{x}) = \mathbf{0}$.

(3) For an $m \times n$ matrix A and an $n \times \ell$ matrix B,

$$\begin{aligned} \mathcal{C}(AB) &= \{AB\mathbf{x} : \mathbf{x} \in \mathbb{R}^{\ell}\} \\ &\subseteq \{A\mathbf{y} : \mathbf{y} \in \mathbb{R}^{n}\} = \mathcal{C}(A), \end{aligned}$$

because $B\mathbf{x} \in \mathbb{R}^n$ for any $\mathbf{x} \in \mathbb{R}^{\ell}$. (See Example 43.)

(4) $\mathcal{R}(AB) = \mathcal{C}((AB)^T) = \mathcal{C}(B^T A^T) \subseteq \mathcal{C}(B^T) = \mathcal{R}(B)$. □

Corollary 3.24.1 rank $(AB) \leq \min\{\text{rank } A, \text{rank } B\}$.

The equality is also possible. In fact, it will be shown in Theorem 3.26 that for any $m \times n$ (square or non-square) matrix A, rank $(A^T A) =$ rank $A =$ rank (AA^T), implying that *two homogeneous systems* $A\mathbf{x} = \mathbf{0}$ *and* $A^T A\mathbf{x} = \mathbf{0}$ *have the same solution set for any matrix* A. Another such case will be shown in Exercise **3.22**.

Theorem 3.25 *Let A be an $m \times n$ matrix of rank r. Then*

(1) *for every submatrix C of A,* rank $C \leq$ rank A, *and*

(2) *the matrix A has at least one $r \times r$ submatrix of rank r, that is, A has an invertible submatrix of order r.*

Proof: (1) Consider an intermediate matrix B which is obtained from A by removing the rows that are not wanted in C. Then clearly $\mathcal{R}(B) \subseteq \mathcal{R}(A)$ and hence rank $B \leq$ rank A. Moreover, since the columns of C are taken from those of B, $\mathcal{C}(C) \subseteq \mathcal{C}(B)$ and rank $C \leq$ rank B.

(2) If rank $A = r$, one can find r linearly independent row vectors of A, which form a basis for the row space of A. Let B be the submatrix of A consisting of such r linearly independent row vectors of A. Clearly, rank $B = r$ and the column space of B should be of dimension r. By taking r linearly independent column vectors of B, one can find an $r \times r$ submatrix C of A with rank r. $\square$

Problem 70 Show that the rank of a matrix is equal to the order of the largest invertible submatrices.

Problem 71 For each of the matrices given in Problem 69, find an invertible submatrix of the largest order.

Problem 72 For any long matrix A which is not square, show that $\det(AA^T) = 0$.
Hint: Use Corollary 3.24.1.

3.9 Applications

3.9.1 One-sided or two-sided invertibility

In Chapter 1, we have seen that a non-square matrix A may have only one-sided (right or left) inverses.

In this section, it will be shown that the existence of a one-sided inverse (right or left) of A implies the existence or the uniqueness of the solutions of a system $A\mathbf{x} = \mathbf{b}$. Furthermore, this happens when A has full rank.

First, we need the following theorem.

Theorem 3.26 *For any $m \times n$ matrix A, A^TA is a symmetric $n \times n$ square matrix. Furthermore,*

$$\mathcal{N}(A^TA) = \mathcal{N}(A) \quad \textit{and} \quad \text{rank}(A^TA) = \text{rank}\, A.$$

Equivalently, two homogeneous systems $A\mathbf{x} = \mathbf{0}$ and $A^TA\mathbf{x} = \mathbf{0}$ have the same solution set for any (square or non-square) matrix A.

Consequently,

$$\text{rank}\, A = \text{rank}\, A^T = \text{rank}\,(A^TA) = \text{rank}\,(AA^T).$$

Proof: Clearly, A^TA is square and symmetric. First, show that $\mathcal{N}(A) = \mathcal{N}(A^TA)$. Since $A\mathbf{x} = \mathbf{0}$ implies $A^TA\mathbf{x} = \mathbf{0}$, one gets $\mathcal{N}(A) \subseteq \mathcal{N}(A^TA)$. To show the converse, let $\mathbf{x} \in \mathcal{N}(A^TA)$. Then $A^TA\mathbf{x} = \mathbf{0}$ and

$$(A\mathbf{x})^T(A\mathbf{x}) = \mathbf{x}^T(A^TA\mathbf{x}) = \mathbf{x}^T\mathbf{0} = 0.$$

Noting $A\mathbf{x} \in \mathbb{R}^m$, let $A\mathbf{x} = (b_1, b_2, \ldots, b_m)$ as a column vector in $\mathbb{R}^m$. Then

$$(A\mathbf{x})^T(A\mathbf{x}) = [b_1\ b_2\ \cdots b_m]\begin{bmatrix} b_1 \\ b_2 \\ \vdots \\ b_m \end{bmatrix} = b_1^2 + b_2^2 + \cdots + b_m^2.$$

Hence $(A\mathbf{x})^T(A\mathbf{x}) = 0$ implies $b_1 = b_2 = \cdots = b_m = 0$, which implies $A\mathbf{x} = \mathbf{0}$ and $\mathbf{x} \in \mathcal{N}(A)$.

Since both A and A^TA have n columns, the Rank-Nullity Theorem 3.23 gives

$$\text{rank}\, A + \dim \mathcal{N}(A) \ = \ n \ = \ \text{rank}\,(A^TA) + \dim \mathcal{N}(A^TA).$$

Since $\mathcal{N}(A) = \mathcal{N}(A^TA)$ is already shown, it follows $\text{rank}\, A = \text{rank}(A^TA)$.

By applying the same argument to A^T and using $\text{rank}\, A = \text{rank}\, A^T$, one can get the last equality. □

Corollary 3.26.1 *For any $\mathbf{x} \neq \mathbf{0}$ in $\mathbb{R}^n$, both products $\mathbf{x}\mathbf{x}^T$ and $\mathbf{x}^T\mathbf{x}$ have rank* 1.

The next theorem shows that the existence of a right inverse of A implies the existence of a solution of a system $A\mathbf{x} = \mathbf{b}$ for any $\mathbf{b}$. This can happen only to a wide rectangular matrix of full rank.

Theorem 3.27 (Right-sided Invertibility) *Let A be an $m \times n$ matrix. Then the following statements are equivalent.*

(1) *For each* $\mathbf{b} \in \mathbb{R}^m$, $A\mathbf{x} = \mathbf{b}$ *has* at least *one solution* $\mathbf{x}$ *in* $\mathbb{R}^n$.
(2) *The column vectors of A span* $\mathbb{R}^m$, *i.e.,* $\mathcal{C}(A) = \mathbb{R}^m$.
(3) $\mathcal{N}(A^T) = \{\mathbf{0}\}$.
(4) rank $A = m$ *(hence $m \leq n$). (A is a* 'wide' matrix of full rank.)
(5) *The row vectors of A are linearly independent.*
(6) AA^T *is square and two-sided invertible.*
(7) *A has a right inverse, (i.e., there exists B such that* $AB = I_m$*).*

Proof: (1) $\Leftrightarrow$ (2) is clear.

(2) $\Leftrightarrow$ (3) $\Leftrightarrow$ (4) comes from the Rank-Nullity Theorem:

$$\dim \mathcal{C}(A) + \dim \mathcal{N}(A^T) = \text{rank } A + \text{nullity of } A^T = m.$$

(4) $\Leftrightarrow$ (5) comes from Corollary 3.19.1 with $\dim \mathcal{C}(A) = \text{rank } A = \dim \mathcal{R}(A)$.

(4) $\Leftrightarrow$ (6) comes from Theorem 3.26, rank A = rank (AA^T), and Corollary 3.23.1: Any square matrix is invertible if and only if it has full rank.

(6) $\Rightarrow$ (7). By (6), $AA^T(AA^T)^{-1} = I_m$. So, A has a right inverse.

(7) $\Rightarrow$ (1). If B is a right inverse of A, then for any $\mathbf{b} \in \mathbb{R}^m$, $\mathbf{x} = B\mathbf{b}$ is a solution of $A\mathbf{x} = \mathbf{b}$. □

The next theorem shows that the existence of a left inverse of A implies the uniqueness of the solutions of $A\mathbf{x} = \mathbf{0}$. This can happen only to a long rectangular matrix of full rank.

Theorem 3.28 (Left-sided Invertibility) *Let A be an $m \times n$ matrix. Then the following statements are equivalent.*

(1) *For each* $\mathbf{b} \in \mathbb{R}^m$, $A\mathbf{x} = \mathbf{b}$ *has* at most *one solution* $\mathbf{x}$ *in* $\mathbb{R}^n$.
(2) $\mathcal{N}(A) = \{\mathbf{0}\}$.
(3) *The row vectors of A span* $\mathbb{R}^n$, *that is,* $\mathcal{R}(A) = \mathbb{R}^n$.
(4) rank $A = n$ *(hence $n \leq m$). (A is a* 'long' matrix of full rank.)
(5) *The column vectors of A are linearly independent.*
(6) A^TA *is square and two-sided invertible.*
(7) *A has a left inverse, (i.e., there exists C such that* $CA = I_n$*).*

Proof: (1) $\Leftrightarrow$(2) is clear.

(2) $\Leftrightarrow$ (3) $\Leftrightarrow$ (4) comes from the Rank-Nullity theorem:

$$\dim \mathcal{R}(A) + \dim \mathcal{N}(A) = \text{rank } A + \text{nullity of } A = n.$$

(4) $\Leftrightarrow$ (5). Use Corollary 3.19.1 with the identity $\dim \mathcal{C}(A) = \dim \mathcal{R}(A)$.
(4) $\Leftrightarrow$ (6) comes from Theorem 3.26, rank $A =$ rank $(A^T A)$, and Corollary 3.23.1: Any square matrix is invertible if and only if it has full rank.

(6) $\Rightarrow$ (7). By (6), $(A^T A)^{-1} A^T A = I_n$. So, A has a left inverse.

(7) $\Rightarrow$ (1). Let C be a left inverse of A and let $A\mathbf{x} = \mathbf{b}$ has a solution. Then $\mathbf{x} = CA\mathbf{x} = C\mathbf{b}$ is a unique solution. □

Example 59 (*Find a one-sided inverse*) The matrix $A = \begin{bmatrix} 2 & 0 & 0 \\ 0 & 3 & 0 \end{bmatrix}$ is wide of full rank, hence A has a right inverse, say C, by Theorems 3.27. In fact,

$$AC = \begin{bmatrix} 2 & 0 & 0 \\ 0 & 3 & 0 \end{bmatrix} \begin{bmatrix} 1/2 & 0 \\ 0 & 1/3 \\ x & y \end{bmatrix} = \begin{bmatrix} 1 & 0 \\ 0 & 1 \end{bmatrix},$$

where x, y are any real. But it cannot have any left inverse, by Theorem 3.28.

On the other hand, the matrix $\begin{bmatrix} 2 & 0 \\ 0 & 3 \\ 0 & 0 \end{bmatrix}$ is long of full rank, hence it has a left inverse, by Theorem 3.28. In fact,

$$\begin{bmatrix} 1/2 & 0 & x \\ 0 & 1/3 & y \end{bmatrix} \begin{bmatrix} 2 & 0 \\ 0 & 3 \\ 0 & 0 \end{bmatrix} = \begin{bmatrix} 1 & 0 \\ 0 & 1 \end{bmatrix},$$

where x, y are any real. But it cannot have any right inverse, by Theorems 3.27. □

Remark: (1) An $m \times n$ matrix A has a right inverse if and only if rank $A = m$, while A has a left inverse if and only if rank $A = n$. Therefore, if $m \neq n$, A cannot have any two-sided inverse. In fact, for an $m \times n$ matrix A of full rank,

if rank $A = n \leq m$, A is long and has a left inverse: $(A^T A)^{-1} A^T A = I_n$,
if rank $A = m \leq n$, A is wide and has a right inverse: $AA^T (AA^T)^{-1} = I_m$.

Another way of finding a one-sided inverse will be shown later in Subsection 7.8.3 as an application of a singular value decomposition.

(2) If A is a square matrix of full rank, "Right-sided Invertibility = Left-sided Invertibility", and the fourteen statements listed in Theorems 3.27 and 3.28 are all equivalent. In particular, for the invertibility of a square matrix it is enough to show the existence of any one-sided inverse. Moreover, in this case all of the one-sided (both left and right) inverses are the same (see Theorem 1.22).

Collecting all results obtained so far about solvability of a system, one can get lots of characterizations of the invertibility of a square matrix.

Theorem 3.29 (Both-sided Invertibility) *For a square matrix A of order n, the following statements are equivalent.*

(1) *A is (both-sided) invertible, (equivalently A^T is (both-sided) invertible).*
(2) $\det A \neq 0$, *(equivalently* $\det A^T \neq 0$*).*
(3) *A is row-equivalent to I_n, (equivalently A is column-equivalent to I_n).*
(4) *A is a product of elementary matrices.*
(5) *Elimination can be completed: $PA = LDU$, with all $d_i \neq 0$.*
(6) *$A\mathbf{x} = \mathbf{b}$ has a solution for every $\mathbf{b} \in \mathbb{R}^n$.*
(7) *$A\mathbf{x} = \mathbf{0}$ has only the trivial solution, i.e., $\mathcal{N}(A) = \{\mathbf{0}\}$.*
(8) *The columns of A are linearly independent.*
(9) *The columns of A span $\mathbb{R}^n$, i.e., $\mathcal{C}(A) = \mathbb{R}^n$.*
(10) *AA^T is invertible.*
(11) *A has a right inverse, (equivalently A^T has a left inverse).*
(12) $\operatorname{rank} A = n$, *(equivalently* $\operatorname{rank} A^T = n$*).*
(13) *The rows of A are linearly independent.*
(14) *The rows of A span $\mathbb{R}^n$, i.e., $\mathcal{R}(A) = \mathbb{R}^n$.*
(15) *$A^T A$ is invertible.*
(16) *A has a left inverse, (equivalently A^T has a right inverse).*
(17)* *The linear transformation $A : \mathbb{R}^n \to \mathbb{R}^n$ via $\mathbf{x} \mapsto A\mathbf{x}$ is injective.*
(18)* *The linear transformation $A : \mathbb{R}^n \to \mathbb{R}^n$ is surjective.*
(19)* *Zero is not an eigenvalue of A.*

Proof: Exercise; where have we proved which claim? Try to prove any not covered yet. The numbers with asterisks will be explained later: (17) and (18) in the Remark on page 177 and (19) in Theorem 6.2. □

Problem 73 For each of the following matrices, find all vectors $\mathbf{b}$ such that the system of linear equations $A\mathbf{x} = \mathbf{b}$ has at least one solution. Also, discuss the uniqueness of the solution.

$$(1)\ A = \begin{bmatrix} 1 & 3 & 0 & 2 & 4 \\ 1 & 4 & -1 & 3 & 5 \\ 2 & 2 & 3 & 2 & 2 \end{bmatrix}, \quad (2)\ A = \begin{bmatrix} 1 & 2 \\ 3 & 4 \\ 5 & 6 \end{bmatrix}, \quad (3)\ A = \begin{bmatrix} 2 & 3 & 4 \\ 0 & 1 & 6 \\ 0 & 0 & 7 \end{bmatrix}.$$

Problem 74 In Theorem 3.29, choose any two statements (i) and (j) from (1)-(16), and try to prove $(i) \Leftrightarrow (j)$ directly. Good and fun challenge!

3.9.2 Rank decompositions

Let

$$A = [\mathbf{c}_1\, \mathbf{c}_2\, \cdots\, \mathbf{c}_n]$$

be an $m \times n$ matrix with columns $\{\mathbf{c}_1, \mathbf{c}_2, \ldots, \mathbf{c}_n\}$ of rank r, and let

$\{\mathbf{q}_1, \mathbf{q}_2, \ldots, \mathbf{q}_r\}$ be any (ordered) basis for the column space $\mathcal{C}(A)$.

Then, each column $\mathbf{c}_j$ is a linear combination of the basis $\mathbf{q}_1, \mathbf{q}_2, \ldots, \mathbf{q}_r$, say

$$\mathbf{c}_j = t_{1j}\mathbf{q}_1 + t_{2j}\mathbf{q}_2 + \cdots + t_{rj}\mathbf{q}_r.$$

Write

$$\mathbf{t}_j = [t_{1j}\, t_{2j}\, \cdots\, t_{rj}]^T, \quad j = 1, 2, \ldots, n,$$

a column vector in $\mathbb{R}^r$. Then the matrix A is decomposed into a product of two matrices of rank r:

$$A = [\mathbf{c}_1\, \mathbf{c}_2\, \cdots\, \mathbf{c}_n]_{m\times n} = [\mathbf{q}_1\, \mathbf{q}_2\, \cdots\, \mathbf{q}_r]_{m\times r}[\mathbf{t}_1\, \mathbf{t}_2\, \cdots\, \mathbf{t}_n]_{r\times n}, \quad r = \text{rank}\, A.$$

This is called a **rank factorization** or a **rank decomposition** of a matrix A. Certainly, a rank decomposition of A is not unique.

3.10 Exercises

3.1. *A vector space is defined to be a set with two algebraic operations: a sum and a scalar multiplication.* The xy-plane $\mathbb{R}^2$ is the set of all pairs (x, y) of real numbers. On the plane $\mathbb{R}^2$, define two operations, a *sum* $\dot{+}$ and a *scalar multiplication* $\cdot$, as

$$\begin{aligned} (x_1, y_1)\dot{+}(x_2, y_2) &= (2x_1 + x_2, 2y_1 + y_2) \\ k \cdot (x, y) &= (kx, y). \end{aligned}$$

The operations on the right-hand sides are the usual ones. Which of the '$4+4$' rules in the definition of a vector space (Definition 3.2) are satisfied for the set $(\mathbb{R}^2, \dot{+}, \cdot)$ with two operations $\dot{+}$ and $\cdot$?

3.2. *A vector space is defined to be a set with two algebraic operations: a sum and a scalar multiplication.* The set $\mathbb{R}^3$ is the set of all triples (x, y, z) of real numbers. On the set $\mathbb{R}^3$, define two operations, a *sum* $+$ and a *scalar multiplication* $\cdot$, as

$$\begin{aligned} (x_1, y_1, z_1)+(x_2, y_2, z_2) &= (x_1 + x_2, y_1 + y_2, z_1 + z_2) \\ k \cdot (x, y, z) &= (kx, 2ky, 3kz), \end{aligned}$$

instead of the standard scalar multiplication defined in Page 102. Here, the operations on the right-hand sides are the usual ones. Is this a vector space?

3.3. Determine whether the given set is a vector space with the usual addition and scalar multiplication of functions.

(1) The set of all continuous functions f defined on the interval $[-1, 1]$ such that $f(0) = 1$.
(2) The set of all continuous functions f defined on the real line $\mathbb{R}$ such that $\lim_{x\to\infty} f(x) = 0$.
(3) The set of all twice differentiable functions f defined on $\mathbb{R}$ such that $f''(x) + f(x) = 0$.

3.4. Let $C[-1, 1]$ be the vector space of all continuous functions defined on the interval $[-1, 1]$. See Example 36(3). Which of the following subsets of $C[-1, 1]$ are subspaces of the vector space $C[-1, 1]$?

(1) $W = \{f(x) \in C[-1, 1] : f(-1) = -f(1)\}$.
(2) $W = \{f(x) \in C[-1, 1] : f(x) \geq 0 \text{ for all } x \in [-1, 1]\}$.
(3) $W = \{f(x) \in C[-1, 1] : f(-1) = -2 \text{ and } f(1) = 2\}$.
(4) $W = \{f(x) \in C[-1, 1] : f(1) = 0\}$.

3.5. Show that the set of matrices of the form $AB - BA$ cannot span the vector space $M_{n\times n}(\mathbb{R})$.

Hint: The sum of diagonal entries of $AB - BA$ is zero.

3.6. Express the polynomial $p(x) = -1 - 3x + 3x^2$ as a linear combination of $p_1(x) = 1 + 2x + x^2$, $p_2(x) = 2 + 5x$, and $p_3(x) = 3 + 8x - 2x^2$.

3.7. In the n-space $\mathbb{R}^n$, determine whether or not the set

$$\{\mathbf{e}_1 - \mathbf{e}_2,\ \mathbf{e}_2 - \mathbf{e}_3, \ldots, \mathbf{e}_{n-1} - \mathbf{e}_n,\ \mathbf{e}_n - \mathbf{e}_1\}$$

is linearly dependent.

3.8. Are the vectors

$$\mathbf{v}_1 = (1, 2, 3, 4),\ \ \mathbf{v}_2 = (2, 3, 4, 1),\ \ \mathbf{v}_3 = (3, 4, 1, 2),\ \ \mathbf{v}_4 = (4, 1, 2, 3)$$

linearly dependent or linearly independent in the 4-space $\mathbb{R}^4$?

3.9. For two matrices A and B for which AB can be defined, prove the following statements.

(1) If both A and B have linearly independent column vectors, then the column vectors of AB are also linearly independent.
(2) If both A and B have linearly independent row vectors, then the row vectors of AB are also linearly independent.
(3) If the column vectors of B are linearly dependent, then the column vectors of AB are also linearly dependent.
(4) If the row vectors of A are linearly dependent, then the row vectors of AB are also linearly dependent.

Hint: Use Lemma 3.11: *The column vectors of A are linearly independent if and only if $A\mathbf{x} = \mathbf{0}$ has only the trivial solution.*

3.10. How many 4×4 permutation matrices are there? Are they linearly independent in the vector space $M_{4\times 4}(\mathbb{R})$? Do they span the vector space $M_{4\times 4}(\mathbb{R})$?

Hint: $\dim M_{4\times 4}(\mathbb{R}) = 4 \times 4$, see Exercise 52(3). Use Lemma 3.16 .

3.11. In the 3-space $\mathbb{R}^3$, let W be the set of all vectors (x, y, z) such that $x - 2y - 3z = 0$. Prove that W is a subspace of $\mathbb{R}^3$. Find a basis for the subspace W.

3.12. Let A and B be two $n \times n$ matrices. Show that $AB = \mathbf{0}$ if and only if the column space of B is a subspace of the null space of A.

Hint: Recall that the column space of an $m \times n$ matrix A is just the image of the function $A : \mathbb{R}^n \to \mathbb{R}^m; \mathbf{x} \mapsto A\mathbf{x}$. See Remark in Page 113.

3.13. Let $\{\mathbf{v}_1, \mathbf{v}_2, \ldots, \mathbf{v}_n\}$ be a basis for a vector space V and let $W_i = \{r\mathbf{v}_i : r \in \mathbb{R}\}$ be the subspace of V spanned by $\mathbf{v}_i$. Show that $V = W_1 \oplus W_2 \oplus \cdots \oplus W_n$.

3.14. Let V and W be any two subspaces of the n-space $\mathbb{R}^n$. Show that

$$\dim(V + W) + \dim(V \cap W) = \dim V + \dim W.$$

Hint: Any basis for $V \cap W$ can be extended to a basis for V, also to a basis for W.

3.15. Show that

$$\begin{aligned} V &= \{(x, y, z, u) \in \mathbb{R}^4 : y + z + u = 0\}, \\ W &= \{(x, y, z, u) \in \mathbb{R}^4 : x + y = 0, z = 2u\} \end{aligned}$$

are both subspaces of $\mathbb{R}^4$. Find bases for V, W, $V + W$, and $V \cap W$.

3.16. Find a basis and the dimension of each of the following subspaces of $M_{n\times n}(\mathbb{R})$ of all $n \times n$ matrices.

(1) The subspace of all $n \times n$ diagonal matrices whose traces are zero. (Here, a *trace* means the sum of diagonal entries.)
(2) The subspace of all $n \times n$ symmetric matrices.
(3) The subspace of all $n \times n$ skew-symmetric matrices.

First, try it when $n = 2$ or $n = 3$.

3.17. Find bases for the row space, the column space, and the null space for each of the following matrices.

$$A = \begin{bmatrix} 1 & 0 & 0 \\ 0 & 1 & 0 \\ 0 & 0 & 0 \end{bmatrix}, \quad B = \begin{bmatrix} 1 & 2 & 0 & 0 \\ 0 & 0 & 1 & 1 \\ 0 & 0 & 0 & 0 \end{bmatrix}, \quad C = \begin{bmatrix} 1 & 3 & 0 & 4 \\ 0 & 1 & 0 & 1 \\ 0 & 0 & 1 & 2 \end{bmatrix}.$$

Hint: A and B are in reduced row-echelon form; C is in row-echelon form.

3.18. Find bases for the row space, the column space, and the null space for each of the following matrices.

$$A = \begin{bmatrix} 1 & 3 & 2 \\ 2 & 6 & 4 \\ 3 & 9 & 6 \end{bmatrix}, \quad B = \begin{bmatrix} 1 & 2 & 1 & 0 \\ 2 & 4 & 3 & 0 \\ 1 & 2 & 1 & 1 \end{bmatrix}, \quad C = \begin{bmatrix} 0 & -2 & 2 & 0 \\ 1 & 3 & 6 & 8 \\ 1 & 6 & 4 & 4 \end{bmatrix}.$$

3.19. Find the rank of A as a function of x when $A = \begin{bmatrix} 1 & 2 & 3 & 4 \\ 5 & 6 & 7 & 8 \\ 9 & 10 & 11 & x \end{bmatrix}$.

3.20. Find the rank and the largest invertible submatrix of each of the following matrices.

$$A = \begin{bmatrix} 1 & 2 & 3 \\ 4 & 5 & 6 \\ 7 & 8 & 9 \end{bmatrix}, \quad B = \begin{bmatrix} 1 & 1 & 0 & 1 \\ 2 & 1 & 1 & 2 \\ 1 & 1 & 1 & 4 \end{bmatrix}, \quad C = \begin{bmatrix} 1 & 4 & 9 & 2 \\ 1 & 5 & 0 & 2 \\ 1 & 7 & 7 & 6 \end{bmatrix}.$$

3.21. Find the rank and the nullity of A^k, $k = 1, 2, \ldots$ for each of the following matrices.

$$(1)\ A = \begin{bmatrix} 0 & 1 & 0 & 0 & 0 & 0 \\ 0 & 0 & 1 & 0 & 0 & 0 \\ 0 & 0 & 0 & 1 & 0 & 0 \\ 0 & 0 & 0 & 0 & 1 & 0 \\ 0 & 0 & 0 & 0 & 0 & 1 \\ 0 & 0 & 0 & 0 & 0 & 0 \end{bmatrix}, \quad (2)\ A = \begin{bmatrix} 0 & 1 & 0 & 0 & 0 & 0 \\ 0 & 0 & 1 & 0 & 0 & 0 \\ 0 & 0 & 0 & 0 & 0 & 0 \\ 0 & 0 & 0 & 0 & 1 & 0 \\ 0 & 0 & 0 & 0 & 0 & 1 \\ 0 & 0 & 0 & 0 & 0 & 0 \end{bmatrix}.$$

3.22. For any two matrices A and B for which AB can be defined, show that $\text{rank}(AB) = \text{rank}\, A$ if the rows of B are linearly independent, while $\text{rank}(CA) = \text{rank}\, A$ if the columns of C are linearly independent. In particular, if A is an invertible square matrix, then $\text{rank}\,(AB) = \text{rank}\, B = \text{rank}\,(BA)$ for any matrix B.

3.23. For any two nonzero column vectors $\mathbf{u}$ and $\mathbf{v}$ in the n-space $\mathbb{R}^n$, show that the matrix $A = \mathbf{u}\mathbf{v}^T$ has rank one. Conversely, every matrix of rank one can be written as $\mathbf{u}\mathbf{v}^T$ for some column vectors $\mathbf{u}$ and $\mathbf{v}$.

Hint: A has rank one if and only if the column space of A is spanned by a single column of A.

3.24. Show that the solution set of a linear equation $a_1x_1 + a_2x_2 + \cdots + a_nx_n = 0, (a_1, a_2, \ldots, a_n) \neq (0, 0, \ldots, 0)$ is an $(n-1)$-dimensional subspace of the n-space $\mathbb{R}^n$. This subspace is called a *hyperplane* of $\mathbb{R}^n$.

Hint: The equation has only one basic variable and so $n - 1$ free variables.

3.25. Find a one-sided inverse of each of the following matrices if there exists.

$$(1) \begin{bmatrix} 1 & 0 & 0 & 1 \\ 0 & 2 & 0 & 2 \\ 0 & 0 & 3 & 3 \end{bmatrix}, \quad (2) \begin{bmatrix} 1 & 0 & 0 \\ 0 & 2 & 0 \\ 0 & 0 & 3 \\ 0 & 0 & 5 \end{bmatrix}, \quad (3) \begin{bmatrix} 1 & 2 & 0 \\ 1 & 0 & 1 \\ 0 & 2 & 1 \end{bmatrix}.$$

3.26. For each matrix A in Exercise **3.25** (1),(2),(3), evaluate the determinants of AA^T and A^TA.

Hint: Use Problem 72. Note that $\det(AB) = \det A \det B$ is not true for non-square matrices A, B.

3.27. Determine whether the following statements are true or false, and justify your answers.

(1) The real line $\mathbb{R}^1$ is a vector space with the usual addition and multiplication.
(2) The set $[0, \infty)$ of all nonnegative real numbers is a subspace of $\mathbb{R}^1$.
(3) The set of all 3×3 matrices A satisfying $A^T = A^{-1}$ is a subspace of the vector space $M_{3\times 3}(\mathbb{R})$.
(4) For any two linearly independent vectors $\mathbf{x}$ and $\mathbf{y}$ in a vector space V, the vectors $\mathbf{x}, \mathbf{y}, \mathbf{x} - \mathbf{y}$ are also linearly independent.
(5) Any four vectors in the 3-space $\mathbb{R}^3$ are always linearly dependent.
(6) Any two vectors in the 3-space $\mathbb{R}^3$ are always linearly independent.
(7) If U and W are subspaces of a vector space V with bases α and β, respectively, then the intersection $\alpha \cap \beta$ is a basis for $U \cap W$.
(8) Let U be the row-echelon form of a square matrix A. If the first r columns of U are linearly independent, so are the first r columns of A.
(9) Let U be the row-echelon form of a square matrix A. If the first r rows of U are linearly independent, so are the first r rows of A.
(10) Any two row-equivalent matrices have the same row and column spaces.
(11) Let A be an $m \times n$ matrix with rank m. Then the column vectors of A span $\mathbb{R}^m$.
(12) Let A be an $m \times n$ matrix with rank n. Then $A\mathbf{x} = \mathbf{b}$ has at most one solution.
(13) If U is a subspace of V and $\mathbf{x}, \mathbf{y}$ are vectors in V such that $\mathbf{x} + \mathbf{y}$ is contained in U, then $\mathbf{x} \in U$ and $\mathbf{y} \in U$.
(14) For any two vector spaces U and V, U is a subspace of V if and only if $\dim U \le \dim V$.
(15) For any square matrix A, $A\mathbf{x} = \mathbf{b}$ has at least one solution for every $\mathbf{b}$ if and only if $A\mathbf{x} = \mathbf{b}$ has at most one solution for every $\mathbf{b}$.
(16) For any symmetric matrix A, $\mathcal{R}(A) = \mathcal{C}(A)$ and $\mathcal{N}(A) = \mathcal{N}(A^T)$.
(17) For any $m \times n$ matrix A, $\dim \mathcal{C}(A^T) + \dim \mathcal{N}(A^T) = m$.
(18) For any rectangular matrix A, the null space $\mathcal{N}(A)$ of A and the null space $\mathcal{N}(A^T)$ of A^T have the same dimension.
(19) Every rectangular matrix has at least one one-sided inverse matrix.
(20) For any $m \times n$ matrix A, $\text{rank}(A^TAA^T) = \text{rank}\,(AA^TA)$.

Hint: Exactly nine of them are true.

Chapter 4

Linear Transformations; Linearity preserving functions

Given any two vector spaces, one might ask how to compare one vector space with the other, or whether two vector spaces have the 'same' structure as vector spaces or not.

To answer this question, one has to compare them first as sets, and then see whether their two arithmetic rules, the vector sums and the scalar multiplications, are the same or not.

A natural way of comparing two sets is to define a *function* between them. When a function f is given between the underlying sets of the vector spaces, one can compare the arithmetic rules of the vector spaces by examining whether the function f preserves the two algebraic operations:

$$f(\mathbf{x}+\mathbf{y}) = f(\mathbf{x}) + f(\mathbf{y}) \quad \text{and} \quad f(k\mathbf{x}) = kf(\mathbf{x})$$

for any vectors $\mathbf{x}, \mathbf{y}$ and any scalar k, that is, preserves the linearity.

In this chapter we discuss these kinds of functions between two vector spaces.

4.1 Examples and properties of linear transformations

Definition 4.1 Let V and W be vector spaces. A function $T : V \to W$ is called a **linear transformation** or a **linear operator** from V to W if for any two vectors $\mathbf{x},\ \mathbf{y} \in V$ and any scalar $k \in \mathbb{R}$ the following two conditions are satisfied:

(1) $T(\mathbf{x}+\mathbf{y}) = T(\mathbf{x}) + T(\mathbf{y})$,
(2) $T(k\mathbf{x}) = kT(\mathbf{x})$.

The space V is called the **domain (space)** of T, and the space W is the **codomain (space)** or the **target space** of T. We often call T simply **linear**.

When $V = W$, T is called a **linear transformation** or a **linear operator** on the space V. [1]

It is easy to see that the two conditions for a linear transformation can be combined into a single requirement

$$T(\mathbf{x} + k\mathbf{y}) = T(\mathbf{x}) + kT(\mathbf{y})$$

for any $\mathbf{x}$, $\mathbf{y} \in V$ and for any scalar $k \in \mathbb{R}$.

Geometrically, the linearity is just the requirement for a straight line to be transformed to a straight line, since $\mathbf{x} + k\mathbf{y}$ represents a straight line through $\mathbf{x}$ in the direction $\mathbf{y}$ in V, and its image $T(\mathbf{x}) + kT(\mathbf{y})$ represents a straight line through $T(\mathbf{x})$ in the direction of $T(\mathbf{y})$ in W.

Example 60 (*Linear or not*) Consider the following functions:

(1) $f : \mathbb{R} \to \mathbb{R}$ defined by $f(x) = ax$, a any scalar,
(2) $g : \mathbb{R} \to \mathbb{R}$ defined by $g(x) = ax + b$, a, b any scalars, $b \neq 0$,
(3) $h : \mathbb{R}^2 \to \mathbb{R}^2$ defined by $h(x, y) = (x - y, 2x)$,
(4) $k : \mathbb{R}^2 \to \mathbb{R}^2$ defined by $k(x, y) = (xy,\ x^2)$.

One can easily see that g and k are not linear, while f and h are linear. Moreover, on the 1-space $\mathbb{R}$, all polynomials of degree greater than one cannot be linear. For example, $f(x) = x^n, n > 1$ is not linear. □

Example 61 (*Every matrix A is a linear transformation*)

(1) For any $m \times n$ matrix A, the transformation $T : \mathbb{R}^n \to \mathbb{R}^m$ defined by the matrix product

$$T(\mathbf{x}) = A\mathbf{x}$$

is a linear transformation by the distributive law $A(\mathbf{x} + k\mathbf{y}) = A\mathbf{x} + kA\mathbf{y}$ for any $\mathbf{x}$, $\mathbf{y} \in \mathbb{R}^n$ and for any scalar $k \in \mathbb{R}$. Therefore, a matrix A, identified with the

[1] The functions are commonly denoted by f or g. However we use a notation T or S for the linear transformations.

transformation T, can be considered as a linear transformation $\mathbf{x} \mapsto A\mathbf{x}$ from $\mathbb{R}^n$ to $\mathbb{R}^m$.

(2) For a vector space V, the **identity transformation** $id : V \to V$ is defined by $id(\mathbf{x}) = \mathbf{x}$ for all $\mathbf{x} \in V$. If W is another vector space, the **zero transformation** $T_0 : V \to W$ is defined by $T_0(\mathbf{x}) = \mathbf{0}$ (the zero vector) for all $\mathbf{x} \in V$. Clearly, both transformations are linear. $\square$

The following theorem is a direct consequence of the definition, and the proof is left as an exercise.

Theorem 4.2 *Let $T : V \to W$ be a linear transformation. Then*

(1) $T(\mathbf{0}) = \mathbf{0}$.

(2) *For any $\mathbf{x}_1, \mathbf{x}_2, \ldots, \mathbf{x}_n \in V$ and scalars $k_1, k_2, \ldots, k_n$,*

$$T(k_1\mathbf{x}_1 + k_2\mathbf{x}_2 + \cdots + k_n\mathbf{x}_n) = k_1T(\mathbf{x}_1) + k_2T(\mathbf{x}_2) + \cdots + k_nT(\mathbf{x}_n).$$

Nontrivial important examples of linear transformations are the rotations, reflections, and projections in geometry.

Example 62 (*The rotations, reflections, and projections in geometry are linear*)

(1) Let θ denote an angle in radians in the Euclidean plane $\mathbb{R}^2$. Then the matrix

$$R_\theta = \begin{bmatrix} \cos\theta & -\sin\theta \\ \sin\theta & \cos\theta \end{bmatrix}$$

defines a linear transformation of $\mathbb{R}^2$ that rotates any vector in $\mathbb{R}^2$ through the angle θ about the origin. It is called a **rotation** by the angle θ.

(2) The **projection** of $\mathbb{R}^2$ onto the x-axis is the linear transformation $P : \mathbb{R}^2 \to \mathbb{R}^2$ defined by, for $\mathbf{x} = (x, y) \in \mathbb{R}^2$,

$$P(\mathbf{x}) = \begin{bmatrix} 1 & 0 \\ 0 & 0 \end{bmatrix} \begin{bmatrix} x \\ y \end{bmatrix} = \begin{bmatrix} x \\ 0 \end{bmatrix}.$$

(3) The linear transformation $T : \mathbb{R}^2 \to \mathbb{R}^2$ defined by, for $\mathbf{x} = (x, y)$,

$$T(\mathbf{x}) = \begin{bmatrix} 1 & 0 \\ 0 & -1 \end{bmatrix} \begin{bmatrix} x \\ y \end{bmatrix} = \begin{bmatrix} x \\ -y \end{bmatrix}$$

is the **reflection** across the x-axis in $\mathbb{R}^2$. The image of a figure (like $\coprod$) by the reflection is its *mirror image* (like $\prod$). $\square$

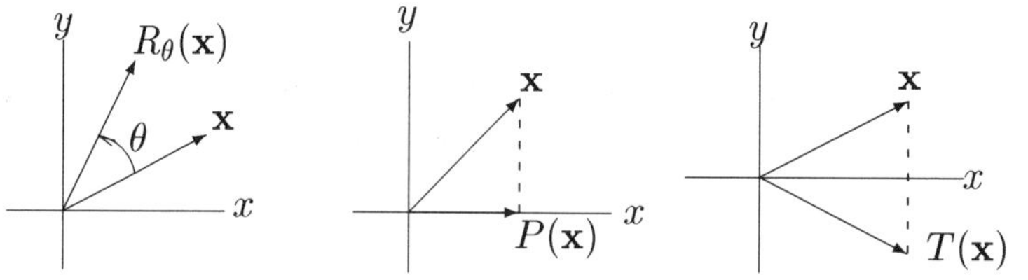

Figure 4.1: A rotation, a projection and a reflection on the plane $\mathbb{R}^2$

Problem 75 Find the matrix of the reflection across the line $y = x$ in the plane $\mathbb{R}^2$, and also the matrix of the reflection through the plane $y = x$ in the 3-space $\mathbb{R}^3$. In both cases, the image of a figure by a reflection is its mirror image. The set $y = x$ of fixed points is its mirror in each case.

Example 63 (*Differentiations and integrations are linear*) In calculus, it is well known that two transformations

$$D : P_n(\mathbb{R}) \to P_{n-1}(\mathbb{R}), \quad \mathcal{I} : P_{n-1}(\mathbb{R}) \to P_n(\mathbb{R})$$

defined by differentiation and integration,

$$D(f)(x) = f'(x), \quad \mathcal{I}(f)(x) = \int_0^x f(t)dt,$$

preserve the linearity. Hence, they are linear transformations. Many problems related to differential and integral equations can be reformulated in terms of linear transformations or matrices. □

As shown in Example 60, on the real line $\mathbb{R}$, only the polynomial functions $f(x) = ax$ of degree one are linear among algebraic functions, and no polynomials of higher degree on $\mathbb{R}$ can be linear. However, all rotations, projections, and reflections are linear on the plane $\mathbb{R}^2$ as important geometric functions.

In future, lots of interesting linear transformations on an (abstract) vector space will appear.

Definition 4.3 Let V and W be two vector spaces, and let $T : V \to W$ be a linear transformation from V to W.

(1) $\text{Ker}(T) = \{\mathbf{v} \in V : T(\mathbf{v}) = \mathbf{0}\} \subseteq V$ is called the **kernel** of T.

(2) $\text{Im}(T) = \{T(\mathbf{v}) : \mathbf{v} \in V\} = T(V) \subseteq W$ is called the **image** of T.

Example 64 Let V and W be vector spaces and let $id : V \to V$ and $T_0 : V \to W$ be the identity and the zero transformations, respectively. Then it is easy to see that

$$\mathrm{Ker}(id) = \{\mathbf{0}\}, \ \mathrm{Im}(id) = V, \ \ \mathrm{Ker}(T_0) = V, \ \text{ and } \ \mathrm{Im}(T_0) = \{\mathbf{0}\}. \qquad \square$$

Theorem 4.4 *Let $T : V \to W$ be a linear transformation from a vector space V to a vector space W. Then the kernel $\mathrm{Ker}(T)$ and the image $\mathrm{Im}(T)$ are subspaces of V and W, respectively.*

Proof: Since $T(\mathbf{0}) = \mathbf{0}$, both Ker($T$) and Im($T$) are always nonempty.

(1) For any $\mathbf{x}, \mathbf{y} \in \mathrm{Ker}(T)$ and for any scalar k,

$$T(\mathbf{x} + k\mathbf{y}) = T(\mathbf{x}) + kT(\mathbf{y}) = \mathbf{0} + k\mathbf{0} = \mathbf{0}.$$

Hence $\mathbf{x} + k\mathbf{y} \in \mathrm{Ker}(T)$, so that Ker($T$) is a subspace of V.

(2) If $\mathbf{v}, \mathbf{w} \in$ Im(T), then there exist $\mathbf{x}$ and $\mathbf{y}$ in V such that $T(\mathbf{x}) = \mathbf{v}$ and $T(\mathbf{y}) = \mathbf{w}$. Thus, for any scalar k,

$$\mathbf{v} + k\mathbf{w} = T(\mathbf{x}) + kT(\mathbf{y}) = T(\mathbf{x} + k\mathbf{y}).$$

Thus $\mathbf{v} + k\mathbf{w} \in$ Im(T), so that Im(T) is a subspace of W. $\square$

Example 65 (Ker(A) $= \mathcal{N}(A)$ *and* Im(A) $= \mathcal{C}(A)$ *for any matrix* A) Let $A : \mathbb{R}^n \to \mathbb{R}^m$ be the linear transformation defined by an $m \times n$ matrix A as in Example 61(1).

The kernel Ker(A) of A consists of all solutions to the homogeneous system $A\mathbf{x} = \mathbf{0}$. Hence, Ker(A) is just the null space $\mathcal{N}(A)$ of the matrix A.

Also, the image Im(A) is just the column space $\mathcal{C}(A) \subseteq \mathbb{R}^m$. See Remark in Page 113 and recall that $A\mathbf{x}$ is a linear combination of the column vectors of A. $\square$

Example 66 (*The trace is a linear transformation*) A **trace** is a function $\mathrm{tr} : M_{n\times n}(\mathbb{R}) \to \mathbb{R}$ defined by the sum of diagonal entries

$$\mathrm{tr}(A) = a_{11} + a_{22} + \cdots + a_{nn} = \sum_{i=1}^{n} a_{ii}$$

for $A = [a_{ij}] \in M_{n\times n}(\mathbb{R})$, and $\mathrm{tr}(A)$ is called the **trace** of the matrix A. It is easy to show that

$$\mathrm{tr}(A+B) = \mathrm{tr}(A) + \mathrm{tr}(B),\ \ \mathrm{tr}(kA) = k\,\mathrm{tr}(A),\ \ \text{and}\ \ \mathrm{tr}(AB) = \mathrm{tr}(BA)$$

for any two matrices A and B in $M_{n\times n}(\mathbb{R})$, which means that 'tr' is a linear transformation from $M_{n\times n}(\mathbb{R})$ to the 1-space $\mathbb{R}$.

In addition, one can show that the set of all $n\times n$ matrices with trace 0 is a subspace of the vector space $M_{n\times n}(\mathbb{R})$. □

Problem 76 Show that the function $T : \mathbb{R}^n \to \mathbb{R}$ defined by $(x_1, x_2, \ldots, x_n) \mapsto x_1 + x_2 + \cdots + x_n$ is linear, but the function $S : \mathbb{R}^n \to \mathbb{R}$ defined by $(x_1, x_2, \ldots, x_n) \mapsto x_1^2 + x_2^2 + \cdots + x_n^2$ is *not* linear.

Problem 77 Let $W = \{A \in M_{n\times n}(\mathbb{R}) : \mathrm{tr}(A) = 0\}$. Show that W is a subspace, and find a basis for W. Compare this with Exercise **3.16**(1).

Problem 78 Show that, for any two matrices A and B in $M_{n\times n}(\mathbb{R})$, $\mathrm{tr}(AB) = \mathrm{tr}(BA)$.

4.2 Constructing linear transformations

One of the most important properties of linear transformations is that they are completely determined by their values on a basis.

Theorem 4.5 *Let V and W be vector spaces. Let $\{\mathbf{v}_1, \mathbf{v}_2, \ldots, \mathbf{v}_n\}$ be any basis for V and let $\mathbf{w}_1, \mathbf{w}_2, \ldots, \mathbf{w}_n$ be any n vectors (possibly repeated) in W. Then there exists a unique linear transformation $T : V \to W$ such that $T(\mathbf{v}_i) = \mathbf{w}_i$ for $i = 1, 2, \ldots, n$.*

In other words, the function $\mathbf{v}_i \mapsto \mathbf{w}_i$ can be extended to a linear transformation in a unique way.

Proof: Let $\mathbf{x} \in V$. Then it has a unique expression: $\mathbf{x} = \sum_{i=1}^n a_i\mathbf{v}_i$ for some scalars $a_1, a_2, \ldots, a_n$.

Extension: Define

$$T : V \to W \quad \text{by} \quad T(\mathbf{x}) = \sum_{i=1}^n a_i T(\mathbf{v}_i) = \sum_{i=1}^n a_i\mathbf{w}_i.$$

In particular, $T(\mathbf{v}_i) = \mathbf{w}_i$ for $i = 1, 2, \ldots, n$.

Linearity: For any two $\mathbf{x} = \sum_{i=1}^{n} a_i\mathbf{v}_i$, $\mathbf{y} = \sum_{i=1}^{n} b_i\mathbf{v}_i \in V$ and any scalar k, we have $\mathbf{x} + k\mathbf{y} = \sum_{i=1}^{n}(a_i + kb_i)\mathbf{v}_i$. Hence,

$$T(\mathbf{x} + k\mathbf{y}) = \sum_{i=1}^{n}(a_i + kb_i)\mathbf{w}_i = \sum_{i=1}^{n} a_i\mathbf{w}_i + k\sum_{i=1}^{n} b_i\mathbf{w}_i = T(\mathbf{x}) + kT(\mathbf{y}).$$

Uniqueness: Suppose that $S : V \to W$ is linear and $S(\mathbf{v}_i) = \mathbf{w}_i$ for $i = 1, 2, \ldots, n$. Then for any $\mathbf{x} \in V$ with $\mathbf{x} = \sum_{i=1}^{n} a_i\mathbf{v}_i$, one can get

$$S(\mathbf{x}) = \sum_{i=1}^{n} a_i S(\mathbf{v}_i) = \sum_{i=1}^{n} a_i\mathbf{w}_i = T(\mathbf{x}).$$

It gives $S = T$. □

The uniqueness in Theorem 4.5 may be rephrased as the following corollary.

Corollary 4.5.1 *Let V and W be vector spaces, and let $\{\mathbf{v}_1, \mathbf{v}_2, \ldots, \mathbf{v}_n\}$ be any basis for V. If $S, T : V \to W$ are linear transformations and $S(\mathbf{v}_i) = T(\mathbf{v}_i)$ for $i = 1, 2, \ldots, n$, then $S = T$, i.e., $S(\mathbf{x}) = T(\mathbf{x})$ for all $\mathbf{x} \in V$.*

Corollary 4.5.2 *Let $\{\mathbf{e}_1, \mathbf{e}_2, \ldots, \mathbf{e}_n\}$ be the standard basis for the n-space $\mathbb{R}^n$ and let $\mathbf{w}_1, \mathbf{w}_2, \ldots, \mathbf{w}_n$ be any n vectors (possibly repeated) in the m-space $\mathbb{R}^m$. Then the function $\mathbf{e}_i \mapsto \mathbf{w}_i$ can be extended uniquely to a linear transformation $\mathbf{x} \mapsto A\mathbf{x}$, where A is an $m \times n$ matrix whose columns are $\mathbf{w}_1, \mathbf{w}_2, \ldots, \mathbf{w}_n$ in order.*

Proof: Trivial because for any $m \times n$ matrix A, $\mathbf{x} \mapsto A\mathbf{x}$ is a linear transformation from $\mathbb{R}^n$ to $\mathbb{R}^m$, and *the matrix product $A\mathbf{e}_i$ is just the i-th column of A* for $i = 1, 2, \ldots, n$. See Problem 54. □

Example 67 (*Linear extension of a function defined on a basis*) Let $\mathbf{w}_1 = (1, 2)$, $\mathbf{w}_2 = (3, 4)$, $\mathbf{w}_3 = (5, 6)$ be three vectors in the plane $\mathbb{R}^2$.

(1) Let $\alpha = \{\mathbf{e}_1, \mathbf{e}_2, \mathbf{e}_3\}$ be the standard basis for the 3-space $\mathbb{R}^3$, and let $T : \mathbb{R}^3 \to \mathbb{R}^2$ be the linear transformation defined by

$$T(\mathbf{e}_1) = \mathbf{w}_1, \qquad T(\mathbf{e}_2) = \mathbf{w}_2, \qquad T(\mathbf{e}_3) = \mathbf{w}_3.$$

Find a formula for $T(x_1, x_2, x_3)$, and then use it to compute $T(2, 0, 1)$.

(2) Let $\beta = \{\mathbf{v}_1, \mathbf{v}_2, \mathbf{v}_3\}$ be another basis for $\mathbb{R}^3$, where $\mathbf{v}_1 = (1,1,1)$, $\mathbf{v}_2 = (1,1,0)$, $\mathbf{v}_3 = (1,0,0)$, and let $T : \mathbb{R}^3 \to \mathbb{R}^2$ be the linear transformation defined by

$$T(\mathbf{v}_1) = \mathbf{w}_1, \qquad T(\mathbf{v}_2) = \mathbf{w}_2, \qquad T(\mathbf{v}_3) = \mathbf{w}_3.$$

Find a formula for $T(x_1, x_2, x_3)$, and then use it to compute $T(2,0,1)$.

Solution: (1) For $\mathbf{x} = (x_1, x_2, x_3) = x_1\mathbf{e}_1 + x_2\mathbf{e}_2 + x_3\mathbf{e}_3 \in \mathbb{R}^3$,

$$\begin{aligned} T(\mathbf{x}) &= \sum_{i=1}^{3} x_i T(\mathbf{e}_i) = \sum_{i=1}^{3} x_i \mathbf{w}_i \\ &= x_1(1,2) + x_2(3,4) + x_3(5,6) \\ &= (x_1 + 3x_2 + 5x_3,\ 2x_1 + 4x_2 + 6x_3). \end{aligned}$$

Thus, $T(2,0,1) = (7,10)$. In matrix notation, this can be expressed as

$$A\mathbf{x} = \begin{bmatrix} 1 & 3 & 5 \\ 2 & 4 & 6 \end{bmatrix} \begin{bmatrix} x_1 \\ x_2 \\ x_3 \end{bmatrix} = \begin{bmatrix} x_1 + 3x_2 + 5x_3 \\ 2x_1 + 4x_2 + 6x_3 \end{bmatrix}$$

with $A = [T(\mathbf{e}_1)\ T(\mathbf{e}_2)\ T(\mathbf{e}_3)] = [\mathbf{w}_1\ \mathbf{w}_2\ \mathbf{w}_3]$. as stated in Corollary 4.5.2.

(2) First, need to express $\mathbf{x} = (x_1, x_2, x_3)$ as a linear combination of the basis vectors $\mathbf{v}_1, \mathbf{v}_2, \mathbf{v}_3$, *i.e.*,

$$\begin{aligned} (x_1, x_2, x_3) = \sum_{i=1}^{3} k_i \mathbf{v}_i &= k_1(1,1,1) + k_2(1,1,0) + k_3(1,0,0) \\ &= (k_1 + k_2 + k_3)\mathbf{e}_1 + (k_1 + k_2)\mathbf{e}_2 + k_1\mathbf{e}_3. \end{aligned}$$

By equating corresponding components one can get a system of linear equations

$$\begin{cases} k_1 + k_2 + k_3 = x_1 \\ k_1 + k_2 = x_2 \\ k_1 = x_3 \,. \end{cases}$$

The solution is $k_1 = x_3,\ k_2 = x_2 - x_3,\ k_3 = x_1 - x_2$. Therefore,

$$\begin{aligned} (x_1, x_2, x_3) &= x_3\mathbf{v}_1 + (x_2 - x_3)\mathbf{v}_2 + (x_1 - x_2)\mathbf{v}_3, \quad \text{and then} \\ T(x_1, x_2, x_3) &= x_3 T(\mathbf{v}_1) + (x_2 - x_3)T(\mathbf{v}_2) + (x_1 - x_2)T(\mathbf{v}_3) \\ &= x_3(1,2) + (x_2 - x_3)(3,4) + (x_1 - x_2)(5,6) \\ &= (5x_1 - 2x_2 - 2x_3,\ 6x_1 - 2x_2 - 2x_3). \end{aligned}$$

From this formula, it follows $T(2,0,1) = (8,10)$. □

Problem 79 Is there a linear transformation $T : \mathbb{R}^3 \to \mathbb{R}^2$ such that $T(1,2,3) = (0,1)$ and $T(2,4,6) = (1,0)$?

Hint: Clear? If not, see the next Problem 80.

Problem 80 Let $T : V \to W$ be any linear transformation and let $\{\mathbf{w}_1, \mathbf{w}_2, \ldots, \mathbf{w}_k\}$ be any linearly independent subset of the image $\text{Im}(T) \subseteq W$. If $\mathbf{v}_1, \mathbf{v}_2, \ldots, \mathbf{v}_k$ are chosen in V so that $T(\mathbf{v}_i) = \mathbf{w}_i$ for $i = 1, 2, \ldots, k$, then $\{\mathbf{v}_1, \mathbf{v}_2, \ldots, \mathbf{v}_k\}$ is also linearly independent. Prove it.

4.3 Which linear transformation has the inverse linear transformation?

A function f from a set X to a set Y (not necessarily vector spaces) is said to be **invertible** if there is a function g from Y to X such that their composition [2] $g \circ f = id_X$, the identity function on X and $f \circ g = id_Y$, the identity function on Y. Such a function g is unique, called the **inverse** function of f and denoted by $g = f^{-1}$.

If there exists an invertible function from a set X to another set Y, then it gives a one-to-one correspondence between these two sets so that they have the same number of elements. In this case, a one-to-one correspondence gives a *relabeling* of the elements of a set along the elements of the other set and hence the sets X and Y can be identified sometimes as sets.

A useful criterion for a function from a set to another to be invertible is that it is injective and surjective. Recall that a function $f : X \to Y$ is **one-to-one** (or **injective**) if $f(u) = f(v)$ in Y implies $u = v$ in X, and is **onto** (or **surjective**) if for each element y in Y there exists an element x in X such that $f(x) = y$.

A function is **bijective** if it is both one-to-one and onto, that is, if for each element y in Y there is a *unique* element x in X such that $f(x) = y$.

Lemma 4.6 *A function $f : X \to Y$ is invertible if and only if it is bijective (or, one-to-one and onto).*

Proof: Let $f : X \to Y$ be invertible, and let $g : Y \to X$ be its inverse. If $f(u) = f(v)$, then $u = g(f(u)) = g(f(v)) = v$. Thus f is injective. For each $y \in Y$, let $g(y) = x$ in X. Then $f(x) = f(g(y)) = y$. Thus it is surjective.

---

[2]The *composition* of two functions: If $f : \mathbb{R} \to \mathbb{R}$ is given by $f(x) = 2x + 1$ and $g : \mathbb{R} \to \mathbb{R}$ is given by $g(x) = \sin x$, then the compositions $(f \circ g)(x) = f(g(x)) = f(\sin x) = 2\sin x + 1$ and $(g \circ f)(x) = g(f(x)) = g(2x + 1) = \sin(2x + 1)$.

Conversely, if f is bijective, then, for each $y \in Y$, there is a unique $x \in X$ such that $f(x) = y$. Now for each $y \in Y$, we define $g(y) = x$. Then it is easy check that $g : Y \to X$ is well defined, and that $f \circ g = id_Y$ and $g \circ f = id_X$. □

Example 68 (*Which function has an inverse?*) The exponential function $f(x) = e^x$ on the real line $\mathbb{R}^1$ is injective, but not surjective. By restricting the target space to $(0, \infty)$, the exponential function $f(x) = e^x : \mathbb{R}^1 \to (0, \infty)$ is bijective, and its inverse function is the (natural) logarithm function $\ln x : (0, \infty) \to \mathbb{R}^1$. □

Example 69 (*Which linear transformation is injective or surjective?*) For an $m \times n$ matrix A,

(1) the linear transformation $A : \mathbb{R}^n \to \mathbb{R}^m$ is injective if and only if $\mathrm{Ker}(A) = \mathcal{N}(A)$ is trivial $\{\mathbf{0}\}$;

(2) the linear transformation $A : \mathbb{R}^n \to \mathbb{R}^m$ is surjective if and only if $\mathrm{Im}(A) = \mathcal{C}(A)$ is the full space $\mathbb{R}^m$. Hence,

(3) $A : \mathbb{R}^n \to \mathbb{R}^m$ is bijective if and only if $\mathcal{N}(A) = \{\mathbf{0}\}$ and $\mathcal{C}(A) = \mathbb{R}^m$.

See Example 65. Also, see Theorems 3.27 and 3.28 for one-sided invertibility. □

If $T : V \to W$ and $S : W \to Z$ are linear transformations, then it is easy to show that their composition $(S \circ T)(\mathbf{v}) = S(T(\mathbf{v}))$ is also a linear transformation. In particular, if two linear transformations are defined by matrices $A : \mathbb{R}^n \to \mathbb{R}^m$ and $B : \mathbb{R}^m \to \mathbb{R}^k$ as in Example 61(1), then their composition is just the matrix product BA, [3] *i.e.*, $(B \circ A)(\mathbf{x}) = B(A\mathbf{x}) = (BA)\mathbf{x}$.

Lemma 4.7 *Let V and W be vector spaces. If $T : V \to W$ is an invertible linear transformation, then its inverse $T^{-1} : W \to V$ is also linear.*

Proof: Let $\mathbf{w}_1, \mathbf{w}_2 \in W$ be any, and let k be any scalar. Since T is invertible, there exist unique $\mathbf{v}_1$ and $\mathbf{v}_2$ in V such that $T(\mathbf{v}_1) = \mathbf{w}_1$ and $T(\mathbf{v}_2) = \mathbf{w}_2$. Then

$$\begin{aligned} T^{-1}(\mathbf{w}_1 + k\mathbf{w}_2) &= T^{-1}(T(\mathbf{v}_1) + kT(\mathbf{v}_2)) \\ &= T^{-1}(T(\mathbf{v}_1 + k\mathbf{v}_2)) \\ &= \mathbf{v}_1 + k\mathbf{v}_2 \\ &= T^{-1}(\mathbf{w}_1) + kT^{-1}(\mathbf{w}_2). \end{aligned}$$

Hence, T^{-1} is linear. □

[3] This is a main reason why the product of two matrices is defined as in Section 1.6 and not defined to be the entrywise multiplications for the Hadamard product.

Definition 4.8 A linear transformation $T : V \to W$ from a vector space V to another vector space W is called an **isomorphism** if it is invertible (or, one-to-one and onto). In this case, we say that V and W are **isomorphic** to each other. Write $V \cong W$ to mean V and W are isomorphic.

In this sense of "isomorphic", we sometimes say that the two vector spaces have the same structure as vector spaces, and hence a vector space can be identified with the other one.

Example 70 (*The vector space $P_n(\mathbb{R})$ is isomorphic to $\mathbb{R}^{n+1}$: $P_n(\mathbb{R}) \cong \mathbb{R}^{n+1}$*) Consider the vector space $P_2(\mathbb{R}) = \{a+bx+cx^2 : a, b, c \in \mathbb{R}\}$ of all polynomials of degree ≤ 2 with real coefficients. To a polynomial $a + bx + cx^2$ in $P_2(\mathbb{R})$, one can assign the column vector $[a\ b\ c]^T$ in the 3-space $\mathbb{R}^3$. Then it is not hard to see that it is an isomorphism from the vector space $P_2(\mathbb{R})$ to the 3-space $\mathbb{R}^3$, by which one can *identify* the polynomial $a + bx + cx^2$ with the column vector $[a\ b\ c]^T$.

This means that the space $P_2(\mathbb{R})$ can be identified with the 3-space $\mathbb{R}^3$ through the isomorphism.

In general, the space $P_n(\mathbb{R})$ is isomorphic to the $(n+1)$-space $\mathbb{R}^{n+1}$ and hence one can be identified with the other as the same vector space. □

More generally, one can get the following theorem.

Theorem 4.9 *Two vector spaces V and W are isomorphic if and only if* $\dim V = \dim W$.

Proof: Let $T : V \to W$ be an isomorphism, and let $\{\mathbf{v}_1, \mathbf{v}_2, \ldots, \mathbf{v}_n\}$ be a basis for V. We show that its image $\beta = \{T(\mathbf{v}_1), T(\mathbf{v}_2), \ldots, T(\mathbf{v}_n)\}$ is a basis for W so that $\dim W = n = \dim V$.

(1) *β is linearly independent:* Since T is injective, the equation

$$\mathbf{0} = c_1T(\mathbf{v}_1) + c_2T(\mathbf{v}_2) + \cdots + c_nT(\mathbf{v}_n) = T(c_1\mathbf{v}_1 + c_2\mathbf{v}_2 + \cdots + c_n\mathbf{v}_n)$$

implies that $\mathbf{0} = c_1\mathbf{v}_1+c_2\mathbf{v}_2+\cdots+c_n\mathbf{v}_n$. Since the $\mathbf{v}_i$'s are linearly independent, we have $c_i = 0$ for all $i = 1, 2, \ldots, n$.

(2) *β spans W:* Since T is surjective, for any $\mathbf{y} \in W$ there exists an $\mathbf{x} \in V$ such that $T(\mathbf{x}) = \mathbf{y}$. Write $\mathbf{x} = \sum_{i=1}^n a_i\mathbf{v}_i$. Then

$$\mathbf{y}=T(\mathbf{x})=T(a_1\mathbf{v}_1+a_2\mathbf{v}_2+\cdots+a_n\mathbf{v}_n)=a_1T(\mathbf{v}_1)+a_2T(\mathbf{v}_2)+\cdots+a_nT(\mathbf{v}_n),$$

i.e., $\mathbf{y}$ is a linear combination of $T(\mathbf{v}_1),\ T(\mathbf{v}_2), \ldots,\ T(\mathbf{v}_n)$.

Conversely, if $\dim V = \dim W = n$, then one can choose bases $\{\mathbf{v}_1, \mathbf{v}_2, \ldots, \mathbf{v}_n\}$ and $\{\mathbf{w}_1, \mathbf{w}_2, \ldots, \mathbf{w}_n\}$ for V and W, respectively. By Theorem 4.5, there exist linear transformations $T : V \to W$ and $S : W \to V$ such that $T(\mathbf{v}_i) = \mathbf{w}_i$ and $S(\mathbf{w}_i) = \mathbf{v}_i$ for $i = 1, 2, \ldots, n$. Clearly, $(S \circ T)(\mathbf{v}_i) = \mathbf{v}_i$ and $(T \circ S)(\mathbf{w}_i) = \mathbf{w}_i$ for $i = 1, 2, \ldots, n$, which implies that $S \circ T$ and $T \circ S$ are the identity transformations on V and on W, respectively, by the uniqueness of extended linear functions in Corollary 4.5.1. Hence, T and S are isomorphisms, and consequently V and W are isomorphic. □

Corollary 4.9.1 *Let V and W be vector spaces.*

(1) *If $\dim V = n$, then V is isomorphic to the n-space $\mathbb{R}^n$.*
(2) *If $\dim V = \dim W$, any bijective function from a basis for V to a basis for W can be extended to an isomorphism from V to W.*

It is clear from Lemma 4.7 that if T is an isomorphism, then its inverse T^{-1} is also an isomorphism. In particular, if a linear transformation $A : \mathbb{R}^n \to \mathbb{R}^n$ is defined by an invertible $n \times n$ matrix A as in Example 61(1), then the inverse matrix A^{-1} plays the role of the inverse linear transformation, so that it is also an isomorphism on $\mathbb{R}^n$.

That is, a linear transformation $A : \mathbb{R}^n \to \mathbb{R}^n$ defined by an $n \times n$ matrix A is an isomorphism if and only if A is invertible as a matrix.

Problem 81 Suppose that S and T are linear transformations whose composition $S \circ T$ is well defined. Show that

(1) if $S \circ T$ is injective, so is T,
(2) if $S \circ T$ is surjective, so is S,
(3) if S and T are isomorphisms, so is $S \circ T$,
(4) if A and B are two $n \times n$ matrices of rank n, so is AB.

Problem 82 Let $T : V \to W$ be a linear transformation. Prove that

(1) T is injective if and only if $\mathrm{Ker}(T) = \{\mathbf{0}\}$.
(2) If $V = W$, then T is injective if and only if T is surjective. Hence, if any linear transformation on V is injective or surjective, it is an isomorphism.

4.4 Natural isomorphisms and coordinate systems for vector spaces

The standard basis $\{\mathbf{e}_1, \mathbf{e}_2, \mathbf{e}_3\}$ for the Euclidean 3-space $\mathbb{R}^3$ defines the standard xyz-coordinate system for the space $\mathbb{R}^3$. In general, the standard basis $\{\mathbf{e}_1, \mathbf{e}_2, \dots, \mathbf{e}_n\}$ for the Euclidean n-space $\mathbb{R}^n$ defines the Cartesian coordinate system [4] for the space $\mathbb{R}^n$.

As known already, every vector space has at least one basis, in fact, infinitely many in the nontrivial cases, and every basis for a vector space induces a coordinate system for the vector space. See Section 3.4.

In this section, we introduce a natural way to define a coordinate system for a vector space.

Every n-dimensional vector space V is isomorphic to the n-space $\mathbb{R}^n$ (as shown in Corollary 4.9.1). Hence, one may also try to define a coordinate system for the vector space V by choosing an isomorphism between two spaces.

However, such an isomorphism depends on the choices of basis for the space V which are supposed to be mapped entrywisely to the standard basis $\{\mathbf{e}_1, \mathbf{e}_2, \dots, \mathbf{e}_n\}$ for $\mathbb{R}^n$. In this case, the order of each basis vectors is important and is supposed to be fixed.

An **ordered basis** for a vector space is a basis endowed with a linear order.

For example, in the 3-space $\mathbb{R}^3$, two bases $\{\mathbf{e}_1, \mathbf{e}_2, \mathbf{e}_3\}$ with the order $\mathbf{e}_1, \mathbf{e}_2, \mathbf{e}_3$ and $\{\mathbf{e}_2, \mathbf{e}_1, \mathbf{e}_3\}$ with the order $\mathbf{e}_2, \mathbf{e}_1, \mathbf{e}_3$ are different as ordered bases, but the same as unordered ones.

The basis $\{\mathbf{e}_1, \mathbf{e}_2, \dots, \mathbf{e}_n\}$ with the order $\mathbf{e}_1, \mathbf{e}_2, \dots, \mathbf{e}_n$ is called the **standard ordered basis** for $\mathbb{R}^n$. However, we often say simply basis for an ordered basis if there is no ambiguity in the context.

Let V be a vector space of dimension n with an ordered basis $\alpha = \{\mathbf{v}_1, \mathbf{v}_2, \dots, \mathbf{v}_n\}$, and let $\beta = \{\mathbf{e}_1, \mathbf{e}_2, \dots, \mathbf{e}_n\}$ be the standard ordered basis for $\mathbb{R}^n$. Then, by Theorem 4.5, the one-to-one correspondence $\mathbf{v}_i \leftrightarrow \mathbf{e}_i$ between two ordered bases α and β extends to an isomorphism

$$\Phi : V \to \mathbb{R}^n \quad \text{given by} \quad \Phi(\mathbf{v}_i) = \mathbf{e}_i,$$

[4]A Cartesian coordinate system is also called a *rectangular coordinate system* or an *orthogonal coordinate system.*

called the **natural isomorphism** or the **coordinate isomorphism** with respect to the (ordered) basis α.

By this coordinate isomorphism, a vector in V can be identified with a column vector in $\mathbb{R}^n$. In fact, for any $\mathbf{x} = \sum_{i=1}^n a_i\mathbf{v}_i \in V$, the image of $\mathbf{x}$ under the coordinate isomorphism is

$$\Phi(\mathbf{x}) = \sum_{i=1}^n a_i\Phi(\mathbf{v}_i) = \sum_{i=1}^n a_i\mathbf{e}_i = (a_1, \ldots, a_n) = \begin{bmatrix} a_1 \\ \vdots \\ a_n \end{bmatrix} \in \mathbb{R}^n,$$

which is called the **coordinates** of $\mathbf{x}$ or the **coordinate vector** of $\mathbf{x}$ with respect to the ordered basis α (as introduced in Theorem 3.15). Sometimes, we simply call it the α-**coordinates** of $\mathbf{x}$. It is denoted by $[\mathbf{x}]_\alpha$. Clearly $[\mathbf{v}_i]_\alpha = \mathbf{e}_i$ for each i.

Example 71 **(1)** (Example 62, again) The rotation by the angle θ of $\mathbb{R}^2$ is given by the matrix

$$R_\theta = \begin{bmatrix} \cos\theta & -\sin\theta \\ \sin\theta & \cos\theta \end{bmatrix}.$$

Clearly, it is invertible and hence is an isomorphism. In fact, the inverse R_θ^{-1} is another rotation $R_{-\theta}$.

(2) Let $\alpha = \{\mathbf{e}_1, \mathbf{e}_2\}$ be the standard ordered basis, and let $\beta = \{\mathbf{v}_1, \mathbf{v}_2\}$, where $\mathbf{v}_i = R_\theta\mathbf{e}_i$, $i = 1, 2$. Then β is also a basis for $\mathbb{R}^2$. The α-coordinates of $\mathbf{v}_i$ are

$$[\mathbf{v}_1]_\alpha = \begin{bmatrix} \cos\theta \\ \sin\theta \end{bmatrix}, \qquad [\mathbf{v}_2]_\alpha = \begin{bmatrix} -\sin\theta \\ \cos\theta \end{bmatrix},$$

(as the standard rectangular coordinates in the xy-plane), while the β-coordinates of $\mathbf{v}_i$ are

$$[\mathbf{v}_1]_\beta = \begin{bmatrix} 1 \\ 0 \end{bmatrix}, \qquad [\mathbf{v}_2]_\beta = \begin{bmatrix} 0 \\ 1 \end{bmatrix},$$

because $\mathbf{v}_1 = 1\mathbf{v}_1 + 0\mathbf{v}_2$ and $\mathbf{v}_2 = 0\mathbf{v}_1 + 1\mathbf{v}_2$.

If we choose $\alpha' = \{\mathbf{e}_2, \mathbf{e}_1\}$ as a different ordered basis for $\mathbb{R}^2$, then the α'-coordinates of $\mathbf{v}_i$ are

$$[\mathbf{v}_1]_{\alpha'} = \begin{bmatrix} \sin\theta \\ \cos\theta \end{bmatrix}, \qquad [\mathbf{v}_2]_{\alpha'} = \begin{bmatrix} \cos\theta \\ -\sin\theta \end{bmatrix}.$$

□

Problem 83 Find the coordinate vector of $5 + 2x + 3x^2$ with respect to the following ordered basis α for $P_2(\mathbb{R})$:

$$(1)\ \ \alpha = \{1, x, x^2\}, \qquad (2)\ \ \alpha = \{1 + x, 1 + x^2, x + x^2\}.$$

Here, in (1), 1 denotes the constant function $p(x) = 1$, x denotes the identity function $p(x) = x$, x^2 does the quadratic function $p(x) = x^2$, as in Problem 59.

4.5 Matrix representations of linear transformations

Let $T : V \to W$ be a linear transformation from an n-dimensional vector space V to an m-dimensional vector space W, and let $\alpha = \{\mathbf{v}_1, \ldots, \mathbf{v}_n\}$ and $\beta = \{\mathbf{w}_1, \ldots, \mathbf{w}_m\}$ be any ordered bases for V and W, respectively.

The domain space V is isomorphic to $\mathbb{R}^n$ by the coordinate isomorphism Φ, and the target space W is isomorphic to $\mathbb{R}^m$ by the coordinate isomorphism Ψ defined in Section 4.3, respectively.

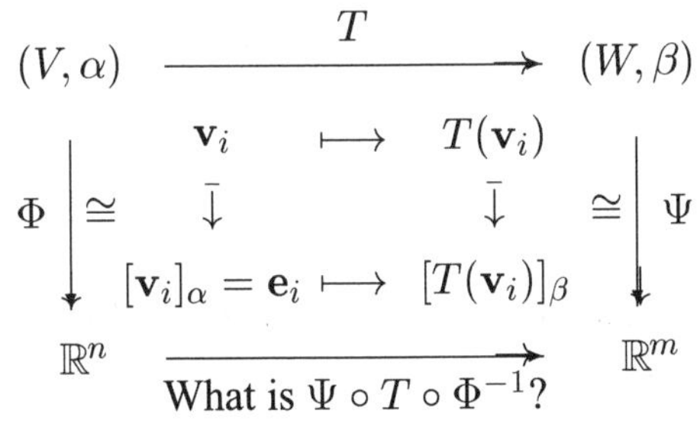

Figure 4.2: What is $\Psi \circ T \circ \Phi^{-1}$?

By chasing the commutative diagram in Figure 4.2, one can see that the linear transformation $\Psi \circ T \circ \Phi^{-1}$ maps the standard basic vector $\mathbf{e}_i$ to the column vector $[T(\mathbf{v}_i)]_\beta$. Hence, by Corollary 4.5.2, $\Psi \circ T \circ \Phi^{-1}$ is just the linear transformation $\mathbf{x} \mapsto A\mathbf{x}$ with the matrix being

$$A = [[T(\mathbf{v}_1)]_\beta\ [T(\mathbf{v}_2)]_\beta\ \cdots\ [T(\mathbf{v}_n)]_\beta] \tag{4.1}$$

with column vectors $[T(\mathbf{v}_j)]_\beta$.

Definition 4.10 The matrix A is called the **associated matrix** for T (or the **matrix representation** of T) with respect to the ordered bases α and β, and denoted by $A = [T]_\alpha^\beta$. When $V = W$ and $\alpha = \beta$, we simply write $[T]_\alpha$ for $[T]_\alpha^\alpha$.

As mentioned before for the coordinate of a vector, whenever we say an associated matrix $[T]_\alpha^\beta$, the bases α and β are supposed to be *ordered* ones even if we say simply the bases α and β.

The following examples illustrate the computation of the associated matrices.

Example 72 (*The associated matrix* $[id]_\alpha$ *is the identity matrix*) Let $id : V \to V$ be the identity transformation on a vector space V. Then for any ordered basis α for V, the matrix $[id]_\alpha = I$, the identity matrix, because if $\alpha = \{\mathbf{v}_1, \mathbf{v}_2, \dots, \mathbf{v}_n\}$, then

$$\begin{cases} id(\mathbf{v}_1) &= 1\mathbf{v}_1 + 0\mathbf{v}_2 + \cdots + 0\mathbf{v}_m \\ id(\mathbf{v}_2) &= 0\mathbf{v}_1 + 1\mathbf{v}_2 + \cdots + 0\mathbf{v}_m \\ \quad\vdots & \qquad\qquad\qquad\qquad\qquad \vdots \\ id(\mathbf{v}_n) &= 0\mathbf{v}_1 + 0\mathbf{v}_2 + \cdots + 1\mathbf{v}_m. \end{cases}$$

Hence, $[id(\mathbf{v}_i)]_\alpha = \mathbf{e}_i$ for each i and

$$[id]_\alpha = [\,[id(\mathbf{v}_1)]_\alpha \;\; [id(\mathbf{v}_2)]_\alpha \;\; \cdots \;\; [id(\mathbf{v}_n)]_\alpha\,] = [\,\mathbf{e}_1 \;\; \mathbf{e}_2 \;\; \cdots \;\; \mathbf{e}_n\,]$$

is the identity matrix. □

Example 73 (*The associated matrix* $[T]_\alpha^\beta$) Let $T : P_1(\mathbb{R}) \to P_2(\mathbb{R})$ be the linear transformation defined by

$$ax + b \mapsto x(ax + b).$$

In another notation, for a polynomial $p(x) \in P_1(\mathbb{R})$,

$$\big(T(p)\big)(x) = xp(x).$$

Find the associated matrix $[T]_\alpha^\beta$ with respected to ordered bases $\alpha = \{1, x\}$ and $\beta = \{1, x, x^2\}$ for $P_1(\mathbb{R})$ and $P_2(\mathbb{R})$, respectively. Here, 1 denotes the constant function $p(x) = 1$, x denotes the identity function $p(x) = x$, x^2 does the quadratic function $p(x) = x^2$, as before. (See Problem 59).

Solution: Clearly,

$$\begin{cases} (T(1))(x) &= x \cdot 1 &= x &= 0 \cdot 1 + 1x + 0x^2 \\ (T(x))(x) &= x \cdot x &= x^2 &= 0 \cdot 1 + 0x + 1x^2. \end{cases}$$

Hence, the associated matrix for T is the transpose of the coefficient matrix in this expression, that is,

$$[T]_\alpha^\beta = [[T(1)]_\beta \ [T(x)]_\beta] = \begin{bmatrix} 0 & 0 \\ 1 & 0 \\ 0 & 1 \end{bmatrix}.$$

□

Example 74 (*The associated matrices* $[T]_\alpha^\beta$ *and* $[T]_\alpha^{\beta'}$) Let $T : \mathbb{R}^2 \to \mathbb{R}^3$ be the linear transformation defined by $T(x, y) = (x+2y,\ 4x+5y,\ 8x+9y)$ with respect to the standard bases α and β for $\mathbb{R}^2$ and $\mathbb{R}^3$, respectively. Then

$$\begin{cases} T(\mathbf{e}_1) & = & T(1,0) & = & (1,4,8) & = & 1\mathbf{e}_1 & + & 4\mathbf{e}_2 & + & 8\mathbf{e}_3 \\ T(\mathbf{e}_2) & = & T(0,1) & = & (2,5,9) & = & 2\mathbf{e}_1 & + & 5\mathbf{e}_2 & + & 9\mathbf{e}_3. \end{cases}$$

Hence,

$$[T]_\alpha^\beta = \begin{bmatrix} 1 & 2 \\ 4 & 5 \\ 8 & 9 \end{bmatrix}.$$

If $\beta' = \{\mathbf{e}_3, \mathbf{e}_2, \mathbf{e}_1\}$, then

$$[T]_\alpha^{\beta'} = \begin{bmatrix} 8 & 9 \\ 4 & 5 \\ 1 & 2 \end{bmatrix}.$$

□

Example 75 (*The associated matrix* $[T]_\alpha$ *for the standard basis* α) Let $T : \mathbb{R}^2 \to \mathbb{R}^2$ be a linear transformation given by $T(1, 1) = (0, 1)$ and $T(-1, 1) = (10, 11)$. Find the matrix representation $[T]_\alpha$ of T with respect to the standard basis $\alpha = \{\mathbf{e}_1, \mathbf{e}_2\}$.

Solution: Note $(a, b) = a\mathbf{e}_1 + b\mathbf{e}_2$ for any $(a, b) \in \mathbb{R}^2$. Thus, the definition of T shows

$$\begin{array}{rclclclcl} T(\mathbf{e}_1) + T(\mathbf{e}_2) & = & T(\mathbf{e}_1 + \mathbf{e}_2) & = & T(1,1) & = & (0,1) & = & \mathbf{e}_2, \\ -T(\mathbf{e}_1) + T(\mathbf{e}_2) & = & T(-\mathbf{e}_1 + \mathbf{e}_2) & = & T(-1,1) & = & (10,11) & = & 10\mathbf{e}_1 + 11\mathbf{e}_2. \end{array}$$

By solving these equations,

$$\begin{array}{rclcl} T(\mathbf{e}_1) & = & -5\mathbf{e}_1 & - & 5\mathbf{e}_2, \\ T(\mathbf{e}_2) & = & 5\mathbf{e}_1 & + & 6\mathbf{e}_2. \end{array}$$

Therefore,

$$[T]_\alpha = [[T(\mathbf{e}_1)]_\alpha \ [T(\mathbf{e}_2)]_\alpha] = \begin{bmatrix} -5 & 5 \\ -5 & 6 \end{bmatrix}.$$

□

Remark: (1) Let A be any $m \times n$ matrix, as a linear transformation $\mathbf{x} \mapsto A\mathbf{x}$ from the n-space $\mathbb{R}^n$ to the m-space $\mathbb{R}^m$ via $\mathbf{x} \mapsto A\mathbf{x}$. Clearly, its matrix representation with respect to the standard bases α for $\mathbb{R}^n$ and β for $\mathbb{R}^m$ is the matrix A itself, *i.e.*, $A = [A]_\alpha^\beta$. (Recall that $A\mathbf{e}_j$ is just the j-th column vector of A.) If A is an invertible $n \times n$ matrix, then its columns $\mathbf{c}_1, \mathbf{c}_2, \ldots, \mathbf{c}_n$ form another basis γ for $\mathbb{R}^n$ and then the matrix representation $[A]_\alpha^\gamma$ is the identity matrix, because $A\mathbf{e}_j = 1 \cdot \mathbf{c}_j$.

(2) Let V and W be vector spaces with (ordered) bases α and β, respectively, and let $T : V \to W$ be a linear transformation with the matrix representation $[T]_\alpha^\beta = A$. Then $\text{Ker}(T)$ and $\text{Im}(T)$ are isomorphic to the null space $\mathcal{N}(A)$ and the column space $\mathcal{C}(A)$, respectively, by the coordinate isomorphisms (see Figure 4.2).

In particular, if $V = \mathbb{R}^n$ and $W = \mathbb{R}^m$ with the standard bases, then

$$\text{Ker}(T) = \mathcal{N}(A) \quad \text{and} \quad \text{Im}(T) = \mathcal{C}(A).$$

From the Rank-Nullity Theorem 3.23, we have

$$\dim \text{Ker}(T) + \dim \text{Im}(T) = \dim V.$$

(3) When the vector spaces V and W are given with (ordered) bases α and β, respectively, as in (2), every linear transformation $T : V \to W$ has its matrix representation $[T]_\alpha^\beta = A : \mathbb{R}^n \to \mathbb{R}^m$, where $n = \dim V$ and $m = \dim W$. Conversely, every $m \times n$ matrix $A : \mathbb{R}^n \to \mathbb{R}^m$ is also associated with a linear transformation $T : V \to W$, see Figure 4.2. In fact, a linear transformation $T : (V, \alpha) \to (W, \beta)$ can be identified with its matrix representation $[T]_\alpha^\beta = A : \mathbb{R}^n \to \mathbb{R}^m$ through the coordinate isomorphisms Φ and Ψ, in a sense of the commutativity of the diagram in Figure 4.2.

Problem 84 Let $T : \mathbb{R}^4 \to \mathbb{R}^3$ be the linear transformation defined by

$$T(x, y, z, u) = (2x + 2y + 2u,\ 3x - 3z + 3u,\ 4x + 5y + 6z + 7u).$$

Let α and β be the standard (ordered) bases for $\mathbb{R}^4$ and $\mathbb{R}^3$, respectively. Find $[T]_\alpha^\beta$.

Problem 85 Find the matrix representations $[T]_\alpha$ and $[T]_\beta$ of each of the following linear transformations T on the 3-space $\mathbb{R}^3$ with respect to the standard (ordered) basis $\alpha = \{\mathbf{e}_1, \mathbf{e}_2, \mathbf{e}_3\}$ and the (ordered) basis $\beta = \{\mathbf{e}_3, \mathbf{e}_2, \mathbf{e}_1\}$:

(1) $T(x, y, z) = (2x + 3y + 4z, 5x + 6y + 7z, -8x - 9y)$,

(2) $T(x, y, z) = (2y + z, x - 4y, 3x)$.

Also, find the matrix representation $[T]_\alpha^\beta$ of each of the linear transformations T.

Problem 86 Find the matrix representation of a reflection across the line $y = \sqrt{3}x$ in the plane $\mathbb{R}^2$ with respect to the standard (ordered) basis $\alpha = \{\mathbf{e}_1, \mathbf{e}_2\}$.

4.6 Change of bases; Change of coordinates

Let $\alpha = \{\mathbf{v}_1, \mathbf{v}_2, \ldots, \mathbf{v}_n\}$ and $\beta = \{\mathbf{w}_1, \mathbf{w}_2, \ldots, \mathbf{w}_n\}$ be two ordered bases for an n-dimensional vector space V. Then any vector $\mathbf{x} \in V$ has two coordinate vectors $[\mathbf{x}]_\alpha$ and $[\mathbf{x}]_\beta$, and they are related by

$$[\mathbf{x}]_\alpha = [id]_\beta^\alpha [\mathbf{x}]_\beta,$$

as shown in Figure 4.2 when $V = W$ and $T = id$ is the identity transformation on V. See the commutative diagram in Figure 4.3.

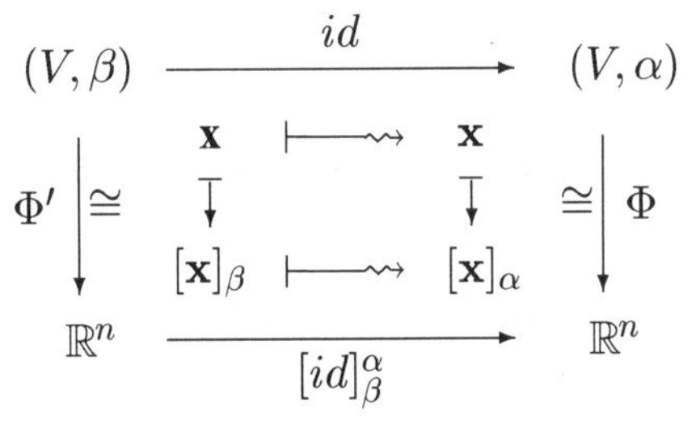

Figure 4.3: The change of basis matrix $[id]_\beta^\alpha$

Definition 4.11 The matrix representation $[id]_\beta^\alpha$ of the identity transformation $id : V \to V$ with respect to any two ordered bases α and β is called the **change of basis matrix** or the **change of coordinate matrix** from β to α.

Since the identity transformation $id : V \to V$ is invertible, the change of basis matrix $[id]_\beta^\alpha$ is also invertible. In fact, $[id]_\beta^\alpha [id]_\alpha^\beta = [id]_\alpha$ is the identity matrix (by chasing Figure 4.3 with Example 72, or see Corollary 4.5.1). Hence,

$$[id]_\alpha^\beta = ([id]_\beta^\alpha)^{-1} \quad \text{and} \quad [\mathbf{x}]_\beta = [id]_\alpha^\beta [\mathbf{x}]_\alpha = ([id]_\beta^\alpha)^{-1} [\mathbf{x}]_\alpha.$$

The following example is well known in the plane $\mathbb{R}^2$.

Example 76 **(1)** The coordinate expression of $\mathbf{x} = (x, y) \in \mathbb{R}^2$ with respect to the standard ordered basis $\alpha = \{\mathbf{e}_1, \mathbf{e}_2\}$ is $\mathbf{x} = x\mathbf{e}_1 + y\mathbf{e}_2$, so that $[\mathbf{x}]_\alpha = \begin{bmatrix} x \\ y \end{bmatrix}$.

(2) Let $\beta = \{\mathbf{e}'_1, \mathbf{e}'_2\}$ be another ordered basis for $\mathbb{R}^2$ obtained by rotating α counterclockwise through an angle θ as in Figure 4.4. Suppose that the coordinate expression of $\mathbf{x} \in \mathbb{R}^2$ with respect to β is written as $\mathbf{x} = x'\mathbf{e}'_1 + y'\mathbf{e}'_2$, or $[\mathbf{x}]_\beta = \begin{bmatrix} x' \\ y' \end{bmatrix}$. Then, the expressions of the vectors in β with respect to α are

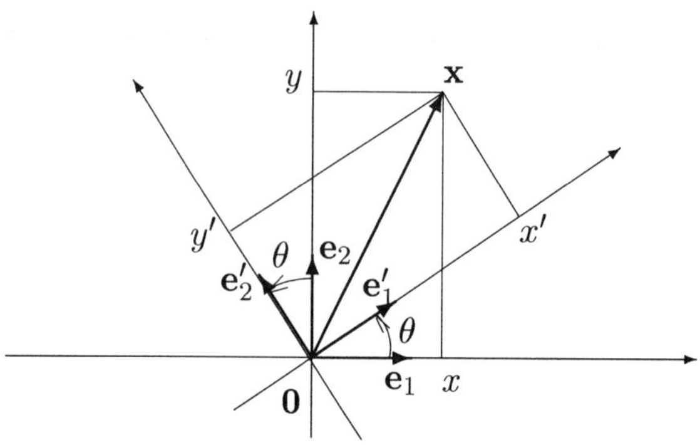

Figure 4.4: Bases $\{\mathbf{e}_1, \mathbf{e}_2\}$ and $\{\mathbf{e}'_1, \mathbf{e}'_2\}$

$$\begin{cases} \mathbf{e}'_1 = id(\mathbf{e}'_1) = \cos\theta\,\mathbf{e}_1 + \sin\theta\,\mathbf{e}_2 \\ \mathbf{e}'_2 = id(\mathbf{e}'_2) = -\sin\theta\,\mathbf{e}_1 + \cos\theta\,\mathbf{e}_2. \end{cases}$$

Therefore,

$$[id]_\beta^\alpha = [[\mathbf{e}'_1]_\alpha\ [\mathbf{e}'_2]_\alpha] = \begin{bmatrix} \cos\theta & -\sin\theta \\ \sin\theta & \cos\theta \end{bmatrix}.$$

Two coordinate vectors $[\mathbf{x}]_\alpha$, $[\mathbf{x}]_\beta$ are related by the equation $[\mathbf{x}]_\alpha = [id]_\beta^\alpha[\mathbf{x}]_\beta$. That is,

$$\begin{bmatrix} x \\ y \end{bmatrix} = \begin{bmatrix} \cos\theta & -\sin\theta \\ \sin\theta & \cos\theta \end{bmatrix} \begin{bmatrix} x' \\ y' \end{bmatrix}.$$

Also,

$$[id]_\alpha^\beta = ([id]_\beta^\alpha)^{-1} = \begin{bmatrix} \cos\theta & \sin\theta \\ -\sin\theta & \cos\theta \end{bmatrix}.$$

(3) In (2), if $\theta = \pi/4$, then

$$\begin{bmatrix} x \\ y \end{bmatrix} = [id]_\beta^\alpha \begin{bmatrix} x' \\ y' \end{bmatrix} = \begin{bmatrix} \cos\frac{\pi}{4} & -\sin\frac{\pi}{4} \\ \sin\frac{\pi}{4} & \cos\frac{\pi}{4} \end{bmatrix} \begin{bmatrix} x' \\ y' \end{bmatrix} = \begin{bmatrix} \frac{1}{\sqrt{2}} & -\frac{1}{\sqrt{2}} \\ \frac{1}{\sqrt{2}} & \frac{1}{\sqrt{2}} \end{bmatrix} \begin{bmatrix} x' \\ y' \end{bmatrix}.$$

Hence, the curve $xy = 1$ is transformed to

$$1 = xy = \left(\frac{1}{\sqrt{2}}x' - \frac{1}{\sqrt{2}}y'\right)\left(\frac{1}{\sqrt{2}}x' + \frac{1}{\sqrt{2}}y'\right) = \frac{(x')^2}{2} - \frac{(y')^2}{2},$$

which is a hyperbola. □

Example 77 (*All nonsingular matrices are change of basis matrices.*) Let Q be a nonsingular matrix with columns $\mathbf{c}_1, \mathbf{c}_2, \ldots, \mathbf{c}_n$. Since Q is nonsingular, the column vectors $\mathbf{c}_i$ are linearly independent and hence form an (ordered) basis for the space $\mathbb{R}^n$, say it β.

Let $\alpha = \{\mathbf{e}_1, \mathbf{e}_2, \ldots, \mathbf{e}_n\}$ be the standard (ordered) basis for the space $\mathbb{R}^n$. Then, for each basis vector $\mathbf{e}_i$,

$$Q\mathbf{e}_i = \mathbf{c}_i = [\mathbf{c}_i]_\alpha = [id]_\beta^\alpha[\mathbf{c}_i]_\beta = [id]_\beta^\alpha \mathbf{e}_i.$$

Therefore, $Q = [id]_\beta^\alpha$ is a change of basis matrix.

Example 78 (*Find a change of basis matrix* $[id]_\beta^\alpha$) On the 3-space $\mathbb{R}^3$, let us take a new $x'y'z'$-coordinate by rotating the standard xyz-coordinate around the x-axis counterclockwise through an angle θ.

In other words, we take a new basis $\beta = \{\mathbf{e}_1', \mathbf{e}_2', \mathbf{e}_3'\}$ by rotating the standard (ordered) basis $\alpha = \{\mathbf{e}_1, \mathbf{e}_2, \mathbf{e}_3\}$ around the x-axis through the angle θ. Then

$$[\mathbf{e}_1']_\alpha = \begin{bmatrix} 1 \\ 0 \\ 0 \end{bmatrix}, \quad [\mathbf{e}_2']_\alpha = \begin{bmatrix} 0 \\ \cos\theta \\ \sin\theta \end{bmatrix}, \quad [\mathbf{e}_3']_\alpha = \begin{bmatrix} 0 \\ -\sin\theta \\ \cos\theta \end{bmatrix},$$

and hence

$$[id]_\beta^\alpha = \begin{bmatrix} 1 & 0 & 0 \\ 0 & \cos\theta & -\sin\theta \\ 0 & \sin\theta & \cos\theta \end{bmatrix} \quad \text{and} \quad [id]_\alpha^\beta = ([id]_\beta^\alpha)^{-1} = \begin{bmatrix} 1 & 0 & 0 \\ 0 & \cos\theta & \sin\theta \\ 0 & -\sin\theta & \cos\theta \end{bmatrix}.$$

Hence,

$$\begin{bmatrix} x' \\ y' \\ z' \end{bmatrix} = [id]_\alpha^\beta \begin{bmatrix} x \\ y \\ z \end{bmatrix} = \begin{bmatrix} 1 & 0 & 0 \\ 0 & \cos\theta & \sin\theta \\ 0 & -\sin\theta & \cos\theta \end{bmatrix} \begin{bmatrix} x \\ y \\ z \end{bmatrix}.$$

□

Problem 87 Let $id : \mathbb{R}^n \to \mathbb{R}^n$ be the identity transformation. Let $\mathbf{x}_k$ denote the vector in the n-space $\mathbb{R}^n$ whose first $k-1$ entries are zero and the last $n-k+1$ entries are 1. Then $\beta = \{\mathbf{x}_1, \mathbf{x}_2, \ldots, \mathbf{x}_n\}$ is a basis for $\mathbb{R}^n$ (see Problem 60). Let $\alpha = \{\mathbf{e}_1, \mathbf{e}_2, \ldots, \mathbf{e}_n\}$ be the standard (ordered) basis for $\mathbb{R}^n$. Show that $[id]_\beta^\alpha = [\mathbf{x}_1\ \mathbf{x}_2\ \cdots\ \mathbf{x}_n]$. Try to find the change of basis matrix $[id]_\alpha^\beta$. Which is easier to find, $[id]_\beta^\alpha$ or $[id]_\alpha^\beta$?

Hint: It is easy to find the change of basis matrix $[id]_\beta^\alpha$ if α is the standard (ordered) basis for $\mathbb{R}^n$ and $[id]_\alpha^\beta = ([id]_\beta^\alpha)^{-1}$.

Problem 88 Take $\beta = \{(1,0,1),(1,1,0),(0,1,1)\}, \gamma = \{(2,3,1),(1,2,0),(2,0,3)\}$ as ordered (ordered) bases for the 3-space $\mathbb{R}^3$ and let $\alpha = \{\mathbf{e}_1, \mathbf{e}_2, \mathbf{e}_3\}$ be the standard basis for $\mathbb{R}^3$. Find the change of basis matrices $[id]_\beta^\alpha$ and $[id]_\gamma^\alpha$. What are the matrices $[id]_\alpha^\beta$ and $[id]_\alpha^\gamma$?

4.7 Which two matrices are similar?

Let V and W be two vector spaces of dimensions n and m with ordered bases α and β, respectively and let $T : V \to W$ be a linear transformation. In Section 4.5, we show how to find the associated matrix $[T]_\alpha^\beta$. If another (ordered) bases α' and β' are taken for V and W, respectively, then another associated matrix $[T]_{\alpha'}^{\beta'}$ will be given.

How two associated matrices $[T]_\alpha^\beta$ and $[T]_{\alpha'}^{\beta'}$ can be related?

In the spaces V and W, each vector has two coordinate vectors

$$\begin{array}{lll} [\mathbf{x}]_\alpha \text{ and } [\mathbf{x}]_{\alpha'} & \text{in } \mathbb{R}^n & \text{for } \mathbf{x} \in V, \\ [T(\mathbf{x})]_\beta \text{ and } [T(\mathbf{x})]_{\beta'} & \text{in } \mathbb{R}^m & \text{for } T(\mathbf{x}) \in W. \end{array}$$

These coordinates are related by the change of basis matrices as follows:

$$[\mathbf{x}]_\alpha = [id_V]_{\alpha'}^\alpha [\mathbf{x}]_{\alpha'} \quad \text{and} \quad [T(\mathbf{x})]_\beta = [id_W]_{\beta'}^\beta [T(\mathbf{x})]_{\beta'},$$

which follow the commutativity of two side rectangular diagrams in Figure 4.2.

Moreover, two $m \times n$ associated matrices $[T]_\alpha^\beta$ and $[T]_{\alpha'}^{\beta'}$ map

$$[T(\mathbf{x})]_\beta = [T]_\alpha^\beta [\mathbf{x}]_\alpha \quad \text{and} \quad [T(\mathbf{x})]_{\beta'} = [T]_{\alpha'}^{\beta'} [\mathbf{x}]_{\alpha'},$$

which follow the commutativity of lower and upper rectangular diagrams in Figure 4.2.

Since the inner rectangular in Figure 4.2 is clearly commutative, the outer rectangular in Figure 4.2 should also be commutative.

Equivalently, it gives, by the equation,

$$\begin{aligned}[T]_{\alpha'}^{\beta'}[\mathbf{x}]_{\alpha'} &= [T(\mathbf{x})]_{\beta'} = [id_W]_{\beta}^{\beta'}[T(\mathbf{x})]_{\beta} = [id_W]_{\beta}^{\beta'}\,[T]_{\alpha}^{\beta}[\mathbf{x}]_{\alpha} \\ &= [id_W]_{\beta}^{\beta'}\,[T]_{\alpha}^{\beta}\,[id_V]_{\alpha'}^{\alpha}[\mathbf{x}]_{\alpha'}\end{aligned}$$

for all $\mathbf{x}$, and it implies

$$[T]_{\alpha'}^{\beta'} = [id_W]_{\beta}^{\beta'}\,[T]_{\alpha}^{\beta}\,[id_V]_{\alpha'}^{\alpha} \tag{4.2}$$

with the change of basis matrices $[id_V]_{\alpha'}^{\alpha}$ and $[id_W]_{\beta}^{\beta'}$.

Later, by using Theorem 4.19, it can also be justified as

$$[T]_{\alpha'}^{\beta'} = [id_W \circ T \circ id_V]_{\alpha'}^{\beta'} = [id_W]_{\beta}^{\beta'}\,[T]_{\alpha}^{\beta}\,[id_V]_{\alpha'}^{\alpha}.$$

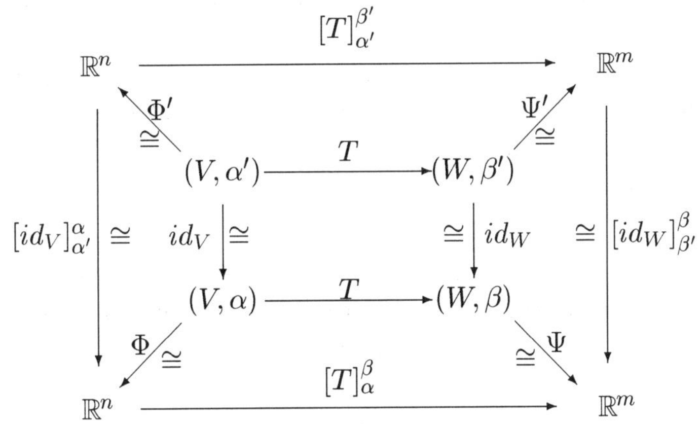

Figure 4.5: Relation between two associated matrices $[T]_{\alpha}^{\beta}$ and $[T]_{\alpha'}^{\beta'}$

In consequence, two associated matrices $[T]_{\alpha}^{\beta}$ and $[T]_{\alpha'}^{\beta'}$ are related as follows.

Theorem 4.12 *Let $T : V \to W$ be a linear transformation from a vector space V with two (ordered) bases α and α' to another vector space W with two (ordered) bases β and β'. Then*

$$[T]_{\alpha'}^{\beta'} = P^{-1}[T]_{\alpha}^{\beta}Q,$$

where $Q = [id_V]_{\alpha'}^{\alpha}$ and $P = [id_W]_{\beta'}^{\beta}$ are the change of basis matrices.

As a special case, if $W = V$, $\alpha = \beta$, and $\alpha' = \beta'$, then $P = Q$ and one can get the following corollary.

Corollary 4.12.1 *Let $T : V \to V$ be a linear transformation on a vector space V and let α and β be two (ordered) bases for V. Then*

(1) *the change of basis matrix $[id]_\beta^\alpha$ is invertible and* $([id]_\beta^\alpha)^{-1} = [id]_\alpha^\beta$.

(2) *For any* $\mathbf{x} \in V$, $[\mathbf{x}]_\alpha = [id]_\beta^\alpha [\mathbf{x}]_\beta$.

(3) $[T]_\beta = ([id]_\beta^\alpha)^{-1}[T]_\alpha[id]_\beta^\alpha = [id]_\alpha^\beta[T]_\alpha[id]_\beta^\alpha$.

The relation (3) of $[T]_\beta$ and $[T]_\alpha$ in Corollary 4.12.1 is called a *similarity*. In general, we have the following definition.

Definition 4.13 For any two square matrices A and B of the same size, A is **similar** to B if there exists a nonsingular matrix Q such that $B = Q^{-1}AQ$.

If A is similar to B, then B is also similar to A. Thus we simply say that A and B are similar. Corollary 4.12.1 says that if A and B are $n \times n$ matrices representing the same linear transformation T on a vector space V with respect to different bases, then A and B are similar.

Example 79 (*Two associated matrices $[T]_\alpha$ and $[T]_\beta$ are similar*) Let $\beta = \{\mathbf{v}_1, \mathbf{v}_2, \mathbf{v}_3\}$ be an (ordered) basis for the 3-space $\mathbb{R}^3$ consisting of $\mathbf{v}_1 = (1, 1, 0)$, $\mathbf{v}_2 = (1, 0, 1)$, and $\mathbf{v}_3 = (0, 1, 1)$. Let T be the linear transformation of $\mathbb{R}^3$ given by the matrix

$$[T]_\beta = \begin{bmatrix} 2 & 3 & 4 \\ 0 & 1 & 6 \\ 0 & 0 & 7 \end{bmatrix}.$$

Let $\alpha = \{\mathbf{e}_1, \mathbf{e}_2, \mathbf{e}_3\}$ be the standard (ordered) basis. Find the change of basis matrix $[id]_\beta^\alpha$ and the associated matrix $[T]_\alpha$.

Solution:

Since $\mathbf{v}_1 = \mathbf{e}_1 + \mathbf{e}_2$, $\mathbf{v}_2 = \mathbf{e}_1 + \mathbf{e}_3$, $\mathbf{v}_3 = \mathbf{e}_2 + \mathbf{e}_3$, we have

$$[id]_\beta^\alpha = \begin{bmatrix} 1 & 1 & 0 \\ 1 & 0 & 1 \\ 0 & 1 & 1 \end{bmatrix}, \quad \text{and} \quad [id]_\alpha^\beta = ([id]_\beta^\alpha)^{-1} = \frac{1}{2}\begin{bmatrix} 1 & 1 & -1 \\ 1 & -1 & 1 \\ -1 & 1 & 1 \end{bmatrix}.$$

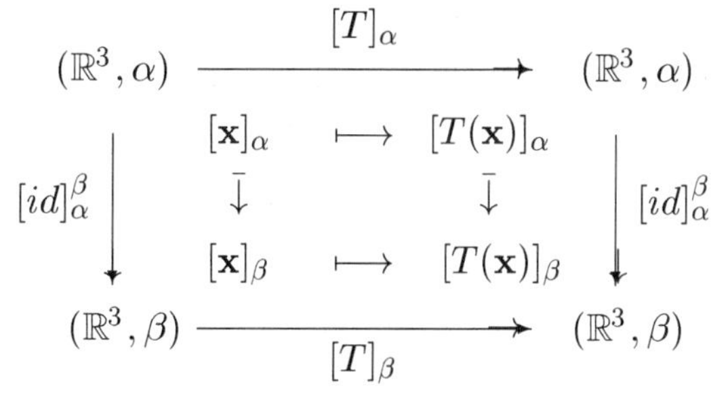

Figure 4.6: Relation between two associated matrices $[T]_\alpha$ and $[T]_\beta$

Therefore, with the given matrix $[T]_\beta$,

$$[T]_\alpha = [id]_\beta^\alpha [T]_\beta [id]_\alpha^\beta = \begin{bmatrix} -2 & 4 & 6 \\ -3 & 5 & 6 \\ -6 & 6 & 7 \end{bmatrix}.$$

This shows that two matrices $[T]_\alpha = \begin{bmatrix} -2 & 4 & 6 \\ -3 & 5 & 6 \\ -6 & 6 & 7 \end{bmatrix}$ and $[T]_\beta = \begin{bmatrix} 2 & 3 & 4 \\ 0 & 1 & 6 \\ 0 & 0 & 7 \end{bmatrix}$ are similar.

Moreover, one can find an explicit formula for $T(x, y, z)$ as a linear combination of $T(\mathbf{e}_1), T(\mathbf{e}_2), T(\mathbf{e}_1)$, which are the column vectors of $[T]_\alpha$. For example, $T(1, 0, 0) = T(\mathbf{e}_1) = (-2, -3, -6)$, as the first column of $[T]_\alpha$. □

Example 80 (*Constructing two similar matrices* $[T]_\alpha$ *and* $[T]_\beta$) Let $T : \mathbb{R}^3 \to \mathbb{R}^3$ be the linear transformation defined by

$$T(x, y, z) = (2x + 3y + 4z,\ 5x + 6y + 7z,\ -8x - 9y).$$

Let $\alpha = \{\mathbf{e}_1, \mathbf{e}_2, \mathbf{e}_3\}$ be the standard (ordered) basis for $\mathbb{R}^3$, and let $\beta = \{\mathbf{v}_1, \mathbf{v}_2, \mathbf{v}_3\}$ be another (ordered) basis consisting of $\mathbf{v}_1 = (1, 0, 0)$, $\mathbf{v}_2 = (1, 1, 0)$, and $\mathbf{v}_3 = (1,\ 1, 1)$. Find the associated matrices $[T]_\alpha$ and $[T]_\beta$ for T.

Solution: One can easily get (by Eq. (4.1) in Page 163)

$$[T]_\alpha = [[T(\mathbf{e}_1)]_\alpha\ [T(\mathbf{e}_2)]_\alpha\ [T(\mathbf{e}_3)]_\alpha] = \begin{bmatrix} 2 & 3 & 4 \\ 5 & 6 & 7 \\ -8 & -9 & 0 \end{bmatrix} \text{ and } [id]_\beta^\alpha = \begin{bmatrix} 1 & 1 & 1 \\ 0 & 1 & 1 \\ 0 & 0 & 1 \end{bmatrix}.$$

Thus, with the inverse $[id]_\alpha^\beta = ([id]_\beta^\alpha)^{-1} = \begin{bmatrix} 1 & -1 & 0 \\ 0 & 1 & -1 \\ 0 & 0 & 1 \end{bmatrix}$,

$$[T]_\beta = [id]_\alpha^\beta [T]_\alpha [id]_\beta^\alpha = \begin{bmatrix} -3 & -6 & -9 \\ 13 & 28 & 35 \\ -8 & -17 & -17 \end{bmatrix}.$$

Clearly, $[T]_\alpha$ and $[T]_\beta$ are similar. □

As the last part of this section, it will be shown that two similar matrices can be the matrix representations of the same linear transformation with respect to different bases.

Let two $n \times n$ matrices A and B be similar, say $B = Q^{-1}AQ$ with a non-singular matrix Q. Let T be a linear transformation on the space $\mathbb{R}^n$ defined by $T(\mathbf{x}) = A\mathbf{x}$ for all $\mathbf{x} \in \mathbb{R}^n$. Then, the matrix representation $[T]_\alpha$ of T with respect to the standard ordered basis $\alpha = \{\mathbf{e}_1, \mathbf{e}_2, \ldots, \mathbf{e}_n\}$ is just the matrix A.

Now, let β be the ordered basis consisting of the column vectors of the nonsingular matrix Q. Then $Q = [id]_\beta^\alpha$, (see Example 77). Therefore, we have

$$B = Q^{-1}AQ = ([id]_\beta^\alpha)^{-1}[T]_\alpha [id]_\beta^\alpha = [T]_\beta.$$

It shows that the matrices $A = [T]_\alpha$ and $B = [T]_\beta$ are the matrix representations of the same linear transformation T with respect to different bases.

As another characterization, it will be shown later in Theorem 8.1 that *two square matrices are similar if and only if their Jordan canonical forms are the same.*

Problem 89 (Problem 88 continued) Take again $\beta = \{(1,0,1),(1,1,0),(0,1,1)\}$ and $\gamma = \{(2,3,1),(1,2,0),(2,0,3)\}$ as (ordered) bases for the 3-space $\mathbb{R}^3$. Find the change of basis matrices $[id]_\beta^\gamma$ and $[id]_\gamma^\beta$.

Hint: Let $\alpha = \{\mathbf{e}_1, \mathbf{e}_2, \mathbf{e}_3\}$ be the standard basis for $\mathbb{R}^3$. Then, $[id]_\beta^\gamma = [id]_\alpha^\gamma [id]_\beta^\alpha = ([id]_\gamma^\alpha)^{-1}[id]_\beta^\alpha$ and $[id]_\gamma^\beta = ([id]_\beta^\gamma)^{-1}$. One can find $[id]_\gamma^\alpha$ and $[id]_\beta^\alpha$ easily.

Problem 90 Let $T : \mathbb{R}^3 \to \mathbb{R}^3$ be the linear transformation defined by

$$T(x_1, x_2, x_3) = (x_1 + 2x_2 + x_3,\ -x_2,\ x_1 + 4x_3).$$

Let α be the standard (ordered) basis, and let $\beta = \{\mathbf{v}_1, \mathbf{v}_2, \mathbf{v}_3\}$ be another (ordered) basis consisting of $\mathbf{v}_1 = (1,0,0)$, $\mathbf{v}_2 = (1,1,0)$, and $\mathbf{v}_3 = (1,1,1)$ for $\mathbb{R}^3$. Find the associated matrix for T with respect to α and the associated matrix for T with respect to β. Are they similar?

Problem 91 Let A and B be square matrices of the same size. Show that if A is similar to B, then A^2 is similar to B^2. In general, A^n is similar to B^n for all $n \geq 2$.

4.8 Applications

4.8.1 Affine spaces and affine transformations

In the 3 space $\mathbb{R}^3$, there are infinitely many lines and planes. Among them, a line or a plane is a vector (sub)space if it passes through the origin $\mathbf{0} = (0, 0, 0)$, and it cannot be a vector space otherwise.

Definition 4.14 Let V be a (real) vector space, $\mathbf{x}_0 \in V$, and let U be a subspace of V. The subset

$$\mathbf{x}_0 + U = \{\mathbf{x}_0 + \mathbf{u} \mid \mathbf{u} \in U\}$$

is called an **affine subspace** of V.

If $\mathbf{x}_0 \in U$, then $\mathbf{x}_0 + U = U$ and it is a (vector) subspace of V. However, if $\mathbf{x}_0 \notin U$, then $\mathbf{x}_0 + U \neq U$ and the set $\mathbf{x}_0 + U$ cannot be a vector (sub)space.

As special cases, all vector spaces and their subspaces are affine spaces. However, all lines and (hyper)planes in the space $\mathbb{R}^n$ not passing through the origin $\mathbf{0}$ are examples of affine spaces which are not vector spaces.

Definition 4.15 Let $\mathbf{x}_0 + U$ be an affine subspace of a vector space V, and let $\mathbf{y}_0 + Z$ be an affine subspace of a vector space W. With any linear transformation $T : U \to Z$, a transformation

$$\mathbf{x}_0 + U \to \mathbf{y}_0 + Z \quad \text{defined by} \quad \mathbf{x}_0 + \mathbf{u} \mapsto \mathbf{y}_0 + T(\mathbf{u})$$

is called an **affine transformation.**

As special cases, all linear transformations from a vector space to another are affine transformations. However, all translations $\mathbf{x} \mapsto \mathbf{x}_0 + \mathbf{x}$ in the plane $\mathbb{R}^2$ or in the space $\mathbb{R}^3$ are affine transformations which are not linear transformations if $\mathbf{x}_0 \neq \mathbf{0}$.

4.8.2 Vector spaces of linear transformations

Let V and W be two vector spaces of dimensions n and m, respectively. Let $\mathcal{L}(V;W)$ denote the set of all linear transformations from V to W, *i.e.*,

$$\mathcal{L}(V;W) = \{T : T \text{ is a linear transformation from } V \text{ to } W\}.$$

For any two linear transformations S and T in $\mathcal{L}(V;W)$ and any $\lambda \in \mathbb{R}$, we define the **sum** $S+T$ and the **scalar multiplication** λS by

$$(S+T)(\mathbf{v}) = S(\mathbf{v}) + T(\mathbf{v}) \quad \text{and} \quad (\lambda S)(\mathbf{v}) = \lambda(S(\mathbf{v}))$$

for any $\mathbf{v} \in V$. Clearly, both $S+T$ and λS are linear and satisfy the '$4+4$' rules of a vector space, so that $\mathcal{L}(V;W)$ becomes a vector space.

Let α and β be two ordered bases for V and W, respectively and let $T : V \to W$ be a linear transformation. Then the associated matrix $[T]_\alpha^\beta$ for T with respect to these bases is uniquely determined. That is, with fixed ordered bases α and β, the function

$$\phi : \mathcal{L}(V;W) \to M_{m\times n}(\mathbb{R}) \quad \text{defined by} \quad \phi(T) = [T]_\alpha^\beta$$

is well defined (see Section 4.5).

The commutative diagram in Figure 4.2 gives the following lemma.

Lemma 4.16 *The function $\phi : \mathcal{L}(V;W) \to M_{m\times n}(\mathbb{R})$ via $T \mapsto [T]_\alpha^\beta$ is injective.*

Furthermore, the following lemma shows that ϕ is linear, so that it becomes in fact an isomorphism from $\mathcal{L}(V;W)$ to $M_{m\times n}(\mathbb{R})$.

Lemma 4.17 *Let V and W be vector spaces with any ordered bases α and β, respectively, and let $S, T : V \to W$ be both linear. Then*

$$[S+T]_\alpha^\beta = [S]_\alpha^\beta + [T]_\alpha^\beta \quad \textit{and} \quad [kS]_\alpha^\beta = k[S]_\alpha^\beta.$$

Proof: Let $\alpha = \{\mathbf{v}_1, \ldots, \mathbf{v}_n\}$ and $\beta = \{\mathbf{w}_1, \ldots, \mathbf{w}_m\}$. Then we have unique expressions $S(\mathbf{v}_j) = \sum_{i=1}^m a_{ji}\mathbf{w}_i$ and $T(\mathbf{v}_j) = \sum_{i=1}^m b_{ji}\mathbf{w}_i$ for each $1 \le j \le n$, so that $[S]_\alpha^\beta = [a_{ij}]$ and $[T]_\alpha^\beta = [b_{ij}]$. Hence

$$(S+T)(\mathbf{v}_j) = \sum_{i=1}^m a_{ji}\mathbf{w}_i + \sum_{i=1}^m b_{ji}\mathbf{w}_i = \sum_{i=1}^m (a_{ji}+b_{ji})\mathbf{w}_i.$$

Thus

$$[S+T]_\alpha^\beta = [a_{ij}+b_{ij}] = [a_{ij}] + [b_{ij}] = [S]_\alpha^\beta + [T]_\alpha^\beta.$$

The proof of the equality $[kS]_\alpha^\beta = k[S]_\alpha^\beta$ is similar and left as an exercise. □

In particular, if $V = \mathbb{R}^n$ and $W = \mathbb{R}^m$, then the vector space $M_{m\times n}(\mathbb{R})$ of $m \times n$ matrices may be identified with the vector space $\mathcal{L}(\mathbb{R}^n; \mathbb{R}^m)$. This is a fact because any $m \times n$ matrix A can be considered as a linear transformation and A is the matrix representation of itself with respect to the standard bases for $\mathbb{R}^n$ and $\mathbb{R}^m$.

One can summarize our discussions in the following theorem.

Theorem 4.18 *For vector spaces V of dimension n and W of dimension m, the vector space $\mathcal{L}(V; W)$ of all linear transformations from V to W is isomorphic to the vector space $M_{m\times n}(\mathbb{R})$ of all $m \times n$ matrices, and*

$$\dim \mathcal{L}(V; W) = \dim M_{m\times n}(\mathbb{R}) = mn = \dim V \dim W.$$

In particular, if $V = W$ then $m = n$ and

$$\dim \mathcal{L}(V; V) = \dim M_{n\times n}(\mathbb{R}) = n^2.$$

Remark: With an isomorphism from the (operator) space $\mathcal{L}(V; V)$ to the (matrix) space $M_{n\times n}(\mathbb{R})$ in Theorem 4.18, one can show that the following statements for a linear transformation $T : V \to V$ are equivalent, as mentioned in Theorem 3.29:

(1) *T is an isomorphism,*
(2) *T is injective,*
(3) *T is surjective.*

Alternatively, one can prove it directly by using a basis for V. See also Problem 82.

Let V, W, and Z be vector spaces. Suppose that $S : V \to W$ and $T : W \to Z$ are linear transformations. Then the composition $T \circ S : V \to Z$ is also linear.

The next theorem shows that the one-to-one correspondence between $\mathcal{L}(V; W)$ and $M_{m\times n}(\mathbb{R})$ preserves not only the vector space structure but also the compositions of linear transformations.

Theorem 4.19 [5] *Let V, W, and Z be vector spaces with ordered bases α, β, and γ, respectively. Suppose that $S : V \to W$ and $T : W \to Z$ are linear transformations. Then*

$$[T \circ S]_\alpha^\gamma = [T]_\beta^\gamma [S]_\alpha^\beta.$$

Proof: Let $\alpha = \{\mathbf{v}_1, \ldots, \mathbf{v}_n\}$, $\beta = \{\mathbf{w}_1, \ldots, \mathbf{w}_m\}$, and $\gamma = \{\mathbf{z}_1, \ldots, \mathbf{z}_\ell\}$. Let $[T]_\beta^\gamma = [a_{ij}]$ and $[S]_\alpha^\beta = [b_{pq}]$. Then, for $1 \le i \le n$,

$$\begin{aligned} (T \circ S)(\mathbf{v}_i) &= T(S(\mathbf{v}_i)) = T\left(\sum_{k=1}^m b_{ki}\mathbf{w}_k\right) = \sum_{k=1}^m b_{ki} T(\mathbf{w}_k) \\ &= \sum_{k=1}^m b_{ki}\left(\sum_{j=1}^\ell a_{jk}\mathbf{z}_j\right) = \sum_{j=1}^\ell \left(\sum_{k=1}^m a_{jk} b_{ki}\right)\mathbf{z}_j. \end{aligned}$$

This shows that $[T \circ S]_\alpha^\gamma = [T]_\beta^\gamma [S]_\alpha^\beta$. □

Theorem 4.20 *Let V and W be vector spaces with ordered bases α and β, respectively, and let $T : V \to W$ be an isomorphism. Then*

$$[T^{-1}]_\beta^\alpha = ([T]_\alpha^\beta)^{-1}.$$

Proof: If T is invertible, then $\dim V = \dim W$, and the matrices $[T]_\alpha^\beta$ and $[T^{-1}]_\beta^\alpha$ are square of the same size. Thus,

$$[T]_\alpha^\beta [T^{-1}]_\beta^\alpha = [T \circ T^{-1}]_\beta = [id]_\beta$$

is the identity matrix (see Example 72). Hence, $[T^{-1}]_\beta^\alpha = ([T]_\alpha^\beta)^{-1}$. □

Problem 92 Let α be the standard (ordered) basis for the 3-space $\mathbb{R}^3$, and let $S, T : \mathbb{R}^3 \to \mathbb{R}^3$ be two linear transformations defined by

$$\begin{aligned} S(\mathbf{e}_1) &= (2, 2, 1), \quad S(\mathbf{e}_2) = (0, 1, 2), \quad S(\mathbf{e}_3) = (-1, 2, 1), \\ T(\mathbf{e}_1) &= (1, 0, 1), \quad T(\mathbf{e}_2) = (0, 1, 1), \quad T(\mathbf{e}_3) = (1, 1, 2). \end{aligned}$$

Compute $[S + T]_\alpha$, $[2T - S]_\alpha$, and $[T \circ S]_\alpha$.

---

[5] **Arthur Cayley** (August 16, 1821– January 26, 1895) studied compositions of linear transformations in 1855 and was led to define the matrix product to satisfy this theorem.

4.8.3 Homogeneous coordinate systems and graphics

One of the applications of linear transformation is to animation or graphical display of images through computer graphics.

As an example, consider the capital letter 'L' in the plane $\mathbb{R}^2$ as in Figure 4.7. This letter 'L' can be represented by a matrix with coordinates of the vertices. The

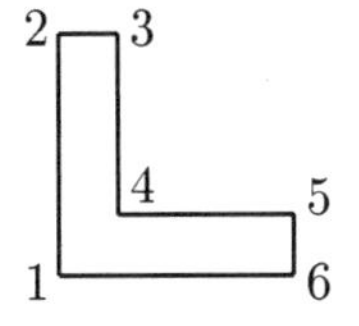

Figure 4.7: Letter "L" on a screen

coordinates of the six vertices of the letter 'L' form a matrix (say it L):

$$\begin{array}{r} \textit{vertices} \\ x\textit{-coordinate} \\ y\textit{-coordinate} \end{array} \begin{array}{c} \begin{matrix} 1 & 2 & 3 & 4 & 5 & 6 \end{matrix} \\ \begin{bmatrix} 0.0 & 0.0 & 0.5 & 0.5 & 2.0 & 2.0 \\ 0.0 & 2.0 & 2.0 & 0.5 & 0.5 & 0.0 \end{bmatrix} \end{array} = L.$$

Let's assume that the computer knows which vertices are connected to which by lines via some algorithm. It is known that line segments are transformed to other line segments by a matrix, considered as a linear transformation. Thus, by multiplying L on the left by a (rotation or reflection) matrix, the vertices are transformed to the other set of vertices, and the line segments connecting the vertices are preserved.

For example, the matrix $B = \begin{bmatrix} 1 & 0.25 \\ 0 & 1 \end{bmatrix}$ transforms the matrix L to the following form, which represents new coordinates of the vertices:

$$\begin{array}{r} \textit{vertices} \\ BL = \end{array} \begin{array}{c} \begin{matrix} 1 & 2 & 3 & 4 & 5 & 6 \end{matrix} \\ \begin{bmatrix} 0.0 & 0.5 & 1.0 & 0.625 & 2.125 & 2.0 \\ 0.0 & 2.0 & 2.0 & 0.5 & 0.5 & 0.0 \end{bmatrix} \end{array}.$$

Now, the computer connects these vertices properly by lines according to the given algorithm and displays on the screen a new image as shown in the left-hand side of Figure 4.8. The multiplication of the matrix $C = \begin{bmatrix} 0.5 & 0 \\ 0 & 1 \end{bmatrix}$ to BL shrinks its width by half producing the image in the right-hand side of Figure 4.8.

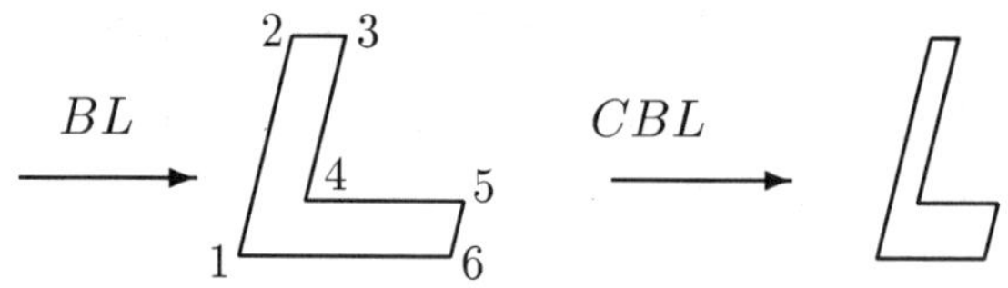

Figure 4.8: Tilting and shrinking

Now, it is suggested that the readers try to find various matrices such as reflections, rotations, or any other ones, and multiply L by them to see how new shapes of the letter 'L' can happen.

Problem 93 For the given matrices L and $B = \begin{bmatrix} 1 & 0.25 \\ 0 & 1 \end{bmatrix}$ above, use the matrix $B^T L$ instead of BL. What kind of shape can you have?

Example 81 It is easy to see in the 3-space $\mathbb{R}^3$ that the matrices

$$R_{(x,\alpha)} = \begin{bmatrix} 1 & 0 & 0 \\ 0 & \cos\alpha & -\sin\alpha \\ 0 & \sin\alpha & \cos\alpha \end{bmatrix}, \quad R_{(y,\beta)} = \begin{bmatrix} \cos\beta & 0 & -\sin\beta \\ 0 & 1 & 0 \\ \sin\beta & 0 & \cos\beta \end{bmatrix},$$

$$R_{(z,\gamma)} = \begin{bmatrix} \cos\gamma & -\sin\gamma & 0 \\ \sin\gamma & \cos\gamma & 0 \\ 0 & 0 & 1 \end{bmatrix}$$

are the rotations around the x, y, z-axes by the angles α, β, and γ, respectively. □

Problem 94 Find the matrix $R_{(\mathbf{u},\pi/2)}$ for the rotation of the space $\mathbf{R}^3$ around the line determined by $\mathbf{u} = (1,1,1)$ by $\pi/2$.

Hint: It can be a composition of basic rotations such as $R_{(x,\alpha)}$, $R_{(y,\beta)}$, and $R_{(z,\gamma)}$. First rotate the $\mathbf{u}$-axis (a line passing through $\mathbf{u}$ and the origin $\mathbf{0}$) about the z-axis into the xz-plane by $R_{(z,-\pi/4)}$ and then into the x-axis by the rotation $R_{(y,-\pi/4)}$ around the y-axis. Then the rotation around the $\mathbf{u}$-axis is the same as the rotation around the x-axis followed by the inverses of the above rotations, *i.e.*, take the rotation $R_{(x,\pi/2)}$ around the x-axis, and then get back to the rotation around the $\mathbf{u}$-axis via $R_{(y,\pi/4)}$ and $R_{(z,\pi/4)}$. In summary, $R_{(\mathbf{u},\pi/4)} = R_{(z,\pi/4)}R_{(y,\alpha)}R_{(x,\pi/2)}R_{(y,-\pi/4)}R_{(z,-\pi/4)}$.

The most common linear transformations for computer graphics are rotations, reflections, tilting (or say shear), scaling (shrinking or stretching), or their compositions on the plane $\mathbb{R}^2$ or on the space $\mathbb{R}^3$.

However, another indispensable transformation for computer graphics is a translation.

A **translation** is by definition a transformation $T : \mathbb{R}^n \to \mathbb{R}^n$ defined by $T(\mathbf{x}) = \mathbf{x} + \mathbf{x}_0$ for any $\mathbf{x} \in \mathbb{R}^n$, where $\mathbf{x}_0$ is a fixed vector in $\mathbb{R}^n$.

Very unfortunately, *a translation is not linear if* $\mathbf{x}_0 \neq \mathbf{0}$ and hence it cannot be represented by a matrix in a rectangular coordinate system.

To escape this disadvantage, we introduce a new coordinate system, called a **homogeneous coordinate**. [6] For brevity, we consider it only in the 3-space $\mathbb{R}^3$.

Definition 4.21 We identify a point $\mathbf{x} = (x, y, z) \in \mathbb{R}^3$ in a rectangular coordinate with the one-dimensional subspace spanned by $(x, y, z, 1)$ in the 4-space $\mathbb{R}^4$, which is a line $\{(hx, hy, hz, h) : h \in \mathbb{R}\}$ in $\mathbb{R}^4$.

The set of vectors $(hx, hy, hz, h), h \in \mathbb{R}$ (which is a line) or any representative $(hx, hy, hz, h), h \neq 0$ is called the **homogeneous coordinate** of a point $\mathbf{x} = (x, y, z) \in \mathbb{R}^3$. Most times, we use $(x, y, z, 1)$ as a representative of the set.

For example, a point $\mathbf{x} = (1, 2, 3)$ in the space $\mathbb{R}^3$ has the rectangular coordinate $(1, 2, 3)$ but its homogeneous coordinate can be (represented by) $(1, 2, 3, 1)$ or $(2, 4, 6, 2)$, actually any of $(h, 2h, 3h, h), h \neq 0$.

In general, any vector (x, y, z, h), $h \neq 0$ in $\mathbb{R}^4$ can be viewed as the homogeneous coordinate of the point $(x/h, y/h, z/h)$ in $\mathbb{R}^3$.

Always, a rectangular coordinate can be recovered from a homogeneous coordinate by dividing the first three entries by the fourth.

However, unlike rectangular coordinates, even a single point can be represented by infinitely many different homogeneous coordinates.

Now, it is possible to represent all of our transformations including translations as 4×4 matrices by using homogeneous coordinates. It will be shown on a case by case in the following.

---

[6]Homogeneous coordinates were introduced by August Ferdinand Möbius in 1827. **August Ferdinand Möbius** (November 17, 1790– September 26, 1868) was a German mathematician and theoretical astronomer. He is famous for the Möbius strip and the Möbius transformation.

(1) Translations: A translation $T : \mathbb{R}^3 \to \mathbb{R}^3$ defined by $T(\mathbf{x}) = \mathbf{x} + \mathbf{x}_0$, where $\mathbf{x}_0 = (x_0, y_0, z_0)$, can be represented by a matrix multiplication in the homogeneous coordinate space as

$$\begin{bmatrix} x \\ y \\ z \\ 1 \end{bmatrix} \mapsto \begin{bmatrix} 1 & 0 & 0 & x_0 \\ 0 & 1 & 0 & y_0 \\ 0 & 0 & 1 & z_0 \\ 0 & 0 & 0 & 1 \end{bmatrix} \begin{bmatrix} x \\ y \\ z \\ 1 \end{bmatrix}.$$

(2) Rotations: With the notations in Example 81,

$$R_{(x,\alpha)} = \begin{bmatrix} 1 & 0 & 0 & 0 \\ 0 & \cos\alpha & -\sin\alpha & 0 \\ 0 & \sin\alpha & \cos\alpha & 0 \\ 0 & 0 & 0 & 1 \end{bmatrix}, \quad R_{(y,\beta)} = \begin{bmatrix} \cos\beta & 0 & -\sin\beta & 0 \\ 0 & 1 & 0 & 0 \\ \sin\beta & 0 & \cos\beta & 0 \\ 0 & 0 & 0 & 1 \end{bmatrix},$$

$$R_{(z,\gamma)} = \begin{bmatrix} \cos\gamma & -\sin\gamma & 0 & 0 \\ \sin\gamma & \cos\gamma & 0 & 0 \\ 0 & 0 & 1 & 0 \\ 0 & 0 & 0 & 1 \end{bmatrix}$$

are the rotations around the x, y, z-axes by the angles α, β, and γ in the homogeneous coordinate space, respectively.

(3) Reflections: The reflection through the xy-plane (say an xy-reflection) is represented by a matrix multiplication in the homogeneous coordinate space as

$$\begin{bmatrix} x \\ y \\ z \\ 1 \end{bmatrix} \mapsto \begin{bmatrix} 1 & 0 & 0 & 0 \\ 0 & 1 & 0 & 0 \\ 0 & 0 & -1 & 0 \\ 0 & 0 & 0 & 1 \end{bmatrix} \begin{bmatrix} x \\ y \\ z \\ 1 \end{bmatrix}.$$

Similarly, one can get matrices for an xz-reflection and a yz-reflection.

(4) Shearing (transvections or **tiltings**): In the plane $\mathbb{R}^2$ with the rectangular coordinate, the transformation which maps a point (x, y) to the point $(x + 2y, y)$ is a shearing. In $\mathbb{R}^3$, an xy-shear can be represented by a matrix multiplication in a homogeneous coordinate as

$$\begin{bmatrix} x \\ y \\ z \\ 1 \end{bmatrix} \mapsto \begin{bmatrix} 1 & 0 & a & 0 \\ 0 & 1 & b & 0 \\ 0 & 0 & 1 & 0 \\ 0 & 0 & 0 & 1 \end{bmatrix} \begin{bmatrix} x \\ y \\ z \\ 1 \end{bmatrix}, \quad a, b \text{ are positive.}$$

Similarly, one can get an xz-shear and a yz-shear with matrices of the form

$$\begin{bmatrix} 1 & a & 0 & 0 \\ 0 & 1 & 0 & 0 \\ 0 & b & 1 & 0 \\ 0 & 0 & 0 & 1 \end{bmatrix} \quad \text{and} \quad \begin{bmatrix} 1 & 0 & 0 & 0 \\ a & 1 & 0 & 0 \\ b & 0 & 1 & 0 \\ 0 & 0 & 0 & 1 \end{bmatrix},$$

respectively.

(5) Scaling (dilation, stretching, or **shrinking)**: A scaling is represented by a matrix multiplication in the homogeneous coordinate as

$$\begin{bmatrix} x \\ y \\ z \\ 1 \end{bmatrix} \mapsto \begin{bmatrix} a & 0 & 0 & 0 \\ 0 & b & 0 & 0 \\ 0 & 0 & c & 0 \\ 0 & 0 & 0 & 1 \end{bmatrix} \begin{bmatrix} x \\ y \\ z \\ 1 \end{bmatrix}, \quad a, b, c \text{ are positive.}$$

In summary, all of the transformations can be represented by matrix multiplications in the homogeneous coordinate space and also their compositions are done by the corresponding matrix multiplications.

Problem 95 (*The homogeneous coordinate in the plane* $\mathbf{R}^2$) We identify a point $\mathbf{x} = (x, y) \in \mathbb{R}^2$ in the rectangular coordinate with the one-dimensional subspace spanned by $(x, y, 1)$ in the 3-space $\mathbb{R}^3$, which is a line $\{(hx, hy, h) : h \in \mathbb{R}\}$ in $\mathbb{R}^3$. The set of vectors $(hx, hy, h), h \in \mathbb{R}$ (which is a line) or any representative $(hx, hy, h), h \neq 0$ is called the **homogeneous coordinate** of a point $\mathbf{x} = (x, y)$ in the plane $\mathbb{R}^2$.

The homogeneous coordinate $(x_1, x_2, 0)$ describes the point at infinity (say, ∞) in the direction with slope x_2/x_1. Show that two points (a_1, a_2, a_3) and (b_1, b_2, b_3) in the homogeneous coordinate are identical if and only if

$$\det \begin{bmatrix} a_2 & a_3 \\ b_2 & b_3 \end{bmatrix} = \det \begin{bmatrix} a_3 & a_1 \\ b_3 & b_1 \end{bmatrix} = \det \begin{bmatrix} a_1 & a_2 \\ b_1 & b_2 \end{bmatrix} = 0.$$

Problem 96 Find the (4×4) matrix of the reflection through the plane $y = x$ in the 3-space $\mathbb{R}^3$ in the homogeneous coordinate. Compare it with the (3×3) matrix of the reflection mentioned in Problem 75.

4.9 Exercises

4.1. Which of the following functions $f : \mathbb{R} \to \mathbb{R}$ are injective, surjective, or bijective?

(1) $f(x) = x + 1$,

(2) $f(x) = 2x^2 + 1$,
(3) $f(x) = 2x^3 + 1$,
(4) $f(x) = e^x$,
(5) $f(x) = |x|$.

4.2. Let $f : X \to Y$ be a function from a set X to a set Y. Show that

(1) f is injective if and only if f has a left inverse $g : Y \to X$ such that $g \circ f = id_X$, the identity function on X,
(2) f is surjective if and only if f has a right inverse $g : Y \to X$ such that $f \circ g = id_Y$, the identity function on Y,
(3) f is bijective if and only if f has a two-sided inverse $g : Y \to X$ such that $g \circ f = id_X$ and $f \circ g = id_Y$.

4.3. *Every linear transformation preserves two algebraic operations; the vector sum and the scalar multiplication:* $f(\mathbf{x}+\mathbf{y}) = f(\mathbf{x}) + f(\mathbf{y})$ *and* $f(k\mathbf{x}) = kf(\mathbf{x})$. Which of the following functions T are linear transformations?

(1) $T(x, y) = (x + 1, y + 1)$,
(2) $T(x, y, z) = (x + y, y, 2x + z)$,
(3) $T(x, y, z) = (y, z, x)$,
(4) $T(x, y, z) = x + y + z$.

4.4. Let $\{\mathbf{e}_1, \mathbf{e}_2, \mathbf{e}_3\}$ be the standard basis for the 3-space $\mathbb{R}^3$ and let $T : \mathbb{R}^3 \to P_3(\mathbb{R})$ be a linear transformation defined on basic elements by $T(\mathbf{e}_1) = x+1, T(\mathbf{e}_2) = x^2$ and $T(\mathbf{e}_3) = x^3 + x + 1$. Find $T(a, b, c)$. Is T invertible?

4.5. Let $T : P_2(\mathbb{R}) \to P_3(\mathbb{R})$ be a linear transformation defined by $T(1) = x, T(x) = x^2$ and $T(x^2) = x^3 + x$. Find $T(ax^2 + bx + c)$.

4.6. Find compositions $S \circ T$ and/or $T \circ S$ whenever it is defined.

(1) $T(x, y, z) = (y, z, x)$, $S(x, y, z) = (x + y, y + z, z + x)$,
(2) $T(x, y, z) = (x + y + z, x + z)$, $S(x, y) = (x, x + y, y)$,
(3) $T(x, y) = (x, 3y + x, 2x - 4y, y)$, $S(x, y, z) = (2x, y)$.

4.7. For a linear transformation $T : \mathbb{R}^3 \to \mathbb{R}^3$ defined by $T(x, y, z) = (2y + z, x - 2y, x + z)$,

(1) determine the kernel $\text{Ker}(T)$ of T, and find a basis for $\text{Ker}(T)$,
(2) determine whether T is injective,
(3) determine whether T is surjective.

4.8. Let T be a linear transformation of a vector space V such that $T^2 = id_V$ but $T \neq id_V$, the identity function on V. Let $U = \{\mathbf{v} \in V : T(\mathbf{v}) = \mathbf{v}\}$ and $W = \{\mathbf{v} \in V : T(\mathbf{v}) = -\mathbf{v}\}$. Show that

(1) both U and W are subspaces of V,
(2) at least one of U and W is a nonzero subspace,

(3) $U \cap W = \{\mathbf{0}\}$,
(4) $V = U + W$.

4.9. Show that each of the following linear transformations T on the 3-space $\mathbb{R}^3$ is invertible, and find a formula for $T^{-1}(r, s, t)$.

(1) $T(x, y, z) = (y, z, x)$,
(2) $T(x, y, z) = (3x, x - 2y,\ 2x - 2y + z)$,
(3) $T(x, y, z) = (2x, 4x - y, 2x + 3y - z)$.

4.10. Show that if $T : V \to W$ and $S : W \to Z$ are invertible linear transformations, then their composition $S \circ T$ is also invertible, and $(S \circ T)^{-1} = T^{-1} \circ S^{-1}$.

4.11. Let T be any linear transformation from $\mathbb{R}^3$ to $\mathbb{R}^2$, and let S be any linear transformation from $\mathbb{R}^2$ to $\mathbb{R}^3$. Show that the composition $S \circ T$ cannot be invertible.

4.12. Let T be a linear transformation of a vector space V satisfying $T + T^2 = id_V$. Show that T is invertible.

4.13. With respect to the ordered basis $\alpha = \{1, x, x^2\}$ for the vector space $P_2(\mathbb{R})$, find the coordinate vector of each of the following polynomials:

(1) $f(x) = 3x^2 + x + 1$, (2) $f(x) = x^2 + 4x - 5$, (3) $f(x) = 2x + 5$.

4.14. Let $\beta = \{\mathbf{v}_1, \mathbf{v}_2, \mathbf{v}_3\}$ be an ordered basis consisting of $\mathbf{v}_1 = (1, 0, 0)$, $\mathbf{v}_2 = (1, 1, 0)$, and $\mathbf{v}_3 = (1, 1, 1)$ for the 3-space $\mathbb{R}^3$. With respect to the ordered basis β, find the coordinate vector of the vector $\mathbf{x} = (x, y, z)$.

4.15. Find the matrix representation of each of the following linear transformations T on $\mathbb{R}^3$ with respect to the standard (ordered) basis $\{\mathbf{e}_1, \mathbf{e}_2, \mathbf{e}_3\}$.

(1) $T(x, y, z) = (x + 2y,\ y + 2z,\ z + 2x)$,
(2) $T(x, y, z) = (2y,\ 3x - y,\ x + 2z)$,
(3) $T(x, y, z) = (3x - 4y + z,\ x + 5y + z,\ y + z)$.

4.16. Let $T : P_2(\mathbb{R}) \to P_2(\mathbb{R})$ be the linear transformation defined by

$$Tf(x) = f''(x) - 4f'(x) + f(x).$$

Find the matrix $[T]_\alpha$ for the (ordered) basis $\alpha = \{x, 1 + x, x + x^2\}$.

4.17. Let T be the linear transformation of $\mathbb{R}^2$ defined by $T(x, y) = (-y, x)$.

(1) Find the matrix representation of T with respect to the standard (ordered) basis $\alpha = \{\mathbf{e}_1, \mathbf{e}_2\}$.
(2) Find the matrix representation of T with respect to an ordered basis $\beta = \{\mathbf{v}_1, \mathbf{v}_2\}$, where $\mathbf{v}_1 = (1, 2)$, $\mathbf{v}_2 = (1, -1)$.
(3) Show that for every real number k, the linear transformation $T - k\, id_{\mathbb{R}^2}$ is invertible.

4.18. Let $A = \begin{bmatrix} 4 & 2 & 1 \\ 0 & 1 & 3 \end{bmatrix}$.

(1) Find the linear transformation $T : \mathbb{R}^3 \to \mathbb{R}^2$ so that $A = [T]_\alpha^\beta$, where

$$\alpha = \left\{ \begin{bmatrix} 1 \\ 0 \\ 0 \end{bmatrix}, \begin{bmatrix} 1 \\ 1 \\ 0 \end{bmatrix}, \begin{bmatrix} 1 \\ 1 \\ 1 \end{bmatrix} \right\}, \qquad \beta = \left\{ \begin{bmatrix} 1 \\ 0 \end{bmatrix}, \begin{bmatrix} 1 \\ 1 \end{bmatrix} \right\}.$$

(2) Find $T(x, y, z)$.

4.19. Find the change of basis matrices $[id]_\alpha^\beta$ and $[id]_\beta^\alpha$, when

(1) $\alpha = \{(1,0), (0,1)\}$, $\beta = \{(3,2), (2,4)\}$,
(2) $\alpha = \{(3,1), (1,2)\}$, $\beta = \{(1,0), (0,1)\}$,
(3) $\alpha = \{(1,1,1), (1,1,0), (1,0,0)\}$, $\beta = \{(1,0,1), (1,1,0), (0,1,1)\}$.

4.20. Let $\alpha = \{\mathbf{e}_1, \mathbf{e}_2, \mathbf{e}_3\}$ and $\beta = \{(1,1,1), (1,1,0), (1,0,0)\}$ be given two ordered bases for the 3-space $\mathbb{R}^3$ and let $T : \mathbb{R}^3 \to \mathbb{R}^3$ be the linear transformation defined by $T(x, y, z) = (x + 2y, x + y + z, 3x + z)$.

(1) Find the change of basis matrix Q from β to α.
(2) Find the change of basis matrix P from α to β.
(3) Find the associated matrices $[T]_\alpha$ and $[T]_\beta$. You can use $[T]_\beta = Q^{-1}[T]_\alpha Q$.

4.21. Any $n \times n$ matrix A can be considered as a linear transformation $\mathbf{x} \mapsto A\mathbf{x}$ on the n-space $\mathbb{R}^n$. Suppose that $\mathbf{c}_1, \mathbf{c}_2, \ldots, \mathbf{c}_n$ are linearly independent vectors in $\mathbb{R}^n$ spanning a parallelepiped. Then A transforms this parallelepiped into another parallelepiped spanned by $A\mathbf{c}_1, A\mathbf{c}_2, \ldots, A\mathbf{c}_n$. Let $B = \begin{bmatrix} \mathbf{c}_1 & \mathbf{c}_2 & \cdots & \mathbf{c}_n \end{bmatrix}$ be the $n \times n$ matrix with columns $\mathbf{c}_j$, and let $C = \begin{bmatrix} A\mathbf{c}_1 & A\mathbf{c}_2 & \cdots & A\mathbf{c}_n \end{bmatrix}$ be the $n \times n$ matrix with columns $A\mathbf{c}_j$. Show that

$$\mathrm{vol}(\mathcal{P}(C)) = |\det A|\ \mathrm{vol}(\mathcal{P}(B)).$$

This means that the absolute value of the determinant of A is the ratio between the volumes of a parallelepiped $\mathcal{P}(B)$ and its image parallelepiped $\mathcal{P}(C)$ under the transformation by A. If $\det A = 0$, then the image $\mathcal{P}(C)$ is a parallelepiped in a subspace of dimension less than n.

Hint: For any square matrix A, $\mathrm{vol}(\mathcal{P}(A)) = |\det(A^T)| = |\det A|$. See Corollary 2.19.1 on page 92.

4.22. Let $T : \mathbb{R}^3 \to \mathbb{R}^3$ be the linear transformation given by

$$T(x, y, z) = (x + 2y,\ y + 2z,\ z + 2x).$$

Let Q denote the unit cube in $\mathbb{R}^3$ determined by the standard basis vectors $\mathbf{e}_1, \mathbf{e}_2, \mathbf{e}_3$. Find the volume of the image parallelepiped $T(Q)$ of Q under T.

Hint: $\mathrm{vol}(T(Q)) = |\det A|\mathrm{vol}(Q)$ for the matrix representation A of T.

4.23. (1) Are there matrices A and B in $M_{n\times n}(\mathbb{R})$ such that $AB + BA = I_n$?
(2) Are there matrices A and B in $M_{n\times n}(\mathbb{R})$ such that $AB - BA = I_n$?

4.24. Show that

(1) the matrix $B_\theta = \begin{bmatrix} \cos\theta & \sin\theta \\ \sin\theta & -\cos\theta \end{bmatrix}$ represents the reflection $T_{\theta/2}$ across the line passing through the origin that forms an angle $\theta/2$ with the positive x-axis,
(2) the composition of any two reflections B_θ and B_η is a rotation.

Hint: For (1), show that $T_{\theta/2}(\mathbf{e}_1)$ and $T_{\theta/2}(\mathbf{e}_2)$ are the first and the second columns of the matrix B_θ, respectively. For (2), compute $(B_\eta \circ B_\theta)(\mathbf{e}_1)$ and $(B_\eta \circ B_\theta)(\mathbf{e}_2)$.

4.25. For each of

$$(1)\ A = \begin{bmatrix} 4 & 1 & 0 & 0 \\ 0 & 4 & 1 & 0 \\ 0 & 0 & 4 & 1 \\ 0 & 0 & 0 & 4 \end{bmatrix} \qquad (2)\ A = \begin{bmatrix} \lambda & 1 & 0 & 0 & 0 \\ 0 & \lambda & 1 & 0 & 0 \\ 0 & 0 & \lambda & 1 & 0 \\ 0 & 0 & 0 & \lambda & 1 \\ 0 & 0 & 0 & 0 & \lambda \end{bmatrix},$$

show that A is similar to its transpose A^T. It is a fact that *every square matrix is similar to its transpose.* See Exercise **8.22**.

Hint: Use the change of basis matrix $Q = [\mathbf{e}_4\ \mathbf{e}_3\ \mathbf{e}_2\ \mathbf{e}_1]$.

4.26. For any two similar $n \times n$ matrices A and B, show that

(1) $\det A = \det B$, (2) $\operatorname{tr}(A) = \operatorname{tr}(B)$, (3) $\operatorname{rank} A = \operatorname{rank} B$.

4.27. Show that the matrix $A = \begin{bmatrix} 1 & 1 \\ 0 & 1 \end{bmatrix}$ cannot be similar to a diagonal matrix.

Hint: Use Exercise **4.26**. If A would be similar to a diagonal matrix D, $\det A = \det D$ and $\operatorname{tr}(A) = \operatorname{tr}(D)$.

4.28. Determine whether or not the following statements are true in general, and justify your answers.

(1) None of real-valued functions $\sin x, e^x, \ln x, x + 1$ are linear.
(2) For every linear transformation $T : \mathbb{R}^2 \to \mathbb{R}^3$, $\operatorname{Ker}(T) = \{\mathbf{0}\}$.
(3) For every linear transformation $T : \mathbb{R}^3 \to \mathbb{R}^2$, $\operatorname{Ker}(T) \neq \{\mathbf{0}\}$.
(4) A linear transformation $T : \mathbb{R}^n \to \mathbb{R}^m$ is injective if and only if the null space of $[T]_\alpha^\beta$ is $\{\mathbf{0}\}$ for any basis α for $\mathbb{R}^n$ and any basis β for $\mathbb{R}^m$.
(5) For every linear transformation T on $\mathbb{R}^n$, the dimension of the image of T is equal to that of the row space of $[T]_\alpha$ for any basis α for $\mathbb{R}^n$.
(6) For any two linear transformations $T : V \to W$ and $S : W \to Z$, if $\operatorname{Ker}(S \circ T) = \mathbf{0}$, then $\operatorname{Ker}(T) = \mathbf{0}$.

(7) Any polynomial $p(x)$ is linear if and only if the degree of $p(x)$ is less than or equal to 1.

(8) Let $T : \mathbb{R}^3 \to \mathbb{R}^2$ be a function given as $T(\mathbf{x}) = (T_1(\mathbf{x}), T_2(\mathbf{x}))$ for any $\mathbf{x} \in \mathbb{R}^3$. Then T is linear if and only if their coordinate functions T_i, $i = 1, 2$, are linear.

(9) For a linear transformation $T : \mathbb{R}^n \to \mathbb{R}^n$, if $[T]_\alpha^\beta = I_n$ for some bases α and β of $\mathbb{R}^n$, then T must be the identity transformation.

(10) If a linear transformation $T : \mathbb{R}^n \to \mathbb{R}^n$ is injective, then every matrix representation of T is nonsingular.

(11) Every $m \times n$ matrix A can be a matrix representation of a linear transformation $T : \mathbb{R}^n \to \mathbb{R}^m$.

(12) Every change of basis matrix is invertible.

(13) Any two $n \times n$ diagonal matrices are similar to each other.

(14) A matrix similar to a change of basis matrix is also a change of basis matrix.

(15) $\det : M_{n\times n}(\mathbb{R}) \to \mathbb{R}$ is a linear transformation.

(16) The trace $\mathrm{tr} : M_{n\times n}(\mathbb{R}) \to \mathbb{R}$ is a linear transformation.

(17) Every rotation on $\mathbb{R}^3$ is a linear transformation.

(18) Every translation in $\mathbb{R}^n$ is a linear transformation.

(19) For any two $n \times n$ matrices A and B, $\mathrm{tr}(AB) = \mathrm{tr}(A)\mathrm{tr}(B)$.

(20) Each of the nontrivial translations cannot be represented by a matrix in the rectangular coordinate system, but it is possible in the homogeneous coordinate system.

Hint: Exactly seven of them are false.

Chapter 5

Inner Product Spaces; Abstraction of Euclidean spaces

Most readers would be already familiar with the dot product formula $\mathbf{x} \cdot \mathbf{y} = \|\mathbf{x}\|\|\mathbf{y}\| \cos\theta$, or equivalently $\mathbf{x} \cdot \mathbf{y} = x_1y_1 + x_2y_2 + x_3y_3$ for two vectors $\mathbf{x}$, $\mathbf{y}$ in the Euclidean 3-space $\mathbb{R}^3$. This formula allows us to define the length of a vector and the angle between two vectors in the space $\mathbb{R}^3$.

As a generalization of the space $\mathbb{R}^3$ with the dot product, an inner product space is a vector space with an additional structure called an *inner product*. This additional structure associates each pair of vectors in the vector space with a scalar quantity known as the inner product of the vectors.

An inner product allows us to define some geometrical notions such as the length of a vector and the angle between two vectors in any inner product space.

5.1 Inner products; Abstraction of the dot product

The **dot product** (or **Euclidean inner product**) of two vectors $\mathbf{x} = (x_1, x_2, x_3)$ and $\mathbf{y} = (y_1, y_2, y_3)$ in the 3-space $\mathbb{R}^3$ is a real number defined by the formula

$$\mathbf{x} \cdot \mathbf{y} = x_1y_1 + x_2y_2 + x_3y_3 = [x_1\ x_2\ x_3] \begin{bmatrix} y_1 \\ y_2 \\ y_3 \end{bmatrix} = \mathbf{x}^T\mathbf{y},$$

where $\mathbf{x}^T\mathbf{y}$ is the matrix product of $\mathbf{x}^T$ and $\mathbf{y}$. This is also a number identified with the 1×1 matrix $\mathbf{x}^T\mathbf{y}$.

Using the dot product, the **Euclidean length** (or **magnitude**) of a vector $\mathbf{x} = (x_1, x_2, x_3)$ is defined by

$$\|\mathbf{x}\| = (\mathbf{x} \cdot \mathbf{x})^{\frac{1}{2}} = \sqrt{x_1^2 + x_2^2 + x_3^2}\,,$$

and the **Euclidean distance** between two vectors $\mathbf{x}$ and $\mathbf{y}$ in $\mathbb{R}^3$ is defined by

$$d(\mathbf{x}, \mathbf{y}) = \|\mathbf{x} - \mathbf{y}\|.$$

In this way, the dot product can be considered to be a ruler for measuring the length of a line segment in the 3-space $\mathbb{R}^3$. Furthermore, it can be used to measure the angle between two nonzero vectors.

The **Euclidean angle** θ between two nonzero vectors $\mathbf{x}$ and $\mathbf{y}$ in $\mathbb{R}^3$ is defined by the formula

$$\cos\theta = \frac{\mathbf{x} \cdot \mathbf{y}}{\|\mathbf{x}\|\|\mathbf{y}\|}, \quad 0 \le \theta \le \pi,$$

since the dot product satisfies the formula

$$\mathbf{x} \cdot \mathbf{y} = \|\mathbf{x}\|\|\mathbf{y}\| \cos\theta.$$

In particular, two vectors $\mathbf{x}$ and $\mathbf{y}$ are **orthogonal** (*i.e.*, they form a right angle $\theta = \pi/2$) if and only if the Pythagorean theorem holds:

$$\|\mathbf{x}\|^2 + \|\mathbf{y}\|^2 = \|\mathbf{x} + \mathbf{y}\|^2.$$

By rewriting this formula in terms of the dot product, one can obtain another equivalent condition:

$$\mathbf{x} \cdot \mathbf{y} = x_1y_1 + x_2y_2 + x_3y_3 = 0.$$

In fact, this dot product is one of the most important structures defined on the Euclidean 3-space $\mathbb{R}^3$. Euclidean geometry begins with the vector space $\mathbb{R}^3$ together with the dot product, because the Euclidean distance can be defined by the dot product.

The dot product in the 3-space $\mathbb{R}^3$ has a direct extension to the n-space $\mathbb{R}^n$ of any dimension n.

For any two vectors $\mathbf{x} = (x_1, x_2, \ldots, x_n)$ and $\mathbf{y} = (y_1, y_2, \ldots, y_n)$ in $\mathbb{R}^n$, their **dot product**, also called the **Euclidean inner product**, and the **Euclidean length** (or **magnitude**) of a vector are defined similarly as

$$\begin{aligned} \mathbf{x} \cdot \mathbf{y} &= x_1 y_1 + x_2 y_2 + \cdots + x_n y_n = \mathbf{x}^T \mathbf{y}, \\ \|\mathbf{x}\| &= (\mathbf{x} \cdot \mathbf{x})^{\frac{1}{2}} = \sqrt{x_1^2 + x_2^2 + \cdots + x_n^2}, \end{aligned}$$

respectively. Also, the **Euclidean distance** between two vectors $\mathbf{x}$ and $\mathbf{y}$ in $\mathbb{R}^n$ is defined by

$$d(\mathbf{x}, \mathbf{y}) = \|\mathbf{x} - \mathbf{y}\|.$$

Problem 97 In the plane $\mathbb{R}^2$ (or the space $\mathbb{R}^3$) with the dot product, prove the *law of cosines*:

$$\begin{aligned} \|\mathbf{x} - \mathbf{y}\|^2 &= \|\mathbf{x}\|^2 + \|\mathbf{y}\|^2 - 2\|\mathbf{x}\|\|\mathbf{y}\| \cos\theta \quad \text{and} \\ \|\mathbf{x} + \mathbf{y}\|^2 &= \|\mathbf{x}\|^2 + \|\mathbf{y}\|^2 + 2\|\mathbf{x}\|\|\mathbf{y}\| \cos\theta \end{aligned}$$

for any two nonzero vectors $\mathbf{x}$ and $\mathbf{y}$ with the angle θ between them. From this equation, derive the equality $\mathbf{x} \cdot \mathbf{y} = \mathbf{x}^T \mathbf{y} = \|\mathbf{x}\|\|\mathbf{y}\| \cos\theta$.

Hint: Compute the side lengths of a triangle with the cosine of one of its angles.

To extend the notion of the dot product to a (real) vector space, we extract the most essential properties that the dot product on the n-space $\mathbb{R}^n$ satisfies and take them as axioms for an inner product on a vector space V.

Definition 5.1 An **inner product** on a real vector space V is a function that associates a real number $\langle \mathbf{x}, \mathbf{y} \rangle$ to each ordered pair of vectors $\mathbf{x}$ and $\mathbf{y}$ in V in such a way that the following rules are satisfied: For any vectors $\mathbf{x}, \mathbf{y}$, and $\mathbf{z}$ in V and any scalar k in $\mathbb{R}$,

(1) $\langle \mathbf{x}, \mathbf{y} \rangle = \langle \mathbf{y}, \mathbf{x} \rangle$ (symmetry),

(2) $\langle \mathbf{x} + \mathbf{y}, \mathbf{z} \rangle = \langle \mathbf{x}, \mathbf{z} \rangle + \langle \mathbf{y}, \mathbf{z} \rangle$ (additivity),

(3) $\langle k\mathbf{x}, \mathbf{y} \rangle = k\langle \mathbf{x}, \mathbf{y} \rangle$ (homogeneity),

(4) $\langle \mathbf{x}, \mathbf{x} \rangle \geq 0$ always, and $\langle \mathbf{x}, \mathbf{x} \rangle = 0 \Leftrightarrow \mathbf{x} = \mathbf{0}$ (positive-definiteness).

A pair $(V, \langle \ , \rangle)$ of a (real) vector space V and an inner product $\langle \ , \rangle$ is called a **(real) inner product space**.

In particular, the pair $(\mathbb{R}^n, \cdot)$ with the dot product is called the **Euclidean n-space**.

By the symmetry (1), the additivity (2) and the homogeneity (3) also hold for the second variable:

(2′) $\langle \mathbf{x}, \mathbf{y}+\mathbf{z}\rangle = \langle \mathbf{x}, \mathbf{y}\rangle + \langle \mathbf{x}, \mathbf{z}\rangle$,
(3′) $\langle \mathbf{x}, k\mathbf{y}\rangle = k\langle \mathbf{x}, \mathbf{y}\rangle$.

It is easy to show that $\langle \mathbf{0}, \mathbf{y}\rangle = 0$ and also $\langle \mathbf{x}, \mathbf{0}\rangle = 0$.

Remark: In Definition 5.1, the rules (2) and (3) mean that the inner product is linear for the first variable, and the rules (2′) and (3′) above mean that the inner product is also linear for the second variable. In this sense, the inner product is said to be **bilinear** [1]. Hence, an inner product on a real vector space is often said to be *positive-definite, symmetric, and bilinear.*

Example 82 (*There are infinitely many non-Euclidean inner products on* $\mathbb{R}^n$)

(1) For any two vectors $\mathbf{x} = (x_1, x_2)$ and $\mathbf{y} = (y_1, y_2)$ in $\mathbb{R}^2$, define

$$\begin{aligned}\langle \mathbf{x}, \mathbf{y}\rangle &= ax_1y_1 + c(x_1y_2 + x_2y_1) + bx_2y_2 \\ &= [x_1\ x_2]\begin{bmatrix} a & c \\ c & b\end{bmatrix}\begin{bmatrix} y_1 \\ y_2\end{bmatrix} \\ &= \mathbf{x}^T A\mathbf{y},\end{aligned}$$

where a, b and c are arbitrary real numbers. Then, this function $\langle\ ,\ \rangle$ clearly satisfies the first three rules of the inner product, *i.e.*, $\langle\ ,\ \rangle$ is symmetric and bilinear (by the properties of the matrix product).

If $a > 0$ and $\det A = ab - c^2 > 0$, both positive, then it also satisfies the rule (4), the positive-definiteness of the inner product. (Hint: The equation $\langle \mathbf{x}, \mathbf{x}\rangle = ax_1^2 + 2cx_1x_2 + bx_2^2 \geq 0$ if and only if either $x_2 = 0$ or the discriminant of $\langle \mathbf{x}, \mathbf{x}\rangle / x_2^2$ is nonpositive.)

In the case of $c = 0$, this reduces to $\langle \mathbf{x}, \mathbf{y}\rangle = ax_1y_1 + bx_2y_2 = \mathbf{x}^T A\mathbf{y}$ with $A = \begin{bmatrix} a & 0 \\ 0 & b\end{bmatrix}$. Notice also that $a = \langle \mathbf{e}_1, \mathbf{e}_1\rangle$, $b = \langle \mathbf{e}_2, \mathbf{e}_2\rangle$, and $c = \langle \mathbf{e}_1, \mathbf{e}_2\rangle = \langle \mathbf{e}_2, \mathbf{e}_1\rangle$.

(2) For any diagonal matrix $A = \begin{bmatrix} d_1 & 0 & 0 \\ 0 & d_2 & 0 \\ 0 & 0 & d_3\end{bmatrix}$ with all $d_i > 0$,

$$\langle \mathbf{x}, \mathbf{y}\rangle = \mathbf{x}^T A\mathbf{y} = d_1x_1y_1 + d_2x_2y_2 + d_3x_3y_3$$

[1] Bi-, a number prefix denoting two and semi-, numerical prefix meaning $\frac{1}{2}$. Sesqui- is used in Latin combinations for $1\frac{1}{2}$.

defines an inner product on $\mathbb{R}^3$. Thus, there are infinitely many different inner products on the 3-space $\mathbb{R}^3$. If $d_1 = d_2 = d_3 = 1$, it is the Euclidean inner product, and all the others are non-Euclidean inner products on $\mathbb{R}^3$. $\square$

Lots of different types of inner products on the 3-space $\mathbb{R}^3$ may play the role of different kinds of rulers and protractors in our physical world $\mathbb{R}^3$.

By a subspace W of an inner product space V, we always mean a subspace of the vector space V together with the inner product that is the restriction [2] of the inner product on V to W.

Example 83 (*A subspace of an inner product space*) The Euclidean 2-space $(\mathbb{R}^2, \cdot)$ (with the dot product) is clearly a subspace of the Euclidean 3-space $(\mathbb{R}^3, \cdot)$ (with the dot product).

However, any non-Euclidean 2-space $\mathbb{R}^2$, for instance the 2-space $\mathbb{R}^2$ with a non-Euclidean inner product

$$\langle \mathbf{x}, \mathbf{y} \rangle = [x_1 \; x_2] \begin{bmatrix} 1 & 0 \\ 0 & 2 \end{bmatrix} \begin{bmatrix} y_1 \\ y_2 \end{bmatrix},$$

defined in Example 82, is not a subspace of the Euclidean 3-space $(\mathbb{R}^3, \cdot)$. $\square$

Remark: In vector calculus, the dot product (or the inner product) and the cross product (or the outer product) are introduced in the 3-space $\mathbb{R}^3$. The dot product in the space $\mathbb{R}^3$ is naturally extended to the Euclidean inner product in a higher dimensional space $\mathbb{R}^n$, as introduced already.

It is known in advanced mathematics that the cross product in the 3-space $\mathbb{R}^3$ cannot be extended to a higher dimensional Euclidean space $\mathbb{R}^n$. In fact, it is known that if there is a bilinear function $f : \mathbb{R}^n \times \mathbb{R}^n \to \mathbb{R}^n$ satisfying the property: $f(\mathbf{x}, \mathbf{y})$ is perpendicular to $\mathbf{x}$ and $\mathbf{y}$, and $\|f(\mathbf{x}, \mathbf{y})\|^2 = \|\mathbf{x}\|^2\|\mathbf{y}\|^2 - (\mathbf{x} \cdot \mathbf{y})^2$, then $n = 3$ or 7. That is, it is possible for only the spaces $\mathbb{R}^3$ and $\mathbb{R}^7$.

Hence it is impossible to introduce the cross product on the vector spaces. See [Prasolov], Section 41.7.3.

Problem 98 In Example 82(1), the converse is also true: Show that if $\langle \mathbf{x}, \mathbf{y} \rangle = \mathbf{x}^T A \mathbf{y}$ is an inner product on $\mathbb{R}^2$, then $a > 0$ and $\det A = ab - c^2 > 0$.

Hint: $\langle \mathbf{x}, \mathbf{x} \rangle = \mathbf{x}^T A \mathbf{x} > 0$ for all nonzero vectors $\mathbf{x}$. Check this with $\mathbf{x} = (1, 0)$ or $\mathbf{x} = (x, 1)$.

[2] Restriction of a function: Let $f : X \to Y$ be any function from a set X to a set Y. For any nonempty subset A of X, the *restriction* of the function $f : X \to Y$ to A is the function $g : A \to Y$ defined by $g(a) = f(a)$ for any $a \in A$. For example, the function $f(x) = \tan x$ is not injective. But the restriction of $f(x) = \tan x$ to $(-\pi/2, \pi/2)$ is bijective, so it has its inverse.

Problem 99 In Example 82(2), show that $\mathbf{x}^T A\mathbf{y}$ cannot be an inner product if A is a diagonal matrix having a nonpositive diagonal entry $d_i \leq 0$.

5.2 Geometry of vectors; Lengths, angles, orthogonality

In this section we study the geometry of an inner product space by introducing a length, an angle, and a distance between two vectors. This generalizes the Euclidean length, the Euclidean angle, and the Euclidean distance.

The following inequality will enable us to define an angle between any two nonzero vectors in an inner product space V.

Theorem 5.2 (Cauchy-Schwarz inequality) [3] *For any two vectors* $\mathbf{x}$ *and* $\mathbf{y}$ *in an inner product space* V*,*

$$\langle \mathbf{x}, \mathbf{y} \rangle^2 \leq \langle \mathbf{x}, \mathbf{x} \rangle \langle \mathbf{y}, \mathbf{y} \rangle.$$

Or equivalently

$$|\langle \mathbf{x}, \mathbf{y} \rangle| \leq \|\mathbf{x}\| \|\mathbf{y}\|.$$

Proof: If $\mathbf{x} = \mathbf{0}$, it is clear. Assume $\mathbf{x} \neq \mathbf{0}$. For any scalar t, we have

$$0 \leq \langle t\mathbf{x} + \mathbf{y}, t\mathbf{x} + \mathbf{y} \rangle = \langle \mathbf{x}, \mathbf{x} \rangle t^2 + 2\langle \mathbf{x}, \mathbf{y} \rangle t + \langle \mathbf{y}, \mathbf{y} \rangle.$$

This inequality implies that the quadratic polynomial $\langle \mathbf{x}, \mathbf{x} \rangle t^2 + 2\langle \mathbf{x}, \mathbf{y} \rangle t + \langle \mathbf{y}, \mathbf{y} \rangle$ in t has either no real roots or a repeated real root. Therefore, its discriminant should be nonpositive:

$$\langle \mathbf{x}, \mathbf{y} \rangle^2 - \langle \mathbf{x}, \mathbf{x} \rangle \langle \mathbf{y}, \mathbf{y} \rangle \leq 0,$$

which is the wanted inequality. □

Problem 100 Prove that the equality in the Cauchy-Schwarz inequality holds if and only if the vectors $\mathbf{x}$ and $\mathbf{y}$ are linearly dependent.

[3]**Baron Augustin-Louis Cauchy** (August 21, 1789– May 23, 1857) was a French mathematician and physicist who made pioneering contributions to analysis. **Karl Hermann Amandus Schwarz** (January 25, 1843– November 30, 1921) was a German mathematician known for his work in complex analysis.

The lengths of vectors and angles between two nonzero vectors in an inner product space are defined in a similar way to the case of the Euclidean n-space.

Definition 5.3 Let V be an inner product space.

(1) The **magnitude** $\|\mathbf{x}\|$ (or the **length**) of a vector $\mathbf{x}$ is defined by

$$\|\mathbf{x}\| = \sqrt{\langle \mathbf{x}, \mathbf{x} \rangle}\,.$$

(2) The **distance** $d(\mathbf{x}, \mathbf{y})$ between two vectors $\mathbf{x}$ and $\mathbf{y}$ is defined by

$$d(\mathbf{x}, \mathbf{y}) = \|\mathbf{x} - \mathbf{y}\|.$$

(3) The Cauchy-Schwarz inequality says that $-1 \le \frac{\langle \mathbf{x}, \mathbf{y} \rangle}{\|\mathbf{x}\|\|\mathbf{y}\|} \le 1$ for any two nonzero vectors $\mathbf{x}$ and $\mathbf{y}$. Hence, there is a unique $\theta \in [0, \pi]$ such that

$$\cos\theta = \frac{\langle \mathbf{x}, \mathbf{y} \rangle}{\|\mathbf{x}\|\|\mathbf{y}\|} \quad \text{or} \quad \langle \mathbf{x}, \mathbf{y} \rangle = \|\mathbf{x}\|\|\mathbf{y}\|\cos\theta.$$

Such a number θ is called the **angle** between $\mathbf{x}$ and $\mathbf{y}$.

Different types of inner products on a vector space give rise to different measures of lengths, angles, and distances as shown in the following example.

Example 84 Let $\mathbf{x} = (1, 1)$ and $\mathbf{y} = (1, 0)$ be two vectors in the plane $\mathbb{R}^2$.

(1) In the Euclidean 2-space $\mathbb{R}^2$ with the dot product, $\|\mathbf{x}\| = \sqrt{2}$ and the angle between $\mathbf{x}$ and $\mathbf{y}$ is clearly $\pi/4$ and $\cos(\pi/4) = 1/\sqrt{2} = 0.7071\cdots$.

(2) In a (non-Euclidean) $\mathbb{R}^2$ equipped with an inner product $\langle \mathbf{x}, \mathbf{y} \rangle = 2x_1y_1 + 3x_2y_2$, $\|\mathbf{x}\| = \sqrt{5}$ and the angle between $\mathbf{x}$ and $\mathbf{y}$ is computed as

$$\cos\theta = \frac{\langle \mathbf{x}, \mathbf{y} \rangle}{\|\mathbf{x}\|\|\mathbf{y}\|} = \frac{2}{\sqrt{5}\cdot\sqrt{2}} = 0.6324\cdots.$$

Thus, $\theta = \cos^{-1}(2/\sqrt{10})$.

The above two cases show that the length of a vector, the angle between two vectors, and the distance between two vectors actually depend on an inner product on a vector space. □

Problem 101 Prove the following properties of length in an inner product space V. For any two vectors $\mathbf{x}, \mathbf{y} \in V$,

(1) $||\mathbf{x}|| \geq 0$,
(2) $||\mathbf{x}|| = 0$ if and only if $\mathbf{x} = \mathbf{0}$,
(3) $||k\mathbf{x}|| = |k|\,||\mathbf{x}||$,
(4) $||\mathbf{x} + \mathbf{y}|| \leq ||\mathbf{x}|| + ||\mathbf{y}||$ (triangle inequality).

Hint for (4): Compute the square of both sides and use the Cauchy-Schwarz inequality.

Problem 102 Let V be an inner product space. By using Problem 101, show that for any vectors $\mathbf{x}, \mathbf{y}$, and $\mathbf{z}$ in V,

(1) $d(\mathbf{x}, \mathbf{y}) \geq 0$,
(2) $d(\mathbf{x}, \mathbf{y}) = 0$ if and only if $\mathbf{x} = \mathbf{y}$,
(3) $d(\mathbf{x}, \mathbf{y}) = d(\mathbf{y}, \mathbf{x})$,
(4) $d(\mathbf{x}, \mathbf{y}) \leq d(\mathbf{x}, \mathbf{z}) + d(\mathbf{z}, \mathbf{y})$ (triangle inequality).

Definition 5.4 Two vectors $\mathbf{x}$ and $\mathbf{y}$ in an inner product space are said to be **orthogonal** (or **perpendicular**) if $\langle \mathbf{x}, \mathbf{y} \rangle = 0$.

For any nonzero vectors $\mathbf{x}$ and $\mathbf{y}$, $\langle \mathbf{x}, \mathbf{y} \rangle = 0$ if and only if the angle between $\mathbf{x}$ and $\mathbf{y}$ is $\theta = \pi/2$.

Example 85 (*Law of cosines and the Pythagorean theorem in an inner product space*) Let V be any inner product space, not just Euclidean space, and let $\mathbf{x}$ and $\mathbf{y}$ be any two nonzero vectors in V with the angle θ between them. Then, $\langle \mathbf{x}, \mathbf{y} \rangle = ||\mathbf{x}||\,||\mathbf{y}|| \cos\theta$ gives the equality

$$\begin{aligned} ||\mathbf{x} - \mathbf{y}||^2 &= ||\mathbf{x}||^2 + ||\mathbf{y}||^2 - 2||\mathbf{x}||\,||\mathbf{y}|| \cos\theta \quad \text{and} \\ ||\mathbf{x} + \mathbf{y}||^2 &= ||\mathbf{x}||^2 + ||\mathbf{y}||^2 + 2||\mathbf{x}||\,||\mathbf{y}|| \cos\theta, \end{aligned}$$

called the *law of cosines* in the inner product space V. Moreover, it implies the *Pythagorean theorem* in an inner product space:

$$||\mathbf{x} + \mathbf{y}||^2 = ||\mathbf{x}||^2 + ||\mathbf{y}||^2$$

for any orthogonal vectors $\mathbf{x}$ and $\mathbf{y}$. □

Next two theorems show some properties of the orthogonality of vectors.

Theorem 5.5 *Let V be an inner product space and let $\mathbf{x} \in V$. Then the following are equivalent.*

(1) $\mathbf{x}$ *is orthogonal to every vector* $\mathbf{y}$ *in* V, *i.e.,* $\langle \mathbf{x}, \mathbf{y} \rangle = 0$ *for all* $\mathbf{y}$ *in* V,

(2) $\mathbf{x}$ *is orthogonal to every vector* $\mathbf{v}_i$ *in a basis* $\alpha = \{\mathbf{v}_1, \mathbf{v}_2, \ldots, \mathbf{v}_n\}$ *for* V,

(3) $\mathbf{x} = \mathbf{0}$.

Proof: (3) $\Rightarrow$ (1) $\Rightarrow$ (2) is clear. To show (2) $\Rightarrow$ (3), let $\alpha = \{\mathbf{v}_1, \mathbf{v}_2, \ldots, \mathbf{v}_n\}$ be any basis for V and let

$$\mathbf{x} = x_1\mathbf{v}_1 + x_2\mathbf{v}_2 + \cdots + x_n\mathbf{v}_n = \sum_{i=1}^{n} x_i\mathbf{v}_i$$

in V. Then $\langle \mathbf{x}, \mathbf{x}\rangle = \sum_{i=1}^{n} x_i\langle \mathbf{x}, \mathbf{v}_i\rangle = 0$. Hence, $\mathbf{x} = \mathbf{0}$ by the positive-definiteness. □

Theorem 5.6 *If* $\mathbf{x}_1, \mathbf{x}_2, \ldots, \mathbf{x}_k$ *are nonzero mutually orthogonal vectors in an inner product space* V *(i.e., each vector is orthogonal to every other vector), then they are linearly independent.*

Proof: Suppose $c_1\mathbf{x}_1 + c_2\mathbf{x}_2 + \cdots + c_k\mathbf{x}_k = \mathbf{0}$. Then for each $i = 1, 2, \ldots, k$,

$$\begin{aligned} 0 &= \langle \mathbf{0}, \mathbf{x}_i\rangle = \langle c_1\mathbf{x}_1 + \cdots + c_k\mathbf{x}_k, \mathbf{x}_i\rangle \\ &= c_1\langle \mathbf{x}_1, \mathbf{x}_i\rangle + \cdots + c_i\langle \mathbf{x}_i, \mathbf{x}_i\rangle + \cdots + c_k\langle \mathbf{x}_k, \mathbf{x}_i\rangle \\ &= c_i\|\mathbf{x}_i\|^2, \end{aligned}$$

because $\mathbf{x}_1, \mathbf{x}_2, \ldots, \mathbf{x}_k$ are mutually orthogonal. Since each $\mathbf{x}_i$ is not the zero vector, $\|\mathbf{x}_i\| \neq 0$ which implies $c_i = 0$ for $i = 1, 2, \ldots, k$. □

Problem 103 Show that $|\ \|\mathbf{x}\| - \|\mathbf{y}\|\ | \leq \|\mathbf{x} - \mathbf{y}\|$ for any two vectors $\mathbf{x}$, $\mathbf{y}$ in an inner product space.

Hint: Use the triangle inequality in Problem 101.

Problem 104 **[Cauchy-Schwarz inequality for sums]** For any two vectors $\mathbf{x} = (x_1, x_2, \ldots, x_n)$ and $\mathbf{y} = (y_1, y_2, \ldots, y_n)$ in the Euclidean n-space $\mathbb{R}^n$, show that

(1) $$\left[\sum_{i=1}^{n} x_i y_i\right]^2 \leq \left[\sum_{i=1}^{n} x_i^2\right]\left[\sum_{i=1}^{n} y_i^2\right]$$ (Cauchy-Schwarz inequality).

(2) $$\left[\sum_{i=1}^{n} (x_i + y_i)^2\right]^{\frac{1}{2}} \leq \left[\sum_{i=1}^{n} x_i^2\right]^{\frac{1}{2}} + \left[\sum_{i=1}^{n} y_i^2\right]^{\frac{1}{2}}$$ (triangle inequality).

Hint: (1) Use the Cauchy-Schwarz inequality, Theorem 5.2. For (2), use the triangle inequality in Problem 101.

Problem 105 **[Cauchy-Schwarz inequality for integrals]** [4] Let $C[0,1]$ be the inner product space of all real-valued continuous functions defined on $[0,1]$ equipped with the inner product

$$\langle f, g\rangle = \int_0^1 f(x)g(x)\,dx \quad \text{for any } f \text{ and } g.$$

For any continuous real-valued functions $f(x)$ and $g(x)$ on $[0,1]$, show that

(1) $\left[\int_0^1 f(x)g(x)dx\right]^2 \le \left[\int_0^1 f^2(x)dx\right]\left[\int_0^1 g^2(x)dx\right]$ (Cauchy-Schwarz inequality).

(2) $\left[\int_0^1 (f(x)+g(x))^2dx\right]^{\frac{1}{2}} \le \left[\int_0^1 f^2(x)dx\right]^{\frac{1}{2}} + \left[\int_0^1 g^2(x)dx\right]^{\frac{1}{2}}$ (triangle inequality).

5.3 Matrix representations of inner products

Let A be an $n \times n$ diagonal matrix with positive diagonal entries. Then, $\langle \mathbf{x}, \mathbf{y}\rangle = \mathbf{x}^T A\mathbf{y}$ defines an inner product on the n-space $\mathbb{R}^n$ as shown in Example 82(2). The converse is also true: Every inner product on a vector space can be expressed as a matrix product $\mathbf{x}^T A\mathbf{y}$ with a symmetric (or diagonal) matrix A.

Let $(V, \langle\ ,\ \rangle)$ be an inner product space and let $\alpha = \{\mathbf{v}_1, \mathbf{v}_2, \dots, \mathbf{v}_n\}$ be a fixed ordered basis for V. Then for any vectors

$$\begin{aligned}\mathbf{x} &= x_1\mathbf{v}_1 + x_2\mathbf{v}_2 + \cdots + x_n\mathbf{v}_n &= \sum_{i=1}^n x_i\mathbf{v}_i,\\ \mathbf{y} &= y_1\mathbf{v}_1 + y_2\mathbf{v}_2 + \cdots + y_n\mathbf{v}_n &= \sum_{j=1}^n y_j\mathbf{v}_j\end{aligned}$$

in V,

$$\langle \mathbf{x}, \mathbf{y}\rangle = \sum_{i=1}^n\sum_{j=1}^n x_i y_j \langle \mathbf{v}_i, \mathbf{v}_j\rangle$$

holds. If we set $a_{ij} = \langle \mathbf{v}_i, \mathbf{v}_j\rangle$ for $i, j = 1, 2, \dots, n$, then these numbers constitute a symmetric matrix $A = [a_{ij}]$, since $\langle \mathbf{v}_i, \mathbf{v}_j\rangle = \langle \mathbf{v}_j, \mathbf{v}_i\rangle$. Thus, in matrix notation, the inner product can be written as

$$\langle \mathbf{x}, \mathbf{y}\rangle = \sum_{i=1}^n\sum_{j=1}^n x_i y_j a_{ij} = [\mathbf{x}]_\alpha^T A[\mathbf{y}]_\alpha. \tag{5.1}$$

[4] The inequality *for sums* was published by Augustin-Louis Cauchy (1821), while the corresponding inequality *for integrals* was first proved by Viktor Bunyakovsky (1859). The modern proof of the integral inequality was given by Hermann Amandus Schwarz (1888).

The matrix A is called the **matrix representation** of the inner product $\langle\ ,\ \rangle$ with respect to the basis α. Moreover, this matrix A can be a diagonal matrix by taking a suitable basis α. This will be shown later in Section 7.6 or in Section 9.10 more explicitly.

Example 86 (*Matrix representation of an inner product*)

(1) (*The matrix representation of the dot product is the identity matrix*) With respect to the standard (ordered) basis $\{\mathbf{e}_1, \mathbf{e}_2, \ldots, \mathbf{e}_n\}$ for $\mathbb{R}^n$, the dot product $\mathbf{e}_i \cdot \mathbf{e}_j = \delta_{ij}$ for all $i, j = 1, 2, \ldots, n$. Thus, the matrix representation of the dot product is the identity matrix and the dot product is just the matrix product $\mathbf{x}^T\mathbf{y}$:

$$\mathbf{x} \cdot \mathbf{y} = [x_1 \ \cdots \ x_n] \begin{bmatrix} 1 & & \mathbf{0} \\ & \ddots & \\ \mathbf{0} & & 1 \end{bmatrix} \begin{bmatrix} y_1 \\ \vdots \\ y_n \end{bmatrix} = [x_1 \ \cdots \ x_n] \begin{bmatrix} y_1 \\ \vdots \\ y_n \end{bmatrix} = \mathbf{x}^T\mathbf{y}.$$

(2) On $V = P_2([0,1])$, we define an inner product on V as

$$\langle f, g \rangle = \int_0^1 f(x)g(x)dx.$$

Then for an (ordered) basis $\alpha = \{f_1(x) = 1, f_2(x) = x,\ f_3(x) = x^2\}$ for V, one can easily find its matrix representation $A = [a_{ij}]$. For instance,

$$a_{23} = \langle f_2, f_3 \rangle = \int_0^1 f_2(x)f_3(x)dx = \int_0^1 x \cdot x^2 dx = \frac{1}{4},$$

which also equals a_{32}. □

The expression of an inner product as a matrix product is very useful in stating or proving theorems in an inner product space.

For any symmetric matrix A and for a fixed (ordered) basis α, the formula $\langle \mathbf{x}, \mathbf{y} \rangle = [\mathbf{x}]_\alpha^T A [\mathbf{y}]_\alpha$ seems to give rise to an inner product on V. In fact, the formula clearly is symmetric and bilinear, but does not necessarily satisfy the fourth rule, the *positive-definiteness*.

Some necessary and sufficient conditions for a symmetric matrix A to give rise to an inner product will be discussed in Chapter 9 Section 9.6. One of them is that there exists a nonsingular matrix W such that $A = W^TW$. See Theorem 9.7. It implies that $\det A = (\det W)^2 > 0$.

Also, recall that the conditions $a > 0$ and $\det A = ab - c^2 > 0$ in Example 82 are necessary for a 2×2 matrix A to give rise to an inner product on $\mathbb{R}^2$. Later, it will be shown in Theorem 9.7 that it is also a sufficient condition for the positive-definiteness.

Problem 106 In Example 86(2), determine all the other entries of $A = [a_{ij}]$. Also show that $\det A > 0$.

5.4 Gram-Schmidt orthogonalization for rectangular coordinates

The standard basis for the Euclidean n-space $\mathbb{R}^n$ has a special property: The basis vectors are mutually orthogonal and are of length 1. For this reason, it is called the **standard rectangular coordinate system** or the **Cartesian coordinate system** for the Euclidean n-space $\mathbb{R}^n$.[5]

In an inner product space, a vector of length 1 is called a **unit vector**. For every nonzero vector $\mathbf{x}$, the vector $\mathbf{x}/\|\mathbf{x}\|$ is a unit vector. To derive a unit vector from a nonzero vector by dividing by its length is called **normalization**.

If there is a set of mutually orthogonal vectors (or a basis) in an inner product space, then the vectors can be converted into unit vectors by normalizing them without losing their mutual orthogonality.

Definition 5.7 A set of vectors $\mathbf{x}_1, \mathbf{x}_2, \ldots, \mathbf{x}_k$ in an inner product space V is said to be **orthonormal** if

$$\langle \mathbf{x}_i, \mathbf{x}_j \rangle = \delta_{ij} = \begin{cases} 0 & \text{if } i \neq j \quad \text{(orthogonality)}, \\ 1 & \text{if } i = j \quad \text{(normality)}. \end{cases}$$

A set $\{\mathbf{x}_1, \mathbf{x}_2, \ldots, \mathbf{x}_n\}$ of vectors is called an **orthonormal basis** for V if it is a basis and orthonormal.

Problem 107 Determine whether each of the following sets of vectors in $\mathbb{R}^2$ or in $\mathbb{R}^3$ is orthogonal, orthonormal, or neither with respect to the Euclidean inner product.

$$(1)\ \left\{\begin{bmatrix}10\\0\end{bmatrix}, \begin{bmatrix}0\\20\end{bmatrix}\right\} \qquad (2)\ \left\{\begin{bmatrix}10\\1\end{bmatrix}, \begin{bmatrix}0\\20\end{bmatrix}\right\} \qquad (3)\ \left\{\begin{bmatrix}10\\1\end{bmatrix}, \begin{bmatrix}-1\\20\end{bmatrix}\right\}$$

$$(4)\ \left\{\begin{bmatrix}1/\sqrt{2}\\1/\sqrt{2}\end{bmatrix}, \begin{bmatrix}-1/\sqrt{2}\\1/\sqrt{2}\end{bmatrix}\right\} \quad (5)\ \left\{\begin{bmatrix}0\\0\\\pm1\end{bmatrix}, \begin{bmatrix}0\\\pm1\\0\end{bmatrix}, \begin{bmatrix}\pm1\\0\\0\end{bmatrix}\right\}.$$

[5] A Cartesian coordinate is also called a *rectangular coordinate* or an *orthogonal coordinate.*

Clearly, there are infinitely many orthonormal bases for the Euclidean 3-space $\mathbb{R}^3$, including the standard basis $\{\mathbf{e}_1, \mathbf{e}_2, \mathbf{e}_3\}$. Later, it will be shown in Theorem 5.8 that every inner product space has at least one (actually infinitely many) orthonormal basis, just like the standard orthonormal basis for the Euclidean 3-space $\mathbb{R}^3$.

The following example illustrates how to construct such an orthonormal basis.

Example 87 (*How to construct an orthonormal basis?*) For the matrix

$$A = \begin{bmatrix} 1 & 1 & 1 \\ 1 & 0 & 1 \\ 1 & 1 & 0 \\ 1 & 0 & 0 \end{bmatrix},$$

find an orthonormal basis for the column space $\mathcal{C}(A)$ of A.

Solution: Let $\mathbf{c_1}$, $\mathbf{c_2}$, and $\mathbf{c_3}$ be the column vectors of A in the order from left to right. It is easily verified that they are linearly independent, so they form a basis for the column space $\mathcal{C}(A)$ of dimension 3 in the 4-space $\mathbb{R}^4$.

For notational convenience, we denote by $\mathbf{Span}\{\mathbf{x}_1, \mathbf{x}_2, \ldots, \mathbf{x}_k\}$ the subspace spanned by $\{\mathbf{x}_1, \mathbf{x}_2, \ldots, \mathbf{x}_k\}$.

(1) First normalize $\mathbf{c}_1$ to get

$$\mathbf{u}_1 = \frac{\mathbf{c}_1}{\|\mathbf{c}_1\|} = \frac{\mathbf{c}_1}{2} = \left(\frac{1}{2}, \frac{1}{2}, \frac{1}{2}, \frac{1}{2}\right),$$

which is a unit vector. Then $\mathbf{Span}\{\mathbf{u}_1\} = \mathbf{Span}\{\mathbf{c}_1\}$, because one is a (nonzero) scalar multiple of the other.

(2) Noting that the vector $\mathbf{c}_2 - \langle \mathbf{u}_1, \mathbf{c}_2 \rangle \mathbf{u}_1 = \mathbf{c}_2 - \mathbf{u}_1 = (\frac{1}{2}, -\frac{1}{2}, \frac{1}{2}, -\frac{1}{2})$ is a nonzero (unit) vector which is orthogonal to $\mathbf{u}_1$, normalize it to get

$$\mathbf{u}_2 = \frac{\mathbf{c}_2 - \langle \mathbf{u}_1, \mathbf{c}_2 \rangle \mathbf{u}_1}{\|\mathbf{c}_2 - \langle \mathbf{u}_1, \mathbf{c}_2 \rangle \mathbf{u}_1\|} = \left(\frac{1}{2}, -\frac{1}{2}, \frac{1}{2}, -\frac{1}{2}\right).$$

Then, $\{\mathbf{u}_1, \mathbf{u}_2\}$ is orthonormal and $\mathbf{Span}\{\mathbf{u}_1, \mathbf{u}_2\} = \mathbf{Span}\{\mathbf{c}_1, \mathbf{c}_2\}$, because each $\mathbf{u}_i$ is a linear combination of $\mathbf{c}_1$ and $\mathbf{c}_2$, and the converse is also true.

(3) Finally, note that $\mathbf{c}_3 - \langle \mathbf{u}_1, \mathbf{c}_3 \rangle \mathbf{u}_1 - \langle \mathbf{u}_2, \mathbf{c}_3 \rangle \mathbf{u}_2 = \mathbf{c}_3 - \mathbf{u}_1 - 0\mathbf{u}_2 = (\frac{1}{2}, \frac{1}{2}, -\frac{1}{2}, -\frac{1}{2})$ is also a nonzero vector which is orthogonal to both $\mathbf{u}_1$ and $\mathbf{u}_2$. By the normalization, the vector

$$\mathbf{u}_3 = \frac{\mathbf{c}_3 - \langle \mathbf{u}_1, \mathbf{c}_3 \rangle \mathbf{u}_1 - \langle \mathbf{u}_2, \mathbf{c}_3 \rangle \mathbf{u}_2}{\|\mathbf{c}_3 - \langle \mathbf{u}_1, \mathbf{c}_3 \rangle \mathbf{u}_1 - \langle \mathbf{u}_2, \mathbf{c}_3 \rangle \mathbf{u}_2\|} = \left(\frac{1}{2}, \frac{1}{2}, -\frac{1}{2}, -\frac{1}{2}\right)$$

is a unit vector, and one can also show that $\mathbf{Span}\{\mathbf{u}_1, \mathbf{u}_2, \mathbf{u}_3\} = \mathbf{Span}\{\mathbf{c}_1, \mathbf{c}_2, \mathbf{c}_3\} = \mathcal{C}(A)$.

Consequently, $\{\mathbf{u}_1, \mathbf{u}_2, \mathbf{u}_3\}$ is orthonormal and linearly independent, and hence it is an orthonormal basis for the column space $\mathcal{C}(A)$. See Theorem 5.6 for the linear independence. □

The orthonormalization process in Example 87 indicates how to prove the general case, called the **Gram-Schmidt orthogonalization.** [6]

Theorem 5.8 *Every inner product space has an orthonormal basis.*

Proof: [Gram-Schmidt orthogonalization process] Let $\{\mathbf{x}_1, \mathbf{x}_2, \ldots, \mathbf{x}_n\}$ be a basis for an n-dimensional inner product space V. Let

$$\mathbf{u}_1 = \frac{\mathbf{x}_1}{\|\mathbf{x}_1\|}, \qquad \mathbf{u}_2 = \frac{\mathbf{x}_2 - \langle \mathbf{u}_1, \mathbf{x}_2 \rangle \mathbf{u}_1}{\|\mathbf{x}_2 - \langle \mathbf{u}_1, \mathbf{x}_2 \rangle \mathbf{u}_1\|}.$$

Of course, $\mathbf{x}_2 - \langle \mathbf{u}_1, \mathbf{x}_2 \rangle \mathbf{u}_1 \neq \mathbf{0}$, because $\{\mathbf{x}_1, \mathbf{x}_2\}$ is linearly independent. Generally, one can define by induction on $k = 1, 2, \ldots, n$,

$$\mathbf{u}_k = \frac{\mathbf{x}_k - \langle \mathbf{u}_1, \mathbf{x}_k \rangle \mathbf{u}_1 - \langle \mathbf{u}_2, \mathbf{x}_k \rangle \mathbf{u}_2 - \cdots - \langle \mathbf{u}_{k-1}, \mathbf{x}_k \rangle \mathbf{u}_{k-1}}{\|\mathbf{x}_k - \langle \mathbf{u}_1, \mathbf{x}_k \rangle \mathbf{u}_1 - \langle \mathbf{u}_2, \mathbf{x}_k \rangle \mathbf{u}_2 - \cdots - \langle \mathbf{u}_{k-1}, \mathbf{x}_k \rangle \mathbf{u}_{k-1}\|}.$$

Then, one can show inductively that $\mathbf{Span}\{\mathbf{u}_1, \mathbf{u}_2, \ldots, \mathbf{u}_k\} = \mathbf{Span}\{\mathbf{c}_1, \mathbf{c}_2, \ldots, \mathbf{c}_k\}$ for all k. Finally, as in Example 87, the vectors $\mathbf{u}_1, \mathbf{u}_2, \ldots, \mathbf{u}_n$ are orthonormal in the n-dimensional vector space V. Since every orthonormal set is linearly independent by Theorem 5.6, it is an orthonormal basis for V. □

Remark: For a better understanding of the Gram-Schmidt orthogonalization process, recall a fact in geometry that the orthogonal projection of the Euclidean space $\mathbb{R}^3$ onto the xy-plane $\mathbb{R}^2$ equipped with their dot products is defined by

$$\mathbf{x} = (x, y, z) \mapsto (x, y, 0) = (\mathbf{e}_1 \cdot \mathbf{x})\mathbf{e}_1 + (\mathbf{e}_2 \cdot \mathbf{x})\mathbf{e}_2.$$

Such a formula can be extended to any inner product space: If V is an inner product space with an inner product $\langle \ , \rangle$ and W is a subspace with an *orthonormal*

[6] The method is named after Jørgen Pedersen Gram and Erhard Schmidt but it appeared earlier in the work of Laplace and Cauchy. **Jørgen Pedersen Gram** (June 27, 1850 – April 29, 1916) was a Danish actuary and mathematician. **Erhard Schmidt** (January 13, 1876– December 6, 1959) was a Baltic German mathematician.

basis $\{\mathbf{u}_1, \mathbf{u}_2, \ldots, \mathbf{u}_{k-1}\}$, then the *orthogonal projection* of the space V onto the subspace W maps a vector $\mathbf{x}$ in V to

$$\langle \mathbf{u}_1, \mathbf{x} \rangle \mathbf{u}_1 + \langle \mathbf{u}_2, \mathbf{x} \rangle \mathbf{u}_2 + \cdots + \langle \mathbf{u}_{k-1}, \mathbf{x} \rangle \mathbf{u}_{k-1}$$

in W, which will be proved later in Section 5.6. Therefore, for any vector $\mathbf{x}$ in V but not in W,

$$\mathbf{x} - \langle \mathbf{u}_1, \mathbf{x} \rangle \mathbf{u}_1 - \langle \mathbf{u}_2, \mathbf{x} \rangle \mathbf{u}_2 - \cdots - \langle \mathbf{u}_{k-1}, \mathbf{x} \rangle \mathbf{u}_{k-1}$$

is nonzero and orthogonal to the subspace W. This is a key idea of constructing the vectors $\mathbf{u}_k$ inductively in the Gram-Schmidt orthogonalization.

Problem 108 Use the Gram-Schmidt orthogonalization on the Euclidean space $\mathbb{R}^4$ to transform the basis

$$\alpha = \{(0,1,1,0),\ (-1,1,0,0),\ (2,4,0,-2),\ (-6,0,0,-6)\}$$

into an orthonormal basis $\{\mathbf{u}_1, \mathbf{u}_2, \mathbf{u}_3, \mathbf{u}_4\}$. With your answer, express each vector in α as a linear combination of the orthonormal basis vectors $\mathbf{u}_1, \mathbf{u}_2, \mathbf{u}_3, \mathbf{u}_4$.

Problem 109 Find an orthonormal basis for the subspace W of the Euclidean space $\mathbb{R}^3$ given as the solution set of $x + 2y - z = 0$.

Let $\{\mathbf{e}_1, \mathbf{e}_2, \ldots, \mathbf{e}_n\}$ be the standard basis for the Euclidean n-space $\mathbb{R}^n$. Then any vector $\mathbf{x}$ in $\mathbb{R}^n$ can be written as

$$\begin{aligned}\mathbf{x} &= (x_1, x_2, \ldots, x_n) \\ &= x_1\mathbf{e}_1 + x_2\mathbf{e}_2 + \cdots + x_n\mathbf{e}_n \\ &= \langle \mathbf{e}_1, \mathbf{x} \rangle \mathbf{e}_1 + \langle \mathbf{e}_2, \mathbf{x} \rangle \mathbf{e}_2 + \cdots + \langle \mathbf{e}_n, \mathbf{x} \rangle \mathbf{e}_n.\end{aligned}$$

The next theorem shows that an orthonormal basis acts just like the standard basis for the Euclidean n-space $\mathbb{R}^n$.

Theorem 5.9 *Let* $\{\mathbf{u}_1, \mathbf{u}_2, \ldots, \mathbf{u}_k\}$ *be an orthonormal basis for a subspace* U *in an inner product space* V*. Then, for any vector* $\mathbf{x}$ *in* U*,*

$$\mathbf{x} = \langle \mathbf{u}_1, \mathbf{x} \rangle \mathbf{u}_1 + \langle \mathbf{u}_2, \mathbf{x} \rangle \mathbf{u}_2 + \cdots + \langle \mathbf{u}_k, \mathbf{x} \rangle \mathbf{u}_k.$$

Proof: For any vector $\mathbf{x} \in U$, one can write $\mathbf{x} = x_1\mathbf{u}_1 + x_2\mathbf{u}_2 + \cdots + x_k\mathbf{u}_k$, as a linear combination of the basis vectors. However, for each $i = 1, 2, \ldots, n$,

$$\begin{aligned} \langle \mathbf{u}_i, \mathbf{x} \rangle &= \langle \mathbf{u}_i,\ x_1\mathbf{u}_1 + \cdots + x_k\mathbf{u}_k \rangle \\ &= x_1\langle \mathbf{u}_i, \mathbf{u}_1 \rangle + \cdots + x_i\langle \mathbf{u}_i, \mathbf{u}_i \rangle + \cdots + x_k\langle \mathbf{u}_i, \mathbf{u}_k \rangle \\ &= x_i, \end{aligned}$$

because $\{\mathbf{u}_1, \mathbf{u}_2, \ldots, \mathbf{u}_k\}$ is orthonormal. □

In an inner product space V, if an (ordered) orthonormal basis $\alpha = \{\mathbf{u}_1, \mathbf{u}_2, \ldots, \mathbf{u}_n\}$ for V is given, then it gives geometrically a coordinate system, called a **rectangular coordinate system** for V, in which any vector $\mathbf{x}$ in V has a unique coordinate expression as

$$\mathbf{x} = \langle \mathbf{u}_1, \mathbf{x} \rangle \mathbf{u}_1 + \langle \mathbf{u}_2, \mathbf{x} \rangle \mathbf{u}_2 + \cdots + \langle \mathbf{u}_n, \mathbf{x} \rangle \mathbf{u}_n,$$

and its **coordinate** is a column vector

$$[\mathbf{x}]_\alpha = \begin{bmatrix} \langle \mathbf{u}_1, \mathbf{x} \rangle \\ \langle \mathbf{u}_2, \mathbf{x} \rangle \\ \vdots \\ \langle \mathbf{u}_n, \mathbf{x} \rangle \end{bmatrix}.$$

Hence, any orthonormal basis for a vector space V gives a *rectangular coordinate system* for V, while a non-orthonormal basis for V gives generally an *oblique coordinate system* for V.

The Gram-Schmidt orthogonalization process is nothing but a process of constructing a rectangular coordinate system from a given oblique coordinate system.

Moreover, any rectangular coordinate system given by an orthonormal basis for an n-dimensional inner product space V can be identified with (or be isomorphic to) the Cartesian coordinate system for the Euclidean n-space $\mathbb{R}^n$ by the coordinate isomorphism.

Let $\alpha = \{\mathbf{u}_1, \mathbf{u}_2, \ldots, \mathbf{u}_n\}$ be any (ordered) orthonormal basis for the space V. Then, the coordinate isomorphism

$$\Phi : V \to \mathbb{R}^n \quad \text{defined by} \quad \Phi(\mathbf{u}_i) = [\mathbf{u}_i]_\alpha = \mathbf{e}_i,\ i = 1, 2, \ldots, n$$

maps the rectangular coordinate system $\alpha = \{\mathbf{u}_1, \mathbf{u}_2, \ldots, \mathbf{u}_n\}$ for V to the Cartesian coordinate system $\{\mathbf{e}_1, \mathbf{e}_2, \ldots, \mathbf{e}_n\}$ for $\mathbb{R}^n$.

Furthermore, such isomorphism Φ preserves the inner product on vectors: For a vector $\mathbf{x} = \sum_{i=1}^{n} x_i\mathbf{u}_i$ in V with $x_i = \langle \mathbf{u}_i, \mathbf{x}\rangle$, the coordinate vector of $\mathbf{x}$ with respect to α is the column vector

$$\Phi(\mathbf{x}) = [\mathbf{x}]_\alpha = \begin{bmatrix} x_1 \\ x_2 \\ \vdots \\ x_n \end{bmatrix}.$$

For another vector $\mathbf{y} = \sum_{i=1}^{n} y_i\mathbf{u}_i$ in V,

$$\langle \mathbf{x}, \mathbf{y}\rangle = \left\langle \sum_{i=1}^{n} x_i\mathbf{u}_i, \sum_{j=1}^{n} y_j\mathbf{u}_j \right\rangle = \sum_{i=1}^{n}\sum_{j=1}^{n} x_i y_j \langle \mathbf{u}_i, \mathbf{u}_j\rangle = \sum_{i=1}^{n} x_i y_i = [\mathbf{x}]_\alpha^T[\mathbf{y}]_\alpha.$$

The right-hand side of this equation is just the dot product of vectors in the Euclidean space $\mathbb{R}^n$. That is,

$$\langle \mathbf{x}, \mathbf{y}\rangle = [\mathbf{x}]_\alpha^T[\mathbf{y}]_\alpha = \Phi(\mathbf{x}) \cdot \Phi(\mathbf{y})$$

for any $\mathbf{x}, \mathbf{y} \in V$.

Consequently, *the coordinate isomorphism Φ preserves the inner product,* and it gives the following theorem (compare it with Corollary 4.9.1(1)) on page 160.

Theorem 5.10 *Any n-dimension inner product space V with an inner product $\langle\ ,\ \rangle$ is isomorphic to the Euclidean n-space $\mathbb{R}^n$ with the dot product.*

Furthermore, if an orthonormal basis α for V is given, then the space V is isomorphic to $\mathbb{R}^n$ by the coordinate isomorphism $\Phi : V \to \mathbb{R}^n$ defined by $\mathbf{x} \mapsto [\mathbf{x}]_\alpha$, which preserves even the inner product.

For this reason, sometimes one restricts the study of an inner product space to the case of the Euclidean n-space $\mathbb{R}^n$ with the dot product.

A special kind of linear transformation that preserves the inner product such as the coordinate isomorphism from V to $\mathbb{R}^n$ given in Theorem 5.10 also plays an important role in linear algebra, and it will be studied in detail in Section 5.9.

Problem 110 Let W be the diagonal line in the Euclidean plane $\mathbb{R}^2$:

$$W = \{(x, x) \in \mathbb{R}^2 : x \in \mathbb{R}\},$$

and let Proj : $\mathbb{R}^2 \to \mathbb{R}^2$ be the orthogonal projection of the plane $\mathbb{R}^2$ onto the subspace W. Determine the image Proj$(\mathbf{x})$ for an $\mathbf{x} = (x, y) \in \mathbb{R}^2$.

Problem 111 In Example 87, we construct an orthonormal basis $\{\mathbf{u}_1, \mathbf{u}_2, \mathbf{u}_3\}$ for the column space $\mathcal{C}(A)$ by showing $\textbf{Span}\{\mathbf{u}_1, \mathbf{u}_2, \mathbf{u}_3\} = \textbf{Span}\{\mathbf{c}_1, \mathbf{c}_2, \mathbf{c}_3\} = \mathcal{C}(A)$. Express each column $\mathbf{c}_i$ as a linear combination of the orthonormal vectors $\mathbf{u}_1, \mathbf{u}_2, \mathbf{u}_3$.

5.5 Projections of a vector space and their algebraic characterization

In the three-dimensional space of Euclidean geometry, the orthogonal projection of the 3-space $\mathbb{R}^3$ onto the xy-plane $\mathbb{R}^2$ is defined by

$$\mathbf{x} = (x, y, z) \mapsto (x, y, 0) = (\mathbf{e}_1 \cdot \mathbf{x})\mathbf{e}_1 + (\mathbf{e}_2 \cdot \mathbf{x})\mathbf{e}_2.$$

These kinds of (orthogonal) projections of the 3-space can be extended to those of any vector space (not necessarily only an inner product space).

Let U be a subspace of any vector space V. Then, there is another subspace W of V such that $V = U \oplus W$ (see Corollary 3.19.2), so that any $\mathbf{x} \in V$ has a unique expression as $\mathbf{x} = \mathbf{u} + \mathbf{w}$ for $\mathbf{u} \in U$ and $\mathbf{w} \in W$.

As an easy exercise, one can see that a function

$$T : V \to V \quad \text{defined by} \quad T(\mathbf{x}) = T(\mathbf{u} + \mathbf{w}) = \mathbf{u}$$

is a linear transformation, whose image $\text{Im}(T)$ is the subspace U and the kernel $\text{Ker}(T)$ is the subspace W.

Definition 5.11 Let U and W be subspaces of a vector space V. A linear transformation $T : V \to V$ is called the **projection** of V onto the subspace U along W if $V = U \oplus W$ and $T(\mathbf{x}) = \mathbf{u}$ for $\mathbf{x} = \mathbf{u} + \mathbf{w} \in U \oplus W$.

Example 88 (*There are infinitely many different projections of $\mathbb{R}^2$ onto the x-axis*) Let X, Y, and Z be the one-dimensional subspaces of the Euclidean 2-space $\mathbb{R}^2$ spanned by the vectors $\mathbf{e}_1$, $\mathbf{e}_2$, and $\mathbf{d} = -\mathbf{e}_1 + \mathbf{e}_2 = (-1, 1)$, respectively:

$$\begin{aligned} X &= \{r\mathbf{e}_1 : r \in \mathbb{R}\}, \text{ the } x\text{-axis}, \\ Y &= \{r\mathbf{e}_2 : r \in \mathbb{R}\}, \text{ the } y\text{-axis}, \\ Z &= \{r(-\mathbf{e}_1 + \mathbf{e}_2) : r \in \mathbb{R}\}, \text{ the line } y = -x. \end{aligned}$$

Since the pairs $\{\mathbf{e}_1, \mathbf{e}_2\}$ and $\{\mathbf{e}_1, \mathbf{d}\}$ are both linearly independent, the space $\mathbb{R}^2$ can be expressed as a direct sum in two different ways:

$$\mathbb{R}^2 = X \oplus Y = X \oplus Z.$$

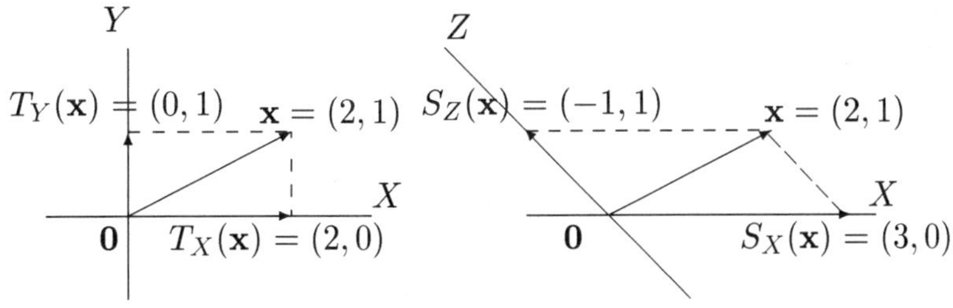

Figure 5.1: Two different projections of $\mathbb{R}^2$ onto the x-axis

Thus, a vector $\mathbf{x} = (2,1) \in \mathbb{R}^2$ is written in two different ways:

$$\mathbf{x} = (2,1) = \begin{cases} 2(1,0) + (0,1) & \in \; X \oplus Y \; = \mathbb{R}^2 \quad \text{and} \\ (3,0) + (-1,1) & \in \; X \oplus Z \; = \mathbb{R}^2, \end{cases}$$

as shown in Figure 5.1, where the left one has a rectangular coordinate system with basis $\{\mathbf{e}_1, \mathbf{e}_2\}$ and the right one has an oblique coordinate system with basis $\{\mathbf{e}_1, \mathbf{d}\}$.

Let T_X and S_X denote the projections of $\mathbb{R}^2$ onto X along Y and onto X along Z, respectively. Then for $\mathbf{x} = (2,1)$,

$$T_X(\mathbf{x}) = 2(1,0) = (2,0) \quad \text{and} \quad S_X(\mathbf{x}) = (3,0).$$

This shows that a projection of $\mathbb{R}^2$ onto the x-axis depends on a choice of complementary subspace of the x-axis.

Instead of Z, by choosing $Z_n = \{r(n,1) : r \in \mathbb{R}\}$ for any integer n as a complementary subspace of the x-axis, one can construct infinitely many different projections of $\mathbb{R}^2$ onto the x-axis. □

By definition, for a given subspace U of V and for any choice of a complementary subspace W of U, a projection T of V onto U along W is defined by $T(\mathbf{u} + \mathbf{w}) = \mathbf{u}$ and hence $T(\mathbf{u}) = T(\mathbf{u} + \mathbf{0}) = \mathbf{u}$ for any $\mathbf{u} \in U$ and $\mathbf{w} \in W$. Therefore, every projection T of V is *idempotent*, that is

$$T \circ T = T.$$

It gives an algebraic characterization of a linear transformation to be a projection.

Theorem 5.12 *A linear transformation $T : V \to V$ is a projection if and only if it is idempotent, i.e., $T = T^2$ ($= T \circ T$ by definition).*

Proof: ($\Rightarrow$) has already been shown.

($\Leftarrow$): Suppose $T^2 = T$. It suffices to show that $V = \text{Im}(T) \oplus \text{Ker}(T)$ and $T(\mathbf{u} + \mathbf{w}) = \mathbf{u}$ for any $\mathbf{u} + \mathbf{w} \in \text{Im}(T) \oplus \text{Ker}(T)$.

First, show that $\text{Im}(T) \cap \text{Ker}(T) = \{\mathbf{0}\}$ and $V = \text{Im}(T) + \text{Ker}(T)$ for the direct decomposition. Indeed, if $\mathbf{y} \in \text{Im}(T) \cap \text{Ker}(T)$, then there exists $\mathbf{x} \in V$ such that $T(\mathbf{x}) = \mathbf{y}$ and $T(\mathbf{y}) = \mathbf{0}$. It implies

$$\mathbf{y} = T(\mathbf{x}) = T^2(\mathbf{x}) = T(T(\mathbf{x})) = T(\mathbf{y}) = \mathbf{0}.$$

The identity $T^2 = T$ also gives that $T(\mathbf{v}) \in \text{Im}(T)$ and $\mathbf{v} - T(\mathbf{v}) \in \text{Ker}(T)$ for any $\mathbf{v} \in V$. It implies

$$V = \text{Im}(T) + \text{Ker}(T).$$

Now, it is clear that $T(\mathbf{u}+\mathbf{w}) = T(\mathbf{u})+T(\mathbf{w}) = T(\mathbf{u}) = \mathbf{u}$ for any $\mathbf{u}+\mathbf{w} \in \text{Im}(T) \oplus \text{Ker}(T)$. □

For any projection $T : V \to V$, it is shown that $V = \text{Im}(T) \oplus \text{Ker}(T)$ and T is a projection of V onto the subspace $\text{Im}(T)$ along $\text{Ker}(T)$.

Next corollary shows that $id_V - T$ is a projection of V onto $\text{Ker}(T)$ along $\text{Im}(T)$, where id_V is the identity transformation on V.

Corollary 5.12.1 *A linear transformation $T : V \to V$ is a projection if and only if $id_V - T$ is a projection. Moreover, if T is the projection of V onto a subspace U along W, then $id_V - T$ is the projection of V onto W along U.*

Proof: For the identity transformation id_V, $id_V - T$ is a projection because

$$(id_V - T) \circ (id_V - T) = (id_V - T) - (T - T^2) = id_V - T.$$

Moreover, $\text{Im}(id_V - T) = \text{Ker}(T)$ and $\text{Ker}(id_V - T) = \text{Im}(T)$. □

Problem 112 Which of the following matrices are projections as linear transformations?

$$A = \begin{bmatrix} 1 & 0 \\ 0 & 1 \end{bmatrix}, \ B = \begin{bmatrix} 1 & 1 \\ 0 & 0 \end{bmatrix}, \ C = \begin{bmatrix} 1 & 0 \\ 1 & 0 \end{bmatrix}, \ D = \begin{bmatrix} 0 & 1 \\ 0 & 1 \end{bmatrix}, \ E = \begin{bmatrix} 0 & 1 \\ 1 & 0 \end{bmatrix}.$$

Hint: Check each matrix whether it is idempotent or not.

Problem 113 For $V = U \oplus W$, let T_U denote the projection of V onto U along W, and let T_W denote the projection of V onto W along U. Prove the following.

(1) For any $\mathbf{x} \in V$, $\mathbf{x} = T_U(\mathbf{x}) + T_W(\mathbf{x})$.
(2) $T_U \circ (id_V - T_U) = \mathbf{0}$.
(3) $T_U \circ T_W = T_W \circ T_U = \mathbf{0}$.
(4) For any projection $T : V \to V$, $\mathrm{Im}(id_V - T) = \mathrm{Ker}(T)$ and $\mathrm{Ker}(id_V - T) = \mathrm{Im}(T)$.

5.6 Orthogonal and oblique projections of an inner product space

Let U be a subspace of a vector space V. As shown in Example 88, there may exist infinitely many different projections of V onto U, depending on a choice of complementary subspace of U.

However, if V is an inner product space, there is a natural choice of complementary subspace W, called the *orthogonal complement* of U, along which the projection of V onto U is called the *orthogonal projection.*

To show this, we first extend the orthogonality of two vectors to an orthogonality of two subspaces.

Definition 5.13 Let U and W be subspaces of an inner product space V.

(1) Two subspaces U and W are said to be **orthogonal**, written $U \perp W$, if $\langle \mathbf{u}, \mathbf{w} \rangle = 0$ for every $\mathbf{u} \in U$ and every $\mathbf{w} \in W$.
(2) The set of vectors in V which are orthogonal to every vector in U is called the **orthogonal complement** of U, denoted by $U^\perp$, *i.e.,*

$$U^\perp = \{\mathbf{v} \in V : \langle \mathbf{v}, \mathbf{u} \rangle = 0 \text{ for all } \mathbf{u} \in U\}.$$

Obviously, $U^\perp$ is a subspace of V, and $\mathbf{v} \in U^\perp$ if and only if $\langle \mathbf{v}, \mathbf{u} \rangle = 0$ for every $\mathbf{u} \in \beta$, where β is any basis for U. Moreover, $W \perp U$ if and only if $W \subseteq U^\perp$.

Theorem 5.14 *Let U be a subspace of an inner product space V. Then*

(1) $(U^\perp)^\perp = U$.

(2) $V = U \oplus U^{\perp}$: *that is, for each* $\mathbf{x} \in V$, *there exist unique* $\mathbf{x}_U \in U$ *and unique* $\mathbf{x}_{U^\perp} \in U^{\perp}$ *such that* $\mathbf{x} = \mathbf{x}_U + \mathbf{x}_{U^\perp}$. *This is called the* **orthogonal decomposition** *of* V *(or of* $\mathbf{x}$*) by* U.

Proof: Let $\dim U = m$. Take an orthonormal basis for U, say $\alpha = \{\mathbf{u}_1, \mathbf{u}_2, \ldots, \mathbf{u}_m\}$, and extend it to an orthonormal basis for V, say $\beta = \{\mathbf{u}_1, \mathbf{u}_2, \ldots, \mathbf{u}_m, \mathbf{u}_{m+1}, \ldots, \mathbf{u}_n\}$. This is always possible by the Gram-Schmidt orthogonalization. Then, $\gamma = \{\mathbf{u}_{m+1}, \ldots, \mathbf{u}_n\}$ forms an (orthonormal) basis for $U^{\perp}$, which gives $(U^{\perp})^{\perp} = U$ and $V = U \oplus U^{\perp}$.

In this case, for any $\mathbf{x} \in V$,

$$\begin{aligned} \mathbf{x}_U &= \langle \mathbf{u}_1, \mathbf{x}\rangle \mathbf{u}_1 + \langle \mathbf{u}_2, \mathbf{x}\rangle \mathbf{u}_2 + \cdots + \langle \mathbf{u}_m, \mathbf{x}\rangle \mathbf{u}_m \quad \text{and} \\ \mathbf{x}_{U^\perp} &= \langle \mathbf{u}_{m+1}, \mathbf{x}\rangle \mathbf{u}_{m+1} + \langle \mathbf{u}_{m+2}, \mathbf{x}\rangle \mathbf{u}_{m+2} + \cdots + \langle \mathbf{u}_n, \mathbf{x}\rangle \mathbf{u}_n, \end{aligned}$$

giving $\mathbf{x} = \mathbf{x}_U + \mathbf{x}_{U^\perp}$ by Theorem 5.9. The uniqueness comes from these identities. □

Definition 5.15 Let U be a subspace of an inner product space V. The **orthogonal projection** of V onto the subspace U is defined to be a projection of V onto the subspace U along its orthogonal complement $U^{\perp}$ and is denoted by Proj_U.

A projection which is not orthogonal is called an **oblique projection**.

Because of the *uniqueness of orthogonal projection* of V onto the subspace U, it is called *the* orthogonal projection of V onto U, and denote it by Proj_U to distinguish it from the others.

Let $\{\mathbf{u}_1, \mathbf{u}_2, \ldots, \mathbf{u}_m\}$ be any orthonormal basis for U. Then the (unique) orthogonal projection Proj_U of V onto the subspace U is given by

$$\mathrm{Proj}_U(\mathbf{x}) = \langle \mathbf{u}_1, \mathbf{x}\rangle \mathbf{u}_1 + \langle \mathbf{u}_2, \mathbf{x}\rangle \mathbf{u}_2 + \cdots + \langle \mathbf{u}_m, \mathbf{x}\rangle \mathbf{u}_m$$

for any $\mathbf{x} \in V$, by the orthogonal decomposition of $\mathbf{x}$ in Theorem 5.14.

Example 89 (*The orthogonal projection of* $\mathbb{R}^3$ *onto the* xy*-plane is unique*) In the Euclidean 3-space $\mathbb{R}^3$, let U be the xy-plane with the orthonormal basis $\alpha = \{\mathbf{e}_1, \mathbf{e}_2\}$. Then, $\mathrm{Proj}_U(\mathbf{x}) = (\mathbf{e}_1 \cdot \mathbf{x})\mathbf{e}_1 + (\mathbf{e}_2 \cdot \mathbf{x})\mathbf{e}_2$ is the orthogonal projection of $\mathbb{R}^3$ onto the xy-plane in a usual sense in geometry, and $\mathbf{x} - \mathrm{Proj}_U(\mathbf{x}) \in U^{\perp}$, which is the z-axis. It actually means that

$$\mathrm{Proj}_U(x_1, x_2, x_3) = (x_1, x_2, 0) \quad \text{and} \quad \mathbf{x} - \mathrm{Proj}_U(\mathbf{x}) = (0, 0, x_3)$$

for any $\mathbf{x} = (x_1, x_2, x_3)$ in $\mathbb{R}^3$. □

Example 90 (*The orthogonal projection of $\mathbb{R}^2$ onto the x-axis is unique*) As in Example 88, let X, Y, and Z be the one-dimensional subspaces of the Euclidean 2-space $\mathbb{R}^2$ spanned by the vectors $\mathbf{e}_1$, $\mathbf{e}_2$, and $\mathbf{d} = -\mathbf{e}_1 + \mathbf{e}_2 = (-1, 1)$, respectively. Then clearly $Y = X^\perp$ and $Z \neq X^\perp$.

And, for the projections T_X and S_X of $\mathbb{R}^2$ given in Example 88, T_X is the *orthogonal projection*, but S_X is not, so that $T_X = \mathrm{Proj}_X$ and $S_X \neq \mathrm{Proj}_X$. In fact, S_X is one of *oblique projections*. □

Theorem 5.16 *Let U be a subspace of an inner product space V and let $\mathbf{x} \in V$. Then, the orthogonal projection $\mathrm{Proj}_U(\mathbf{x})$ of $\mathbf{x}$ onto U satisfies*

$$\|\mathbf{x} - \mathrm{Proj}_U(\mathbf{x})\| \leq \|\mathbf{x} - \mathbf{y}\|$$

for all $\mathbf{y} \in U$. The equality holds if and only if $\mathbf{y} = \mathrm{Proj}_U(\mathbf{x})$.

Proof: By the definition of Proj_U, for any vector $\mathbf{x} \in V$, we have $\mathrm{Proj}_U(\mathbf{x}) \in U$ and $\mathbf{x} - \mathrm{Proj}_U(\mathbf{x}) \in U^\perp$. Thus, for all $\mathbf{y} \in U$,

$$\begin{aligned} \|\mathbf{x} - \mathbf{y}\|^2 &= \|(\mathbf{x} - \mathrm{Proj}_U(\mathbf{x})) + (\mathrm{Proj}_U(\mathbf{x}) - \mathbf{y})\|^2 \\ &= \|\mathbf{x} - \mathrm{Proj}_U(\mathbf{x})\|^2 + \|\mathrm{Proj}_U(\mathbf{x}) - \mathbf{y}\|^2 \\ &\geq \|\mathbf{x} - \mathrm{Proj}_U(\mathbf{x})\|^2, \end{aligned}$$

where the second equality comes from the Pythagorean theorem for the orthogonality

$$(\mathbf{x} - \mathrm{Proj}_U(\mathbf{x})) \perp (\mathrm{Proj}_U(\mathbf{x}) - \mathbf{y})$$

(because $\mathbf{x} - \mathrm{Proj}_U(\mathbf{x}) \in U^\perp$ and $\mathrm{Proj}_U(\mathbf{x}) - \mathbf{y} \in U$). See Figure 5.2. □

It follows from Theorem 5.16 that *the orthogonal projection $\mathrm{Proj}_U(\mathbf{x})$ of $\mathbf{x}$ is the unique vector in U that is closest to $\mathbf{x}$* in the sense that it minimizes the distance from $\mathbf{x}$ to the vectors in U. It also shows that the vector $\mathrm{Proj}_U(\mathbf{x})$ is independent of the choice of an orthonormal basis for the subspace U. See Figure 5.2 again.

The orthogonal projection Proj_U of the Euclidean space $\mathbb{R}^n$ onto a subspace U appears quite often in applied science and engineering problems.

Problem 114 Let U and W be subspaces of an inner product space V. Show that

(1) If $U \perp W$, then $U \cap W = \{\mathbf{0}\}$. (2) $U \subseteq W$ if and only if $W^\perp \subseteq U^\perp$.

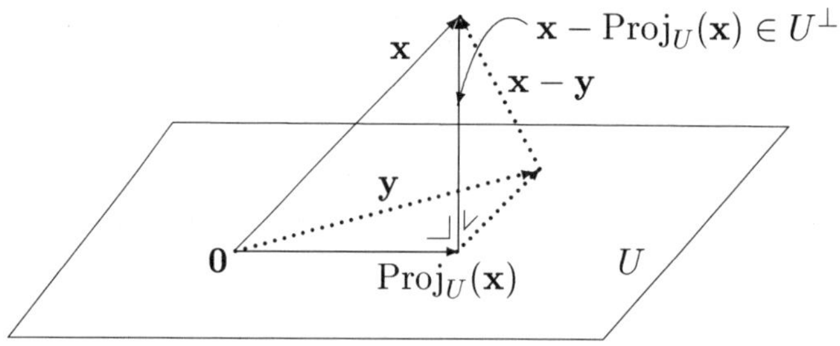

Figure 5.2: The orthogonal projection Proj_U

Problem 115 For a subspace U of an inner product space V, show that

(1) Proj_U is a linear map,
(2) $\mathrm{Proj}_U \circ \mathrm{Proj}_U = \mathrm{Proj}_U$,
(3) $\mathrm{Proj}_{U^{\perp}} = id_V - \mathrm{Proj}_U$.

Hint: (2) $\mathrm{Proj}_U(\mathbf{u}_i) = \mathbf{u}_i$ for every basic vector $\mathbf{u}_i$ in U. For (3), use Corollary 5.12.1.

Problem 116 Find the point on the plane $x - y - z = 0$ that is closest to $\mathbf{x} = (1, 2, 0)$.
Hint: Find an orthonormal basis for the plane and use the orthogonal projection.

5.7 Orthogonal projection matrices and Oblique projection matrices

Every projection of a vector space V onto a subspace is one of linear transformations, hence it has a matrix representation if a basis for the space V is given.

If V is an inner product space, every orthogonal or oblique projection of V also has its matrix representation if a basis for the space V is given.

In this section, we restrict our discussion to the Euclidean n-space $\mathbb{R}^n$ with the dot product and with the standard (ordered) basis α even though the entire discussion in this section applies in the same way to any inner product space.

Definition 5.17 The matrix representation of a projection $T_U : \mathbb{R}^n \to \mathbb{R}^n$ of the Euclidean space $\mathbb{R}^n$ onto a subspace U (along a complementary subspace W) with respect to the standard (ordered) basis is called a **projection matrix** of a projection T_U. This is also sometimes called a **projection matrix** of $\mathbb{R}^n$ onto U.

A projection matrix of an orthogonal (oblique, respectively) projection is called an **orthogonal projection matrix** (an **oblique projection matrix**, respectively).

By Theorem 5.12, *a square matrix A is a projection matrix if and only if it is idempotent, i.e., $A^2 = A$.*

Example 91 (*Orthogonal projection matrices and Oblique projection matrices*)

(1) $\begin{bmatrix} 1 & 0 \\ 1 & 0 \end{bmatrix}$ is idempotent and hence it is a projection matrix. This projects a vector of $\mathbb{R}^2$ onto the diagonal line $y = x$ along the y-axis; while $\begin{bmatrix} 0 & 1 \\ 0 & 1 \end{bmatrix}$ is the projection matrix which projects a vector of $\mathbb{R}^2$ onto the same diagonal line $y = x$ along the x-axis.

Both projections *are not orthogonal projections* and both matrices are not symmetric matrices.

(2) $A = \begin{bmatrix} 1/2 & 1/2 \\ 1/2 & 1/2 \end{bmatrix}$ is idempotent and hence it is a projection matrix. The matrix A projects *orthogonally* a vector of $\mathbb{R}^2$ onto the diagonal line $y = x$. Hence, A is an orthogonal projection matrix and it is clearly a symmetric matrix.

(3) Let $\alpha = \{\mathbf{e}_1, \mathbf{e}_2, \ldots, \mathbf{e}_n\}$ be the standard basis for the n-space $\mathbb{R}^n$. Then, the $n \times n$ matrix $\mathbf{e}_1\mathbf{e}_1^T + \mathbf{e}_2\mathbf{e}_2^T + \mathbf{e}_3\mathbf{e}_3^T$ is idempotent and hence it is a projection matrix. In fact, for any $\mathbf{x} = (x_1, x_2, x_3, \ldots, x_n)$ in $\mathbb{R}^n$,

$$\begin{aligned} (\mathbf{e}_1\mathbf{e}_1^T + \mathbf{e}_2\mathbf{e}_2^T + \mathbf{e}_3\mathbf{e}_3^T)\mathbf{x} &= (\mathbf{e}_1^T\mathbf{x})\mathbf{e}_1 + (\mathbf{e}_2^T\mathbf{x})\mathbf{e}_2 + (\mathbf{e}_3^T\mathbf{x})\mathbf{e}_3 \\ &= (x_1, x_2, x_3, 0, \ldots, 0) \\ &= \mathrm{Proj}_{\mathbb{R}^3}(\mathbf{x}), \end{aligned}$$

which is the *orthogonal projection* of the n-space $\mathbb{R}^n$ onto the subspace $\mathbb{R}^3$, Hence, the $n \times n$ matrix $\mathbf{e}_1\mathbf{e}_1^T + \mathbf{e}_2\mathbf{e}_2^T + \mathbf{e}_3\mathbf{e}_3^T$ is an orthogonal projection matrix, and it is also symmetric. □

In Example 91(2),(3), the orthogonal projection matrices are symmetric and idempotent. It will be shown later in Subsection 5.10.4 that *a square matrix A is an orthogonal projection matrix if and only if it is symmetric and idempotent.*

For a better understanding of orthogonal projection matrices, let $\mathbf{u} \in \mathbb{R}^n$ be a unit vector in the Euclidean n-space, and let $U = \{r\mathbf{u} \ : \ r \in \mathbb{R}\}$ be a one-dimensional subspace spanned by a unit vector $\mathbf{u}$. Then for a vector $\mathbf{x}$ in $\mathbb{R}^n$, the orthogonal projection of $\mathbf{x}$ onto U is

$$\mathrm{Proj}_U(\mathbf{x}) = (\mathbf{u} \cdot \mathbf{x})\mathbf{u} = (\mathbf{u}^T\mathbf{x})\mathbf{u} = \mathbf{u}(\mathbf{u}^T\mathbf{x}) = (\mathbf{u}\mathbf{u}^T)\mathbf{x}.$$

(Here, the last two equalities come from the facts that $\mathbf{u} \cdot \mathbf{x} = \mathbf{u}^T\mathbf{x}$ is a scalar and the associativity of the matrix product $\mathbf{u}\mathbf{u}^T\mathbf{x}$, respectively.)

This equation shows that the matrix representation of the orthogonal projection Proj_U with respect to the standard (ordered) basis α is

$$[\mathrm{Proj}_U]_\alpha = \mathbf{u}\mathbf{u}^T = \begin{bmatrix} u_1 \\ \vdots \\ u_n \end{bmatrix} [u_1 \ \cdots \ u_n] = \begin{bmatrix} |u_1|^2 & \cdots & u_1 u_n \\ \vdots & \ddots & \vdots \\ u_n u_1 & \cdots & |u_n|^2 \end{bmatrix},$$

which is symmetric and idempotent.

If U is an m-dimensional subspace of $\mathbb{R}^n$ with an orthonormal basis $\{\mathbf{u}_1, \mathbf{u}_2, \ldots, \mathbf{u}_m\}$, then for any $\mathbf{x} \in \mathbb{R}^n$,

$$\begin{aligned} \mathrm{Proj}_U(\mathbf{x}) &= (\mathbf{u}_1 \cdot \mathbf{x})\mathbf{u}_1 + (\mathbf{u}_2 \cdot \mathbf{x})\mathbf{u}_2 + \cdots + (\mathbf{u}_m \cdot \mathbf{x})\mathbf{u}_m \\ &= \mathbf{u}_1(\mathbf{u}_1^T\mathbf{x}) + \mathbf{u}_2(\mathbf{u}_2^T\mathbf{x}) + \cdots + \mathbf{u}_m(\mathbf{u}_m^T\mathbf{x}) \\ &= (\mathbf{u}_1\mathbf{u}_1^T + \mathbf{u}_2\mathbf{u}_2^T + \cdots + \mathbf{u}_m\mathbf{u}_m^T)\mathbf{x}. \end{aligned}$$

Thus, the matrix representation of the orthogonal projection Proj_U with respect to the standard (ordered) basis α is

$$[\mathrm{Proj}_U]_\alpha = \mathbf{u}_1\mathbf{u}_1^T + \mathbf{u}_2\mathbf{u}_2^T + \cdots + \mathbf{u}_m\mathbf{u}_m^T, \tag{5.2}$$

which is also symmetric, idempotent, and a sum of rank one matrices. (See Exercise **3.23** for the rank).

Now, the reader should be noticed that an orthogonal projection matrix is *NOT* one of the orthogonal matrices. This is a fact for a square matrix A.

- A is a *projection matrix* if and only if it is idempotent,
- A is an *orthogonal projection matrix* if and only if it is symmetric and idempotent,
- A is an *oblique projection matrix* if and only if it is idempotent but not symmetric, and
- A is an *orthogonal matrix* if and only if it is invertible and $A^{-1} = A^T$.

Example 92 (*Distance from a point to a line in* $\mathbb{R}^2$) For a given line $L : ax + by + c = 0$, $(a, b) \neq (0, 0)$ and for a given point $P = (x_0, y_0)$ in the plane $\mathbb{R}^2$, determine

(1) the distance d from the point P to the line L,

(2) the orthogonal projection matrix $[\mathrm{Proj}_U]_\alpha$, where U is the subspace of $\mathbb{R}^2$ consisting of all vectors perpendicular to the line L, and

(3) the orthogonal projection matrix $[\mathrm{Proj}_W]_\alpha$, where W is the subspace of $\mathbb{R}^2$ consisting of all vectors parallel to the line L, where α is the standard (ordered) basis for $\mathbb{R}^2$.

Solution: Firstly, note that the line L is not a subspace of $\mathbb{R}^2$ if $c \neq 0$, and $U \perp W$.

(1) For any two distinct points $Q_1 = (x_1, y_1)$ and $Q_2 = (x_2, y_2)$ on the line L, the vector $\overrightarrow{Q_1Q_2}$ is parallel to the line L and the equality $a(x_2 - x_1) + b(y_2 - y_1) = 0$ implies that the nonzero vector $\mathbf{n} = (a, b)$ is perpendicular to L, *i.e.*, $\overrightarrow{Q_1Q_2} \perp \mathbf{n}$.

For a given point $P = (x_0, y_0)$ in $\mathbb{R}^2$, the distance d from the point P to the line L is simply the length of the orthogonal projection of $\overrightarrow{Q_1P}$ onto the subspace $\langle \mathbf{n} \rangle$ spanned by $\mathbf{n}$. Thus,

$$\begin{aligned} d &= \|\mathrm{Proj}_{\langle \mathbf{n} \rangle}(\overrightarrow{Q_1P})\| \\ &= \left| \left\langle \overrightarrow{Q_1P}, \frac{\mathbf{n}}{\|\mathbf{n}\|} \right\rangle \right| \qquad \text{(the dot product)} \\ &= \frac{|a(x_0 - x_1) + b(y_0 - y_1)|}{\sqrt{a^2 + b^2}} \\ &= \frac{|ax_0 + by_0 + c|}{\sqrt{a^2 + b^2}}. \end{aligned}$$

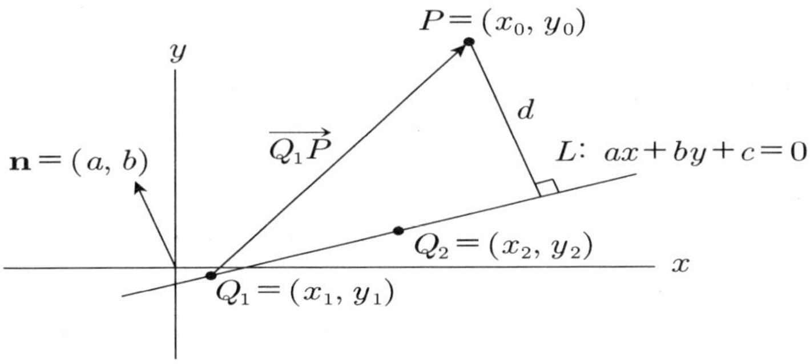

Figure 5.3: Distance d from a point P to a line L in the plane

Here, the last equality is due to the fact that the point Q_1 is on the line, *i.e.*, $ax_1 + by_1 + c = 0$.

(2) It is clear that the subspace U of $\mathbb{R}^2$ consisting of all vectors perpendicular to the line L is spanned by the vector $\mathbf{n} = (a, b)$. By normalizing it, let $\mathbf{u} = \frac{\mathbf{n}}{\|\mathbf{n}\|} = \frac{1}{\sqrt{a^2+b^2}}(a, b)$. Then the projection matrix $[\mathrm{Proj}_U]_\alpha$ onto U is

$$\mathbf{u}\mathbf{u}^T = \frac{1}{a^2+b^2}\begin{bmatrix} a \\ b \end{bmatrix}[a\ b] = \frac{1}{a^2+b^2}\begin{bmatrix} a^2 & ab \\ ab & b^2 \end{bmatrix}.$$

(3) Clearly, the subspace W consisting of all vectors parallel to the line L is the line $ax + by = 0$ passing through the origin. One can choose $\mathbf{w} = \frac{1}{\sqrt{a^2+b^2}}(-b, a)$ as an orthonormal basis vector for W. Then the projection matrix $[\mathrm{Proj}_W]_\alpha$ is

$$\mathbf{w}\mathbf{w}^T = \frac{1}{a^2+b^2}\begin{bmatrix} -b \\ a \end{bmatrix}[-b\ a] = \frac{1}{a^2+b^2}\begin{bmatrix} b^2 & -ab \\ -ab & a^2 \end{bmatrix}.$$

The orthogonal projection matrices $\mathbf{u}\mathbf{u}^T$ and $\mathbf{w}\mathbf{w}^T$ are both idempotent and symmetric. □

Example 93 (*The volume of the parallelepiped $\mathcal{P}(A)$*) In Theorem 2.19, it is stated that *for any square or non-square (but wide rectangular) $m \times n$ matrix A, the volume of the parallelepiped $\mathcal{P}(A)$ spanned by the row vectors $\mathbf{r}_1, \mathbf{r}_2, \ldots, \mathbf{r}_m$ of A is*

$$\mathrm{vol}(\mathcal{P}(A)) = \sqrt{\det(AA^T)} \quad \textit{or equivalently} \quad \mathrm{vol}(\mathcal{P}(A))^2 = \det(AA^T).$$

We verify this formula here as an application case of the orthogonal projection.

Proof: By the Cauchy-Binet formula, $\det(AA^T) \geq 0$ always, and $\sqrt{\det(AA^T)}$ is well defined. (See Corollary 2.18.1).

We use mathematical induction on m. In the case $m = 1$, the parallelepiped is reduced to a single row vector $\mathbf{r}$ and its volume is just its length $\|\mathbf{r}\|$. Clearly, $\|\mathbf{r}\|^2 = x_1^2 + x_2^2 + \cdots + x_n^2 = \mathbf{r}\mathbf{r}^T$ as usual.

With the assumption that it is true for any wide matrix B having $m - 1$ row vectors, let A be an $m \times n$ matrix with row vectors $\mathbf{r}_1, \mathbf{r}_2, \ldots, \mathbf{r}_m$, but still $m \leq n$. Then, the height h of $\mathbf{r}_1$ from the (sub)space W spanned by the row vectors $\mathbf{r}_2, \ldots, \mathbf{r}_m$ is just $\|\mathbf{r}_1 - \mathrm{Proj}_W(\mathbf{r}_1)\|$.

Now, let B be the submatrix of A obtained from A by deleting the first row $\mathbf{r}_1$. Then, by the definition of the volume, $\mathrm{vol}(\mathcal{P}(A))$ is h times the volume $\mathrm{vol}(\mathcal{P}(B))$

of the base parallelepiped $\mathcal{P}(B)$. Hence, one can replace the first row $\mathbf{r}_1$ by $\mathbf{r}_1 - \mathrm{Proj}_W(\mathbf{r}_1)$ in the matrix A without variation in the value $\mathrm{vol}(\mathcal{P}(A))$.

Also, in the Cauchy-Binet formula

$$\det(AA^T) = \sum_{1 \le k_1 < k_2 < \cdots < k_n \le m} (\det A_{k_1,k_2,\ldots,k_n})^2,$$

one can also replace the first row $\mathbf{r}_1$ by $\mathbf{r}_1 - \mathrm{Proj}_W(\mathbf{r}_1)$ without variation in the value $\det(AA^T)$ because the (type III) elementary row operation that adds a constant multiple of one row to another row leaves the determinant unchanged (in each $\det A_{k_1,k_2,\ldots,k_n}$).

As a consequence, one can assume from the beginning that the first row vector $\mathbf{r}_1$ is orthogonal to every other row vector $\mathbf{r}_2, \ldots, \mathbf{r}_m$ without any loss of generality. Then $h = \|\mathbf{r}_1\|$ and

$$\begin{aligned} AA^T &= \begin{bmatrix} -- & \mathbf{r}_1 & -- \\ -- & \mathbf{r}_2 & -- \\ & \vdots & \\ -- & \mathbf{r}_m & -- \end{bmatrix} \begin{bmatrix} | & | & & | \\ \mathbf{r}_1 & \mathbf{r}_2 & \cdots & \mathbf{r}_m \\ | & | & & | \end{bmatrix} \\ &= \left[\begin{array}{c|ccc} \mathbf{r}_1\mathbf{r}_1^T & 0 & \cdot & 0 \\ \hline 0 & & & \\ \vdots & & C & \\ 0 & & & \end{array}\right] = \begin{bmatrix} \mathbf{r}_1\mathbf{r}_1^T & \mathbf{0} \\ \mathbf{0} & C \end{bmatrix}, \end{aligned}$$

where

$$C = BB^T = \begin{bmatrix} -- & \mathbf{r}_2 & -- \\ & \vdots & \\ -- & \mathbf{r}_m & -- \end{bmatrix} \begin{bmatrix} | & & | \\ \mathbf{r}_2 & \cdots & \mathbf{r}_m \\ | & & | \end{bmatrix}.$$

Now, by the induction hypothesis, $\mathrm{vol}(\mathcal{P}(B)^2 = \det(BB^T) = \det C$, and then

$$\mathrm{vol}(\mathcal{P}(A))^2 = h^2 \cdot \mathrm{vol}(\mathcal{P}(B))^2 = \mathbf{r}_1\mathbf{r}_1^T \cdot \det C = \det(AA^T). \qquad \square$$

Problem 117 In Example 92, two projection matrices $\mathbf{u}\mathbf{u}^T$ and $\mathbf{w}\mathbf{w}^T$ are given. Show that both matrices are symmetric and idempotent.

Problem 118 Let $\{\mathbf{u}_1, \mathbf{u}_2, \ldots, \mathbf{u}_n\}$ be an orthonormal basis for $\mathbb{R}^n$. Then $\mathbf{u}_i\mathbf{u}_i^T$ is idempotent and hence is a projection for each i. Write $P_i = \mathbf{u}_i\mathbf{u}_i^T$ for each i. Show that

(1) for any $\mathbf{x} \in \mathbb{R}^n$ and $i, j = 1, 2, \ldots, n$,

$$\begin{aligned} P_i\mathbf{x} &= \mathbf{u}_i\mathbf{u}_i^T\mathbf{x} = \langle \mathbf{u}_i, \mathbf{x}\rangle \mathbf{u}_i, \\ P_iP_j &= \mathbf{u}_i\mathbf{u}_i^T\mathbf{u}_j\mathbf{u}_j^T = \langle \mathbf{u}_i, \mathbf{u}_j\rangle \mathbf{u}_i\mathbf{u}_j^T = \begin{cases} 1\mathbf{u}_i\mathbf{u}_i^T = P_i & \text{if } i = j, \\ 0\mathbf{u}_i\mathbf{u}_j^T = \mathbf{0} & \text{if } i \neq j. \end{cases} \end{aligned}$$

(2) $(P_1 + \cdots + P_n)\mathbf{x} = P_1\mathbf{x} + \cdots + P_n\mathbf{x} = \sum_{i=1}^n \langle \mathbf{u}_i, \mathbf{x}\rangle \mathbf{u}_i = \mathbf{x} = id(\mathbf{x})$. That is, $\mathbf{u}_1\mathbf{u}_1^T + \mathbf{u}_2\mathbf{u}_2^T + \cdots + \mathbf{u}_n\mathbf{u}_n^T = I_n$, the identity matrix.

Problem 119 Let $V = P_3(\mathbb{R})$ be the vector space of polynomials of degree ≤ 3 equipped with the inner product $\langle f, g\rangle = \int_0^1 f(x)g(x)\, dx$. Let W be the subspace of V spanned by $\{1, x\}$ and let $f(x) = x^2$. Find the orthogonal projection $\mathrm{Proj}_W(f)$ of f onto W.

Hint: $\{1, 2\sqrt{3}(x - 1/2)\}$ is an orthonormal basis for the subspace W.

5.8 Orthogonal complements of the row and the column spaces

For any $m \times n$ matrix A, the row space $\mathcal{R}(A)$ and the column space $\mathcal{C}(A)$ are subspaces of the domain space $\mathbb{R}^n$ and the target space $\mathbb{R}^m$, respectively.

What is the orthogonal projection of the domain space $\mathbb{R}^n$ onto the subspace $\mathcal{R}(A)$? What is the orthogonal projection of the target space $\mathbb{R}^m$ onto the subspace $\mathcal{C}(A)$?

To answer these questions, one needs to find the orthogonal complement of each subspace in $\mathbb{R}^n$, or in $\mathbb{R}^m$. It will give orthogonal decompositions of the domain space $\mathbb{R}^n$ and the target space $\mathbb{R}^m$, and from these decompositions, one can completely determine the solution set of a consistent system $A\mathbf{x} = \mathbf{b}$.

Lemma 5.18 *For an $m \times n$ matrix A, the null space $\mathcal{N}(A)$ and the row space $\mathcal{R}(A)$ are orthogonal: i.e., $\mathcal{N}(A) \perp \mathcal{R}(A)$ in the Euclidean space $\mathbb{R}^n$.*

Similarly, $\mathcal{N}(A^T) \perp \mathcal{C}(A)$ in $\mathbb{R}^m$.

Proof: Clearly, $\mathbf{w} \in \mathcal{N}(A)$ if and only if $A\mathbf{w} = \mathbf{0}$, or equivalently, for every row vector $\mathbf{r}$ of A, $\mathbf{r} \cdot \mathbf{w} = 0$. For the second statement, do the same with A^T. □

From Lemma 5.18, it follows that

$$\begin{aligned} \mathcal{N}(A) &\subseteq \mathcal{R}(A)^\perp \quad (\text{ or } \mathcal{R}(A) \subseteq \mathcal{N}(A)^\perp) \quad \text{and} \\ \mathcal{N}(A^T) &\subseteq \mathcal{C}(A)^\perp \quad (\text{ or } \mathcal{C}(A) \subseteq \mathcal{N}(A^T)^\perp). \end{aligned}$$

By comparing the dimensions of these subspaces, one can have

$$\begin{aligned}\dim \mathcal{R}(A) + \dim \mathcal{N}(A) &= n &= \dim \mathcal{R}(A) + \dim \mathcal{R}(A)^{\perp},\\ \dim \mathcal{C}(A) + \dim \mathcal{N}(A^T) &= m &= \dim \mathcal{C}(A) + \dim \mathcal{C}(A)^{\perp}.\end{aligned}$$

Here, in each equation, the first equality comes from the Rank-Nullity Theorem and the second equality comes from the orthogonal decomposition mentioned in Theorem 5.14. It concludes that

$$\dim \mathcal{N}(A) = \dim \mathcal{R}(A)^{\perp} \quad \text{and} \quad \dim \mathcal{N}(A^T) = \dim \mathcal{C}(A)^{\perp},$$

which give actually

$$\mathcal{N}(A) = \mathcal{R}(A)^{\perp} \quad \text{and} \quad \mathcal{N}(A^T) = \mathcal{C}(A)^{\perp}.$$

By using the orthogonal decomposition mentioned in Theorem 5.14 again, one can get the following theorem.

Theorem 5.19 *For any $m \times n$ matrix A,*

$$\mathcal{N}(A) \oplus \mathcal{R}(A) = \mathbb{R}^n \ \text{ and } \ \mathcal{N}(A^T) \oplus \mathcal{C}(A) = \mathbb{R}^m.$$

Moreover, both are orthogonal decompositions; $\mathcal{R}(A) \perp \mathcal{N}(A)$, $\mathcal{C}(A) \perp \mathcal{N}(A^T)$.

If rank $A = r$, that is, $\dim \mathcal{R}(A) = r = \dim \mathcal{C}(A)$, then $\dim \mathcal{N}(A) = n - r$ and $\dim \mathcal{N}(A^T) = m - r$. Considering the matrix A as a linear transformation $A : \mathbb{R}^n \to \mathbb{R}^m$, Figure 5.4 depicts Theorem 5.19.

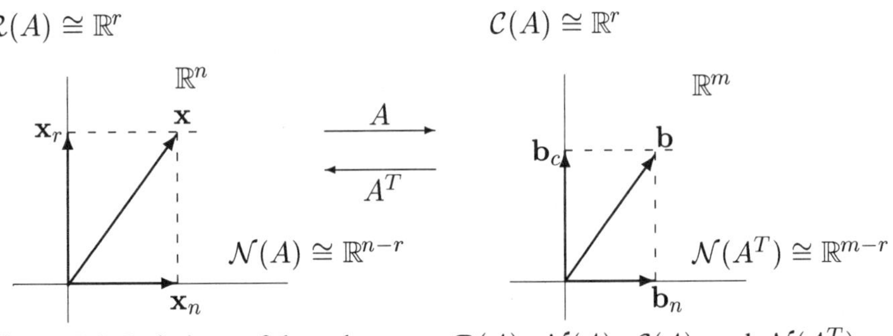

Figure 5.4: Relations of the subspaces $\mathcal{R}(A)$, $\mathcal{N}(A)$, $\mathcal{C}(A)$, and $\mathcal{N}(A^T)$

Corollary 5.19.1 *The set of solutions of a consistent system $A\mathbf{x} = \mathbf{b}$ is precisely $\mathbf{x}_0 + \mathcal{N}(A)$, where $\mathbf{x}_0$ is any solution of $A\mathbf{x} = \mathbf{b}$.*

Proof: Let $\mathbf{x}_0 \in \mathbb{R}^n$ be a solution of a system $A\mathbf{x} = \mathbf{b}$. Now consider the set $\mathbf{x}_0 + \mathcal{N}(A)$, which is just a translation of $\mathcal{N}(A)$ by $\mathbf{x}_0$.

(1) Any vector $\mathbf{x}_0 + \mathbf{n}$ in $\mathbf{x}_0 + \mathcal{N}(A)$ is also a solution; $A(\mathbf{x}_0 + \mathbf{n}) = A\mathbf{x}_0 = \mathbf{b}$.

(2) If $\mathbf{x}$ is another solution, then clearly $\mathbf{x} - \mathbf{x}_0$ belongs to the null space $\mathcal{N}(A)$ so that $\mathbf{x} = \mathbf{x}_0 + \mathbf{n}$ for some $\mathbf{n} \in \mathcal{N}(A)$, *i.e.*, $\mathbf{x} \in \mathbf{x}_0 + \mathcal{N}(A)$. □

As a special case, if rank $A = m$ (so that $m \leq n$), then $\mathcal{C}(A) = \mathbb{R}^m$. Thus, for any $\mathbf{b} \in \mathbb{R}^m$, the system $A\mathbf{x} = \mathbf{b}$ has at least one solution in $\mathbb{R}^n$. (This is the case of Theorem 3.27).

On the other hand, if rank $A = n$ (so that $n \leq m$), then $\mathcal{N}(A) = \{\mathbf{0}\}$ and $\mathcal{R}(A) = \mathbb{R}^n$. Therefore the system $A\mathbf{x} = \mathbf{b}$ has at most one solution, that is, it has a unique solution $\mathbf{x}$ in $\mathcal{R}(A)$ if $\mathbf{b} \in \mathcal{C}(A)$, and has no solutions if $\mathbf{b} \notin \mathcal{C}(A)$. (This is the case of Theorem 3.28). The latter case may occur when $m > n =$ rank A; that is, $\mathcal{N}(A^T)$ is a nontrivial subspace of $\mathbb{R}^m$.

Even if a system $A\mathbf{x} = \mathbf{b}$ is inconsistent and has no solutions, one may try to determine the *best approximate solution* set of $A\mathbf{x} = \mathbf{b}$. This will be discussed later in Subsection 5.10.1 separately.

Problem 120 Let A be any $m \times n$ matrix and let rank $A = r$. Show that for any basis $\{\mathbf{v}_1, \mathbf{v}_2, \ldots, \mathbf{v}_r\}$ for the row space $\mathcal{R}(A)$ of A, their images $A\mathbf{v}_1, A\mathbf{v}_2, \ldots, A\mathbf{v}_r$ are linearly independent in $\mathbb{R}^m$. Similarly, if $\{\mathbf{w}_1, \mathbf{w}_2, \ldots, \mathbf{w}_r\}$ is a basis for the column space of A, then the vectors $A^T\mathbf{w}_1, A^T\mathbf{w}_2, \ldots, A^T\mathbf{w}_r$ are linearly independent in $\mathbb{R}^n$.

Conclude that

the transformation form the row space $\mathcal{R}(A)$ to the column space $\mathcal{C}(A)$ defined by $\mathbf{v} \to A\mathbf{v}$ is an isomorphism, and

the transformation form the column space $\mathcal{C}(A)$ to the row space $\mathcal{R}(A)$ defined by $\mathbf{w} \to A^T\mathbf{w}$ is an isomorphism.

Hint: If $c_1A\mathbf{v}_1 + c_2A\mathbf{v}_2 + \cdots + c_rA\mathbf{v}_r = \mathbf{0}$, then $c_1\mathbf{v}_1 + c_2\mathbf{v}_2 + \cdots + c_r\mathbf{v}_r \in \mathcal{N}(A) \cap \mathcal{R}(A)$.

Problem 121 Prove the following statements.

(1) If $A\mathbf{x} = \mathbf{b}$ and $A^T\mathbf{y} = \mathbf{0}$, then $\mathbf{y}^T\mathbf{b} = 0$, *i.e.*, $\mathbf{y} \perp \mathbf{b}$.
(2) If $A\mathbf{x} = \mathbf{0}$ and $A^T\mathbf{y} = \mathbf{c}$, then $\mathbf{x}^T\mathbf{c} = 0$, *i.e.*, $\mathbf{x} \perp \mathbf{c}$.

Problem 122 Given two vectors $(1, 2, 1, 2)$ and $(0, -1, -1, 1)$ in the Euclidean 4-space $\mathbb{R}^4$, find all vectors in $\mathbb{R}^4$ that are perpendicular to both of them. Do they form a subspace of $\mathbb{R}^4$? If yes, find a basis for this subspace.
Hint: Consider a 2×4 matrix whose row vectors are the given two vectors.

5.9 Orthogonal matrices preserve the lengths, so are isometries

In Chapter 4, it has been shown that a linear transformation can be represented by a matrix, and a matrix is also a kind of linear transformation.

Moreover, in Theorem 5.10, any n-dimension inner product space V can be isomorphic to the Euclidean n-space $\mathbb{R}^n$ by an inner product preserving coordinate isomorphism.

In this section we are mainly interested in matrices A or linear transformations T on the Euclidean n-space $\mathbb{R}^n$ that *preserve the Euclidean inner product*: $A\mathbf{x} \cdot A\mathbf{y} = \mathbf{x} \cdot \mathbf{y}$ for all vectors $\mathbf{x}, \mathbf{y} \in \mathbb{R}^n$.

Let $A = [\mathbf{c}_1 \ \cdots \ \mathbf{c}_n]$ be an $n \times n$ matrix with columns $\mathbf{c}_1, \dots, \mathbf{c}_n$. Then, a simple computation shows that

$$A^T A = \begin{bmatrix} -- & \mathbf{c}_1^T & -- \\ & \vdots & \\ -- & \mathbf{c}_n^T & -- \end{bmatrix} \begin{bmatrix} | & & | \\ \mathbf{c}_1 & \cdots & \mathbf{c}_n \\ | & & | \end{bmatrix} = [\mathbf{c}_i^T \mathbf{c}_j] = [\mathbf{c}_i \cdot \mathbf{c}_j].$$

Hence, if the column vectors are orthonormal, $\mathbf{c}_i^T \mathbf{c}_j = \delta_{ij}$, then $A^T A = I_n$, that is, A^T is a left inverse of A. The converse is also true. Since A is a square matrix, the left inverse is also the right inverse of A, *i.e.*, $AA^T = I_n$. Equivalently, the row vectors of A are orthonormal.

This argument can be characterized as follows.

Lemma 5.20 *For an $n \times n$ matrix A, the following are equivalent.*

(1) *The column vectors of A are orthonormal,*
(2) $A^T A = I_n$,
(3) $A^T = A^{-1}$,
(4) $AA^T = I_n$,
(5) *The row vectors of A are orthonormal.*

Definition 5.21 A square matrix A is called an **orthogonal** matrix if A satisfies any one (and hence all) of the statements in Lemma 5.20.

Clearly, the identity matrix and hence every permutation matrix is an orthogonal matrix, and A is orthogonal if and only if A^T is orthogonal.

Example 94 (*Rotations and reflections are orthogonal*) The matrices

$$A = \begin{bmatrix} \cos\theta & -\sin\theta \\ \sin\theta & \cos\theta \end{bmatrix} \quad \text{and} \quad B = \begin{bmatrix} \cos\theta & \sin\theta \\ \sin\theta & -\cos\theta \end{bmatrix}$$

are orthogonal, and satisfy

$$A^{-1} = A^T = \begin{bmatrix} \cos\theta & \sin\theta \\ -\sin\theta & \cos\theta \end{bmatrix} \quad \text{and} \quad B^{-1} = B^T = \begin{bmatrix} \cos\theta & \sin\theta \\ \sin\theta & -\cos\theta \end{bmatrix}.$$

In fact, the linear transformation $T : \mathbb{R}^2 \to \mathbb{R}^2$ defined by $T(\mathbf{x}) = A\mathbf{x}$ is a rotation through the angle θ, while $S : \mathbb{R}^2 \to \mathbb{R}^2$ defined by $S(\mathbf{x}) = B\mathbf{x}$ is the reflection across the line passing through the origin that forms an angle $\theta/2$ with the positive x-axis. See Exercise **4.24**.

Note that $\det A = 1$ but $\det B = -1$. □

Example 95 (*The determinants of orthogonal matrices are* 1 *or* -1) Let A be an orthogonal matrix and let $\det A = \delta$ be its determinant. Then, since $A^T = A^{-1}$, $\delta = \det A = \det(A^T) = \det(A^{-1}) = 1/\delta$, giving $\delta^2 = 1$. So, $\delta = 1$ or -1. □

Example 96 (*All* 2×2 *orthogonal matrices are rotations or reflections*) Show that every 2×2 orthogonal matrix should be one of the forms

$$\begin{bmatrix} \cos\theta & -\sin\theta \\ \sin\theta & \cos\theta \end{bmatrix} \quad \text{or} \quad \begin{bmatrix} \cos\theta & \sin\theta \\ \sin\theta & -\cos\theta \end{bmatrix}.$$

That is, *any* 2×2 *orthogonal matrix is either a rotation or a reflection.*

Solution: Suppose that $A = \begin{bmatrix} a & b \\ c & d \end{bmatrix}$ is an orthogonal matrix, so that

$$AA^T = I_2 = A^T A.$$

The first equality gives

$$a^2 + b^2 = 1, \quad ac + bd = 0, \quad \text{and} \quad c^2 + d^2 = 1.$$

The second equality $A^T A = I_2$ gives

$$a^2 + c^2 = 1, \; ab + cd = 0, \quad \text{and} \quad b^2 + d^2 = 1.$$

Thus, $b = \pm c$. If $b = -c$, then one can get $a = d$. If $b = c$, then one can get $a = -d$. Now,

$$\text{choose } \theta \text{ so that } a = \cos\theta \text{ and } b = \sin\theta.$$

Actually, it is a fact that any $n \times n$ orthogonal matrix is either a rotation or a reflection of the Euclidean space $\mathbb{R}^n$. □

Problem 123 Find the inverse of each of the following matrices.

$$(1)\ \begin{bmatrix} 1 & 0 & 0 \\ 0 & \cos\theta & \sin\theta \\ 0 & -\sin\theta & \cos\theta \end{bmatrix}, \qquad (2)\ \begin{bmatrix} 1/\sqrt{2} & -1/\sqrt{2} & 0 \\ -1/\sqrt{2} & -1/\sqrt{2} & 0 \\ 0 & 0 & 1 \end{bmatrix}.$$

What are they as linear transformations on $\mathbb{R}^3$; rotations, reflections, or other?

Problem 124 Find all 2×2 orthogonal matrices which transform the square $-1 \le x, y \le 1$ onto itself. How many are there? Compute the determinant of each of them.

Hint: There are four rotations and four reflections.

As shown in Examples 94 and 96, all rotations and reflections on the Euclidean 2-space $\mathbb{R}^2$ are orthogonal, and they intuitively, preserve both the lengths of vectors and the angle between two vectors.

In fact, every orthogonal matrix A preserves the lengths of vectors:

$$\|A\mathbf{x}\|^2 = A\mathbf{x} \cdot A\mathbf{x} = (A\mathbf{x})^T(A\mathbf{x}) = \mathbf{x}^T A^T A\mathbf{x} = \mathbf{x}^T\mathbf{x} = \|\mathbf{x}\|^2.$$

Definition 5.22 Let V and W be two inner product spaces. A linear transformation $T : V \to W$ is called an **isometry**, or an **orthogonal transformation**, if it preserves the lengths of vectors, that is, for every vector $\mathbf{x} \in V$

$$\|T(\mathbf{x})\| = \|\mathbf{x}\|.$$

If $T : V \to W$ is an isometry, then T is injective, since the kernel of T is trivial: $T(\mathbf{x}) = \mathbf{0}$ implies $\|\mathbf{x}\| = \|T(\mathbf{x})\| = 0$.

An isometry needs not to be bijective. But, if $\dim V = \dim W$, then an isometry is bijective and hence becomes an isomorphism. For example, every orthogonal matrix is a bijective isometry and hence an isomorphism as a linear transformation.

The following theorem gives an interesting characterization of an isometry.

Theorem 5.23 *Let $T : V \to W$ be a linear transformation from an inner product space V to another inner product space W. Then, T is an isometry, that is, preserves the lengths if and only if T preserves inner products, that is,*

$$\langle T(\mathbf{x}), T(\mathbf{y})\rangle = \langle \mathbf{x}, \mathbf{y}\rangle$$

for any vectors $\mathbf{x}, \mathbf{y}$ in V.

Proof: Let T be an isometry. Then $\|T(\mathbf{x})\|^2 = \|\mathbf{x}\|^2$ for any $\mathbf{x} \in V$. Hence,

$$\langle T(\mathbf{x}+\mathbf{y}), T(\mathbf{x}+\mathbf{y})\rangle = \|T(\mathbf{x}+\mathbf{y})\|^2 = \|\mathbf{x}+\mathbf{y}\|^2 = \langle \mathbf{x}+\mathbf{y}, \mathbf{x}+\mathbf{y}\rangle$$

for any $\mathbf{x}, \mathbf{y} \in V$. On the other hand,

$$\begin{aligned} \langle T(\mathbf{x}+\mathbf{y}), T(\mathbf{x}+\mathbf{y})\rangle &= \langle T(\mathbf{x}), T(\mathbf{x})\rangle + 2\langle T(\mathbf{x}), T(\mathbf{y})\rangle + \langle T(\mathbf{y}), T(\mathbf{y})\rangle, \\ \langle \mathbf{x}+\mathbf{y}, \mathbf{x}+\mathbf{y}\rangle &= \langle \mathbf{x}, \mathbf{x}\rangle + 2\langle \mathbf{x}, \mathbf{y}\rangle + \langle \mathbf{y}, \mathbf{y}\rangle, \end{aligned}$$

from which one can get $\langle T(\mathbf{x}), T(\mathbf{y})\rangle = \langle \mathbf{x}, \mathbf{y}\rangle$.

The converse is clear by choosing $\mathbf{y} = \mathbf{x}$. □

Since $d(\mathbf{x}, \mathbf{y}) = \|\mathbf{x} - \mathbf{y}\|$ for any $\mathbf{x}$ and $\mathbf{y}$ in V, one can get the following.

Corollary 5.23.1 *A linear transformation $T : V \to W$ is an isometry if and only if it preserves the distance,*

$$d(T(\mathbf{x}), T(\mathbf{y})) = d(\mathbf{x}, \mathbf{y})$$

for any $\mathbf{x}$ and $\mathbf{y}$ in V.

If θ is the angle between two nonzero vectors $\mathbf{x}$ and $\mathbf{y}$ in an inner product space V, then for any isometry $T : V \to V$,

$$\cos\theta = \frac{\langle \mathbf{x}, \mathbf{y}\rangle}{\|\mathbf{x}\|\|\mathbf{y}\|} = \frac{\langle T\mathbf{x}, T\mathbf{y}\rangle}{\|T\mathbf{x}\|\|T\mathbf{y}\|}.$$

It gives the following.

Corollary 5.23.2 *An isometry preserves angles.*

The converse of Corollary 5.23.2 is not true in general. A linear transformation $T(\mathbf{x}) = 2\mathbf{x}$ on the Euclidean space $\mathbb{R}^n$ preserves the angle but does not preserve the lengths of vectors (*i.e.,* it's not an isometry).

Such a linear transformation is called a **dilation** or a **scaling**.

Going back to a matrix, one has further characterizations of an orthogonal matrix.

Theorem 5.24 *For an $n \times n$ matrix A, the following are equivalent.*

- **(1)** *A is an orthogonal matrix,*
- **(2)** *A is an isometry: $\|A\mathbf{x}\| = \|\mathbf{x}\|$ for any $\mathbf{x} \in \mathbb{R}^n$,*
- **(3)** *A preserves the dot product: $A\mathbf{x} \cdot A\mathbf{y} = \mathbf{x} \cdot \mathbf{y}$ for any $\mathbf{x}, \mathbf{y} \in \mathbb{R}^n$,*
- **(4)** *A maps an orthonormal bases α to another orthonormal bases. In other words, A maps a rectangular coordinate system to another rectangular coordinate system.*

Proof: (1) $\Rightarrow$ (2) $\Rightarrow$ (3) have been shown already.
(3) $\Rightarrow$ (4): Let $\alpha = \{\mathbf{v}_1, \mathbf{v}_2, \ldots, \mathbf{v}_n\}$ be an orthonormal basis for $\mathbb{R}^n$. Then $A\mathbf{v}_1, A\mathbf{v}_2, \ldots, A\mathbf{v}_n$ are also orthonormal by (3).
(4) $\Rightarrow$ (1): Since the standard basis $\{\mathbf{e}_1, \mathbf{e}_2, \ldots, \mathbf{e}_n\}$ for $\mathbb{R}^n$ is orthonormal and $A\mathbf{e}_i = \mathbf{c}_i$ is the i-th column of A, the columns of A are orthonormal by (4). □

It is shown that any orthogonal matrix is a bijective isometry as the linear transformation $T(\mathbf{x}) = A\mathbf{x}$. Now, one might ask whether every isometry $T : V \to V$ has an orthogonal matrix representation $[T]_\alpha^\beta$ or not.

The following theorem shows that the answer is affirmative if α and β are both orthonormal bases.

Theorem 5.25 *Let $T : V \to W$ be a linear transformation from an inner product space V to another inner product space W of the same dimension. Let α and β be orthonormal bases for V and W, respectively. Then, T is an isometry if and only if the matrix representation $[T]_\alpha^\beta$ of T is an orthogonal matrix.*

Proof: We visit the commutative diagram in Figure 4.2 on page 163 again: If V and W are inner product spaces of the same dimension, say n, and α and β are orthonormal bases for V and W, respectively, then two coordinate isomorphisms $\Phi : V \to \mathbb{R}^n$ and $\Psi : W \to \mathbb{R}^n$ both preserve the inner product by Theorem 5.10, that is, both of them are bijective isometries if α and β are orthonormal bases.

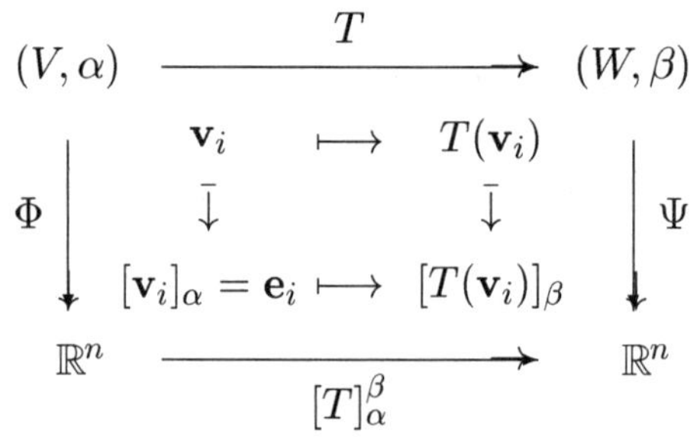

Figure 5.5: Are Φ and Ψ isometries?

Hence, by the commutativity of the diagram, the transformation $T : V \to W$ preserves the inner product if and only if the matrix $[T]_\alpha^\beta$ preserves the inner product.

In other words, $T : V \to W$ is an isometry if and only if the matrix $[T]_\alpha^\beta$ is an orthogonal matrix. □

In Theorem 5.25, if α or β is not orthonormal, then it cannot be true anymore. As an example, let $id : \mathbb{R}^n \to \mathbb{R}^n$ be the identity transformation. Then id is clearly an isometry. Take two bases as $\alpha = \{\mathbf{e}_1, \mathbf{e}_2, \ldots, \mathbf{e}_n\}$ and $\beta = \{2\mathbf{e}_1, 2\mathbf{e}_2, \ldots, 2\mathbf{e}_n\}$ for $\mathbb{R}^n$. Then, $[id]_\beta^\alpha = 2I_n$ which is not an orthogonal matrix.

Problem 125 Determine whether the following linear transformations on the Euclidean space $\mathbb{R}^3$ are orthogonal or not.

(1) $T(x, y, z) = (z, y, x)$,
(2) $T(x, y, z) = (x, y, (y + z)/2)$,
(3) $T(x, y, z) = (z, (\sqrt{3}x + y)/2,\ (x - \sqrt{3}y)/2)$,
(4) $T(x, y, z) = ((5x + 11z)/13,\ (12y - 5z)/13,\ x)$.

Hint: Find $T(\mathbf{e}_i)$ in each case.

Problem 126 (**Bessel's Inequality**) Let V be an inner product space, and let $\{\mathbf{v}_1, \mathbf{v}_2, \ldots, \mathbf{v}_m\}$ be a set of orthonormal vectors in V (not necessarily a basis for V). Show that for any $\mathbf{x}$ in V, $||\mathbf{x}||^2 \geq \sum_{i=1}^m |\langle \mathbf{x}, \mathbf{v}_i \rangle|^2$.

Hint: Extend $\{\mathbf{v}_1, \ldots, \mathbf{v}_m\}$ to an orthonormal basis $\{\mathbf{v}_1, \ldots, \mathbf{v}_m, \ldots, \mathbf{v}_n\}$. Then $||\mathbf{x}||^2 = \sum_{i=1}^m |\langle \mathbf{x}, \mathbf{v}_i \rangle|^2 + \sum_{j=m+1}^n |\langle \mathbf{x}, \mathbf{v}_j \rangle|^2$.

5.10 Applications

5.10.1 Least squares solutions of a system $A\mathrm{x} = \mathrm{b}$ always exist

We start with an example of inconsistent system of linear equations.

Example 97 (*The best approximate solutions of an inconsistent system* $A\mathbf{x} = \mathbf{b}$)
Clearly, the system

$$\begin{cases} x+y &= 2 \\ x+y &= 0 \end{cases} \quad \text{or equivalently} \quad A\mathbf{x} = \begin{bmatrix} 1 & 1 \\ 1 & 1 \end{bmatrix} \begin{bmatrix} x \\ y \end{bmatrix} = \begin{bmatrix} x+y \\ x+y \end{bmatrix} = \begin{bmatrix} 2 \\ 0 \end{bmatrix} = \mathbf{b}$$

has no solutions.

The row space $\mathcal{R}(A)$ is the diagonal line $y = x$ spanned by the row $\begin{bmatrix} 1 & 1 \end{bmatrix}$ and the null space $\mathcal{N}(A)$ is the line $y = -x$ in the domain space $\mathbb{R}^2$. While, the column space $\mathcal{C}(A)$ is the diagonal line $y = x$ in the target space $\mathbb{R}^2$, and $\mathbf{b} \notin \mathcal{C}(A)$.

The orthogonal projection $\mathrm{Proj}_{\mathcal{C}(A)}(\mathbf{b})$ of $\mathbf{b} = \begin{bmatrix} 2 \\ 0 \end{bmatrix}$ is the unique vector in $\mathcal{C}(A)$ that is closest to $\mathbf{b}$. It is easy to see that

$$\mathrm{Proj}_{\mathcal{C}(A)}(\mathbf{b}) = \begin{bmatrix} 1 \\ 1 \end{bmatrix}.$$

Now, $A\mathbf{x} = \begin{bmatrix} 1 \\ 1 \end{bmatrix}$ has a solution since $\mathrm{Proj}_{\mathcal{C}(A)}(\mathbf{b}) \in \mathcal{C}(A)$. In fact, there are infinitely many solutions;

$$\mathbf{x} = (t, 1-t), \quad t \text{ is any.}$$

Each solution $\mathbf{x}_0$ of them satisfies

$$\|A\mathbf{x}_0 - \mathbf{b}\| = \sqrt{(1-2)^2 + (1-0)^2} \le \|A\mathbf{x} - \mathbf{b}\|$$

for any $\mathbf{x}$ in $\mathbb{R}^2$.

In this sense, such an $\mathbf{x}_0$ is called a *best approximate solution* of $A\mathbf{x} = \mathbf{b}$. Certainly, this is not a 'true' solution. Moreover, $\mathbf{x}_0 + \mathcal{N}(A)$ is the set of all best approximate solutions of $A\mathbf{x} = \mathbf{b}$. □

Let A be an $m \times n$ matrix and let $A\mathbf{x} = \mathbf{b}$ be any inconsistent system, where $\mathbf{b} \notin \mathcal{C}(A) \subseteq \mathbb{R}^m$.

For every vector $\mathbf{x}$ in $\mathbb{R}^n$, $A\mathbf{x} \in \mathcal{C}(A)$ which is the image of $A : \mathbb{R}^n \to \mathbb{R}^m$. As shown in Example 97, the best one can do for an inconsistent system $A\mathbf{x} = \mathbf{b}$ is to find an (*approximate*) vector $\mathbf{x}_0 \in \mathbb{R}^n$ whose image $A\mathbf{x}_0$ is the *closest* to the given vector $\mathbf{b} \in \mathbb{R}^m$.

The *closest* means that the '$error$' $\|A\mathbf{x}_0 - \mathbf{b}\|$ is as small as possible in $\mathbb{R}^m$. Such a vector $\mathbf{x}_0$ is called

a **least squares solution** of $A\mathbf{x} = \mathbf{b}$.

Here, *least squares* means that the sum of the squares of the errors $(a_{i1}x_1 + a_{i2}x_2 + \cdots + a_{in}x_n) - b_i$ occurring in the results of every single equation in $\|A\mathbf{x}_0 - \mathbf{b}\|$ is minimized, or equivalently $\|A\mathbf{x} - \mathbf{b}\|^2$ and hence the length $\|A\mathbf{x} - \mathbf{b}\|$ is minimized.

From the orthogonal decomposition $\mathbb{R}^m = \mathcal{C}(A) \oplus \mathcal{N}(A^T)$ and $\mathcal{C}(A) \perp \mathcal{N}(A^T)$, any $\mathbf{b} \in \mathbb{R}^m$ has the unique orthogonal decomposition as

$$\mathbf{b} = \mathbf{b}_c + \mathbf{b}_n \in \mathcal{C}(A) \oplus \mathcal{N}(A^T) = \mathbb{R}^m,$$

where $\mathbf{b}_c = \mathrm{Proj}_{\mathcal{C}(A)}(\mathbf{b}) \in \mathcal{C}(A)$ and $\mathbf{b}_n = \mathbf{b} - \mathbf{b}_c \in \mathcal{N}(A^T)$. Here, the vector $\mathbf{b}_c = \mathrm{Proj}_{\mathcal{C}(A)}(\mathbf{b}) \in \mathcal{C}(A)$ has two important properties:

(1) There exists at least one solution $\mathbf{x}_0 \in \mathbb{R}^n$ of $A\mathbf{x} = \mathbf{b}_c$, since $\mathbf{b}_c \in \mathcal{C}(A)$,

(2) $\mathbf{b}_c$ is the closest vector to $\mathbf{b}$ among the vectors in $\mathcal{C}(A)$ (see Theorem 5.16).

Therefore, *a least squares solution* $\mathbf{x}_0 \in \mathbb{R}^n$ *of* $A\mathbf{x} = \mathbf{b}$ *always exists* that is just a solution of $A\mathbf{x} = \mathbf{b}_c = \mathrm{Proj}_{\mathcal{C}(A)}(\mathbf{b})$. Furthermore, if $\mathbf{x}_0 \in \mathbb{R}^n$ is any least squares solution, then the set of all least squares solutions is $\mathbf{x}_0 + \mathcal{N}(A)$ by Corollary 5.19.1. Hence a least squares solution is unique if $\mathcal{N}(A) = \mathbf{0}$.

In particular, if $\mathbf{b} \in \mathcal{C}(A)$, then $\mathbf{b} = \mathbf{b}_c$ and the least squares solutions are just the 'true' solutions of $A\mathbf{x} = \mathbf{b}$.

In summary, to have a least squares solution of $A\mathbf{x} = \mathbf{b}$, first it is needed to find the coordinate vector of the orthogonal projection $\mathbf{b}_c = \mathrm{Proj}_{\mathcal{C}(A)}(\mathbf{b}) \in \mathcal{C}(A) \subseteq \mathbb{R}^m$. If it is done, then with the coordinate vector of $\mathbf{b}_c$, one can solve the system $A\mathbf{x} = \mathbf{b}_c$ as usual.

Fortunately, there is another efficient way to find a least squares solution without determining the coordinate vector of $\mathbf{b}_c = \mathrm{Proj}_{\mathcal{C}(A)}(\mathbf{b})$ or even without determining the vector $\mathbf{b}_c$.

For the given system $A\mathbf{x} = \mathbf{b}$, if $\mathbf{x}_0 \in \mathbb{R}^n$ is a least squares solution of $A\mathbf{x} = \mathbf{b}$, then

$$A\mathbf{x}_0 - \mathbf{b} = A\mathbf{x}_0 - (\mathbf{b}_c + \mathbf{b}_n) = -\mathbf{b}_n \in \mathcal{N}(A^T)$$

holds since $A\mathbf{x}_0 = \mathbf{b}_c$. Thus, $A^T(A\mathbf{x}_0 - \mathbf{b}) = A^T(-\mathbf{b}_n) = \mathbf{0}$ or equivalently $A^T A\mathbf{x}_0 = A^T\mathbf{b}$, that is, a least squares solution $\mathbf{x}_0$ is a (true) solution of the equation

$$A^T A\mathbf{x} = A^T\mathbf{b}. \tag{5.3}$$

This equation is very interesting and important because it is also a sufficient condition for a least squares solution. We call it the **normal equation** of the given system $A\mathbf{x} = \mathbf{b}$.

Theorem 5.26 *Let A be any $m \times n$ matrix, and let $\mathbf{b} \in \mathbb{R}^m$ be any vector. Then,*

(1) *every normal equation $A^T A\mathbf{x} = A^T\mathbf{b}$ always has at least one solution,*

(2) *a vector $\mathbf{x}_0 \in \mathbb{R}^n$ is a least squares solution of $A\mathbf{x} = \mathbf{b}$ if and only if $\mathbf{x}_0$ is a solution of the normal equation $A^T A\mathbf{x} = A^T\mathbf{b}$.* [7]

Proof: (1) is shown already. For (2), we only need to show the sufficiency. Let $\mathbf{x}_0$ be a solution of $A^T A\mathbf{x} = A^T\mathbf{b}$. Then, $A^T(A\mathbf{x}_0 - \mathbf{b}) = \mathbf{0}$, so $A\mathbf{x}_0 - \mathbf{b} \in \mathcal{N}(A^T)$. Say, $A\mathbf{x}_0 - \mathbf{b} = \mathbf{n}$ in $\mathcal{N}(A^T)$, and let $\mathbf{b} = \mathbf{b}_c + \mathbf{b}_n \in \mathcal{C}(A) \oplus \mathcal{N}(A^T)$. Then, $A\mathbf{x}_0 - \mathbf{b}_c = \mathbf{n} + \mathbf{b}_n \in \mathcal{C}(A) \cap \mathcal{N}(A^T)$. Since $\mathcal{C}(A) \cap \mathcal{N}(A^T) = \{\mathbf{0}\}$, $A\mathbf{x}_0 = \mathbf{b}_c = \mathrm{Proj}_{\mathcal{C}(A)}(\mathbf{b})$, *i.e.*, $\mathbf{x}_0$ is a least squares solution of $A\mathbf{x} = \mathbf{b}$. □

Consequently, for any (square or non-square) matrix A and for any (consistent or inconsistent) system $A\mathbf{x} = \mathbf{b}$, one can get the true solutions or the best approximate solutions of $A\mathbf{x} = \mathbf{b}$ by solving the normal equation $A^T A\mathbf{x} = A^T\mathbf{b}$, where the coefficient matrix $A^T A$ is square and symmetric.

In this reason, we are sometimes interested in systems of linear equations with only *square* coefficient matrices. (For example, see Chapter 6.)

[7] The least-squares method is usually credited to Carl Friedrich Gauss (1795), but it was first published by Adrien-Marie Legendre in 1805. The method of least squares grew out of the fields of astronomy and geodesy for fitting navigating curves in the Earth's oceans during the Age of Exploration.

Example 98 (*The best approximate solutions of an over-determined system* $A\mathbf{x} = \mathbf{b}$) Let

$$A = \begin{bmatrix} 0 & -1 & 1 \\ 1 & 0 & 1 \\ 1 & 1 & 0 \\ 1 & -1 & 2 \end{bmatrix}, \quad \mathbf{b} = \begin{bmatrix} 0 \\ 0 \\ 1 \\ 1 \end{bmatrix}.$$

Find all least squares solutions of the (over-determined) system $A\mathbf{x} = \mathbf{b}$ and determine the orthogonal projection $\mathbf{b}_c$ of $\mathbf{b}$ into the column space $\mathcal{C}(A)$.

Here, '*over-determined*' means a system has more equations than unknowns.

Solution: (The reader may check that $A\mathbf{x} = \mathbf{b}$ is inconsistent). Solve the normal equation $A^T A\mathbf{x} = A^T\mathbf{b}$ with

$$A^T A = \begin{bmatrix} 0 & 1 & 1 & 1 \\ -1 & 0 & 1 & -1 \\ 1 & 1 & 0 & 2 \end{bmatrix} \begin{bmatrix} 0 & -1 & 1 \\ 1 & 0 & 1 \\ 1 & 1 & 0 \\ 1 & -1 & 2 \end{bmatrix} = \begin{bmatrix} 3 & 0 & 3 \\ 0 & 3 & -3 \\ 3 & -3 & 6 \end{bmatrix}$$

which is not invertible (rank A = rank $(A^T A) = 2$ and hence the nullity of A is 1) and

$$A^T\mathbf{b} = \begin{bmatrix} 0 & 1 & 1 & 1 \\ -1 & 0 & 1 & -1 \\ 1 & 1 & 0 & 2 \end{bmatrix} \begin{bmatrix} 0 \\ 0 \\ 1 \\ 1 \end{bmatrix} = \begin{bmatrix} 2 \\ 0 \\ 2 \end{bmatrix}.$$

A least squares solution of $A\mathbf{x} = \mathbf{b}$ is a solution of $A^T A\mathbf{x} = A^T\mathbf{b}$, *i.e.*,

$$\begin{bmatrix} 3 & 0 & 3 \\ 0 & 3 & -3 \\ 3 & -3 & 6 \end{bmatrix} \begin{bmatrix} x_1 \\ x_2 \\ x_3 \end{bmatrix} = \begin{bmatrix} 2 \\ 0 \\ 2 \end{bmatrix}.$$

By solving it (possibly by the Gauss-Jordan elimination), one can get (infinitely many) the least squares solutions which are

$$\mathbf{x}_0 = \begin{bmatrix} x_1 \\ x_2 \\ x_3 \end{bmatrix} = \frac{1}{3} \begin{bmatrix} 2 \\ 0 \\ 0 \end{bmatrix} + t \begin{bmatrix} -1 \\ 1 \\ 1 \end{bmatrix}$$

for any number $t \in \mathbb{R}$ (but not true solutions). Moreover,

$$\text{Proj}_{\mathcal{C}(A)}(\mathbf{b}) = \mathbf{b}_c = A\mathbf{x}_0 = \frac{1}{3} \begin{bmatrix} 0 \\ 2 \\ 2 \\ 2 \end{bmatrix} \in \mathcal{C}(A).$$

The sum of the squares of the errors is

$$\|A\mathbf{x}_0 - \mathbf{b}\|^2 = (0-0)^2 + \left(\frac{2}{3} - 0\right)^2 + \left(\frac{2}{3} - 1\right)^2 + \left(\frac{2}{3} - 1\right)^2 = \frac{6}{9}.$$

Hence, the error of the best approximation $A\mathbf{x}$ to $\mathbf{b}$ is $\|A\mathbf{x}_0 - \mathbf{b}\| = \sqrt{6}/3$. □

Problem 127 Find the least squares solutions $\mathbf{x}_0$ in $\mathbb{R}^3$ of $A\mathbf{x} = \mathbf{b}$, where

$$A = \begin{bmatrix} 1 & 0 & 2 \\ 0 & 2 & 2 \\ -1 & 1 & -1 \\ -1 & 2 & 0 \end{bmatrix}, \quad \mathbf{b} = \begin{bmatrix} 3 \\ -3 \\ 0 \\ -3 \end{bmatrix}.$$

Hint: rank $A = 2$ and the columns of A are linearly dependent. Hence, it has infinitely many least squares solutions.

5.10.2 Least squares solutions of $A\mathrm{x} = \mathrm{b}$ when A is of full rank

Let A be any $m \times n$ square or rectangular matrix, and let $A\mathbf{x} = \mathbf{b}$ be any consistent or inconsistent system of linear equations. Then, the normal equation $A^T A\mathbf{x} = A^T\mathbf{b}$ has always at least one solution, which is a true solution or a best approximate solution of $A\mathbf{x} = \mathbf{b}$.

In this section, we consider the case that A is of full rank; rank $A = \min\{m, n\}$.

Firstly, let rank $A = m$. Then, A is wide and $\mathcal{C}(A) = \mathbb{R}^m$. Hence, $A\mathbf{x} = \mathbf{b}$ has always at least one (true) solution for any $\mathbf{b}$.

Secondly, let rank $A = n$. Then, A is long, and $\mathcal{N}(A) = \mathbf{0}$, or equivalently, $A\mathbf{x} = \mathbf{b}$ has at most one (true) solution. However, since rank A = rank $(A^T A) = n$, see Theorem 3.26, the $n \times n$ matrix $A^T A$ is invertible and

$$\mathbf{x}_0 = (A^T A)^{-1} A^T \mathbf{b}$$

is the unique solution of the normal equation $A^T A\mathbf{x} = A^T\mathbf{b}$, and hence the unique least squares solution of the equation $A\mathbf{x} = \mathbf{b}$.

It gives the following theorem.

Theorem 5.27 *Let A be an $m \times n$ matrix. If* rank $A = n$, *or equivalently the columns of A are linearly independent, then*

(1) *A^TA is invertible so that $(A^TA)^{-1}A^T$ is a left inverse of A,*

(2) *the vector $\mathbf{x}_0 = (A^TA)^{-1}A^T\mathbf{b}$ is the unique least squares solution of a system $A\mathbf{x} = \mathbf{b}$ and $A\mathbf{x}_0 = A(A^TA)^{-1}A^T\mathbf{b} = \mathbf{b}_c = \mathrm{Proj}_{\mathcal{C}(A)}(\mathbf{b})$,*

(3) *$A(A^TA)^{-1}A^T$ is the (orthogonal) projection matrix of the orthogonal projection $\mathrm{Proj}_{\mathcal{C}(A)}$ of $\mathbb{R}^m$ onto $\mathcal{C}(A)$.*

In particular, if $A^TA = I_n$, or equivalently the columns of A are orthonormal, then AA^T is the (orthogonal) projection matrix of $\mathrm{Proj}_{\mathcal{C}(A)}$ and the least squares solution is unique which is $\mathbf{x}_0 = A^T\mathbf{b}$.

Remark: (1) Let A be an $m \times n$ rectangular matrix. If $A^TA = I_n$, or equivalently the columns of A are orthonormal, then the columns $\mathbf{u}_1, \ldots, \mathbf{u}_n$ of A form an orthonormal basis for the column space $\mathcal{C}(A)$, so that for any $\mathbf{b} \in \mathbb{R}^m$,

$$\mathrm{Proj}_{\mathcal{C}(A)}\mathbf{b} = \mathbf{b}_c = (\mathbf{u}_1 \cdot \mathbf{b})\mathbf{u}_1 + \cdots + (\mathbf{u}_n \cdot \mathbf{b})\mathbf{u}_n = (\mathbf{u}_1\mathbf{u}_1^T + \cdots + \mathbf{u}_n\mathbf{u}_n^T)\mathbf{b}.$$

Hence, the orthogonal projection matrix $A(A^TA)^{-1}A^T$ of $\mathrm{Proj}_{\mathcal{C}(A)}$ becomes

$$AA^T = \mathbf{u}_1\mathbf{u}_1^T + \cdots + \mathbf{u}_n\mathbf{u}_n^T,$$

and the least squares solution is

$$\mathbf{x}_0 = A^T\mathbf{b} = \begin{bmatrix} -- & \mathbf{u}_1^T & -- \\ & \vdots & \\ -- & \mathbf{u}_n^T & -- \end{bmatrix} \mathbf{b} = \begin{bmatrix} \mathbf{u}_1 \cdot \mathbf{b} \\ \vdots \\ \mathbf{u}_n \cdot \mathbf{b} \end{bmatrix},$$

which is the coordinate vector of $A\mathbf{x}_0 = \mathbf{b}_c = \mathrm{Proj}_{\mathcal{C}(A)}(\mathbf{b})$ with respect to the orthonormal basis $\{\mathbf{u}_1, \ldots, \mathbf{u}_n\}$ for $\mathcal{C}(A)$.

(2) Let rank $A = r < n$, not of full rank. Then one can reduce the columns of A to be a basis for the column space and work with this reduced matrix $\bar{A}$ (thus, $\bar{A}^T\bar{A}$ is invertible) to find the projection matrix of $\mathrm{Proj}_{\mathcal{C}(A)}$ of $\mathbb{R}^m$, and one may expect that the projection matrix of $\mathrm{Proj}_{\mathcal{C}(A)}$ would be $\bar{A}(\bar{A}^T\bar{A})^{-1}\bar{A}^T$. However, the least squares solutions of $A\mathbf{x} = \mathbf{b}$ should be found from the original normal equation directly, since the least squares solution $\bar{\mathbf{x}}_0 = (\bar{A}^T\bar{A})^{-1}\bar{A}^T\mathbf{b}$ of $\bar{A}\mathbf{x} = \mathbf{b}$ has only r components so that it cannot be a solution of $A\mathbf{x} = \mathbf{b}_c$.

Example 99 (*Solving an over-determined system $A\mathbf{x} = \mathbf{b}$ by the normal equation*) Let

$$A = \begin{bmatrix} 0 & -1 & 1 \\ 1 & 0 & 1 \\ 1 & 0 & 2 \\ 1 & -1 & 0 \end{bmatrix}, \quad \mathbf{b} = \begin{bmatrix} 0 \\ 0 \\ 1 \\ 1 \end{bmatrix}.$$

Find the least squares solutions of the system $A\mathbf{x} = \mathbf{b}$ and determine the orthogonal projection $\mathbf{b}_c$ of $\mathbf{b}$ into the column space $\mathcal{C}(A)$.

Solution: Before solving it, first notice that this example looks the same as Example 98. Not true; the columns of A in Example 98 are linearly dependent, while the columns of A here are linearly independent.

The system $A\mathbf{x} = \mathbf{b}$ is still inconsistent. Solve the normal equation $A^TA\mathbf{x} = A^T\mathbf{b}$ with

$$A^TA = \begin{bmatrix} 0 & 1 & 1 & 1 \\ -1 & 0 & 0 & -1 \\ 1 & 1 & 2 & 0 \end{bmatrix} \begin{bmatrix} 0 & -1 & 1 \\ 1 & 0 & 1 \\ 1 & 0 & 2 \\ 1 & -1 & 0 \end{bmatrix} = \begin{bmatrix} 3 & -1 & 3 \\ -1 & 2 & -1 \\ 3 & -1 & 6 \end{bmatrix}$$

which is invertible (rank A = rank $(A^TA) = 3$ and hence $\mathcal{N}(A) = \mathbf{0}$) and

$$(A^TA)^{-1} = \frac{1}{15} \begin{bmatrix} 11 & 3 & -5 \\ 3 & 9 & 0 \\ -5 & 0 & 5 \end{bmatrix}.$$

From the normal equation, the unique least squares solution is

$$\mathbf{x}_0 = (A^TA)^{-1}A^T\mathbf{b} = \frac{1}{15} \begin{bmatrix} 11 & 3 & -5 \\ 3 & 9 & 0 \\ -5 & 0 & 5 \end{bmatrix} \begin{bmatrix} 2 \\ -1 \\ 2 \end{bmatrix} = \frac{1}{5} \begin{bmatrix} 3 \\ -1 \\ 0 \end{bmatrix},$$

where

$$A^T\mathbf{b} = \begin{bmatrix} 0 & 1 & 1 & 1 \\ -1 & 0 & 0 & -1 \\ 1 & 1 & 2 & 0 \end{bmatrix} \begin{bmatrix} 0 \\ 0 \\ 1 \\ 1 \end{bmatrix} = \begin{bmatrix} 2 \\ -1 \\ 2 \end{bmatrix}.$$

The orthogonal projection of $\mathbf{b}$ onto the space $\mathcal{C}(A)$ is

$$\mathrm{Proj}_{\mathcal{C}(A)}(\mathbf{b}) = \mathbf{b}_c = A\mathbf{x}_0 = \begin{bmatrix} 0 & -1 & 1 \\ 1 & 0 & 1 \\ 1 & 0 & 2 \\ 1 & -1 & 0 \end{bmatrix} \frac{1}{5} \begin{bmatrix} 3 \\ -1 \\ 0 \end{bmatrix} = \frac{1}{5} \begin{bmatrix} 1 \\ 3 \\ 3 \\ 4 \end{bmatrix}.$$

□

Problem 128 Find all of the true or the best approximate solutions of

$$A\mathbf{x} = \begin{bmatrix} 1 & 2 \\ 1 & 5 \\ 0 & 0 \end{bmatrix} \begin{bmatrix} x \\ y \end{bmatrix} = \begin{bmatrix} 4 \\ 3 \\ 9 \end{bmatrix} = \mathbf{b}.$$

Determine also the orthogonal projection $\mathbf{b}_c$ of $\mathbf{b}$ onto the column space $\mathcal{C}(A)$.

Hint: rank $A = 2$ and the columns of A are linearly independent. Hence, it has only one least squares solution.

Problem 129 Find the true or the best approximate solutions of $\begin{cases} x + y = 5 \\ 2x + 2y = 5 \\ 3x + 4y = 0. \end{cases}$

Hint: Two coefficient columns are linearly independent.

5.10.3 Best polynomial curve fitting using least squares method

In this section we go back to the polynomial curve fitting problem discussed in Subsection 2.7.2.

There is a (unique) polynomial $y = ax^2 + bx + c$ of degree ≤ 2 that passes exactly through (exact fits to) given three points (x_0, y_0), (x_1, y_1), and (x_2, y_2). However, if we require the degree of a polynomial to be one, there may not exist a straight line $y = ax + b$ that passes exactly through the given three points. If there does not exist, can one find a line $y = ax + b$ that passes closest to the given three points?

In general, one needs sometimes to find a polynomial of degree k that may not pass exactly through (exact fit to) but passes closest to (say, *best fit to*) given $n + 1$ experimental input/output data.

As an example, one may ask a straight line $y = a + bx$ that fits to (*i.e.*, passes through) three different points (x_0, y_0), (x_1, y_1), and (x_2, y_2). It gives an over-determined system $A\mathbf{x} = \mathbf{b}$ of three linear equations $y_i = a + bx_i$ of two unknowns a, b. In most cases, $A\mathbf{x} = \mathbf{b}$ has no solutions but its least squares solutions can be the best approximate solutions, that *best fit to the given data.*

The following two examples illustrate how to use the least squares method for a polynomial of degree k that *best fits* to given $n + 1$ experimental input/output data.

Example 100 (*A straight line that passes closest to give four points*) Find a straight line $y = a+bx$ that best fits to the following four experimental input/output data

input x_i	0	1	2	3
output y_i	0	2	3	2

Solution: The given four data (x_i, y_i) give an over-determined system $A\mathbf{x} = \mathbf{b}$ of four linear equations $y_i = a + bx_i$ of two unknowns a, b, where

$$A = \begin{bmatrix} 1 & x_0 \\ 1 & x_1 \\ 1 & x_2 \\ 1 & x_3 \end{bmatrix} = \begin{bmatrix} 1 & 0 \\ 1 & 1 \\ 1 & 2 \\ 1 & 3 \end{bmatrix}, \quad \mathbf{x} = \begin{bmatrix} a \\ b \end{bmatrix} \quad \text{and} \quad \mathbf{b} = \begin{bmatrix} 0 \\ 2 \\ 3 \\ 2 \end{bmatrix}.$$

Since the columns of A are linearly independent, the normal equation $A^T A\mathbf{x} = A^T\mathbf{b}$ has the unique solution $\mathbf{x} = (A^T A)^{-1}A^T\mathbf{b}$. With

$$A^T A = \begin{bmatrix} 4 & 6 \\ 6 & 14 \end{bmatrix}, \ (A^T A)^{-1} = \frac{1}{10}\begin{bmatrix} 7 & -3 \\ -3 & 2 \end{bmatrix} \quad \text{and} \quad A^T\mathbf{b} = \begin{bmatrix} 7 \\ 14 \end{bmatrix},$$

one can get $\mathbf{x} = (A^T A)^{-1}A^T\mathbf{b} = \begin{bmatrix} 0.7 \\ 0.7 \end{bmatrix}$. Hence, the line $y = 0.7 + 0.7x$ best fits to the given four data. □

Example 101 (*A quadratic curve that passes closest to give four points*) Find a quadratic curve $y = a + bx + cx^2$ that best fits to the given experimental data

input x_i	0	1	2	3
output y_i	0	10	20	40

Solution: By adopting matrix notation for four linear equations $y_i = a + bx_i + cx_i^2$, one can get $A\mathbf{x} = \mathbf{b}$, where

$$A = \begin{bmatrix} 1 & x_0 & x_0^2 \\ 1 & x_1 & x_1^2 \\ 1 & x_2 & x_2^2 \\ 1 & x_3 & x_3^2 \end{bmatrix} = \begin{bmatrix} 1 & 0 & 0^2 \\ 1 & 1 & 1^2 \\ 1 & 2 & 2^2 \\ 1 & 3 & 3^2 \end{bmatrix}, \quad \mathbf{x} = \begin{bmatrix} a \\ b \\ c \end{bmatrix}, \quad \text{and} \quad \mathbf{b} = \begin{bmatrix} y_0 \\ y_1 \\ y_2 \\ y_3 \end{bmatrix} = \begin{bmatrix} 0 \\ 10 \\ 20 \\ 40 \end{bmatrix}.$$

Since the columns of A are linearly independent, the normal equation $A^T A\mathbf{x} = A^T\mathbf{b}$ has the unique solution $\mathbf{x} = (A^T A)^{-1}A^T\mathbf{b}$. With

$$A^T A = \begin{bmatrix} 4 & 6 & 14 \\ 6 & 14 & 36 \\ 14 & 36 & 92 \end{bmatrix}, \ (A^T A)^{-1} = \frac{1}{20}\begin{bmatrix} 19 & -21 & 5 \\ -21 & 49 & -15 \\ 5 & -15 & 5 \end{bmatrix}, \ A^T\mathbf{b} = \begin{bmatrix} 70 \\ 170 \\ 450 \end{bmatrix},$$

one can get $\mathbf{x} = (A^T A)^{-1}A^T\mathbf{b} = \begin{bmatrix} 0.5 \\ 5.5 \\ 2.5 \end{bmatrix}$. Hence, $y = 0.5 + 5.5x + 2.5x^2$ is the best fitting quadratic curve. □

Problem 130 Find a straight line $y = a + bx$ that best fits to the given experimental data

input x_i	0	1	2	3
output y_i	0	10	20	40

5.10.4 What is the entire class of orthogonal projection matrices?

In Section 5.7, it is mentioned that *a matrix is an orthogonal projection matrix if and only if it is symmetric and idempotent.*

In this section, its proof will be given.

Let U be any subspace of the Euclidean space $\mathbb{R}^m$ and let Proj_U be the orthogonal projection of $\mathbb{R}^m$ onto U. Say $\dim U = n$. To find the orthogonal projection matrix of Proj_U, take any basis for U and construct an $m \times n$ matrix A with these basis vectors as columns. Then, $U = \mathcal{C}(A)$ and the least squares method gives (see Theorem 5.27) that the (orthogonal) projection matrix of $\text{Proj}_U = \text{Proj}_{\mathcal{C}(A)}$ with respect to the standard (ordered) basis α is

$$[\text{Proj}_U]_\alpha = A(A^TA)^{-1}A^T,$$

which is always symmetric and idempotent.

Theorem 5.28 *A square matrix P is an orthogonal projection matrix if and only if it is symmetric and idempotent, i.e., $P^T = P$ and $P^2 = P$.*

Proof: ($\Rightarrow$) is shown already.

($\Leftarrow$): By Theorem 5.19, one has an orthogonal decomposition $\mathbb{R}^m = \mathcal{C}(P) \oplus \mathcal{N}(P^T)$. But, $\mathcal{N}(P^T) = \mathcal{N}(P)$ since $P^T = P$. Thus,

$$\text{for any } \mathbf{u} + \mathbf{n} \in \mathcal{C}(P) \oplus \mathcal{N}(P^T) = \mathbb{R}^m,\ P(\mathbf{u} + \mathbf{n}) = P\mathbf{u} + P\mathbf{n} = P\mathbf{u} = \mathbf{u},$$

because $P^2 = P$ implies $P\mathbf{u} = \mathbf{u}$ for $\mathbf{u} \in \mathcal{C}(P)$. This shows that P is a projection matrix of an orthogonal projection $P = \text{Proj}_{\mathcal{C}(P)}$. □

If a projection matrix P is symmetric, then its image is $\mathcal{C}(P) = \mathcal{R}(P^T) = \mathcal{R}(P)$ and its kernel is $\mathcal{N}(P) = \mathcal{N}(P^T)$. By Corollary 5.12.1, $I - P$ is a projection matrix on the null space $\mathcal{N}(P)$ $(= \mathcal{C}(I - P))$, which is orthogonal to $\mathcal{C}(P)$ $(= \mathcal{N}(I - P))$.

Example 102 For any unit vector $\mathbf{u}$ in $\mathbb{R}^m$, it is clear that $(\mathbf{u}\mathbf{u}^T)^T = \mathbf{u}\mathbf{u}^T$ and $(\mathbf{u}\mathbf{u}^T)^2 = \mathbf{u}\mathbf{u}^T$. Hence $\mathbf{u}\mathbf{u}^T$ is the orthogonal projection of $\mathbb{R}^m$ onto the column space $\mathcal{C}(\mathbf{u}\mathbf{u}^T)$, which is the one-dimensional subspace spanned by $\mathbf{u}$, and $I_m - \mathbf{u}\mathbf{u}^T$ is the orthogonal projection of $\mathbb{R}^m$ onto the null space $\mathcal{N}(\mathbf{u}\mathbf{u}^T)$. □

Example 103 Let $P_i : \mathbb{R}^m \to \mathbb{R}^m$ be defined by

$$P_i(x_1, \ldots, x_m) = (0, \ldots, 0, x_i, 0, \ldots, 0),$$

for $i = 1, 2, \ldots, m$. Then, each P_i is the orthogonal projection of $\mathbb{R}^m$ onto the i-th axis, whose associated matrix is

$$P_i = \begin{bmatrix} \ddots & & & & \mathbf{0} \\ & 0 & & & \\ & & 1 & & \\ & & & 0 & \\ \mathbf{0} & & & & \ddots \end{bmatrix}, \quad \text{and} \quad I - P_i = \begin{bmatrix} \ddots & & & & \mathbf{0} \\ & 1 & & & \\ & & 0 & & \\ & & & 1 & \\ \mathbf{0} & & & & \ddots \end{bmatrix}.$$

Both are symmetric and idempotent. □

Problem 131 Show that any square matrix P that satisfies $P^TP = P$ is an orthogonal projection matrix.

Hint: First, show that P is symmetric.

Problem 132 Let $\mathbf{u} = (\frac{1}{\sqrt{2}}, \frac{1}{\sqrt{2}})$ be a unit vector in the plane $\mathbb{R}^2$, which spans the diagonal line in the plane $\mathbb{R}^2$. Show that the matrix

$$P = \mathbf{u}\mathbf{u}^T = \frac{1}{2}\begin{bmatrix} 1 \\ 1 \end{bmatrix}[1\ 1] = \frac{1}{2}\begin{bmatrix} 1 & 1 \\ 1 & 1 \end{bmatrix}$$

is the orthogonal projection matrix of $\mathbb{R}^2$ onto the diagonal line.

Problem 133 (Problem 112 continued) Which of the following matrices are the orthogonal projection matrices?

$$A = \begin{bmatrix} 1 & 0 \\ 0 & 1 \end{bmatrix}, \quad B = \begin{bmatrix} 1 & 1 \\ 0 & 0 \end{bmatrix}, \quad C = \begin{bmatrix} 1 & 0 \\ 1 & 0 \end{bmatrix}, \quad D = \begin{bmatrix} 0 & 1 \\ 0 & 1 \end{bmatrix}, \quad E = \begin{bmatrix} 0 & 1 \\ 1 & 0 \end{bmatrix}.$$

Hint: Check each matrix whether it is both symmetric and idempotent or not.

5.10.5 QR decompositions from the Gram-Schmidt process

Let A be any square or rectangular $m \times n$ matrix of full rank.

QR decomposition: Firstly, let rank $A = n$ with $n \leq m$. Or equivalently A is a long matrix whose columns are linearly independent. In this case, $A^T A$ is invertible and A has a left inverse, see Theorem 3.26. Let $A = [\mathbf{c}_1 \ \mathbf{c}_2 \ \cdots \ \mathbf{c}_n]$ with columns $\mathbf{c}_i$'s.

By applying the Gram-Schmidt orthogonalization to the columns $\{\mathbf{c}_1, \mathbf{c}_2, \ldots, \mathbf{c}_n\}$, one can derive an orthonormal basis $\{\mathbf{u}_1, \mathbf{u}_2, \ldots, \mathbf{u}_n\}$ for the column space $\mathcal{C}(A)$. Then each column $\mathbf{c}_i$ can be written (uniquely) as a linear combination of the orthonormal basis vectors $\mathbf{u}_1, \mathbf{u}_2, \ldots, \mathbf{u}_n$. This can be shown as the following steps.

(1) [Gram-Schmidt process] Orthogonalize the columns $\{\mathbf{c}_1, \mathbf{c}_2, \ldots, \mathbf{c}_n\}$ by the process

$$\begin{aligned}
\mathbf{u}_1 &= \frac{\mathbf{c}_1}{\|\mathbf{c}_1\|}, \\
\mathbf{u}_2 &= \frac{\mathbf{c}_2 - \langle \mathbf{u}_1, \mathbf{c}_2 \rangle \mathbf{u}_1}{\|\mathbf{c}_2 - \langle \mathbf{u}_1, \mathbf{c}_2 \rangle \mathbf{u}_1\|}, \\
&\vdots \\
\mathbf{u}_n &= \frac{\mathbf{c}_n - \langle \mathbf{u}_1, \mathbf{c}_n \rangle \mathbf{u}_1 - \langle \mathbf{u}_2, \mathbf{c}_n \rangle \mathbf{u}_2 - \cdots - \langle \mathbf{u}_{n-1}, \mathbf{c}_n \rangle \mathbf{u}_{n-1}}{\|\mathbf{c}_n - \langle \mathbf{u}_1, \mathbf{c}_n \rangle \mathbf{u}_1 - \langle \mathbf{u}_2, \mathbf{c}_n \rangle \mathbf{u}_2 - \cdots - \langle \mathbf{u}_{n-1}, \mathbf{c}_n \rangle \mathbf{u}_{n-1}\|}.
\end{aligned}$$

(2) From the equations in (1), each column $\mathbf{c}_i$ can be written as a linear combination of the orthonormal basis vectors $\mathbf{u}_1, \mathbf{u}_2, \ldots, \mathbf{u}_n$ as follows:

$$\begin{aligned}
\mathbf{c}_1 &= b_{11}\mathbf{u}_1, \\
\mathbf{c}_2 &= b_{12}\mathbf{u}_1 + b_{22}\mathbf{u}_2, \\
&\vdots \\
\mathbf{c}_n &= b_{1n}\mathbf{u}_1 + \cdots + b_{nn}\mathbf{u}_n,
\end{aligned}$$

where $b_{ij} = \langle \mathbf{u}_i, \mathbf{c}_j \rangle$ for $i \leq j$, which is just the coordinate component of $\mathbf{c}_j$ in $\mathbf{u}_i$ direction. See Theorem 5.9.

(3) In matrix notation,

$$A = [\mathbf{c}_1 \ \mathbf{c}_2 \ \cdots \ \mathbf{c}_n] = [\mathbf{u}_1 \ \mathbf{u}_2 \ \cdots \ \mathbf{u}_n] \begin{bmatrix} b_{11} & b_{12} & \cdots & b_{1n} \\ 0 & b_{22} & \cdots & b_{2n} \\ \vdots & & \ddots & \\ 0 & \cdots & 0 & b_{nn} \end{bmatrix} = QR,$$

where $Q = [\mathbf{u}_1\ \mathbf{u}_2\ \cdots\ \mathbf{u}_n]$ is the $m \times n$ matrix with orthonormal columns, which are obtained from the $\mathbf{c}_i$'s by the Gram-Schmidt orthogonalization, and R is an $n \times n$ upper triangular (invertible) matrix with entry $b_{ij} = \langle \mathbf{u}_i, \mathbf{c}_j \rangle$.

The matrix $Q = [\mathbf{u}_1\ \mathbf{u}_2\ \cdots\ \mathbf{u}_n]$ is called the **orthogonal part** of A, and the matrix $R = [b_{ij}]$ is called the **upper triangular part** of A.

(4) Let U_{j-1} denote the subspace of $\mathbb{R}^m$ spanned by $\{\mathbf{c}_1, \mathbf{c}_2, \ldots, \mathbf{c}_{j-1}\}$ (or equivalently, by $\{\mathbf{u}_1, \mathbf{u}_2, \ldots, \mathbf{u}_{j-1}\}$) for each j. Then,

$$1 = \langle \mathbf{u}_j, \mathbf{u}_j \rangle = \left\langle \mathbf{u}_j, \frac{\mathbf{c}_j - \mathrm{Proj}_{U_{j-1}}(\mathbf{c}_j)}{\|\mathbf{c}_j - \mathrm{Proj}_{U_{j-1}}(\mathbf{c}_j)\|} \right\rangle = \frac{1}{\|\mathbf{c}_j - \mathrm{Proj}_{U_{j-1}}(\mathbf{c}_j)\|} \langle \mathbf{u}_j, \mathbf{c}_j \rangle.$$

Hence, each diagonal entry $b_{jj} = \langle \mathbf{u}_j, \mathbf{c}_j \rangle$ of the matrix R is equal to $\|\mathbf{c}_j - \mathrm{Proj}_{U_{j-1}}(\mathbf{c}_j)\|$, which is not zero.

Therefore, the matrix R is an invertible upper triangular matrix.

(5) Since R is invertible, rank A = rank $Q = n \leq m$ and $\mathcal{C}(Q) = \mathcal{C}(A)$.

As a special case, if A is square, then Q becomes an orthogonal matrix. It gives $\det A = \det Q \det R = \pm \det R = \pm b_{11} b_{22} \cdots b_{nn}$.

LQ decomposition: Let rank $A = m$ with $m \leq n$. Or equivalently A is a wide matrix whose row vectors are linearly independent. In this case, AA^T is invertible, and A has a right inverse. The QR decomposition of A^T gives $A = R^T Q = LQ$, where Q is an $m \times n$ matrix whose rows are orthonormal and $L = R^T$ is an $n \times n$ invertible lower triangular matrix.

In summary, one can get the following theorem.

Theorem 5.29 **(1)** *Any $m \times n$ matrix of rank n can be factored into a product QR, where Q is an $m \times n$ matrix whose columns are orthonormal, or equivalently $Q^T Q = I$, and R is an $n \times n$ invertible upper triangular matrix.*

(2) *Any $m \times n$ matrix of rank m can be factored into a product LQ, where Q is an $m \times n$ matrix whose rows are orthonormal, or equivalently $QQ^T = I$, and L is an $m \times m$ invertible lower triangular matrix.*

(3) *In particular, if A is an invertible square matrix, then Q is an orthogonal matrix in $A = QR$ or in $A = LQ$.*

One of the applications of QR decomposition is to find least squares solutions of over-determined systems.

The (orthogonal) projection matrix P of $\mathrm{Proj}_{C(A)}$ and the least squares solution of $A\mathbf{x} = \mathbf{b}$ for $\mathbf{b} \in \mathbb{R}^m$ can be calculated easily as

$$\begin{aligned} P &= A(A^TA)^{-1}A^T = QR(R^TQ^TQR)^{-1}R^TQ^T = QQ^T, \\ \mathbf{x}_0 &= R^{-1}Q^T\mathbf{b}, \end{aligned}$$

because $R^{-1}Q^T$ is a left inverse of $A = QR$. In fact, the (orthogonal) projection matrix of $\mathrm{Proj}_{C(A)} = \mathrm{Proj}_{C(Q)}$ is

$$QQ^T,$$

as shown in Theorem 5.27.

Corollary 5.29.1 *Let A be an $m \times n$ (long) matrix of rank n and let $A = QR$ be its QR factorization. Then,*

(1) *the projection matrix of $\mathrm{Proj}_{C(A)}$ on the column space of A is QQ^T.*

(2) *The least squares solution of the system $A\mathbf{x} = \mathbf{b}$ is given by $\mathbf{x}_0 = R^{-1}Q^T\mathbf{b}$, which can be solved by using backward substitution to the system $R\mathbf{x} = Q^T\mathbf{b}$.*

For an $m \times n$ (long) matrix $A = QR$ of rank n, notice that Q^TQ *is the identity matrix I_n, while QQ^T is the projection matrix of* $\mathrm{Proj}_{C(A)}$. In this case, the matrix R can be computed as a solution of $A = QR$, that is, $R = Q^TA$.

Example 104 (*QR decomposition of A*) **(1)** Find a QR decomposition of a long matrix

$$A = [\mathbf{c}_1\ \mathbf{c}_2\ \mathbf{c}_3] = \begin{bmatrix} 1 & 1 & 0 \\ 1 & 0 & 1 \\ 0 & 1 & 0 \\ 0 & 0 & 1 \end{bmatrix}.$$

(2) Find the (orthogonal) projection matrix of $\mathrm{Proj}_{C(A)}$.

(3) Solve the system $A\mathbf{x} = \mathbf{b}$, where $\mathbf{b} = (1, 0, 0, 0)$.

Solution: (1) Following the Gram-Schmidt orthogonalization process, we first orthogonalize $\mathbf{c}_1, \mathbf{c}_2, \mathbf{c}_3$ to get

$$\begin{aligned}
\mathbf{u}_1 &= \frac{\mathbf{c}_1}{\|\mathbf{c}_1\|} = \left(\frac{1}{\sqrt{2}}, \frac{1}{\sqrt{2}}, 0, 0\right), \\
\mathbf{u}_2 &= \frac{\mathbf{c}_2 - \langle \mathbf{u}_1, \mathbf{c}_2\rangle \mathbf{u}_1}{\|\mathbf{c}_2 - \langle \mathbf{u}_1, \mathbf{c}_2\rangle \mathbf{u}_1\|} = \left(\frac{1}{\sqrt{6}}, -\frac{1}{\sqrt{6}}, \frac{\sqrt{2}}{\sqrt{3}}, 0\right), \\
\mathbf{u}_3 &= \frac{\mathbf{c}_3 - \langle \mathbf{u}_1, \mathbf{c}_3\rangle \mathbf{u}_1 - \langle \mathbf{u}_2, \mathbf{c}_3\rangle \mathbf{u}_2}{\|\mathbf{c}_3 - \langle \mathbf{u}_1, \mathbf{c}_3\rangle \mathbf{u}_1 - \langle \mathbf{u}_2, \mathbf{c}_3\rangle \mathbf{u}_2\|} = \left(-\frac{1}{2\sqrt{3}}, \frac{1}{2\sqrt{3}}, \frac{1}{2\sqrt{3}}, \frac{\sqrt{3}}{2}\right).
\end{aligned}$$

Now write each $\mathbf{c}_j$ as a linear combination of $\mathbf{u}_1, \mathbf{u}_2, \mathbf{u}_3$ as

$$\begin{aligned}
\mathbf{c}_1 &= b_{11}\mathbf{u}_1 \\
\mathbf{c}_2 &= b_{12}\mathbf{u}_1 + b_{22}\mathbf{u}_2 \\
\mathbf{c}_3 &= b_{13}\mathbf{u}_1 + b_{23}\mathbf{u}_2 + b_{33}\mathbf{u}_3,
\end{aligned}$$

where $b_{ij} = \langle \mathbf{u}_i, \mathbf{c}_j\rangle$. Therefore, $Q = [\mathbf{u}_1\ \mathbf{u}_2\ \mathbf{u}_3]$ and $R = [b_{ij}]$ gives a factorization

$$\begin{aligned}
A &= \begin{bmatrix} 1 & 1 & 0 \\ 1 & 0 & 1 \\ 0 & 1 & 1 \\ 0 & 0 & 1 \end{bmatrix} = [\mathbf{u}_1\ \mathbf{u}_2\ \mathbf{u}_3] \begin{bmatrix} \sqrt{2} & \frac{1}{\sqrt{2}} & \frac{1}{\sqrt{2}} \\ 0 & \frac{\sqrt{3}}{\sqrt{2}} & -\frac{1}{\sqrt{6}} \\ 0 & 0 & \frac{2}{\sqrt{3}} \end{bmatrix} \\
&= \begin{bmatrix} \frac{1}{\sqrt{2}} & \frac{1}{\sqrt{6}} & -\frac{1}{2\sqrt{3}} \\ \frac{1}{\sqrt{2}} & -\frac{1}{\sqrt{6}} & \frac{1}{2\sqrt{3}} \\ 0 & \frac{\sqrt{2}}{\sqrt{3}} & \frac{1}{2\sqrt{3}} \\ 0 & 0 & \frac{\sqrt{3}}{2} \end{bmatrix} \begin{bmatrix} \sqrt{2} & \frac{1}{\sqrt{2}} & \frac{1}{\sqrt{2}} \\ 0 & \frac{\sqrt{3}}{\sqrt{2}} & -\frac{1}{\sqrt{6}} \\ 0 & 0 & \frac{2}{\sqrt{3}} \end{bmatrix} = QR.
\end{aligned}$$

In another way, R can be computed by the product $Q^T A$.

(2) The projection matrix of $\mathrm{Proj}_{\mathcal{C}(A)} = \mathrm{Proj}_{\mathcal{C}(Q)}$ is

$$QQ^T = \frac{1}{4}\begin{bmatrix} 3 & 1 & 1 & -1 \\ 1 & 3 & -1 & 1 \\ 1 & -1 & 3 & 1 \\ -1 & 1 & 1 & 3 \end{bmatrix}.$$

(3) The given vector $\mathbf{b}$ is not contained in the column space $\mathcal{C}(A)$ because it cannot be a linear combination of $\mathbf{u}_1, \mathbf{u}_2, \mathbf{u}_3$. The least squares solution of $A\mathbf{x} = \mathbf{b}$

is the solution of $QR\mathbf{x} = A\mathbf{x} = \mathrm{Proj}_{C(A)}\mathbf{b} = QQ^T\mathbf{b}$. Multiply this equation by Q^T on the left to get $R\mathbf{x} = Q^T\mathbf{b}$, that is

$$\begin{bmatrix} \sqrt{2} & \frac{1}{\sqrt{2}} & \frac{1}{\sqrt{2}} \\ 0 & \frac{\sqrt{3}}{\sqrt{2}} & -\frac{1}{\sqrt{6}} \\ 0 & 0 & \frac{2}{\sqrt{3}} \end{bmatrix} \begin{bmatrix} x_1 \\ x_2 \\ x_3 \end{bmatrix} = \frac{1}{\sqrt{6}} \begin{bmatrix} \sqrt{3} \\ 1 \\ -\frac{\sqrt{2}}{2} \end{bmatrix}.$$

By using backward substitution, one gets $x_1 = 1/2, x_2 = 1/4, x_3 = -1/4$. □

Problem 134 Find the projection matrix P of the Euclidean 3-space $\mathbb{R}^3$ onto the column space $\mathcal{C}(A)$ for a long matrix

$$A = \begin{bmatrix} 1 & 0 \\ 1 & 1 \\ 0 & 1 \end{bmatrix}.$$

Hint: $P = A(A^TA)^{-1}A^T$ or $P = QQ^T$.

Problem 135 (Revisit Problem 111) Find a QR factorization of the matrix

$$A = \begin{bmatrix} 1 & 1 & 1 \\ 1 & 0 & 1 \\ 1 & 1 & 0 \\ 1 & 0 & 0 \end{bmatrix}$$

which is given in Example 87 on page 201.

Hint: Use the solutions of Example 87 on page 201 and Problem 111 on page 206.

5.11 Exercises

5.1. *Every inner product is positive-definite, symmetric, and bilinear.* Decide which of the following functions defined on the 2-space $\mathbb{R}^2$ are inner products and which are not. For any two $\mathbf{x} = (x_1, x_2)$ and $\mathbf{y} = (y_1, y_2)$ in $\mathbb{R}^2$,

(1) $\langle \mathbf{x}, \mathbf{y} \rangle = x_1x_2 + y_1y_2$,
(2) $\langle \mathbf{x}, \mathbf{y} \rangle = x_1y_1 + x_2y_2$,
(3) $\langle \mathbf{x}, \mathbf{y} \rangle = 4x_1y_1 + 2x_2y_2 - x_1y_2 - x_2y_1$,
(4) $\langle \mathbf{x}, \mathbf{y} \rangle = x_1y_2 - x_2y_1$,
(5) $\langle \mathbf{x}, \mathbf{y} \rangle = x_1y_1 + 3x_2y_2$,
(6) $\langle \mathbf{x}, \mathbf{y} \rangle = x_1y_1 - x_1y_2 - x_2y_1 + 3x_2y_2$.

5.2. *Every inner product is positive-definite, symmetric, and bilinear.* Show that the function $\langle A, B\rangle = \mathrm{tr}(A^T B)$ for $A, B \in M_{n\times n}(\mathbb{R})$ defines an inner product on the vector space $M_{n\times n}(\mathbb{R})$. Compute the length of the identity matrix I_n in this inner product space.

5.3. Let $\mathbf{x} = (1, 0, 0)$, $\mathbf{y} = (1, 1, 1)$ be two vectors in the 3-space $\mathbb{R}^3$.

(1) In the Euclidean 3-space $\mathbb{R}^3$ with the dot product, find the length $||\mathbf{y}||$ and $\cos\theta$ for the angle θ between $\mathbf{x}$ and $\mathbf{y}$.

(2) In a (non-Euclidean) $\mathbb{R}^3$ equipped with an inner product $\langle \mathbf{x}, \mathbf{y}\rangle = x_1y_1 + 2x_2y_2 + 3x_3y_3$, find the length $||\mathbf{y}||$ and $\cos\theta$ for the angle θ between $\mathbf{x}$ and $\mathbf{y}$.

5.4. Let $\mathbf{x} = (1, 1, 1, 1, 1)$, $\mathbf{y} = (1, 2, 3, 4, 5)$ be two vectors in the Euclidean 5-space $\mathbb{R}^5$ (with the dot product). Find the length $||\mathbf{y}||$ and $\cos\theta$ for the angle θ between $\mathbf{x}$ and $\mathbf{y}$.

5.5. Determine the values of k so that the given vectors are orthogonal with respect to the Euclidean inner product defined on $\mathbb{R}^4$.

$$(1)\left\{\begin{bmatrix}1\\1\\1\\1\end{bmatrix}, \begin{bmatrix}k\\2k\\3k\\4k\end{bmatrix}\right\}, \qquad (2)\left\{\begin{bmatrix}1\\1\\1\\1\end{bmatrix}, \begin{bmatrix}k\\k+1\\k+2\\k+3\end{bmatrix}\right\}.$$

5.6. Prove that $(a + b)^2 \le 2(a^2 + b^2)$ for any real numbers a, b. In general,

$$(a_1 + a_2 + \cdots + a_n)^2 \le n(a_1^2 + a_2^2 + \cdots + a_n^2)$$

for any real numbers $a_1, a_2, \ldots, a_n$. When does the equality hold?

Hint: Use the Cauchy-Schwarz inequality and Problem 100 with $\mathbf{x} = (a_1, \ldots, a_n)$ and $\mathbf{y} = (1, \ldots, 1)$ in the Euclidean space $\mathbb{R}^n$.

5.7. Let V be an inner product space. For any vectors $\mathbf{x}$ and $\mathbf{y}$ in V, establish the following identities:

(1) $\langle \mathbf{x}, \mathbf{y}\rangle = \frac{1}{4}\left(||\mathbf{x}+\mathbf{y}||^2 - ||\mathbf{x}-\mathbf{y}||^2\right)$ (polarization identity),

(2) $\langle \mathbf{x}, \mathbf{y}\rangle = \frac{1}{2}\left(||\mathbf{x}+\mathbf{y}||^2 - ||\mathbf{x}||^2 - ||\mathbf{y}||^2\right)$ (polarization identity),

(3) $||\mathbf{x}+\mathbf{y}||^2 + ||\mathbf{x}-\mathbf{y}||^2 = 2(||\mathbf{x}||^2 + ||\mathbf{y}||^2)$ (parallelogram equality).

Hint: $||\mathbf{x}+\mathbf{y}||^2 = ||\mathbf{x}||^2 + 2\langle \mathbf{x}, \mathbf{y}\rangle + ||\mathbf{y}||^2$ and $||\mathbf{x}-\mathbf{y}||^2 = ||\mathbf{x}||^2 - 2\langle \mathbf{x}, \mathbf{y}\rangle + ||\mathbf{y}||^2$.

5.8. *The matrix representation of an inner product $\langle\ ,\ \rangle$ depends on the choice of a basis.* Let $\alpha = \{\mathbf{e}_1, \mathbf{e}_2, \ldots, \mathbf{e}_n\}$ be the standard basis for the Euclidean space $\mathbb{R}^n$ and let $\beta = \{\mathbf{v}_1, \mathbf{v}_2, \ldots, \mathbf{v}_n\}$ be another basis for $\mathbb{R}^n$, where $\mathbf{v}_1 = \mathbf{e}_1 + \mathbf{e}_2 + \cdots + \mathbf{e}_n$, $\mathbf{v}_2 = \mathbf{e}_2 + \cdots + \mathbf{e}_n, \ldots, \mathbf{v}_n = \mathbf{e}_n$. (See Example 50). With respect to the (ordered) basis β, determine the matrix representation of the dot product on the space $\mathbb{R}^n$.

5.9. *The Gram-Schmidt orthogonalization process is constructing a rectangular coordinate system from a given oblique coordinate system.*
Find an orthonormal basis for the Euclidean 3-space $\mathbb{R}^3$ by applying the Gram-Schmidt orthogonalization to the three vectors $\mathbf{x} = (1, 0, 1)$, $\mathbf{x}_2 = (1, 0, -1)$, and $\mathbf{x}_3 = (0, 3, 4)$.

5.10. Show that if $\mathbf{u}$ is orthogonal to $\mathbf{v}$, then every scalar multiple of $\mathbf{u}$ is also orthogonal to $\mathbf{v}$. Find a unit vector which is orthogonal to $\mathbf{v}_1 = (1, 1, 2)$ and to $\mathbf{v}_2 = (0,\ 1, 3)$ in the Euclidean 3-space $\mathbb{R}^3$.

5.11. Determine the orthogonal projection of $\mathbf{v}_1$ onto $\mathbf{v}_2$ for the following vectors in the relevant Euclidean n-space $\mathbb{R}^n$ (with the dot product).

(1) $\mathbf{v}_1 = (1, 2, 3),\ \mathbf{v}_2 = (1, 1, 2)$,
(2) $\mathbf{v}_1 = (1, 2, 1),\ \mathbf{v}_2 = (2, 1, -1)$,
(3) $\mathbf{v}_1 = (1, 0, 1, 0),\ \mathbf{v}_2 = (0, 2, 2, 0)$.

5.12. With respect to the standard (ordered) basis for the relevant Euclidean space $\mathbb{R}^n$, find the projection matrix of each the orthogonal projections stated in Exercise **5.11**.

Hint: For each Proj_U, $\mathrm{Proj}_U(\mathbf{e}_j)$ is the j-th column of the projection matrix. Or, $[\mathrm{Proj}_U]_\alpha = \mathbf{u}\mathbf{u}^T$, if an orthonormal basis $\{\mathbf{u}\}$ for the one-dimensional space U is given.

5.13. Let S be the set of vectors $\mathbf{v}_i$, where $\mathbf{v}_i$'s are given below. For each S, find a basis for the subspace $S^\perp$ in the relevant Euclidean space $\mathbb{R}^n$.

(1) $\mathbf{v}_1 = (0, 1, 0),\ \mathbf{v}_2 = (0, 0, 1)$,
(2) $\mathbf{v}_1 = (1, 1, 0),\ \mathbf{v}_2 = (1, 1, 1)$,
(3) $\mathbf{v}_1 = (1, 0, 1, 2),\ \mathbf{v}_2 = (1, 1, 1, 1),\ \mathbf{v}_3 = (2, 2, 0, 1)$.

5.14. In any inner product space V, show that $\mathbf{x} + \mathbf{y}$ is perpendicular to $\mathbf{x} - \mathbf{y}$ if and only if $||\mathbf{x}|| = ||\mathbf{y}||$.

5.15. Let U and W be subspaces of an inner product space V. Show that

(1) $(U + W)^\perp = U^\perp \cap W^\perp$, (2) $(U \cap W)^\perp = U^\perp + W^\perp$.

5.16. Let U be the subspace of the Euclidean 4-space $\mathbb{R}^4$ spanned by $(1, 1, 0, 0)$ and $(1, 0, 1,\ 0)$, and let W be the subspace of $\mathbb{R}^4$ spanned by $(0, 1, 0, 1)$ and $(0, 0, 1, 1)$. Find a basis for and the dimension of each of the following subspaces:

(1) $U + W$, (2) $U^\perp$, (3) $U^\perp + W^\perp$, (4) $U \cap W$.

5.17. (*Distance from a point to a line in* $\mathbb{R}^3$) For a given line $L : x = 4 - t, y = -2 + 3t, z = 1 + 2t, (t \in \mathbb{R})$ and for a given point $P = (x_0, y_0, z_0)$ in the space $\mathbb{R}^3$, determine

(1) the projection matrix of Proj_W, where W is the one-dimensional subspace of $\mathbb{R}^3$ consisting of all vectors parallel to the line L,

(2) the distance d from the point P to the line L,

(3) the projection matrix of Proj_U, where U is the two-dimensional subspace of $\mathbb{R}^3$ consisting of all vectors perpendicular to the line L.

Hint: Review Example 92 on page 214.

5.18. (*Distance from a point to a plane in* $\mathbb{R}^3$) For a given plane $Y : 2x + y + z = 2$ and for a given point $P = (1, 1, 1)$ in the space $\mathbb{R}^3$, determine

(1) the projection matrix of Proj_W, where W is the two-dimensional subspace of $\mathbb{R}^3$ consisting of all vectors parallel with the plane Y,

(2) the distance d from the point P to the plane Y,

(3) the projection matrix of Proj_U, where U is the one-dimensional subspace of $\mathbb{R}^3$ consisting of all vectors perpendicular to the plane Y.

Hint: Review Example 92 on page 214.

5.19. Find the volume of

(1) the three-dimensional tetrahedron in the Euclidean 4-space $\mathbb{R}^4$ whose vertices are at $(0, 0, 0, 0)$, $(1, 0, 0, 0)$, $(0, 1, 2, 2)$, and $(0, 0, 1, 2)$, (see Problem 46),

(2) the three-dimensional tetrahedron in the Euclidean 5-space $\mathbb{R}^5$ whose vertices are at $(0, 0, 0, 0, 0)$, $(1, 0, 0, 0, 0)$, $(0, 1, 2, 2, 2)$, and $(0, 0, 1, 2, 2)$.

Hint: Review Example 93 on page 216.

5.20. Which of the following matrices are orthogonal?

$$(1) \begin{bmatrix} 1/2 & -1/3 \\ -1/2 & 1/3 \end{bmatrix}, \qquad (2) \begin{bmatrix} 4/5 & -3/5 \\ -3/5 & 4/5 \end{bmatrix},$$

$$(3) \begin{bmatrix} 1/\sqrt{2} & 0 & -1/\sqrt{2} \\ 0 & -1/\sqrt{2} & 1/\sqrt{2} \\ -1/\sqrt{2} & 1/\sqrt{2} & 0 \end{bmatrix}, \quad (4) \begin{bmatrix} 1/\sqrt{2} & 1/\sqrt{3} & -1/\sqrt{6} \\ 1/\sqrt{2} & -1/\sqrt{3} & 1/\sqrt{6} \\ 0 & 1/\sqrt{3} & 2/\sqrt{6} \end{bmatrix}.$$

5.21. For any orthogonal matrix A, show that $\det A = \pm 1$. Give an example of an orthogonal matrix A for which $\det A = -1$.

5.22. Find orthonormal bases for the row space $\mathcal{R}(A)$ and the null space $\mathcal{N}(A)$ of each of the following matrices.

$$(1)\ A = \begin{bmatrix} 1 & 1 & 0 \\ 1 & -2 & 1 \\ 1 & 0 & 2 \end{bmatrix}, \quad (2)\ A = \begin{bmatrix} 2 & 4 & 3 \\ 1 & 1 & 1 \\ 2 & 0 & 1 \end{bmatrix}, \quad (3)\ A = \begin{bmatrix} 1 & 1 & 1 & 1 \\ 1 & 0 & 1 & 0 \\ 1 & 0 & 0 & 1 \\ 1 & 0 & 0 & 0 \end{bmatrix}.$$

Hint: $\mathcal{N}(A) = \mathcal{N}(U)$ for reduced row-echelon form U of A.

5.23. Let W be the subspace of the Euclidean 3-space $\mathbb{R}^3$ spanned by the vectors $\mathbf{x}_1$ and $\mathbf{x}_2$ given in each of the following cases.

(1) $\mathbf{x}_1 = (1, 1, 1), \mathbf{x}_2 = (1, 0, 2),$ and $\mathbf{b} = (1, 1, 0)$,
(2) $\mathbf{x}_1 = (1, 1, 2), \mathbf{x}_2 = (1, 1, -1),$ and $\mathbf{b} = (1, 3, -2)$.

Find the projection matrix of the orthogonal projection Proj_W onto the subspace W and compute $\mathrm{Proj}_W(\mathbf{b})$ for $\mathbf{b}$ given in each case.

5.24. Find the (orthogonal) projection matrices of the orthogonal projections onto the row space, onto the null space, and onto the column space of each of the following matrices.

$$(1)\ \begin{bmatrix} 1 & 2 \\ 3 & 4 \end{bmatrix}, \qquad (2)\ \begin{bmatrix} 2 & 4 & 1 \\ 1 & 1 & 1 \end{bmatrix}, \qquad (3)\ \begin{bmatrix} 1 & 4 & 0 \\ 0 & 0 & 2 \\ 2 & 3 & -1 \end{bmatrix}.$$

Hint: For the orthogonal projection $\mathrm{Proj}_{C(A)}$ onto the column space $C(A)$, see Subsection 5.10.4.

5.25. Consider the vector space $C[-1, 1]$ equipped with the inner product defined by

$$\langle f, g \rangle = \int_{-1}^{1} f(x)g(x)dx.$$

A function $f \in C[-1, 1]$ is *even* if $f(-x) = f(x)$, or *odd* if $f(-x) = -f(x)$. Let U and V be the sets of all even functions and of all odd functions in $C[-1, 1]$, respectively.

(1) Show that both U and V are subspaces of $C[-1, 1]$ and $C[-1, 1] = U \oplus V$.
(2) Show that $U \perp V$.

5.26. Find the equation of the straight line that best fits to the four points $(0, 1)$, $(1, 3)$, $(2, 4)$, and $(3, 4)$.

5.27. Find the quadratic curve that best fits to the five points $(-1, -14)$, $(0, -5)$, $(1, -4)$, $(2, 1)$, and $(3, 22)$.

Compare it with 'the exact fitting polynomial problem' in Problem 42 on page 87.

5.28. Find all least squares solutions of the following systems of linear equations,

$$(1)\ \begin{cases} x + y = 5 \\ x - y = 2 \\ 2x - y = 7, \end{cases} \qquad (2)\ \begin{cases} 2x - 3y = 8 \\ 2x - y + z = 12 \\ 4x - 5y + z = 15 \\ 2x + 4z = 1. \end{cases}$$

5.29. Find a QR decomposition of each of the following matrices.

$$A = \begin{bmatrix} 1 & 2 & 3 \\ 0 & 2 & 3 \\ 0 & 0 & 3 \end{bmatrix}, \qquad B = \begin{bmatrix} 1 & 2 \\ 1 & 2 \\ 0 & 2 \end{bmatrix}.$$

5.30. Determine whether the following statements are true or false, in general, and justify your answers.

(1) On any vector space, one can define at least one inner product.

(2) For any $n \times n$ symmetric matrix A, $\langle \mathbf{x}, \mathbf{y} \rangle = \mathbf{x}^T A\mathbf{y}$ defines an inner product on $\mathbb{R}^n$.

(3) For a diagonal matrix $D = \begin{bmatrix} d_1 & 0 \\ 0 & d_2 \end{bmatrix}$, $\langle \mathbf{x}, \mathbf{y} \rangle = \mathbf{x}^T D\mathbf{y}$ defines an inner product on $\mathbb{R}^2$ if and only if the diagonal entries d_1, d_2 are both positive.

(4) Two nonzero vectors $\mathbf{x}$ and $\mathbf{y}$ in an inner product space are linearly independent if and only if the angle between $\mathbf{x}$ and $\mathbf{y}$ is not zero.

(5) If U is perpendicular to W, then $U^\perp$ is perpendicular to $W^\perp$.

(6) Let V be an inner product space. Then $||\mathbf{x} - \mathbf{y}|| \geq ||\mathbf{x}|| - ||\mathbf{y}||$ for any vectors $\mathbf{x}$ and $\mathbf{y}$ in V.

(7) The Pythagorean theorem holds in any inner product space.

(8) Every inner product space has at least one rectangular coordinate system like the xyz-coordinates in the Euclidean space $\mathbb{R}^3$.

(9) Every projection matrix is invertible.

(10) A square matrix A is a projection matrix if and only if A^T is a projection matrix.

(11) For a linear transformation $T : \mathbb{R}^n \to \mathbb{R}^n$, T is an orthogonal projection if and only if $id_{\mathbb{R}^n} - T$ is an orthogonal projection.

(12) For any $m \times n$ matrix A, the row space $\mathcal{R}(A)$ and the column space $\mathcal{C}(A)$ are orthogonal.

(13) Every projection is a linear transformation and every orthogonal projection is an orthogonal transformation.

(14) A linear transformation T on an inner product space preserves the lengths of vectors if and only if it preserves the inner product.

(15) A linear transformation on an inner product space preserves the lengths of vectors if and only if it preserves the angle between any two nonzero vectors.

(16) A linear transformation T on an inner product space is an isomorphism if and only if it is an isometry.

(17) Every permutation matrix is an orthogonal matrix.

(18) Every matrix representation of an orthogonal projection is an orthogonal matrix.

(19) For any $m \times n$ matrix A and $\mathbf{b} \in \mathbb{R}^m$, $A^T A\mathbf{x} = A^T\mathbf{b}$ always has a solution.

(20) The least squares solution of $A\mathbf{x} = \mathbf{b}$ is unique for any matrix A.

Hint: Exactly ten of them are true.

Chapter 6

Diagonalization makes it easy, as in Powers of Matrices

In solving a system of linear equations $A\mathbf{x} = \mathbf{b}$, or in solving the normal equation $A^T A\mathbf{x} = A^T\mathbf{b}$, if the matrix A is diagonal, then both can be solved quite easily.

With keeping this fact in mind, several other matrix problems will be challenged in this chapter by the following procedures:

- First convert a given problem to matrix form with a square matrix A,
- Next try to use a decomposition $A = QDQ^{-1}$ if A is similar to a diagonal matrix D.

Possible challenging problems are

(1) solving a system $A\mathbf{x} = \mathbf{b}$ or $A^T A\mathbf{x} = A^T\mathbf{b}$ of linear equations,
(2) checking the invertibility of A or estimating $\det A$,
(3) computing the matrix power A^n or the limit of a matrix series $\sum_{n=1}^{\infty} A^n$, [1]
(4) computing a matrix exponential $e^A = \sum_{k=0}^{\infty} \frac{A^k}{k!} = I + A + \frac{A^2}{2!} + \frac{A^3}{3!} + \cdots$,
(5) solving a system of linear differential equations or difference equations, and
(6) finding a simple form of the matrix representation of a linear transformation, etc.

The readers are suggested to try the problems (1)-(4) when the matrix A is diagonal. In fact, all are not difficult if A is diagonal.

---

[1] Infinite sum: $\sum_{n=1}^{\infty} A^n = \lim_{N\to\infty} \sum_{n=1}^{N} A^n$ by definition.

In this chapter we mainly discuss how to solve the same problems when the matrix A is diagonal or is similar to a diagonal matrix D. However, it is strongly related with *eigenvalues* and *eigenvectors* of the matrix A.

For this reason, we start with introducing an *eigenvalue* and an *eigenvector* of a square matrix. These play an important role in finding a decomposition $A = QDQ^{-1}$. They also have far-reaching applications, not only in mathematics but also in many other fields besides mathematics.

6.1 Eigenvalues and eigenvectors

Definition 6.1 Let A be an $n \times n$ square matrix. A nonzero vector $\mathbf{x}$ in the n-space $\mathbb{R}^n$ is called an **eigenvector** (or **characteristic vector**) of A if there is a scalar λ in $\mathbb{R}$ such that [2]

$$A\mathbf{x} = \lambda\mathbf{x}.$$

The scalar λ is called an **eigenvalue** (or **characteristic value**) of A, and the vector $\mathbf{x}$ is said to **belong** to λ, or λ is **associated** with an eigenvector $\mathbf{x}$.

Geometrically, an eigenvector of a matrix A is a nonzero vector $\mathbf{x}$ in the n-space $\mathbb{R}^n$ such that the vectors $\mathbf{x}$ and $A\mathbf{x}$ are collinear. In other words, the subspace W spanned by $\mathbf{x}$ is **invariant** under the linear transformation $A : \mathbb{R}^n \to \mathbb{R}^n$ in the sense $A(W) \subset W$. Or, W is called an **invariant subspace** of the matrix A.

Algebraically, an eigenvector $\mathbf{x}$ is any nontrivial solution of the homogeneous system $(\lambda I - A)\mathbf{x} = \mathbf{0}$ of linear equations, that is, $\mathbf{x}$ is any nonzero vector in the null space $\mathcal{N}(\lambda I - A)$. This null space is called the **eigenspace** of A belonging to λ and is sometimes denoted by $E(\lambda)$.

As a trivial example, let A be a diagonal matrix, say

$$A = \begin{bmatrix} d_1 & 0 & \cdots & 0 \\ 0 & d_2 & & \vdots \\ \vdots & & \ddots & 0 \\ 0 & \cdots & 0 & d_n \end{bmatrix}.$$

Then $A\mathbf{e}_i = d_i\mathbf{e}_i$ is the i-th column of A for $i = 1, 2, \ldots, n$. Thus, each diagonal entry d_i is an eigenvalue of A associated with the eigenvector $\mathbf{e}_i$ and each (Euclidean) coordinate axis is an invariant subspace of A. Each i-th coordinate axis is the eigenspace $E(d_i)$.

[2] λ is a letter lambda in Greek

There are two unknowns in the system $(\lambda I - A)\mathbf{x} = \mathbf{0}$;

$$\text{an eigenvalue } \lambda \quad \text{and} \quad \text{an eigenvector } \mathbf{x}.$$

To find these two unknowns, first determine the eigenvalues λ by using the fact that the system $(\lambda I - A)\mathbf{x} = \mathbf{0}$ has a nontrivial solution $\mathbf{x}$ if and only if λ satisfies the equation

$$\det(\lambda I - A) = 0.$$

This is called the **characteristic equation** of A. Here, $\det(\lambda I - A)$ is a polynomial of degree n in λ, called the **characteristic polynomial** of A. Thus, the eigenvalues λ are just the roots of the characteristic equation $\det(\lambda I - A) = 0$.

Next, the eigenvectors of A can be determined by solving the homogeneous system $(\lambda I - A)\mathbf{x} = \mathbf{0}$ for each eigenvalue λ.

In summary, by referring to Theorem 3.29, one can have the following theorem.

Theorem 6.2 *For any square matrix A, the following are equivalent.*

(1) *λ is an eigenvalue of A,*
(2) $\det(\lambda I - A) = 0$ *(or equivalently* $\det(A - \lambda I) = 0$*),*
(3) *$\lambda I - A$ is singular, i.e., non-invertible,*
(4) *the homogeneous system $(\lambda I - A)\mathbf{x} = \mathbf{0}$ has a nontrivial solution.*

Example 105 (*Matrix having distinct eigenvalues*) Find the eigenvalues and eigenvectors of

$$A = \begin{bmatrix} 2 & 1 \\ 0 & 1 \end{bmatrix}.$$

Solution: *First determine the eigenvalues.* The characteristic polynomial of A is

$$\det(\lambda I - A) = \det \begin{bmatrix} \lambda - 2 & -1 \\ 0 & \lambda - 1 \end{bmatrix} = (\lambda - 1)(\lambda - 2).$$

Thus the eigenvalues are

$$\lambda_1 = 1 \quad \text{and} \quad \lambda_2 = 2.$$

To determine the eigenvectors belonging to each λ_i, it is needed to solve the homogeneous system of equations $(\lambda_i I - A)\mathbf{x} = \mathbf{0}$.

Let us take $\lambda_1 = 1$ first. Then the system $(\lambda_1 I - A)\mathbf{x} = \mathbf{0}$ becomes

$$\begin{cases} -x_1 & - & x_2 & = & 0, \\ 0x_1 & + & 0x_2 & = & 0. \end{cases}$$

Hence, $x_1 = -x_2$ and $\mathbf{x}_1 = (1, -1)$ is an eigenvector belonging to $\lambda_1 = 1$. The eigenspace is $E(1) = \{t\mathbf{x}_1 : t \in \mathbb{R}\}$. (Here, one can choose any nonzero solution $(t, -t)$ as an eigenvector $\mathbf{x}_1$ belonging to $\lambda_1 = 1$.)

For $\lambda_2 = 2$, the system $(\lambda_2 I - A)\mathbf{x} = \mathbf{0}$ becomes

$$\begin{cases} 0x_1 & - & x_2 & = & 0, \\ 0x_1 & + & x_2 & = & 0. \end{cases}$$

Thus, $\mathbf{x}_2 = (1, 0)$ is an eigenvector belonging to λ_2 and $E(2) = \{t\mathbf{x}_2 : t \in \mathbb{R}\}$.

Note that the eigenvectors $\mathbf{x}_1$ and $\mathbf{x}_2$ belonging to the (distinct) eigenvalues λ_1 and λ_2, respectively are linearly independent. □

Example 106 (3×3 *matrix having a repeated eigenvalue with three linearly independent eigenvectors*) Find bases for the eigenspaces of

$$A = \begin{bmatrix} 3 & 2 & 0 \\ 2 & 3 & 0 \\ 0 & 0 & 5 \end{bmatrix}.$$

Solution: *First determine the eigenvalues.* The characteristic polynomial of A is $\det(\lambda I - A) = (\lambda - 1)(\lambda - 5)^2$, so that the eigenvalues of A are

$$\lambda_1 = 1 \quad \text{and} \quad \lambda_2 = 5 \text{ of multiplicity } 2.$$

Thus, there are two eigenspaces of A; $E(1)$ and $E(5)$.

If $\lambda_1 = 1$, then the system $(\lambda I - A)\mathbf{x} = \mathbf{0}$ becomes

$$\begin{bmatrix} -2 & -2 & 0 \\ -2 & -2 & 0 \\ 0 & 0 & -4 \end{bmatrix} \begin{bmatrix} x_1 \\ x_2 \\ x_3 \end{bmatrix} = \begin{bmatrix} 0 \\ 0 \\ 0 \end{bmatrix}.$$

Solving this system yields $x_1 = t$, $x_2 = -t$, $x_3 = 0$ for any $t \in \mathbb{R}$. Thus, the eigenvectors belonging to $\lambda_1 = 1$ are nonzero vectors of the form

$$\mathbf{x} = \begin{bmatrix} t \\ -t \\ 0 \end{bmatrix} = t \begin{bmatrix} 1 \\ -1 \\ 0 \end{bmatrix}, \quad t \in \mathbb{R},$$

so that $(1, -1, 0)$ is a basis for the eigenspace $E(\lambda_1)$ belonging to $\lambda_1 = 1$.

If $\lambda_2 = 5$, then the system $(\lambda I - A)\mathbf{x} = \mathbf{0}$ becomes

$$\begin{bmatrix} 2 & -2 & 0 \\ -2 & 2 & 0 \\ 0 & 0 & 0 \end{bmatrix} \begin{bmatrix} x_1 \\ x_2 \\ x_3 \end{bmatrix} = \begin{bmatrix} 0 \\ 0 \\ 0 \end{bmatrix}.$$

Its solutions are $x_1 = s$, $x_2 = s$, $x_3 = t$ for any $s, t \in \mathbb{R}$. Thus, the eigenvectors of A belonging to $\lambda_2 = 5$ are nonzero vectors of the form

$$\mathbf{x} = \begin{bmatrix} s \\ s \\ t \end{bmatrix} = \begin{bmatrix} s \\ s \\ 0 \end{bmatrix} + \begin{bmatrix} 0 \\ 0 \\ t \end{bmatrix} = s \begin{bmatrix} 1 \\ 1 \\ 0 \end{bmatrix} + t \begin{bmatrix} 0 \\ 0 \\ 1 \end{bmatrix}$$

for any $s, t \in \mathbb{R}$. Since $(1, 1, 0)$ and $(0, 0, 1)$ are linearly independent, they form a basis for the eigenspace $E(\lambda_2)$ belonging to $\lambda_2 = 5$.

Consequently, the 3×3 matrix A has three linearly independent eigenvectors $(1, -1, 0)$, $(1, 1, 0)$, and $(0, 0, 1)$. □

As shown in Examples 105 and 106, every 2×2 or 3×3 matrix A can have at most 2 or 3 eigenvalues λ because the characteristic polynomial of A is of degree 2 or 3 in λ, depending on the size of A. However, for each eigenvalue λ, there are infinitely many different choices of an eigenvector $\mathbf{x}$ belonging to λ.

For each eigenvalue λ of A in Examples 105 and 106, one can see that the dimension of the eigenspace $E(\lambda)$ is equal to the multiplicity of λ as a root of the equation $\det(\lambda I - A) = 0$.

But, in general it is not true as shown in the next example.

Example 107 (5×5 *matrix having a repeated eigenvalue with only one linearly independent eigenvector*) Consider the matrix

$$A = \begin{bmatrix} k & 1 & 0 & 0 & 0 \\ 0 & k & 1 & 0 & 0 \\ 0 & 0 & k & 1 & 0 \\ 0 & 0 & 0 & k & 1 \\ 0 & 0 & 0 & 0 & k \end{bmatrix}.$$

A simple computation shows that the characteristic polynomial of the matrix A is $(\lambda - k)^5$ so that the eigenvalue of A is

$$\lambda = k \quad \text{of multiplicity 5.}$$

Clearly, the rank of

$$kI - A = \begin{bmatrix} 0 & -1 & 0 & 0 & 0 \\ 0 & 0 & -1 & 0 & 0 \\ 0 & 0 & 0 & -1 & 0 \\ 0 & 0 & 0 & 0 & -1 \\ 0 & 0 & 0 & 0 & 0 \end{bmatrix}$$

is 4 and by the Rank-Nullity theorem, $\dim E(k) = \dim \mathcal{N}(kI - A) = 1$. Hence there is only one linearly independent eigenvector belonging to $\lambda = k$ and it can be chosen to be

$$\mathbf{e}_1 = (1, 0, 0, 0, 0),$$

because $A\mathbf{e}_1 = k\mathbf{e}_1$ is the first column of A.

This example shows that $\dim E(k) = \dim \mathcal{N}(kI - A)$ can be strictly less than the multiplicity of λ. □

The characteristic equation $\det(\lambda I - A) = 0$ may have complex roots, which then are called **complex eigenvalues**. However, the complex numbers are not scalars for a real vector space. In many cases, it is necessary to allow those complex roots as eigenvalues. It means, it is necessary to expand the set of scalars to the set of complex numbers.

This expansion of the set of scalars to the set of complex numbers leads us to work with complex vector spaces. These will be treated in Chapter 7 separately.

In this chapter we restrict our discussion to the case of real eigenvalues even though the entire discussion in this chapter applies in the same way to complex vector spaces.

Example 108 (*Matrix having no real eigenvalues*) The characteristic polynomial of the matrix

$$A = \begin{bmatrix} \cos\theta & -\sin\theta \\ \sin\theta & \cos\theta \end{bmatrix}$$

is

$$\lambda^2 - 2\cos\theta\lambda + (\cos^2\theta + \sin^2\theta).$$

Thus, the eigenvalues are

$$\lambda = \cos\theta \pm i\sin\theta,$$

which are complex numbers, so this matrix as a rotation of $\mathbb{R}^2$ has no real eigenvalues unless $\theta = n\pi, n = 0, \pm 1, \pm 2, \ldots$.

Clearly, no one-dimension subspaces are *invariant* under a nontrivial rotation if the rotation angle is neither 0 nor π. □

Problem 136 For any 3×3 matrix $A = [a_{ij}]$, show that $\det(\lambda I - A) = \lambda^3 - (a_{11} + a_{22} + a_{33})\lambda^2 + \cdots$, which is a polynomial of degree 3 in λ. In general, for any $n \times n$ matrix A, $\det(\lambda I - A) = \lambda^n - \operatorname{tr}(A)\lambda^{n-1} + \cdots$, which is a polynomial of degree n in λ.

Problem 137 Let λ be an eigenvalue of A and let $\mathbf{x}$ be an eigenvector belonging to λ. Use mathematical induction to show that λ^m is an eigenvalue of A^m and $\mathbf{x}$ is an eigenvector of A^m belonging to λ^m for each $m = 1, 2, \ldots$.

Hint: $A^2\mathbf{x} = A(A\mathbf{x}) = A(\lambda\mathbf{x}) = \lambda A\mathbf{x} = \lambda^2\mathbf{x}$.

6.2 Properties of eigenvalues and eigenvectors

We introduce some basic properties of eigenvalues and eigenvectors.

Lemma 6.3 **(1)** *Two similar matrices have the same characteristic polynomial.*

(2) *If A is a triangular matrix, then the diagonal entries are exactly the eigenvalues of A.*

Proof: (1) Let two matrices A and B be similar and let Q be a nonsingular matrix such that $B = Q^{-1}AQ$. Then

$$\begin{aligned}\det(\lambda I - B) &= \det\left(Q^{-1}(\lambda I)Q - Q^{-1}AQ\right) \\ &= \det\left(Q^{-1}(\lambda I - A)Q\right) \\ &= \det Q^{-1} \det(\lambda I - A) \det Q \\ &= \det(\lambda I - A).\end{aligned}$$

(2) The characteristic equation of an upper triangular matrix is

$$\begin{aligned}\det(\lambda I - A) &= \det\begin{bmatrix}\lambda - a_{11} & * & \cdots & * \\ & \ddots & \ddots & \vdots \\ & & \ddots & * \\ \mathbf{0} & & & \lambda - a_{nn}\end{bmatrix} \\ &= (\lambda - a_{11}) \cdots (\lambda - a_{nn}) = 0.\end{aligned}$$

Hence, the diagonal entries are the eigenvalues of A. □

By Lemma 6.3(1), two similar matrices have the same eigenvalues, *i.e., the eigenvalues are invariant under the similarity* . However, their eigenvectors might be different: In fact, when $B = Q^{-1}AQ$, $\mathbf{x}$ is an eigenvector of B belonging to λ if and only if $Q\mathbf{x}$ is an eigenvector of A belonging to λ, because

$$(Q^{-1}AQ)\mathbf{x} = B\mathbf{x} = \lambda\mathbf{x} \;\Leftrightarrow\; A(Q\mathbf{x}) = \lambda(Q\mathbf{x}).$$

Theorem 6.4 *Let an $n \times n$ matrix A have n eigenvalues $\lambda_1, \lambda_2, \ldots, \lambda_n$ possibly with repetition. Then,*

(1) $\det A = \lambda_1 \lambda_2 \cdots \lambda_n$, (*the product of the n eigenvalues*),
(2) $\mathrm{tr}(A) = \lambda_1 + \lambda_2 + \cdots + \lambda_n$, (*the sum of the n eigenvalues*).

Proof: (1) Since eigenvalues $\lambda_1, \lambda_2, \ldots, \lambda_n$ are the roots of the characteristic polynomial of A, we have

$$\det(\lambda I - A) = (\lambda - \lambda_1)\,(\lambda - \lambda_2)\,\cdots\,(\lambda - \lambda_n).$$

By taking $\lambda = 0$ in both sides, one can get

$$(-1)^n \det A = (-1)^n \lambda_1 \lambda_2 \cdots \lambda_n.$$

(2) On the other hand,

$$\begin{aligned}(\lambda - \lambda_1)(\lambda - \lambda_2)\cdots(\lambda - \lambda_n) &= \det(\lambda I - A) \\ &= \det \begin{bmatrix} \lambda - a_{11} & -a_{12} & \cdots & -a_{1n} \\ -a_{21} & \lambda - a_{22} & \cdots & -a_{2n} \\ \vdots & \vdots & & \vdots \\ -a_{n1} & -a_{n2} & \cdots & \lambda - a_{nn} \end{bmatrix},\end{aligned}$$

which is a polynomial of the form $p(\lambda) = \lambda^n + c_{n-1}\lambda^{n-1} + \cdots + c_1\lambda + c_0$ in λ. One can compute the coefficient c_{n-1} of λ^{n-1} in two different ways by expanding both sides. This gives

$$\lambda_1 + \lambda_2 + \cdots + \lambda_n = a_{11} + a_{22} + \cdots + a_{nn} = \mathrm{tr}(A).$$

See Problem 136. □

Problem 138 Show that

(1) for any 2×2 matrix A, $\det(\lambda I - A) = \lambda^2 - \mathrm{tr}(A)\lambda + \det A$,
(2) for any 3×3 matrix A,

$$\begin{aligned}\det(\lambda I - A) &= \lambda^3 - \mathrm{tr}(A)\lambda^2 + \frac{1}{2}\sum_{i \neq j}(a_{ii}a_{jj} - a_{ij}a_{ji})\,\lambda - \det A \\ &= \lambda^3 - \mathrm{tr}(A)\lambda^2 + \frac{1}{2}\left((\mathrm{tr}(A))^2 - \mathrm{tr}(A^2)\right)\lambda - \det A.\end{aligned}$$

Hint: Compare the coefficients of λ^i on both sides. See Problem 136 for (2).

In Theorem 6.4, it is assumed that the matrix A has n (real) eigenvalues counting multiplicities. But, by allowing the scalars to be complex numbers, every $n \times n$ matrix has n eigenvalues counting multiplicities, so that Theorem 6.4 remains true for any square matrix, even it has a complex eigenvalue. [3]

Corollary 6.4.1 *Two similar matrices have the same determinant and the same trace.*

Any square matrix A is singular if and only if $\det A = 0$. However, $\det A$ is the product of its n eigenvalues. Thus a square matrix A is singular if and only if zero is an eigenvalue of A, equivalently

A is invertible if and only if zero is not an eigenvalue of A.

This gives a proof of Theorem 3.29(19).

The following two corollaries are easy consequences of this fact.

Corollary 6.4.2 *For any $n \times n$ matrices A and B, the following are equivalent.*

(1) *Zero is an eigenvalue of AB,*
(2) *A or B is singular,*
(3) *Zero is an eigenvalue of BA.*

Corollary 6.4.3 *For any $n \times n$ matrices A and B, the matrices AB and BA have the same set of eigenvalues.*

---

[3]By allowing complex scalars, *any polynomial of degree n (even with complex coefficients) has n complex roots counting multiplicities.* This is well known as *the fundamental theorem of algebra.*

Proof: By Corollary 6.4.2, zero is an eigenvalue of AB if and only if it is an eigenvalue of BA. Let λ be a nonzero eigenvalue of AB with $(AB)\mathbf{x} = \lambda\mathbf{x}$ for a nonzero vector $\mathbf{x}$. Then the vector $B\mathbf{x}$ is not zero because $\lambda \neq 0$, and

$$(BA)(B\mathbf{x}) = B(\lambda\mathbf{x}) = \lambda(B\mathbf{x}).$$

Hence, $B\mathbf{x}$ is an eigenvector of BA belonging to the eigenvalue λ, and hence λ is an eigenvalue of BA.

By a similar argument, any nonzero eigenvalue of BA is also an eigenvalue of AB. □

Problem 139 Show that A and A^T have the same eigenvalues of the same multiplicities. Do they necessarily have the same eigenvectors?

Hint: $\det(\lambda I - A) = \det(\lambda I - A)^T = \det(\lambda I - A^T)$. It is a fact that *every square matrix is similar to its transpose.* See Exercise **4.25** and Exercise **8.22**.)

Problem 140 Let $\lambda_1, \lambda_2, \ldots, \lambda_n$ be the eigenvalues of an $n \times n$ matrix A. If A is invertible, then the inverse A^{-1} has the eigenvalues $1/\lambda_1, 1/\lambda_2, \ldots, 1/\lambda_n$.

Problem 141 Give an example of two matrices A and B such that $\det A = \det B$, $\operatorname{tr}(A) = \operatorname{tr}(B)$, but A is not similar to B.

6.3 Diagonalizations and Eigenvalue decompositions(EVD)

In this section it will be shown what kinds of square matrices are similar to diagonal matrices. That is, given a square matrix A, it will be determined whether there exists an invertible matrix Q such that $Q^{-1}AQ$ is diagonal, and if so how one can find such a matrix Q.

Definition 6.5 A square matrix A is said to be **diagonalizable** if there exists an invertible matrix Q such that $Q^{-1}AQ$ is diagonal (*i.e., A* is similar to a diagonal matrix).

For an $n \times n$ matrix A, every eigenvector of A is a nonzero vector in $\mathbb{R}^n$. Hence, A can have at most n linearly independent eigenvectors.

When an $n \times n$ matrix A has n linearly independent eigenvectors, the matrix A is said to have a full set of linearly independent eigenvectors.

The next theorem characterizes a diagonalizable matrix, and the proof shows a practical way of diagonalizing a matrix.

Theorem 6.6 *Let A be an $n \times n$ matrix. Then A is diagonalizable if and only if A has n linearly independent eigenvectors, that is, A has a full set of linearly independent eigenvectors.*

Proof: ($\Rightarrow$) Suppose that A is diagonalizable. Then there is an invertible matrix Q such that $Q^{-1}AQ$ is diagonal, say

$$Q^{-1}AQ = D = \begin{bmatrix} \lambda_1 & 0 & \cdots & 0 \\ 0 & \lambda_2 & \cdots & 0 \\ \vdots & \vdots & \ddots & \vdots \\ 0 & 0 & \cdots & \lambda_n \end{bmatrix},$$

or, equivalently, $AQ = QD$. Let $\mathbf{x}_1, \mathbf{x}_2, \ldots, \mathbf{x}_n$ denote the column vectors of Q. Since

$$\begin{aligned} AQ &= [A\mathbf{x}_1 \; A\mathbf{x}_2 \; \cdots \; A\mathbf{x}_n], \\ QD &= [\lambda_1\mathbf{x}_1 \; \lambda_2\mathbf{x}_2 \; \cdots \; \lambda_n\mathbf{x}_n], \end{aligned}$$

the matrix equation $AQ = QD$ implies

$$A\mathbf{x}_i = \lambda_i\mathbf{x}_i \quad \text{for } i = 1, 2, \ldots, n.$$

Moreover, since Q is invertible, its column vectors cannot be zero and are linearly independent, that is, the $\mathbf{x}_i$'s are n linearly independent eigenvectors of A.

($\Leftarrow$) Assume that A has n linearly independent eigenvectors $\mathbf{x}_1, \mathbf{x}_2, \ldots, \mathbf{x}_n$, say $A\mathbf{x}_i = \lambda_i\mathbf{x}_i$ for $i = 1, 2, \ldots, n$. Let

$$Q = [\mathbf{x}_1 \; \mathbf{x}_2 \; \cdots \; \mathbf{x}_n]$$

with the eigenvectors $\mathbf{x}_j$ as the columns. Then the equations $A\mathbf{x}_i = \lambda_i\mathbf{x}_i$ can be written as $AQ = QD$, where D is the diagonal matrix having the eigenvalues $\lambda_1, \lambda_2, \ldots, \lambda_n$ on the diagonal. Since the column vectors of Q are assumed to be linearly independent, Q is invertible, so $Q^{-1}AQ = D$. □

For a diagonalizable matrix A, $A = QDQ^{-1}$ is called an **eigenvalue decomposition(EVD)**, or an **eigenvalue factorization** of a diagonalizable matrix A.

Remark: (1) The proof of Theorem 6.6 reveals how to diagonalize an $n \times n$ (diagonalizable) matrix A, giving the following algorithm for diagonalizing a matrix.

Step 1 Find n linearly independent eigenvectors $\mathbf{x}_1, \mathbf{x}_2, \ldots, \mathbf{x}_n$ of A.
Step 2 Form the matrix $Q = [\mathbf{x}_1\ \mathbf{x}_2\ \cdots\ \mathbf{x}_n]$.
Step 3 The matrix $Q^{-1}AQ$ will be a diagonal matrix with $\lambda_1, \lambda_2, \ldots, \lambda_n$ as its successive diagonal entries, where λ_j is the eigenvalue associated with the eigenvector $\mathbf{x}_j$, $j = 1, 2, \ldots, n$.

(2) Let α denote the standard basis for $\mathbb{R}^n$ and let $\beta = \{\mathbf{x}_1, \mathbf{x}_2, \ldots, \mathbf{x}_n\}$ be a basis for $\mathbb{R}^n$ consisting of n linearly independent eigenvectors of A. Then,

(i) the matrix

$$Q = [\mathbf{x}_1\ \mathbf{x}_2\ \cdots\ \mathbf{x}_n] = [[\mathbf{x}_1]_\alpha\ [\mathbf{x}_2]_\alpha\ \cdots\ [\mathbf{x}_n]_\alpha] = [id]_\beta^\alpha$$

is the change of basis matrix from β to α, and

(ii) *the matrix representation of A with respect to β is diagonal:*

$$[A]_\beta = [id]_\alpha^\beta [A]_\alpha [id]_\beta^\alpha = Q^{-1}AQ = \begin{bmatrix} \lambda_1 & & \mathbf{0} \\ & \ddots & \\ \mathbf{0} & & \lambda_n \end{bmatrix},$$

where the diagonal entries λ_i's are the eigenvalues of A.

Example 109 (*Not all matrices are diagonalizable*) As a simple example, consider

$$A = \begin{bmatrix} 0 & 1 \\ 0 & 0 \end{bmatrix}.$$

Its eigenvalues are $\lambda_1 = \lambda_2 = 0$. If A were diagonalizable, then

$$Q^{-1}AQ = \begin{bmatrix} \lambda_1 & 0 \\ 0 & \lambda_2 \end{bmatrix} = \mathbf{0}$$

for some invertible matrix Q, and then A should be the zero matrix; $A = Q\mathbf{0}Q^{-1} = \mathbf{0}$. Since A is not the zero matrix, no invertible matrices Q exist so that $Q^{-1}AQ$ is diagonal.

As another example, the matrix

$$A = \begin{bmatrix} k & 1 & 0 & 0 & 0 \\ 0 & k & 1 & 0 & 0 \\ 0 & 0 & k & 1 & 0 \\ 0 & 0 & 0 & k & 1 \\ 0 & 0 & 0 & 0 & k \end{bmatrix}$$

shown in Example 107 has only one linearly independent eigenvector and hence cannot be diagonalizable for any k. □

Example 110 (*Several different ways to diagonalize a matrix*) Diagonalize the matrix

$$A = \begin{bmatrix} 1 & 1 & 0 \\ 0 & 2 & 1 \\ 0 & 0 & 3 \end{bmatrix}.$$

Solution: Since A is upper triangular, the eigenvalues of A are the diagonal entries

$$\lambda_1 = 1, \quad \lambda_2 = 2, \quad \text{and} \quad \lambda_3 = 3,$$

and their associated eigenvectors can be chosen to be

$$\mathbf{x}_1 = (1, 0, 0), \quad \mathbf{x}_2 = (1, 1, 0), \quad \text{and} \quad \mathbf{x}_3 = (1, 2, 2),$$

respectively. They are linearly independent. Thus, the matrix

$$P = [\mathbf{x}_1 \ \mathbf{x}_2 \ \mathbf{x}_3] = \begin{bmatrix} 1 & 1 & 1 \\ 0 & 1 & 2 \\ 0 & 0 & 2 \end{bmatrix}$$

diagonalizes A. In fact, one can verify that

$$\begin{aligned} P^{-1}AP &= \begin{bmatrix} 1 & -1 & 1/2 \\ 0 & 1 & -1 \\ 0 & 0 & 1/2 \end{bmatrix} \begin{bmatrix} 1 & 1 & 0 \\ 0 & 2 & 1 \\ 0 & 0 & 3 \end{bmatrix} \begin{bmatrix} 1 & 1 & 1 \\ 0 & 1 & 2 \\ 0 & 0 & 2 \end{bmatrix} \\ &= \begin{bmatrix} 1 & 0 & 0 \\ 0 & 2 & 0 \\ 0 & 0 & 3 \end{bmatrix} = \begin{bmatrix} \lambda_1 & 0 & 0 \\ 0 & \lambda_2 & 0 \\ 0 & 0 & \lambda_3 \end{bmatrix}. \end{aligned}$$

What would happen if one chooses different eigenvectors belonging to the eigenvalues $1, 2$, and 3? According to the proof of Theorem 6.6, nothing would happen: Any matrix whose columns are linearly independent eigenvectors will diagonalize A. For example,

$$-\mathbf{x}_1 = (-1, 0, 0), \quad -\mathbf{x}_2 = (-1, -1, 0), \quad \text{and} \quad 2\mathbf{x}_3 = (2, 4, 4)$$

are also eigenvectors belonging to the eigenvalues

$$\lambda_1 = 1, \quad \lambda_2 = 2, \quad \text{and} \quad \lambda_3 = 3,$$

respectively, and the matrix

$$Q = [-\mathbf{x}_1 \ -\mathbf{x}_2 \ 2\mathbf{x}_3] = \begin{bmatrix} -1 & -1 & 2 \\ 0 & -1 & 4 \\ 0 & 0 & 4 \end{bmatrix}$$

diagonalizes A as

$$Q^{-1}AQ = \begin{bmatrix} 1 & 0 & 0 \\ 0 & 2 & 0 \\ 0 & 0 & 3 \end{bmatrix} = \begin{bmatrix} \lambda_1 & 0 & 0 \\ 0 & \lambda_2 & 0 \\ 0 & 0 & \lambda_3 \end{bmatrix}.$$

A change in the order of the eigenvectors in constructing a matrix Q does not affect the diagonalizability of A, but the eigenvalues on the diagonal of the resulting diagonal matrix should appear in accordance with the order of the eigenvectors in the change of basis matrix.

For example, a matrix

$$S = [\mathbf{x}_1\ \mathbf{x}_3\ \mathbf{x}_2] = \begin{bmatrix} 1 & 1 & 1 \\ 0 & 2 & 1 \\ 0 & 2 & 0 \end{bmatrix}$$

also diagonalizes A, because its columns are linearly independent eigenvectors. In fact, one can show that

$$S^{-1} = \begin{bmatrix} 1 & -1 & 1/2 \\ 0 & 0 & 1/2 \\ 0 & 1 & -1 \end{bmatrix} \quad \text{and} \quad S^{-1}AS = \begin{bmatrix} 1 & 0 & 0 \\ 0 & 3 & 0 \\ 0 & 0 & 2 \end{bmatrix} = \begin{bmatrix} \lambda_1 & 0 & 0 \\ 0 & \lambda_3 & 0 \\ 0 & 0 & \lambda_2 \end{bmatrix},$$

with a different order of the eigenvalues.

As an additional consequence, an EVD of A is not unique. □

Problem 142 Show that the following matrices are not diagonalizable.

$$(1)\ A = \begin{bmatrix} \lambda & 1 & 0 \\ 0 & \lambda & 1 \\ 0 & 0 & \lambda \end{bmatrix}, \qquad (2)\ B = \begin{bmatrix} \lambda & 0 & 0 \\ 1 & \lambda & 0 \\ 0 & 1 & \lambda \end{bmatrix}, \qquad \lambda \text{ is any scalar.}$$

Problem 143 Construct a 2×2 matrix A whose eigenvalues are 2 and 3, and whose eigenvectors are $(2, 1)$ and $(3, 2)$, respectively.

From Theorem 6.6, we learn how to diagonalize a matrix and what the diagonal matrix is when the matrix has a full set of linearly independent eigenvectors.

The next question is which square matrix A can have a full set of linearly independent eigenvectors.

The following theorem shows that it can happen when an $n \times n$ matrix has n distinct eigenvalues.

Theorem 6.7 *Let $\lambda_1, \lambda_2, \ldots, \lambda_k$ be distinct eigenvalues of a matrix A and let $\mathbf{x}_1, \mathbf{x}_2, \ldots, \mathbf{x}_k$ be any eigenvectors belonging to them, respectively. Then $\{\mathbf{x}_1, \mathbf{x}_2, \ldots, \mathbf{x}_k\}$ is linearly independent.*

Proof: Let r be the largest integer such that $\{\mathbf{x}_1, \ldots, \mathbf{x}_r\}$ is linearly independent. If $r = k$, then there is nothing to prove. Suppose not, *i.e.*, $1 \le r < k$. Then $\{\mathbf{x}_1, \mathbf{x}_2, \ldots, \mathbf{x}_{r+1}\}$ is linearly dependent. Thus, there exist scalars $c_1, c_2, \ldots, c_{r+1}$ with $c_{r+1} \neq 0$ such that

$$c_1\mathbf{x}_1 + c_2\mathbf{x}_2 + \cdots + c_{r+1}\mathbf{x}_{r+1} = \mathbf{0}. \tag{1}$$

Multiplying both sides by A and using

$$A\mathbf{x}_1 = \lambda_1\mathbf{x}_1, \; A\mathbf{x}_2 = \lambda_2\mathbf{x}_2, \; \ldots, \; A\mathbf{x}_{r+1} = \lambda_{r+1}\mathbf{x}_{r+1},$$

one can get

$$c_1\lambda_1\mathbf{x}_1 + c_2\lambda_2\mathbf{x}_2 + \cdots + c_{r+1}\lambda_{r+1}\mathbf{x}_{r+1} = \mathbf{0}. \tag{2}$$

Multiplying both sides of (1) by λ_{r+1} and subtracting the resulting equation from (2) yields

$$c_1(\lambda_1 - \lambda_{r+1})\mathbf{x}_1 + c_2(\lambda_2 - \lambda_{r+1})\mathbf{x}_2 + \cdots + c_r(\lambda_r - \lambda_{r+1})\mathbf{x}_r = \mathbf{0}.$$

Since $\{\mathbf{x}_1, \mathbf{x}_2, \ldots, \mathbf{x}_r\}$ is linearly independent and $\lambda_1, \lambda_2, \ldots, \lambda_{r+1}$ are all distinct, it follows that $c_1 = c_2 = \cdots = c_r = 0$. Substituting these values in (1) yields $c_{r+1} = 0$, which is a contradiction to the assumption. □

As a consequence of Theorems 6.6 and 6.7, one can get the following.

Corollary 6.7.1 *If an $n \times n$ matrix A has n distinct eigenvalues, then A is diagonalizable.*

It follows from Corollary 6.7.1 that if $\mathbf{x}_1, \mathbf{x}_2, \ldots, \mathbf{x}_n$ are eigenvectors of an $n \times n$ matrix A belonging to n distinct eigenvalues $\lambda_1, \lambda_2, \ldots, \lambda_n$, respectively, then they form a basis for $\mathbb{R}^n$ and the matrix representation of A with respect to this basis is a diagonal matrix as shown in Remark (2) on page 260.

Of course, some matrices can have eigenvalues with multiplicities greater than 1 so that the number of distinct eigenvalues is strictly less than n. In this case, if such a matrix still has n linearly independent eigenvectors, then it is also diagonalizable (see Example 106 or the matrix λI_n).

In some cases, such a matrix cannot have n linearly independent eigenvectors (see Example 109), so a diagonalization is impossible. This latter case will be discussed separately in Chapter 8.

The next example shows a simple application of the diagonalization of a matrix to compute the power A^n of a matrix A.

Example 111 Compute A^{100} for $A = \begin{bmatrix} 1 & 1 \\ 0 & 2 \end{bmatrix}$.

Solution: The eigenvalues of A are 1 and 2 with associated eigenvectors $(1, 0)$ and $(1, 1)$, respectively. Hence $Q = \begin{bmatrix} 1 & 1 \\ 0 & 1 \end{bmatrix}$ diagonalizes A, *i.e.*,

$$Q^{-1}AQ = \begin{bmatrix} 1 & 0 \\ 0 & 2 \end{bmatrix} \quad \text{or} \quad A = Q\begin{bmatrix} 1 & 0 \\ 0 & 2 \end{bmatrix}Q^{-1}.$$

Therefore,

$$\begin{aligned} A^{100} &= Q\begin{bmatrix} 1 & 0 \\ 0 & 2 \end{bmatrix}^{100}Q^{-1} = \begin{bmatrix} 1 & 1 \\ 0 & 1 \end{bmatrix}\begin{bmatrix} 1^{100} & 0 \\ 0 & 2^{100} \end{bmatrix}\begin{bmatrix} 1 & -1 \\ 0 & 1 \end{bmatrix} \\ &= \begin{bmatrix} 1^{100} & 2^{100} - 1^{100} \\ 0 & 2^{100} \end{bmatrix} = \begin{bmatrix} 1 & 2^{100} - 1 \\ 0 & 2^{100} \end{bmatrix} \end{aligned}$$

as a product of only three matrices. □

Problem 144 Compute A^{100} for $A = \begin{bmatrix} 1 & 4 \\ 3 & 2 \end{bmatrix}$.

Problem 145 For $A = \begin{bmatrix} 5 & -4 & 4 \\ 12 & -11 & 12 \\ 4 & -4 & 5 \end{bmatrix}$, (1) diagonalize the matrix A, (2) find the eigenvalues and eigenvectors of $A^{10} + A^7 + 5A$, and (3) compute A^{100}.

6.4 Applications

If a square matrix A is diagonalizable, then an EVD $A = QDQ^{-1}$ gives an easy way to solve many problems related to the matrix A, like (1)–(6) on page 249.

As an example, let's try to solve a system $A\mathbf{x} = \mathbf{b}$ with any (square or non-square) coefficient matrix A. It can be done by solving the normal equation $A^T A\mathbf{x} = A^T\mathbf{b}$, where the coefficient matrix $A^T A$ is square and symmetric.

Now, one can assume that the coefficient matrix A is square and symmetric from the beginning in the system $A\mathbf{x} = \mathbf{b}$. Later, it will be shown (in Section 7.6) that every symmetric matrix A is diagonalizable, say $Q^{-1}AQ = D$, with D diagonal. Then $A\mathbf{x} = QDQ^{-1}\mathbf{x} = \mathbf{b}$, equivalently $DQ^{-1}\mathbf{x} = Q^{-1}\mathbf{b}$. Since D is diagonal, this can be solved easily to get the solution $\mathbf{y} = Q^{-1}\mathbf{x}$, and then $\mathbf{x} = Q\mathbf{y}$. Notice that $D\mathbf{y} = \mathbf{c}$ with $\mathbf{c} = Q^{-1}\mathbf{b}$ can be solved immediately when $Q^{-1}\mathbf{b}$ is computed.

As the first application of the diagonalization of a matrix, one may consider linear recurrence relations.

6.4.1 Fibonacci sequence

Early in the thirteenth century, Fibonacci [4] posed the following problem:

Suppose that a newly born pair of rabbits produces no offspring during the first month of their lives, but each pair gives birth to a new pair once a month from the second month onward.
Starting with one newly born pair in the first month, how many pairs of rabbits can be bred in a given time, assuming no rabbit dies?

Initially, there is one pair. After one month there is still one pair, but two months later it gives a birth, so there are two pairs. If at the end of n months there are x_n pairs, then after $n+1$ months the number will be the x_n pairs plus the number of offspring of the x_{n-1} pairs who were alive at $n-1$ months.

As an equation, for $n \geq 2$,

$$x_{n+1} = x_n + x_{n-1}.$$

Here, if two beginning values are given as $x_0 = 0$ and $x_1 = 1$, then the first several terms of the sequence become

$$0, 1, 1, 2, 3, 5, 8, 13, 21, 34, 55, \ldots .$$

[4]Fibonacci (1175 - 1250?) was a mathematician from the Republic of Pisa in what is now Italy. He is considered to be the most talented Western mathematician of the Middle Ages. He is famous for the Fibonacci sequence. He is also known as Leonardo Bonacci, Leonardo da Pisa, Leonardo Pisano Bigollo, or Leonardo Fibonacci. Fibonacci is his nickname.

This sequence is called the **Fibonacci sequence** and each term is called a **Fibonacci number**.

Example 112 Find the 2000th Fibonacci number.

Solution: A standard trick is to add a trivial extra equation $x_n = x_n$ together with the given equation:

$$\begin{cases} x_{n+1} &= x_n + x_{n-1} \\ x_n &= x_n. \end{cases}$$

Equivalently in matrix notation,

$$\begin{bmatrix} x_{n+1} \\ x_n \end{bmatrix} = \begin{bmatrix} 1 & 1 \\ 1 & 0 \end{bmatrix} \begin{bmatrix} x_n \\ x_{n-1} \end{bmatrix},$$

which is of the form

$$\mathbf{x}_n = A\mathbf{x}_{n-1} = AA\mathbf{x}_{n-2} = \cdots = A^n\mathbf{x}_0, \quad n = 1, 2, \ldots,$$

where $\mathbf{x}_n = \begin{bmatrix} x_{n+1} \\ x_n \end{bmatrix}$, $\mathbf{x}_0 = \begin{bmatrix} 1 \\ 0 \end{bmatrix}$, and $A = \begin{bmatrix} 1 & 1 \\ 1 & 0 \end{bmatrix}$. Thus, the problem is reduced to computing the power A^n.

The characteristic polynomial of A is $\lambda^2 - \lambda - 1$ with the eigenvalues

$$\lambda_1 = \frac{1}{2}(1 + \sqrt{5}), \qquad \lambda_2 = \frac{1}{2}(1 - \sqrt{5})$$

and their associated eigenvectors

$$\mathbf{v}_1 = \begin{bmatrix} \lambda_1 \\ 1 \end{bmatrix}, \qquad \mathbf{v}_2 = \begin{bmatrix} \lambda_2 \\ 1 \end{bmatrix},$$

respectively. Moreover, the change of basis matrix and its inverse are found to be

$$Q = [\mathbf{v}_1\ \mathbf{v}_2] = \begin{bmatrix} \lambda_1 & \lambda_2 \\ 1 & 1 \end{bmatrix}, \qquad Q^{-1} = \frac{1}{\sqrt{5}} \begin{bmatrix} 1 & -\lambda_2 \\ -1 & \lambda_1 \end{bmatrix}.$$

With $D = \begin{bmatrix} \lambda_1 & 0 \\ 0 & \lambda_2 \end{bmatrix}$,

$$A^n = QD^nQ^{-1} = Q \begin{bmatrix} \lambda_1^n & 0 \\ 0 & \lambda_2^n \end{bmatrix} Q^{-1}.$$

For instance, if $n = 2000$, then

$$\begin{bmatrix} x_{2001} \\ x_{2000} \end{bmatrix} = \mathbf{x}_{2000} = A^{2000} \begin{bmatrix} 1 \\ 0 \end{bmatrix} = QD^{2000}Q^{-1} \begin{bmatrix} 1 \\ 0 \end{bmatrix} = \frac{1}{\sqrt{5}} \begin{bmatrix} \lambda_1^{2001} - \lambda_2^{2001} \\ \lambda_1^{2000} - \lambda_2^{2000} \end{bmatrix}.$$

By taking its second entry, it follows

$$x_{2000} = \frac{1}{\sqrt{5}} \left(\lambda_1^{2000} - \lambda_2^{2000} \right) = \frac{1}{\sqrt{5}} \left(\left(\frac{1+\sqrt{5}}{2} \right)^{2000} - \left(\frac{1-\sqrt{5}}{2} \right)^{2000} \right).$$

In general, the Fibonacci numbers are

$$x_n = \frac{1}{\sqrt{5}} \left(\left(\frac{1+\sqrt{5}}{2} \right)^{n} - \left(\frac{1-\sqrt{5}}{2} \right)^{n} \right)$$

for $n \geq 0$. This shows that the n-th term x_n is a linear combination of λ_1^n and λ_2^n.

Since x_{2000} is an integer, we look for the nearest integer to the huge number $\frac{1}{\sqrt{5}} \left(\frac{1+\sqrt{5}}{2} \right)^{2000}$ because $\left(\frac{1-\sqrt{5}}{2} \right)^{k}$ is actually very small for large k. Historically, the number $\frac{1+\sqrt{5}}{2}$, which is very close to the ratio $\frac{x_{2001}}{x_{2000}}$, is called the **golden mean** or the **golden ratio**. □

Remark: The golden mean is one of the mysterious naturally occurring numbers, like $e = 2.71828182\cdots$ or $\pi = 3.14159265\cdots$ and it is denoted by ϕ. Its decimal representation is $\phi = 1.61803398\cdots$. It is also described as $\phi = \frac{\ell}{s}$ for $0 < s < \ell$ satisfying $\frac{\ell}{s} = \frac{\ell+s}{\ell}$. [5]

Problem 146 For $A = \begin{bmatrix} 1 & 1 \\ 1 & 0 \end{bmatrix}$, show that A has eigenvalues $\lambda_1 = \frac{1}{2}(1+\sqrt{5})$, $\lambda_2 = \frac{1}{2}(1-\sqrt{5})$ with associated eigenvectors $\mathbf{v}_1 = \begin{bmatrix} \lambda_1 \\ 1 \end{bmatrix}$, $\mathbf{v}_2 = \begin{bmatrix} \lambda_2 \\ 1 \end{bmatrix}$, respectively.

Problem 147 For $n \geq 1$, let a_n denote the number of subsets of $\{1, 2, \ldots, n\}$ that contain no consecutive integers. Find the number a_n for all $n \geq 1$.

Hint: First, try to find the beginning numbers a_1, a_2, a_3.

[5] The golden mean is also called the **golden number**; $\phi = \frac{\ell}{s} = \frac{\ell+s}{\ell} = 1 + \frac{s}{\ell} = 1 + \frac{1}{\phi}$, which gives $\phi = \frac{1}{2}(1+\sqrt{5})$. It is a root of a polynomial with integer coefficients, so ϕ is an *algebraic number*. ϕ is a letter phi in Greek.

6.4.2 Linear recurrence relations; LRR

The construction of the Fibonacci sequence can be generalized as follows.

Definition 6.8 A sequence $\{x_n : n \geq 0\}$ of numbers is said to satisfy a **linear recurrence relation (LRR) of order** k (with constant coefficients) if there exist k constants $a_1, a_2, \ldots, a_k$ with a_1 and a_k nonzero such that

$$x_n = a_1 x_{n-1} + a_2 x_{n-2} + \cdots + a_k x_{n-k} \quad \text{for all } n \geq k, \tag{6.1}$$

i.e., each x_n is a linear combination of the preceding k terms $x_{n-1}, x_{n-2}, \ldots, x_{n-k}$.

Example 113 (*LRR of low order*) **(1)** An LRR of order 1 is $x_n = a x_{n-1}$ for $n = 1, 2, \ldots$. It gives a geometric sequence $x_n = a^n x_0$.

(2) An LRR of order 2 is $x_n = a_1 x_{n-1} + a_2 x_{n-2}$ for $n = 1, 2, \ldots$. If $a_1 = a_2 = 1$ and $x_0 = 0,\ x_1 = 1$, then this gives the Fibonacci sequence. □

A **solution** to an LRR is any sequence $\{x_n : n \geq 0\}$ of numbers that satisfies the LRR. Of course, a solution can be found by simply writing out enough terms of the sequence whenever k beginning values $x_0, x_1, \ldots, x_{k-1}$, called the **initial values**, are given.

As in the case of the Fibonacci sequence, an LRR of order k

$$x_n = a_1 x_{n-1} + a_2 x_{n-2} + \cdots + a_k x_{n-k} \quad \text{for all } \ n \geq k,$$

or equivalently,

$$x_{n+k-1} = a_1 x_{n+k-2} + a_2 x_{n+k-3} + \ \cdots \ + a_k x_{n-1} \quad \text{for all } \ n \geq 1$$

can be written in matrix form with some extra trivial equations $x_j = x_j$ as follows:

$$\mathbf{x}_n = \begin{bmatrix} x_{n+k-1} \\ x_{n+k-2} \\ \vdots \\ x_{n+1} \\ x_n \end{bmatrix} = \begin{bmatrix} a_1 & a_2 & a_3 & \cdots & a_{k-1} & a_k \\ 1 & 0 & 0 & \cdots & 0 & 0 \\ 0 & 1 & 0 & \cdots & 0 & 0 \\ \vdots & \vdots & \vdots & & \vdots & \vdots \\ 0 & 0 & 0 & \cdots & 1 & 0 \end{bmatrix} \begin{bmatrix} x_{n+k-2} \\ x_{n+k-3} \\ \vdots \\ x_n \\ x_{n-1} \end{bmatrix} = A\mathbf{x}_{n-1}$$

for $n \geq 1$, or simply $\mathbf{x}_n = A\mathbf{x}_{n-1}$. The matrix A is called the **companion matrix** of an LRR $\mathbf{x}_n = A\mathbf{x}_{n-1}$.

To solve the LRR $\mathbf{x}_n = A\mathbf{x}_{n-1}$, first compute the eigenvalues and the eigenvectors of a companion matrix A.

Lemma 6.9 *For a companion matrix*

$$A = \begin{bmatrix} a_1 & a_2 & a_3 & \cdots & a_{k-1} & a_k \\ 1 & 0 & 0 & \cdots & 0 & 0 \\ 0 & 1 & 0 & \cdots & 0 & 0 \\ \vdots & \vdots & \vdots & & \vdots & \vdots \\ 0 & 0 & 0 & \cdots & 1 & 0 \end{bmatrix} \quad \text{with } a_1 \text{ and } a_k \text{ nonzero,}$$

(1) *the characteristic polynomial of A is* $\lambda^k - a_1\lambda^{k-1} - \cdots - a_{k-1}\lambda - a_k$.
(2) $\det A \neq 0$, *hence all eigenvalues of A are nonzero.*
(3) *For each eigenvalue* λ *of A,* $x_n = \lambda^n$ *is a solution to the LRR in Eq. (6.1).*

Proof: (1) Use mathematical induction on k. Clearly true for $k = 1$. With the assumption of the equality for $k-1$, let's prove it for any $k \times k$ companion matrix A. By taking the cofactor expansion of $\det(\lambda I - A)$ along the last column, the induction hypothesis gives

$$\begin{aligned} \det(\lambda I - A) &= \lambda(\lambda^{k-1} - a_1\lambda^{k-2} - \cdots - a_{k-1}) + (-1)^{2k-1}a_k \\ &= \lambda^k - a_1\lambda^{k-1} - \cdots - a_{k-1}\lambda - a_k. \end{aligned}$$

(2) By the cofactor expansion along the last column, $\det A = (-1)^{k+1}a_k \neq 0$. Hence, all eigenvalues are not zero.

(3) By (1), for any eigenvalue λ of A, $x_n = \lambda^n$ satisfies the LRR $x_n = a_1x_{n-1} + a_2x_{n-2} + \cdots + a_kx_{n-k}$. □

Example 114 (*LRR of low orders*) (Example 113 again) In matrix notation,

(1) an LRR of order 1 is $[x_n] = [a][x_{n-1}]$ for $n = 1, 2, \ldots$. Its companion matrix is $[a]$ with eigenvalue $\lambda = a$. Clearly, $\lambda^n = a^n$ is a geometric sequence with $x_0 = 1$.

(2) An LRR of order 2: The companion matrix of the Fibonacci sequence is $A = \begin{bmatrix} 1 & 1 \\ 1 & 0 \end{bmatrix}$. The characteristic polynomial of A is $\lambda^2 - \lambda - 1$, with eigenvalues

$$\lambda_1 = \tfrac{1}{2}(1 + \sqrt{5}) \quad \text{and} \quad \lambda_2 = \tfrac{1}{2}(1 - \sqrt{5}).$$

Hence, for each λ_i, $x_n = \lambda_i^n$ is a solution to the LRR $x_n = x_{n-1} + x_{n-2}$. Furthermore, any of their linear combinations $c_1\lambda_1^n + c_2\lambda_2^n$ is also a solution because it satisfies the LRR $x_n = x_{n-1} + x_{n-2}$. Finally, the constants c_1, c_2 can be determined by the two initial values $x_0 = 1$ and $x_1 = 1$.

As shown in the two cases of LRRs of orders 1 and 2, the reader might guess that for any LRR of order 3, if the companion matrix has three distinct eigenvalues $\lambda_i, \lambda_2, \lambda_3$, then each $x_n = \lambda_i^n$ is a solution to the LRR and any of their linear combinations is also a solution. In fact, this is true for any LRR of higher order. It will be shown later in Theorem 6.13. □

Remark: (1) From Lemma 6.9 (1), one can see that if an LRR

$$x_n = a_1x_{n-1} + a_2x_{n-2} + \cdots + a_kx_{n-k}$$

is given, then the characteristic equation of the associated companion matrix A can be derived *directly* from the LRR by replacing x_i with λ^i and dividing the resulting equation by λ^{n-k}. This relation between the LRR and the characteristic equation of the companion matrix A gives a reason why $\{\lambda^n : n \geq 0\}$ is a solution of the LRR for each eigenvalue λ.

2) As an alternative way of defining the companion matrix A of the LRR $x_n = a_1x_{n-1} + a_2x_{n-2} + \cdots + a_kx_{n-k}$ for all $n \geq k$ with a_1 and a_k nonzero, one can write as

$$\mathbf{x}_n = \begin{bmatrix} x_n \\ x_{n+1} \\ \vdots \\ x_{n+k-2} \\ x_{n+k-1} \end{bmatrix} = \begin{bmatrix} 0 & 1 & 0 & 0 & \cdots & 0 \\ 0 & 0 & 1 & 0 & \cdots & 0 \\ \vdots & \vdots & \vdots & & \vdots & \vdots \\ 0 & 0 & 0 & 0 & \cdots & 1 \\ a_k & a_{k-1} & a_{k-2} & \cdots & a_2 & a_1 \end{bmatrix} \begin{bmatrix} x_{n-1} \\ x_n \\ \vdots \\ x_{n+k-3} \\ x_{n+k-2} \end{bmatrix} = A\mathbf{x}_{n-1}$$

for $n \geq 1$. In this case, one can show that Lemma 6.9 also holds for this companion matrix A. In fact, the two different companion matrices of an LRR do not induce any essential difference in solving the LRR.

Next, it is needed to find the eigenvectors of a companion matrix A.

Lemma 6.10 *If λ_0 is an eigenvalue of the companion matrix A of an LRR of order k, then the eigenspace $E(\lambda_0)$ is always a one-dimensional (sub)space, which is spanned by the eigenvector $[\lambda_0^{k-1} \; \cdots \; \lambda_0 \; 1]^T$.*

Proof: An entrywise comparison of $A\mathbf{x} = \lambda_0\mathbf{x}$,

$$\begin{bmatrix} a_1 & a_2 & \cdots & a_{k-1} & a_k \\ 1 & 0 & \cdots & 0 & 0 \\ \vdots & \vdots & & \vdots & \vdots \\ 0 & 0 & \cdots & 1 & 0 \end{bmatrix} \begin{bmatrix} x_k \\ \vdots \\ x_2 \\ x_1 \end{bmatrix} = \lambda_0 \begin{bmatrix} x_k \\ \vdots \\ x_2 \\ x_1 \end{bmatrix},$$

gives that $x_i = \lambda_0 x_{i-1} = \lambda_0^{i-1} x_1$ for $i = 2, \ldots, k$. Now, take $x_1 = 1$. □

An LRR $\mathbf{x}_n = A\mathbf{x}_{n-1}$ can be easily solved if A has k *distinct* eigenvalues, where k is the order of A.

Example 115 (*LRR* $\mathbf{x}_n = A\mathbf{x}_{n-1}$ *with* k *distinct eigenvalues*) Solve the LRR

$$x_n = 9x_{n-1} - 23x_{n-2} + 15x_{n-3} \quad \text{for } n \geq 3$$

with initial values $x_0 = 0,\ x_1 = 1,\ x_2 = -1$.

Solution: In matrix notation, the LRR is

$$\mathbf{x}_n = \begin{bmatrix} x_{n+2} \\ x_{n+1} \\ x_n \end{bmatrix} = \begin{bmatrix} 9 & -23 & 15 \\ 1 & 0 & 0 \\ 0 & 1 & 0 \end{bmatrix} \begin{bmatrix} x_{n+1} \\ x_n \\ x_{n-1} \end{bmatrix} = A\mathbf{x}_{n-1} \quad \text{for } n \geq 1.$$

By Lemma 6.9, the characteristic polynomial of A is

$$\det(\lambda I - A) = \lambda^3 - 9\lambda^2 + 23\lambda - 15 = (\lambda - 1)(\lambda - 3)(\lambda - 5),$$

giving three distinct eigenvalues. Furthermore,

$$\text{for each eigenvalue } \lambda_1 = 1, \lambda_2 = 3, \lambda_3 = 5, \quad x_n = \lambda_i^n \text{ is a solution}$$

to the LRR $\mathbf{x}_n = A\mathbf{x}_{n-1}$, and any of their linear combinations

$$x_n = c_1 \cdot 1^n + c_2 \cdot 3^n + c_3 \cdot 5^n \quad \text{is also a solution}$$

because it satisfies the LRR. From the three initial values, one can determine the constants c_1, c_2, c_3.

As an alternative method, one can solve the LRR $\mathbf{x}_n = A\mathbf{x}_{n-1}$ by an EVD of the matrix A. For the three eigenvalues $\lambda_1 = 1,\ \lambda_2 = 3,\ \lambda_3 = 5$ of A, their associated eigenvectors can be chosen to be

$$\mathbf{v}_1 = \begin{bmatrix} 1 \\ 1 \\ 1 \end{bmatrix}, \qquad \mathbf{v}_2 = \begin{bmatrix} 9 \\ 3 \\ 1 \end{bmatrix}, \qquad \mathbf{v}_3 = \begin{bmatrix} 25 \\ 5 \\ 1 \end{bmatrix},$$

respectively, by Lemma 6.10. Then the change of basis matrix and its inverse are

$$Q = [\mathbf{v}_1\ \mathbf{v}_2\ \mathbf{v}_3] = \begin{bmatrix} 1 & 9 & 25 \\ 1 & 3 & 5 \\ 1 & 1 & 1 \end{bmatrix}, \quad Q^{-1} = \frac{1}{8} \begin{bmatrix} 1 & -8 & 15 \\ -2 & 12 & -10 \\ 1 & -4 & 3 \end{bmatrix}.$$

With

$$D = \begin{bmatrix} 1 & 0 & 0 \\ 0 & 3 & 0 \\ 0 & 0 & 5 \end{bmatrix} \quad \text{and} \quad Q^{-1}\mathbf{x}_0 = Q^{-1}\begin{bmatrix} -1 \\ 1 \\ 0 \end{bmatrix} = \begin{bmatrix} -9 \\ 14 \\ -5 \end{bmatrix},$$

one can get

$$\mathbf{x}_n = A^n\mathbf{x}_0 = QD^nQ^{-1}\mathbf{x}_0 = \begin{bmatrix} 1 & 9 & 25 \\ 1 & 3 & 5 \\ 1 & 1 & 1 \end{bmatrix}\begin{bmatrix} 1^n & 0 & 0 \\ 0 & 3^n & 0 \\ 0 & 0 & 5^n \end{bmatrix}\begin{bmatrix} -9 \\ 14 \\ -5 \end{bmatrix},$$

or

$$\begin{bmatrix} x_{n+2} \\ x_{n+1} \\ x_n \end{bmatrix} = \begin{bmatrix} -9 & +14 \times 3^{n+2} & -5^{n+3} \\ -9 & +14 \times 3^{n+1} & -5^{n+2} \\ -9 & +14 \times 3^n & -5^{n+1} \end{bmatrix}.$$

This implies that the solution is $x_n = -9 + 14 \times 3^n - 5^{n+1}$. So the n-th term x_n is a linear combination of λ_1^n, λ_2^n, and λ_3^n. □

Problem 148 Let $\{x_n\}$ be a sequence with $x_0 = 1$, $x_1 = 2$, $x_2 = 0$, and let $x_n = 2x_{n-1} + x_{n-2} - 2x_{n-3}$ for $n \geq 3$. Find the n-th term x_n.

Problem 149 Let $\{x_n\}$ be a sequence with $x_0 = 0$, $x_1 = 1$, $x_2 = -1$ and let $x_n = 6x_{n-1} - 11x_{n-2} + 6x_{n-3}$ for $n \geq 3$. Find the n-th term x_n.

6.4.3 Linear difference equations; LdE

Before discussing an LRR $\mathbf{x}_n = A\mathbf{x}_{n-1}$ with a companion matrix A having repeated eigenvalues, let us consider more general problems of an LRR.

Define a sequence $\{\mathbf{x}_n : n \geq 0\}$ of vectors in $\mathbb{R}^k$ by a matrix equation

$$\mathbf{x}_n = A\mathbf{x}_{n-1}, \quad n \geq 1, \quad \text{with an } \textit{arbitrary } k \times k \text{ square matrix } A$$

(not necessarily a companion matrix). Such an equation $\mathbf{x}_n = A\mathbf{x}_{n-1}$, $n \geq 1$ is called a **linear difference equation (LdE) of order** k.

A **solution** to an LdE $\mathbf{x}_n = A\mathbf{x}_{n-1}$ is a sequence $\{\mathbf{x}_n \in \mathbb{R}^k : n \geq 0\}$ of vectors that satisfies the equation. In fact, solving an LdE is reduced to a simple computation of A^n if the starting vector $\mathbf{x}_0$, called the **initial value** or the **initial vector**, is given. This is because $\mathbf{x}_n = A^n\mathbf{x}_0$ *is a unique solution with initial value* $\mathbf{x}_0$. Hence, every LdE $\mathbf{x}_n = A\mathbf{x}_{n-1}$ has a solution, (in fact, infinitely many depending on the choice of initial vector $\mathbf{x}_0$).

If $\mathbf{v}$ is an eigenvector of A belonging to λ, then $\lambda^n\mathbf{v} = A(\lambda^{n-1}\mathbf{v})$ for all $n \geq 1$. Hence, a sequence

$$\{\mathbf{x}_n\} = \{\lambda^n\mathbf{v}\} \text{ is a solution to an LdE } \mathbf{x}_n = A\mathbf{x}_{n-1}.$$

Notice that from the definition, every companion matrix A of an LRR $\mathbf{x}_n = A\mathbf{x}_{n-1}$ is invertible (see Lemma 6.9 (2)), but the matrix A in an LdE $\mathbf{x}_n = A\mathbf{x}_{n-1}$ can be any matrix, singular or nonsingular.

We first examine the set of solutions to an LRR or to an LdE.

Theorem 6.11 (Dimension of the solution space of an LdE or of an LRR) *Let $\mathbf{x}_n = A\mathbf{x}_{n-1}$ be an LdE, where A is any $k \times k$ matrix.*

(1) *If λ is an eigenvalue of A with an associated eigenvector $\mathbf{v}$, then $\mathbf{x}_n = \lambda^n\mathbf{v}$ is a solution to $\mathbf{x}_n = A\mathbf{x}_{n-1}$.*
(2) *The set of all solutions to the LdE $\mathbf{x}_n = A\mathbf{x}_{n-1}$ forms a k-dimensional vector space.*
(3) *The set of solutions to an LRR of order k,*

$$x_n = a_1x_{n-1} + a_2x_{n-2} + \cdots + a_kx_{n-k}, \quad n \geq k$$

with nonzero a_1 and a_k, is a k-dimensional vector space.

Proof: Since (1) has already been shown, we prove (2). Clearly, a sum of two solutions and any scalar multiplication of a solution are also solutions. Hence, the set of solutions forms a vector space (as is shown in Example 36(4) on page 104.)

Let $\{\mathbf{e}_1, \mathbf{e}_2, \ldots, \mathbf{e}_k\}$ be the standard basis for $\mathbb{R}^k$. For each $\mathbf{e}_i$, there exists the unique solution $\{\mathbf{x}_n \in \mathbb{R}^k : n \geq 0\}$ of vectors $\mathbf{x}_n = A^n\mathbf{e}_i$ to the LdE $\mathbf{x}_n = A\mathbf{x}_{n-1}$ with initial value $\mathbf{e}_i$.[6] All such solutions are clearly linearly independent. Moreover, they generate the vector space of all solutions, and hence

$$\{A^n\mathbf{e}_i \in \mathbb{R}^k : n \geq 0\}, \quad i = 1, 2, \ldots, k$$

is a basis for the vector space of all solutions. Indeed, if $\{\mathbf{x}_n \in \mathbb{R}^k : n \geq 0\}$ is a solution, then the initial value $\mathbf{x}_0$ can be written as a linear combination of the standard basis vectors: say, $\mathbf{x}_0 = c_1\mathbf{e}_1 + c_2\mathbf{e}_2 + \cdots + c_k\mathbf{e}_k$. Then,

$$\mathbf{x}_n = A^n\mathbf{x}_0 = c_1A^n\mathbf{e}_1 + c_2A^n\mathbf{e}_2 + \cdots + c_kA^n\mathbf{e}_k.$$

[6]The vector $\mathbf{x}_n = A^n\mathbf{e}_i$ is just the i-th column of the power A^n.

In fact, the space of all solutions to the LdE $\mathbf{x}_n = A\mathbf{x}_{n-1}$ is isomorphic to the space $\mathbb{R}^k$ via $\{\mathbf{x}_n\} \mapsto \mathbf{x}_0$.

(3) comes from (2). However, as a direct proof, every solution $\{x_n : n \geq 0\}$ to an LRR $\mathbf{x}_n = A\mathbf{x}_{n-1}$ is completely determined by the initial vector $\mathbf{x}_0 = (x_{k-1}, \ldots, x_1, x_0)$ in $\mathbb{R}^k$ and hence the solution space has dimension k. □

Definition 6.12 A basis for the solution space of an LdE or an LRR is called a **fundamental set** of solutions. The solution expressed as an arbitrary linear combination of a fundamental set is called a **general solution** to the LdE or the LRR. The solution determined by a given initial value is called a **particular solution**.

By Theorem 6.11, it is enough to find k linearly independent solutions to solve an LdE or an LRR of order k, and its general solution is just a linear combination of those linearly independent solutions.

As a favorite case, let the square matrix A be *diagonalizable* with k linearly independent eigenvectors $\mathbf{v}_1, \mathbf{v}_2, \ldots, \mathbf{v}_k$ belonging to the eigenvalues $\lambda_1, \lambda_2, \ldots, \lambda_k$, respectively. Since $\{\mathbf{v}_1, \mathbf{v}_2, \ldots, \mathbf{v}_k\}$ forms a basis for $\mathbb{R}^k$, any initial value $\mathbf{x}_0$ can be written as

$$\mathbf{x}_0 = c_1\mathbf{v}_1 + c_2\mathbf{v}_2 + \cdots + c_k\mathbf{v}_k.$$

Since $A\mathbf{v}_i = \lambda_i\mathbf{v}_i$, for all $n = 1, 2, \ldots,$

$$\mathbf{x}_n = A^n\mathbf{x}_0 = c_1\lambda_1^n\mathbf{v}_1 + c_2\lambda_2^n\mathbf{v}_2 + \cdots + c_k\lambda_k^n\mathbf{v}_k$$

which gives a general solution to the LdE.

In particular, if the companion matrix A of an LRR

$$x_n = a_1x_{n-1} + a_2x_{n-2} + \cdots + a_kx_{n-k}$$

of order k has k *distinct* eigenvalues $\lambda_1, \lambda_2, \ldots, \lambda_k$, then their associated eigenvectors $\mathbf{v}_i = [\lambda_i^{k-1} \ \cdots \ \lambda_i \ 1]^T$, $i = 1, 2, \ldots, k$ chosen in Lemma 6.10 are linearly independent. Now, the last k-th coordinate of

$$\mathbf{x}_n = A^n\mathbf{x}_0 = c_1\lambda_1^n\mathbf{v}_1 + c_2\lambda_2^n\mathbf{v}_2 + \cdots + c_k\lambda_k^n\mathbf{v}_k$$

is

$$x_n = c_1\lambda_1^n + c_2\lambda_2^n + \cdots + c_k\lambda_k^n \quad \text{with constants } c_i\text{'s},$$

which is a general solution to the given LRR. If initial k values $x_0, x_1, \ldots, x_{k-1}$ are given, the constants c_i's can be determined.

This can be summarized in the following theorem.

Theorem 6.13 *Let A be a $k \times k$ diagonalizable matrix with k linearly independent eigenvectors $\mathbf{v}_1, \mathbf{v}_2, \ldots, \mathbf{v}_k$ belonging to the eigenvalues $\lambda_1, \lambda_2, \ldots, \lambda_k$, respectively. Then, a general solution to an LdE $\mathbf{x}_n = A\mathbf{x}_{n-1}$ is*

$$\mathbf{x}_n = c_1\lambda_1^n\mathbf{v}_1 + c_2\lambda_2^n\mathbf{v}_2 + \cdots + c_k\lambda_k^n\mathbf{v}_k \quad \textit{with constants } c_i\text{'s}. \tag{6.2}$$

If an initial vector $\mathbf{x}_0$ is given, the constants $c_1, c_2, \ldots, c_k$ can be determined for a particular solution.

In particular, for an LRR

$$x_n = a_1x_{n-1} + a_2x_{n-2} + \cdots + a_kx_{n-k}, \quad n \geq k$$

with nonzero a_1 and a_k, if the associated companion matrix A has k distinct eigenvalues $\lambda_1, \lambda_2, \ldots, \lambda_k$, then its general solution is

$$x_n = c_1\lambda_1^n + c_2\lambda_2^n + \cdots + c_k\lambda_k^n \quad \textit{with constants } c_i\text{'s}. \tag{6.3}$$

If initial k values $x_0, x_1, \ldots, x_{k-1}$ are given, the constants c_i's can be determined for a particular solution.

The next example gives a hint on how to solve an LRR when its associated companion matrix A has a repeated eigenvalue.

Example 116 (*LRR $\mathbf{x}_n = A\mathbf{x}_{n-1}$ when A has a repeated eigenvalue*) Solve the LRR

$$x_n = -2x_{n-1} - x_{n-2} \quad \text{for } n \geq 2.$$

Also find its particular solution having initial values $x_0 = 1,\ x_1 = 2$.

Solution: The characteristic polynomial of the associated companion matrix is

$$\lambda^2 + 2\lambda + 1 = (\lambda + 1)^2,$$

and $\lambda = -1$ is an eigenvalue of multiplicity 2. Hence, the geometric sequence

$$\{x_n\} = \{(-1)^n\} \quad \text{is a solution to the LRR.}$$

Since the dimension of the solution space is 2 by Theorem 6.11, there is another solution which is linearly independent of $\{x_n\} = \{(-1)^n\}$. But, in this case

$$\{x_n\} = \{n(-1)^n\}$$

is also a solution to the LRR. In fact, for $n \geq 2$,

$$-2x_{n-1} - x_{n-2} = -2(n-1)(-1)^{n-1} - (n-2)(-1)^{n-2} = n(-1)^n = x_n.$$

Clearly, the two solutions $\{(-1)^n\}$ and $\{n(-1)^n\}$ are linearly independent, and so

$$x_n = c_1(-1)^n + c_2 n(-1)^n$$

is a general solution.

The given initial conditions determine $c_1 = 1, c_2 = -3$, and then $x_n = (-1)^n - 3n(-1)^n$ is a particular solution. □

In Example 116 it is shown that $\lambda = -1$ is an eigenvalue of multiplicity 2 and the two sequences $\{(-1)^n\}$ and $\{n(-1)^n\}$ are linearly independent solutions to the LRR. In this case, the second solution $\{n(-1)^n\}$ is not a geometric sequence.

As a general case, let us consider an LRR

$$x_n = a_1 x_{n-1} + a_2 x_{n-2} + \cdots + a_k x_{n-k}, \quad n \geq k$$

with nonzero a_1 and a_k, and let the associated companion matrix A have t distinct eigenvalues $\lambda_1, \lambda_2, \ldots, \lambda_t$ with multiplicities $m_1, m_2, \ldots, m_t$, respectively.

For each eigenvalue λ_i with multiplicity $m_i > 1$, the characteristic polynomial $f(\lambda)$ of A can be written as

$$f(\lambda) = \lambda^k - a_1\lambda^{k-1} - a_2\lambda^{k-2} - \cdots - a_{k-1}\lambda - a_k = (\lambda - \lambda_i)^{m_i} g(\lambda),$$

where $g(\lambda_i) \neq 0$. Hence, $f(\lambda_i) = f'(\lambda_i) = \cdots = f^{(m_i-1)}(\lambda_i) = 0$. [7] Since $f(\lambda_i) = 0$,

$$\lambda^{n-k} f(\lambda) = \lambda^n - a_1\lambda^{n-1} - a_2\lambda^{n-2} - \cdots - a_{k-1}\lambda^{n-k+1} - a_k\lambda^{n-k} = 0 \quad (3)$$

at $\lambda = \lambda_i$ for $n \geq k$. It shows that

$$\{x_n\} = \{\lambda_i^n\} \text{ is a solution to the given LRR.}$$

With $f(\lambda_i) = f'(\lambda_i) = 0$, by differentiating Eq (3), $(\lambda^{n-k} f(\lambda))' = 0$ at $\lambda = \lambda_i$ for $n \geq k$. By multiplying it by λ, $\lambda(\lambda^{n-k} f(\lambda))'$ is still zero at $\lambda = \lambda_i$. In an expansion form,

$$\begin{aligned} n\lambda_i^n \quad - \quad & a_1(n-1)\lambda_i^{n-1} - a_2(n-2)\lambda_i^{n-2} - \cdots \\ & -a_{k-1}(n-k+1)\lambda_i^{n-k+1} - a_k(n-k)\lambda_i^{n-k} = 0 \end{aligned} \quad (4)$$

[7] $f'(\lambda_i), f''(\lambda_i), \ldots, f^{(m_i-1)}(\lambda_i)$ are the successive derivatives of a function $f(\lambda)$ with respect to λ at λ_i.

for $n \geq k$. This shows that

$$\{x_n\} = \{n\lambda_i^n\} \quad \text{is also a solution to the LRR.}$$

As a similar process with $f(\lambda_i) = f'(\lambda_i) = f''(\lambda_i) = 0$, by differentiating Eq (4) and multiplying it by λ again, one can get $\lambda(\lambda(\lambda^{n-k}f(\lambda))')' = 0$ at $\lambda = \lambda_i$, which gives

$$\text{one more additional solution} \quad \{x_n\} = \{n^2\lambda_i^n\}.$$

By repeating it, one can conclude that the sequence

$$\{x_n\} = \{\lambda_i^n\},\ \{n\lambda_i^n\},\ \ldots,\ \{n^{m_i-1}\lambda_i^n\}$$

give m_i linearly independent solutions to the LRR.

Collecting such m_i linearly independent solutions for each eigenvalue λ_i with multiplicity $m_i > 1$, one can get a fundamental set of solutions to the LRR.

In summary, one has the following theorem.

Theorem 6.14 *For any given LRR*

$$x_n = a_1x_{n-1} + a_2x_{n-2} + \cdots + a_kx_{n-k}, \quad n \geq k$$

with nonzero a_1 and a_k, if the associated companion matrix A has t distinct eigenvalues $\lambda_1, \lambda_2, \ldots, \lambda_t$ with multiplicities $m_1, m_2, \ldots, m_t$, respectively, where $m_1 + m_2 + \cdots + m_t = k$, then

$$\{\{\lambda_i^n\}, \{n\lambda_i^n\}, \ldots, \{n^{m_i-1}\lambda_i^n\} \mid i = 1, 2, \ldots, t\} \tag{6.4}$$

forms a fundamental set of solutions, and a general solution is a linear combination of these solutions with constants c_i's.

Some other kinds of applications of LdE $\mathbf{x}_n = A\mathbf{x}_{n-1}$ will be continuously discussed in the coming sections.

Problem 150 Solve the LRR

$$x_n = x_{n-1} + 7x_{n-2} - x_{n-3} - 6x_{n-4} \quad \text{for } n \geq 4.$$

Also find its particular solution having initial values $x_0 = 0,\ x_1 = 1,\ x_2 = -1,\ x_3 = 2.$

Problem 151 As a direct proof of Theorem 6.14, show that each of $\{\lambda_i^n\}, \{n\lambda_i^n\}, \ldots,$ $\{n^{m_i-1}\lambda_i^n\}$ in Eq. (6.4) is a solution of the given LRR and that they are linearly independent.

Hint: Try it first when the multiplicity m_i is 2 or 3.

Problem 152 Solve the LRR $x_n = 3x_{n-1} - 4x_{n-3}$ for $n \geq 3$. What is a particular solution if $x_0 = 1, x_1 = x_2 = -1$?

6.4.4 Discrete dynamical systems

An LdE $\mathbf{x}_n = A\mathbf{x}_{n-1}$ sometimes represents a mathematical model of a dynamic process that changes over time.

This is widely used in several areas such as economics, electrical engineering, and ecology, etc. In this case, a vector $\mathbf{x}_n$ gives information about a dynamic process at time t_n.

For such a model, an LdE

$$\mathbf{x}_n = A\mathbf{x}_{n-1},\ n = 1, 2, \ldots$$

with a square matrix A is sometimes called a **discrete dynamical system**.

Throughout this section, we are concerned with an LdE $\mathbf{x}_n = A\mathbf{x}_{n-1},\ n = 1, 2, \ldots$ for only a diagonalizable matrix A, and the non-diagonalizable case will be discussed later in Chapter 8 separately.

Let A be a $k \times k$ diagonalizable matrix with k linearly independent eigenvectors $\mathbf{v}_1, \mathbf{v}_2, \ldots, \mathbf{v}_k$ belonging to the eigenvalues $\lambda_1, \lambda_2, \ldots, \lambda_k$, respectively. Then, by Theorem 6.13, a general solution to $\mathbf{x}_n = A\mathbf{x}_{n-1}$ is

$$\mathbf{x}_n = c_1\lambda_1^n\mathbf{v}_1 + c_2\lambda_2^n\mathbf{v}_2 + \cdots + c_k\lambda_k^n\mathbf{v}_k.$$

Hence, if $|\lambda_i| < 1$ for all i, then the vector $\mathbf{x}_n$ approaches the zero vector as n increases. On the other hand, if there exists an eigenvalue λ_i with $|\lambda_i| > 1$, this vector $\mathbf{x}_n$ may grow exponentially in magnitude.

Therefore, there are three possible cases for a dynamic process given by $\mathbf{x}_n = A\mathbf{x}_{n-1}, n = 1, 2, \ldots$. The process is said to be

(1) unstable if A has an eigenvalue λ with $|\lambda| > 1$,
(2) stable if $|\lambda| < 1$ for all eigenvalues of A,
(3) neutrally stable if $|\lambda| \leq 1$ for all eigenvalues of A, but it is not stable.

To determine the stability of a dynamic process, it is necessary sometimes to estimate the (upper) bound for the absolute values of eigenvalues of a matrix A.

To do this, for any square matrix $A = [a_{ij}]$ of order k, let $D(a_{ii}, r_i)$ denote the closed disc in the (complex) plane centered at the diagonal entry a_{ii} with radius $r_i = \sum_{j \neq i} |a_{ij}|$ for $i = 1, 2, \ldots, k$. Such a disc is called a *Gershgorin (row) disc*. In fact, r_i is the sum of the absolute values of the off-diagonal entries in the i-th row of A.

Theorem 6.15 (Gershgorin Theorem) [8] *Let A be any square matrix of order k. Then each eigenvalue of A lies in a Gershgorin disc $D(a_{ii}, r_i)$ for some $1 \le i \le k$.*

This theorem says that if the off-diagonal entries of a square matrix have sufficiently small absolute values, then its eigenvalues are quite close in length to the diagonal entries of the matrix.

Proof: Let λ be an eigenvalue with eigenvector $\mathbf{x} = [x_1\ x_2\ \cdots\ x_k]^T$. Then the i-th coordinate of $A\mathbf{x} = \lambda\mathbf{x}$ is $\sum_{j=1}^{k} a_{ij}x_j = \lambda x_i$ for $i = 1, 2, \ldots, k$. Take a coordinate x_ℓ of $\mathbf{x}$ with the largest absolute value. Then clearly $x_\ell \neq 0$ because $\mathbf{x} \neq \mathbf{0}$, and

$$|\lambda - a_{\ell\ell}||x_\ell| = |\lambda x_\ell - a_{\ell\ell}x_\ell| = \left|\sum_{j=1}^{k} a_{\ell j}x_j - a_{\ell\ell}x_\ell\right| = \left|\sum_{j\neq\ell} a_{\ell j}x_j\right| \le \sum_{j\neq\ell} |a_{\ell j}||x_\ell|$$

Since $|x_\ell| > 0$, $|\lambda - a_{\ell\ell}| \le \sum_{j\neq\ell} |a_{\ell j}| = r_\ell$ and hence $\lambda \in D(a_{\ell\ell}, r_\ell)$. □

Because A and A^T have the same eigenvalues, it follows.

Corollary 6.15.1 **(1)** *For a diagonal matrix, the Gershgorin discs coincide with the set of the eigenvalues with multiplicity.*

(2) *Every eigenvalue λ of a $k \times k$ matrix A lies in a* Gershgorin column *disc $D(a_{jj}, c_j)$ for some $1 \le j \le k$, where $D(a_{jj}, c_j)$ is the closed disc centered at the diagonal entry a_{jj} with radius $c_j = \sum_{i\neq j} |a_{ij}|$; the sum of the absolute values of the off-diagonal entries in the j-th column of A.*

Example 117 **(1)** Use the Gershgorin theorem to estimate the eigenvalues of

$$A = \begin{bmatrix} 0.4 & 0.0 & 0.1 & 0.0 \\ 0.1 & 0.6 & 0.2 & 0.1 \\ 0.1 & 0.1 & 0.0 & 0.0 \\ 0.0 & 0.1 & 0.1 & 0.8 \end{bmatrix}.$$

For each row $\mathbf{r}_i$, take the diagonal entry a_{ii} as the center for the Gershgorin disc and take the sum of the absolute values of the off-diagonal entries in the row

$$\sum_{j\neq i} |a_{ij}| = r_i$$

[8] **Semyon Aronovich Gershgorin** (August 24, 1901– May 30, 1933) was a Soviet (born in Pruzhany, Belarus, Russian Empire) mathematician.

as the radius to obtain the following four discs:

$$D(0.4, 0.1),\ D(0.6, 0.4),\ D(0.0, 0.2),\ \text{and } D(0.8, 0.2).$$

However, one can improve the accuracy of the second and the last discs by applying the formula to the corresponding columns of the matrix, obtaining $D(0.6, 0.2)$ and $D(0.8, 0.1)$, respectively.

The true eigenvalues of A are approximated by (use a calculator for computing $\lambda_i{}^9$)

$$\lambda_1 \approx 0.850182,\ \ \lambda_2 \approx 0.590377,\ \ \lambda_3 \approx 0.407807,\ \ \lambda_4 \approx -0.0483659.$$

(2) Use the Gershgorin theorem to estimate the eigenvalues of the symmetric matrix

$$A = \begin{bmatrix} 100 & 0 & 1 & 0 \\ 0 & 200 & 2 & 1 \\ 1 & 2 & 300 & 0 \\ 0 & 1 & 0 & 400 \end{bmatrix}.$$

From the rows (or columns) of A, we obtain the following four Gershgorin discs:

$$D(100, 1),\ D(200, 3),\ D(300, 3),\ \text{and } D(400, 1).$$

The true eigenvalues of A are approximated by

$$\lambda_1 \approx 99.995,\quad \lambda_2 \approx 199.955,\quad \lambda_3 \approx 300.045,\quad \lambda_4 \approx 400.005,$$

which are quite close in length to the diagonal entries of A in order. □

Example 118 (*Stable or unstable*) Solve a discrete dynamical system $\mathbf{x}_n = A\mathbf{x}_{n-1}$, $n = 1, 2, \ldots,$ where

$$(1)\ A = \begin{bmatrix} 0.8 & 0.0 \\ 0.0 & 0.4 \end{bmatrix},\ (2)\ A = \begin{bmatrix} 2 & 1 \\ 1 & 2 \end{bmatrix},\ (3)\ A = \begin{bmatrix} 1 & 0 \\ 0 & -1 \end{bmatrix}.$$

Solution: (1) Clearly, the eigenvalues of A are 0.8 and 0.4 with eigenvectors $\mathbf{v}_1 = \begin{bmatrix} 1 \\ 0 \end{bmatrix}$ and $\mathbf{v}_2 = \begin{bmatrix} 0 \\ 1 \end{bmatrix}$, respectively. Hence, its general solution is

$$\mathbf{x}_n = c_1(0.8)^n \begin{bmatrix} 1 \\ 0 \end{bmatrix} + c_2(0.4)^n \begin{bmatrix} 0 \\ 1 \end{bmatrix} = \begin{bmatrix} c_1(0.8)^n \\ c_2(0.4)^n \end{bmatrix}.$$

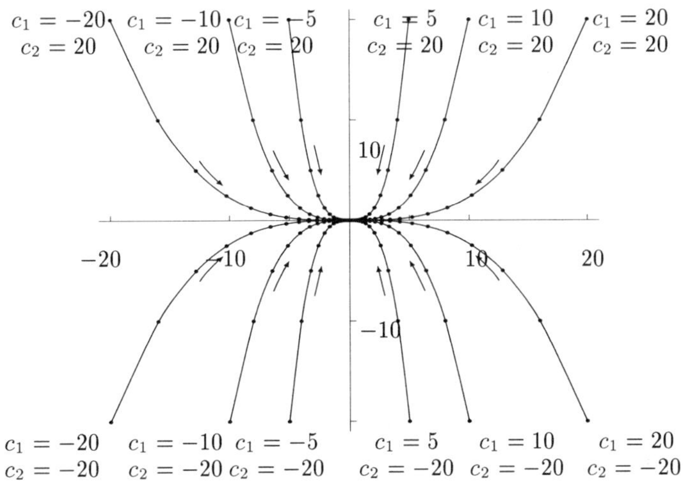

Figure 6.1: A stable dynamical system

This system $\mathbf{x}_n = A\mathbf{x}_{n-1}$ is stable. See Figure 6.1.

(2) By the Gershgorin theorem, the eigenvalues of A are all ≥ 1, and so the system cannot be stable. In fact, the eigenvalues of A are 1 and 3 with eigenvectors $\mathbf{v}_1 = \begin{bmatrix} -1 \\ 1 \end{bmatrix}$ and $\mathbf{v}_2 = \begin{bmatrix} 1 \\ 1 \end{bmatrix}$, respectively. A general solution to $\mathbf{x}_n = A\mathbf{x}_{n-1}$ is

$$\mathbf{x}_n = c_1\, 1^n \begin{bmatrix} -1 \\ 1 \end{bmatrix} + c_2\, 3^n \begin{bmatrix} 1 \\ 1 \end{bmatrix}.$$

This is unstable if $c_2 \neq 0$. For example, if $c_1 = -1, c_2 = 1$, then

$$\mathbf{x}_0 = \begin{bmatrix} 2 \\ 0 \end{bmatrix},\ \mathbf{x}_1 = \begin{bmatrix} 4 \\ 2 \end{bmatrix},\ \mathbf{x}_2 = \begin{bmatrix} 10 \\ 8 \end{bmatrix},\ \mathbf{x}_3 = \begin{bmatrix} 28 \\ 26 \end{bmatrix},\ \mathbf{x}_4 = \begin{bmatrix} 82 \\ 80 \end{bmatrix},\ \ldots$$

are the beginning vectors.

(3) In this case,

$$\mathbf{x}_n = c_1(1)^n \begin{bmatrix} 1 \\ 0 \end{bmatrix} + c_2(-1)^n \begin{bmatrix} 0 \\ 1 \end{bmatrix} = \begin{bmatrix} c_1(1)^n \\ c_2(-1)^n \end{bmatrix}.$$

This is neutrally stable and oscillates between $\begin{bmatrix} c_1 \\ c_2 \end{bmatrix}$ and $\begin{bmatrix} c_1 \\ -c_2 \end{bmatrix}$ if $c_2 \neq 0$. □

[9]One might try to use softwares like WolframAlpha, Maple, Matlab, or Mathematica, etc.

Problem 153 (*Stable or unstable*) Solve a discrete dynamical system $\mathbf{x}_n = A\mathbf{x}_{n-1}$, $n = 1, 2, \ldots$, where

$$(1)\ A = \begin{bmatrix} 1.2 & 0 \\ 0 & 0.8 \end{bmatrix}, \qquad (2)\ A = \begin{bmatrix} 0 & 1.2 \\ 0.8 & 0 \end{bmatrix}.$$

Hint: (2) A has two eigenvalues whose absolute values are less than 1.

6.4.5 Markov processes

The following example is a special type of a discrete dynamical system, called a *Markov process.* [10]

Example 119 (*Markov process with two states for residence*) There are two possible states for residence in a county; *a mountainous area and a coastal area.* Suppose that the population of the county starts with x_0 people in the mountainous area and y_0 people in the coastal area.

Suppose that each year 10% of the people in the mountainous area move to the coastal area, and 20% of the people in the coastal area move to the mountainous area.

What is the distribution of the population of the two residence areas after 10 years? What is the 'eventual' distribution of the population of the two residence areas?

Solution: At the end of the first year, the distribution of the population will be

$$\begin{cases} x_1 = 0.9\, x_0 + 0.2\, y_0 \\ y_1 = 0.1\, x_0 + 0.8\, y_0. \end{cases}$$

Or, in matrix form,

$$\mathbf{x}_1 = \begin{bmatrix} x_1 \\ y_1 \end{bmatrix} = \begin{bmatrix} 0.9 & 0.2 \\ 0.1 & 0.8 \end{bmatrix} \begin{bmatrix} x_0 \\ y_0 \end{bmatrix} = A\mathbf{x}_0.$$

Thus if $\mathbf{x}_n = (x_n, y_n)$ denotes the distribution of the population in the county after n years, then clearly $\mathbf{x}_n = A^n\mathbf{x}_0$.

In this formulation, the matrix A can be characterized as follows:

(1) The entries of A are all nonnegative, and

[10]Markov processes are based on the work of the Russian mathematician Andrei Markov (June 14, 1856– July 20, 1922) in 1907.

(2) the entries of each column of A add up to 1,

because the entries of a column of A represent the probabilities of residing in each of the two residence areas in the next year and the total population of the county remains constant.

Now, to solve the problem, first find the eigenvalues and eigenvectors of A. They are

$$\lambda_1 = 1,\ \lambda_2 = 0.7 \quad \text{and} \quad \mathbf{v}_1 = (2, 1),\ \mathbf{v}_2 = (-1,\ 1),$$

respectively, so that its general solution is

$$\mathbf{x}_n = c_1 1^n \begin{bmatrix} 2 \\ 1 \end{bmatrix} + c_2(0.7)^n \begin{bmatrix} -1 \\ 1 \end{bmatrix} = \begin{bmatrix} 2c_1 1^n - c_2(0.7)^n \\ c_1 1^n + c_2(0.7)^n \end{bmatrix}.$$

The initial condition $\mathbf{x}_0 = (x_0, y_0)$ gives $c_1 = \frac{x_0}{3} + \frac{y_0}{3}$ and $c_2 = \frac{-x_0}{3} + \frac{2y_0}{3}$, so that

$$\begin{aligned} \mathbf{x}_{10} &= \left(\frac{x_0}{3} + \frac{y_0}{3}\right) 1^{10} \begin{bmatrix} 2 \\ 1 \end{bmatrix} + \left(\frac{-x_0}{3} + \frac{2y_0}{3}\right) (0.7)^{10} \begin{bmatrix} -1 \\ 1 \end{bmatrix} \\ &= \frac{1}{3} \begin{bmatrix} 2(x_0 + y_0) - (0.7)^{10}(-x_0 + 2y_0) \\ (x_0 + y_0) + (0.7)^{10}(-x_0 + 2y_0) \end{bmatrix}, \end{aligned}$$

and eventually

$$\mathbf{x}_n \to \left(\frac{x_0}{3} + \frac{y_0}{3}\right) \begin{bmatrix} 2 \\ 1 \end{bmatrix}, \quad \text{as } n \to \infty.$$

Thus, for any given initial population $\mathbf{x}_0 = (x_0, y_0)$, the eventual ratio $x_n : y_n$ of the populations tends to $2 : 1$, which is independent of the initial distribution. □

Recall that the matrix A in Example 119 satisfies the following two conditions:

(1) all entries of A are nonnegative,
(2) the entries of each column of A add up to 1.

Such a matrix characterizes transitions for a finite Markov chain. Hence, the matrix A is called a **Markov matrix**, a **stochastic matrix**, or a **probability matrix**. It is also called a **transition matrix** or a **probability transition matrix**, depending on the cases of states in question. In general, a dynamical system $\mathbf{x}_n = A\mathbf{x}_{n-1}$ with a Markov matrix A is called a **Markov process** (or a **stochastic process**).

The next corollary follows directly from Gershgorin Theorem 6.15.

Corollary 6.15.2 *If λ is an eigenvalue of any Markov matrix A, then $|\lambda| \le 1$.*

In fact, every Markov matrix has an eigenvalue $\lambda = 1$.

To show this, let A be any Markov matrix. Then the entries of each column of A add up to 1. This means that the sum of each column of $A - I$ is 0, or equivalently $\mathbf{r}_1 + \mathbf{r}_2 + \cdots + \mathbf{r}_n = \mathbf{0}$ for the row vectors $\mathbf{r}_i$ of $A - I$. Hence, the row vectors are linearly dependent and $\det(A - I) = 0$. Consequently, $\lambda = 1$ is an eigenvalue of A.

If $\mathbf{x}$ is an eigenvector of A belonging to $\lambda = 1$, then $A\mathbf{x} = \mathbf{x}$. This is called the **equilibrium state** in the Markov process $\mathbf{x}_n = A\mathbf{x}_{n-1}$.

Theorem 6.16 *If A is a Markov matrix, then*

(1) *$\lambda = 1$ is an eigenvalue of A,*

(2) *there exists a vector $\mathbf{x}$ that remains fixed by the Markov process, that is, $A\mathbf{x} = \mathbf{x}$ is the equilibrium state.*

Problem 154 (*Markov process of weather with three possible states*) Suppose that there are three possible states for weather: *sunny, cloudy or rainy,* with the (probability) transition matrix

$$A = \begin{bmatrix} 0.7 & 0.1 & 0.1 \\ 0.2 & 0.7 & 0.2 \\ 0.1 & 0.2 & 0.7 \end{bmatrix}.$$

Here, for example, the third row $(0.1, 0.2, 0.7)$ means that if it is rainy on a day, then there is a 0.1 probability that it is sunny on the next day, there is a 0.2 probability that it is cloudy on the next day, and a 0.7 probability of rain again on the next day.

Find the distribution of the probabilities of the weather states *sunny, cloudy or rainy* after 10 days. What is the eventual steady-state of the weather?

Hint: If $\mathbf{x}_n = (x_n, y_n, z_n)$ denotes the distribution of the probabilities of the weather states *sunny, cloudy or rainy,* after n days, $\mathbf{x}_n = A\mathbf{x}_{n-1}$ in a matrix equation.

Problem 155 (*Random walk*) As an elementary example of a random walk, consider a random walk on the following integer number line

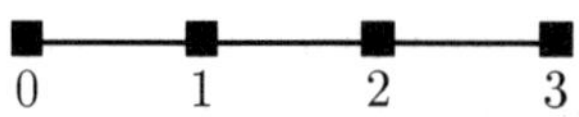

At integers 1 and 2, a step moves +1 (one step forward) or -1 (one step backward) with equal probability, but at the end integers 0 and 3, a step moves +1 and -1 always, respectively. That is, when a step reaches 0 or 3, the step has to turn back. See Figure 6.2. When the random walk starts at 0, find the probability of landing at each integer after 10 steps. How about if the random walk starts at 1?

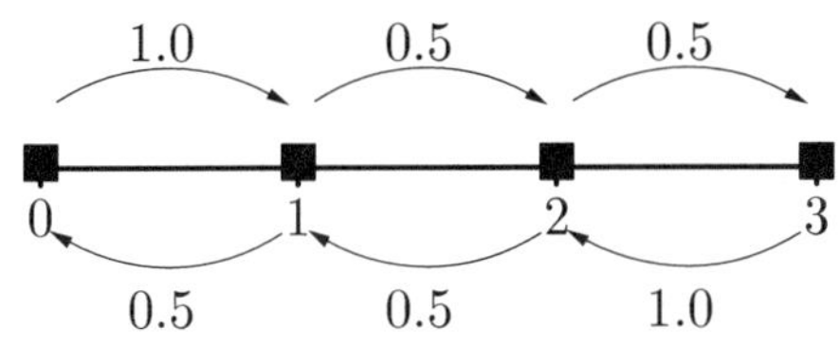

Figure 6.2: A random walk on the integer number line

Hint: For the (probability) matrix $A = \begin{bmatrix} 0.0 & 0.5 & 0.0 & 0.0 \\ 1.0 & 0.0 & 0.5 & 0.0 \\ 0.0 & 0.5 & 0.0 & 1.0 \\ 0.0 & 0.0 & 0.5 & 0.0 \end{bmatrix}$, what are the columns of the power A^{10}? The eigenvalues and eigenvectors of A are $\lambda_1 = -1, \lambda_2 = 1, \lambda_3 = -0.5, \lambda_4 = 0.5$ and $\mathbf{v}_1 = (-1, 2, -2, 1)$, $\mathbf{v}_2 = (1, 2, 2, 1), \mathbf{v}_3 = (1, -1, -1, 1)$, $\mathbf{v}_4 = (-1, -1, 1, 1)$, respectively.

6.4.6 Linear differential equations; LDE

A first-order linear differential equation is an equation of the form

$$y' + P(t)y = Q(t),$$

where $y = f(t)$, $P(t)$, and $Q(t)$ are real-valued differentiable functions of (a time) t. The equation is called *first order* because it only involves the function $y = f(t)$ and its first derivative $y' = \frac{df(t)}{dt}$. [11]

A (general) *solution* to the linear differential equation $y' + P(t)y = Q(t)$ is a function $y = f(t)$ which satisfies the equation.

As a trivial case, if $P(t) = \mathbf{0}$, then $y'(t) = Q(t)$ which is integrable [12] and

$y'(t) = Q(t)$ has a *general solution* $y(t) = \int Q(t)dt + c$,

where c is an arbitrary constant.

---

[11] A **linear differential equation of order** n is a differential equation of the form $a_0(t)y + a_1(t)y' + a_2(t)y'' + \cdots + a_n(t)y^{(n)} = b(t)$, where $a_0(t), a_1(t), \ldots, a_n(t)(\neq 0)$ and $b(t)$ are arbitrary real-valued differentiable functions that do not need to be linear, and $y', y'', \ldots, y^{(n)}$ are the successive derivatives of an unknown function y of the variable t.

[12] In calculus, it is well known that every real-valued differentiable function is continuous and every real-valued continuous function is integrable.

As another trivial case, if $Q(t) = \mathbf{0}$ and $P(t) = k$ is a constant function, then

$$y'(t) = -ky(t) \text{ has a } \textit{general solution } \ y(t) = ce^{-kt},$$

where c is an arbitrary constant. If an additional condition $y(0) = 3$, called an *initial condition*, is given, then the solution is $y = 3e^{-kt}$, which is called a *particular solution*.

Most of other cases seem much difficult to solve, like as $y' + e^t\, y = t^n$.

The second trivial example can be generalized to a **system of n linear differential equations** (we call it simply an **LDE** in this book) with constant coefficients and with $Q(t) = \mathbf{0}$, which is by definition of the form

$$\begin{cases} y_1' &= a_{11}y_1 + a_{12}y_2 + \cdots + a_{1n}y_n \\ y_2' &= a_{21}y_1 + a_{22}y_2 + \cdots + a_{2n}y_n \\ &\vdots \\ y_n' &= a_{n1}y_1 + a_{n2}y_2 + \cdots + a_{nn}y_n, \end{cases} \tag{6.5}$$

where a_{ij}'s are scalars and $y_i = f_i(t)$ for $i = 1, 2, \ldots, n$ are real-valued differentiable functions defined on an interval $I = (a, b)$. In most cases, one assumes that the interval I contains 0 and initial conditions are given as $f_i(0) = d_i$ at $0 \in I$.

To have a matrix form of an LDE, let $\mathbf{y}(t) = [f_1(t)\ f_2(t)\ \cdots\ f_n(t)]^T$, or simply $\mathbf{y} = [f_1\ f_2\ \cdots\ f_n]^T$, denote the vector whose entries are the differentiable functions $y_i = f_i(t)$ defined on an interval $I = (a, b)$. Thus, for each $t \in I$, $\mathbf{y}(t)$ is a vector in $\mathbb{R}^n$. Its derivative is defined by

$$\mathbf{y}'(t) = \begin{bmatrix} f_1'(t) \\ f_2'(t) \\ \vdots \\ f_n'(t) \end{bmatrix} \quad \text{or} \quad \mathbf{y}' = \begin{bmatrix} f_1' \\ f_2' \\ \vdots \\ f_n' \end{bmatrix} \quad \text{simply}.$$

If A denotes the coefficient matrix of the LDE, then the matrix form of the LDE given in Eq. (6.5) is

$$\mathbf{y}' = A\mathbf{y}, \text{ or } \mathbf{y}'(t) = A\mathbf{y}(t) \ \text{ for all } t \in I.$$

An initial condition can be given by $\mathbf{y}_0 = \mathbf{y}(0) = [d_1\ d_2\ \cdots\ d_n]^T \in \mathbb{R}^n$.

A differentiable vector function $\mathbf{y}(t)$ is called a **solution** to the LDE $\mathbf{y}'(t) = A\mathbf{y}(t)$ if it satisfies the equation.

In general, the entries a_{ij}'s of the coefficient matrix A can be (differentiable) functions of t. However, in this book, we restrict our discussion to an LDE with constant coefficients a_{ij}'s.

Example 120 Consider the following three systems:

$$\left\{ \begin{array}{rcl} y_1' & = & 2y_1 - 3y_2 \\ y_2' & = & 2y_1 + \ y_2 \end{array} \right., \quad \left\{ \begin{array}{rcl} y_1' & = & ty_1 + t^2y_2 \\ y_2' & = & 2y_1 + t^3y_2 \end{array} \right., \quad \left\{ \begin{array}{rcl} y_1' & = & 2y_1 - 3y_2^2 \\ y_2' & = & y_1y_2 + 5y_2 \end{array} \right..$$

The first two cases are systems of linear differential equations even some coefficients of the second one are not constant. The third one is not linear because of the terms y_2^2 and y_1y_2. □

Example 121 (*An exponentially increasing population* [13]) Let $p(t)$ be the population of a given species, like bacteria or virus, at time t and let $r(t,p)$ be the difference between its birth rate and its death rate at time t.

Assume that $r(t,p)$ is independent of time t and p. This means that $r(t,p)$ is a constant. Then the rate of change of the population is $\frac{dp(t)}{dt} = rp(t)$ and its general solution is $p(t) = p(0)e^{rt}$, an exponential function. □

Let A be any $n \times n$ matrix. Some basic facts about the solutions to an LDE $\mathbf{y}'(t) = A\mathbf{y}(t)$ defined on $I = (a,b)$ are listed below. (A reader may skip reading their proofs if not interested in the details.)

(I) (*Existence and uniqueness of solutions to an LDE*) Every LDE $\mathbf{y}'(t) = A\mathbf{y}(t)$ always has a solution. In addition, if an initial condition $\mathbf{y}_0$ is given, then there is a unique solution $\mathbf{y}(t)$ on I which satisfies the initial condition.

Unlike an LdE, the existence of such a solution to an LDE is not clear. It will be shown later in Theorem 6.26 on page 300.

(II) (*Linear independence of solutions*) Let $\{\mathbf{y}_1, \mathbf{y}_2, \ldots, \mathbf{y}_n\}$ be a set of n solutions to an LDE $\mathbf{y}' = A\mathbf{y}$ on I. The linear independence of $\mathbf{y}_1, \mathbf{y}_2, \ldots, \mathbf{y}_n$ on

---

[13] The Swiss mathematician Jacob Bernoulli (January 6, 1655– August 16, 1705) was the first to reveal the value of e by exploring the nature of compound interest in the late 17th century. If an account starts with \$1.00 and pays 100% interest per year, then the value of the account will be \$2.00 at the end of the year. But, if 100% interest splits every six months, then the value of the account will be $\$1.00 \times 1.5 \times 1.5 = \$2.25$ at the end of the year. Compounding monthly yields $\$1.00 \times (1 + 1/12)^{12} = \$2.613\cdots$ at the end of the year. Bernoulli calculated what the account would be if the interest was calculated continuously. The answer was it grew at the annual rate of $2.7182818\cdots$ or the number e. This number e is often allied to growth or decay, such as accumulation of capital, radioactivity, bacterial infections, or epidemics. The number e is sometimes called *Euler's number* after the Swiss mathematician Leonhard Euler who first used the letter e.

I is defined as usual: if $c_1\mathbf{y}_1 + c_2\mathbf{y}_2 + \cdots + c_n\mathbf{y}_n = \mathbf{0}$ has only the trivial solution $c_1 = c_2 = \cdots = c_n = 0$. Or, equivalently, they are linearly dependent if and only if one of them can be expressed as a linear combination of all the others.

Define

$$Y(t) = [\mathbf{y}_1(t)\ \mathbf{y}_2(t)\ \cdots\ \mathbf{y}_n(t)] = \begin{bmatrix} y_{11}(t) & y_{12}(t) & \cdots & y_{1n}(t) \\ y_{21}(t) & y_{22}(t) & \cdots & y_{2n}(t) \\ \vdots & \vdots & & \vdots \\ y_{n1}(t) & y_{n2}(t) & \cdots & y_{nn}(t) \end{bmatrix} \quad \text{for } t \in I,$$

where $\mathbf{y}_i(t)$ is the i-th column of $Y(t)$. If the n solutions are linearly dependent, then $\det Y(t) = 0$ for all $t \in I$.

Or, equivalently, *if* $\det Y(t) \neq 0$ *for at least one point* $t \in I$, *then the solutions are linearly independent.*

The determinant of $Y(t)$ is called the **Wronskian** of the solutions, which is a real-valued differentiable function on I.

Lemma 6.17 $\frac{d}{dt}\det Y(t) = \operatorname{tr}(A)\det Y(t)$ *and hence* $\det Y(t) = ce^{\operatorname{tr}(A)t}$.

Proof:

$$\begin{aligned}
\frac{d}{dt}\det Y(t) &= \sum_{\sigma\in S_n} \operatorname{sign}(\sigma)(y_{1\sigma(1)}\cdots y_{n\sigma(n)})' \\
&= \sum_{\sigma} \operatorname{sign}(\sigma)y'_{1\sigma(1)}\cdots y_{n\sigma(n)} + \cdots + \sum_{\sigma} \operatorname{sign}(\sigma)y_{1\sigma(1)}\cdots y'_{n\sigma(n)} \\
&= (y'_{11}Y_{11} + \cdots + y'_{1n}Y_{1n}) + \cdots + (y'_{n1}Y_{n1} + \cdots + y'_{nn}Y_{nn}) \\
&= \sum_i^n \sum_j^n y'_{ij}Y_{ij} = \sum_i^n \left(\sum_j^n y'_{ij}[\operatorname{adj} Y]_{ji} \right) = \sum_i^n [Y' \cdot \operatorname{adj} Y]_{ii} \\
&= \operatorname{tr}(Y' \cdot \operatorname{adj} Y) = \operatorname{tr}(A \cdot (Y \cdot \operatorname{adj} Y)) \\
&= \operatorname{tr}(\det Y(t) A) = \operatorname{tr}(A)\det Y(t),
\end{aligned}$$

where $Y_{ij}(t)$ is the cofactor of y_{ij}, and we use the following two identities:

$$\begin{aligned}
Y'(t) = [\mathbf{y}'_1(t)\ \cdots\ \mathbf{y}'_n(t)] &= A[\mathbf{y}_1(t)\ \cdots\ \mathbf{y}_n(t)] = AY(t), \\
Y(t)\operatorname{adj} Y(t) &= \det Y(t) I_n.
\end{aligned}$$

For the second identity, see Eq. (2.6) on page 81. □

From Lemma 6.17, one can see that *the Wronskian* $\det Y(t)$ *is an exponential function of the form* $\det Y(t) = ce^{\mathrm{tr}(A)t}$ *with an initial condition* $\det Y(0) = c$, and hence the Wronskian $\det Y(t) \neq 0$ for all $t \in I$ if and only if $c = \det Y(0) \neq 0$. Or equivalently

$$\det Y(t) \neq 0 \textit{ for all } t \in I \textit{ if and only if } \det Y(t) \neq 0 \textit{ at one point } t \in I.$$

It gives the following theorem.

Theorem 6.18 *Let* $\{\mathbf{y}_1, \mathbf{y}_2, \ldots, \mathbf{y}_n\}$ *be a set of* n *solutions to an LDE* $\mathbf{y}' = A\mathbf{y}$ *on* I*, where* A *is any* $n \times n$ *matrix. Then the following are equivalent.*

(1) *The vectors* $\mathbf{y}_1, \mathbf{y}_2, \ldots, \mathbf{y}_n$ *are linearly independent,*

(2) $\det Y(t) \neq 0$ *for some* t*, that is,* $\mathbf{y}_1(t), \mathbf{y}_2(t), \ldots, \mathbf{y}_n(t)$ *are linearly independent in* $\mathbb{R}^n$ *for some* t*,*

(3) $\det Y(t) \neq 0$ *for all* t*, that is,* $\mathbf{y}_1(t), \mathbf{y}_2(t), \ldots, \mathbf{y}_n(t)$ *are linearly independent in* $\mathbb{R}^n$ *for all* t*.*

(III) (*Dimension of the solution space of an LDE*) Just like the case of an LdE, one can show that the set of all solutions to an LDE $\mathbf{y}'(t) = A\mathbf{y}(t)$ is a vector space. Moreover, for each standard basis vector $\mathbf{e}_i$, there exists a unique solution $\mathbf{y}_i$ of $\mathbf{y}'(t) = A\mathbf{y}(t)$ satisfying the initial condition $\mathbf{y}_i(0) = \mathbf{e}_i$, by (I). All such solutions $\mathbf{y}_1, \mathbf{y}_2, \ldots, \mathbf{y}_n$ are linearly independent by Theorem 6.18. Moreover, they span the vector space of solutions.

This gives the following theorem (in a similar way as in Theorem 6.11).

Theorem 6.19 (Dimension of the solution space of an LDE) *For any* $n \times n$ *matrix* A*, the set of solutions to an LDE* $\mathbf{y}' = A\mathbf{y}$ *on* I *is an* n*-dimensional vector space.*

If $\mathbf{y} = [y_1\ y_2\ \cdots\ y_n]^T$ is a solution on I, then it draws a curve in $\mathbb{R}^n$ passing through the initial vector $\mathbf{y}(0) = [d_1\ d_2\ \cdots\ d_n]^T$ as t varies in the interval I.

Like the LdE case, one can define a **fundamental set** of solutions, a **general solution** and a **particular solution** to an LDE.

6.4.7 LDE $\mathbf{y}' = A\mathbf{y}$ for a diagonalizable matrix A

By Theorem 6.19, it is enough to find n linearly independent solutions to solve an LDE $\mathbf{y}' = A\mathbf{y}$ on I, and then its general solution is just a linear combination of those linearly independent solutions.

This is also considered in three separate cases: (1) A is diagonal, (2) A is diagonalizable, and finally (3) A is any square matrix (in a later section).

(1) Firstly, let A be a diagonal matrix D. Then $\mathbf{y}'(t) = D\mathbf{y}(t)$ is

$$\begin{bmatrix} y_1'(t) \\ y_2'(t) \\ \vdots \\ y_n'(t) \end{bmatrix} = \begin{bmatrix} \lambda_1 & & & \mathbf{0} \\ & \lambda_2 & & \\ & & \ddots & \\ \mathbf{0} & & & \lambda_n \end{bmatrix} \begin{bmatrix} y_1(t) \\ y_2(t) \\ \vdots \\ y_n(t) \end{bmatrix} = \begin{bmatrix} \lambda_1 y_1(t) \\ \lambda_2 y_2(t) \\ \vdots \\ \lambda_n y_n(t) \end{bmatrix}.$$

This system is just n simple linear differential equations of the first order:

$$y_i'(t) = \lambda_i y_i(t), \quad i = 1, 2, \ldots, n,$$

and its solution is trivial: $y_i(t) = c_i e^{\lambda_i t}$ with a constant c_i for $i = 1, 2, \ldots, n$.

In other words, the diagonal matrix D has n linearly independent eigenvectors $\mathbf{e}_1, \mathbf{e}_2, \ldots, \mathbf{e}_n$ belonging to the eigenvalues $\lambda_1, \lambda_2, \ldots, \lambda_n$, respectively. One can see that $\mathbf{y}_i(t) = e^{\lambda_i t}\mathbf{e}_i$ is a solution to the LDE $\mathbf{y}'(t) = D\mathbf{y}(t)$ for $i = 1, 2, \ldots, n$.

Moreover, at $t = 0$, the solution set $\{\mathbf{y}_1(t), \mathbf{y}_2(t), \ldots, \mathbf{y}_n(t)\} = \{\mathbf{e}_1, \mathbf{e}_2, \ldots, \mathbf{e}_n\}$ is linearly independent.

By Theorem 6.18, a general solution to the LDE $\mathbf{y}'(t) = D\mathbf{y}(t)$ is

$$\begin{aligned} \mathbf{y}(t) &= c_1\mathbf{y}_1(t) + c_2\mathbf{y}_2(t) + \cdots + c_n\mathbf{y}_n(t) \\ &= c_1 e^{\lambda_1 t}\mathbf{e}_1 + c_2 e^{\lambda_2 t}\mathbf{e}_2 + \cdots + c_n e^{\lambda_n t}\mathbf{e}_n. \end{aligned}$$

In matrix notation,

$$\mathbf{y}(t) = \begin{bmatrix} y_1(t) \\ y_2(t) \\ \vdots \\ y_n(t) \end{bmatrix} = \begin{bmatrix} e^{\lambda_1 t} & & & \mathbf{0} \\ & e^{\lambda_2 t} & & \\ & & \ddots & \\ \mathbf{0} & & & e^{\lambda_n t} \end{bmatrix} \begin{bmatrix} c_1 \\ c_2 \\ \vdots \\ c_n \end{bmatrix} = e^{tD}\mathbf{y}_0,$$

where e^{tD} is by definition,

$$e^{tD} = \begin{bmatrix} e^{\lambda_1 t} & & & \mathbf{0} \\ & e^{\lambda_2 t} & & \\ & & \ddots & \\ \mathbf{0} & & & e^{\lambda_n t} \end{bmatrix}.$$

Example 122 (*A* **predator-prey** *problem as a model of* $\mathbf{y}' = A\mathbf{y}$) One of the fundamental problems of mathematical ecology is the predator-prey problem. As an example, we consider sharks (predators) and small fishes (preys) in a restricted region of the ocean.

Without the small fishes (preys) the population of the sharks (predators) will decrease, and without the sharks the population of the fishes will increase. Let

$x(t)$ denote the population of preys (small fishes) at time t, and

$y(t)$ denote the population of predators (sharks) at time t.

Then, the growth rates of the two populations $x(t)$ and $y(t)$ over time t can be written as the following system of (nonlinear) differential equations:

$$\begin{cases} x'(t) &= \quad a\,x(t) \quad - \quad b\,x(t)y(t) \\ y'(t) &= \, -c\,y(t) \quad + \quad d\,x(t)y(t), \end{cases}$$

called the **predator-prey equation** or the **Lotka-Volterra equation**. In this equation, a, b, c, d are all positive real constants, and the coefficients a and c are the birth rate of x (preys) and the death rate of y (predators), respectively.

The nonlinear $x(t)y(t)$ terms in the two equations give the interaction of the two species such as the number of contacts per unit time between predators and prey. So the coefficients b and d are the measures of the effect of the interaction between them.

A study of this general system of differential equations leads to very interesting developments in the theory of ecological systems and it is interested in determining whether an ecological balance, *i.e.,* an equilibrium state exists or not.

Here, we restrict our study to the case of $x(t)$ and $y(t)$ being very small. In this case, the nonlinear term $x(t)y(t)$ becomes sufficiently small and it can be neglected in each of the two equations, so the system can be assumed to be an LDE as follows: In a matrix form,

$$\begin{bmatrix} x'(t) \\ y'(t) \end{bmatrix} = \begin{bmatrix} a & 0 \\ 0 & -c \end{bmatrix} \begin{bmatrix} x(t) \\ y(t) \end{bmatrix}.$$

The eigenvalues of its matrix are $\lambda_1 = a$ and $\lambda_2 = -c$ with their associated eigenvectors $\mathbf{e}_1$ and $\mathbf{e}_2$, respectively. Since $\mathbf{e}_1$ and $\mathbf{e}_2$ are linearly independent, its general solution is

$$\begin{bmatrix} x(t) \\ y(t) \end{bmatrix} = \begin{bmatrix} c_1 e^{at} \\ c_2 e^{-ct} \end{bmatrix} = \begin{bmatrix} e^{at} & 0 \\ 0 & e^{-ct} \end{bmatrix} \begin{bmatrix} c_1 \\ c_2 \end{bmatrix} = c_1 e^{at}\mathbf{e}_1 + c_2 e^{-ct}\mathbf{e}_2.$$

An ecological balance occurs when neither of the population levels is changing, *i.e.*, when $x'(t) = y'(t) = 0$. By solving $x(t)(a - by(t)) = -y(t)(c - dx(t)) = 0$ from the predator-prey equation, one can get $\{y(t) = 0, x(t) = 0\}$ and $\{y(t) = a/b, x(t) = c/d\}$, giving two cases of the balance. □

(2) Secondly, let A be diagonalizable in the LDE $\mathbf{y}'(t) = A\mathbf{y}(t)$ and let $\mathbf{v}_1, \mathbf{v}_2, \ldots, \mathbf{v}_n$ be n linearly independent eigenvectors of A belonging to the eigenvalues $\lambda_1, \lambda_2, \ldots, \lambda_n$, respectively. Then, the matrix $Q = [\mathbf{v}_1\ \mathbf{v}_2\ \cdots\ \mathbf{v}_n]$ diagonalizes A and

$$A = QDQ^{-1} = Q \begin{bmatrix} \lambda_1 & & & \mathbf{0} \\ & \lambda_2 & & \\ & & \ddots & \\ \mathbf{0} & & & \lambda_n \end{bmatrix} Q^{-1}.$$

Thus the given LDE becomes $Q^{-1}\mathbf{y}' = DQ^{-1}\mathbf{y}$. Now, a change of variables $\mathbf{x} = Q^{-1}\mathbf{y}$ (or $\mathbf{y} = Q\mathbf{x}$) gives a new LDE

$$\mathbf{x}' = D\mathbf{x},$$

with an initial condition $\mathbf{x}_0 = Q^{-1}\mathbf{y}_0 = (c_1, c_2, \ldots, c_n)$. Since D is diagonal, its general solution is

$$\mathbf{x} = e^{tD}\mathbf{x}_0 = c_1 e^{\lambda_1 t}\mathbf{e}_1 + c_2 e^{\lambda_2 t}\mathbf{e}_2 + \cdots + c_n e^{\lambda_n t}\mathbf{e}_n.$$

Now, a general solution to the original LDE $\mathbf{y}' = A\mathbf{y}$ is

$$\begin{aligned} \mathbf{y} = Q\mathbf{x} &= Qe^{tD}Q^{-1}\mathbf{y}_0 \\ &= [\mathbf{v}_1\ \mathbf{v}_2\ \cdots\ \mathbf{v}_n] \begin{bmatrix} e^{\lambda_1 t} & & & \mathbf{0} \\ & e^{\lambda_2 t} & & \\ & & \ddots & \\ \mathbf{0} & & & e^{\lambda_n t} \end{bmatrix} \begin{bmatrix} c_1 \\ c_2 \\ \vdots \\ c_n \end{bmatrix} \\ &= c_1 e^{\lambda_1 t}\mathbf{v}_1 + c_2 e^{\lambda_2 t}\mathbf{v}_2 + \cdots + c_n e^{\lambda_n t}\mathbf{v}_n. \end{aligned}$$

Clearly, the n solutions

$$e^{\lambda_1 t}\mathbf{v}_1,\ e^{\lambda_2 t}\mathbf{v}_2,\ \ldots,\ e^{\lambda_n t}\mathbf{v}_n$$

are linearly independent, by Theorem 6.18.

Remark: One can check directly that each vector function $\mathbf{y}_i(t) = e^{\lambda_i t}\mathbf{v}_i$ is a particular solution to the LDE with initial condition $\mathbf{y}_i(0) = \mathbf{v}_i$ for $i = 1, 2, \ldots, n$. Since $\mathbf{y}_i(0) = \mathbf{v}_i$ for $i = 1, 2, \ldots, n$ are linearly independent, $e^{\lambda_i t}\mathbf{v}_i$ for $i = 1, 2, \ldots, n$ form a fundamental set of solutions.

As a consequence, one can have the following theorem.

Theorem 6.20 *Let A be a diagonalizable $n \times n$ matrix with n linearly independent eigenvectors $\mathbf{v}_1, \mathbf{v}_2, \ldots, \mathbf{v}_n$ belonging to the eigenvalues $\lambda_1, \lambda_2, \ldots, \lambda_n$, respectively. Then, a general solution to an LDE $\mathbf{y}'(t) = A\mathbf{y}(t)$ is*

$$\mathbf{y}(t) = c_1 e^{\lambda_1 t}\mathbf{v}_1 + c_2 e^{\lambda_2 t}\mathbf{v}_2 + \cdots + c_n e^{\lambda_n t}\mathbf{v}_n. \tag{6.6}$$

If an initial condition $\mathbf{y}(0)$ is given, the constants $c_1, c_2, \ldots, c_n$ can be determined for a particular solution.

Example 123 ($\mathbf{y}' = A\mathbf{y}$ *with a diagonalizable matrix* A) Solve the LDE

$$\begin{cases} y_1' &= 5y_1 - 4y_2 + 6y_3 \\ y_2' &= 12y_1 - 9y_2 + 12y_3 \\ y_3' &= 6y_1 - 4y_2 + 5y_3. \end{cases}$$

Solution: In matrix notation, the LDE is $\mathbf{y}' = A\mathbf{y}$ with

$$A = \begin{bmatrix} 5 & -4 & 6 \\ 12 & -9 & 12 \\ 6 & -4 & 5 \end{bmatrix}.$$

The eigenvalues of A and their associated eigenvectors are

$$\lambda_1 = \lambda_2 = -1,\ \lambda_3 = 3, \quad \text{and} \quad \mathbf{v}_1 = (2,3,0),\ \mathbf{v}_2 = (-1,0,1),\ \mathbf{v}_3 = (1,2,1),$$

respectively, which are linearly independent (see Problem 145). Hence, by Theorem 6.20, its general solution is

$$\begin{aligned} \begin{bmatrix} y_1 \\ y_2 \\ y_3 \end{bmatrix} &= c_1 e^{-t}\begin{bmatrix} 2 \\ 3 \\ 0 \end{bmatrix} + c_2 e^{-t}\begin{bmatrix} -1 \\ 0 \\ 1 \end{bmatrix} + c_3 e^{3t}\begin{bmatrix} 1 \\ 2 \\ 1 \end{bmatrix} \\ &= c_1 e^{-t}\mathbf{v}_1 + c_2 e^{-t}\mathbf{v}_2 + c_3 e^{3t}\mathbf{v}_3. \end{aligned}$$

If an initial condition is given, the constants c_i can be determined. □

(3) An LDE $\mathbf{y}' = A\mathbf{y}$ with a non-diagonalizable matrix A will be discussed later in Section 6.6 and again in Subsection 8.8.5.

Problem 156 Solve the LDE

$$\begin{bmatrix} y_1'(t) \\ y_2'(t) \end{bmatrix} = \begin{bmatrix} 0 & 1 \\ 1 & 0 \end{bmatrix} \begin{bmatrix} y_1(t) \\ y_2(t) \end{bmatrix} \quad \text{or} \quad \begin{cases} y_1'(t) &= y_2(t) \\ y_2'(t) &= y_1(t) \end{cases}$$

with initial conditions $y_1(0) = 1, y_2(0) = 0$. This will be repeated again in Example 128.

Problem 157 Solve the LDE $\mathbf{y}' = A\mathbf{y}$, where

$$A = \begin{bmatrix} 3 & 2 & 0 \\ 2 & 3 & 0 \\ 0 & 0 & 5 \end{bmatrix},$$

which is shown to be diagonalizable in Example 106.

Hint: Use the eigenvalues and eigenvectors of A given in Example 106.

6.5 Exponential matrices

Just as a real-valued function $y(t) = ce^{at}$ satisfies the differential equation $y'(t) = ay(t)$, a vector-valued function $\mathbf{y}(t) = e^{tD}\mathbf{y}_0$ satisfies an LDE $\mathbf{y}'(t) = D\mathbf{y}(t)$ with a diagonal matrix D, as the first case in Subsection 6.4.7.

With the expectation of expanding it to any square matrix A, *the exponent of a square matrix* A is defined in this section.

Following the real-valued exponential function

$$e^x = \sum_{k=0}^{\infty} \frac{x^k}{k!} = 1 + x + \frac{x^2}{2!} + \frac{x^3}{3!} + \cdots,$$

the matrix-valued exponential function is defined as follows.

Definition 6.21 For any square matrix A, the **exponential matrix** of A is defined as the series

$$e^A = \sum_{k=0}^{\infty} \frac{A^k}{k!} = I + A + \frac{A^2}{2!} + \frac{A^3}{3!} + \cdots.$$

That is, the exponential matrix e^A is defined to be the (entrywise) limit of the sequence: $\left[e^A\right]_{ij} = \lim\limits_{N\to\infty}\left[\sum_{k=0}^{N}\frac{A^k}{k!}\right]_{ij}$ for all i, j. Sometimes e^A is also written as $\exp A$.

Clearly, $e^{\mathbf{0}}$ is the identity matrix for any zero matrix $\mathbf{0}$.

Practically, the computation of e^A involves the computation of the powers A^k for all $k \geq 1$, which is not easy in general. Nevertheless, one can show that the limit e^A always exists for every square matrix A.

Theorem 6.22 *For any square matrix A, the matrix e^A is well defined. In fact, every (i, j)-entry $\lim\limits_{N\to\infty}\left[\sum_{k=0}^{N}\frac{A^k}{k!}\right]_{ij}$ of e^A is absolutely convergent, so is convergent.*

Proof: Since any $n \times n$ matrix A has only n^2 entries $[A]_{ij}$, there is a number M such that $|[A]_{ij}| \leq M$ for all (i, j)-entries $[A]_{ij}$ of A. Hence, one can easily show that $|[A^k]_{ij}| \leq n^{k-1}M^k$ for all k and i, j, which gives

$$|[e^A]_{ij}| \leq \sum_{k=0}^{\infty}\frac{1}{k!}n^{k-1}M^k = \frac{1}{n}e^{nM}.$$

By the comparison test, each entry of $e^A = \sum_{n=0}^{\infty}\frac{A^k}{k!}$ is absolutely convergent for any square matrix A. It is well known (in calculus) that every absolutely convergent series is always convergent, as one of the series convergence tests. □

Example 124 (e^D *for a diagonal matrix* D) If

$$D = \begin{bmatrix} \lambda_1 & & \mathbf{0} \\ & \ddots & \\ \mathbf{0} & & \lambda_n \end{bmatrix}, \quad \text{then} \quad D^k = \begin{bmatrix} \lambda_1^k & & \mathbf{0} \\ & \ddots & \\ \mathbf{0} & & \lambda_n^k \end{bmatrix} \quad \text{for any } k \geq 0.$$

Thus, the exponential matrix e^D is

$$e^D = \sum_{k=0}^{\infty}\frac{D^k}{k!} = \begin{bmatrix} \sum\limits_{k=0}^{\infty}\frac{\lambda_1^k}{k!} & & \mathbf{0} \\ & \ddots & \\ \mathbf{0} & & \sum\limits_{k=0}^{\infty}\frac{\lambda_n^k}{k!} \end{bmatrix} = \begin{bmatrix} e^{\lambda_1} & & \mathbf{0} \\ & \ddots & \\ \mathbf{0} & & e^{\lambda_n} \end{bmatrix},$$

which coincides with the definition given on page 291. □

Example 125 (e^A *from the definition*) If $A = \begin{bmatrix} 0 & 1 \\ 1 & 0 \end{bmatrix}$, then $A^2 = I$. Hence

$$A^0 = A^2 = A^4 = \cdots = I \quad \text{and} \quad A^1 = A^3 = A^5 = \cdots = A.$$

This gives,

$$e^A = \sum_{k=0}^{\infty} \frac{A^k}{k!} = \begin{bmatrix} \displaystyle\sum_{k=\text{even}} \frac{1}{k!} & \displaystyle\sum_{k=\text{odd}} \frac{1}{k!} \\ \displaystyle\sum_{k=\text{odd}} \frac{1}{k!} & \displaystyle\sum_{k=\text{even}} \frac{1}{k!} \end{bmatrix} = \frac{1}{2} \begin{bmatrix} e^1 + e^{-1} & e^1 - e^{-1} \\ e^1 - e^{-1} & e^1 + e^{-1} \end{bmatrix}.$$

□

The following theorem is sometimes helpful for computing e^A.

Theorem 6.23 *Let A_1, A_2, A_3, ... be a sequence of $m \times n$ matrices such that $\lim_{k\to\infty} A_k$ exists. Then*

$$\lim_{k\to\infty} BA_k = B\left(\lim_{k\to\infty} A_k\right) \quad \textit{and} \quad \lim_{k\to\infty} A_kC = \left(\lim_{k\to\infty} A_k\right)C$$

for any matrices B and C for which the products can be defined.

Proof: Let $\lim_{k\to\infty} A_k = L$. By comparing the (i,j)-entries on both sides

$$\begin{aligned} \lim_{k\to\infty} [BA_k]_{ij} &= \lim_{k\to\infty} \left(\sum_{\ell=1}^{m} [B]_{i\ell}[A_k]_{\ell j} \right) = \sum_{\ell=1}^{m} [B]_{i\ell} \lim_{k\to\infty} [A_k]_{\ell j} \\ &= \sum_{\ell=1}^{m} [B]_{i\ell}[L]_{\ell j} = [BL]_{ij}, \end{aligned}$$

we get $\lim_{k\to\infty} BA_k = BL$. Similarly $\lim_{k\to\infty} A_kC = LC$. □

For example, if A is a diagonalizable matrix and $Q^{-1}AQ = D$ is diagonal for some invertible matrix Q, then, for each integer $k \geq 0$, $A^k = QD^kQ^{-1}$ and by Theorem 6.23,

$$\begin{aligned} e^A = e^{QDQ^{-1}} &= I + QDQ^{-1} + \frac{(QDQ^{-1})^2}{2!} + \frac{(QDQ^{-1})^3}{3!} + \cdots \\ &= Q\left(I + D + \frac{D^2}{2!} + \frac{D^3}{3!} + \cdots\right)Q^{-1} \\ &= Qe^DQ^{-1}, \end{aligned}$$

as a product of three matrices.

Example 126 (e^A *for a diagonalizable matrix* A) (Example 125 again)
Let $A = \begin{bmatrix} 0 & 1 \\ 1 & 0 \end{bmatrix}$, as in Example 125. It is a diagonalizable matrix: its eigenvalues are 1 and -1 with associated eigenvectors $\mathbf{u}_1 = \begin{bmatrix} 1 \\ 1 \end{bmatrix}$ and $\mathbf{u}_2 = \begin{bmatrix} -1 \\ 1 \end{bmatrix}$, respectively. Thus $A = QDQ^{-1}$ with $D = \begin{bmatrix} 1 & 0 \\ 0 & -1 \end{bmatrix}$ and $Q = \begin{bmatrix} 1 & -1 \\ 1 & 1 \end{bmatrix}$. Hence, by Theorem 6.23,

$$e^A = Qe^DQ^{-1} = \frac{1}{2}\begin{bmatrix} 1 & -1 \\ 1 & 1 \end{bmatrix}\begin{bmatrix} e^1 & 0 \\ 0 & e^{-1} \end{bmatrix}\begin{bmatrix} 1 & 1 \\ -1 & 1 \end{bmatrix} = \frac{1}{2}\begin{bmatrix} e^1 + e^{-1} & e^1 - e^{-1} \\ e^1 - e^{-1} & e^1 + e^{-1} \end{bmatrix},$$

as shown in Example 125. □

The following theorem shows some basic properties of the exponential matrices, whose proofs are easy and left for exercise.

Theorem 6.24 **(1)** $e^{A+B} = e^A e^B$ *if* $AB = BA$.

(2) e^A *is invertible for any square matrix* A *and* $(e^A)^{-1} = e^{-A}$.

(3) $e^{Q^{-1}AQ} = Q^{-1}e^AQ$ *for any invertible matrix* Q.

(4) *If* $\lambda_1, \lambda_2, \ldots, \lambda_n$ *are the eigenvalues of a matrix* A *with their associated eigenvectors* $\mathbf{v}_1, \mathbf{v}_2, \ldots, \mathbf{v}_n$*, then* e^{λ_i}*'s are the eigenvalues of* e^A *with the same associated eigenvectors* $\mathbf{v}_i$*'s for* $i = 1, 2, \ldots, n$*. Moreover,* $\det e^A = e^{\lambda_1} \cdots e^{\lambda_n} = e^{\mathrm{tr}(A)} \neq 0$ *for any square matrix* A.

In general, the computation of e^A is not easy at all if A is not diagonalizable. However, every square matrix is similar to a simpler triangular matrix called the Jordan canonical form, and this makes the computation of e^A relatively easy.

The following example shows a computation of e^A for a triangular matrix A even if it is not diagonalizable. The general case will be treated again in Chapter 8.

Example 127 (e^A *for a triangular of the form* $A = \lambda I + N$) Compute e^A for

$$A = \begin{bmatrix} 2 & 3 \\ 0 & 2 \end{bmatrix}.$$

Solution: Write $A = 2I + N$ with $N = \begin{bmatrix} 0 & 3 \\ 0 & 0 \end{bmatrix}$. Since $(2I)N = N(2I)$, by Theorem 6.24(1), $e^A = e^{2I}e^N$. From the direct computation of the series expansion, we get $e^{2I} = e^2 I$. Moreover, since $N^k = \mathbf{0}$ for $k \geq 2$,

$$e^N = I + N + \frac{N^2}{2!} + \cdots = I + N = \begin{bmatrix} 1 & 3 \\ 0 & 1 \end{bmatrix}.$$

Thus,

$$e^A = e^2 e^N = e^2(I + N) = e^2 \begin{bmatrix} 1 & 3 \\ 0 & 1 \end{bmatrix} = \begin{bmatrix} e^2 & 3e^2 \\ 0 & e^2 \end{bmatrix},$$

which is also a triangular matrix. □

It will be shown in the next section that the columns of the exponential matrix e^{tA} form a fundamental set of solutions to the LDE $\mathbf{y}' = A\mathbf{y}$.

Problem 158 Let $A = \begin{bmatrix} 1 & 1 \\ 0 & 1 \end{bmatrix}$. Then $A = I + N$ with $N = \begin{bmatrix} 0 & 1 \\ 0 & 0 \end{bmatrix}$. (1) Find $\lim_{n \to \infty} A^n$ if there exists. (2) Compute e^A. (3) Show that A is not diagonalizable.

Problem 159 Prove Theorem 6.24.

6.6 Applications continued

6.6.1 Existence and uniqueness of solutions to an LDE

One of the most prominent applications of exponential matrices is to show the existence of solutions to an LDE.

In this section it will be shown that a general solution to $\mathbf{y}'(t) = A\mathbf{y}(t)$ is $\mathbf{y}(t) = e^{tA}\mathbf{y}_0$ for any square (diagonalizable or not) matrix A.

For any matrix whose entries are differentiable function of t, its derivative is defined by the entrywise derivation. It means, for example,

$$\frac{d}{dt}\begin{bmatrix} 3 & t^n \\ e^t & e^{5t} \end{bmatrix} = \begin{bmatrix} \frac{d}{dt}3 & \frac{d}{dt}t^n \\ \frac{d}{dt}e^t & \frac{d}{dt}e^{5t} \end{bmatrix} = \begin{bmatrix} 0 & nt^{n-1} \\ e^t & 5e^{5t} \end{bmatrix},$$

or

$$\text{the derivative of } \begin{bmatrix} f_1(t) \\ f_2(t) \\ \vdots \\ f_n(t) \end{bmatrix} \quad \text{is} \quad \mathbf{y}'(t) = \begin{bmatrix} f_1'(t) \\ f_2'(t) \\ \vdots \\ f_n'(t) \end{bmatrix}.$$

Lemma 6.25 *For any $t \in \mathbb{R}$ and any square matrix A, the exponential matrix*

$$e^{tA} = I + tA + \frac{t^2}{2!}A^2 + \frac{t^3}{3!}A^3 + \cdots$$

is a differentiable function of t and $\frac{d}{dt}e^{tA} = Ae^{tA}$.

Proof: Because of the absolute convergence of the series expansion of e^{tA}, one can differentiate term by term for $\frac{d}{dt}e^{tA}$, *i.e.,*

$$\begin{aligned} \frac{d}{dt}e^{tA} &= \frac{d}{dt}\left(I + tA + \frac{t^2}{2!}A^2 + \frac{t^3}{3!}A^3 + \cdots\right) \\ &= A + tA^2 + \frac{t^2}{2!}A^3 + \cdots = Ae^{tA}, \end{aligned}$$

as expected. □

As a direct consequence of Lemma 6.25, one can see that

$$\mathbf{y}(t) = e^{tA}\mathbf{y}_0 \quad \text{is a solution to an LDE } \mathbf{y}' = A\mathbf{y}$$

with initial vector $\mathbf{y}_0$.

By taking the standard basis vectors $\mathbf{e}_i$, $1 \le i \le n$, as the initial vector $\mathbf{y}_0$, $e^{tA}\mathbf{e}_1, e^{tA}\mathbf{e}_2, \ldots, e^{tA}\mathbf{e}_n$ are linearly independent n solutions to an LDE $\mathbf{y}' = A\mathbf{y}$. It means that

the n *columns of* e^{tA} *are (particular) solutions to the LDE* $\mathbf{y}' = A\mathbf{y}$

and they are linearly independent and form a fundamental set of solutions.

It gives the following theorem.

Theorem 6.26 (Existence and uniqueness of solutions to an LDE) *For any $n \times n$ matrix A, an LDE $\mathbf{y}' = A\mathbf{y}$ always has a general solution. In fact, the columns of e^{tA} form a fundamental set of solutions to the LDE $\mathbf{y}' = A\mathbf{y}$.*

If an initial vector $\mathbf{y}_0$ is given, then

$$\mathbf{y}(t) = e^{tA}\mathbf{y}_0$$

is the unique (particular) solution to the LDE which satisfies the initial condition.

As a special case, let A be diagonalizable and let $Q^{-1}AQ = D$ be diagonal, where $Q = [\mathbf{v}_1 \cdots \mathbf{v}_n]$ is a change of basis matrix consisting of n linearly independent eigenvectors of A belonging to the eigenvalues λ_i's. Hence, as a repetition of Theorem 6.20, a general solution to an LDE $\mathbf{y}' = A\mathbf{y}$ is

$$\begin{aligned}
\mathbf{y}(t) = e^{tA}\mathbf{y}_0 &= e^{tQDQ^{-1}}\mathbf{y}_0 = Qe^{tD}Q^{-1}\mathbf{y}_0 \\
&= [\mathbf{v}_1 \ \cdots \ \mathbf{v}_n] \begin{bmatrix} e^{\lambda_1 t} & & \mathbf{0} \\ & \ddots & \\ \mathbf{0} & & e^{\lambda_n t} \end{bmatrix} \begin{bmatrix} c_1 \\ \vdots \\ c_n \end{bmatrix} \\
&= c_1 e^{\lambda_1 t}\mathbf{v}_1 + c_2 e^{\lambda_2 t}\mathbf{v}_2 + \cdots + c_n e^{\lambda_n t}\mathbf{v}_n.
\end{aligned}$$

Example 128 ($\mathbf{y}' = A\mathbf{y}$ *with diagonalizable matrix A*) Solve the LDE

$$\begin{bmatrix} y_1'(t) \\ y_2'(t) \end{bmatrix} = \begin{bmatrix} 0 & 1 \\ 1 & 0 \end{bmatrix} \begin{bmatrix} y_1(t) \\ y_2(t) \end{bmatrix}, \quad \text{or} \quad \begin{cases} y_1'(t) &= y_2(t) \\ y_2'(t) &= y_1(t) \end{cases}$$

with initial conditions $y_1(0) = 1$, $y_2(0) = 0$.

Solution: The matrix $A = \begin{bmatrix} 0 & 1 \\ 1 & 0 \end{bmatrix}$ is given in Example 126. Its eigenvalues are 1 and -1 with associated eigenvectors $\mathbf{u}_1 = \begin{bmatrix} 1 \\ 1 \end{bmatrix}$ and $\mathbf{u}_2 = \begin{bmatrix} -1 \\ 1 \end{bmatrix}$, respectively. By Theorem 6.20, or Theorem 6.26, its general solution $\mathbf{y}(t) = e^{tA}\mathbf{y}_0$ is

$$\begin{bmatrix} y_1 \\ y_2 \end{bmatrix} = c_1 e^t \begin{bmatrix} 1 \\ 1 \end{bmatrix} + c_2 e^{-t} \begin{bmatrix} -1 \\ 1 \end{bmatrix}.$$

The initial condition $y_1(0) = 1$, $y_2(0) = 0$ gives $c_1 = 1/2$ and $c_2 = -1/2$. □

The next example shows another possible computation of e^{tA} for some triangular matrices A, as in Example 127. The general case will be treated again in Chapter 8.

Example 129 ($\mathbf{y}' = A\mathbf{y}$ *with a non-diagonalizable triangular matrix* $A = \lambda I + N$) Solve the LDE $\mathbf{y}' = A\mathbf{y}$ with initial condition $\mathbf{y}(0) = \mathbf{y}_0$, where

$$A = \begin{bmatrix} \lambda & 1 \\ 0 & \lambda \end{bmatrix}, \qquad \mathbf{y}_0 = \begin{bmatrix} a \\ b \end{bmatrix}.$$

Solution: The matrix A has an eigenvalue λ of multiplicity 2 and is not diagonalizable. One can rewrite A as

$$A = \begin{bmatrix} \lambda & 0 \\ 0 & \lambda \end{bmatrix} + \begin{bmatrix} 0 & 1 \\ 0 & 0 \end{bmatrix} = \lambda I + N.$$

Then, by a similar argument as in Example 127,

$$e^{tA} = e^{t(\lambda I + N)} = e^{\lambda t} e^{tN} = e^{\lambda t} \begin{bmatrix} 1 & t \\ 0 & 1 \end{bmatrix}.$$

Therefore, the solution is

$$\mathbf{y} = e^{tA}\mathbf{y}_0 = e^{\lambda t} \begin{bmatrix} 1 & t \\ 0 & 1 \end{bmatrix} \begin{bmatrix} a \\ b \end{bmatrix} = \begin{bmatrix} (a+bt)e^{\lambda t} \\ be^{\lambda t} \end{bmatrix} = e^{\lambda t} \begin{bmatrix} a \\ b \end{bmatrix} + te^{\lambda t} \begin{bmatrix} b \\ 0 \end{bmatrix}.$$

In terms of components, $y_1 = (a + bt)e^{\lambda t}$ and $y_2 = be^{\lambda t}$. □

Example 130 ($\mathbf{y}' = A\mathbf{y}$ *with* A *having complex eigenvalues*) Find a general solution to an LDE $\mathbf{y}' = A\mathbf{y}$, where

$$A = \begin{bmatrix} a & -b \\ b & a \end{bmatrix}, \quad b \neq 0.$$

Solution: The eigenvalues of A are $a \pm ib$, which are not real. However, one can compute e^{tA} directly without using diagonalization. We first write A as

$$A = \begin{bmatrix} a & -b \\ b & a \end{bmatrix} = a \begin{bmatrix} 1 & 0 \\ 0 & 1 \end{bmatrix} + b \begin{bmatrix} 0 & -1 \\ 1 & 0 \end{bmatrix} = aI + bJ.$$

Then clearly $IJ = JI$ and $e^{tA} = e^{atI + btJ} = e^{at}e^{btJ}$. Since

$$J^2 = \begin{bmatrix} -1 & 0 \\ 0 & -1 \end{bmatrix} = -I, \quad J^3 = \begin{bmatrix} 0 & 1 \\ -1 & 0 \end{bmatrix} = -J, \quad J^4 = \begin{bmatrix} 1 & 0 \\ 0 & 1 \end{bmatrix} = I,$$

one can deduce $J^k = J^{k+4}$ for all $k = 1, 2, \ldots$, and

$$\begin{aligned} e^{btJ} &= I + \frac{btJ}{1!} + \frac{(bt)^2J^2}{2!} + \frac{(bt)^3J^3}{3!} + \frac{(bt)^4J^4}{4!} + \cdots \\ &= \begin{bmatrix} 1 - \frac{(bt)^2}{2!} + \frac{(bt)^4}{4!} - \cdots & -(bt) + \frac{(bt)^3}{3!} - \frac{(bt)^5}{5!} + \cdots \\ (bt) - \frac{(bt)^3}{3!} + \frac{(bt)^5}{5!} - \cdots & 1 - \frac{(bt)^2}{2!} + \frac{(bt)^4}{4!} - \cdots \end{bmatrix} \\ &= \begin{bmatrix} \cos bt & -\sin bt \\ \sin bt & \cos bt \end{bmatrix} \end{aligned}$$

for any t and constant b. Thus, a general solution to $\mathbf{y}' = A\mathbf{y}$ is

$$\mathbf{y} = e^{tA}\mathbf{c} = e^{at}e^{btJ}\mathbf{c} = e^{at}\begin{bmatrix} \cos bt & -\sin bt \\ \sin bt & \cos bt \end{bmatrix}\begin{bmatrix} c_1 \\ c_2 \end{bmatrix}.$$

That is, $y_1(t) = e^{at}(c_1 \cos bt - c_2 \sin bt)$, $y_2(t) = e^{at}(c_1 \sin bt + c_2 \cos bt)$. □

Problem 160 Solve the LDE $\begin{cases} y_1' = y_1 + y_2 \\ y_2' = 4y_1 - 2y_2 . \end{cases}$

Problem 161 Solve the LDE $\mathbf{y}' = A\mathbf{y}$ with initial condition $\mathbf{y}(0) = \mathbf{y}_0$ by computing $e^{tA}\mathbf{y}_0$ for

(1) $A = \begin{bmatrix} 1 & -2 \\ 0 & -1 \end{bmatrix}$, $\mathbf{y}_0 = \begin{bmatrix} 1 \\ 1 \end{bmatrix}$, (2) $A = \begin{bmatrix} 1 & 1 & 1 \\ 0 & 0 & 1 \\ 0 & 0 & -1 \end{bmatrix}$, $\mathbf{y}_0 = \begin{bmatrix} 1 \\ 1 \\ 1 \end{bmatrix}$.

Hint: In both cases, A are diagonalizable.

6.7 Exercises

6.1. *An $n \times n$ matrix A is diagonalizable if and only if A has n linearly independent eigenvectors.* Find the characteristic polynomial, eigenvalues, and eigenvectors of each of the following matrices, if there exist.

(1) $\begin{bmatrix} 2 & 0 \\ 0 & 3 \end{bmatrix}$, (2) $\begin{bmatrix} 1 & 2 & 3 \\ 0 & 2 & 3 \\ 0 & 0 & 3 \end{bmatrix}$, (3) $\begin{bmatrix} -2 & 0 & 0 \\ 3 & 2 & 3 \\ 4 & -1 & 6 \end{bmatrix}$,

(4) $\begin{bmatrix} 0 & 1 & 0 & 1 \\ 1 & 0 & 1 & 0 \\ 0 & 1 & 0 & 1 \\ 1 & 0 & 1 & 0 \end{bmatrix}$, (5) $\begin{bmatrix} 1 & 1 & 1 & 1 \\ 1 & 1 & 1 & 1 \\ 1 & 1 & 1 & 1 \\ 1 & 1 & 1 & 1 \end{bmatrix}$, (6) $\begin{bmatrix} 0 & -5 & 0 & 0 \\ 5 & 0 & 0 & 0 \\ 0 & 0 & 0 & -2 \\ 0 & 0 & 2 & 0 \end{bmatrix}$.

Hint: (1) is diagonal. (2) is triangular. (4) and (5) are symmetric. (6) has two nonzero (2 × 2) blocks, each of which is skew-symmetric.

6.2. *The eigenvalues are just the roots of the characteristic equation* $\det(\lambda I - A) = 0$. For a 2 × 2 matrix $A = \begin{bmatrix} a & b \\ c & d \end{bmatrix}$, show that the characteristic polynomial of A is

$$f(\lambda) = \lambda^2 - (a+d)\lambda + (ad - bc) = \lambda^2 - \mathrm{tr}(A)\lambda + \det A,$$

and hence the eigenvalues of A are $\frac{1}{2}\left(a + d \pm \sqrt{(a-d)^2 + 4bc}\right)$.

By using the discriminant of $f(\lambda) = 0$, show that the matrix A has

(1) two distinct real eigenvalues if $(a-d)^2 + 4bc > 0$,
(2) one eigenvalue if $(a-d)^2 + 4bc = 0$,
(3) no real eigenvalues if $(a-d)^2 + 4bc < 0$,
(4) only real eigenvalues if it is symmetric (*i.e.*, $b = c$).

6.3. Find a 3 × 3 matrix that has the eigenvalues $\lambda_1 = 1, \lambda_2 = 2, \lambda_3 = 3$ with associated eigenvectors $\mathbf{x}_1 = (2, -1, 0),\ \mathbf{x}_2 = (-1, 2, -1),\ \mathbf{x}_3 = (0, -1, 2)$, respectively.

6.4. Let a 3 × 3 matrix A have the eigenvalues $-1, 0, 1$ with associated eigenvectors $\mathbf{u}, \mathbf{v}, \mathbf{w}$, respectively. Describe the null space $\mathcal{N}(A)$, the row space $\mathcal{R}(A)$, and the column space $\mathcal{C}(A)$. Find their bases.

Hint: Three different eigenvalues, so their associated eigenvectors are linearly independent.

6.5. Show that every 2 × 2 matrix A such that $\mathrm{tr}(A) = 1$ and $\det A = -2$ should be diagonalizable.

Hint: Show that A has two distinct eigenvalues.

6.6. Show that if A is a 3 × 3 matrix having three eigenvalues $1, 2, 3$, then $(A - I)(A - 2I)(A - 3I) = \mathbf{0}$, the zero matrix.

6.7. Corollary 6.4.3 says that for any two $n \times n$ matrices A and B, their products AB and BA have the same set of eigenvalues. Show that AB and BA are similar if A or B is nonsingular. Is this still true for any two singular matrices A and B?

6.8. Show that every 2 × 2 skew-symmetric nonzero matrix has no real eigenvalues. However, every 3 × 3 skew-symmetric matrix has at least one real eigenvalue (see page 313 and also Theorem 7.7), which should be zero.

Hint: For any 3 × 3 skew-symmetric matrix A, $\det A = 0$ which is the product of three eigenvalues.

6.9. Let W be a subspace of $\mathbb{R}^n$ and let P be the projection matrix of the orthogonal projection Proj_W. Find all eigenvalues of P and the eigenspace of each eigenvalue of P.

Hint: Every eigenspace E should be invariant under the projection; $P(E) \subset E$.

6.10. Let $\mathbf{u}, \mathbf{v}$ be $n \times 1$ column vectors, and let $A = \mathbf{u}\mathbf{v}^T$. Show that $\mathbf{u}$ is an eigenvector of A. Find all eigenvalues and their associated eigenvectors of A.

Hint: $A\mathbf{x} = \mathbf{u}\mathbf{v}^T\mathbf{x} = \mathbf{v}^T\mathbf{x}\mathbf{u} = \langle \mathbf{v}, \mathbf{x} \rangle \mathbf{u} = \langle \mathbf{x}, \mathbf{v} \rangle \mathbf{u}$, and rank $A = 1$.

6.11. Let $A = [a_{ij}]$ be an $n \times n$ matrix with eigenvalues $\lambda_1, \lambda_2, \ldots, \lambda_n$. Show that

$$\lambda_j = a_{jj} + \sum_{i \neq j} (a_{ii} - \lambda_i) \qquad \text{for } j = 1, 2, \ldots, n.$$

Hint: Use $\mathrm{tr}(A) = \lambda_1 + \cdots + \lambda_n = a_{11} + \cdots + a_{nn}$.

6.12. *An $n \times n$ matrix A is diagonalizable if and only if A has n linearly independent eigenvectors.* Which of the following matrices are diagonalizable?

$$(1) \begin{bmatrix} 2 & 0 \\ 0 & 3 \end{bmatrix}, \qquad (2) \begin{bmatrix} 1 & 1 & 0 \\ 0 & 2 & 1 \\ 0 & 0 & 3 \end{bmatrix}, \qquad (3) \begin{bmatrix} 1 & 1 & 0 \\ 0 & 2 & 0 \\ 0 & 0 & 2 \end{bmatrix},$$

$$(4) \begin{bmatrix} 2 & 1 & 0 \\ 0 & 2 & 1 \\ 0 & 0 & 2 \end{bmatrix}, \qquad (5) \begin{bmatrix} 1 & 1 & 1 & 1 \\ 0 & 1 & 1 & 1 \\ 0 & 0 & 1 & 1 \\ 0 & 0 & 0 & 1 \end{bmatrix}, \qquad (6) \begin{bmatrix} 1 & 1 & 0 & 0 \\ 1 & 1 & 0 & 0 \\ 0 & 0 & 2 & 2 \\ 0 & 0 & 2 & 2 \end{bmatrix}.$$

Hint: (1) is diagonal. (2) is triangular with three distinct diagonal entries. For (3)-(6), use the Rank-Nullity theorem for $A - \lambda I$. (6) is symmetric.

6.13. Find an invertible matrix Q and a diagonal matrix D such that $Q^{-1}AQ = D$ for each of the following symmetric matrices.

$$(1)\ A = \begin{bmatrix} 1 & 0 & 0 \\ 0 & 1 & 1 \\ 0 & 1 & 1 \end{bmatrix}, \qquad (2)\ A = \begin{bmatrix} 1 & 2 & 0 \\ 2 & 2 & 2 \\ 0 & 2 & 3 \end{bmatrix}, \qquad (3)\ A = \begin{bmatrix} -3 & 2 & 4 \\ 2 & -6 & 2 \\ 4 & 2 & -3 \end{bmatrix}.$$

Hint: It will be shown later in Theorem 7.19 that *every symmetric matrix is diagonalizable.* See also Exercise **7.20**.

6.14. For each matrix A in Exercise **6.13**, compute A^{100}.

6.15. Let $f(\lambda) = \det(\lambda I - A)$ be the characteristic polynomial of A. Evaluate the matrix polynomial $f(A)$ for

$$(1)\ A = \begin{bmatrix} 0 & 1 & 1 \\ 1 & 0 & 1 \\ 1 & 1 & 0 \end{bmatrix}, \qquad (2)\ A = \begin{bmatrix} 0 & 1 & 2 \\ 1 & 0 & 2 \\ 1 & 2 & 0 \end{bmatrix}.$$

See Exercise **1.10** for the definition of a matrix polynomial $f(A)$. In fact, *for any square matrix A and for its characteristic polynomial $f(\lambda)$, we always have* $f(A) = \mathbf{0}$. (This is the **Cayley-Hamilton theorem**, see Section 8.6).

6.16. Calculate $A^{10}\mathbf{x}$ for $A = \begin{bmatrix} 1 & 2 & -1 \\ 0 & 5 & -2 \\ 0 & 6 & -2 \end{bmatrix}, \qquad \mathbf{x} = \begin{bmatrix} 2 \\ 4 \\ 7 \end{bmatrix}.$

6.17. Find a general solution to each of the following LRRs.

(1) $x_n = 2x_{n-1} + x_{n-2}, \quad n \geq 2,$
(2) $x_n = 2x_{n-1} + x_{n-2} - 2x_{n-3}, \quad n \geq 3,$
(3) $x_n = 5x_{n-1} - 5x_{n-2} - 5x_{n-3} + 6x_{n-4}, \quad n \geq 4.$

With the companion matrix A for each case of the above (1)-(3), find a general solution to an LdE $\mathbf{x}_n = A\mathbf{x}_{n-1}$.

Hint: $x_n = \lambda_i^n$ for LRR; $\mathbf{x}_n = \lambda_i^n \mathbf{v}_i$ for LdE.

6.18. Solve the LdE $\mathbf{x}_n = A\mathbf{x}_{n-1}$ for $A = \begin{bmatrix} 3 & 1 \\ 1 & 3 \end{bmatrix}$ with $\mathbf{x}_0 = \begin{bmatrix} 2 \\ 0 \end{bmatrix}$.

6.19. Solve the LdE $\mathbf{x}_n = A\mathbf{x}_{n-1}$ for $A = \begin{bmatrix} -6 & 24 & 8 \\ -1 & 8 & 4 \\ 2 & -12 & -6 \end{bmatrix}$ with $\mathbf{x}_0 = \begin{bmatrix} 2 \\ 1 \\ 0 \end{bmatrix}$.

6.20. Let $A = \begin{bmatrix} 0 & 0.3 \\ 0.6 & x \end{bmatrix}$. Find a value of x so that A has an eigenvalue $\lambda = 1$. For such A and $\mathbf{x}_0 = (1, 1)$, calculate $\lim_{k\to\infty} A^k\mathbf{x}_0$.

Hint: Compute A^k with the eigenvalues and the eigenvectors of A.

6.21. Let $A = \begin{bmatrix} 1 & 1 \\ 0 & 2 \end{bmatrix}$.

(1) Compute e^A directly from the series expansion.
(2) Compute e^A by diagonalizing A.

6.22. Solve the LDE $\mathbf{y}' = A\mathbf{y}$, where

$$(1)\ A = \begin{bmatrix} 3 & 1 \\ 1 & 3 \end{bmatrix}, \ \mathbf{y}(0) = \begin{bmatrix} 2 \\ 0 \end{bmatrix}, \qquad (2)\ A = \begin{bmatrix} 1 & -1 \\ -1 & 3 \end{bmatrix}, \ \mathbf{y}(0) = \begin{bmatrix} 1 \\ 0 \end{bmatrix}.$$

Hint: Both A are symmetric. First, find the eigenvalues and eigenvectors of each A. If $A\mathbf{v} = \lambda\mathbf{v}$, then $ce^{\lambda t}\mathbf{v}$ is a solution of LDE $\mathbf{y}' = A\mathbf{y}$. See Theorem 6.20.

6.23. Solve the LDE $\mathbf{y}' = A\mathbf{y}$, where

$$(1)\ A = \begin{bmatrix} -6 & 24 & 8 \\ -1 & 8 & 4 \\ 2 & -12 & -6 \end{bmatrix} \quad \text{and} \quad \mathbf{y}(0) = \begin{bmatrix} 2 \\ 1 \\ 0 \end{bmatrix},$$

$$(2)\ A = \begin{bmatrix} 1 & 1 & 0 \\ 0 & 2 & 1 \\ 0 & 0 & 3 \end{bmatrix} \quad \text{and} \quad \mathbf{y}(0) = \begin{bmatrix} 1 \\ 1 \\ 0 \end{bmatrix}.$$

Hint: First, find the eigenvalues and eigenvectors of each A. If $A\mathbf{v} = \lambda\mathbf{v}$, then $ce^{\lambda t}\mathbf{v}$ is a solution of LDE $\mathbf{y}' = A\mathbf{y}$.

6.24. Prove that if A is skew-symmetric, then e^A is orthogonal.

6.25. Compute e^A for

$$(1)\ A = \begin{bmatrix} 1 & 1 \\ 0 & 0 \end{bmatrix}, \qquad (2)\ A = \begin{bmatrix} 3 & 1 \\ 1 & 3 \end{bmatrix}, \qquad (3)\ A = \begin{bmatrix} 3 & t \\ 0 & 3 \end{bmatrix},$$

$$(4)\ A = \begin{bmatrix} 2 & 3 & 0 \\ 0 & 2 & 3 \\ 0 & 0 & 2 \end{bmatrix}, \qquad (5)\ A = \begin{bmatrix} \lambda & 1 & 0 & 0 \\ 0 & \lambda & 1 & 0 \\ 0 & 0 & \lambda & 1 \\ 0 & 0 & 0 & \lambda \end{bmatrix}.$$

Hint: In (3),(4),(5), A can be written as $A = \lambda I + N$.

6.26. Solve the LDE $\mathbf{y}' = A\mathbf{y}$ with initial condition $\mathbf{y}(0) = \mathbf{y}_0$ for $A = \begin{bmatrix} 2 & 3 & 0 \\ 0 & 2 & 3 \\ 0 & 0 & 2 \end{bmatrix}$.

Use the solution of Exercise **6.25**.

6.27. Let A be a square matrix of full rank and let $A = QR$ be its QR decomposition, where Q is the orthogonal part and R is the upper triangular part of A. Show that the matrix $A = QR$ and RQ have the same set of eigenvalues.

Hint: The matrices $A = QR$ and $Q^{-1}AQ = Q^{-1}(QR)Q = RQ$ are similar.

6.28. For any two (invertible or not) matrices A and B, show that

(1) $\text{adj } AB = \text{adj } B \cdot \text{adj } A$,
(2) $\text{adj } QAQ^{-1} = Q(\text{adj } A)Q^{-1}$ for any invertible matrix Q,
(3) if $AB = BA$, then $(\text{adj } A)B = B(\text{adj } A)$.

Hint: This is mentioned for any two invertible matrices A and B in Exercise **2.23**. A little difficult to prove it. Assume that it is true for invertible matrices. In each of the equations (1)–(3) both sides depend continuously on the entries of A and B. Any matrix A can be approximated by matrices of the form $A_\varepsilon = A + \varepsilon I$ which are invertible for sufficiently small nonzero ε. (Actually, if $\lambda_1, \ldots, \lambda_n$ are all of the eigenvalues of A, then A_ε is invertible for all $\varepsilon \neq -\lambda_i$.) Also if $AB = BA$, then $A_\varepsilon B = BA_\varepsilon$.

6.29. Determine whether the following statements are true or false, in general, and justify your answers.

(1) Every (real) square matrix is similar to at least one diagonal matrix.
(2) Every square matrix has at least one real eigenvalue.
(3) Each eigenvalue of $A + B$ is a sum of an eigenvalue of A and an eigenvalue of B.
(4) If λ is an eigenvalue of A of multiplicity k, then there exist k linearly independent eigenvectors belonging to λ.
(5) If A has two distinct eigenvalues, then A has at least two linearly independent eigenvectors.

(6) A sum of two eigenvectors of A is also an eigenvector of A.
(7) Any two similar matrices have the same eigenvalues.
(8) Any two similar matrices have the same eigenvectors.
(9) If B is obtained from A by interchanging two rows, then A is diagonalizable if and only if B is diagonalizable.
(10) If A and B are diagonalizable, so is AB.
(11) Every orthogonal matrix is similar to a real diagonal matrix.
(12) Every diagonalizable matrix is invertible.
(13) For an invertible matrix A, A is diagonalizable if and only if A^{-1} is diagonalizable.
(14) For a square matrix A, A is diagonalizable if and only if A^T is diagonalizable.
(15) For any square matrix A and for an eigenvalue λ of A, $\lambda + k$ is an eigenvalue of $A + kI$.
(16) Every monic polynomial can be the characteristic polynomial of a matrix.
(17) For any square matrix A, $\det e^A = e^{\det A}$.
(18) For any identity matrix I, $e^I = eI$.
(19) A companion matrix of an LRR is always invertible.
(20) Every LdE has at least one solution.

Hint: Exactly nine of them are true

Chapter 7

Complex Vector Spaces; Abstraction of the complex plane

Up to the last chapter, we have been dealing with matrices having only real entries and vector spaces with only real scalars. Also, in a system of linear (difference or differential) equations, all coefficients of the equations are assumed to be real.

However, for many applications of linear algebra, it is desirable to extend the scalars to be complex numbers. For example, by allowing complex scalars, any polynomial of degree n (even with complex coefficients) has n complex roots counting multiplicities. (This is well known as *the fundamental theorem of algebra*[1]).

By applying it to the characteristic polynomial of a matrix, one can say that all square matrices of order n always have n eigenvalues counting multiplicities. For instance, a (skew-symmetric) matrix $A = \begin{bmatrix} 0 & -1 \\ 1 & 0 \end{bmatrix}$ has no real eigenvalues, but it has two complex eigenvalues $\lambda = \pm i$. Thus, it is indispensable to work with complex numbers to find the full set of eigenvalues and eigenvectors for any square matrix.

The definition of a *complex vector space* is the same as the real case, but the scalars are now allowed to be complex numbers. We naturally extend the properties of real vector spaces to those of complex vector spaces, and also examine what are the differences between the real and complex vector spaces.

[1]In 1799, German mathematician **Carl Friedrich Gauss** (April 30, 1777– February 23, 1855) gave a proof of this theorem at the young age of 22, but his proof had some gaps. In 1816, Gauss published a complete proof building on the earlier works of several others.

7.1 How are complex vector spaces different from real vector spaces?

The complex n-space $\mathbb{C}^n$ is the set of all ordered n-tuples $(z_1, z_2, \ldots, z_n)$ of complex numbers:

$$\mathbb{C}^n = \{(z_1, z_2, \ldots, z_n) : z_i \in \mathbb{C},\ i = 1, 2, \ldots, n\}.$$

Clearly, it is a complex vector space with sum and scalar multiplication defined by

$$\begin{aligned}(z_1, z_2, \ldots, z_n) + (z_1', z_2', \ldots, z_n') &= (z_1 + z_1', z_2 + z_2', \ldots, z_n + z_n')\\ k(z_1, z_2, \ldots, z_n) &= (kz_1, kz_2, \ldots, kz_n) \text{ for } k \in \mathbb{C}.\end{aligned}$$

The standard basis for the complex vector space $\mathbb{C}^n$ is again $\{\mathbf{e}_1, \mathbf{e}_2, \ldots, \mathbf{e}_n\}$ but *the scalars are now complex numbers* so that any vector $\mathbf{z}$ in $\mathbb{C}^n$ is of the form $\mathbf{z} = \sum_{k=1}^n z_k\mathbf{e}_k$ with $z_k = x_k + iy_k \in \mathbb{C}$. Hence, $\mathbf{z} = \mathbf{x} + i\mathbf{y}$ with $\mathbf{x}, \mathbf{y} \in \mathbb{R}^n$. [2]

In a complex vector space, linear combinations are defined in the same way as the real case except that scalars are allowed to be complex numbers. Thus the same is true for linear independence, spanning spaces, basis, dimension, and subspaces.

For complex matrices, whose entries are complex numbers, the matrix sum and the matrix product follow the same rules as real matrices. The same is true for the concept of a linear transformation $T : V \to W$ from a complex vector space V to a complex vector space W. The definitions of the kernel and the image of a linear transformation remain the same as those in the real case, as well as the facts about row spaces, null spaces, column spaces, matrix representations of linear transformations, similarity, and so on.

A significant difference from the real case is happened in the inner product, and it should be a modification from the real case. If one would take the dot product on $\mathbb{C}^n$ as $\mathbf{u} \cdot \mathbf{v} = u_1v_1 + u_2v_2 + \cdots + u_nv_n$ following the real case, then the nonzero vector $(1, i)$ in $\mathbb{C}^2$ would have zero length, $1^2 + i^2 = 0$, terribly bad. To modify this problem, recall that the *absolute value* (or *modulus*) of a complex number $z = x + iy$ is defined as the nonnegative real number

$$|z| = (\bar{z}z)^{\frac{1}{2}} = \sqrt{x^2 + y^2},$$

[2] The *imaginary unit* i is defined as a solution to the quadratic equation $z^2 + 1 = 0$, that is, $i^2 = -1$. With this number i, a *complex number* can be expressed in the form $z = x + iy$, where x and y are real numbers. We call x the *real part* of z, denoted by $\mathrm{Re}(z)$, and y the *imaginary part* of z, denoted by $\mathrm{Im}(z)$.

where $\bar{z}$ is the complex conjugate of z. [3] Naturally, a modified definition of the inner product should coincide with the real case when the vectors are real.

The dot product on the complex n-space $\mathbb{C}^n$ is defined as follows.

Definition 7.1 For two vectors $\mathbf{u} = [u_1\ u_2\ \cdots\ u_n]^T$ and $\mathbf{v} = [v_1\ v_2\ \cdots\ v_n]^T$ in $\mathbb{C}^n$, the **dot** (or **Euclidean inner**) **product** $\mathbf{u} \cdot \mathbf{v}$ of $\mathbf{u}$ and $\mathbf{v}$ is defined by

$$\mathbf{u} \cdot \mathbf{v} = \overline{u_1}v_1 + \overline{u_2}v_2 + \cdots + \overline{u_n}v_n = \overline{\mathbf{u}}^T\mathbf{v},$$

where $\overline{\mathbf{u}} = [\overline{u}_1\ \overline{u}_2\ \cdots\ \overline{u}_n]^T$ is the complex conjugate of $\mathbf{u}$.

Note that the dot product on $\mathbb{C}^n$ is not symmetric. In fact, $\mathbf{v} \cdot \mathbf{u} = \overline{\mathbf{u} \cdot \mathbf{v}}$ for any $\mathbf{u}$ and $\mathbf{v}$ in $\mathbb{C}^n$.

The **Euclidean length** (or **magnitude**) of a vector $\mathbf{u}$ in $\mathbb{C}^n$ is defined by

$$\|\mathbf{u}\| = (\mathbf{u} \cdot \mathbf{u})^{\frac{1}{2}} = \sqrt{\overline{u_1}u_1 + \overline{u_2}u_2 + \cdots + \overline{u_n}u_n} = \sqrt{|u_1|^2 + |u_2|^2 + \cdots + |u_n|^2},$$

where $|u_k|^2 = \overline{u}_k u_k$.

The (**Euclidean**) **distance** between two vectors $\mathbf{u}$ and $\mathbf{v}$ in $\mathbb{C}^n$ is

$$d(\mathbf{u}, \mathbf{v}) = \|\mathbf{u} - \mathbf{v}\|.$$

To see more differences between the real vector spaces and the complex vector spaces, let's consider the complex n-space $\mathbb{C}^n$ as a real vector space by restricting the scalar multiplication to only real scalars;

$$r(z_1, z_2, \ldots, z_n) \quad = \quad (rz_1, rz_2, \ldots, rz_n) \ \text{ for } \ r \in \mathbb{R}.$$

Two vectors $\mathbf{e}_1 = (1, 0, \ldots, 0)$ and $i\mathbf{e}_1 = (i, 0, \ldots, 0)$ are *linearly dependent* in the complex vector space $\mathbb{C}^n$. However, they are *linearly independent* if $\mathbb{C}^n$ is considered as a real vector space with only real scalars. Therefore,

$$\{\mathbf{e}_1, \mathbf{e}_2, \ldots, \mathbf{e}_n, i\mathbf{e}_1, i\mathbf{e}_2, \ldots, i\mathbf{e}_n\}$$

forms a basis for $\mathbb{C}^n$ considered as a real vector space. In this way, $\mathbb{C}^n$ can be naturally identified with the $2n$-dimensional real vector space $\mathbb{R}^{2n}$ as real vector

[3]The *complex conjugate* of a complex number $z = x + iy$ is $\bar{z} = x - iy$. Note that $z = \bar{z}$ if and only if z is real, while $z = -\bar{z}$ if and only if z is purely imaginary.

spaces. That is, $\dim \mathbb{C}^n = n$ as a complex vector space, but $\dim \mathbb{C}^n = 2n$ as a real vector space.

When $\mathbb{C}^n$ is considered as a $2n$-dimensional real vector space, the space $\mathbb{R}^n = \{(x_1, x_2, \ldots, x_n) : x_i \in \mathbb{R}\}$ is a subspace of $\mathbb{C}^n$, but not when $\mathbb{C}^n$ is considered as an n-dimensional complex vector space.

The definitions of eigenvalues and eigenvectors in a complex vector space are the same as the real case, but the eigenvalues are now allowed to be complex numbers. Hence, for any $n \times n$ (real or complex) matrix A, the characteristic polynomial $\det(\lambda I - A)$ has always n complex roots (*i.e.*, eigenvalues) counting multiplicities.

For example, consider a rotation matrix

$$A = \begin{bmatrix} \cos\theta & -\sin\theta \\ \sin\theta & \cos\theta \end{bmatrix}$$

with real entries. This matrix has two complex eigenvalues for any $\theta \in \mathbb{R}$ which are $e^{i\theta}$ and $e^{-i\theta}$, see Exercise **7.12**, but no real eigenvalues unless $\theta = k\pi$ for an integer k.

Consequently, all theorems and corollaries in Chapter 6 regarding eigenvalues and eigenvectors remain true without requiring the existence of n eigenvalues explicitly, and exactly the same proofs as the real case are valid since the arguments in the proofs are not concerned with what the scalars are.

For example, one has a theorem like *'for an $n \times n$ matrix A, the eigenvectors belonging to distinct eigenvalues are linearly independent'*, and *'if the n eigenvalues of A are all distinct, then the eigenvectors belonging to them form a basis for $\mathbb{C}^n$ so that A is diagonalizable'*.

As another difference, an $n \times n$ real matrix A can be considered as a linear transformation $\mathbf{x} \mapsto A\mathbf{x}$ in two different ways. On the space $\mathbb{R}^n$ and also on the space $\mathbb{C}^n$,

$$\begin{aligned} T : \mathbb{R}^n \to \mathbb{R}^n \quad & \text{defined by} \quad T(\mathbf{x}) = A\mathbf{x}, \\ S : \mathbb{C}^n \to \mathbb{C}^n \quad & \text{defined by} \quad S(\mathbf{x}) = A\mathbf{x}. \end{aligned}$$

Since the entries of A are all real, the coefficients of the characteristic polynomial $f(\lambda) = \det(\lambda I - A)$ of A are all real. Thus, if λ is a root of $f(\lambda) = 0$, then $f(\bar{\lambda}) = \overline{f(\lambda)} = 0$ and hence its conjugate $\bar{\lambda}$ is also a root. In other words,

if λ is an eigenvalue of a real matrix A, then $\bar{\lambda}$ is also an eigenvalue of A.

In particular, any $n \times n$ real matrix A has at least one real eigenvalue $\lambda = \bar{\lambda}$ if n is odd.

Furthermore,

if $\mathbf{x}$ *is an eigenvector belonging to a complex eigenvalue* λ *of a real matrix* A, *then the complex conjugate* $\bar{\mathbf{x}}$ *is an eigenvector belonging to* $\bar{\lambda}$.

In fact, if $A\mathbf{x} = \lambda\mathbf{x}$ with $\mathbf{x} \neq \mathbf{0}$, then

$$A\bar{\mathbf{x}} = \overline{A\mathbf{x}} = \overline{\lambda\mathbf{x}} = \bar{\lambda}\bar{\mathbf{x}},$$

where $\bar{\mathbf{x}}$ denotes the complex conjugate of the vector $\mathbf{x}$.

Using this fact, the following example shows that the 2×2 real matrices having no real eigenvalues are (nontrivial) rotations and their scalar multiples.

Example 131 If A is a 2×2 real matrix having no real eigenvalues, then A is similar to a matrix of the form

$$rR_\theta = \begin{bmatrix} r\cos\theta & -r\sin\theta \\ r\sin\theta & r\cos\theta \end{bmatrix} \quad \text{or} \quad rR_{-\theta} = \begin{bmatrix} r\cos\theta & r\sin\theta \\ -r\sin\theta & r\cos\theta \end{bmatrix}$$

for some $r > 0$ and $\theta \neq 0$.

In particular, if $\theta = \pi/2$, $rR_\theta = \begin{bmatrix} 0 & -r \\ r & 0 \end{bmatrix}$ is skew-symmetric which has no real eigenvalues.

Hence, among the 2×2 real matrices, only the (nontrivial) rotations and their (nontrivial) scalar multiplications can have no real eigenvalues.

Solution: Let A be a 2×2 real matrix having no real eigenvalues. As a complex matrix, A has two complex eigenvalues

$$\lambda = a + ib \quad \text{and} \quad \bar{\lambda} = a - ib,$$

$(a, b \in \mathbb{R}$ and $b \neq 0)$ with associated complex eigenvectors

$$\mathbf{x} = \mathbf{u} + i\mathbf{v} \quad \text{and} \quad \bar{\mathbf{x}} = \mathbf{u} - i\mathbf{v},$$

$(\mathbf{u}, \mathbf{v} \in \mathbb{R}^2$ and $\mathbf{v} \neq \mathbf{0})$, respectively.

Then, their real and imaginary parts $\mathbf{u}, \mathbf{v}, a,$ and b are given as

$$\begin{aligned} \mathbf{u} &= \tfrac{1}{2}(\mathbf{x} + \bar{\mathbf{x}}), & \mathbf{v} &= -\tfrac{i}{2}(\mathbf{x} - \bar{\mathbf{x}}), \\ a &= \frac{1}{2}(\lambda + \bar{\lambda}), & b &= -\frac{i}{2}(\lambda - \bar{\lambda}), \end{aligned}$$

which are all real.

Since $\lambda \neq \bar{\lambda}$, the eigenvectors $\mathbf{x}$ and $\bar{\mathbf{x}}$ are nonzero and linearly independent in the complex vector space $\mathbb{C}^2$, as they are also linearly independent when $\mathbb{C}^2$ is considered as a real vector space. This implies that the vectors $\mathbf{x}+\bar{\mathbf{x}}$ and $\mathbf{x}-\bar{\mathbf{x}}$ are nonzero real vectors and linearly independent in the space $\mathbb{C}^2$, (see Problem 164 below), so that the real vectors $\mathbf{u}$ and $\mathbf{v}$ are both nonzero and linearly independent in the subspace $\mathbb{R}^2$ of the real vector space $\mathbb{C}^2$.

Thus, $\alpha = \{\mathbf{u}, \mathbf{v}\}$ is a basis for the real vector space $\mathbb{R}^2$. Their images $A\mathbf{u}$ and $A\mathbf{v}$ can be computed as follows:

$$\begin{aligned} A\mathbf{u} &= \frac{1}{2}(A\mathbf{x} + A\bar{\mathbf{x}}) = \frac{1}{2}(\lambda\mathbf{x} + \bar{\lambda}\bar{\mathbf{x}}) \\ &= \lambda\left(\frac{\mathbf{u}+i\mathbf{v}}{2}\right) + \bar{\lambda}\left(\frac{\mathbf{u}-i\mathbf{v}}{2}\right) = a\mathbf{u} - b\mathbf{v}, \end{aligned}$$

and also

$$A\mathbf{v} = b\mathbf{u} + a\mathbf{v}.$$

Hence, the matrix representation of the linear transformation $A : \mathbb{R}^2 \to \mathbb{R}^2$ with respect to the basis $\alpha = \{\mathbf{u}, \mathbf{v}\}$ is

$$[A]_\alpha = \begin{bmatrix} a & b \\ -b & a \end{bmatrix}.$$

Now, by setting $r = \sqrt{a^2 + b^2} > 0$, one can get

$$a = r\cos\theta \quad \text{and} \quad b = r\sin\theta$$

for some $\theta \in \mathbb{R}$, so

$$[A]_\alpha = \begin{bmatrix} r\cos\theta & r\sin\theta \\ -r\sin\theta & r\cos\theta \end{bmatrix}.$$

By taking $-\theta$ instead of θ, one can get the second matrix $rR_{-\theta}$. □

In contrast to Example 131, there exist infinitely many 2×2 (not real) complex matrices having no real eigenvalues that are not scalar multiples of rotations. The 2×2 diagonal matrices with complex diagonal entries can be such examples.

As the last remark, the Gauss-Jordan elimination can certainly be applied to solve a system of linear equations with complex coefficients a_{ij}. It gives an LU or LDU decomposition of a complex matrix $A = [a_{ij}]$ or of PA if a permutation matrix P is needed.

Problem 162 Find the eigenvalues and the eigenvectors of complex matrices

$$(1)\ \begin{bmatrix} i & 0 \\ 1 & -i \end{bmatrix}, \qquad (2)\ \begin{bmatrix} i & 0 & 1 \\ 0 & 2 & 0 \\ 1 & 0 & i \end{bmatrix}.$$

Problem 163 Prove that an $n \times n$ complex matrix A is diagonalizable if and only if A has n linearly independent eigenvectors in the complex vector space $\mathbb{C}^n$.

Hint: Refer to the real case. See Theorem 6.6.

Problem 164 Let $\mathbf{x}$ and $\mathbf{y}$ be two vectors in a (real or complex) vector space V. Show that $\mathbf{x}$ and $\mathbf{y}$ are linearly independent if and only if $\mathbf{x}+\mathbf{y}$ and $\mathbf{x}-\mathbf{y}$ are linearly independent.

7.2 Hermitian inner products

In an (abstract) complex vector space, one can also define an inner product by adopting the basic properties of the Euclidean inner product on $\mathbb{C}^n$ as axioms.

Definition 7.2 A (complex) **inner product** (or **Hermitian inner product**) on a complex vector space V is a function that associates a complex number $\langle \mathbf{u}, \mathbf{v} \rangle$ to each ordered pair of vectors $\mathbf{u}$ and $\mathbf{v}$ in V in such a way that the following rules are satisfied: For all vectors $\mathbf{u}, \mathbf{v}$, and $\mathbf{w}$ in V and all scalars k in $\mathbb{C}$,

(1) $\langle \mathbf{u}, \mathbf{v} \rangle = \overline{\langle \mathbf{v}, \mathbf{u} \rangle}$ (Hermitian, not symmetric),
(2) $\langle \mathbf{u}+\mathbf{v}, \mathbf{w} \rangle = \langle \mathbf{u}, \mathbf{w} \rangle + \langle \mathbf{v}, \mathbf{w} \rangle$ (additivity),
(3) $\langle k\mathbf{u}, \mathbf{v} \rangle = \bar{k}\langle \mathbf{u}, \mathbf{v} \rangle$ (antilinear, not linear),
(4) $\langle \mathbf{v}, \mathbf{v} \rangle \geq 0$ always, and $\langle \mathbf{v}, \mathbf{v} \rangle = 0 \Leftrightarrow \mathbf{v} = \mathbf{0}$ (positive-definiteness).

A complex vector space together with an inner product is called a **complex inner product space** or a **unitary space**. In particular, the n-space $\mathbb{C}^n$ with the dot product is called the **Euclidean** (complex) n**-space**.

The following properties are immediate from the definition of an inner product:

(5) $\langle \mathbf{0}, \mathbf{v} \rangle = \langle \mathbf{v}, \mathbf{0} \rangle = 0$,
(6) $\langle \mathbf{u}, \mathbf{v}+\mathbf{w} \rangle = \langle \mathbf{u}, \mathbf{v} \rangle + \langle \mathbf{u}, \mathbf{w} \rangle$,
(7) $\langle \mathbf{u}, k\mathbf{v} \rangle = k\langle \mathbf{u}, \mathbf{v} \rangle$ (linear in 2nd variable).

Consequently, every complex inner product is a *positive-definite Hermitian sesquilinear form.* [4] (See also Definition 9.12 on page 451.)

Remark: There is another way to define an inner product on a complex vector space. If we redefine the dot product $\mathbf{u} \cdot \mathbf{v}$ on the n-space $\mathbb{C}^n$ by

$$\mathbf{u} \cdot \mathbf{v} = u_1\overline{v_1} + u_2\overline{v_2} + \cdots + u_n\overline{v_n},$$

then the third rule in Definition 7.2 should be modified to be

(3′) $\langle k\mathbf{u}, \mathbf{v}\rangle = k\langle \mathbf{u}, \mathbf{v}\rangle$, and $\langle \mathbf{u}, k\mathbf{v}\rangle = \overline{k}\langle \mathbf{u}, \mathbf{v}\rangle$.

But these two different definitions do not induce any essential difference in a complex inner product space.

In a complex inner product space, as in the real case, the **length** (or **magnitude**) of a vector $\mathbf{u}$ and the **distance** between two vectors $\mathbf{u}$ and $\mathbf{v}$ are defined by

$$\|\mathbf{u}\| = \langle \mathbf{u}, \mathbf{u}\rangle^{\frac{1}{2}}, \qquad d(\mathbf{u}, \mathbf{v}) = \|\mathbf{u} - \mathbf{v}\|,$$

respectively. Two vectors $\mathbf{u}$, $\mathbf{v}$ are **orthogonal** (or **perpendicular**) if $\langle \mathbf{u}, \mathbf{v}\rangle = 0$.

Example 132 (*A complex inner product on a function space*) Let $C_{\mathbb{C}}[a, b]$ denote the set of all complex-valued continuous functions defined on $[a, b]$. Thus an element in $C_{\mathbb{C}}[a, b]$ is of the form $\mathbf{f}(x) = f_1(x) + if_2(x)$, where $f_1(x)$ and $f_2(x)$ are real-valued and continuous on $[a, b]$. It is known in calculus that $\mathbf{f}$ is continuous if and only if each component function f_i is continuous.

Clearly, the set $C_{\mathbb{C}}[a, b]$ is a complex vector space under the sum and scalar multiplication of functions. For a vector $\mathbf{f}(x) = f_1(x) + if_2(x)$ in $C_{\mathbb{C}}[a, b]$, its integral is defined by

$$\int_a^b \mathbf{f}(x)dx = \int_a^b [f_1(x) + if_2(x)]\, dx = \int_a^b f_1(x)dx + i\int_a^b f_2(x)dx.$$

It is an elementary exercise to show that, for vectors $\mathbf{f}(x) = f_1(x) + if_2(x)$ and $\mathbf{g}(x) = g_1(x) + ig_2(x)$ in the complex vector space $C_{\mathbb{C}}[a, b]$, the following formula

[4] Bi-, a number prefix denoting two and semi-, numerical prefix meaning $\frac{1}{2}$. Sesqui- is used in Latin combinations for $1\frac{1}{2}$. Here, sesquilinear means that a complex inner product is $\frac{1}{2}$ linear for the first variable and is linear for the second variable.

defines an inner product on $C_{\mathbb{C}}[a,b]$:

$$\begin{aligned}\langle \mathbf{f}, \mathbf{g}\rangle &= \int_a^b \overline{\mathbf{f}(x)}\mathbf{g}(x)dx \\ &= \int_a^b [f_1(x) - if_2(x)]\,[g_1(x) + ig_2(x)]\,dx \\ &= \int_a^b [f_1(x)g_1(x) + f_2(x)g_2(x)]\,dx + i\int_a^b [f_1(x)g_2(x) - f_2(x)g_1(x)]\,dx\end{aligned}$$

as a complex number. □

In a complex inner product space, an inner product $\langle \mathbf{u}, \mathbf{v}\rangle$ is not necessarily a real number. Hence, it is also needed to modify the (real) Cauchy-Schwarz inequality for the case of a complex inner product space.

Theorem 7.3 (Cauchy-Schwarz inequality, the complex case) *For any two vectors* $\mathbf{u}$ *and* $\mathbf{v}$ *in a complex inner product space* V*,*

$$|\langle \mathbf{u}, \mathbf{v}\rangle|^2 \le \langle \mathbf{u}, \mathbf{u}\rangle\langle \mathbf{v}, \mathbf{v}\rangle,$$

or equivalently

$$|\langle \mathbf{u}, \mathbf{v}\rangle| \le \|\mathbf{u}\|\|\mathbf{v}\|.$$

Proof: If $\mathbf{u} = \mathbf{0}$, it is clear. Assume $\mathbf{u} \neq \mathbf{0}$. Then the orthogonal projection of the vector $\mathbf{v}$ onto the one-dimensional subspace $\langle \mathbf{u}\rangle$ spanned by $\mathbf{u}$ is

$$\mathrm{Proj}_{\langle \mathbf{u}\rangle}\mathbf{v} = \left\langle \frac{\mathbf{u}}{\|\mathbf{u}\|}, \mathbf{v}\right\rangle \frac{\mathbf{u}}{\|\mathbf{u}\|} = \frac{\langle \mathbf{u}, \mathbf{v}\rangle}{\|\mathbf{u}\|^2}\mathbf{u}.$$

(This projection of the complex case will be described as in the real case in the next section.) Let $\lambda = \langle \mathbf{u}, \mathbf{v}\rangle/\|\mathbf{u}\|^2$. Then the distance from the vector $\mathbf{v}$ to the subspace $\langle \mathbf{u}\rangle$ is $\|\mathbf{v} - \mathrm{Proj}_{\langle \mathbf{u}\rangle}(\mathbf{v})\| = \|\mathbf{v} - \lambda\mathbf{u}\|$. But

$$\begin{aligned}0 &\le \|\mathbf{v} - \lambda\mathbf{u}\|^2 \\ &= \langle \mathbf{v} - \lambda\mathbf{u}, \mathbf{v} - \lambda\mathbf{u}\rangle \\ &= \langle \mathbf{v}, \mathbf{v}\rangle - \langle \mathbf{v}, \lambda\mathbf{u}\rangle - \langle \lambda\mathbf{u}, \mathbf{v}\rangle + \langle \lambda\mathbf{u}, \lambda\mathbf{u}\rangle \\ &= \langle \mathbf{v}, \mathbf{v}\rangle - \lambda\overline{\langle \mathbf{u}, \mathbf{v}\rangle} - \overline{\lambda}\langle \mathbf{u}, \mathbf{v}\rangle + \overline{\lambda}\lambda\langle \mathbf{u}, \mathbf{u}\rangle \\ &= \|\mathbf{v}\|^2 - \frac{|\langle \mathbf{u}, \mathbf{v}\rangle|^2}{\|\mathbf{u}\|^2} - \frac{|\langle \mathbf{u}, \mathbf{v}\rangle|^2}{\|\mathbf{u}\|^2} + \frac{|\langle \mathbf{u}, \mathbf{v}\rangle|^2}{\|\mathbf{u}\|^2} \\ &= \|\mathbf{v}\|^2 - \frac{|\langle \mathbf{u}, \mathbf{v}\rangle|^2}{\|\mathbf{u}\|^2},\end{aligned}$$

which implies the (Cauchy-Schwarz) inequality $|\langle \mathbf{u}, \mathbf{v}\rangle| \le ||\mathbf{u}||||\mathbf{v}||$. □

The **angle** θ between two nonzero vectors $\mathbf{u}$ and $\mathbf{v}$ in a complex inner product space V can be defined as the (real) angle θ by considering two vectors $\mathbf{u}$ and $\mathbf{v}$ as real vectors (in the space V considered as a real vector space). In fact, it is a unique $\theta \in [0, \pi]$ such that $\cos\theta = \frac{|\langle \mathbf{u},\mathbf{v}\rangle|}{||\mathbf{u}||||\mathbf{v}||}$.

Problem 165 Show that the Euclidean inner product on the complex n-space $\mathbb{C}^n$ satisfies all of the rules in the definition of a complex inner product space.

Problem 166 For any two vectors $\mathbf{u}$ and $\mathbf{v}$ in a complex inner product space V, prove

(1) $||\mathbf{u}+\mathbf{v}|| \le ||\mathbf{u}|| + ||\mathbf{v}||$ (triangle inequality),
(2) $||\mathbf{u}+\mathbf{v}||^2 = ||\mathbf{u}||^2 + ||\mathbf{v}||^2$ if $\mathbf{u}$ and $\mathbf{v}$ are orthogonal, (Pythagorean theorem).

Hint for (1): Expand $||\mathbf{u}+\mathbf{v}||^2 = \langle \mathbf{u}+\mathbf{v}, \mathbf{u}+\mathbf{v}\rangle$ and use the Cauchy-Schwarz inequality, Theorem 7.3 for $|\mathrm{Re}(\langle \mathbf{u}, \mathbf{v}\rangle)| \le |\langle \mathbf{u}, \mathbf{v}\rangle| \le ||\mathbf{u}||||\mathbf{v}||$.

Problem 167 [**Cauchy-Schwarz inequality for sums**] For any two vectors $\mathbf{z} = (z_1, z_2, \ldots, z_n)$ and $\mathbf{w} = (w_1, w_2, \ldots, w_n)$ in the Euclidean complex n-space $\mathbb{C}^n$, prove

(1) $$\left|\sum_{i=1}^{n} \bar{z}_i w_i\right|^2 \le \left[\sum_{i=1}^{n} |z_i|^2\right]\left[\sum_{i=1}^{n} |w_i|^2\right]$$ (Cauchy-Schwarz inequality).

(2) $$\left[\sum_{i=1}^{n} |z_i + w_i|^2\right]^{\frac{1}{2}} \le \left[\sum_{i=1}^{n} |z_i|^2\right]^{\frac{1}{2}} + \left[\sum_{i=1}^{n} |w_i|^2\right]^{\frac{1}{2}}$$ (triangle inequality).

The Cauchy-Schwarz inequality for integrals of complex-valued continuous functions defined on $[0, 1]$ can be derived in a similar way as the real case in Problem 105 with the inner product given in Example 132.

Hint: Use the Cauchy-Schwarz inequality in the Euclidean complex n-space $\mathbb{C}^n$, or use Problem 166.

7.3 Orthogonal projections, the complex case

In complex inner product spaces, the definitions of such terms as orthogonal sets, orthonormal sets, orthogonal complements, and orthonormal basis remain the same as in the real cases.

In the same way, the *orthogonal projection* Proj_U from a complex inner product space V onto a subspace U is also defined to be a projection of the space V onto the subspace U along its orthogonal complement $U^\perp$.

For example, on the Euclidean complex n-space $\mathbb{C}^n$ with the dot product, let $\langle \mathbf{u} \rangle$ be a one-dimensional subspace spanned by a unit vector $\mathbf{u}$. Then for a vector $\mathbf{v}$ in $\mathbb{C}^n$, the orthogonal projection of $\mathbf{v}$ onto $\langle \mathbf{u} \rangle$ is

$$\mathrm{Proj}_{\langle \mathbf{u} \rangle}(\mathbf{v}) = (\mathbf{u} \cdot \mathbf{v})\mathbf{u} = (\overline{\mathbf{u}}^T \mathbf{v})\mathbf{u} = \mathbf{u}(\overline{\mathbf{u}}^T \mathbf{v}) = (\mathbf{u}\overline{\mathbf{u}}^T)\mathbf{v}.$$

In general, on any complex inner product space V with an inner product $\langle \ , \rangle$, the orthogonal projection of $\mathbf{v} \in V$ onto the one-dimensional subspace $\langle \mathbf{u} \rangle$ spanned by a unit vector $\mathbf{u}$ is

$$\mathrm{Proj}_{\langle \mathbf{u} \rangle}(\mathbf{v}) = \langle \mathbf{u}, \mathbf{v} \rangle \mathbf{u}.$$

The Gram-Schmidt orthogonalization is still valid in complex inner product spaces, and can be used to convert an arbitrary basis into an orthonormal basis. If V is an n-dimensional complex inner product space, then by taking an orthonormal basis for V, one can define a natural isometry from V to the Euclidean complex n-space $\mathbb{C}^n$ that preserves the inner product, as in the real case. It gives a rectangular (complex) coordinate system for a complex inner product space.

Example 133 (*Gram-Schmidt orthogonalization, the complex case*) On the Euclidean complex 3-space $\mathbb{C}^3$, apply the Gram-Schmidt orthogonalization to convert

$$\mathbf{x}_1 = (i, i, i), \quad \mathbf{x}_2 = (0, i, \ i), \quad \mathbf{x}_3 = (0, 0, i)$$

into an orthonormal basis.

Solution: Step 1: Set

$$\mathbf{u}_1 = \frac{\mathbf{x}_1}{\|\mathbf{x}_1\|} = \frac{(i, i, i)}{\sqrt{3}} = \left(\frac{i}{\sqrt{3}}, \frac{i}{\sqrt{3}}, \frac{i}{\sqrt{3}} \right).$$

Step 2: Let W_1 denote the subspace spanned by $\mathbf{u}_1$. Then

$$\begin{aligned} \mathbf{x}_2 - \mathrm{Proj}_{W_1}\mathbf{x}_2 &= \mathbf{x}_2 - \langle \mathbf{u}_1, \mathbf{x}_2 \rangle \mathbf{u}_1 \\ &= (0, i, i) - \frac{2}{\sqrt{3}} \left(\frac{i}{\sqrt{3}}, \frac{i}{\sqrt{3}}, \frac{i}{\sqrt{3}} \right) \\ &= \left(-\frac{2i}{3}, \frac{i}{3}, \frac{i}{3} \right). \end{aligned}$$

Therefore,

$$\mathbf{u}_2 = \frac{\mathbf{x}_2 - \mathrm{Proj}_{W_1}\mathbf{x}_2}{\|\mathbf{x}_2 - \mathrm{Proj}_{W_1}\mathbf{x}_2\|} = \frac{3}{\sqrt{6}} \left(-\frac{2i}{3}, \frac{i}{3}, \frac{i}{3} \right) = \left(-\frac{2i}{\sqrt{6}}, \frac{i}{\sqrt{6}}, \frac{i}{\sqrt{6}} \right).$$

Step 3: Let W_2 denote the subspace spanned by $\{\mathbf{u}_1, \mathbf{u}_2\}$. Then

$$\begin{aligned}
&\mathbf{x}_3 - \text{Proj}_{W_2}\mathbf{x}_3 \\
&= \mathbf{x}_3 - \langle \mathbf{u}_1, \mathbf{x}_3 \rangle \mathbf{u}_1 - \langle \mathbf{u}_2, \mathbf{x}_3 \rangle \mathbf{u}_2 \\
&= (0, 0, i) - \frac{1}{\sqrt{3}}\left(\frac{i}{\sqrt{3}}, \frac{i}{\sqrt{3}}, \frac{i}{\sqrt{3}}\right) - \frac{1}{\sqrt{6}}\left(-\frac{2i}{\sqrt{6}}, \frac{i}{\sqrt{6}}, \frac{i}{\sqrt{6}}\right) \\
&= \left(0, -\frac{i}{2}, \frac{i}{2}\right).
\end{aligned}$$

Therefore,

$$\mathbf{u}_3 = \frac{\mathbf{x}_3 - \text{Proj}_{W_2}\mathbf{x}_3}{\|\mathbf{x}_3 - \text{Proj}_{W_2}\mathbf{x}_3\|} = \sqrt{2}\left(0, -\frac{i}{2}, \frac{i}{2}\right) = \left(0, -\frac{i}{\sqrt{2}}, \frac{i}{\sqrt{2}}\right).$$

Thus,

$$\mathbf{u}_1 = \left(\frac{i}{\sqrt{3}}, \frac{i}{\sqrt{3}}, \frac{i}{\sqrt{3}}\right), \quad \mathbf{u}_2 = \left(-\frac{2i}{\sqrt{6}}, \frac{i}{\sqrt{6}}, \frac{i}{\sqrt{6}}\right), \quad \mathbf{u}_3 = \left(0, -\frac{i}{\sqrt{2}}, \frac{i}{\sqrt{2}}\right)$$

form an orthonormal basis for the complex vector space $\mathbb{C}^3$. □

Example 134 (*An orthonormal set in the function space $C_{\mathbb{C}}[0, 2\pi]$*) Let $C_{\mathbb{C}}[0, 2\pi]$ be the complex vector space of all complex-valued continuous functions defined on $[a, b]$. This is a complex inner product space with the inner product given in Example 132. In this space, the set of vectors of the form

$$e^{ikx} = \cos kx + i \sin kx,$$

where k is an integer, is orthogonal. In fact, for any two

$$\mathbf{g}_k(x) = e^{ikx} \quad \text{and} \quad \mathbf{g}_\ell(x) = e^{i\ell x},$$

we have

$$\begin{aligned}
\langle \mathbf{g}_k, \mathbf{g}_\ell \rangle &= \int_0^{2\pi} \overline{e^{ikx}} e^{i\ell x} dx = \int_0^{2\pi} e^{-ikx} e^{i\ell x} dx = \int_0^{2\pi} e^{i(\ell - k)x} dx \\
&= \int_0^{2\pi} \cos(\ell - k)x dx + i \int_0^{2\pi} \sin(\ell - k)x dx \\
&= \begin{cases} \left[\frac{1}{\ell - k}\sin(\ell - k)x\right]_0^{2\pi} + i\left[\frac{-1}{\ell - k}\cos(\ell - k)x\right]_0^{2\pi} & \text{if } k \neq \ell, \\ \int_0^{2\pi} dx & \text{if } k = \ell \end{cases} \\
&= \begin{cases} 0 & \text{if } k \neq \ell, \\ 2\pi & \text{if } k = \ell. \end{cases}
\end{aligned}$$

Therefore, the vectors

$$\mathbf{f}_k(x) = \frac{1}{\sqrt{2\pi}} e^{ikx}, \quad k = 0, \pm 1, \pm 2, \ldots$$

form an orthonormal set in the complex inner product space $C_{\mathbb{C}}[0, 2\pi]$. □

As the real case, for any $m \times n$ complex matrix A of rank n, one can apply the Gram-Schmidt orthogonalization to convert the column vectors $\mathbf{c}_i$'s of A into an orthonormal basis $\beta = \{\mathbf{u}_1, \mathbf{u}_2, \ldots, \mathbf{u}_n\}$ for the column space $\mathcal{C}(A)$. It gives a QR factorization $A = QR$, where $Q = [\mathbf{u}_1\ \mathbf{u}_2\ \cdots\ \mathbf{u}_n]$ is the $m \times n$ complex matrix with columns $\mathbf{u}_i$ satisfying $\overline{Q}^T Q = I_n$, and R is an $n \times n$ upper triangular complex matrix. In fact, the j-th column of the matrix R is simply the coordinate vector of $\mathbf{c}_j$ with respect to the orthonormal basis $\beta = \{\mathbf{u}_1, \mathbf{u}_2, \ldots, \mathbf{u}_n\}$ for the column space $\mathcal{C}(A)$, that is, the j-th column is the coordinate vector in

$$\mathbf{c}_j = \langle \mathbf{u}_1, \mathbf{c}_j \rangle \mathbf{u}_1 + \langle \mathbf{u}_2, \mathbf{c}_j \rangle \mathbf{u}_2 + \cdots + \langle \mathbf{u}_j, \mathbf{c}_j \rangle \mathbf{u}_j.$$

Problem 168 Let U be an m-dimensional subspace of the Euclidean complex n-space $\mathbb{C}^n$ with an orthonormal basis $\{\mathbf{u}_1, \mathbf{u}_2, \ldots, \mathbf{u}_m\}$. Show that for any $\mathbf{z} \in \mathbb{C}^n$,

$$\text{Proj}_U(\mathbf{z}) \quad = \quad (\mathbf{u}_1 \overline{\mathbf{u}_1}^T + \mathbf{u}_2 \overline{\mathbf{u}_2}^T + \cdots + \mathbf{u}_m \overline{\mathbf{u}_m}^T)\, \mathbf{z}.$$

Hint: Extend the orthonormal basis $\{\mathbf{u}_1, \mathbf{u}_2, \ldots, \mathbf{u}_m\}$ for U to an orthonormal basis $\{\mathbf{u}_1, \mathbf{u}_2, \ldots, \mathbf{u}_m, \mathbf{u}_{m+1}, \ldots, \mathbf{u}_n\}$ for the n-space $\mathbb{C}^n$. Then for any $\mathbf{z} \in \mathbb{C}^n$,

$$\mathbf{z} = (\mathbf{u}_1 \cdot \mathbf{z})\mathbf{u}_1 + (\mathbf{u}_2 \cdot \mathbf{z})\mathbf{u}_2 + \cdots + (\mathbf{u}_m \cdot \mathbf{z})\mathbf{u}_m + \cdots + (\mathbf{u}_n \cdot \mathbf{z})\mathbf{u}_n \in U \oplus U^{\perp}.$$

7.4 Hermitian, skew-Hermitian, and unitary matrices

The dot product of real vectors $\mathbf{x}, \mathbf{y} \in \mathbb{R}^n$ is given by $\mathbf{x} \cdot \mathbf{y} = \mathbf{x}^T \mathbf{y}$ in matrix form. For complex vectors $\mathbf{u}, \mathbf{v} \in \mathbb{C}^n$, the dot product is defined by $\mathbf{u} \cdot \mathbf{v} = \bar{u}_1 v_1 + \cdots + \bar{u}_n v_n = \bar{\mathbf{u}}^T \mathbf{v}$, which involves the conjugate transpose, not just the transpose.

Let $\overline{A}$ denote the matrix whose entries are the complex conjugates of the corresponding entries in A.

Definition 7.4 For a complex matrix A, its complex conjugate transpose, $A^H = \overline{A}^T$, is called the **Hermitian adjoint**, or simply **adjoint**, of A. Thus, $[a_{ij}]^H = [\overline{a_{ji}}]$.

One can easily see that

$$(A+B)^H = A^H + B^H, \ (kA)^H = \overline{k}A^H, \ (A^H)^H = A, \text{ and } (AB)^H = B^H A^H.$$

Also, the Euclidean inner product on $\mathbb{C}^n$ can be written as

$$\mathbf{u} \cdot \mathbf{v} = \overline{\mathbf{u}}^T \mathbf{v} = \mathbf{u}^H \mathbf{v}.$$

For complex matrices, the symmetric and skew-symmetric real matrices are extended to Hermitian and skew-Hermitian matrices, respectively. [5]

Definition 7.5 A complex square matrix A is said to be **Hermitian** or **self-adjoint** if $A^H = A$, while is **skew-Hermitian** if $A^H = -A$.

For example,

$$\begin{bmatrix} 2 & 4+i \\ 4-i & 3 \end{bmatrix} \text{ is Hermitian, and } \begin{bmatrix} i & 1+i \\ -1+i & -i \end{bmatrix} \text{ is skew-Hermitian.}$$

As shown in these two matrices, the diagonal entries of any Hermitian matrix should be always real numbers, while the diagonal entries of any skew-Hermitian matrix should be always purely imaginary numbers. Hence, all 2×2 Hermitian or skew-Hermitian matrices are of the types; with real r_i,

$$\text{all Hermitians are of } \begin{bmatrix} r_1 & z \\ \bar{z} & r_2 \end{bmatrix}, \text{ and all skew-Hermitians are of } \begin{bmatrix} ir_1 & z \\ -\bar{z} & ir_2 \end{bmatrix}.$$

A (skew-) Hermitian matrix with real entries is just a real (skew-) symmetric matrix, and conversely, every real (skew-) symmetric matrix is (skew-) Hermitian.

Like real matrices, any $m \times n$ complex matrix A can be considered as a linear transformation $\mathbf{x} \mapsto A\mathbf{x}$ from $\mathbb{C}^n$ to $\mathbb{C}^m$, and

$$(A\mathbf{x}) \cdot \mathbf{y} = (A\mathbf{x})^H \mathbf{y} = \mathbf{x}^H A^H \mathbf{y} = \mathbf{x} \cdot (A^H \mathbf{y})$$

for any $\mathbf{x} \in \mathbb{C}^n$ and $\mathbf{y} \in \mathbb{C}^m$.

Problem 169 Prove that if A is invertible, so is A^H, and $(A^H)^{-1} = (A^{-1})^H$.

[5] Hermitian matrices are named after Charles Hermite (December 24, 1822– January 14, 1901). He was a French mathematician who did research concerning number theory, quadratic forms, invariant theory, and algebra.

Problem 170 Let A and B be either both Hermitian or both skew-Hermitian. Show that

(1) AB is Hermitian if and only if $AB = BA$,
(2) AB is skew-Hermitian if and only if $AB = -BA$.

The following theorem lists some basic properties of Hermitian matrices.

Theorem 7.6 *Every $n \times n$ Hermitian matrix A satisfies the following.*

(1) *For any (complex) vector $\mathbf{x} \in \mathbb{C}^n$, $\mathbf{x}^H A\mathbf{x}$ is real.*
(2) *All (complex) eigenvalues of A are real. In particular, every $n \times n$ real symmetric matrix has precisely n real eigenvalues.*
(3) *The eigenvectors of A belonging to distinct eigenvalues are mutually orthogonal, (so are clearly linearly independent). In particular, eigenvectors belonging to distinct eigenvalues of a real symmetric matrix are mutually orthogonal.*

Proof: (1) Since $\mathbf{x}^H A\mathbf{x}$ is a 1×1 matrix,

$$\overline{(\mathbf{x}^H A\mathbf{x})} = (\mathbf{x}^H A\mathbf{x})^H = \mathbf{x}^H A\mathbf{x}.$$

(2) If $A\mathbf{x} = \lambda\mathbf{x}$ with $\mathbf{x} \neq \mathbf{0}$, then

$$\mathbf{x}^H A\mathbf{x} = \mathbf{x}^H \lambda\mathbf{x} = \lambda\mathbf{x}^H\mathbf{x} = \lambda\|\mathbf{x}\|^2.$$

The left-hand side is real and $\|\mathbf{x}\|^2$ is positive, because $\mathbf{x} \neq \mathbf{0}$. So, λ should be real.

(3) Let $\mathbf{x}$ and $\mathbf{y}$ be eigenvectors of A belonging to eigenvalues λ and μ, respectively. Let $\lambda \neq \mu$. Because $A = A^H$ and λ is real, it follows that

$$\lambda(\mathbf{x} \cdot \mathbf{y}) = (\lambda\mathbf{x}) \cdot \mathbf{y} = A\mathbf{x} \cdot \mathbf{y} = \mathbf{x} \cdot A\mathbf{y} = \mu(\mathbf{x} \cdot \mathbf{y}).$$

Since $\lambda \neq \mu$, this gives that $\mathbf{x} \cdot \mathbf{y} = \mathbf{x}^H\mathbf{y} = 0$, *i.e.*, $\mathbf{x}$ is orthogonal to $\mathbf{y}$. □

For a (complex) vector $\mathbf{x} \in \mathbb{C}^n$,

$$\mathbf{x}^H A\mathbf{x} = [\overline{x_1}\ \overline{x_2}\ \cdots\ \overline{x_n}][a_{ij}]\begin{bmatrix} x_1 \\ x_2 \\ \vdots \\ x_n \end{bmatrix} = \sum_{i=1}^{n}\sum_{j=1}^{n} a_{ij}\overline{x_i}x_j$$

is a (homogeneous) polynomial [6] of degree two in n variables $x_1, x_2, \ldots, x_n$, called a **quadratic form** on $\mathbb{C}^n$. Similarly, one can define a **quadratic form** on $\mathbb{R}^n$ to be $\mathbf{x}^T A\mathbf{x} = \sum_{i=1}^n \sum_{j=1}^n a_{ij}x_i x_j$ for $\mathbf{x} \in \mathbb{R}^n$. Quadratic forms will be studied later in Chapter 9. A dot product $\mathbf{x}^T\mathbf{x} = x_1^2 + x_2^2 + \cdots + x_n^2$ with $A = I_n$ is a standard example of a quadratic form on $\mathbb{R}^n$.

It is easy to see that if A is Hermitian, then iA is skew-Hermitian. Conversely, if A is skew-Hermitian, then iA is Hermitian. Therefore, the following theorem is a direct consequence of this fact and Theorem 7.6. The proof is left as an exercise.

Theorem 7.7 *Every $n \times n$ skew-Hermitian matrix A satisfies the following.*

(1) *For any vector $\mathbf{x}$ in $\mathbb{C}^n$, a quadratic form $\mathbf{x}^H A\mathbf{x}$ is purely imaginary.*
(2) *All eigenvalues of A are purely imaginary. In particular, every $n \times n$ real skew-symmetric matrix has n purely imaginary eigenvalues.*
(3) *The eigenvectors of A belonging to distinct eigenvalues are mutually orthogonal, (so are clearly linearly independent). In particular, eigenvectors belonging to distinct eigenvalues of a real skew-symmetric matrix are mutually orthogonal.*

Problem 171 Prove Theorem 7.7 by using Theorem 7.6.

Problem 172 True or False: (1) The determinant of any Hermitian matrix is always real. (2) The determinant of any skew-Hermitian matrix is always purely imaginary.

Problem 173 Show that $A = B + iC$ (B and C real matrices) is skew-Hermitian if and only if B is skew-symmetric and C is symmetric.

A square matrix Q with real entries is orthogonal if $Q^{-1} = Q^T$, *i.e.*, the column vectors are orthonormal. It can be stated for complex matrices as follows. (Compare it with Lemma 5.20).

Lemma 7.8 *For a complex square matrix U, the following are equivalent.*

(1) *the column vectors of U are orthonormal,*
(2) $U^H U = I$,
(3) $U^{-1} = U^H$,
(4) $UU^H = I$,
(5) *the row vectors of U are orthonormal.*

[6] A (multivariate) polynomial $f(x_1, x_2, \ldots, x_n)$ is said to be *homogeneous* of degree k if all of its nonzero terms have the same degree k, or equivalently $f(\alpha x_1, \alpha x_2, \ldots, \alpha x_n) = \alpha^k f(x_1, x_2, \ldots, x_n)$ for a fixed integer k. Commonly, it is called a *k-form*.

An orthogonal matrix is extended to a unitary matrix in a complex case.

Definition 7.9 A complex square matrix U is said to be **unitary** if it satisfies any one (and hence, all) of the conditions in Lemma 7.8.

Like a real orthogonal matrix, any unitary matrix preserves the lengths and the dot product of vectors.

Theorem 7.10 *Every $n \times n$ unitary matrix U satisfies the following.*

(1) *U preserves the dot product on $\mathbb{C}^n$: For all $\mathbf{x}$, $\mathbf{y}$ in $\mathbb{C}^n$, $(U\mathbf{x})^H(U\mathbf{y}) = \mathbf{x}^H\mathbf{y}$.*

(2) *All eigenvalues of U are of unit length. In particular, every $n \times n$ real orthogonal matrix has n (real or complex) eigenvalues of unit length and hence its real eigenvalues are equal to 1 or -1.*

(3) *The eigenvectors of U belonging to distinct eigenvalues are mutually orthogonal, (so are clearly linearly independent). In particular, eigenvectors belonging to distinct eigenvalues of a real orthogonal matrix are mutually orthogonal.*

Proof: (1) $(U\mathbf{x})^H(U\mathbf{y}) = \mathbf{x}^H U^H U\mathbf{y} = \mathbf{x}^H\mathbf{y}$.

(2) If $U\mathbf{x} = \lambda\mathbf{x}$ with $\mathbf{x} \neq \mathbf{0}$, then by (1) $\mathbf{x}^H\mathbf{x} = (U\mathbf{x})^H(U\mathbf{x}) = |\lambda|^2\mathbf{x}^H\mathbf{x}$. Since $\mathbf{x}^H\mathbf{x} \neq 0$, it should be $|\lambda| = 1$.

(3) Let $U\mathbf{x} = \lambda\mathbf{x}$, $U\mathbf{y} = \mu\mathbf{y}$ with $\lambda \neq \mu$. By (2), we have $\lambda\bar{\lambda} = 1 = \mu\bar{\mu}$ and $U^{-1}\mathbf{y} = \mu^{-1}\mathbf{y} = \bar{\mu}\mathbf{y}$. Therefore,

$$\bar{\lambda}\mathbf{x}^H\mathbf{y} = (\lambda\mathbf{x})^H\mathbf{y} = (U\mathbf{x})^H\mathbf{y} = \mathbf{x}^H U^{-1}\mathbf{y} = \mathbf{x}^H(\bar{\mu}\mathbf{y}) = \bar{\mu}\mathbf{x}^H\mathbf{y}$$

holds, and $\lambda \neq \mu$ implies $\mathbf{x}^H\mathbf{y} = 0$. □

From the same argument as in the proof of Theorem 5.23, U preserves the dot product if and only if it preserves the lengths of vectors: $\|U\mathbf{x}\| = \|\mathbf{x}\|$ for all $\mathbf{x}$ in $\mathbb{C}^n$. Thus,

every unitary matrix is an isometry.

Moreover, in a real inner product space, a change of basis matrix from an orthonormal basis to another orthonormal basis is an orthogonal matrix. See Theorem 5.25. This fact in the real case can be extended to the complex case.

Theorem 7.11 *A change of basis matrix from an orthonormal basis to another orthonormal basis in a complex inner product space is a unitary matrix.*

Proof: Let $\alpha = \{\mathbf{v}_1, \ldots, \mathbf{v}_n\}$ and $\beta = \{\mathbf{w}_1, \ldots, \mathbf{w}_n\}$ be two orthonormal bases for a complex inner product space, and let $Q = [q_{ij}]$ be the change of basis matrix from the basis β to the basis α. By definition,

$$\mathbf{w}_j = \sum_{\ell=1}^{n} q_{\ell j}\mathbf{v}_\ell.$$

Thus,

$$\begin{aligned}\delta_{ij} = \langle \mathbf{w}_i, \mathbf{w}_j\rangle &= \left\langle \sum_{k=1}^{n} q_{ki}\mathbf{v}_k, \sum_{\ell=1}^{n} q_{\ell j}\mathbf{v}_\ell \right\rangle \\ &= \sum_{k=1}^{n} \overline{q_{ki}} \sum_{\ell=1}^{n} q_{\ell j}\langle \mathbf{v}_k, \mathbf{v}_\ell\rangle \\ &= \sum_{k=1}^{n} \overline{q_{ki}} q_{kj} = \sum_{k=1}^{n} [Q^H]_{ik}[Q]_{kj} = [Q^H Q]_{ij}.\end{aligned}$$

That is, the matrix Q is unitary. □

Just as in the real case, it is true that two matrices representing the same linear transformation of a complex vector space with respect to different bases are similar.

Problem 174 Show that $|\det U| = 1$ for any unitary matrix U.
Hint: $\det U$ is the product of all eigenvalues.

Problem 175 Show that the matrices

$$\begin{bmatrix} 1 & 0 \\ 0 & i \end{bmatrix} \quad \text{and} \quad \begin{bmatrix} \frac{1+i}{2} & \frac{1+i}{2} \\ \frac{1-i}{2} & \frac{-1+i}{2} \end{bmatrix}$$

are unitary but neither Hermitian nor skew-Hermitian.

Problem 176 Show that (1) the product of any two unitary matrices is also unitary, and (2) the adjugate adj U of a unitary matrix U is also unitary.

Problem 177 Describe all 3×3 matrices that are simultaneously Hermitian, unitary, and diagonal. How many such matrices are there?

Hint: Hermitian means that the diagonal entries are all real, and the diagonality implies that all off-diagonal entries are zero. Now, the unitarity gives that the diagonal entries must be ± 1.

7.5 Hermitian, skew-Hermitian, and unitary matrices are all unitarily diagonalizable

It is already shown that if an $n \times n$ square matrix A is Hermitian, skew-Hermitian, or unitary, then the eigenvectors belonging to distinct eigenvalues are mutually orthogonal.

Hence, if such a matrix A has n distinct eigenvalues, then there exists an orthonormal basis α for $\mathbb{C}^n$ consisting of eigenvectors of A so that the matrix representation $[A]_\alpha$ is diagonal, *i.e.,* A is diagonalizable by a unitary matrix.

In this section, it will be shown that every Hermitian, skew-Hermitian, or unitary matrix has a full set of orthonormal eigenvectors even if the eigenvalues are not all distinct. In fact, they are always diagonalizable by unitary matrices.

Definition 7.12 **(1)** Two real matrices A and B are **orthogonally similar** if there exists an orthogonal matrix P such that $P^{-1}AP = B$. A matrix is **orthogonally diagonalizable** if it is orthogonally similar to a diagonal matrix.

(2) Two complex matrices A and B are **unitarily similar** if there exists a unitary matrix U such that $U^{-1}AU = B$. A matrix is **unitarily diagonalizable** if it is unitarily similar to a diagonal matrix.

We begin with Schur's triangulation theorem concerning orthogonal and unitary similarity between matrices.

Theorem 7.13 (Schur's triangulation Theorem) [7]

(1) *If an $n \times n$ real matrix A has only real eigenvalues, then A is orthogonally similar to an upper triangular matrix.*

(2) *Every $n \times n$ complex matrix is unitarily similar to an upper triangular matrix.*

Proof: We prove only the second assertion (2) by mathematical induction on n, because (1) can be done in a similar way.

Clearly, it is true for $n = 1$. Assume that the assertion (2) holds for $n = r - 1$ as an inductive hypothesis. Let A be any $r \times r$ complex matrix and let λ_1 be an

[7]**Issai Schur** (January 10, 1875 - January 10, 1941) was a Jewish mathematician who worked in Germany for most of his life. He studied at Berlin. He obtained his doctorate in 1901, became lecturer in 1903 and, after a stay at Bonn, professor in 1919. As a student of Frobenius, he worked on group representations, but also on combinatorics and number theory and even theoretical physics.

eigenvalue of A with a normalized eigenvector $\mathbf{u}$. Extend it to an orthonormal basis by the Gram-Schmidt orthogonalization, say $\{\mathbf{u}, \mathbf{u}_2, \ldots, \mathbf{u}_r\}$ for $\mathbb{C}^r$. Let

$$U_1 = [\mathbf{u}\ \mathbf{u}_2 \cdots\ \mathbf{u}_r]$$

be the unitary matrix with these basis vectors as the columns. A direct computation of the product $U_1^{-1}AU_1$ shows

$$\begin{aligned} U_1^{-1}AU_1 &= U_1^H AU_1 = U_1^H[A\mathbf{u}\ A\mathbf{u}_2\ \cdots\ A\mathbf{u}_r] \\ &= \begin{bmatrix} -- & \bar{\mathbf{u}}^T & -- \\ -- & \bar{\mathbf{u}}_2^T & -- \\ & \vdots & \\ -- & \bar{\mathbf{u}}_r^T & -- \end{bmatrix} \begin{bmatrix} | & | & & | \\ \lambda_1\mathbf{u} & A\mathbf{u}_2 & \cdots & A\mathbf{u}_r \\ | & | & & | \end{bmatrix} \\ &= \left[\begin{array}{c|c} \lambda_1 & * \\ \hline 0 & \\ \vdots & B \\ 0 & \end{array}\right], \end{aligned}$$

where B is an $(r-1)\times(r-1)$ matrix. By the induction hypothesis there exists an $(r-1)\times(r-1)$ unitary matrix U_2 such that $U_2^{-1}BU_2$ is an upper triangular matrix with diagonal entries $\lambda_2, \lambda_3, \ldots, \lambda_r$. Define

$$U = U_1 \begin{bmatrix} 1 & \mathbf{0} \\ \mathbf{0} & U_2 \end{bmatrix}.$$

Then it is easy to check that U is also a unitary matrix, and

$$U^{-1}AU = U^H AU = \begin{bmatrix} \lambda_1 & * \\ \mathbf{0} & U_2^H BU_2 \end{bmatrix} = \begin{bmatrix} \lambda_1 & * & \cdots & * \\ 0 & \lambda_2 & & \vdots \\ \vdots & & \ddots & * \\ 0 & \cdots & 0 & \lambda_r \end{bmatrix}.$$

This completes the proof. □

A decomposition $A = UTU^H$ with an upper triangular matrix T is called a **Schur factorization** or a **Schur decomposition** of A. Every square matrix has a Schur decomposition, but in general this decomposition $A = UTU^H = U(D + N)U^H$ is not unique, where D is diagonal and N is a nilpotent (upper triangular) matrix, (see Exercise **1.31**).

Schur's triangulation Theorem is a key for diagonalizing real and complex matrices.

Theorem 7.14 *All Hermitian, skew-Hermitian, and unitary matrices are unitarily diagonalizable.*

Proof: By Schur's triangulation Theorem, $U^H AU = B$ is an upper triangular matrix for some unitary matrix U. However,

$$B^H = U^H A^H U = \begin{cases} U^H(\pm A)U = \pm B & \text{if } A \text{ is (skew-) Hermitian} \\ U^{-1}A^{-1}U = B^{-1} & \text{if } A \text{ is unitary.} \end{cases}$$

Hence, if A is (skew-) Hermitian, then the equality $\bar{B}^T = B^H = \pm B$ implies that B is both upper triangular and lower triangular, implying that B is diagonal.

Similarly, if A is unitary, $B^H = B^{-1}$ is lower triangular, and $B^H B = I = BB^H$ implies that B is diagonal. Hence, A is unitarily diagonalizable, in each case. □

In the similarity condition $U^{-1}AU(= U^H AU) = D$, the equation $AU = UD$ shows that the column vectors of U constitute a set of n orthonormal eigenvectors of A while the diagonal entries of D are eigenvalues of A as shown in Theorem 6.6.

Furthermore, by Theorems 7.6, 7.7, and 7.10, all of the diagonal entries of D are real, purely imaginary, or of unit length depending on the type (Hermitian, skew-Hermitian, or unitary, respectively) of the matrix A.

Example 135 (*All Hermitian matrices are unitarily diagonalizable*) Diagonalize the Hermitian matrix

$$A = \begin{bmatrix} 1 & i \\ -i & 1 \end{bmatrix}$$

by a unitary matrix.

Solution: Since A is Hermitian, it is unitarily diagonalizable. One can show that A has the eigenvalues

$\lambda_1 = 0$ and $\lambda_2 = 2$ with associated eigenvectors $\mathbf{x}_1 = (i, -1)$ and $\mathbf{x}_2 = (i, 1)$,

respectively, (which are orthogonal to each other). Let

$$\begin{aligned} \mathbf{u}_1 &= \frac{\mathbf{x}_1}{\|\mathbf{x}_1\|} = \frac{1}{\sqrt{2}}(i, -1), \\ \mathbf{u}_2 &= \frac{\mathbf{x}_2}{\|\mathbf{x}_2\|} = \frac{1}{\sqrt{2}}(i, 1). \end{aligned}$$

Then, they are orthonormal. Now, the matrix

$$U = [\mathbf{u}_1\ \mathbf{u}_2] = \frac{1}{\sqrt{2}}\begin{bmatrix} i & i \\ -1 & 1 \end{bmatrix}$$

is a unitary matrix and diagonalizes A:

$$U^H AU = \frac{1}{2}\begin{bmatrix} -i & -1 \\ -i & 1 \end{bmatrix}\begin{bmatrix} 1 & i \\ -i & 1 \end{bmatrix}\begin{bmatrix} i & i \\ -1 & 1 \end{bmatrix} = \begin{bmatrix} 0 & 0 \\ 0 & 2 \end{bmatrix},$$

whose diagonal entries are all real. □

Example 136 (*Diagonal, but neither Hermitian, skew-Hermitian, nor unitary*) A matrix

$$A = \begin{bmatrix} 1 & 0 & 0 \\ 0 & i & 0 \\ 0 & 0 & x \end{bmatrix}, \qquad x \in \mathbb{R} \text{ with } |x| \neq 1,$$

is neither Hermitian, skew-Hermitian, nor unitary, but is a diagonal matrix. Hence, there are infinitely many unitarily diagonalizable matrices which are neither Hermitian, skew-Hermitian, nor unitary. □

Remark: In Theorem 6.7, it is shown that *the eigenvectors belonging to distinct eigenvalues of a matrix are linearly independent.* However, for Hermitian, skew-Hermitian, or unitary matrices, a stronger result holds: *the eigenvectors belonging to distinct eigenvalues are mutually orthogonal.*

Theorem 7.15 **(1)** *Any real square matrix A can be expressed as the sum of symmetric part $B = (A+A^T)/2$ and skew-symmetric part $C = (A-A^T)/2$. Moreover, this expression is unique, which means that if $B_1 + C_1 = A = B_2 + C_2$ with symmetric B_i and skew-symmetric C_i, then $B_1 = B_2$ and $C_1 = C_2$.*

(2) *Any complex square matrix A can be expressed as the sum of Hermitian part $B = (A + A^H)/2$ and skew-Hermitian part $C = (A - A^H)/2$. This expression is also unique.*

Consequently, the vector space $M_{n\times n}(\mathbb{R})$ of all $n\times n$ real matrices is a direct sum of the subspace of all $n\times n$ symmetric matrices and the subspace of all $n\times n$ skew-symmetric matrices.

Similarly, the complex vector space $M_{n\times n}(\mathbb{C})$ of all $n\times n$ complex matrices is a direct sum of the subspace of all $n\times n$ Hermitian matrices and the subspace of all $n\times n$ skew-Hermitian matrices.

Proof: A proof is not difficult and left as an exercise. Also, see Problem 10. □

In the complex case, the complex matrix A is also expressed as follows:

$$A = B + C = B + i(-iC) = \left[\frac{1}{2}(A + A^H)\right] + i\left[-\frac{i}{2}(A - A^H)\right] = H_1 + iH_2,$$

where $H_1 = \frac{1}{2}(A + A^H)$, $H_2 = -\frac{i}{2}(A - A^H)$ are both Hermitian.

Theorem 7.16 *For any unitarily diagonalizable matrix A, the following are equivalent.*

(1) *A is Hermitian,*
(2) *the quadratic form* $\mathbf{x}^H A\mathbf{x}$ *is real for every* $\mathbf{x} \in \mathbb{C}^n$,
(3) *all eigenvalues of A are real.*

Proof: (1) $\Rightarrow$ (2) $\Rightarrow$ (3) has been proved already in Theorem 7.6.

(3) $\Rightarrow$ (1) : Let A be unitarily diagonalizable, say $U^H AU = D$, a diagonal matrix. By (3), all diagonal entries of D are assumed to be real. Then, $U^H AU = D = D^H = (U^H AU)^H = U^H A^H U$, which implies $A = A^H$, Hermitian. □

In a similar way, one can prove the following theorem.

Theorem 7.17 *For any unitarily diagonalizable matrix A, the following are equivalent.*

(1) *A is skew-Hermitian,*
(2) *the quadratic form* $\mathbf{x}^H A\mathbf{x}$ *is purely imaginary for every* $\mathbf{x} \in \mathbb{C}^n$,
(3) *all eigenvalues of A are purely imaginary.*

Theorem 7.18 *For any unitarily diagonalizable matrix A, the following are equivalent.*

(1) *A is unitary,*
(2) *all eigenvalues of A are of unit length;* $|\lambda| = 1$.

Proof: Let A be unitarily diagonalizable, say $U^H AU = D$ is a diagonal matrix with diagonal entries λ_i.

(1) $\Rightarrow$ (2) has been proved already in Theorem 7.10.

To show (2) $\Rightarrow$ (1), let all λ_i be of unit length, *i.e.*, each $\overline{\lambda_i}\lambda_i = 1$. This gives $D^H D = I_n$ and then $A^H A = (UDU^H)^H(UDU^H) = U(D^H D)U^H = I_n$. □

Problem 178 For each of matrices

$$(1)\ \begin{bmatrix} 1 & i \\ i & 1 \end{bmatrix} \qquad (2)\ \begin{bmatrix} 2i & 0 & 0 \\ i & -1 & -i \\ -1 & 0 & 2i \end{bmatrix}$$

find a unitary matrix U such that $U^{-1}AU$ is an upper triangular matrix.

Hint: First find an eigenvalue λ with a normalized eigenvector $\mathbf{x}$. Next, extend it to an orthonormal basis.

7.6 Which real matrices are all of orthogonally diagonalizable?

All real symmetric matrices are Hermitian, so are unitarily diagonalizable by Theorem 7.14. However, the following theorem gives a more accurate and beautiful characterization of symmetric matrices.

Theorem 7.19 *For any $n \times n$ real matrix A, the following are equivalent.*

(1) *A is symmetric,*
(2) *A is orthogonally diagonalizable,*
(3) *A has a full set of n orthonormal real eigenvectors.*

Proof: $(1) \Rightarrow (2)$: If A is real and symmetric, then it is Hermitian, so it has only real eigenvalues. By Schur's triangulation Theorem 7.13, A is orthogonally similar to an upper triangular matrix, which must be already diagonal if A is symmetric. Hence it is orthogonally diagonalizable.

$(2) \Rightarrow (3)$: If A is diagonalized by an orthogonal matrix Q, then the columns of Q are eigenvectors of A. Hence A has a full set of n orthonormal real eigenvectors.

$(3) \Rightarrow (1)$: If A has a full set of n orthonormal eigenvectors, then these eigenvectors form an orthogonal change of basis matrix Q such that $AQ = QD$. It is now trivial to show that $A = QDQ^{-1} = QDQ^T$ is symmetric. □

Corollary 7.19.1 *Let A be a real symmetric matrix, and let λ be an eigenvalue of A of multiplicity m_λ. Then*

$$\dim E(\lambda) = \dim \mathcal{N}(\lambda I - A) = m_\lambda.$$

Corollary 7.19.2 *For any (square or non-square) real matrix A, A^TA is symmetric and hence has a full set of orthonormal real eigenvectors. Moreover, the eigenvalues of A^TA are all nonnegative real.*

The same is true for the matrix AA^T.

Proof: Let λ be an eigenvalue of A^TA, with $(A^TA)\mathbf{x} = \lambda\mathbf{x}$. Then

$$\lambda\|\mathbf{x}\|^2 = \langle\lambda\mathbf{x}, \mathbf{x}\rangle = \langle(A^TA)\mathbf{x}, \mathbf{x}\rangle = \langle A\mathbf{x}, A\mathbf{x}\rangle = \|A\mathbf{x}\|^2,$$

implying that λ is nonnegative. A similar proof is possible for AA^T. □

By Theorem 7.19, all real symmetric matrices are diagonalizable, even more they are orthogonally diagonalizable. Moreover, they are all of the matrices that can be orthogonally diagonalized.

Even though not all matrices are diagonalizable, certain non-symmetric matrices may still have a full set of linearly independent eigenvectors so that they are diagonalizable, but in this case the eigenvectors cannot be orthogonal. That is, the change of basis matrix Q cannot be an orthogonal matrix.

For example, any triangular matrix having all distinct diagonal entries is diagonalizable because their eigenvalues are all distinct, but it cannot be orthogonally diagonalizable if it is not diagonal (*i.e.,* not symmetric).

Example 137 (*All symmetric matrix are orthogonally diagonalizable*) Find an orthogonal matrix Q that diagonalizes a symmetric matrix

$$A = \begin{bmatrix} 0 & 2 & 2 \\ 2 & 0 & 2 \\ 2 & 2 & 0 \end{bmatrix}.$$

Solution: The characteristic polynomial of A is

$$\det(\lambda I - A) = \det \begin{bmatrix} \lambda & -2 & -2 \\ -2 & \lambda & -2 \\ -2 & -2 & \lambda \end{bmatrix} = (\lambda+2)^2(\lambda-4).$$

Thus, the eigenvalues are

$$\lambda_1 = \lambda_2 = -2 \text{ of multiplicity 2} \quad \text{and} \quad \lambda_3 = 4 \text{ of multiplicity 1}.$$

As in Example 106, it can be shown that

$$\mathbf{x}_1 = (-1, 0, 1) \quad \text{and} \quad \mathbf{x}_2 = (-1, 1, 0)$$

form a basis for the eigenspace belonging to $\lambda = -2$. Applying the Gram-Schmidt orthogonalization to $\{\mathbf{x}_1, \mathbf{x}_2\}$, one can get orthonormal eigenvectors (verify)

$$\mathbf{u}_1 = \frac{1}{\sqrt{2}}(-1, 0, 1) \quad \text{and} \quad \mathbf{u}_2 = \frac{1}{\sqrt{6}}(-1, 2, -1).$$

The eigenspace belonging to $\lambda = 4$ has $\mathbf{x}_3 = (1, 1, 1)$ as a basis. The normalization of $\mathbf{x}_3$ yields

$$\mathbf{u}_3 = \frac{1}{\sqrt{3}}(1, 1, 1).$$

Finally, using $\mathbf{u}_1, \mathbf{u}_2$, and $\mathbf{u}_3$ as column vectors, one can get

$$Q = \begin{bmatrix} -\frac{1}{\sqrt{2}} & -\frac{1}{\sqrt{6}} & \frac{1}{\sqrt{3}} \\ 0 & \frac{2}{\sqrt{6}} & \frac{1}{\sqrt{3}} \\ \frac{1}{\sqrt{2}} & -\frac{1}{\sqrt{6}} & \frac{1}{\sqrt{3}} \end{bmatrix},$$

which orthogonally diagonalizes A. (The reader is suggested to verify that $Q^T AQ$ is actually a diagonal matrix.)

In this example, in order to get the third vector $\mathbf{u}_3$, it is not necessary to use the Gram-Schmidt orthogonalization

$$\mathbf{u}_3 = \frac{\mathbf{x}_3 - \langle \mathbf{u}_1, \mathbf{x}_3 \rangle \mathbf{u}_1 - \langle \mathbf{u}_2, \mathbf{x}_3 \rangle \mathbf{u}_2}{\|\mathbf{x}_3 - \langle \mathbf{u}_1, \mathbf{x}_3 \rangle \mathbf{u}_1 - \langle \mathbf{u}_2, \mathbf{x}_3 \rangle \mathbf{u}_2\|},$$

because the vector $\mathbf{x}_3$ is already orthogonal to both $\mathbf{u}_1$ and $\mathbf{u}_2$ by Theorem 7.6(3), (also see Remark on page 330). □

Problem 179 Show that the non-symmetric matrices

$$(1) \begin{bmatrix} 1 & 0 & -1 \\ 0 & 1 & 0 \\ 0 & 0 & 2 \end{bmatrix} \quad \text{and} \quad (2) \begin{bmatrix} x & 0 & 1 \\ 0 & x+1 & 0 \\ 0 & 0 & x+2 \end{bmatrix}, \ x \in \mathbb{R}$$

are both diagonalizable, but not orthogonally.

7.7 Which complex matrices are all of unitarily diagonalizable?

As shown in the previous section, the real symmetric matrices constitute the entire class of matrices that are orthogonally diagonalizable.

For complex matrices, Hermitian, skew-Hermitian, and unitary matrices are all unitarily diagonalizable. However, they do not constitute the entire class of unitarily diagonalizable matrices. There are infinitely many unitarily diagonalizable matrices that are neither Hermitian, skew-Hermitian, nor unitary. See Example 136.

In fact, the entire class of unitarily diagonalizable matrices is the following class of matrices, called *normal* matrices.

Definition 7.20 A complex square matrix A is called **normal** if

$$AA^H = A^H A.$$

All of the Hermitian, skew-Hermitian, and unitary matrices are clearly normal. All diagonal matrices are also normal, hence there are infinitely many normal matrices that are neither Hermitian, skew-Hermitian, nor unitary, as shown in Example 136. Moreover, there exist examples of such matrices which are not even diagonal.

Example 138 (*Normal, but neither Hermitian, skew-Hermitian, unitary, nor diagonal*) The 2×2 matrix

$$A = \begin{bmatrix} 1 & i \\ i & 1 \end{bmatrix}$$

is normal. One can easily see that this is unitarily diagonalizable. In fact,

$$U^{-1}AU = \left(\frac{1}{\sqrt{2}}\begin{bmatrix} 1 & 1 \\ 1 & -1 \end{bmatrix}\right)\begin{bmatrix} 1 & i \\ i & 1 \end{bmatrix}\left(\frac{1}{\sqrt{2}}\begin{bmatrix} 1 & 1 \\ 1 & -1 \end{bmatrix}\right) = \begin{bmatrix} 1+i & 0 \\ 0 & 1-i \end{bmatrix} = D.$$

However, A is neither Hermitian, skew-Hermitian, unitary, nor diagonal. □

Problem 180 Which of following matrices are Hermitian, skew-Hermitian, unitary, or normal?

$$(1)\ \begin{bmatrix} 1 & 1+i \\ 1-i & 0 \end{bmatrix},\quad (2)\ \begin{bmatrix} 1 & 1 & 1 \\ 1 & 1 & 1 \\ 1 & 1 & 1 \end{bmatrix},\quad (3)\ \begin{bmatrix} i & 1 & -i \\ -1 & 2i & 0 \\ -i & 0 & -i \end{bmatrix},$$

$$(4)\ \begin{bmatrix} -i & 3 \\ 3 & 2 \end{bmatrix},\quad (5)\ \begin{bmatrix} 0 & 0 & i \\ 0 & i & 0 \\ i & 0 & 0 \end{bmatrix},\quad (6)\ \begin{bmatrix} 3 & 1+i & i \\ 1-i & 1 & 3 \\ -i & 3 & 1 \end{bmatrix}.$$

To show that the normal matrices constitute the entire class of unitarily diagonalizable matrices, we begin with a lemma.

Lemma 7.21 *If an upper triangular matrix T is normal, then it is diagonal.*

Proof: Let T be normal; $TT^H = T^HT$. In an expanded form,

$$\begin{bmatrix} t_{11} & & t_{1n} \\ & \ddots & \\ 0 & & t_{nn} \end{bmatrix} \begin{bmatrix} \overline{t_{11}} & & 0 \\ & \ddots & \\ \overline{t_{1n}} & & \overline{t_{nn}} \end{bmatrix} = \begin{bmatrix} \overline{t_{11}} & & 0 \\ & \ddots & \\ \overline{t_{1n}} & & \overline{t_{nn}} \end{bmatrix} \begin{bmatrix} t_{11} & & t_{1n} \\ & \ddots & \\ 0 & & t_{nn} \end{bmatrix}.$$

By comparing their $(1,1)$-entries on both sides, one can see that

$$[TT^H]_{11} = |t_{11}|^2 + \cdots + |t_{1n}|^2 \quad \text{and} \quad [T^HT]_{11} = |t_{11}|^2,$$

which implies $t_{12} = \cdots = t_{1n} = 0$. With these zeros, compare the $(2,2)$-entries on both sides to get

$$[TT^H]_{22} = |t_{22}|^2 + \cdots + |t_{2n}|^2 \quad \text{and} \quad [T^HT]_{22} = |t_{22}|^2,$$

which implies $t_{23} = \cdots = t_{2n} = 0$. Continuing to compare their diagonal entries $[TT^H]_{ii}$ and $[T^HT]_{ii}$ for $i = 3, 4, \ldots$, one can show that all off-diagonal entries of T should be zero, *i.e.*, T is diagonal. □

Theorem 7.22 *For any $n \times n$ complex matrix A, the following are equivalent.*

(1) *A is normal,*
(2) *A is unitarily diagonalizable,*
(3) *A has a full set of n orthonormal complex eigenvectors.*

Proof: (1) $\Rightarrow$ (2): Suppose that A is normal. By Schur's triangulation Theorem, there exists a unitary matrix U such that $T = U^HAU$ is upper triangular. Then T is also normal, since

$$\begin{aligned} TT^H &= U^HAUU^HA^HU = U^HAA^HU = U^HA^HAU \\ &= U^HA^HUU^HAU = T^HT. \end{aligned}$$

Thus, by Lemma 7.21, T is already diagonal so that A is unitarily diagonalizable.

(2) $\Rightarrow$ (3): It is clear from a fact that the columns of the change of basis matrix U are n orthonormal complex eigenvectors of A.

(3) $\Rightarrow$ (1): Let U be the unitary matrix whose columns are the n orthonormal eigenvectors of A. Then $AU = UD$ or $A = UDU^H$, and

$$\begin{aligned} AA^H &= UDU^HUD^HU^H = UDD^HU^H = UD^HDU^H \\ &= UD^HU^HUDU^H = A^HA. \end{aligned}$$

That is, A is normal. □

Corollary 7.22.1 *For any (square or non-square) complex matrix A, A^HA is Hermitian and hence has a full set of orthonormal complex eigenvectors. Moreover, the eigenvalues of A^HA are all nonnegative real.*

The same is true for the matrix AA^H.

Proof: Let λ be an eigenvalue of A^HA, with $(A^HA)\mathbf{x} = \lambda\mathbf{x}$. Then

$$\lambda\|\mathbf{x}\|^2 = \langle\lambda\mathbf{x}, \mathbf{x}\rangle = \langle(A^HA)\mathbf{x}, \mathbf{x}\rangle = \langle A\mathbf{x}, A\mathbf{x}\rangle = \|A\mathbf{x}\|^2,$$

implying that λ is nonnegative. A similar proof is possible for AA^H. □

There exist infinitely many non-normal complex matrices that are still diagonalizable, but of course not unitarily. One can find such examples among the triangular matrices having distinct diagonal entries.

Problem 181 Show that the matrix $A = \begin{bmatrix} x & 0 \\ 1 & i \end{bmatrix}$, $x \in \mathbb{R}$, is not normal, so it cannot be unitarily diagonalizable. But it is diagonalizable.

Hint: A has two distinct eigenvalues.

Problem 182 Determine whether or not the matrix

$$A = \begin{bmatrix} 1 & i & i \\ i & 1 & i \\ i & i & 1 \end{bmatrix}$$

is unitarily diagonalizable. If it is, try to find a unitary matrix U that diagonalizes A.

Hint: $1 - i$ is one of the eigenvalues of A.

Problem 183 Let H_1 and H_2 be two Hermitian matrices. Show that $A = H_1 + iH_2$ is normal if and only if $H_1H_2 = H_2H_1$.

Problem 184 Which two matrices have the same set of eigenvalues? Try to justify the following statements as much as you can:

(1) For any real square matrix A, A and A^T have the same real and complex eigenvalues of the same multiplicity, (see Problem 139 on page 258).
(2) For any square matrices A and B of the same size, AB and BA have the same set of real and complex eigenvalues, (see Corollary 6.4.3 on page 257).
(3) For any (non-square) rectangular real or complex matrices A and B, AB and BA cannot have the same set of eigenvalues even if both are well defined.
(4) For any (non-square) rectangular real or complex matrix A, both A^HA and AA^H are both Hermitian and they have the same set of *nonzero* eigenvalues. Moreover, all of their eigenvalues are nonnegative real numbers. In particular, $\mathbf{x}^H\mathbf{x}$ and $\mathbf{x}\mathbf{x}^H$ have the same set of *nonzero* eigenvalues.

Hint: (4) See Corollaries 7.19.2 and 7.22.1. Also, see Problem 189.

7.8 Applications

7.8.1 The eigenvalue decompositions(EVD) of normal matrices and computing the power A^k

It is shown that the normal matrices constitute the entire class of unitarily diagonalizable matrices. A decomposition $A = UDU^{-1}$ of a normal matrix A can be further decomposed as follows.

Theorem 7.23 (EVD) *Let A be an $n \times n$ normal matrix with a decomposition $A = UDU^{-1} = UDU^H$, where $U = [\mathbf{u}_1\ \mathbf{u}_2\ \cdots\ \mathbf{u}_n]$ is a unitary matrix and $U^{-1}AU = D$ is diagonal with diagonal entries λ_i. Then A can be decomposed into a linear combination of rank one matrices as*

$$A = UDU^H = \lambda_1\mathbf{u}_1\mathbf{u}_1^H + \lambda_2\mathbf{u}_2\mathbf{u}_2^H + \cdots + \lambda_n\mathbf{u}_n\mathbf{u}_n^H \tag{7.1}$$

and $\mathbf{u}_i\mathbf{u}_i^H$ is the orthogonal projection matrix of the complex space $\mathbb{C}^n$ onto the one-dimensional subspace spanned by the eigenvector $\mathbf{u}_i$ for $i = 1, 2, \ldots, n$.

A decomposition in Eq. (7.1) is called an **eigenvalue decomposition(EVD)**, or an **eigenvalue factorization** of a normal matrix A.

Proof: By a direct computation of $A = UDU^{-1} = UDU^H$,

$$\begin{aligned} A &= UDU^H = [\lambda_1\mathbf{u}_1\ \lambda_2\mathbf{u}_2\ \cdots\ \lambda_n\mathbf{u}_n] \begin{bmatrix} \mathbf{u}_1^H \\ \mathbf{u}_2^H \\ \vdots \\ \mathbf{u}_n^H \end{bmatrix} \\ &= \lambda_1\mathbf{u}_1\mathbf{u}_1^H + \lambda_2\mathbf{u}_2\mathbf{u}_2^H + \cdots + \lambda_n\mathbf{u}_n\mathbf{u}_n^H, \end{aligned}$$

as a linear combination of rank one matrices. □

Let $P_i = \mathbf{u}_i\mathbf{u}_i^H$, as a projection. Then the equation $P_1 + P_2 + \cdots + P_n = id$ shows that if one restricts the image of the P_i to be the subspace spanned by $\mathbf{u}_i$ which is isomorphic to $\mathbb{C}$, then $(P_1, P_2, \ldots, P_n)$ defines another rectangular coordinate system with respect to the orthonormal basis $\{\mathbf{u}_1, \mathbf{u}_2, \ldots, \mathbf{u}_n\}$ just like the Cartesian coordinate system $(z_1, z_2, \ldots, z_n)$ of the space $\mathbb{C}^n$, (see Sections 5.6, 5.7, and 5.10.4 for the real case).

Let the normal matrix A have t distinct eigenvalues $\lambda_1, \lambda_2, \ldots, \lambda_t$ with multiplicities $m_1, m_2, \ldots, m_t$, respectively.

For each λ_i of the t distinct eigenvalues, say $\lambda_i = \lambda_{i_1} = \cdots = \lambda_{im_i}$ with associated m_i orthonormal eigenvectors $\mathbf{u}_{i_1}, \ldots, \mathbf{u}_{im_i}$. Then, they form an orthonormal basis for the eigenspace $E(\lambda_i)$, and

$$P_{\lambda_i} = P_{i_1} + \cdots + P_{im_i} = \mathbf{u}_{i_1}\mathbf{u}_{i_1}^H + \cdots + \mathbf{u}_{im_i}\mathbf{u}_{im_i}^H$$

is the projection matrix of the space $\mathbb{C}^n$ onto $E(\lambda_i) = \mathcal{N}(\lambda_i I - A)$.

Now, the decomposition of A in Eq. (7.1) can be

$$A = \lambda_1 P_{\lambda_1} + \lambda_2 P_{\lambda_2} + \cdots + \lambda_t P_{\lambda_t}, \tag{7.2}$$

where λ_i's are all distinct.

A decomposition of A in Eq. (7.2) is also called an **eigenvalue decomposition(EVD)**, or an **eigenvalue factorization** of a normal matrix A.

Corollary 7.23.1 *Let A be a normal matrix. Then*

(1) *the eigenspaces of A belonging to distinct eigenvalues are mutually orthogonal.*

(2) *If an eigenvalue λ of A has multiplicity k, then the dimension of the eigenspace $\mathcal{N}(A-\lambda I)$ belonging to λ is k.*

Corollary 7.23.2 *Let A be a normal matrix with EVD $A = \lambda_1 P_{\lambda_1} + \lambda_2 P_{\lambda_2} + \cdots + \lambda_t P_{\lambda_t}$. Then, for any positive integer m,*

$$A^m = \lambda_1^m P_{\lambda_1} + \lambda_2^m P_{\lambda_2} + \cdots + \lambda_t^m P_{\lambda_t}.$$

This is also true for any negative integer m if A is invertible. That is, for any positive integer ℓ,

$$A^{-\ell} = \frac{1}{\lambda_1^\ell} P_{\lambda_1} + \frac{1}{\lambda_2^\ell} P_{\lambda_2} + \cdots + \frac{1}{\lambda_t^\ell} P_{\lambda_t}.$$

Example 139 (*EVD of a symmetric matrix*) Find an EVD of the symmetric matrix

$$A = \begin{bmatrix} 0 & 2 & 2 \\ 2 & 0 & 2 \\ 2 & 2 & 0 \end{bmatrix}$$

as a sum of projection matrices onto the eigenspaces. By using the decomposition, compute the power A^{100} and A^{-100}.

Solution: As shown in Example 137, A has eigenvalues

$$\lambda_1 = \lambda_2 = -2 \quad \text{and} \quad \lambda_3 = 4$$

with associated orthonormal eigenvectors

$$\mathbf{u}_1 = \frac{1}{\sqrt{2}}(-1, 0, 1), \quad \mathbf{u}_2 = \frac{1}{\sqrt{6}}(-1, 2, -1) \quad \text{and} \quad \mathbf{u}_3 = \frac{1}{\sqrt{3}}(1, 1, 1),$$

respectively.

An EVD of A is $A = -2(P_1 + P_2) + 4P_3$, where the projections are

$$\begin{aligned}
P_1 &= \mathbf{u}_1\mathbf{u}_1^H = \frac{1}{2}\begin{bmatrix} -1 \\ 0 \\ 1 \end{bmatrix}[-1\ 0\ 1] = \frac{1}{2}\begin{bmatrix} 1 & 0 & -1 \\ 0 & 0 & 0 \\ -1 & 0 & 1 \end{bmatrix}, \\
P_2 &= \mathbf{u}_2\mathbf{u}_2^H = \frac{1}{6}\begin{bmatrix} -1 \\ 2 \\ -1 \end{bmatrix}[-1\ 2\ -1] = \frac{1}{6}\begin{bmatrix} 1 & -2 & 1 \\ -2 & 4 & -2 \\ 1 & -2 & 1 \end{bmatrix}, \\
P_3 &= \mathbf{u}_3\mathbf{u}_3^H = \frac{1}{3}\begin{bmatrix} 1 \\ 1 \\ 1 \end{bmatrix}[1\ 1\ 1] = \frac{1}{3}\begin{bmatrix} 1 & 1 & 1 \\ 1 & 1 & 1 \\ 1 & 1 & 1 \end{bmatrix}.
\end{aligned}$$

Hence,

$$P_1 + P_2 = \frac{1}{3}\begin{bmatrix} 2 & -1 & -1 \\ -1 & 2 & -1 \\ -1 & -1 & 2 \end{bmatrix}$$

is the projection of $\mathbb{R}^3$ onto the eigenspace $E(-2)$ belonging to $\lambda = -2$, and P_3 is the projection of $\mathbb{R}^3$ onto the eigenspace $E(4)$ belonging to $\lambda = 4$.

Therefore, the EVD of A is

$$A = \begin{bmatrix} 0 & 2 & 2 \\ 2 & 0 & 2 \\ 2 & 2 & 0 \end{bmatrix} = \frac{-2}{3}\begin{bmatrix} 2 & -1 & -1 \\ -1 & 2 & -1 \\ -1 & -1 & 2 \end{bmatrix} + \frac{4}{3}\begin{bmatrix} 1 & 1 & 1 \\ 1 & 1 & 1 \\ 1 & 1 & 1 \end{bmatrix}.$$

Since A has no zero eigenvalues, A is invertible and

$$A^k = \begin{bmatrix} 0 & 2 & 2 \\ 2 & 0 & 2 \\ 2 & 2 & 0 \end{bmatrix}^k = \frac{(-2)^k}{3}\begin{bmatrix} 2 & -1 & -1 \\ -1 & 2 & -1 \\ -1 & -1 & 2 \end{bmatrix} + \frac{4^k}{3}\begin{bmatrix} 1 & 1 & 1 \\ 1 & 1 & 1 \\ 1 & 1 & 1 \end{bmatrix}$$

for any positive or negative integer k. □

Example 140 (*EVD of a complex normal matrix*) Find an EVD of a normal matrix

$$A = \begin{bmatrix} 0 & 0 & 11i \\ 0 & 11i & 0 \\ 11i & 0 & 0 \end{bmatrix}$$

as a sum of projection matrices of the space $\mathbb{C}^3$ onto the eigenspaces. By using the decomposition, compute the power A^{100} and A^{-100} if A is invertible.

Solution: Since A is normal ($AA^H = A^H A$), it is unitarily diagonalizable. The characteristic polynomial of A is

$$\det(\lambda I - A) = \det\begin{bmatrix} \lambda & 0 & -11i \\ 0 & \lambda - 11i & 0 \\ -11i & 0 & \lambda \end{bmatrix} = (\lambda - 11i)^2(\lambda + 11i).$$

Hence, the eigenvalues are

$$\lambda_1 = \lambda_2 = 11i \text{ of multiplicity 2,} \quad \text{and} \quad \lambda_3 = -11i.$$

By a simple computation using the Gram-Schmidt orthogonalization, one can find that

$$\mathbf{u}_1 = \frac{1}{\sqrt{2}}\begin{bmatrix}1\\0\\1\end{bmatrix}, \quad \mathbf{u}_2 = \begin{bmatrix}0\\1\\0\end{bmatrix}, \quad \text{and} \quad \mathbf{u}_3 = \frac{1}{\sqrt{2}}\begin{bmatrix}-1\\0\\1\end{bmatrix}$$

are orthonormal eigenvectors of A belonging to the eigenvalues λ_1, λ_2, and λ_3, respectively. Now, an EVD is $A = 11i(P_1 + P_2) - 11iP_3$, where the projection matrices are

$$P_1 = \mathbf{u}_1\mathbf{u}_1^H = \frac{1}{2}\begin{bmatrix}1\\0\\1\end{bmatrix}[1\ 0\ 1] = \frac{1}{2}\begin{bmatrix}1 & 0 & 1\\0 & 0 & 0\\1 & 0 & 1\end{bmatrix},$$

$$P_2 = \mathbf{u}_2\mathbf{u}_2^H = \begin{bmatrix}0\\1\\0\end{bmatrix}[0\ 1\ 0] = \begin{bmatrix}0 & 0 & 0\\0 & 1 & 0\\0 & 0 & 0\end{bmatrix},$$

$$P_3 = \mathbf{u}_3\mathbf{u}_3^H = \frac{1}{2}\begin{bmatrix}-1\\0\\1\end{bmatrix}[-1\ 0\ 1] = \frac{1}{2}\begin{bmatrix}1 & 0 & -1\\0 & 0 & 0\\-1 & 0 & 1\end{bmatrix}.$$

Hence,

$$A = 11i(P_1 + P_2) - 11iP_3 = \frac{11i}{2}\begin{bmatrix}1 & 0 & 1\\0 & 2 & 0\\1 & 0 & 1\end{bmatrix} - \frac{11i}{2}\begin{bmatrix}1 & 0 & -1\\0 & 0 & 0\\-1 & 0 & 1\end{bmatrix}.$$

Since A has no zero eigenvalues, A is invertible and by Corollary 7.23.2

$$A^k = \frac{(11i)^k}{2}\begin{bmatrix}1 & 0 & 1\\0 & 2 & 0\\1 & 0 & 1\end{bmatrix} + \frac{(-11i)^k}{2}\begin{bmatrix}1 & 0 & -1\\0 & 0 & 0\\-1 & 0 & 1\end{bmatrix}$$

for any positive or negative integer k. □

Problem 185 Find the EVD of each of the following matrices as a linear combination of rank one matrices.

$$(1)\ A = \begin{bmatrix}2 & 3\\3 & 2\end{bmatrix}, \qquad (2)\ B = \begin{bmatrix}2 & 3i\\-3i & 2\end{bmatrix},$$

$$(3)\ C = \begin{bmatrix}0 & 0 & 0 & 1\\0 & 0 & 2 & 0\\0 & 2 & 0 & 0\\4 & 0 & 0 & 0\end{bmatrix}, \qquad (4)\ D = \begin{bmatrix}1+i & 0 & 0 & 0\\0 & 2+i & 0 & 0\\0 & 0 & 2+i & 0\\0 & 0 & 0 & 4+i\end{bmatrix}.$$

Problem 186 Given a symmetric matrix $A = \begin{bmatrix} 0 & 2 & -1 \\ 2 & 3 & -2 \\ -1 & -2 & 0 \end{bmatrix}$, find an orthogonal matrix Q that diagonalizes A, and find the EVD of A as a linear combination of projection matrices onto the eigenspaces. Also compute the power A^n.

7.8.2 Singular value decompositions(SVD); Diagonalizations of rectangular matrices

Normal matrices are unitarily diagonalizable and have EVD (eigenvalue decompositions). For example, every symmetric matrix A is diagonalizable and has an EVD $A = QDQ^{-1}$ with a diagonal matrix D. But there are lots of square matrices which are not diagonalizable. For example, any matrix of the type $\begin{bmatrix} d & 1 \\ 0 & d \end{bmatrix}$ cannot be diagonalized.

In this section, we introduce a similar kind of decomposition of any (square or non-square) matrix into a product of three matrices, one of which is also a diagonal matrix.

To see more clear description of such decomposition, let a (long or wide) rectangular matrix be allowed to be called *rectangular diagonal* or simply *diagonal* if every (i, j)-entry is zero for $i \neq j$. For example,

$$\Sigma = \begin{bmatrix} a & 0 & 0 \\ 0 & b & 0 \\ 0 & 0 & c \\ 0 & 0 & 0 \end{bmatrix} \quad \text{and} \quad \Sigma = \begin{bmatrix} x & 0 & 0 & 0 \\ 0 & y & 0 & 0 \\ 0 & 0 & z & 0 \end{bmatrix}$$

are (rectangular) diagonal matrices. If a square matrix is diagonal, we call it *square diagonal* if needed to be distinguished.

Commonly, a square diagonal matrix and its diagonal entries are written by D and d_i (or λ_i), respectively. To be distinguished from the square case, a rectangular diagonal matrix and its diagonal entries will be written as Σ and σ_i, respectively.

Now, we introduce a decomposition of any real or complex (square or non-square) matrix A, called a *singular value decomposition* (SVD) of A.

A singular value decomposition(SVD) of a matrix A is a decomposition $A = U\Sigma V^T$ ($A = U\Sigma V^H$ in the complex case) of A into a product of three matrices of two orthogonal (unitary in the complex case) matrices U, V and a diagonal matrix Σ (of the same size as A).

In comparison with an EVD of a normal matrix (say, every normal matrix A can be decomposed into $A = UDU^H$ with a diagonal matrix D), an SVD $A = U\Sigma V^H$ says that if we allow the unitary matrices U and U^H in an EVD $A = UDU^H$ to be any two unitary matrices independently, then *every real or complex (even non-square) matrix* A can be decomposed into $A = U\Sigma V^H$ so that $\Sigma = U^H AV$ is diagonal.

To give a proof of an SVD, we first remind the following facts in the real case.

For any (long or wide) $m \times n$ matrix $A \in M_{m\times n}(\mathbb{R})$, the matrices $A^T A$ and AA^T are both square and symmetric, and they are of size $n \times n$ and $m \times m$, respectively.

The following properties are known already.

(I) rank $(A^T A) =$ rank $A =$ rank (AA^T) and $\mathcal{N}(A^T A) = \mathcal{N}(A)$. Equivalently, two homogeneous systems $A\mathbf{x} = \mathbf{0}$ and $A^T A\mathbf{x} = \mathbf{0}$ have the same solution set. (See Theorem 3.26). Here, the domain of the matrix A is $\mathcal{R}(A) \oplus \mathcal{N}(A) = \mathbb{R}^n = \mathcal{R}(A^T A) \oplus \mathcal{N}(A^T A)$. (See Theorem 5.19). Consequently, it follows $\mathcal{R}(A) = \mathcal{R}(A^T A)$.

(II) The symmetric matrix $A^T A$ has a full set of orthonormal real eigenvectors. Moreover, the eigenvalues of $A^T A$ are all nonnegative real. The same is true for the matrix AA^T.

In this reason, the matrices $A^T A$ and AA^T are called **positive-semidefinite**. (See Corollary 7.19.2).

Since A can be any matrix, the properties **(I)**, **(II)** are also true for the matrix A^T instead of A. For example, one can get $\mathcal{C}(A) = \mathcal{R}(A^T) = \mathcal{R}(AA^T)$.

Moreover, for any complex matrix $A \in M_{m\times n}(\mathbb{C})$, the matrices $A^H A$ and AA^H have similar properties as **(I)** and **(II)**. (See Corollary 7.22.1).

Lemma 7.24 *For any real rectangular matrix $A \in M_{m\times n}(\mathbb{R})$,*

(1) *the $n \times n$ matrix $A^T A$ has n nonnegative real eigenvalues, say*

$$\sigma_1 \geq \sigma_2 \geq \cdots \geq \sigma_r > 0, \quad \sigma_{r+1} = \cdots = \sigma_n = 0$$

in descending order, with associated eigenvectors

$$\mathbf{v}_1, \mathbf{v}_2, \ldots, \mathbf{v}_r, \mathbf{v}_{r+1}, \ldots, \mathbf{v}_n,$$

respectively, which are chosen to be orthonormal. Then the eigenvectors $\mathbf{v}_1, \mathbf{v}_2, \ldots, \mathbf{v}_r$ *of* A^TA *belonging to positive eigenvalues form an orthonormal basis for the row space* $\mathcal{R}(A) = \mathcal{R}(A^TA)$ *in the domain space* $\mathbb{R}^n$, *where* rank $A = r$, *and the eigenvectors* $\mathbf{v}_{r+1}, \ldots, \mathbf{v}_n$ *belonging to the zero eigenvalue form an orthonormal basis for the null space* $\mathcal{N}(A)$.

(2) *For positive* σ_i, $i = 1, 2, \ldots, r$, *set* $\mathbf{u}_i = A\mathbf{v}_i/\sqrt{\sigma_i}$. *Then the vectors* $\mathbf{u}_1, \mathbf{u}_2, \ldots, \mathbf{u}_r$ *form an orthonormal basis for the column space* $\mathcal{C}(A)(= \mathcal{R}(A^T) = \mathcal{R}(AA^T))$. *In particular, each* $\mathbf{u}_i = A\mathbf{v}_i/\sqrt{\sigma_i}$ *is a unit vector, and hence*

$$\sqrt{\sigma_i} = \|A\mathbf{v}_i\|. \tag{7.3}$$

That is, $\mathbf{u}_i = A\mathbf{v}_i/\|A\mathbf{v}_i\|$.

(3) *The unit vectors* $\mathbf{v}_i$ *in* $\mathcal{R}(A)$ *and* $\mathbf{u}_i$ *in* $\mathcal{C}(A)$ *are related by the following equations:*

$$A\mathbf{v}_i = \sqrt{\sigma_i}\mathbf{u}_i \quad \textit{and} \quad A^T\mathbf{u}_i = \sqrt{\sigma_i}\mathbf{v}_i, \qquad i = 1, 2, \ldots, r. \tag{7.4}$$

For any square or rectangular complex matrix A, *the same is also true with* A^HA *instead of* A^TA.

Based on a relation between vectors $\mathbf{v}_i$ and $\mathbf{u}_i$ in Eq. (7.4), the square roots $\sqrt{\sigma_1}, \sqrt{\sigma_2}, \ldots, \sqrt{\sigma_r}$ of positive eigenvalues of A^TA are called **singular values** of a matrix A.

The basic vectors $\mathbf{v}_1, \mathbf{v}_2, \ldots, \mathbf{v}_r$ for the row space $\mathcal{R}(A)$ are the eigenvectors of A^TA belonging to positive eigenvalues. These vectors are called the **right singular vectors** of A.

The basic vectors $\mathbf{u}_1, \mathbf{u}_2, \ldots, \mathbf{u}_r$ for the column space $\mathcal{C}(A)$ are the eigenvectors of AA^T belonging to positive eigenvalues. These vectors are called the **left singular vectors** of A.

Proof: (1) follows directly from the properties (I) and (II) above.

(2) Clearly, the vectors $\mathbf{u}_i = A\mathbf{v}_i/\sqrt{\sigma_i}$ are nonzero in the column space $\mathcal{C}(A)$ for $i = 1, \ldots, r$. Furthermore, these r vectors are orthonormal:

$$\begin{aligned}
\langle \mathbf{u}_i, \mathbf{u}_j \rangle &= \langle A\mathbf{v}_i/\sqrt{\sigma_i}, A\mathbf{v}_j/\sqrt{\sigma_j} \rangle = \frac{1}{\sqrt{\sigma_i}\sqrt{\sigma_j}} \langle A\mathbf{v}_i, A\mathbf{v}_j \rangle \\
&= \frac{1}{\sqrt{\sigma_i}\sqrt{\sigma_j}} \langle A^TA\mathbf{v}_i, \mathbf{v}_j \rangle = \frac{1}{\sqrt{\sigma_i}\sqrt{\sigma_j}} \langle \sigma_i\mathbf{v}_i, \mathbf{v}_j \rangle \\
&= \frac{\sigma_i}{\sqrt{\sigma_i}\sqrt{\sigma_j}} \langle \mathbf{v}_i, \mathbf{v}_j \rangle = \delta_{ij}.
\end{aligned}$$

Since rank $A = r$, they form an orthonormal basis for the column space $\mathcal{C}(A)$. Now, Eq. (7.3) is clear.

(3) From the definition $\mathbf{u}_i = A\mathbf{v}_i/\sqrt{\sigma_i}$ of $\mathbf{u}_i$, it follows $A\mathbf{v}_i = \sqrt{\sigma_i}\mathbf{u}_i$, $i = 1, 2, \ldots, r$ and $A^T\mathbf{u}_i = A^TA\mathbf{v}_i/\sqrt{\sigma_i} = \sigma_i\mathbf{v}_i/\sqrt{\sigma_i} = \sqrt{\sigma_i}\mathbf{v}_i$, $i = 1, 2, \ldots, r$. □

Now we state the **singular value decomposition (SVD)** or **singular value factorization** of a real matrix.

Theorem 7.25 (Real SVD) *Let A be any $m \times n$ real matrix with singular values $\sqrt{\sigma_1}, \sqrt{\sigma_2}, \ldots, \sqrt{\sigma_r}$ in descending order, and corresponding right singular vectors $\mathbf{v}_1, \mathbf{v}_2, \ldots, \mathbf{v}_r$, left singular vectors $\mathbf{u}_1, \mathbf{u}_2, \ldots, \mathbf{u}_r$. Then A can be decomposed into a linear combination of rank one matrices, called a* **singular value decomposition (SVD)** *of A, as*

$$A = U\Sigma V^T = \sqrt{\sigma_1}\mathbf{u}_1\mathbf{v}_1^T + \sqrt{\sigma_2}\mathbf{u}_2\mathbf{v}_2^T + \cdots + \sqrt{\sigma_r}\mathbf{u}_r\mathbf{v}_r^T. \qquad (7.5)$$

In other words, A can be factored into a product $U\Sigma V^T$, where V is an $n \times n$ orthogonal matrix whose columns are $\mathbf{v}_1, \mathbf{v}_2, \ldots, \mathbf{v}_r$ and any orthonormal basis vectors $\mathbf{v}_{r+1}, \ldots, \mathbf{v}_n$ for $\mathcal{N}(A)$, Σ is an $m\times n$ rectangular diagonal matrix with diagonal entries $\sqrt{\sigma_1}, \sqrt{\sigma_2}, \ldots, \sqrt{\sigma_r}, 0, \ldots$ in descending order, and U is an $m \times m$ orthogonal matrix whose columns are $\mathbf{u}_1, \mathbf{u}_2, \ldots, \mathbf{u}_r$ and any orthonormal basis vectors $\mathbf{u}_{r+1}, \ldots, \mathbf{u}_m$ for $\mathcal{N}(A^T)$.

Proof: The theorem seems complicated, but the proof is rather easy. For each singular vector $\mathbf{v}_j$, one gets

$$A\mathbf{v}_j = \sqrt{\sigma_j}\mathbf{u}_j = \left(\sum_{i=1}^{r} \sqrt{\sigma_i}\mathbf{u}_i\mathbf{v}_i^T\right)\mathbf{v}_j.$$

Since the domain space is $\mathbb{R}^n = \mathcal{N}(A) \oplus \mathcal{R}(A)$ (see Theorem 5.19) and the vectors $\mathbf{v}_1, \mathbf{v}_2, \ldots, \mathbf{v}_r$ form a basis for $\mathcal{R}(A)$, one can conclude $A = \sum_{i=1}^r \sqrt{\sigma_i}\mathbf{u}_i\mathbf{v}_i^T$, which is just $U\Sigma V^T$ as a matrix product. □

A similar argument gives a decomposition of a complex matrix.

Theorem 7.26 (Complex SVD) *Let A be any $m \times n$ complex matrix with singular values $\sqrt{\sigma_1}, \sqrt{\sigma_2}, \ldots, \sqrt{\sigma_r}$ in descending order, and corresponding right singular vectors $\mathbf{v}_1, \mathbf{v}_2, \ldots, \mathbf{v}_r$, left singular vectors $\mathbf{u}_1, \mathbf{u}_2, \ldots, \mathbf{u}_r$. Then A can* *be decomposed as a linear combination of rank one matrices, called a* **singular value decomposition (SVD)** *of* A*, as*

$$A = U\Sigma V^H = \sqrt{\sigma_1}\mathbf{u}_1\mathbf{v}_1^H + \sqrt{\sigma_2}\mathbf{u}_2\mathbf{v}_2^H + \cdots + \sqrt{\sigma_r}\mathbf{u}_r\mathbf{v}_r^H. \quad (7.6)$$

In other words, A *can be factored into a product* $U\Sigma V^H$*, where* V *is an* $n \times n$ *unitary matrix whose columns are* $\mathbf{v}_1, \mathbf{v}_2, \ldots, \mathbf{v}_r$ *and any orthonormal basis vectors* $\mathbf{v}_{r+1}, \ldots, \mathbf{v}_n$ *for* $\mathcal{N}(A)$*,* Σ *is an* $m \times n$ *rectangular diagonal matrix with diagonal entries* $\sqrt{\sigma_1}, \sqrt{\sigma_2}, \ldots, \sqrt{\sigma_r}, 0, \ldots$ *in descending order, and* U *is an* $m \times m$ *unitary matrix whose columns are* $\mathbf{u}_1, \mathbf{u}_2, \ldots, \mathbf{u}_r$ *and any orthonormal basis vectors* $\mathbf{u}_{r+1}, \ldots, \mathbf{u}_m$ *for* $\mathcal{N}(A^H)$*.*

The following three examples illustrate how to get an SVD $A = U\Sigma V^T$ of a matrix A.

Example 141 (*SVD* $A = U\Sigma V^T$ *of a square but not diagonalizable matrix* A) Find an SVD of a square matrix

$$A = \begin{bmatrix} 0 & 1 \\ 0 & 0 \end{bmatrix},$$

which is not diagonalizable as shown in Example 109.

Solution: Clearly, rank $A = 1$. To find a matrix V first, note that

$$A^T A = \begin{bmatrix} 0 & 0 \\ 1 & 0 \end{bmatrix} \begin{bmatrix} 0 & 1 \\ 0 & 0 \end{bmatrix} = \begin{bmatrix} 0 & 0 \\ 0 & 1 \end{bmatrix}.$$

Its eigenvalues are

$$\sigma_1 = 1, \quad \sigma_2 = 0$$

in descending order (recall that rank A is equal to the number of positive eigenvalues of $A^T A$), with associated *unit* eigenvectors

$$\begin{aligned} \mathbf{v}_1 &= \pm(0, 1), \\ \mathbf{v}_2 &= \pm(1, 0), \end{aligned}$$

respectively, where $\mathbf{v}_2$ can be any unit vector in the null space $\mathcal{N}(A)$. By choosing $\mathbf{v}_1$ and $\mathbf{v}_2$, one can construct an orthogonal matrix V, say

$$V = [\mathbf{v}_1 \ \mathbf{v}_2] = \begin{bmatrix} 0 & 1 \\ 1 & 0 \end{bmatrix}.$$

The matrix Σ is diagonal of the same size as A with diagonal entries $\sqrt{\sigma_1}$ and 0:

$$\Sigma = \begin{bmatrix} \sqrt{\sigma_1} & 0 \\ 0 & 0 \end{bmatrix} = \begin{bmatrix} 1 & 0 \\ 0 & 0 \end{bmatrix}.$$

The matrix U should be a 2×2 matrix with columns

$$\begin{aligned} \mathbf{u}_1 &= \frac{A\mathbf{v}_1}{\|A\mathbf{v}_1\|} = (1,0), \\ \mathbf{u}_2 &= (0,1), \end{aligned}$$

where $\mathbf{u}_2$ can be any unit vector in the null space $\mathcal{N}(A^T)$. Therefore,

$$U = [\mathbf{u}_1 \ \mathbf{u}_2] = \begin{bmatrix} 1 & 0 \\ 0 & 1 \end{bmatrix}.$$

Now, V, U are orthogonal matrices and

$$A = U\Sigma V^T = \begin{bmatrix} 1 & 0 \\ 0 & 1 \end{bmatrix} \begin{bmatrix} 1 & 0 \\ 0 & 0 \end{bmatrix} \begin{bmatrix} 0 & 1 \\ 1 & 0 \end{bmatrix}^T$$

is an SVD of A.

Note that this is NOT an EVD of A.

Clearly, an SVD of A is not unique, because it depends on the choice of vectors $\mathbf{v}_1$ and $\mathbf{v}_2$. □

Example 142 (*SVD $A = U\Sigma V^T$ of a wide matrix A*) Find an SVD of a wide matrix

$$A = \begin{bmatrix} 1 & 0 & 1 \\ 0 & 1 & 1 \end{bmatrix}.$$

Solution: To find a matrix V first, note that

$$A^T A = \begin{bmatrix} 1 & 0 \\ 0 & 1 \\ 1 & 1 \end{bmatrix} \begin{bmatrix} 1 & 0 & 1 \\ 0 & 1 & 1 \end{bmatrix} = \begin{bmatrix} 1 & 0 & 1 \\ 0 & 1 & 1 \\ 1 & 1 & 2 \end{bmatrix}.$$

Its eigenvalues are

$$\sigma_1 = 3, \quad \sigma_2 = 1, \quad \sigma_3 = 0$$

in descending order (so rank $A = 2$), with associated *unit* eigenvectors

$$\begin{aligned} \mathbf{v}_1 &= \pm\left(\frac{1}{\sqrt{6}}, \frac{1}{\sqrt{6}}, \frac{2}{\sqrt{6}}\right), \\ \mathbf{v}_2 &= \pm\left(-\frac{1}{\sqrt{2}}, \frac{1}{\sqrt{2}}, 0\right), \\ \mathbf{v}_3 &= \pm\left(-\frac{1}{\sqrt{3}}, -\frac{1}{\sqrt{3}}, \frac{1}{\sqrt{3}}\right), \end{aligned}$$

respectively, where $\mathbf{v}_3$ can be any unit vector in the null space $\mathcal{N}(A)$. By choosing $\mathbf{v}_i$, one can construct an orthogonal matrix V, say

$$V = [\mathbf{v}_1\ \mathbf{v}_2\ \mathbf{v}_3] = \begin{bmatrix} \frac{1}{\sqrt{6}} & -\frac{1}{\sqrt{2}} & -\frac{1}{\sqrt{3}} \\ \frac{1}{\sqrt{6}} & \frac{1}{\sqrt{2}} & -\frac{1}{\sqrt{3}} \\ \frac{2}{\sqrt{6}} & 0 & \frac{1}{\sqrt{3}} \end{bmatrix}.$$

The matrix Σ is diagonal of the same size as A with diagonal entries $\sqrt{\sigma_1}$ and $\sqrt{\sigma_1}$:

$$\Sigma = \begin{bmatrix} \sqrt{\sigma_1} & 0 & 0 \\ 0 & \sqrt{\sigma_2} & 0 \end{bmatrix} = \begin{bmatrix} \sqrt{3} & 0 & 0 \\ 0 & 1 & 0 \end{bmatrix}.$$

The matrix U should be a 2×2 matrix with columns

$$\begin{aligned} \mathbf{u}_1 &= \frac{A\mathbf{v}_1}{\|A\mathbf{v}_1\|} = \frac{A\mathbf{v}_1}{\sigma_1} = \left(\frac{1}{\sqrt{2}}, \frac{1}{\sqrt{2}}\right), \\ \mathbf{u}_2 &= \frac{A\mathbf{v}_2}{\|A\mathbf{v}_2\|} = \frac{A\mathbf{v}_2}{\sigma_2} = \left(-\frac{1}{\sqrt{2}}, \frac{1}{\sqrt{2}}\right). \end{aligned}$$

That is,

$$U = [\mathbf{u}_1\ \mathbf{u}_2] = \begin{bmatrix} \frac{1}{\sqrt{2}} & -\frac{1}{\sqrt{2}} \\ \frac{1}{\sqrt{2}} & \frac{1}{\sqrt{2}} \end{bmatrix}.$$

Now, V, U are orthogonal matrices and

$$A = U\Sigma V^T = \begin{bmatrix} \frac{1}{\sqrt{2}} & -\frac{1}{\sqrt{2}} \\ \frac{1}{\sqrt{2}} & \frac{1}{\sqrt{2}} \end{bmatrix} \begin{bmatrix} \sqrt{3} & 0 & 0 \\ 0 & 1 & 0 \end{bmatrix} \begin{bmatrix} \frac{1}{\sqrt{6}} & -\frac{1}{\sqrt{2}} & -\frac{1}{\sqrt{3}} \\ \frac{1}{\sqrt{6}} & \frac{1}{\sqrt{2}} & -\frac{1}{\sqrt{3}} \\ \frac{2}{\sqrt{6}} & 0 & \frac{1}{\sqrt{3}} \end{bmatrix}^T$$

is an SVD of A. □

In an SVD $A = U\Sigma V^T$ of an $m \times n$ matrix A, the matrix A can be considered as a linear transformation $\mathbb{R}^n \to \mathbb{R}^m; \mathbf{x} \mapsto A\mathbf{x}$. In this case, the orthogonal matrix V maps the standard orthonormal basic vectors $\mathbf{e}_1, \mathbf{e}_2, \ldots, \mathbf{e}_n$ to the column vectors of V $\mathbf{v}_1, \mathbf{v}_2, \ldots, \mathbf{v}_n$, respectively which form a new orthonormal basis for the domain space $\mathbb{R}^n$. It implies that V is a rotation or a reflection of the Euclidean space $\mathbb{R}^n$, see Example 96. The same is also true for the orthogonal matrix U in the target space $\mathbb{R}^m$.

What is the rectangular diagonal matrix Σ as a linear transformation?

For example,

$$\Sigma = \begin{bmatrix} 2 & 0 & 0 & 0 \\ 0 & 1 & 0 & 0 \\ 0 & 0 & 1/2 & 0 \end{bmatrix}$$

can be considered as a linear transformation $\mathbf{x} \mapsto \Sigma\mathbf{x}$ from the space $\mathbb{R}^4$ to the space $\mathbb{R}^3$. The matrix Σ maps the standard basic vectors $\mathbf{e}_1, \mathbf{e}_2, \mathbf{e}_3, \mathbf{e}_4$ in $\mathbb{R}^4$ to the column vectors $(2,0,0)$, $(0,1,0)$, $(0,0,1/2)$, $(0,0,0)$ of Σ in $\mathbb{R}^3$, respectively. It shows that the matrix Σ stretches the x-coordinate by multiple 2, the y-coordinate by multiple 1, shrinks the z-coordinate by multiple $1/2$, and shrinks the last coordinate to the origin point of the target space $\mathbb{R}^3$. In this sense, Σ is a *scaling (shrinking or stretching) matrix.*

Example 143 (*SVD $A = U\Sigma V^T$ of a long matrix A and its geometrical meaning*) Find an SVD of the long matrix

$$A = \begin{bmatrix} 1 & 0 \\ 0 & 1 \\ 0 & -1 \end{bmatrix}$$

and state its geometrical meaning. Here the matrix A is considered as a linear transformation $\mathbf{x} \mapsto A\mathbf{x}$ from the xy-plane $\mathbb{R}^2$ to the 3-space $\mathbb{R}^3$.

Solution:

$$A^T A = \begin{bmatrix} 1 & 0 & 0 \\ 0 & 1 & -1 \end{bmatrix} \begin{bmatrix} 1 & 0 \\ 0 & 1 \\ 0 & -1 \end{bmatrix} = \begin{bmatrix} 1 & 0 \\ 0 & 2 \end{bmatrix}.$$

Its eigenvalues are

$$\sigma_1 = 2, \quad \sigma_2 = 1$$

in descending order (so rank $A = 2$), with associated *unit* eigenvectors

$$\begin{aligned} \mathbf{v}_1 &= \pm(0,1), \\ \mathbf{v}_2 &= \pm(1,0), \end{aligned}$$

respectively. By choosing $\mathbf{v}_i$, one can construct an orthogonal matrix V, say

$$V = [\mathbf{v}_1\ \mathbf{v}_2] = \begin{bmatrix} 0 & 1 \\ 1 & 0 \end{bmatrix}.$$

Geometrically, this orthogonal matrix V maps the standard orthonormal basic vectors $\mathbf{e}_1, \mathbf{e}_2$ of the plane $\mathbb{R}^2$ to the column vectors $\mathbf{v}_1 = \mathbf{e}_2, \mathbf{v}_2 = \mathbf{e}_1$ of V respectively, which form a new orthonormal basis for the plane $\mathbb{R}^2$. That is, V is an orthogonal matrix which maps the standard xy-rectangular coordinate system to a $\mathbf{v}_1\mathbf{v}_2$-rectangular coordinate system. In fact, V is a reflection of the xy-plane across the line $y = x$,

The matrix Σ is diagonal of the same size as A with diagonal entries $\sqrt{\sigma_1}$ and $\sqrt{\sigma_2}$:

$$\Sigma = \begin{bmatrix} \sqrt{\sigma_1} & 0 \\ 0 & \sqrt{\sigma_2} \\ 0 & 0 \end{bmatrix} = \begin{bmatrix} \sqrt{2} & 0 \\ 0 & 1 \\ 0 & 0 \end{bmatrix}.$$

This 3×2 matrix Σ stretches the x-coordinate by multiple $\sqrt{2}$ and the y-coordinate by multiple 1 so that its image is the xy-plane of the target space $\mathbb{R}^3$.

The matrix U should be a 3×3 matrix with columns

$$\begin{aligned} \mathbf{u}_1 &= \frac{A\mathbf{v}_1}{\|A\mathbf{v}_1\|} = \frac{A\mathbf{v}_1}{\sigma_1} = \left(0, \frac{1}{\sqrt{2}}, -\frac{1}{\sqrt{2}}\right), \\ \mathbf{u}_2 &= \frac{A\mathbf{v}_2}{\|A\mathbf{v}_2\|} = \frac{A\mathbf{v}_2}{\sigma_2} = (1,0,0), \\ \mathbf{u}_3 &= \left(0, \frac{1}{\sqrt{2}}, \frac{1}{\sqrt{2}}\right), \end{aligned}$$

where $\mathbf{u}_3$ can be any unit vector in the null space $\mathcal{N}(A^T)$. Therefore,

$$U = [\mathbf{u}_1\ \mathbf{u}_2\ \mathbf{u}_3] = \begin{bmatrix} 0 & 1 & 0 \\ \frac{1}{\sqrt{2}} & 0 & \frac{1}{\sqrt{2}} \\ -\frac{1}{\sqrt{2}} & 0 & \frac{1}{\sqrt{2}} \end{bmatrix}.$$

Geometrically, U is the composition $R_{-\pi/4} \circ T$ of a rotation $R_{-\pi/4}$ of the yz-plane around the x-axis by $-\pi/4$ in the target space $\mathbb{R}^3$ followed by a reflection T in the space $\mathbb{R}^3$ through the plane $y = x$.

One can check the identity $A = U\Sigma V^T$ by a direct multiplication of these three matrices. The matrices U and V are orthogonal, so each is a rotation or a reflection as a linear transformation, and Σ is rectangular diagonal, so is a stretching/shrinking along each coordinate axis.

It is shown that the matrix $A = U\Sigma V^T$ is a composition of the type "Rotation (or reflection) $\circ$ Stretching $\circ$ Rotation (or reflection)". □

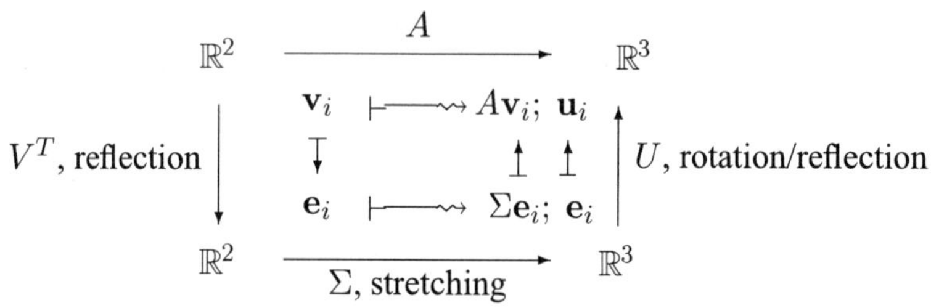

Figure 7.1: SVD, $A = U\Sigma V^T$

An SVD $A = U\Sigma V^T$ of a matrix A is extremely useful in

- determining the rank and finding bases for row space, null space, column space of A,
- finding a one-sided (left or right) inverse or the two-sided inverse of a matrix A if there exists, (see the next section),
- solving a normal equation $A^TA\mathbf{x} = A^T\mathbf{b}$ in least squares fitting, etc by using $A^TA = (U\Sigma V^T)^T U\Sigma V^T = V\Sigma^T\Sigma V^T$.

Problem 187 In an SVD $A = U\Sigma V^T$, show that

(1) if A is square, then A has a zero singular value if and only if A is singular (not invertible),

(2) the number of nonzero singular values counting multiplicities equals the rank of A.

Problem 188 Find an SVD of $A = \begin{bmatrix} 0 & 1 & -1 \\ 1 & 1 & 0 \end{bmatrix}$. By using it, find bases for the row, column, and null spaces of the matrix A.

Problem 189 Using an SVD $A = U\Sigma V^T$ of a matrix A, diagonalize symmetric matrices A^TA and AA^T. From your diagonalizations, show that two matrices A^TA and AA^T have the same nonzero eigenvalues.

Hint:

$$A^TA = V\Sigma^T U^T U\Sigma V^T = V(\Sigma^T\Sigma)V^T$$

$$AA^T = U\Sigma V^T V\Sigma^T U^T = U(\Sigma\Sigma^T)U^T.$$

Problem 190 (Revisit Example 143) With notations in Example 143, show that

(1) $U^TA = \Sigma V^T$ maps the unit circle in $\mathbb{R}^2$ to an ellipse centered at the origin $\mathbf{0}$ in the xy-plane of $\mathbb{R}^3$ with a horizontal major axis and the vertical minor axis; for basic vectors $\mathbf{e}_2 \mapsto \mathbf{e}_1 \mapsto \sqrt{2}\mathbf{e}_1$, and $\mathbf{e}_1 \mapsto \mathbf{e}_2 \mapsto \mathbf{e}_2$.

(2) The matrix U maps $\sqrt{2}\mathbf{e}_1$ and $\mathbf{e}_2$ to $A\mathbf{v}_1$ and $A\mathbf{v}_2$, respectively;

$$\sqrt{2}\mathbf{e}_1 \mapsto A\mathbf{v_1} = A\mathbf{e}_2 = \begin{bmatrix} 0 \\ 1 \\ -1 \end{bmatrix}, \qquad \mathbf{e}_2 \mapsto A\mathbf{v_2} = A\mathbf{e}_1 = \begin{bmatrix} 1 \\ 0 \\ 0 \end{bmatrix},$$

so that it maps the ellipse in the xy-plane to another ellipse in the column space $\mathcal{C}(A)$.

(3) The matrix U is the composition $R_{-\pi/4} \circ T$ of a rotation $R_{-\pi/4}$ of the yz-plane around the x-axis by $-\pi/4$ in the target space $\mathbb{R}^3$ followed by a reflection T in the space $\mathbb{R}^3$ through the plane $y = x$.

7.8.3 How to find a one-sided (left or right) inverse of a matrix

As an application of an SVD, one can use it to find a one-sided inverse (left or right inverse) of a matrix A if there exists. Recall that an $m \times n$ matrix A has a one-sided inverse if and only if it has full rank, that is, rank $A = \min\{m, n\}$, (see Subsection 3.9.1).

First, let's consider any rectangular diagonal matrix of full rank.

(1) If $A = \Sigma$ is an $m \times n$ rectangular diagonal matrix with nonzero diagonal entries, then one can easily find a one-sided inverse of Σ. As an example, if

$$\Sigma = \begin{bmatrix} 1 & 0 & 0 \\ 0 & 2 & 0 \\ 0 & 0 & 3 \\ 0 & 0 & 0 \end{bmatrix}$$

which is long of full rank, then Σ has a left inverse, but no right inverses. Its left inverse is of the form (in fact, infinitely many)

$$\Sigma^{\text{left}} = \begin{bmatrix} 1 & 0 & 0 & x \\ 0 & 1/2 & 0 & y \\ 0 & 0 & 1/3 & z \end{bmatrix}$$

with any x, y, z.

Similarly, if

$$\Sigma = \begin{bmatrix} 1 & 0 & 0 & 0 \\ 0 & 2 & 0 & 0 \\ 0 & 0 & 3 & 0 \end{bmatrix}$$

which is wide of full rank, then Σ has a right inverse, but no left inverses. Its right inverse is of the form (also infinitely many)

$$\Sigma^{\text{right}} = \begin{bmatrix} 1 & 0 & 0 \\ 0 & 1/2 & 0 \\ 0 & 0 & 1/3 \\ x & y & z \end{bmatrix}.$$

Certainly, an $m \times n$ rectangular diagonal matrix cannot have any one-sided inverse if it has a zero diagonal entry (so not of full rank).

(2) Now, let A be any square or rectangular matrix of full rank.

If $A = U\Sigma V^T$ is a square matrix of full rank, then A is invertible and so Σ is also an invertible square matrix. Hence $A^{-1} = V\Sigma^{-1}U^T$.

Let A be any rectangular matrix of full rank. In Theorems 3.27 and 3.28 (on page 140), an $m \times n$ matrix A has a right inverse if rank $A = m$, while it has a left inverse if rank $A = n$.

First, let rank $A = m$. Then $A = U\Sigma V^H$ is a wide matrix. Hence Σ is also an $m \times n$ wide matrix of the type

$$\Sigma = \begin{bmatrix} \sigma_1 & 0 & 0 & 0 & 0 & \cdots & 0 \\ 0 & \sigma_2 & 0 & 0 & 0 & \cdots & 0 \\ \vdots & 0 & \ddots & 0 & 0 & \cdots & 0 \\ 0 & 0 & 0 & \sigma_m & 0 & \cdots & 0 \end{bmatrix}$$

with $\sigma_1 \geq \cdots \geq \sigma_m > 0$. It has a right inverse Σ^{right} by the case (1) mentioned already. Since U and V are unitary,

$$V\Sigma^{\text{right}}U^H \text{ is a right inverse of } A = U\Sigma V^H$$

for any right inverse Σ^{right} of Σ.

Similarly, if $A = U\Sigma V^H$ has rank n, then Σ should be a long and rectangular diagonal matrix with positive diagonal entries. Hence, Σ has a left inverse Σ^{left} and

$$V\Sigma^{\text{left}}U^H \text{ is a left inverse of } A = U\Sigma V^H$$

for any left inverse Σ^{left} of Σ.

Example 144 (*One-sided inverse by an SVD*) As shown in Example 142, the 2×3 matrix

$$A = \begin{bmatrix} 1 & 0 & 1 \\ 0 & 1 & 1 \end{bmatrix}$$

has an SVD

$$A = U\Sigma V^T = \begin{bmatrix} \frac{1}{\sqrt{2}} & -\frac{1}{\sqrt{2}} \\ \frac{1}{\sqrt{2}} & \frac{1}{\sqrt{2}} \end{bmatrix} \begin{bmatrix} \sqrt{3} & 0 & 0 \\ 0 & 1 & 0 \end{bmatrix} \begin{bmatrix} \frac{1}{\sqrt{6}} & -\frac{1}{\sqrt{2}} & -\frac{1}{\sqrt{3}} \\ \frac{1}{\sqrt{6}} & \frac{1}{\sqrt{2}} & -\frac{1}{\sqrt{3}} \\ \frac{2}{\sqrt{6}} & 0 & \frac{1}{\sqrt{3}} \end{bmatrix}^T$$

Hence, the matrix A has a right inverse

$$\begin{aligned} V\Sigma^{\text{right}}U^T &= \begin{bmatrix} \frac{1}{\sqrt{6}} & -\frac{1}{\sqrt{2}} & -\frac{1}{\sqrt{3}} \\ \frac{1}{\sqrt{6}} & \frac{1}{\sqrt{2}} & -\frac{1}{\sqrt{3}} \\ \frac{2}{\sqrt{6}} & 0 & \frac{1}{\sqrt{3}} \end{bmatrix} \begin{bmatrix} \frac{1}{\sqrt{3}} & 0 \\ 0 & 1 \\ 0 & 0 \end{bmatrix} \begin{bmatrix} \frac{1}{\sqrt{2}} & -\frac{1}{\sqrt{2}} \\ \frac{1}{\sqrt{2}} & \frac{1}{\sqrt{2}} \end{bmatrix}^T \\ &= \frac{1}{3}\begin{bmatrix} 2 & -1 \\ -1 & 2 \\ 1 & 1 \end{bmatrix}. \end{aligned}$$

One can show that for any $\Sigma^{\text{right}} = \begin{bmatrix} \frac{1}{\sqrt{3}} & 0 \\ 0 & 1 \\ x & y \end{bmatrix}$ with arbitrary x, y, the product $V\Sigma^{\text{right}}U^T$ is also a right inverse of A. □

Problem 191 Find a one-sided inverse of $A = \begin{bmatrix} 1 & 0 & 1 & 0 \\ 0 & 1 & 0 & 1 \\ -1 & 1 & 1 & -1 \end{bmatrix}$. Try to find one-sided or two-sided inverses of A^TA and AA^T.

Hint: AA^T has the two-sided inverse, but A^TA cannot have any one-sided inverse.

Problem 192 Given an $m \times n$ matrix A, an $n \times m$ matrix B is called a **generalized inverse** of A if B satisfies $ABA = A$. Show that

(1) an inverse or a one-sided inverse of A is a generalized inverse of A. Hence, a generalized inverse might not be unique.

(2) *Every (square or rectangular) matrix A has at least one generalized inverse.* Prove it by using an SVD of A.

Hint for (2). Let $A = U\Sigma V^H$ be an SVD with $\Sigma = \begin{bmatrix} D_r & \mathbf{0} \\ \mathbf{0} & \mathbf{0} \end{bmatrix}$, where D_r is an $r \times r$ diagonal matrix with nonzero diagonal entries σ_i and $r = \operatorname{rank} A$. Then the matrix $V \begin{bmatrix} D_r^{-1} & \mathbf{0} \\ \mathbf{0} & \mathbf{0} \end{bmatrix} U^H$ is a generalized inverse of A.

7.9 Exercises

7.1. *By allowing complex scalars, a complex vector space can be defined.* Solve the system of linear equations

$$(1) \begin{cases} x + iy = 0 \\ ix + 2y = 3i, \end{cases} \qquad (2) \begin{cases} (1-i)x + (1+i)y = 2-i \\ (1+i)x + (1+i)y = 1+3i. \end{cases}$$

7.2. *A complex inner product is positive-definite, Hermitian, and sesquilinear.* Decide which of the following functions defined on the complex 2-space $\mathbb{C}^2$ are complex inner products and which are not. For any $\mathbf{z} = (z_1, z_2)$ and $\mathbf{w} = (w_1, w_2)$ in $\mathbb{C}^2$,

(1) $\langle \mathbf{z}, \mathbf{w} \rangle = z_1w_1 + z_2w_2$,

(2) $\langle \mathbf{z}, \mathbf{w} \rangle = \overline{z_1w_1} + 2z_2w_2$,

(3) $\langle \mathbf{z}, \mathbf{w} \rangle = \overline{z_1}w_1 + 2\overline{z_2}w_2$,

(4) $\langle \mathbf{z}, \mathbf{w} \rangle = \overline{z_1}w_1 - \overline{z_1}w_2 - \overline{z_2}w_1 + 3\overline{z_2}w_2$.

7.3. *A complex inner product space is a complex vector space with a Hermitian inner product.* In the complex Euclidean space $\mathbb{C}^2$ or $\mathbb{C}^3$, calculate $||\mathbf{x}||$ for

$$(1)\ \mathbf{x} = \begin{bmatrix} 1 \\ i \end{bmatrix}, \qquad (2)\ \mathbf{x} = \begin{bmatrix} 1-i \\ 2 \end{bmatrix}, \qquad (3)\ \mathbf{x} = \begin{bmatrix} 2i \\ 1+i \\ 3+i \end{bmatrix}.$$

7.4. Determine the orthogonal projection of $\mathbf{v}_1$ onto $\mathbf{v}_2$ for the following vectors in the complex 3-space $\mathbb{C}^3$ (with the dot product).

(1) $\mathbf{v}_1 = (1, i, 0),\ \mathbf{v}_2 = (1, 1, 2)$,
(2) $\mathbf{v}_1 = (i, i, 1),\ \mathbf{v}_2 = (1, 1, i)$

Hint: With $\mathbf{u}_2 = \mathbf{v}_2/||\mathbf{v}_2||,\ \mathrm{Proj}_{\langle \mathbf{v}_2 \rangle}(\mathbf{v}_1) = (\mathbf{u_2} \cdot \mathbf{v}_1)\mathbf{u_2} = (\mathbf{u}_2{\mathbf{u}_2}^H)\mathbf{v}_1$.

7.5. *A complex inner product space is a complex vector space with a Hermitian inner product.* Construct an orthonormal basis for the Euclidean 2-space $\mathbb{C}^2$ from $\{(i, 2i),\ (1+i, 1)\}$ by the Gram-Schmidt orthogonalization.

7.6. Find a reduced row-echelon form, a basis for the column space and the rank of the matrix

$$(1)\ \begin{bmatrix} i & 1 & 1-i \\ 1-i & 1+i & 1 \end{bmatrix}, \qquad (2)\ \begin{bmatrix} i & 1 & 1-i \\ -i & 1+i & 1+i \\ 3i & 1-i & 2-i \end{bmatrix}.$$

7.7. Diagonalize the (non-symmetric) complex matrix $A = \begin{bmatrix} i & 1 & 0 \\ 1 & 0 & 0 \\ 0 & 0 & 1 \end{bmatrix}$.

7.8. Find an example of real square matrix A such that $A + rI$ is invertible for all real $r \in \mathbb{R}$. Is there a complex square matrix A such that $A + cI$ is invertible for all complex $c \in \mathbb{C}$?

Hint: $\det(A + rI) = 0 \Leftrightarrow -r$ is an eigenvalue of A. Every skew-symmetric matrix cannot have a nonzero real eigenvalue.

7.9. Let $\mathbf{x}$ be any nonzero vector in the n-space $\mathbb{C}^n$ and let $A = \mathbf{x}\mathbf{x}^H$. Show that A is Hermitian, and find the eigenvalues and their eigenspaces for A.

7.10. For a skew-Hermitian matrix A, show that

(1) both $A - I$ and $A + I$ are invertible, (2) e^A is unitary.

7.11. Construct a unitary matrix whose first row is (1) $k(1, 1-i)$ where k is a scalar, or (2) $(1, i, 1-i)\ /||\ (1, i, 1-i)\ ||$.

7.12. Recall that the *complex exponential* $e^{i\theta}$ *(θ real) is* $e^{i\theta} = \cos\theta + i \sin\theta$.

(1) Show that as the 2×2 complex matrix, $\begin{bmatrix} \cos\theta & -\sin\theta \\ \sin\theta & \cos\theta \end{bmatrix}$, (where θ is real) has two (complex) eigenvalues $\lambda_1 = e^{i\theta}$ and $\lambda_2 = e^{-i\theta}$ with associated eigenvectors $\mathbf{v}_1 = (i, 1)$ and $\mathbf{v}_2 = (-i, 1)$, respectively.

(2) Show that the matrix $\begin{bmatrix} e^{i\theta} & 0 \\ 0 & e^{-i\theta} \end{bmatrix}$ has (the same complex) eigenvalues $\lambda_1 = e^{i\theta}$ and $\lambda_2 = e^{-i\theta}$ with associated eigenvectors $\mathbf{e}_1$ and $\mathbf{e}_2$, respectively.

(3) Show that matrices $\begin{bmatrix} \cos\theta & -\sin\theta \\ \sin\theta & \cos\theta \end{bmatrix}$ and $\begin{bmatrix} e^{i\theta} & 0 \\ 0 & e^{-i\theta} \end{bmatrix}$ are unitarily similar.

Hint for (3): Both are unitarily diagonalizable.

7.13. Find the characteristic polynomial and all (real and complex) eigenvalues of each of the following matrices.

$$A = \begin{bmatrix} -2 & 0 & 0 \\ 3 & 2 & 3 \\ 4 & -1 & 6 \end{bmatrix}, \qquad B = \begin{bmatrix} -2 & 0 & 0 \\ 3 & 3 & 2 \\ 4 & 6 & -1 \end{bmatrix}, \qquad C = \begin{bmatrix} 0 & -2 & 0 \\ 3 & 3 & 2 \\ 6 & 4 & -1 \end{bmatrix}.$$

Here, B is obtained from A by interchanging the second and the third columns; C is obtained from A by a cyclic permutation of the columns. Your solutions will show that the eigenvalues or the eigenvectors of a matrix cannot be preserved by elementary row or column operations.

Hint: For each of three matrices, 3 or -3 is one of the eigenvalues.

7.14. For each of the following Hermitian matrices A, find the eigenvalues and eigenvectors of A. Also, find a unitary matrix U such that $U^H AU$ is diagonal.

$$(1) \begin{bmatrix} 0 & i \\ -i & 0 \end{bmatrix}, \quad (2) \begin{bmatrix} 1 & i \\ -i & 2 \end{bmatrix}, \quad (3) \begin{bmatrix} 1 & 2+3i \\ 2-3i & -1 \end{bmatrix}.$$

7.15. For each of the following skew-Hermitian matrices A, find the eigenvalues and eigenvectors. Also, find a unitary matrix U such that $U^H AU$ is diagonal.

$$(1) \begin{bmatrix} 0 & 1 & 2 \\ -1 & 0 & 3 \\ -2 & -3 & 0 \end{bmatrix}, \quad (2) \begin{bmatrix} 0 & i & i \\ i & 0 & i \\ i & i & 0 \end{bmatrix}, \quad (3) \begin{bmatrix} 0 & 1 & i \\ -1 & 0 & 0 \\ i & 0 & 0 \end{bmatrix}.$$

7.16. For each of the following unitary matrices A, find the eigenvalues and eigenvectors. Also, find a unitary matrix U such that $U^H AU$ is diagonal.

$$(1) \begin{bmatrix} 0 & 1 & 0 \\ 0 & 0 & 1 \\ 1 & 0 & 0 \end{bmatrix}, \qquad (2) \begin{bmatrix} 0 & i & 0 \\ 0 & 0 & i \\ 1 & 0 & 0 \end{bmatrix}, \qquad (3) \begin{bmatrix} \cos\theta & 0 & -\sin\theta \\ 0 & i & 0 \\ \sin\theta & 0 & \cos\theta \end{bmatrix}.$$

7.17. Determine whether each of the following matrices is Hermitian, skew-Hermitian, or unitary. For each matrix A, find a diagonal matrix which is unitarily similar to A.

$$(1)\ A = \frac{1}{2}\begin{bmatrix} 1+i & 1-i \\ 1-i & 1+i \end{bmatrix}, \quad (2)\ A = \begin{bmatrix} 0.6 & -0.8 \\ 0.8 & 0.6 \end{bmatrix}, \quad (3)\ A = \begin{bmatrix} 1 & i & 0 \\ -i & 1 & i \\ 0 & -i & 1 \end{bmatrix}.$$

7.18. Verify that $A = \begin{bmatrix} 2 & i \\ i & 2 \end{bmatrix}$ is normal. Diagonalize the matrix A by a unitary matrix U.

7.19. Verify that $A = \begin{bmatrix} 1 & i \\ 1 & 1 \end{bmatrix}$ is normal. Diagonalize A by a unitary matrix U. Also, diagonalize AA^H and $A^H A$ by unitary matrices. Try to do the same for $AA^H A$.

Hint: To find the eigenvalues of A, use the fact: The solutions to the complex quadratic equation $z^2 = i$ are $z = \pm(\frac{1}{\sqrt{2}} + \frac{1}{\sqrt{2}}i)$ because $i = \cos(\pi/2) + i\sin(\pi/2) = e^{i(\pi/2)}$ and $e^{i(\pi/4)} = \frac{1}{\sqrt{2}} + \frac{1}{\sqrt{2}}i$. For the complex exponential function $e^{ix} = \cos x + i\sin x$, see a footnote on page 85.

7.20. Find an orthogonal matrix Q and a diagonal matrix D such that $Q^TAQ = D$ for each of the following symmetric matrices.

$$(1)\ A = \begin{bmatrix} 1 & 0 & 0 \\ 0 & 1 & 1 \\ 0 & 1 & 1 \end{bmatrix}, \qquad (2)\ A = \begin{bmatrix} 1 & 2 & 0 \\ 2 & 2 & 2 \\ 0 & 2 & 3 \end{bmatrix}, \qquad (3)\ A = \begin{bmatrix} -3 & 2 & 4 \\ 2 & -6 & 2 \\ 4 & 2 & -3 \end{bmatrix}.$$

Hint: See Exercise **6.13** on page 304.

7.21. Let A and B be 2×2 symmetric matrices. Show that A and B are similar if and only if $\det A = \det B$ and $\mathrm{tr}(A) = \mathrm{tr}(B)$.

7.22. Show that a matrix A is nilpotent, *i.e.*, $A^n = \mathbf{0}$ for some integer $n \geq 1$, if and only if its eigenvalues are all zero.

Hint: When A is upper triangular, is it clear?

7.23. Find an EVD of a symmetric matrix $A = \begin{bmatrix} 2 & 1 & 1 \\ 1 & 2 & 1 \\ 1 & 1 & 2 \end{bmatrix}$ and use it to compute A^n.

7.24. Find a general solution to each of the following LRRs.

$$\begin{aligned}
&(1)\ x_n = x_{n-1} - 2x_{n-2}, \quad n \geq 2, \\
&(2)\ x_n = x_{n-1} + 4x_{n-2} + 6x_{n-3}, \quad n \geq 3, \\
&(3)\ x_n = 3x_{n-1} - 4x_{n-2} + 2x_{n-3}, \quad n \geq 3.
\end{aligned}$$

Hint: With the companion matrix A of each case, $\lambda = 3$ and $\lambda = 1$ are eigenvalues of A of the cases (2) and (3), respectively.

7.25. Solve the LDE $\mathbf{y}' = A\mathbf{y}$, where

$$(1)\ A = \begin{bmatrix} 1 & -1 \\ 2 & 3 \end{bmatrix},\ \mathbf{y}(0) = \begin{bmatrix} 1 \\ 0 \end{bmatrix}, \quad (2)\ A = \begin{bmatrix} 1 & 1 & 0 \\ 0 & 1 & 1 \\ 1 & 0 & 1 \end{bmatrix},\ \mathbf{y}(0) = \begin{bmatrix} 1 \\ 0 \\ 1 \end{bmatrix}.$$

Hint: See Theorem 6.20.

7.26. Find an LDU decomposition and an SVD of each of complex matrices

$$(1)\ A = \begin{bmatrix} 1 & 0 & i \\ i & 0 & 1 \end{bmatrix}, \qquad (2)\ A = \begin{bmatrix} 1 & 0 & 1 & 0 \\ 0 & 1 & 0 & 1 \\ -1 & 1 & 1 & -1 \end{bmatrix}.$$

7.27. For the matrix A given in Exercise **7.26**, determine its one-sided inverse if there exists.

7.28. For each of the following matrices, determine its one-sided inverse if there exists.

$$(1) \begin{bmatrix} 2 & 0 \\ 0 & i \\ 0 & 0 \end{bmatrix}, \quad (2) \begin{bmatrix} i & 0 & 0 \\ 0 & i & 0 \end{bmatrix}, \quad (3)\ A = U\Sigma V^H \text{ with } \Sigma = \begin{bmatrix} 1+i & 0 \\ 0 & 1-i \\ 0 & 0 \end{bmatrix}.$$

Hint for (3): $V\Sigma^{\text{left}}U^H$ is a left inverse of A.

7.29. If two $n \times n$ (real or complex) diagonalizable matrices A and B have the same set of eigenvectors, or equivalently, there exists an invertible matrix Q such that both $Q^{-1}AQ$ and $Q^{-1}BQ$ are diagonal, then the matrices A and B are said to be **simultaneously diagonalizable**. Show that A and B are simultaneously diagonalizable if and only if $AB = BA$.

7.30. Determine whether the following statements are true or false, in general, and justify your answers.

(1) On any complex vector space, one can define at least one complex inner product.
(2) The Pythagorean theorem holds in any complex inner product space.
(3) Every Hermitian matrix is unitarily similar to a diagonal matrix.
(4) An orthogonal matrix is unitarily similar to a real diagonal matrix.
(5) For any square matrix A, AA^H and A^HA have the same eigenvalues.
(6) If a triangular matrix is similar to a diagonal matrix, it is already diagonal.
(7) If all of the columns of a square matrix A are orthonormal, then A is diagonalizable.
(8) Every permutation matrix is diagonalizable.
(9) Every permutation matrix is Hermitian.
(10) For any unit vector $\mathbf{u}$ in $\mathbb{C}^m$, $\mathbf{u}\mathbf{u}^H$ is Hermitian and idempotent.
(11) A nonzero nilpotent matrix cannot be Hermitian.
(12) Every square matrix is similar to a triangular matrix.
(13) If A is a Hermitian matrix, then $A + iI$ is invertible.
(14) If A is a real matrix, then $A + iI$ is invertible.
(15) If A is an orthogonal matrix, then $A + \frac{1}{2}I$ is invertible.
(16) Every unitarily diagonalizable matrix with real eigenvalues is Hermitian.
(17) Every diagonalizable matrix is normal.
(18) Any two unitary matrices are similar to each other.
(19) If A is a normal matrix, then any two eigenvectors belonging to distinct eigenvalues are mutually orthogonal.
(20) For any complex square matrix A, $\det A = \det A^H$.

Hint: Exactly seven of them are false.

Chapter 8

Jordan Canonical Forms; Block diagonalizations of Matrices

Most problems related to a (real or complex) matrix A can be solved easily if the matrix is diagonal or even diagonalizable. This is true in computing $\det A$, the power A^n, and in solving an LdE $\mathbf{x}_n = A\mathbf{x}_{n-1}$ or an LDE $\mathbf{y}'(t) = A\mathbf{y}(t)$.

In this chapter we discuss how to solve the same problems when the matrix A is not diagonalizable.

Even if a matrix A is not diagonalizable, one may try to find a matrix which is similar to A and has as many zero off-diagonal entries as possible. Schur's triangulation Theorem says that any square matrix is (unitarily) similar to an upper triangular matrix.

But, it is a fact that every square matrix A is similar to a matrix much 'closer' to a diagonal matrix, called a *Jordan canonical matrix*. Its diagonal entries are the eigenvalues of A, each superdiagonal entry (the entry immediately above and to the right of the main diagonal entry) is 0 or 1, and all the other off-diagonal entries are 0.

8.1 Jordan blocks and Jordan matrices

A matrix of the type of

$$J = \begin{bmatrix} \lambda & 1 & & \mathbf{0} \\ & \ddots & \ddots & \\ & & \ddots & 1 \\ \mathbf{0} & & & \lambda \end{bmatrix}$$

is a famous example which is not diagonalizable. All diagonal entries are the same as a single eigenvalue of J. Each superdiagonal entry (the entry immediately above and to the right of the main diagonal entry) is 1, and all the other off-diagonal entries are 0. If the order of J is k, $\operatorname{rank}(J - \lambda I) = k - 1$.

By by the Rank-Nullity theorem, $\dim \mathcal{N}(J - \lambda I) = 1$, and hence J has only one linearly independent eigenvector (belonging to λ).

Such a matrix is called a **Jordan block**, or a **Jordan block** belonging to λ.

A **Jordan matrix** is a block diagonal matrix whose diagonal blocks are all Jordan blocks.

For example,

$$J = \begin{bmatrix} 6 & 1 & 0 & 0 & 0 \\ 0 & 6 & 0 & 0 & 0 \\ 0 & 0 & 2 & 1 & 0 \\ 0 & 0 & 0 & 2 & 0 \\ 0 & 0 & 0 & 0 & 3 \end{bmatrix} = \begin{bmatrix} \begin{bmatrix} 6 & 1 \\ 0 & 6 \end{bmatrix} & & \\ & \begin{bmatrix} 2 & 1 \\ 0 & 2 \end{bmatrix} & \\ & & [3] \end{bmatrix} = \begin{bmatrix} J_1 & & \\ & J_2 & \\ & & J_3 \end{bmatrix}$$

is a Jordan matrix having three Jordan blocks on the diagonal. Clearly, it is not a Jordan block.

Example 145 The following four matrices are upper triangular matrices each of which has the same diagonal entry as its single eigenvalue. Which ones are Jordan blocks? Which ones are Jordan matrices?

$$(1) \begin{bmatrix} 1 & 1 & 0 \\ 0 & 1 & 0 \\ 0 & 0 & 1 \end{bmatrix}, \ (2) \begin{bmatrix} 3 & 1 & 0 \\ 0 & 3 & 1 \\ 0 & 0 & 3 \end{bmatrix}, \ (3) \begin{bmatrix} 2 & 1 & 0 & 0 \\ 0 & 2 & 1 & 0 \\ 0 & 0 & 2 & 1 \\ 0 & 0 & 0 & 2 \end{bmatrix}, \ (4) \begin{bmatrix} 2 & 1 & 0 & 0 \\ 0 & 2 & 0 & 0 \\ 0 & 0 & 2 & 0 \\ 0 & 0 & 0 & 2 \end{bmatrix}.$$

(1) is not a Jordan block because its $(2, 3)$-entry is a superdiagonal one, but it is not 1. (2) and (3) are all Jordan blocks. (4) also is not a Jordan block. However, all of four matrices are Jordan matrices. (1) and (4) have two and three Jordan blocks on the diagonal, respectively.

Problem 193 For each Jordan matrix J listed in Example 145, let its diagonal entries are λ as its single eigenvalue. Find the sequence $\operatorname{rank}(J - \lambda I)^{\ell}$ for $\ell = 1, 2, \ldots$.

Hint: See Exercise **1.30** on page 54.

8.2 Jordan canonical forms and Jordan decompositions

The following theorem shows that *every square matrix is similar to a Jordan matrix.*

Theorem 8.1 *For any square matrix A, A is similar to a matrix J of the following form, called the* **Jordan canonical form** *of A or a* **Jordan canonical matrix**,

$$Q^{-1}AQ = J = \begin{bmatrix} J_1 & & & \mathbf{0} \\ & J_2 & & \\ & & \ddots & \\ \mathbf{0} & & & J_s \end{bmatrix},$$

in which

(1) *s is the number of linearly independent eigenvectors of A, and*
(2) *each J_i on the diagonal of J is a Jordan block.*

A Jordan canonical matrix J is a block diagonal matrix. A factorization $A = QJQ^{-1}$ is called a **Jordan decomposition** or a **Jordan factorization** of A. [1]

The proof of Theorem 8.1 is beyond the scope of this book. Hence, we refer to some advanced algebra books (for example, Artin's Algebra [Artin]) and we are only concerned with how to find the Jordan canonical form J of A and a change of basis matrix Q, and applications of Jordan decomposition in this book.

If an $n \times n$ matrix A has a full set of n linearly independent eigenvectors (that is $s = n$), then there exist n Jordan blocks so that each block should be just a 1×1 matrix.

In this case, the Jordan canonical form J of A is just a diagonal matrix with eigenvalues on the diagonal and a change of basis matrix Q is defined by taking any n linearly independent eigenvectors as its columns.

Hence, a diagonal matrix is a particular case of a Jordan canonical matrix. Moreover, a Jordan block can appear more than once in a Jordan canonical matrix, as for an eigenvalue in a diagonal matrix.

Remark: (1) If a Jordan canonical form J of A is given, one can get another one by permuting the Jordan blocks of J, and this new one is also similar to A, (see

[1] **Marie Ennemond Camille Jordan** (January 5, 1838– January 22, 1922) was a French mathematician. He is remembered now by name in: The Jordan curve theorem, the Jordan canonical form and the Jordan-Holder theorem on composition series of finite groups.

Example 110 as a special case of permuting the eigenvalues). It will be shown in the next section that any two Jordan canonical forms of A cannot be similar except for this possibility.

In other words, any (real or complex) square matrix A is similar to only one Jordan canonical matrix up to permutations of the Jordan blocks. In this sense, it is called *the* Jordan canonical form of a matrix A.

(2) As an alternative way to define the Jordan canonical form of A, one can take the transpose J^T of the Jordan canonical matrix J given in Theorem 8.1. In this case, each Jordan block becomes a lower triangular matrix with a single eigenvalue and with each subdiagonal entry (the entry immediately below and to the left of the main diagonal entry) a value 1.

But this alternative definition does not induce any essential difference from the original one. In fact, every square matrix is similar to its transpose. See Exercise **8.22**.

If J is the Jordan canonical form of a matrix A, then the two matrices A and J have the same eigenvalues and the same number of linearly independent eigenvectors, but not the same set of eigenvectors in general. (Recall that $\mathbf{x}$ is an eigenvector of $J = Q^{-1}AQ$ if and only if $Q\mathbf{x}$ is an eigenvector of A).

The following example illustrates which matrices A can have a given Jordan canonical form J and how the s linearly independent eigenvectors of A correspond to the Jordan blocks in J.

Example 146 (*Jordan blocks $\leftrightarrow$ Eigenvectors in a one-to-one correspondence*) Let J be a Jordan canonical matrix of the form

$$J = \begin{bmatrix} 6 & 1 & 0 & 0 & 0 \\ 0 & 6 & 0 & 0 & 0 \\ 0 & 0 & 2 & 1 & 0 \\ 0 & 0 & 0 & 2 & 0 \\ 0 & 0 & 0 & 0 & 3 \end{bmatrix} = \begin{bmatrix} \begin{bmatrix} 6 & 1 \\ 0 & 6 \end{bmatrix} & & \\ & \begin{bmatrix} 2 & 1 \\ 0 & 2 \end{bmatrix} & \\ & & [3] \end{bmatrix} = \begin{bmatrix} J_1 & & \\ & J_2 & \\ & & J_3 \end{bmatrix}.$$

Find the number of linearly independent eigenvectors of J and determine all matrices whose Jordan canonical forms are the matrix J.

Solution: Each Jordan block should have a single eigenvalue λ as the main diagonal entries and have a value 1 for the superdiagonal entries, with 0 for all the other off-diagonal entries. Hence, the matrix J has three Jordan blocks J_1, J_2, and J_3.

Since J is a triangular matrix, the eigenvalues of J are the diagonal entries $6, 6, 2, 2, 3$.

The eigenspace $E(6)$ has a single linearly independent eigenvector because $\dim E(6) = \dim \mathcal{N}(J - 6I) = 5 - \operatorname{rank}(J - 6I) = 1$, by the Rank-Nullity theorem. In fact, $\mathbf{e}_1 = (1, 0, 0, 0, 0)$ is such a vector, ($J\mathbf{e}_1 = 6\mathbf{e}_1$), and $\lambda = 6$ appears only in a single block J_1.

Similarly, one can see that the eigenspace $E(2)$ has only one linearly independent eigenvector because $\dim E(2) = 5 - \operatorname{rank}(J - 2I) = 1$ and this can be chosen to be $\mathbf{e}_3 = (0, 0, 1, 0, 0)$.

Finally, the eigenspace $E(3)$ has only one linearly independent eigenvector $\mathbf{e}_5 = (0, 0, 0, 0, 1)$.

One can conclude that if the Jordan canonical form of a matrix A is the given matrix J then A should be a 5×5 matrix whose eigenvalues are 6, 2, and 3 with multiplicities 2, 2, and 1, respectively, but for each eigenvalue λ of A, its eigenspace $E(\lambda)$ should have only one linearly independent eigenvector.

Moreover, the converse is also true by Theorem 8.1: *A matrix A and its Jordan canonical matrix J have the same eigenvalues of the same multiplicities and* $\dim \mathcal{N}(\lambda I - A) = \dim \mathcal{N}(\lambda I - J)$ *for each eigenvalue* λ. □

Now, to get a Jordan decomposition $A = QJQ^{-1}$ of any given square matrix A, one needs to find a Jordan canonical matrix J and an invertible matrix Q such that $A = QJQ^{-1}$.

To do this, first find all distinct eigenvalues of a given matrix A and their multiplicities, and then one can try to find a Jordan canonical matrix J whose eigenvalues are the same as the eigenvalues of A with the same multiplicities. It means, one needs to determine the Jordan blocks belonging to each eigenvalue λ of A.

The following example shows all possible types of Jordan blocks belonging to a single eigenvalue λ for a fixed multiplicity m_λ.

Example 147 (*Classifying the Jordan canonical matrices having a single eigenvalue of multiplicity 5*) Classify all possible Jordan canonical forms of a 5×5 matrix A that has a single eigenvalue λ of multiplicity 5 (up to permutations of the Jordan blocks).

Solution: First, Theorem 8.1 says that the number of Jordan blocks in the Jordan canonical form of a matrix A is equal to the number of linearly independent eigenvectors of A.

There are seven possible Jordan canonical forms of order 5 as follows.

(1) Suppose A has only one linearly independent eigenvector belonging to λ. Then the Jordan canonical form of A is of the form

$$J^{(1)} = Q^{-1}AQ = \begin{bmatrix} \lambda & 1 & 0 & 0 & 0 \\ 0 & \lambda & 1 & 0 & 0 \\ 0 & 0 & \lambda & 1 & 0 \\ 0 & 0 & 0 & \lambda & 1 \\ 0 & 0 & 0 & 0 & \lambda \end{bmatrix},$$

which consists of only one Jordan block with eigenvalue λ on the diagonal. And, both A and $J^{(1)}$ have only one linearly independent eigenvector belonging to λ. In this case, rank $(J^{(1)} - \lambda I) = 4$.

(2) Suppose A has two linearly independent eigenvectors belonging to λ. Then the Jordan canonical form of A is either one of the forms

$$J^{(2)} = \begin{bmatrix} \lambda & 1 & & & \\ 0 & \lambda & & & \\ & & \lambda & 1 & 0 \\ & & 0 & \lambda & 1 \\ & & 0 & 0 & \lambda \end{bmatrix} \quad \text{or} \quad J^{(3)} = \begin{bmatrix} \lambda & & & & \\ & \lambda & 1 & 0 & 0 \\ & 0 & \lambda & 1 & 0 \\ & 0 & 0 & \lambda & 1 \\ & 0 & 0 & 0 & \lambda \end{bmatrix},$$

each of which consists of two Jordan blocks belonging to the eigenvalue λ. The matrix $J^{(2)}$ has two linearly independent eigenvectors $\mathbf{e}_1$ and $\mathbf{e}_3$, while $J^{(3)}$ has two eigenvectors $\mathbf{e}_1$ and $\mathbf{e}_2$. These two matrices $J^{(2)}$ and $J^{(3)}$ cannot be similar, because $(J^{(2)} - \lambda I)^3 = \mathbf{0}$, but $(J^{(3)} - \lambda I)^3 \neq \mathbf{0}$. (One can justify this by a direct computation, or see Exercise **1.30**.)

(3) Suppose A has three linearly independent eigenvectors belonging to λ. Then the Jordan canonical form of A is either one of the forms

$$J^{(4)} = \begin{bmatrix} \lambda & & & & \\ & \lambda & 1 & & \\ & 0 & \lambda & & \\ & & & \lambda & 1 \\ & & & 0 & \lambda \end{bmatrix} \quad \text{or} \quad J^{(5)} = \begin{bmatrix} \lambda & & & & \\ & \lambda & & & \\ & & \lambda & 1 & 0 \\ & & 0 & \lambda & 1 \\ & & 0 & 0 & \lambda \end{bmatrix},$$

each of which consists of three Jordan blocks belonging to the eigenvalue λ. The matrix $J^{(4)}$ has three linearly independent eigenvectors $\mathbf{e}_1$, $\mathbf{e}_2$, and $\mathbf{e}_4$, while $J^{(5)}$

has $\mathbf{e}_1$, $\mathbf{e}_2$, and $\mathbf{e}_3$. These two matrices $J^{(4)}$ and $J^{(5)}$ are not similar, because $(J^{(4)} - \lambda I)^2 = \mathbf{0}$, but $(J^{(5)} - \lambda I)^2 \neq \mathbf{0}$.

(4) Suppose A has four linearly independent eigenvectors belonging to λ. Then the Jordan canonical form of A is of the form

$$J^{(6)} = \begin{bmatrix} \lambda & & & & \\ & \lambda & & & \\ & & \lambda & & \\ & & & \lambda & 1 \\ & & & 0 & \lambda \end{bmatrix},$$

which consists of four Jordan blocks with eigenvalue λ. The matrix $J^{(6)}$ has four linearly independent eigenvectors $\mathbf{e}_1, \mathbf{e}_2, \mathbf{e}_3$, and $\mathbf{e}_4$.

(5) Suppose A has five linearly independent eigenvectors belonging to λ. Then the Jordan canonical form of A is diagonal with diagonal entries λ, say $J^{(7)}$. The matrix $J^{(7)}$ has five linearly independent eigenvectors $\mathbf{e}_1, \mathbf{e}_2, \mathbf{e}_3, \mathbf{e}_4$, and $\mathbf{e}_5$, as expected.

In conclusion, all of the seven possible Jordan canonical matrices have the same trace, the same determinant, and the same characteristic polynomial, but any two of them are not similar to each other.

Moreover, if A is any 5×5 matrix having a single eigenvalue λ of multiplicity 5, then the Jordan canonical form of A is one of the seven Jordan canonical matrices $J^{(1)}, J^{(2)}, \ldots, J^{(7)}$. □

Problem 194 For $n = 1, 2, \ldots, 7$, count all possible Jordan canonical matrices of size $n \times n$ which have a single eigenvalue λ of multiplicity n, up to permutations of the Jordan blocks.

Problem 195 Show that for any Jordan block J_λ belonging to λ, the order of J_λ is equal to the smallest positive integer k such that $(J - \lambda I)^k = \mathbf{0}$.

Hint: As an example, let $J = J_\lambda = \begin{bmatrix} \lambda & 1 & 0 & 0 & 0 \\ 0 & \lambda & 1 & 0 & 0 \\ 0 & 0 & \lambda & 1 & 0 \\ 0 & 0 & 0 & \lambda & 1 \\ 0 & 0 & 0 & 0 & \lambda \end{bmatrix}$. Then,

$$J - \lambda I = \begin{bmatrix} 0 & 1 & 0 & 0 & 0 \\ 0 & 0 & 1 & 0 & 0 \\ 0 & 0 & 0 & 1 & 0 \\ 0 & 0 & 0 & 0 & 1 \\ 0 & 0 & 0 & 0 & 0 \end{bmatrix} \quad \text{and} \quad (J - \lambda I)^2 = \begin{bmatrix} 0 & 0 & 1 & 0 & 0 \\ 0 & 0 & 0 & 1 & 0 \\ 0 & 0 & 0 & 0 & 1 \\ 0 & 0 & 0 & 0 & 0 \\ 0 & 0 & 0 & 0 & 0 \end{bmatrix},$$

and $(J - \lambda I)^4 \neq \mathbf{0}$ but $(J - \lambda I)^5 = \mathbf{0}$. See Exercise **1.30**.

Problem 196 Count all possible 5×5 Jordan canonical matrices which have two distinct eigenvalues, say λ of multiplicity 3 and μ of multiplicity 2, up to permutations of the Jordan blocks.

8.3 How to find the Jordan canonical form J

For a given square matrix A, we are interested in determining a Jordan decomposition $A = QJQ^{-1}$. It will be shown how to find the Jordan canonical form J of A in this section, and how to find the change of basis matrix Q in the next section.

In fact, It will be shown that the Jordan canonical form J of a matrix A is completely determined by the sequence of ranks, rank $(A - \lambda I)^\ell$, $\ell = 1, 2, \ldots$ for each eigenvalue λ of A.

Example 148 (*Jordan canonical form of a triangular matrix*) Find the Jordan canonical form J of the matrix

$$A = \begin{bmatrix} 2 & 1 & 4 \\ 0 & 3 & 1 \\ 0 & 0 & 3 \end{bmatrix}.$$

Solution: Since A is triangular, the eigenvalues of A are the diagonal entries

$$\lambda_1 = 2, \quad \lambda_2 = \lambda_3 = 3.$$

Hence, there are two possibilities of the Jordan canonical form of A:

$$J^{(1)} = \begin{bmatrix} 2 & 0 & 0 \\ 0 & 3 & 0 \\ 0 & 0 & 3 \end{bmatrix} = \begin{bmatrix} [2] & & \\ & [3] & \\ & & [3] \end{bmatrix} \quad \text{or } J^{(2)} = \begin{bmatrix} 2 & 0 & 0 \\ 0 & 3 & 1 \\ 0 & 0 & 3 \end{bmatrix} = \begin{bmatrix} [2] & \\ & \begin{bmatrix} 3 & 1 \\ 0 & 3 \end{bmatrix} \end{bmatrix}.$$

But, one can see that rank $(A - 3I) = 2$ (and rank $(A - 3I)^\ell = 1$ for $\ell \geq 2$). This implies that rank $(J - 3I) = 2$ and so the Jordan canonical form of A should be $J^{(2)}$, not $J^{(1)}$. □

Example 149 (*Jordan canonical form of a companion matrix*) Find the Jordan canonical form J of the matrix

$$A = \begin{bmatrix} 4 & -6 & 4 & -1 \\ 1 & 0 & 0 & 0 \\ 0 & 1 & 0 & 0 \\ 0 & 0 & 1 & 0 \end{bmatrix}.$$

Solution: The characteristic polynomial of the matrix A is

$$\det(\lambda I - A) = \lambda^4 - 4\lambda^3 + 6\lambda^2 - 4\lambda + 1 = (\lambda - 1)^4.$$

The eigenvalue of A is $\lambda = 1$ of multiplicity 4. The rank of $A - I$ is

$$\text{rank}\,(A - I) = \text{rank} \begin{bmatrix} 3 & -6 & 4 & -1 \\ 1 & -1 & 0 & 0 \\ 0 & 1 & -1 & 0 \\ 0 & 0 & 1 & -1 \end{bmatrix} = 3$$

by noticing that the last three rows are linearly independent, or the determinant of the 3×3 submatrix of the lower left corner is not zero.

Hence, rank $(J - I)$ should be 3 and $J - I$ should have three 1's as the super-diagonal entries;

$$J = \begin{bmatrix} 1 & 1 & 0 & 0 \\ 0 & 1 & 1 & 0 \\ 0 & 0 & 1 & 1 \\ 0 & 0 & 0 & 1 \end{bmatrix}.$$

On the other hand, one can also check the following equations:

$$(A - I)^3 = \begin{bmatrix} 1 & -3 & 3 & -1 \\ 1 & -3 & 3 & -1 \\ 1 & -3 & 3 & -1 \\ 1 & -3 & 3 & -1 \end{bmatrix} \neq \mathbf{0} \quad \text{but} \quad (A - I)^4 = \mathbf{0},$$

which says that the order of the largest Jordan block is 4. □

Two similar matrices have the same rank. Hence, if J is the Jordan canonical form of a matrix A, then for any eigenvalue λ and for any positive integer ℓ, it should be

$$\text{rank}(A - \lambda I)^\ell = \text{rank}\,(J - \lambda I)^\ell, \quad \ell = 1, 2, \ldots$$

because of the similarity of two matrices. Hence, with the given sequence rank $(A-\lambda I)^\ell$, one can determine the Jordan blocks J_λ belonging to λ so that rank $(A-\lambda I)^\ell = $ rank $(J-\lambda I)^\ell$ for all ℓ.

In particular, the number of Jordan blocks belonging to an eigenvalue λ of A is equal to the number of linearly independent eigenvectors of A belonging to λ, which is the dimension of the eigenspace $E(\lambda) = \mathcal{N}(\lambda I - A)$ (or $= \mathcal{N}(A-\lambda I)$). Moreover, the sum of the orders of all Jordan blocks belonging to λ is equal to the multiplicity m_λ of λ in $\det(\lambda I - A) = 0$.

Example 150 (*The sequence of* rank $(J-\lambda I)^\ell$ *for a Jordan matrix J having a single eigenvalue* λ) **(1)** Let

$$J = J_\lambda = \begin{bmatrix} \lambda & 1 & 0 & 0 & 0 \\ 0 & \lambda & 1 & 0 & 0 \\ 0 & 0 & \lambda & 1 & 0 \\ 0 & 0 & 0 & \lambda & 1 \\ 0 & 0 & 0 & 0 & \lambda \end{bmatrix}$$

as an example of a Jordan block. Then, $(J-\lambda I)^\ell$ for $\ell = 1, 2, \ldots$ are

$$\begin{bmatrix} 0 & 1 & 0 & 0 & 0 \\ 0 & 0 & 1 & 0 & 0 \\ 0 & 0 & 0 & 1 & 0 \\ 0 & 0 & 0 & 0 & 1 \\ 0 & 0 & 0 & 0 & 0 \end{bmatrix}, \begin{bmatrix} 0 & 0 & 1 & 0 & 0 \\ 0 & 0 & 0 & 1 & 0 \\ 0 & 0 & 0 & 0 & 1 \\ 0 & 0 & 0 & 0 & 0 \\ 0 & 0 & 0 & 0 & 0 \end{bmatrix}, \begin{bmatrix} 0 & 0 & 0 & 1 & 0 \\ 0 & 0 & 0 & 0 & 1 \\ 0 & 0 & 0 & 0 & 0 \\ 0 & 0 & 0 & 0 & 0 \\ 0 & 0 & 0 & 0 & 0 \end{bmatrix}, \begin{bmatrix} 0 & 0 & 0 & 0 & 1 \\ 0 & 0 & 0 & 0 & 0 \\ 0 & 0 & 0 & 0 & 0 \\ 0 & 0 & 0 & 0 & 0 \\ 0 & 0 & 0 & 0 & 0 \end{bmatrix},$$

and the zero matrices $\mathbf{0}, \ldots$, respectively (as is shown in Exercise **1.30**). Hence, the sequence of ranks, rank $(J-\lambda I)^\ell$ for $\ell = 1, 2, \ldots$ is $4, 3, 2, 1, 0, 0, \ldots$.

In general, if J is the $n \times n$ Jordan block with an eigenvalue λ, then the sequence rank $(J-\lambda I)^\ell$ for $\ell = 1, 2, \ldots$ is $n-1, n-2, \ldots, 3, 2, 1, 0, 0, \ldots$.

(2) If a Jordan matrix J has two Jordan blocks of size 5×5 and 3×3 with the same eigenvalue λ, then the rank $(J-\lambda I)^\ell$ for $\ell = 1, 2, \ldots$ is $4+2, 3+1, 2+0, 1+0, 0, 0, \ldots$. □

The next example shows a relationship between the orders of the Jordan blocks belonging to an eigenvalue λ of a matrix A and the sequence of rank $(J-\lambda I)^\ell = $ rank $(A-\lambda I)^\ell$ for $\ell = 1, 2, \ldots$.

Example 151 (*Determine the Jordan blocks belonging to an eigenvalue λ of multiplicity m_λ*) Let J be a Jordan canonical matrix with a single eigenvalue λ:

$$J = \begin{bmatrix} J_1 & & & \mathbf{0} \\ & J_2 & & \\ & & J_3 & \\ \mathbf{0} & & & J_4 \end{bmatrix} = \begin{bmatrix} \begin{bmatrix} \lambda & 1 & 0 \\ 0 & \lambda & 1 \\ 0 & 0 & \lambda \end{bmatrix} & & & \\ & \begin{bmatrix} \lambda & 1 \\ 0 & \lambda \end{bmatrix} & & \\ & & \begin{bmatrix} \lambda & 1 \\ 0 & \lambda \end{bmatrix} & \\ & & & [\lambda] \end{bmatrix},$$

whose blocks J_1, J_2, J_3, J_4 have orders $3, 2, 2, 1$, respectively. Then,

$$(J - \lambda I)^2 = \begin{bmatrix} \begin{bmatrix} 0 & 0 & 1 \\ 0 & 0 & 0 \\ 0 & 0 & 0 \end{bmatrix} & & & \\ & \begin{bmatrix} 0 & 0 \\ 0 & 0 \end{bmatrix} & & \\ & & \begin{bmatrix} 0 & 0 \\ 0 & 0 \end{bmatrix} & \\ & & & [0] \end{bmatrix}$$

and

$$(J - \lambda I)^3 = \mathbf{0}.$$

Thus,

$$\text{rank}\,(J - \lambda I) = 4, \quad \text{rank}\,(J - \lambda I)^2 = 1, \quad \text{and} \quad \text{rank}\,(J - \lambda I)^3 = 0.$$

The sequence of rank $(J - \lambda I)^\ell$ *for* $\ell = 1, 2, \ldots$ *is* $4, 1, 0, 0, \ldots$ *and this determines completely the orders of the blocks of the* 8×8 *Jordan matrix* J *belonging to a single* λ.

In fact, one can notice that

(1) $(J - \lambda I)^3 = \mathbf{0}$ but $(J - \lambda I)^2 \neq \mathbf{0}$ implies that the largest block has order 3,

(2) rank $(J - \lambda I)^2 = 1$ is equal to the number of blocks of order 3,

(3) rank $(J - \lambda I) = 4$ is equal to twice the number of blocks of order 3 plus the number of blocks of order 2, so there are two blocks of order 2,

(4) the number of blocks of order 1 is $8 - (2 \times 2) - (3 \times 1) = 1$. □

In a sequence of matrices $J - \lambda I, (J - \lambda I)^2, (J - \lambda I)^3, \ldots$, all Jordan blocks belonging to the eigenvalue λ will terminate at a zero matrix but all the other blocks (belonging to an eigenvalue different from λ) remain as upper triangular matrices with nonzero diagonal entries. Hence, the sequence of rank $(J - \lambda I)^\ell$ must stop decreasing at $n - m_\lambda$. (Note: rank $(J - \lambda I)^\ell = n - m_\lambda$ for $\ell \geq m_\lambda$.)

For simplicity, let $c_\lambda = n - m_\lambda$, which is the number of the eigenvalues different from λ counting multiplicities. Then, the decreasing sequence

$$\{\text{rank}(A - \lambda I)^k - c_\lambda \, : \, k = 1, 2, \ldots, m_\lambda\}$$

completely determines the orders of the blocks of J belonging to λ, as is shown in Example 151:

(1) The order of the largest Jordan block belonging to λ is the smallest positive integer k such that rank $(A - \lambda I)^k - c_\lambda = 0$. Then, the number of such largest blocks is equal to rank $(A - \lambda I)^{k-1} - c_\lambda$. Denote this number by ℓ_1.

(2) The number of blocks of order $k - 1$ is equal to rank $(A - \lambda I)^{k-2} - c_\lambda - 2\ell_1$. Denote this number by ℓ_2.

(3) The number of blocks of order $k - 2$ is equal to rank $(A - \lambda I)^{k-3} - c_\lambda - 3\ell_1 - 2\ell_2$. Denote this number by ℓ_3 and so on.

Inductively, if $\ell_1, \ell_2, \ldots, \ell_j$ are given, then one can determine ℓ_{j+1} as follows:

(4) The number of blocks of order $k - j$ is equal to rank $(A - \lambda I)^{k-(j+1)} - c_\lambda - (j + 1)\ell_1 - j\ell_2 - \cdots - 2\ell_j$, (we denote this number by ℓ_{j+1} with $\ell_0 = 0$) for $j = 0, \ldots, k - 1$.

In summary, one can determine the Jordan canonical form J of an $n \times n$ matrix A by the following procedure.

Step 1 Find all distinct eigenvalues $\lambda_1, \lambda_2, \ldots, \lambda_t$ of A and their multiplicities $m_{\lambda_1}, m_{\lambda_2}, \ldots, m_{\lambda_t}$, so that $m_{\lambda_1} + m_{\lambda_2} + \cdots + m_{\lambda_\ell} = n$. Let $c_{\lambda_i} = n - m_{\lambda_i}$.

Step 2 For each eigenvalue λ, (as one of $\lambda_1, \lambda_2, \dots, \lambda_t$), the Jordan blocks belonging to λ are determined by the following criteria:

(i) The order of the largest block belonging to λ is the smallest positive integer k such that rank $(A - \lambda I)^k - c_\lambda = 0$, and

(ii) the number of blocks of order $k - j$ belonging to λ is inductively determined as

$$\text{rank}\,(A - \lambda I)^{k-(j+1)} - c_\lambda - (j+1)\ell_1 - j\ell_2 - \cdots - 2\ell_j,$$

(we denote this number by ℓ_{j+1} with $\ell_0 = 0$) for $j = 0, \dots, k-1$.

This is a standard algorithm to determine the Jordan canonical form of a matrix. However, for a matrix of large order having several eigenvalues, the evaluation of rank$(A - \lambda I)^k$ might not be easy at all, while, for matrices of lower order or relatively simple matrices, the computations may be possible by hand.

Example 152 (*The sequence of* rank $(A - \lambda I)^\ell$ *determines the Jordan blocks belonging to an eigenvalue* λ) Let A be a 5×5 matrix with a single eigenvalue λ of multiplicity 5. Then, its Jordan canonical form J is one of the seven Jordan matrices $J^{(1)}, J^{(2)}, \dots, J^{(7)}$ listed in Example 147.

Moreover, these seven matrices give seven different sequences of rank$(J - \lambda I)^\ell$ for $\ell = 1, 2, \dots$.

In fact, for $J^{(1)}$, the sequence is $4, 3, 2, 1, 0, 0, \dots$. For $J^{(2)}$, the sequence is $3, 1, 0, 0, \dots$; while for $J^{(3)}$, the sequence is $3, 2, 1, 0, 0, \dots$. For $J^{(4)}$, it is $2, 0, 0, \dots$; while for $J^{(5)}$, it is $2, 1, 0, 0, \dots$. Finally, for $J^{(6)}$ and $J^{(7)}$, one can get $1, 0, 0, \dots$ and $0, 0, 0, \dots$, respectively.

Now, by computing the sequence of rank$(A - \lambda I)^\ell$ for a given matrix A, one can figure out which Jordan matrix J can be the Jordan canonical form of the matrix A, by comparing their sequences with rank $(A - \lambda I)^\ell =$ rank $(J - \lambda I)^\ell$ for all ℓ.

For instance, the matrix

$$A = \begin{bmatrix} \lambda & 1 & 0 & 0 & 3 \\ 0 & \lambda & 0 & 2 & 0 \\ 0 & 0 & \lambda & 1 & 0 \\ 0 & 0 & 0 & \lambda & 0 \\ 0 & 0 & 0 & 0 & \lambda \end{bmatrix},$$

has clearly a single eigenvalue λ of multiplicity 5, and the sequence of rank$(A - \lambda I)^\ell$ for $\ell = 1, 2, \ldots$ is $2, 1, 0, 0, \ldots$. Hence, the Jordan canonical form of the matrix A is the matrix $J^{(5)}$ in Example 147. $\square$

Problem 197 Find the Jordan canonical form of each of the following matrices.

$$(1)\ \begin{bmatrix} i & 0 \\ 1 & i \end{bmatrix}, \qquad (2)\ \begin{bmatrix} 2 & 1 & 0 \\ 0 & 2 & 1 \\ 0 & 0 & 3 \end{bmatrix}, \qquad (3)\ \begin{bmatrix} 4 & 1 & 2 & 0 \\ 0 & 4 & 2 & 0 \\ 0 & 0 & 4 & 2 \\ 0 & 0 & 0 & 2 \end{bmatrix}.$$

8.4 How to find a change of basis matrix Q for a Jordan decomposition

In this section we show how to find a change of basis matrix Q for a Jordan decomposition $A = QJQ^{-1}$ of a (diagonalizable or not) matrix A when J is known.

As shown in Example 146, the standard basis vectors $\mathbf{e}_1$, $\mathbf{e}_3$, and $\mathbf{e}_5$ associated with the first column vectors of the Jordan blocks J_i's of J form linearly independent eigenvectors of the matrix J, and then the vectors $Q\mathbf{e}_1$, $Q\mathbf{e}_3$, and $Q\mathbf{e}_5$ form linearly independent eigenvectors of the matrix A, where $Q^{-1}AQ = J$. Note that $Q\mathbf{e}_1$, $Q\mathbf{e}_3$, and $Q\mathbf{e}_5$ are the first, the third, and the fifth columns of the matrix Q.

The following example shows what the column vectors of a change of basis matrix Q look like when a matrix A and its Jordan canonical form J are given.

Example 153 (*Jordan blocks $\leftrightarrow$ Chains of generalized eigenvectors*) Let A be a 5×5 matrix similar to a Jordan block of the form

$$Q^{-1}AQ = J = \begin{bmatrix} \lambda & 1 & 0 & 0 & 0 \\ 0 & \lambda & 1 & 0 & 0 \\ 0 & 0 & \lambda & 1 & 0 \\ 0 & 0 & 0 & \lambda & 1 \\ 0 & 0 & 0 & 0 & \lambda \end{bmatrix}.$$

Determine a change of basis matrix $Q = [\mathbf{x}_1\ \mathbf{x}_2\ \mathbf{x}_3\ \mathbf{x}_4\ \mathbf{x}_5]$.

Solution: Clearly, two similar matrices A and J have the same eigenvalue λ of multiplicity 5. Since rank$(J - \lambda I) = 4$, it follows $\dim \mathcal{N}(J - \lambda I) = \dim E(\lambda) = 1$ by the Rank-Nullity theorem.

So, J has only one linearly independent eigenvector, which can be $\mathbf{e}_1 = (1, 0, 0, 0, 0)$ (because $J\mathbf{e}_1 = \lambda\mathbf{e}_1$ as the first column of J). Thus A also has only one linearly independent eigenvector, that is $Q\mathbf{e}_1 = \mathbf{x}_1$, the first column of Q.

To see what the other columns of Q are, expand $AQ = QJ$ as

$$[A\mathbf{x}_1\ A\mathbf{x}_2\ A\mathbf{x}_3\ A\mathbf{x}_4\ A\mathbf{x}_5] = [\lambda\mathbf{x}_1\ \ \mathbf{x}_1 + \lambda\mathbf{x}_2\ \ \mathbf{x}_2 + \lambda\mathbf{x}_3\ \ \mathbf{x}_3 + \lambda\mathbf{x}_4\ \ \mathbf{x}_4 + \lambda\mathbf{x}_5].$$

By comparing the column vectors, it follows

$$\begin{array}{lll} A\mathbf{x}_5 = \mathbf{x}_4 + \lambda\mathbf{x}_5, & \text{or} & (A - \lambda I)\mathbf{x}_5 = (A - \lambda I)^1\mathbf{x}_5 = \mathbf{x}_4, \\ A\mathbf{x}_4 = \mathbf{x}_3 + \lambda\mathbf{x}_4, & \text{or} & (A - \lambda I)\mathbf{x}_4 = (A - \lambda I)^2\mathbf{x}_5 = \mathbf{x}_3, \\ A\mathbf{x}_3 = \mathbf{x}_2 + \lambda\mathbf{x}_3, & \text{or} & (A - \lambda I)\mathbf{x}_3 = (A - \lambda I)^3\mathbf{x}_5 = \mathbf{x}_2, \\ A\mathbf{x}_2 = \mathbf{x}_1 + \lambda\mathbf{x}_2, & \text{or} & (A - \lambda I)\mathbf{x}_2 = (A - \lambda I)^4\mathbf{x}_5 = \mathbf{x}_1, \\ A\mathbf{x}_1 = \lambda\mathbf{x}_1, & \text{or} & (A - \lambda I)\mathbf{x}_1 = (A - \lambda I)^5\mathbf{x}_5 = \mathbf{0}. \end{array}$$

The last column vector $\mathbf{x}_5$ of Q satisfies

$$(A - \lambda I)^5\mathbf{x}_5 = \mathbf{0} \quad \text{but} \quad (A - \lambda I)^4\mathbf{x}_5 = \mathbf{x}_1 \neq \mathbf{0}.$$

However, $(A - \lambda I)^5 = Q(J - \lambda I)^5 Q^{-1} = Q\mathbf{0}Q^{-1} = \mathbf{0}$. Hence, if one chooses $\mathbf{x}_5$ as any solution of $(A - \lambda I)^4\mathbf{x} \neq \mathbf{0}$, then all the other $\mathbf{x}_i$'s can be determined easily;

$$\mathbf{x}_4 = (A - \lambda I)\mathbf{x}_5, \quad \mathbf{x}_3 = (A - \lambda I)\mathbf{x}_4, \quad \mathbf{x}_2 = (A - \lambda I)\mathbf{x}_3, \quad \text{etc.}$$

Such a vector $\mathbf{x}_5$ is called a *generalized eigenvector* of rank 5, and the (ordered) set $\{\mathbf{x}_1, \mathbf{x}_2, \ldots, \mathbf{x}_5\}$ is called a *chain of generalized eigenvectors* belonging to λ. Therefore, the (ordered) columns $\mathbf{x}_1, \mathbf{x}_2, \ldots, \mathbf{x}_5$ of the matrix Q are nothing but a chain of generalized eigenvectors.

This is always true when the Jordan matrix J has only one Jordan block. □

Definition 8.2 A nonzero vector $\mathbf{x}$ is said to be a **generalized eigenvector** of A of **rank** k belonging to an eigenvalue λ if

$$(A - \lambda I)^k\mathbf{x} = \mathbf{0} \quad \text{and} \quad (A - \lambda I)^{k-1}\mathbf{x} \neq \mathbf{0}.$$

If $k = 1$, this is just the definition of an eigenvector belonging to an eigenvalue λ.

If a generalized eigenvector $\mathbf{x}$ of rank $k \geq 1$ belonging to an eigenvalue λ is given, then one can define a sequence of nonzero vectors like as in Example 153: say $\mathbf{x} = \mathbf{x}_k$ and define

$$\mathbf{x}_{k-1} = (A - \lambda I)\mathbf{x}_k, \ \mathbf{x}_{k-2} = (A - \lambda I)\mathbf{x}_{k-1}, \ \mathbf{x}_{k-3} = (A - \lambda I)\mathbf{x}_{k-2},$$

and so on, finally stops at $(A - \lambda I)\mathbf{x}_1 = (A - \lambda I)^k \mathbf{x}_k = \mathbf{0}$.

Such a sequence of vectors is named as follows.

Definition 8.3 The (ordered) set of vectors $\{\mathbf{x}_1, \mathbf{x}_2, \ldots, \mathbf{x}_k\}$ is called a **chain of generalized eigenvectors** belonging to the eigenvalue λ. The eigenvector $\mathbf{x}_1$ is called the **initial eigenvector** of the chain.

Except for the initial eigenvector $\mathbf{x}_1$, all the others in the chain $\{\mathbf{x}_1, \mathbf{x}_2, \ldots, \mathbf{x}_k\}$ are not eigenvectors of A belonging to the eigenvalue λ.

As shown in Example 153, one can see that in a Jordan decomposition $A = QJQ^{-1}$ of a matrix A, if a Jordan block J_λ of order k is given, one can get a chain of generalized eigenvectors of A belonging to λ which will be k columns of the change of basis matrix Q.

Moreover, all such chains of generalized eigenvectors derived from the Jordan blocks form the whole columns of the matrix Q.

Finally, the invertibility of a change of basis matrix Q comes from the next three theorems.

Theorem 8.4 *A chain of generalized eigenvectors* $\{\mathbf{x}_1, \mathbf{x}_2, \ldots, \mathbf{x}_k\}$ *belonging to an eigenvalue* λ *is linearly independent.*

Proof: Let $\{\mathbf{x}_1, \mathbf{x}_2, \ldots, \mathbf{x}_k\}$ be a chain of generalized eigenvectors belonging to λ and let $c_1\mathbf{x}_1 + c_2\mathbf{x}_2 + \cdots + c_k\mathbf{x}_k = \mathbf{0}$ with constants c_i. By taking the linear transformation $A - \lambda I$ on both sides of this equation successively, one can get

$$\begin{aligned}
(A - \lambda I)(c_1\mathbf{x}_1 + c_2\mathbf{x}_2 + \cdots + c_k\mathbf{x}_k) &= \mathbf{0} + c_2\mathbf{x}_1 + c_3\mathbf{x}_2 + \cdots + c_k\mathbf{x}_{k-1} = \mathbf{0}, \\
(A - \lambda I)^2(c_1\mathbf{x}_1 + c_2\mathbf{x}_2 + \cdots + c_k\mathbf{x}_k) &= \mathbf{0} + \mathbf{0} + c_3\mathbf{x}_1 + \cdots + c_k\mathbf{x}_{k-2} = \mathbf{0}, \\
&\ \vdots \\
(A - \lambda I)^{k-2}(c_1\mathbf{x}_1 + c_2\mathbf{x}_2 + \cdots + c_k\mathbf{x}_1) &= \mathbf{0} + \mathbf{0} + \cdots + \mathbf{0} + c_{k-1}\mathbf{x}_1 + c_k\mathbf{x}_2 = \mathbf{0}, \\
(A - \lambda I)^{k-1}(c_1\mathbf{x}_1 + c_2\mathbf{x}_2 + \cdots + c_k\mathbf{x}_1) &= \mathbf{0} + \mathbf{0} + \cdots + \mathbf{0} + \mathbf{0} + c_k\mathbf{x}_1 = \mathbf{0}.
\end{aligned}$$

The last equation gives $c_k = 0$, and then the second last one gives $c_{k-1} = 0$. Proceeding successively with known c_j's, it follows all $c_i = 0$. □

Theorem 8.5 *The union of chains of generalized eigenvectors of a square matrix A belonging to distinct eigenvalues is linearly independent.*

Theorem 8.6 *Let $S = \{\mathbf{x}_1, \mathbf{x}_2, \ldots, \mathbf{x}_k\}$ and $T = \{\mathbf{y}_1, \mathbf{y}_2, \ldots, \mathbf{y}_\ell\}$ be two chains of generalized eigenvectors of A belonging to the same eigenvalue λ. If the initial eigenvectors $\mathbf{x}_1$ and $\mathbf{y}_1$ are linearly independent, then the union $S \cup T$ is also linearly independent.*

This can be extended to any finite number of chains of generalized eigenvectors of A belonging to an eigenvalue λ

The proofs of these two theorems will be placed in the end of this section. A reader may skip reading their proofs if not interested in details.

Example 154 (*Change of basis matrix Q for a triangular matrix*) For the matrix

$$A = \begin{bmatrix} 2 & 1 & 4 \\ 0 & 3 & 1 \\ 0 & 0 & 3 \end{bmatrix},$$

find a change of basis matrix Q so that $Q^{-1}AQ$ is the Jordan canonical matrix.

Solution: In Example 148, the Jordan canonical form J of A is determined as

$$J = \begin{bmatrix} 2 & 0 & 0 \\ 0 & 3 & 1 \\ 0 & 0 & 3 \end{bmatrix} = \begin{bmatrix} [2] & \\ & \begin{bmatrix} 3 & 1 \\ 0 & 3 \end{bmatrix} \end{bmatrix}.$$

Thus, a change of basis matrix $Q = [\mathbf{x}_1 \ \mathbf{x}_2 \ \mathbf{x}_3]$ consists of two chains of generalized eigenvectors.

As one of the two chains, an eigenvector belonging to $\lambda_1 = 2$ can be easily chosen as $\mathbf{x}_1 = (1, 0, 0)$.

For the other chain, noting the second 2×2 Jordan block, it is needed to find a generalized eigenvector of rank 2 as a solution $\mathbf{x}$ to the following system:

$$\begin{aligned} (A - 3I)\mathbf{x} &= \begin{bmatrix} -1 & 1 & 4 \\ 0 & 0 & 1 \\ 0 & 0 & 0 \end{bmatrix} \mathbf{x} \neq \mathbf{0}, \\ (A - 3I)^2\mathbf{x} &= \begin{bmatrix} 1 & -1 & -3 \\ 0 & 0 & 0 \\ 0 & 0 & 0 \end{bmatrix} \mathbf{x} = \mathbf{0}. \end{aligned}$$

From the second equation, $\mathbf{x}$ should be of the form $(s + 3t, s, t)$, but the first equation says $t \neq 0$. Let us choose $\mathbf{x} = (4, 1, 1)$, take it $\mathbf{x}_3$ as a generalized eigenvector of rank 2. Then, $\mathbf{x}_2 = (A - 3I)\mathbf{x}_3 = (1, 1, 0)$. Now, one can set

$$Q = [\mathbf{x}_1\ \mathbf{x}_2\ \mathbf{x}_3] = \begin{bmatrix} 1 & 1 & 4 \\ 0 & 1 & 1 \\ 0 & 0 & 1 \end{bmatrix}.$$

One may check by a direct computation

$$Q^{-1}AQ = \begin{bmatrix} 2 & 0 & 0 \\ 0 & 3 & 1 \\ 0 & 0 & 3 \end{bmatrix} = J \qquad \text{with } Q^{-1} = \begin{bmatrix} 1 & -1 & -3 \\ 0 & 1 & -1 \\ 0 & 0 & 1 \end{bmatrix}.$$

The readers are suggested to find a change of basis matrix $Q = [\mathbf{x}_1\ \mathbf{x}_2\ \mathbf{x}_3]$ so that $Q^{-1}AQ = J$ by comparing directly the columns of the both sides of a matrix equation $AQ = QJ$. Can you do it? □

Example 155 (*Find a change of basis matrix Q*) For the matrix

$$A = \begin{bmatrix} 2 & 0 & 0 \\ 0 & 2 & 0 \\ -2 & 1 & 2 \end{bmatrix},$$

find a change of basis matrix Q so that $Q^{-1}AQ$ is the Jordan canonical matrix.

Solution: The matrix A is (lower) triangular and $\lambda = 2$ is an eigenvalue of A of multiplicity 3. The rank of the matrix

$$A - 2I = \begin{bmatrix} 0 & 0 & 0 \\ 0 & 0 & 0 \\ -2 & 1 & 0 \end{bmatrix}$$

is 1. Hence, there are two linearly independent eigenvectors belonging to $\lambda = 2$, and the Jordan canonical form J of A is determined as

$$J = \begin{bmatrix} 2 & 1 & 0 \\ 0 & 2 & 0 \\ 0 & 0 & 2 \end{bmatrix} = \begin{bmatrix} \begin{bmatrix} 2 & 1 \\ 0 & 2 \end{bmatrix} & \\ & [2] \end{bmatrix}.$$

Thus, a change of basis matrix $Q = [\mathbf{x}_1\ \mathbf{x}_2\ \mathbf{x}_3]$ consists of two chains of generalized eigenvectors. Noting the first 2×2 Jordan block, it is needed to find a generalized eigenvector of rank 2 as a solution $\mathbf{x}$ to the following system:

$$\begin{aligned} (A-2I)\mathbf{x} &= \begin{bmatrix} 0 & 0 & 0 \\ 0 & 0 & 0 \\ -2 & 1 & 0 \end{bmatrix} \begin{bmatrix} x \\ y \\ z \end{bmatrix} \neq \mathbf{0}, \\ (A-2I)^2\mathbf{x} &= \mathbf{0}. \end{aligned}$$

But, a direct computation shows that $(A-2I)^2 = \mathbf{0}$. Hence, one can choose any solution of the first equation as a generalized eigenvector of rank 2. Choose

$$\mathbf{x}_2 = (1, 0, 0).$$

Then,

$$\mathbf{x}_1 = (A-2I)\mathbf{x}_2 = \begin{bmatrix} 0 & 0 & 0 \\ 0 & 0 & 0 \\ -2 & 1 & 0 \end{bmatrix} \begin{bmatrix} 1 \\ 0 \\ 0 \end{bmatrix} = \begin{bmatrix} 0 \\ 0 \\ -2 \end{bmatrix},$$

and $\{\mathbf{x}_1, \mathbf{x}_2\}$ gives a chain. Now, by taking another eigenvector $\mathbf{x}_3 = (1, 2, 0)$ so that $\mathbf{x}_1$ and $\mathbf{x}_3$ are linearly independent, one can get

$$Q = \{\mathbf{x}_1, \mathbf{x}_2, \mathbf{x}_3\} = \begin{bmatrix} 0 & 1 & 1 \\ 0 & 0 & 2 \\ -2 & 0 & 0 \end{bmatrix}.$$

□

Example 156 (*Change of basis matrix Q for a companion matrix*) Find a change of basis matrix Q so that $Q^{-1}AQ = J$ is the Jordan canonical form of the matrix

$$A = \begin{bmatrix} 4 & -6 & 4 & -1 \\ 1 & 0 & 0 & 0 \\ 0 & 1 & 0 & 0 \\ 0 & 0 & 1 & 0 \end{bmatrix}.$$

Solution: As is shown in Example 149, the characteristic polynomial of A is

$$\det(A - \lambda I) = \lambda^4 - 4\lambda^3 + 6\lambda^2 - 4\lambda + 1 = (\lambda - 1)^4.$$

The eigenvalue of A is $\lambda = 1$ of multiplicity 4. But, $\dim \mathcal{N}(A - I) = 1$ because the rank of the matrix

$$A - I = \begin{bmatrix} 3 & -6 & 4 & -1 \\ 1 & -1 & 0 & 0 \\ 0 & 1 & -1 & 0 \\ 0 & 0 & 1 & -1 \end{bmatrix}$$

is 3. Thus, a change of basis matrix $Q = [\mathbf{x}_1\ \mathbf{x}_2\ \mathbf{x}_3\ \mathbf{x}_4]$ consists of only one chain of generalized eigenvectors. Find a generalized eigenvector $\mathbf{x}_4$ of rank 4, which is a solution of

$$\begin{aligned} (A-I)^3\mathbf{x} &= \begin{bmatrix} 1 & -3 & 3 & -1 \\ 1 & -3 & 3 & -1 \\ 1 & -3 & 3 & -1 \\ 1 & -3 & 3 & -1 \end{bmatrix} \mathbf{x} \neq \mathbf{0}, \\ (A-I)^4\mathbf{x} &= \mathbf{0}. \end{aligned}$$

But, $(A-I)^4 = \mathbf{0}$. Hence, one can take any solution of the first equation for $\mathbf{x}_4$; take

$$\mathbf{x}_4 = (1, 0, 0, 0).$$

Then,

$$\begin{aligned} \mathbf{x}_3 &= (A-I)\mathbf{x}_4 = \begin{bmatrix} 3 & -6 & 4 & -1 \\ 1 & -1 & 0 & 0 \\ 0 & 1 & -1 & 0 \\ 0 & 0 & 1 & -1 \end{bmatrix} \begin{bmatrix} 1 \\ 0 \\ 0 \\ 0 \end{bmatrix} = \begin{bmatrix} 3 \\ 1 \\ 0 \\ 0 \end{bmatrix}, \\ \mathbf{x}_2 &= (A-I)\mathbf{x}_3 = (3,\ 2,\ 1,\ 0), \\ \mathbf{x}_1 &= (A-I)\mathbf{x}_2 = (1, 1, 1, 1). \end{aligned}$$

Now, one can set

$$Q = [\mathbf{x}_1\ \mathbf{x}_2\ \mathbf{x}_3\ \mathbf{x}_4] = \begin{bmatrix} 1 & 3 & 3 & 1 \\ 1 & 2 & 1 & 0 \\ 1 & 1 & 0 & 0 \\ 1 & 0 & 0 & 0 \end{bmatrix}.$$

The reader may check $Q^{-1}AQ = J$ by a direct computation. □

Remark: (1) In order to get a chain $\{\mathbf{x}_1, \mathbf{x}_2, \ldots, \mathbf{x}_k\}$ of generalized eigenvectors, we first choose a solution $\mathbf{x}_k$ of $(A-\lambda I)^k\mathbf{x} = \mathbf{0}$ but $(A-\lambda I)^{k-1}\mathbf{x} \neq \mathbf{0}$. With this $\mathbf{x}_k$, one can get $\mathbf{x}_{k-1}$ as $\mathbf{x}_{k-1} = (A-\lambda I)\mathbf{x}_k$ and $\mathbf{x}_{k-2} = (A-\lambda I)\mathbf{x}_{k-1}$, and so on, and finally get an eigenvector $\mathbf{x}_1$. This method always works,

However, one may also try to find an initial eigenvector $\mathbf{x}_1$ first in order to get a chain $\{\mathbf{x}_1, \mathbf{x}_2, \ldots, \mathbf{x}_k\}$. After that, try to find $\mathbf{x}_2$ as a solution of $(A-\lambda I)\mathbf{x}_2 = \mathbf{x}_1$, and $\mathbf{x}_3$ as a solution of $(A-\lambda I)\mathbf{x}_3 = \mathbf{x}_2$, and so on. But, this may not work sometimes. This is the case in Example 155. In this example, one can see that the eigenvectors of A belonging to $\lambda = 2$ are of the form $(s, 2s, t)$ for any real s, t.

Take two linearly independent eigenvectors to be $\mathbf{u}_1 = (1, 2, 0)$ and $\mathbf{u}_2 = (1, 2, 1)$. If we take the eigenvector $\mathbf{x}_1$ as $\mathbf{u}_1$ or $\mathbf{u}_2$, then a generalized eigenvector $\mathbf{x}_2$ should be a solution of

$$(A - 2I)\mathbf{x} = \begin{bmatrix} 0 & 0 & 0 \\ 0 & 0 & 0 \\ -2 & 1 & 0 \end{bmatrix} \begin{bmatrix} x \\ y \\ z \end{bmatrix} = \mathbf{u}_1 \text{ or } \mathbf{u}_2.$$

But this system is inconsistent in either case and fail to get a generalized eigenvector $\mathbf{x}_2$ in this way.

(2) By expanding $AQ = QJ$, one can see that the columns of Q which are associated with the first columns of Jordan blocks of J form a maximal set of linearly independent eigenvectors of A, and all of the remaining columns of Q except them *are not eigenvectors, all are generalized eigenvectors.*

Proof of Theorem 8.5 For brevity, we prove this theorem only for two distinct eigenvalues. Let $\{\mathbf{x}_1, \mathbf{x}_2, \ldots, \mathbf{x}_k\}$ and $\{\mathbf{y}_1, \mathbf{y}_2, \ldots, \mathbf{y}_\ell\}$ be chains of generalized eigenvectors of A belonging to the eigenvalues λ and μ, respectively, and let $\lambda \neq \mu$. To show the linearly independence of the set of vectors $\{\mathbf{x}_1, \ldots, \mathbf{x}_k, \mathbf{y}_1, \ldots, \mathbf{y}_\ell\}$, let

$$c_1\mathbf{x}_1 + \cdots + c_k\mathbf{x}_k + d_1\mathbf{y}_1 + \cdots + d_\ell\mathbf{y}_\ell = \mathbf{0}$$

with constants c_i's and d_j's. By taking the linear transformation $(A - \lambda I)^k$ on both sides and by noticing that $(A - \lambda I)^k\mathbf{x}_i = \mathbf{0}$ for all $i = 1, \ldots, k$, one can get

$$(A - \lambda I)^k(d_1\mathbf{y}_1 + d_2\mathbf{y}_2 + \cdots + d_\ell\mathbf{y}_\ell) = \mathbf{0}.$$

Again, multiply this equation by $(A - \mu I)^{\ell-1}$ and note that

$$\begin{aligned} (A - \mu I)^{\ell-1}(A - \lambda I)^k &= (A - \lambda I)^k(A - \mu I)^{\ell-1}, \\ (A - \mu I)^{\ell-1}\mathbf{y}_\ell &= \mathbf{y}_1, \\ (A - \mu I)^{\ell-1}\mathbf{y}_i &= \mathbf{0} \end{aligned}$$

for $i = 1, \ldots, \ell - 1$. Thus we obtain

$$\mathbf{0} = d_\ell(A - \lambda I)^k\mathbf{y}_1.$$

Because $A\mathbf{y}_1 = \mu\mathbf{y}_1$ for an eigenvector $\mathbf{y}_1$, this reduces to

$$\mathbf{0} = d_\ell(\mu - \lambda)^k\mathbf{y}_1,$$

which implies that $d_\ell = 0$ by the assumption $\lambda \neq \mu$ and $\mathbf{y}_1 \neq \mathbf{0}$. Proceeding successively with $(A-\mu I)^{\ell-2}$ next, one can show that $d_i = 0, i = \ell, \ell-1, \ldots, 2, 1$, so we are left with

$$c_1\mathbf{x}_1 + \cdots + c_k\mathbf{x}_k = \mathbf{0}.$$

Since $\{\mathbf{x}_1, \ldots, \mathbf{x}_k\}$ is already linearly independent by Theorem 8.4, $c_i = 0$ for all $i = 1, \ldots, k$. Thus the set of generalized eigenvectors $\{\mathbf{x}_1, \ldots, \mathbf{x}_k, \mathbf{y}_1, \ldots, \mathbf{y}_\ell\}$ is linearly independent. It completes the proof of Theorem 8.5. □

We leave a proof of Theorem 8.6 as an exercise even though it can be proved by induction on the number of vectors in $S \cup T$.

Problem 198 (From Problem 197) Find a change of basis matrix Q so that $Q^{-1}AQ$ is a Jordan canonical matrix for each of the following matrices A.

$$(1)\ \begin{bmatrix} i & 0 \\ 1 & i \end{bmatrix}, \qquad (2)\ \begin{bmatrix} 2 & 1 & 0 \\ 0 & 2 & 1 \\ 0 & 0 & 3 \end{bmatrix}, \qquad (3)\ \begin{bmatrix} 4 & 1 & 2 & 0 \\ 0 & 4 & 2 & 0 \\ 0 & 0 & 4 & 2 \\ 0 & 0 & 0 & 2 \end{bmatrix}.$$

Problem 199 Prove Theorem 8.6 by induction on ℓ when $k = 1$. Next try it when $k = 2$.

Problem 200 Let $\mathbf{x}$ be a generalized eigenvector of A of rank k belonging to an eigenvalue λ and let $\{\mathbf{x}_1, \mathbf{x}_2, \ldots, \mathbf{x}_k\}$ be a chain of generalized eigenvectors with $\mathbf{x} = \mathbf{x}_k$. For each $\mathbf{x}_j$, show that $\mathbf{x}_1 = (A - \lambda I)^{j-1}\mathbf{x}$ is an eigenvector of A belonging to λ. Hence, all of $\mathbf{x}_i$'s except $\mathbf{x}_1$ are not eigenvectors of A belonging to λ.

8.5 Computing J^k and e^J for a Jordan matrix J

If a matrix A is diagonal, it is easy (see Theorem 124) to compute the power A^k and the exponential matrix $e^A (= \exp A)$:

$$\begin{bmatrix} \lambda_1 & & \mathbf{0} \\ & \ddots & \\ \mathbf{0} & & \lambda_n \end{bmatrix}^k = \begin{bmatrix} \lambda_1^k & & \mathbf{0} \\ & \ddots & \\ \mathbf{0} & & \lambda_n^k \end{bmatrix}, \quad \exp \begin{bmatrix} \lambda_1 & & \mathbf{0} \\ & \ddots & \\ \mathbf{0} & & \lambda_n \end{bmatrix} = \begin{bmatrix} e^{\lambda_1} & & \mathbf{0} \\ & \ddots & \\ \mathbf{0} & & e^{\lambda_n} \end{bmatrix}$$

If A is diagonalizable, its EVD $A = QDQ^{-1}$ gives

$$A^k = QD^kQ^{-1} \quad \text{and} \quad e^A = Qe^DQ^{-1},$$

as shown in Sections 6.3 and 6.6.

Now we try to do the same computing for a non-diagonalizable matrix A.

It can be dealt with in two ways. Firstly, one can do it by computing the power J^k and the exponential matrix e^J with a Jordan decomposition $A = QJQ^{-1}$ of A. This will be done in this section.

The second method is based on the Cayley-Hamilton theorem or the minimal polynomial. This will be shown later in Section 8.7 and in Subsections 8.8.2–8.8.3.

Let us start by computing J^k for a single Jordan block J. An $n \times n$ Jordan block J with an eigenvalue λ can be written as

$$\begin{aligned} J = \begin{bmatrix} \lambda & 1 & & 0 \\ 0 & \ddots & \ddots & \\ \vdots & \ddots & \lambda & 1 \\ 0 & \cdots & 0 & \lambda \end{bmatrix} &= \lambda \begin{bmatrix} 1 & 0 & & 0 \\ 0 & \ddots & \ddots & \\ \vdots & \ddots & 1 & 0 \\ 0 & \cdots & 0 & 1 \end{bmatrix} + \begin{bmatrix} 0 & 1 & \cdots & 0 \\ 0 & \ddots & \ddots & 0 \\ \vdots & \ddots & 0 & 1 \\ 0 & \cdots & 0 & 0 \end{bmatrix} \\ &= \lambda I + N. \end{aligned}$$

Since I is the identity matrix, clearly $IN = NI$ and [2]

$$J^k = (\lambda I + N)^k = \sum_{j=0}^{k} \binom{k}{j} (\lambda I)^{k-j} N^j = \sum_{j=0}^{k} \binom{k}{j} \lambda^{k-j} N^j.$$

But, N is nilpotent and $N^k = \mathbf{0}$ for $k \geq n$. Thus, with $\binom{k}{\ell} = 0$ if $k < \ell$ by

[2] The *binomial formula* or *binomial expansion:* For any (real or complex) numbers x, y and for any natural number n,

$$\begin{aligned} (x+y)^n &= \sum_{k=0}^{n} \binom{n}{k} x^k y^{n-k} \\ &= \binom{n}{0} x^0 y^n + \binom{n}{1} x^1 y^{n-1} + \cdots + \binom{n}{n-1} x^{n-1} y^1 + \binom{n}{n} x^n y^0, \end{aligned}$$

where each $\binom{n}{k}$ is a binomial coefficient, and $\binom{n}{0} = x^0 = y^0 = 1$ by definition.

For any two square matrices A, B of the same size and for any natural number n, if $AB = BA$,

$$\begin{aligned} (A+B)^n &= \sum_{k=0}^{n} \binom{n}{k} A^k B^{n-k} \\ &= \binom{n}{0} A^0 B^n + \binom{n}{1} A^1 B^{n-1} + \cdots + \binom{n}{n-1} A^{n-1} B^1 + \binom{n}{n} A^n B^0, \end{aligned}$$

where each $\binom{n}{k}$ is a binomial coefficient, and A^0, B^0 are the identity matrix by definition.

definition,

$$
\begin{aligned}
J^k &= \sum_{j=0}^{n-1} \binom{k}{j} \lambda^{k-j} N^j \\
&= \lambda^k I + \binom{k}{1} \lambda^{k-1} N + \cdots + \binom{k}{n-1} \lambda^{k-(n-1)} N^{n-1} \\
&= \begin{bmatrix}
\lambda^k & \binom{k}{1}\lambda^{k-1} & \binom{k}{2}\lambda^{k-2} & \cdots & \binom{k}{n-1}\lambda^{k-n+1} \\
0 & \lambda^k & \binom{k}{1}\lambda^{k-1} & \cdots & \binom{k}{n-2}\lambda^{k-n+2} \\
\vdots & & \ddots & & \vdots \\
\vdots & & & \lambda^k & \binom{k}{1}\lambda^{k-1} \\
0 & \cdots & \cdots & 0 & \lambda^k
\end{bmatrix} \\
&= \begin{bmatrix}
\lambda^k & \frac{\partial_\lambda \lambda^k}{1!} & \frac{\partial_\lambda^2 \lambda^k}{2!} & \cdots & \frac{\partial_\lambda^{n-1} \lambda^k}{(n-1)!} \\
0 & \lambda^k & \frac{\partial_\lambda \lambda^k}{1!} & \vdots & \frac{\partial_\lambda^{n-2} \lambda^k}{(n-2)!} \\
0 & 0 & \ddots & \ddots & \vdots \\
\vdots & \cdots & \ddots & \lambda^k & \frac{\partial_\lambda \lambda^k}{1!} \\
0 & \cdots & \cdots & 0 & \lambda^k
\end{bmatrix},
\end{aligned}
\tag{8.1}
$$

where ∂_λ denotes the derivative with respect to λ, and $\partial_\lambda^2 = \partial_\lambda \partial_\lambda$ denotes the second derivative with respect to λ, etc.

If the block J is invertible, equivalently $\lambda \neq 0$, then J^{-1} is defined as the inverse of J and $J^{-2} = J^{-1}J^{-1}$, etc. In this case, Eq. (8.1) is still valid for negative integers $k < 0$, (see Exercise **8.3**), with $\binom{k}{j} = \frac{1}{j!}k(k-1)\cdots(k-j+1)$ for $k < 0$.

Now, let $A = QJQ^{-1}$ be a Jordan decomposition with $J = \begin{bmatrix} J_1 & & \\ & \ddots & \\ & & J_s \end{bmatrix}$.

Then, one can compute

$$
A^k = QJ^kQ^{-1} = Q \begin{bmatrix} J_1^k & & \\ & \ddots & \\ & & J_s^k \end{bmatrix} Q^{-1}
$$

for $k = 1, 2, \ldots$, by computing J_i^k for each Jordan block J_i.

Next, to compute the exponential matrix e^A, first note that

$$e^A = e^{QJQ^{-1}} = Qe^J Q^{-1} = Q \begin{bmatrix} e^{J_1} & & & \mathbf{0} \\ & e^{J_2} & & \\ & & \ddots & \\ \mathbf{0} & & & e^{J_s} \end{bmatrix} Q^{-1}$$

Thus, it is enough to compute e^J for a single Jordan block J. Let $J = \lambda I + N$ as before. Then, $N^k = \mathbf{0}$ for $k \geq n$ and

$$\begin{aligned} e^J &= e^{\lambda I} e^N = e^{\lambda} I e^N = e^{\lambda} \sum_{k=0}^{n-1} \frac{N^k}{k!} \\ &= e^{\lambda} \begin{bmatrix} 1 & \frac{1}{1!} & \frac{1}{2!} & \cdots & \frac{1}{(n-1)!} \\ 0 & 1 & \frac{1}{1!} & \cdots & \frac{1}{(n-2)!} \\ 0 & 0 & \ddots & \ddots & \vdots \\ \vdots & \cdots & & 1 & \frac{1}{1!} \\ 0 & \cdots & \cdots & 0 & 1 \end{bmatrix} \\ &= \begin{bmatrix} e^{\lambda} & \frac{\partial_{\lambda} e^{\lambda}}{1!} & \frac{\partial_{\lambda}^2 e^{\lambda}}{2!} & \cdots & \frac{\partial_{\lambda}^{n-1} e^{\lambda}}{(n-1)!} \\ 0 & e^{\lambda} & \frac{\partial_{\lambda} e^{\lambda}}{1!} & \vdots & \frac{\partial_{\lambda}^{n-2} e^{\lambda}}{(n-2)!} \\ 0 & 0 & \ddots & \ddots & \vdots \\ \vdots & \cdots & \ddots & e^{\lambda} & \frac{\partial_{\lambda} e^{\lambda}}{1!} \\ 0 & \cdots & \cdots & 0 & e^{\lambda} \end{bmatrix}. \end{aligned} \tag{8.2}$$

Eqs. (8.1) and (8.2) are frequently used in computing matrix powers and matrix exponents.

Example 157 (*Computing A^k and e^A by using the Jordan canonical form*) Compute the power A^k and the exponential matrix e^A by using the Jordan canonical form of A for

$$(1)\ A = \begin{bmatrix} 2 & 1 & 4 \\ 0 & 3 & 1 \\ 0 & 0 & 3 \end{bmatrix}, \qquad (2)\ A = \begin{bmatrix} 4 & -6 & 4 & -1 \\ 1 & 0 & 0 & 0 \\ 0 & 1 & 0 & 0 \\ 0 & 0 & 1 & 0 \end{bmatrix}.$$

Solution: (1) From Examples 148 and 154, one can see that

$$A = QJQ^{-1} = \begin{bmatrix} 1 & 1 & 4 \\ 0 & 1 & 1 \\ 0 & 0 & 1 \end{bmatrix} \begin{bmatrix} 2 & 0 & 0 \\ 0 & 3 & 1 \\ 0 & 0 & 3 \end{bmatrix} \begin{bmatrix} 1 & -1 & -3 \\ 0 & 1 & -1 \\ 0 & 0 & 1 \end{bmatrix},$$

and by Eqs. (8.1) and (8.2),

$$J^k = \begin{bmatrix} 2^k & 0 & 0 \\ 0 & 3^k & \binom{k}{1}3^{k-1} \\ 0 & 0 & 3^k \end{bmatrix} \quad \text{and} \quad e^J = \begin{bmatrix} e^2 & 0 & 0 \\ 0 & e^3 & e^3 \\ 0 & 0 & e^3 \end{bmatrix}.$$

Hence,

$$\begin{aligned} A^k = QJ^kQ^{-1} &= \begin{bmatrix} 1 & 1 & 4 \\ 0 & 1 & 1 \\ 0 & 0 & 1 \end{bmatrix} \begin{bmatrix} 2^k & 0 & 0 \\ 0 & 3^k & \binom{k}{1}3^{k-1} \\ 0 & 0 & 3^k \end{bmatrix} \begin{bmatrix} 1 & -1 & -3 \\ 0 & 1 & -1 \\ 0 & 0 & 1 \end{bmatrix} \\ &= \begin{bmatrix} 2^k & -2^k + 3^k & 3^{k-1}k - 3 \cdot 2^k + 3^{k+1} \\ 0 & 3^k & 3^{k-1}k \\ 0 & 0 & 3^k \end{bmatrix} \end{aligned}$$

and

$$\begin{aligned} e^A = Qe^JQ^{-1} &= \begin{bmatrix} 1 & 1 & 4 \\ 0 & 1 & 1 \\ 0 & 0 & 1 \end{bmatrix} \begin{bmatrix} e^2 & 0 & 0 \\ 0 & e^3 & e^3 \\ 0 & 0 & e^3 \end{bmatrix} \begin{bmatrix} 1 & -1 & -3 \\ 0 & 1 & -1 \\ 0 & 0 & 1 \end{bmatrix} \\ &= \begin{bmatrix} e^2 & -e^2 + e^3 & -3e^2 + 4e^3 \\ 0 & e^3 & e^3 \\ 0 & 0 & e^3 \end{bmatrix}. \end{aligned}$$

(2) We do the same process as (1). From Examples 149 and 156,

$$A = QJQ^{-1} = \begin{bmatrix} 1 & 3 & 3 & 1 \\ 1 & 2 & 1 & 0 \\ 1 & 1 & 0 & 0 \\ 1 & 0 & 0 & 0 \end{bmatrix} \begin{bmatrix} 1 & 1 & 0 & 0 \\ 0 & 1 & 1 & 0 \\ 0 & 0 & 1 & 1 \\ 0 & 0 & 0 & 1 \end{bmatrix} \begin{bmatrix} 0 & 0 & 0 & 1 \\ 0 & 0 & 1 & -1 \\ 0 & 1 & -2 & 1 \\ 1 & -3 & 3 & -1 \end{bmatrix},$$

and by Eqs. (8.1) and (8.2),

$$J^k = \begin{bmatrix} 1 & \binom{k}{1} & \binom{k}{2} & \binom{k}{3} \\ 0 & 1 & \binom{k}{1} & \binom{k}{2} \\ 0 & 0 & 1 & \binom{k}{1} \\ 0 & 0 & 0 & 1 \end{bmatrix} \quad \text{and} \quad e^J = e \begin{bmatrix} 1 & 1 & \frac{1}{2!} & \frac{1}{3!} \\ 0 & 1 & 1 & \frac{1}{2!} \\ 0 & 0 & 1 & 1 \\ 0 & 0 & 0 & 1 \end{bmatrix}.$$

Now, one can compute $A^k = QJ^kQ^{-1}$ and $e^A = Qe^JQ^{-1}$. For example,

$$\begin{aligned} e^A &= e\begin{bmatrix}1&3&3&1\\1&2&1&0\\1&1&0&0\\1&0&0&0\end{bmatrix}\begin{bmatrix}1&1&\frac{1}{2!}&\frac{1}{3!}\\0&1&1&\frac{1}{2!}\\0&0&1&1\\0&0&0&1\end{bmatrix}\begin{bmatrix}0&0&0&1\\0&0&1&-1\\0&1&-2&1\\1&-3&3&-1\end{bmatrix} \\ &= \frac{e}{6}\begin{bmatrix}34&-63&48&-13\\13&-18&15&-4\\4&-3&6&-1\\1&0&3&2\end{bmatrix}. \end{aligned}$$

□

Example 158 (*Computing A^k and e^A by using the Jordan canonical form*) Compute the power A^k and the exponential matrix e^A by using the Jordan canonical form for

$$A = \begin{bmatrix}1&0&0&-1\\1&2&0&1\\1&2&2&0\\1&0&0&3\end{bmatrix}.$$

Solution: (1) The characteristic polynomial of A is $\det(\lambda I - A) = \lambda^4 - 8\lambda^3 + 24\lambda^2 - 32\lambda + 16 = (\lambda - 2)^4$ and $\lambda = 2$ is an eigenvalue of multiplicity 4. Since

$$\operatorname{rank}(A - 2I) = \operatorname{rank}\begin{bmatrix}-1&0&0&-1\\1&0&0&1\\1&2&0&0\\1&0&0&1\end{bmatrix} = 2$$

and

$$\operatorname{rank}(A - 2I)^2 = \operatorname{rank}\begin{bmatrix}0&0&0&0\\0&0&0&0\\1&0&0&1\\0&0&0&0\end{bmatrix} = 1,$$

the Jordan canonical form J of A should be of the form

$$J = \begin{bmatrix}2&1&0&0\\0&2&1&0\\0&0&2&0\\0&0&0&2\end{bmatrix} = \begin{bmatrix}\begin{bmatrix}2&1&0\\0&2&1\\0&0&2\end{bmatrix} & \mathbf{0}\\ \mathbf{0} & [2]\end{bmatrix}.$$

To find a change of basis matrix Q such that $Q^{-1}AQ = J$, first need to find a generalized eigenvector of rank 3. This is a solution $\mathbf{x}$ to the following system:

$$\begin{aligned} (A-2I)^2\mathbf{x} &= \begin{bmatrix} 0 & 0 & 0 & 0 \\ 0 & 0 & 0 & 0 \\ 1 & 0 & 0 & 1 \\ 0 & 0 & 0 & 0 \end{bmatrix} \begin{bmatrix} x \\ y \\ z \\ w \end{bmatrix} \neq \mathbf{0}, \\ (A-2I)^3\mathbf{x} &= \mathbf{0}. \end{aligned}$$

But, a direct computation shows that $(A-2I)^3 = \mathbf{0}$. Hence, one can choose any solution of the first equation as a generalized eigenvector of rank 3; choose

$$\mathbf{x}_3 = (0, 0, 0, 1).$$

Then,

$$\mathbf{x}_2 = (A-2I)\mathbf{x}_3 = \begin{bmatrix} -1 & 0 & 0 & -1 \\ 1 & 0 & 0 & 1 \\ 1 & 2 & 0 & 0 \\ 1 & 0 & 0 & 1 \end{bmatrix} \begin{bmatrix} 0 \\ 0 \\ 0 \\ 1 \end{bmatrix} = \begin{bmatrix} -1 \\ 1 \\ 0 \\ 1 \end{bmatrix}$$

and

$$\mathbf{x}_1 = (A-2I)\mathbf{x}_2 = \begin{bmatrix} -1 & 0 & 0 & -1 \\ 1 & 0 & 0 & 1 \\ 1 & 2 & 0 & 0 \\ 1 & 0 & 0 & 1 \end{bmatrix} \begin{bmatrix} -1 \\ 1 \\ 0 \\ 1 \end{bmatrix} = \begin{bmatrix} 0 \\ 0 \\ 1 \\ 0 \end{bmatrix}$$

Now, by taking another eigenvector $\mathbf{x}_4 = (-2, 1, 0, 2)$ belonging to $\lambda = 2$, so that $\mathbf{x}_1$ and $\mathbf{x}_4$ are linearly independent, one can get

$$Q = \{\mathbf{x}_1, \mathbf{x}_2, \mathbf{x}_3, \mathbf{x}_4\} = \begin{bmatrix} 0 & -1 & 0 & -2 \\ 0 & 1 & 0 & 1 \\ 1 & 0 & 0 & 0 \\ 0 & 1 & 1 & 2 \end{bmatrix}.$$

One may check by a direct computation

$$Q^{-1}AQ = \begin{bmatrix} 2 & 1 & 0 & 0 \\ 0 & 2 & 1 & 0 \\ 0 & 0 & 2 & 0 \\ 0 & 0 & 0 & 2 \end{bmatrix} = J \quad \text{with } Q^{-1} = \begin{bmatrix} 0 & 0 & 1 & 0 \\ 1 & 2 & 0 & 0 \\ 1 & 0 & 0 & 1 \\ -1 & -1 & 0 & 0 \end{bmatrix}.$$

Therefore,

$$A^k = QJ^kQ^{-1} = Q\begin{bmatrix} \begin{bmatrix} 2 & 1 & 0 \\ 0 & 2 & 1 \\ 0 & 0 & 2 \end{bmatrix}^k & \mathbf{0} \\ \mathbf{0} & [2]^k \end{bmatrix} Q^{-1},$$

and by Eq. (8.1),

$$\begin{bmatrix} 2 & 1 & 0 \\ 0 & 2 & 1 \\ 0 & 0 & 2 \end{bmatrix}^k = \begin{bmatrix} 2^k & \binom{k}{1}2^{k-1} & \binom{k}{2}2^{k-2} \\ 0 & 2^k & \binom{k}{1}2^{k-1} \\ 0 & 0 & 2^k \end{bmatrix}.$$

Hence,

$$\begin{aligned} A^k &= QJ^kQ^{-1} = Q\begin{bmatrix} \begin{bmatrix} 2^k & \binom{k}{1}2^{k-1} & \binom{k}{2}2^{k-2} \\ 0 & 2^k & \binom{k}{1}2^{k-1} \\ 0 & 0 & 2^k \end{bmatrix} & \mathbf{0} \\ \mathbf{0} & 2^k \end{bmatrix} Q^{-1} \\ &= \begin{bmatrix} 2^k - k2^{k-1} & 0 & 0 & -k2^{k-1} \\ k2^{k-1} & 2^k & 0 & k2^{k-1} \\ 2^{k-3}(3k+k^2) & k2^k & 2^k & k(k-1)2^{k-3} \\ k2^{k-1} & 0 & 0 & k2^{k-1} + 2^k \end{bmatrix}. \end{aligned}$$

(2) With the same notation,

$$e^A = e^{QJQ^{-1}} = Qe^JQ^{-1} = e^2Q\begin{bmatrix} 1 & 1 & \frac{1}{2} & 0 \\ 0 & 1 & 1 & 0 \\ 0 & 0 & 1 & 0 \\ 0 & 0 & 0 & 1 \end{bmatrix} Q^{-1} = \begin{bmatrix} 0 & 0 & 0 & -e^2 \\ e^2 & e^2 & 0 & e^2 \\ \frac{3}{2}e^2 & 2e^2 & e^2 & \frac{1}{2}e^2 \\ e^2 & 0 & 0 & 2e^2 \end{bmatrix}$$

by Eq. (8.2). □

Problem 201 Compute N^k and e^N for each of the following nilpotent matrices

$$(1)\ N = \begin{bmatrix} 0 & 1 & 0 \\ 0 & 0 & 1 \\ 0 & 0 & 0 \end{bmatrix}, \qquad (2)\ N = \begin{bmatrix} 0 & 1 & 0 & 0 \\ 0 & 0 & 1 & 0 \\ 0 & 0 & 0 & 1 \\ 0 & 0 & 0 & 0 \end{bmatrix}.$$

Problem 202 Compute J^k and e^J for each of the following Jordan blocks

$$(1)\ J = \begin{bmatrix} 3 & 1 & 0 \\ 0 & 3 & 1 \\ 0 & 0 & 3 \end{bmatrix}, \qquad (2)\ J = \begin{bmatrix} \lambda & 1 & 0 & 0 \\ 0 & \lambda & 1 & 0 \\ 0 & 0 & \lambda & 1 \\ 0 & 0 & 0 & \lambda \end{bmatrix}.$$

Hint: Use Eqs. (8.1) and (8.2).

Problem 203 Compute A^k and e^A by using the Jordan canonical form of

$$(1)\ A = \begin{bmatrix} 3 & 1 & 0 & 0 \\ 0 & 3 & 1 & 0 \\ 0 & 0 & 3 & 0 \\ 0 & 0 & 0 & 1 \end{bmatrix}, \qquad (2)\ A = \begin{bmatrix} 4 & -5 & 3 & -2 \\ 2 & -1 & 1 & -2 \\ 2 & -1 & 1 & -2 \\ 2 & -5 & 3 & 0 \end{bmatrix}.$$

8.6 Cayley-Hamilton theorem

As shown in earlier chapters, the association of the characteristic polynomial with a matrix is very useful in studying the diagonalization of a matrix.

In this section, using this association of characteristic polynomials with matrices, we introduce another useful theorem, called the *Cayley-Hamilton theorem*, which makes the computing of matrix polynomials simple, and has many applications to practical problems.

Let $f(x) = a_m x^m + a_{m-1}x^{m-1} + \cdots + a_1 x + a_0$ be a polynomial, and let A be an $n \times n$ square matrix. The matrix defined by

$$f(A) = a_m A^m + a_{m-1}A^{m-1} + \cdots + a_1 A + a_0 I_n$$

is called a **matrix polynomial** of A.

Theorem 8.7 (Cayley-Hamilton) [3] *For any $n \times n$ matrix A, let $f(\lambda) = \det(\lambda I - A)$ be the characteristic polynomial of A. Then $f(A) = \mathbf{0}$. In other words, every square matrix satisfies its own characteristic equation.*

[3]This was first proved in 1853 by William Rowan Hamilton (August 4, 1805– September 2, 1865) for a special case of 4×4 real or 2×2 complex matrices. Arthur Cayley (August 16, 1821– January 26, 1895) in 1858 stated it for 3×3 and smaller matrices, but only published a proof for the 2×2 case. The general case was first proved by the German mathematician Ferdinand Georg Frobenius (October 26, 1849–August 3, 1917) in 1878.

Arthur Cayley was a British mathematician. He helped found the modern British school of pure mathematics. He was the first to define the concept of a group in the modern way - as a set with a binary operation satisfying certain laws.

William Rowan Hamilton was an Irish physicist, astronomer, and mathematician. His best known contribution to mathematical physics is the reformulation of Newtonian mechanics.

Proof: If A is diagonal, its proof is certainly easy. For example, if $A = \begin{bmatrix} \lambda_1 & 0 \\ 0 & \lambda_2 \end{bmatrix}$, then its characteristic polynomial is $f(\lambda) = (\lambda - \lambda_1)(\lambda - \lambda_2)$. Hence,

$$f(A) = (A - \lambda_1 I)(A - \lambda_2 I) = \begin{bmatrix} \lambda_1 - \lambda_1 & 0 \\ 0 & \lambda_2 - \lambda_1 \end{bmatrix} \begin{bmatrix} \lambda_1 - \lambda_2 & 0 \\ 0 & \lambda_2 - \lambda_2 \end{bmatrix} = \mathbf{0}.$$

For an arbitrary square matrix A, let its Jordan canonical form be

$$J = \begin{bmatrix} J_1 & & \mathbf{0} \\ & \ddots & \\ \mathbf{0} & & J_s \end{bmatrix} = Q^{-1}AQ,$$

so that $f(A) = Qf(J)Q^{-1}$. Since

$$J^k = \begin{bmatrix} J_1^k & & \mathbf{0} \\ & \ddots & \\ \mathbf{0} & & J_s^k \end{bmatrix} \quad \text{and} \quad f(J) = \begin{bmatrix} f(J_1) & & \mathbf{0} \\ & \ddots & \\ \mathbf{0} & & f(J_s) \end{bmatrix},$$

it is sufficient to show that $f(J_i) = \mathbf{0}$ for each Jordan block J_i. Let $J_i = \lambda_0 I + N$ with eigenvalue λ_0 of multiplicity m, for which $N^m = \mathbf{0}$. Since $f(\lambda) = \det(\lambda I - A) = \det(\lambda I - J) = \det(\lambda I - J_i) \prod_{j \neq i} \det(\lambda I - J_j) = (\lambda - \lambda_0)^m g(\lambda)$ for some polynomial $g(\lambda)$, we have

$$f(J_i) = (J_i - \lambda_0 I)^m g(J_i) = (\lambda_0 I + N - \lambda_0 I)^m g(J_i) = N^m g(J_i) = \mathbf{0}\, g(J_i) = \mathbf{0},$$

as wanted. □

Example 159 (*Cayley-Hamilton theorem says every square matrix satisfies its own characteristic equation*) The characteristic polynomial of

$$A = \begin{bmatrix} 2 & 1 & 4 \\ 0 & 3 & 1 \\ 0 & 0 & 3 \end{bmatrix}$$

is $f(\lambda) = (\lambda - 2)(\lambda - 3)^2$ and

$$\begin{aligned} f(A) &= (A - 2I)(A - 3I)^2 \\ &= \begin{bmatrix} 0 & 1 & 4 \\ 0 & 1 & 1 \\ 0 & 0 & 1 \end{bmatrix} \begin{bmatrix} -1 & 1 & 4 \\ 0 & 0 & 1 \\ 0 & 0 & 0 \end{bmatrix}^2 = \begin{bmatrix} 0 & 1 & 4 \\ 0 & 1 & 1 \\ 0 & 0 & 1 \end{bmatrix} \begin{bmatrix} 1 & -1 & 3 \\ 0 & 0 & 0 \\ 0 & 0 & 0 \end{bmatrix} \\ &= \begin{bmatrix} 0 & 0 & 0 \\ 0 & 0 & 0 \\ 0 & 0 & 0 \end{bmatrix}. \end{aligned}$$

□

The Cayley-Hamilton theorem can be used to find the inverse of a nonsingular matrix.

If $f(\lambda) = \lambda^n + a_{n-1}\lambda^{n-1} + \cdots + a_1\lambda + a_0$ is the characteristic polynomial of a matrix A, then

$$\begin{aligned} \mathbf{0} = f(A) &= A^n + a_{n-1}A^{n-1} + \cdots + a_1 A + a_0 I, \\ \text{or} \qquad -a_0 I &= (A^{n-1} + a_{n-1}A^{n-2} + \cdots + a_1 I)A. \end{aligned}$$

Since $a_0 = f(0) = \det(0I - A) = \det(-A) = (-1)^n \det A$, the matrix A is nonsingular if and only if $a_0 = (-1)^n \det A \neq 0$. Therefore, if A is nonsingular,

$$A^{-1} = -\frac{1}{a_0}(A^{n-1} + a_{n-1}A^{n-2} + \cdots + a_1 I). \tag{8.3}$$

From this identity, one can see that A^{-1} (so A^k for every negative integer k) can also be written as a matrix polynomial of A.

Example 160 (*Compute A^{-1} by the Cayley-Hamilton theorem*) With the same matrix

$$A = \begin{bmatrix} 2 & 1 & 4 \\ 0 & 3 & 1 \\ 0 & 0 & 3 \end{bmatrix}$$

given in Example 148, A has the characteristic polynomial

$$f(\lambda) = (\lambda - 2)(\lambda - 3)^2 = \lambda^3 - 8\lambda^2 + 21\lambda - 18.$$

The Cayley-Hamilton theorem yields

$$A^3 - 8A^2 + 21A = 18I_3.$$

Hence

$$\begin{aligned} A^{-1} &= \frac{1}{18}(A^2 - 8A + 21I_3) \\ &= \frac{1}{18}\begin{bmatrix} 4 & 5 & 21 \\ 0 & 9 & 6 \\ 0 & 0 & 9 \end{bmatrix} - \frac{8}{18}\begin{bmatrix} 2 & 1 & 4 \\ 0 & 3 & 1 \\ 0 & 0 & 3 \end{bmatrix} + \frac{21}{18}\begin{bmatrix} 1 & 0 & 0 \\ 0 & 1 & 0 \\ 0 & 0 & 1 \end{bmatrix} \\ &= \frac{1}{18}\begin{bmatrix} 9 & -3 & -11 \\ 0 & 6 & -2 \\ 0 & 0 & 6 \end{bmatrix}. \end{aligned}$$

□

Problem 204 Prove: Let λ be an eigenvalue of a matrix A, and let $f(x)$ be any polynomial. Then, $f(\lambda)$ is an eigenvalue of the matrix polynomial $f(A)$.

Hint: See Problem 137.

Problem 205 Let's try to prove the Cayley-Hamilton Theorem 8.7 as follows: By setting $\lambda = A$, $f(A) = \det(AI - A) = \det \mathbf{0} = 0$. Is this correct or not? If not, which step is incorrect?

Problem 206 Let A and B be square matrices, not necessarily of the same order, and let $f(\lambda) = \det(\lambda I - A)$ be the characteristic polynomial of A. Show that $f(B)$ is invertible if and only if A has no eigenvalues in common with B.

Hint: If $f(\lambda) = (\lambda - \lambda_1)\cdots(\lambda - \lambda_n)$, then $f(B) = (B - \lambda_1 I_m)\cdots(B - \lambda_n I_m)$.

8.7 Computing the power A^k by using the Cayley-Hamilton theorem

The Cayley-Hamilton theorem makes the computing of matrix polynomials simple.

Let $p(\lambda)$ be any polynomial (commonly, of very high degree) and let $f(\lambda)$ be the characteristic polynomial of a square matrix A. The division algorithm for polynomials [4] says that there are polynomials $q(\lambda)$ and $r(\lambda)$ such that

$$p(\lambda) = q(\lambda)f(\lambda) + r(\lambda),$$

where the degree of the remainder $r(\lambda)$ is less than the degree of $f(\lambda)$. Then

$$p(A) = q(A)f(A) + r(A).$$

By the Cayley-Hamilton theorem, $f(A) = \mathbf{0}$ and then

$$p(A) = r(A).$$

Thus, the problem of computing a polynomial of an $n \times n$ matrix A or in particular a power A^k can be reduced to the problem of computing a matrix polynomial of degree less than n.

[4]This is an obvious extension of *the division algorithm for integers*: For any two integers, n, m with $m > 0$, the division n/m gives the quotient q and the remainder $r, 0 \le r < m$ so that $n = mq + r$.

Example 161 (*Computing a matrix polynomial $p(A)$ by the Cayley-Hamilton theorem*) The characteristic polynomial of a matrix $A = \begin{bmatrix} 1 & 2 \\ 2 & 1 \end{bmatrix}$ is

$$f(\lambda) = \lambda^2 - 2\lambda - 3.$$

Let $p(\lambda) = \lambda^4 - 3\lambda^2 + \lambda + 4$ be a given polynomial. A division by $f(\lambda)$ gives

$$p(\lambda) = (\lambda^2 + 2\lambda + 4)f(\lambda) + 15\lambda + 16.$$

Therefore

$$\begin{aligned} A^4 - 3A^2 + A + 4I = p(A) &= (A^2 + 2A + 4I)f(A) + 15A + 16I \\ &= 15A + 16I \\ &= 15\begin{bmatrix} 1 & 2 \\ 2 & 1 \end{bmatrix} + 16\begin{bmatrix} 1 & 0 \\ 0 & 1 \end{bmatrix} = \begin{bmatrix} 31 & 30 \\ 30 & 31 \end{bmatrix}. \end{aligned}$$

Similarly one can compute any power A^n with a division of $p(\lambda) = \lambda^n$ by $f(\lambda) = \lambda^2 - 2\lambda - 3$. For example, if $p(\lambda) = \lambda^5$ then a division by $f(\lambda)$ gives

$$p(\lambda) = (\lambda^3 + 2\lambda^2 + 7\lambda + 20I)f(\lambda) + 61\lambda + 60.$$

Therefore

$$\begin{aligned} A^5 = p(A) &= 61A + 60I \\ &= 61\begin{bmatrix} 1 & 2 \\ 2 & 1 \end{bmatrix} + 60\begin{bmatrix} 1 & 0 \\ 0 & 1 \end{bmatrix} = \begin{bmatrix} 121 & 122 \\ 122 & 121 \end{bmatrix}. \end{aligned}$$ □

Example 162 (*Computing A^k by the Cayley-Hamilton theorem*) Compute the power A^8 by using the Cayley-Hamilton theorem for

$$A = \begin{bmatrix} 1 & 0 & 0 & -1 \\ 1 & 2 & 0 & 1 \\ 1 & 2 & 2 & 0 \\ 1 & 0 & 0 & 3 \end{bmatrix}.$$

Solution: The characteristic polynomial of A is

$$f(\lambda) = \det(\lambda I - A) = \lambda^4 - 8\lambda^3 + 24\lambda^2 - 32\lambda + 16 = (\lambda - 2)^4,$$

see Example 158.

A division of λ^8 by $f(\lambda)$ gives

$$\lambda^8 = q(\lambda)f(\lambda) + r(\lambda)$$

with the quotient $q(\lambda) = \lambda^4 + 8\lambda^3 + 40\lambda^2 + 160\lambda + 560$ and the remainder $r(\lambda) = 1792(\lambda-2)^3 + 1792(\lambda-2)^2 + 1024(\lambda-2) + 256$. Hence,

$$\begin{aligned} A^8 &= r(A) = 1792(A-2I)^3 + 1792(A-2I)^2 + 1024(A-2I) + 256I \\ &= \mathbf{0} + 1792\begin{bmatrix} 0 & 0 & 0 & 0 \\ 0 & 0 & 0 & 0 \\ 1 & 0 & 0 & 1 \\ 0 & 0 & 0 & 0 \end{bmatrix} + 1024\begin{bmatrix} -1 & 0 & 0 & -1 \\ 1 & 0 & 0 & 1 \\ 1 & 2 & 0 & 0 \\ 1 & 0 & 0 & 1 \end{bmatrix} + 256I \\ &= 256\begin{bmatrix} -3 & 0 & 0 & -4 \\ 4 & 1 & 0 & 4 \\ 11 & 8 & 1 & 7 \\ 4 & 0 & 0 & 5 \end{bmatrix}. \end{aligned}$$

(Compare this result with A^k given in Example 158, in which a change of basis matrix Q is needed for the computation.) One may notice that even this simplified computing method for A^n by the Cayley-Hamilton theorem will become increasingly complicated in the division by $f(\lambda)$ when n becomes larger and larger. A further simpler method will be introduced later in Example 169. □

Problem 207 For the matrix $A = \begin{bmatrix} 1 & 0 & 0 \\ 0 & 2 & 0 \\ 5 & 0 & 2 \end{bmatrix}$, compute (1) the inverse A^{-1}, and (2) the matrix polynomial $A^5 + 3A^4 + A^3 - A^2 + 4A + 6I$.

Hint: The characteristic polynomial of A is $f(\lambda) = (\lambda-1)(\lambda-2)^2$. For (2), use $\lambda^5 + 3\lambda^4 + \lambda^3 - \lambda^2 + 4\lambda + 6 = (\lambda^2 + 8\lambda + 33)f(\lambda) + 104\lambda^2 - 228\lambda + 138$.

8.8 Applications

8.8.1 The minimal polynomial of a matrix

Let A be a square matrix of order n and let

$$f(\lambda) = \det(\lambda I - A) = \lambda^n + a_{n-1}\lambda^{n-1} + \cdots + a_1\lambda + a_0 = \prod_{i=1}^{t}(\lambda - \lambda_i)^{m_i}$$

be the characteristic polynomial of A, where m_i is the multiplicity of the eigenvalue λ_i. Then, the Cayley-Hamilton theorem says that

$$A^n + a_{n-1}A^{n-1} + \cdots + a_1 A + a_0 I = \prod_{i=1}^{t}(A - \lambda_i I)^{m_i} = \mathbf{0}.$$

The **minimal polynomial** of a matrix A is the *monic* polynomial $m(\lambda) = \sum_{i=0}^{m} c_i \lambda^i$ of smallest degree m such that

$$m(A) = \sum_{i=0}^{m} c_i A^i = \mathbf{0}.$$

Clearly, the minimal polynomial $m(\lambda)$ divides any polynomial $p(\lambda)$ satisfying $p(A) = \mathbf{0}$. In fact, if $p(\lambda) = q(\lambda)m(\lambda) + r(\lambda)$ by the division algorithm for polynomials, where the degree of $r(\lambda)$ is less than the degree of $m(\lambda)$, then $\mathbf{0} = p(A) = q(A)m(A) + r(A) = r(A)$, but the minimality of the degree of $m(\lambda)$ implies that $r(\lambda) = 0$. In particular, the minimal polynomial $m(\lambda)$ divides the characteristic polynomial so that

$$m(\lambda) = \prod_{i=1}^{t}(\lambda - \lambda_i)^{k_i}$$

for some $k_i \le m_i$.

Clearly, any two similar matrices have the same minimal polynomial because

$$\prod_{i=1}^{t}\left(Q^{-1}AQ - \lambda_i I\right)^{k_i} = Q^{-1}\left(\prod_{i=1}^{t}(A - \lambda_i I)^{k_i}\right)Q$$

for any invertible matrix Q.

Example 163 (*Characteristic polynomials versus Minimal polynomials*)

(1) The characteristic polynomial of the $n \times n$ zero matrix is λ^n, but its minimal polynomial is just λ. The characteristic polynomial of the $n \times n$ identity matrix is $(\lambda - 1)^n$, but its minimal polynomial is just $\lambda - 1$.

(2) For the diagonal matrix

$$A = \begin{bmatrix} 3 & 0 & 0 & 0 & 0 \\ 0 & 3 & 0 & 0 & 0 \\ 0 & 0 & 3 & 0 & 0 \\ 0 & 0 & 0 & 5 & 0 \\ 0 & 0 & 0 & 0 & 5 \end{bmatrix},$$

the characteristic polynomial is $f(\lambda) = (\lambda - 3)^3(\lambda - 5)^2$. However, for the polynomial $m(\lambda) = (\lambda - 3)(\lambda - 5)$, the matrix $m(A)$ is

$$m(A) = (A - 3I)(A - 5I) = \begin{bmatrix} 0 & 0 & 0 & 0 & 0 \\ 0 & 0 & 0 & 0 & 0 \\ 0 & 0 & 0 & 0 & 0 \\ 0 & 0 & 0 & 2 & 0 \\ 0 & 0 & 0 & 0 & 2 \end{bmatrix} \begin{bmatrix} -2 & 0 & 0 & 0 & 0 \\ 0 & -2 & 0 & 0 & 0 \\ 0 & 0 & -2 & 0 & 0 \\ 0 & 0 & 0 & 0 & 0 \\ 0 & 0 & 0 & 0 & 0 \end{bmatrix} = \mathbf{0}.$$

Hence, the minimal polynomial of A is $m(\lambda) = (\lambda - 3)(\lambda - 5)$. □

Example 164 (*The minimal polynomial of a diagonal matrix*)

(1) Any diagonal matrix of the form $\lambda_0 I$ has the minimal polynomial $\lambda - \lambda_0$. In particular, the minimal polynomials of the zero matrix and the identity matrix are the monomials λ and $\lambda - 1$, respectively as mentioned already.

(2) If an $n \times n$ matrix A has n distinct eigenvalues $\lambda_1, \lambda_2, \ldots, \lambda_n$, then its minimal polynomial coincides with the characteristic polynomial $f(\lambda) = \prod_{i=1}^{n}(\lambda - \lambda_i)$. In fact, for the diagonal matrix D having distinct diagonal entries $\lambda_1, \lambda_2, \ldots, \lambda_n$ and for any given $\ell = 1, 2, \ldots, n$, $\prod_{i \neq \ell}(D - \lambda_i I) \neq \mathbf{0}$, because its (ℓ, ℓ)-entry is equal to $\prod_{i \neq \ell}(\lambda_\ell - \lambda_i)$, which is not zero. □

Example 165 (*The minimal polynomial of a Jordan canonical matrix J having one or two blocks*)

(1) For any 5×5 matrix A similar to a Jordan block of the form

$$Q^{-1}AQ = J = \begin{bmatrix} \lambda_0 & 1 & 0 & 0 & 0 \\ 0 & \lambda_0 & 1 & 0 & 0 \\ 0 & 0 & \lambda_0 & 1 & 0 \\ 0 & 0 & 0 & \lambda_0 & 1 \\ 0 & 0 & 0 & 0 & \lambda_0 \end{bmatrix},$$

the minimal polynomial is equal to the characteristic polynomial $f(\lambda) = (\lambda - \lambda_0)^5$, because $(J - \lambda_0 I)^4 \neq \mathbf{0}$ but $(J - \lambda_0 I)^5 = \mathbf{0}$.

(2) For a matrix J having two Jordan blocks belonging to a single eigenvalue λ, say

$$J = \begin{bmatrix} J_1 & \mathbf{0} \\ \mathbf{0} & J_2 \end{bmatrix} = \begin{bmatrix} \begin{bmatrix} \lambda_0 & 1 & 0 \\ 0 & \lambda_0 & 1 \\ 0 & 0 & \lambda_0 \end{bmatrix} & \mathbf{0} \\ \mathbf{0} & \begin{bmatrix} \lambda_0 & 1 \\ 0 & \lambda_0 \end{bmatrix} \end{bmatrix},$$

the minimal polynomial of the smaller block J_2 is a divisor of the minimal polynomial of the larger block J_1 by (1).

In general, *if a matrix A or its Jordan canonical form J has a single eigenvalue λ_0, then its minimal polynomial is $(\lambda-\lambda_0)^k$, where k is the smallest positive integer ℓ such that $(A-\lambda_0 I)^\ell = \mathbf{0}$.* In fact, such a number k is the order of the largest Jordan block belonging to λ_0.

(3) For a matrix J having two Jordan blocks belonging to two distinct eigenvalues $\lambda_0 \neq \lambda_1$, respectively, say

$$J = \begin{bmatrix} J_1 & \mathbf{0} \\ \mathbf{0} & J_2 \end{bmatrix} = \begin{bmatrix} \begin{bmatrix} \lambda_0 & 1 & 0 & 0 & 0 \\ 0 & \lambda_0 & 1 & 0 & 0 \\ 0 & 0 & \lambda_0 & 1 & 0 \\ 0 & 0 & 0 & \lambda_0 & 1 \\ 0 & 0 & 0 & 0 & \lambda_0 \end{bmatrix} & \mathbf{0} \\ \mathbf{0} & \begin{bmatrix} \lambda_1 & 1 & 0 \\ 0 & \lambda_1 & 1 \\ 0 & 0 & \lambda_1 \end{bmatrix} \end{bmatrix},$$

the minimal polynomial of J is a product of the minimal polynomials of J_1 and J_2, which is $(\lambda-\lambda_0)^5(\lambda-\lambda_1)^3$.

In general, for a Jordan canonical matrix J, let J_λ denote the direct sum of Jordan blocks belonging to the eigenvalue λ. Then, $J = \bigoplus_{i=1}^t J_{\lambda_i}$, where t is the number of distinct eigenvalues of J. In this case, *the minimal polynomial of J is the product of the minimal polynomials of the summands J_{λ_i} for $i = 1, 2, \ldots, t$.* □

By a method similar to Example 165(2) and (3) and with Step 2(i) on page 373, one has the following theorem.

Theorem 8.8 *For any $n \times n$ matrix A or its Jordan canonical matrix J, the minimal polynomial is $\prod_{i=1}^t (\lambda-\lambda_i)^{k_i}$, where $\lambda_1, \lambda_2, \ldots, \lambda_t$ are the distinct eigenvalues of A and k_i is the order of the largest Jordan block in J belonging to the eigenvalue λ_i. Or equivalently, k_i is the smallest positive integer ℓ such that $\operatorname{rank}(A-\lambda_i I)^\ell + m_i = n$, where m_i is the multiplicity of the eigenvalue λ_i.*

A square matrix A is diagonalizable if and only if each Jordan block of the Jordan canonical form of A has order one, and a square matrix A is invertible if and only if every eigenvalue of A is not zero.

Hence, one can have the following corollary.

Corollary 8.8.1 **(1)** *A matrix A is diagonalizable if and only if its minimal polynomial is equal to $\prod_{i=1}^{t}(\lambda - \lambda_i)$, where $\lambda_1, \lambda_2, \ldots, \lambda_t$ are the distinct eigenvalues of A.*

(2) *A matrix A is invertible if and only if its minimal polynomial $m(\lambda) = \sum_{i=0}^{m} c_i\lambda^i$ has a nonzero constant term c_0.*

As a result, on can easily find the minimal polynomial of any normal matrix, like symmetric, skew-symmetric, orthogonal, Hermitian, or any unitary matrix.

Example 166 (*The minimal polynomial of a non-diagonalizable matrix A*) Find the minimal polynomial of A for

$$(1)\ A = \begin{bmatrix} 2 & 1 & 4 \\ 0 & 3 & 1 \\ 0 & 0 & 3 \end{bmatrix} \qquad (2)\ A = \begin{bmatrix} 4 & -6 & 4 & -1 \\ 1 & 0 & 0 & 0 \\ 0 & 1 & 0 & 0 \\ 0 & 0 & 1 & 0 \end{bmatrix}.$$

Solution: (1) Since A is triangular, its eigenvalues are $\lambda_1 = 2,\ \lambda_2 = \lambda_3 = 3$. But, rank $(A - 3I) = 2$ and so A is not diagonalizable. Hence, its minimal polynomial is $m(\lambda) = (\lambda - 2)(\lambda - 3)^2$.

(2) Recalling Example 149, we know that the eigenvalue of A is $\lambda = 1$ of multiplicity 4, and rank $(A - I) = 3$. This implies that the Jordan canonical form of A is

$$J = \begin{bmatrix} 1 & 1 & 0 & 0 \\ 0 & 1 & 1 & 0 \\ 0 & 0 & 1 & 1 \\ 0 & 0 & 0 & 1 \end{bmatrix},$$

and the minimal polynomial of A is $m(\lambda) = (\lambda - 1)^4$, by Theorem 8.8. □

Example 167 (*The minimal polynomial of a non-diagonalizable matrix A*) Find the minimal polynomial of A for

$$A = \begin{bmatrix} 1 & 0 & 0 & -1 \\ 1 & 2 & 0 & 1 \\ 1 & 2 & 2 & 0 \\ 1 & 0 & 0 & 3 \end{bmatrix},$$

which is given in Example 158.

Solution: As shown in Example 158, the characteristic polynomial of A is $f(\lambda) = (\lambda-2)^4$, rank $(A-2I)=2$, rank $(A-2I)^2=1$ and the Jordan canonical form of A is

$$J = \begin{bmatrix} 2 & 1 & 0 & 0 \\ 0 & 2 & 1 & 0 \\ 0 & 0 & 2 & 0 \\ 0 & 0 & 0 & 2 \end{bmatrix} = \begin{bmatrix} \begin{bmatrix} 2 & 1 & 0 \\ 0 & 2 & 1 \\ 0 & 0 & 2 \end{bmatrix} & \mathbf{0} \\ \mathbf{0} & [2] \end{bmatrix}.$$

So, the minimal polynomial of A is $m(\lambda) = (\lambda-2)^3$, by Theorem 8.8. □

Problem 208 Find the minimal polynomial of each of the following matrices.

$$(1)\ \begin{bmatrix} 1 & 0 \\ 0 & 1 \end{bmatrix}, \qquad (2)\ \begin{bmatrix} 1 & 0 & 0 \\ 0 & 1 & 0 \\ 0 & 0 & 0 \end{bmatrix}, \qquad (3)\ \begin{bmatrix} 1 & 0 & 0 \\ 0 & 0 & 0 \\ 0 & 0 & 0 \end{bmatrix}.$$

Problem 209 Show that λ is an eigenvalue of a matrix A if and only if λ is a root of the minimal polynomial $m(A)$.

Problem 210 In Example 147 on page 365, we have seen that there are seven possible (nonsimilar) Jordan canonical matrices of order 5 with a single eigenvalue. Find the minimal polynomial of each of them.

8.8.2 Computing A^k by using the minimal polynomial

How to compute the matrix power A^k?

Already two methods are introduced: One is by using the Jordan canonical form J of A with a change of basis matrix Q as is shown in Section 8.5, and the other is by using the Cayley-Hamilton theorem with only the characteristic polynomial of A as is shown in Section 8.6.

In this section we introduce the third method using the minimal polynomial, possibly as the simplest method in most cases.

The following example illustrates how to compute the power A^k and A^{-1} for a diagonalizable matrix A by using the minimal polynomial instead of the characteristic polynomial and also without computing the Jordan canonical form J of A and a change of basis matrix Q.

Example 168 (*Computing A^k by using the minimal polynomial when A is diagonalizable*) Compute the power A^k and A^{-1} by using the minimal polynomial for a symmetric matrix

$$A = \begin{bmatrix} 4 & 0 & 1 & 1 \\ 0 & 4 & 1 & -1 \\ 1 & 1 & 5 & 0 \\ 1 & -1 & 0 & 5 \end{bmatrix}.$$

Solution: The characteristic polynomial of A is $f(\lambda) = (\lambda - 3)^2(\lambda - 6)^2$. Since A is symmetric and so diagonalizable, its minimal polynomial is

$$m(\lambda) = (\lambda - 3)(\lambda - 6),$$

or equivalently $m(A) = A^2 - 9A + 18I = \mathbf{0}$. Hence,

$$A^{-1} = -\frac{1}{18}(A - 9I) = -\frac{1}{18}\begin{bmatrix} -5 & 0 & 1 & 1 \\ 0 & -5 & 1 & -1 \\ 1 & 1 & -4 & 0 \\ 1 & -1 & 0 & -4 \end{bmatrix}.$$

To compute A^k for any natural number k, first note that the power A^k with $k \geq 2$ can be expressed as a linear combination of I and A because $A^2 = 9A - 18I$. In fact, this is also true for any negative integer k because $A^{-1} = -\frac{1}{18}(A - 9I)$. (Much simpler than using the Cayley-Hamilton theorem. See page 393.)

Hence, one can write

$$A^k = x_0 I + x_1 A$$

with unknown coefficients x_i's, (even without the division of λ^k by $m(\lambda)$). Now, by multiplying an eigenvector $\mathbf{x}$ of A belonging to each eigenvalue λ on the right of this equation, that is, by computing $A^k\mathbf{x} = (x_0 I + x_1 A)\mathbf{x}$, we have a system of linear equations

$$\begin{array}{llcccc} \text{for } \lambda = 3, & 3^k & = & x_0 & + & 3x_1, \\ \text{for } \lambda = 6, & 6^k & = & x_0 & + & 6x_1. \end{array}$$

The solution is $x_0 = 2 \cdot 3^k - 6^k$ and $x_1 = \frac{1}{3}(6^k - 3^k)$. Hence,

$$A^k = x_0 I + x_1 A = \frac{1}{3}\begin{bmatrix} 2 \cdot 3^k + 6^k & 0 & 6^k - 3^k & 6^k - 3^k \\ 0 & 2 \cdot 3^k + 6^k & 6^k - 3^k & -6^k + 3^k \\ 6^k - 3^k & 6^k - 3^k & 3^k + 2 \cdot 6^k & 0 \\ 6^k - 3^k & -6^k + 3^k & 0 & 3^k + 2 \cdot 6^k \end{bmatrix}.$$

When $k = -1$, this gives the inverse A^{-1} of A. □

Now we discuss how to compute the power A^k in two separate cases depending on the diagonalizability of A.

(1) Firstly, suppose that A is diagonalizable. Then its minimal polynomial is equal to $m(\lambda) = \prod_{i=1}^{t}(\lambda - \lambda_i)$, where $\lambda_1, \lambda_2, \ldots, \lambda_t$ are the distinct eigenvalues of A (of any multiplicities). Now, by using $m(A) = \prod_{i=1}^{t}(A - \lambda_i I) = \mathbf{0}$ as in Example 168, one can write

$$A^k = x_0 I + x_1 A + \cdots + x_{t-1} A^{t-1}$$

with unknowns x_i's. Now, by multiplying an eigenvector $\mathbf{x}_i$ of A belonging to each eigenvalue λ_i on the right of this equation, one has a system of t linear equations

$$\begin{bmatrix} 1 & \lambda_1 & \cdots & \lambda_1^{t-1} \\ 1 & \lambda_2 & \cdots & \lambda_2^{t-1} \\ \vdots & \vdots & & \vdots \\ 1 & \lambda_t & \cdots & \lambda_t^{t-1} \end{bmatrix} \begin{bmatrix} x_0 \\ x_1 \\ \vdots \\ x_{t-1} \end{bmatrix} = \begin{bmatrix} {\lambda_1}^k \\ {\lambda_2}^k \\ \vdots \\ {\lambda_t}^k \end{bmatrix}.$$

The coefficient matrix $[\lambda_i^j]$ is an invertible Vandermonde matrix of order k because the eigenvalues λ_i are all distinct. Hence, the system is consistent and its unique solution for the x_i's determines the power A^k completely.

(2) Secondly, let A be any $n \times n$ (diagonalizable or not) matrix. By Theorem 8.8, the minimal polynomial of A is $m(\lambda) = \prod_{i=1}^{t}(\lambda - \lambda_i)^{k_i}$, where $\lambda_1, \lambda_2, \ldots, \lambda_t$ are the distinct eigenvalues of A and k_i is the smallest positive integer ℓ such that rank $(A - \lambda_i I)^\ell + m_i = n$ with the multiplicity m_i of the eigenvalue λ_i. Then $s = \sum_{i=1}^{t} k_i$ is the degree of the minimal polynomial $m(\lambda)$. Let

$$A^k = x_0 I + x_1 A + \cdots + x_{s-1} A^{s-1}$$

with unknown x_i's as before.

Now, by multiplying an eigenvector $\mathbf{x}$ of A belonging to each eigenvalue λ_i on the right of this equation, we have

$$\lambda_i^k = x_0 + \lambda_i x_1 + \cdots + \lambda_i^{s-1} x_{s-1}.$$

For each eigenvalue λ_i, let $\{\mathbf{x}_1, \mathbf{x}_2, \ldots, \mathbf{x}_{k_i}\}$ be a chain of generalized eigenvectors belonging to λ_i. Then these vectors satisfy

$$\begin{aligned} A\mathbf{x}_1 &= \lambda_i \mathbf{x}_1, \\ A\mathbf{x}_2 &= \lambda_i \mathbf{x}_2 + \mathbf{x}_1, \\ &\vdots \\ A\mathbf{x}_{k_i} &= \lambda_i \mathbf{x}_{k_i} + \mathbf{x}_{k_i-1}. \end{aligned}$$

By using the first two equations repeatedly, one can get

$$A^k\mathbf{x}_2 = \lambda_i^k\mathbf{x}_2 + k\lambda_i^{k-1}\mathbf{x}_1. \tag{1}$$

On the other hand,

$$\begin{aligned} A^k\mathbf{x}_2 &= \left(x_0 I + x_1 A + \cdots + x_{s-1}A^{s-1}\right)\mathbf{x}_2 \\ &= x_0\mathbf{x}_2 + x_1(\lambda_i\mathbf{x}_2 + \mathbf{x}_1) + \cdots + x_{s-1}\left(\lambda_i^{s-1}\mathbf{x}_2 + (s-1)\lambda_i^{s-2}\mathbf{x}_1\right) \\ &= \left(x_0 + \lambda_i x_1 + \cdots + \lambda_i^{s-1}x_{s-1}\right)\mathbf{x}_2 \\ &\qquad + \left(x_1 + 2\lambda_i x_2 + \cdots + (s-1)\lambda_i^{s-2}x_{s-1}\right)\mathbf{x}_1. \end{aligned} \tag{2}$$

In these two equations (1) and (2) for $A^k\mathbf{x}_2$, by comparing the coefficients of two linearly independent vectors $\mathbf{x}_1$ and $\mathbf{x}_2$, one has

$$\begin{cases} \lambda_i{}^k = x_0 + \lambda_i x_1 + \lambda_i^2 x_2 + \cdots + \lambda_i^{s-1}x_{s-1}, \\ k\lambda_i{}^{k-1} = x_1 + 2\lambda_i x_2 + \cdots + (s-1)\lambda_i^{s-2}x_{s-1}. \end{cases}$$

Here, the second equation is the derivative of the first equation with respect to λ_i.

Similarly, one can write $A^k\mathbf{x}_3$ as a linear combination of linearly independent vectors $\mathbf{x}_1, \mathbf{x}_2, \mathbf{x}_3$ in two different ways (like the equations (1) and (2)) and by comparing the coefficients, one can get the same two equations as above with an additional third equation:

$$\begin{cases} \lambda_i{}^k = x_0 + \lambda_i x_1 + \cdots + \lambda_i^{s-1}x_{s-1}, \\ \binom{k}{1}\lambda_i{}^{k-1} = \binom{1}{1}x_1 + \binom{2}{1}\lambda_i x_2 + \cdots + \binom{s-1}{1}\lambda_i^{s-2}x_{s-1}, \\ \binom{k}{2}\lambda_i{}^{k-2} = \binom{2}{2}x_2 + \binom{3}{2}\lambda_i x_3 + \cdots + \binom{s-1}{2}\lambda_i^{s-3}x_{s-1}. \end{cases}$$

By repeating similar processes for $A^k\mathbf{x}_4, \ldots, A^k\mathbf{x}_{k_i}$, as a linear combination of linearly independent vectors $\mathbf{x}_1, \mathbf{x}_2, \ldots, \mathbf{x}_{k_i}$ in two different ways and then a comparison of their coefficients gives the following k_i equations:

$$\begin{cases} \lambda_i{}^k = x_0 + \lambda_i x_1 + \cdots + \lambda_i^{s-1}x_{s-1}, \\ \binom{k}{1}\lambda_i{}^{k-1} = \binom{1}{1}x_1 + \binom{2}{1}\lambda_i x_2 + \cdots + \binom{s-1}{1}\lambda_i^{s-2}x_{s-1}, \\ \binom{k}{2}\lambda_i{}^{k-2} = \binom{2}{2}x_2 + \binom{3}{2}\lambda_i x_3 + \cdots + \binom{s-1}{2}\lambda_i^{s-3}x_{s-1}, \\ \qquad\vdots \\ \binom{k}{k_i-1}\lambda_i{}^{k-k_i+1} = \binom{k_i-1}{k_i-1}x_{k_i-1} + \binom{k_i}{k_i-1}\lambda_i x_{k_i} + \cdots + \binom{s-1}{k_i-1}\lambda_i^{s-k_i}x_{s-1}. \end{cases}$$

Here, the last $k_i - 1$ equations are equivalent to the successive derivatives of the first equation with respect to λ_i. For example, two times the third equation is just the second derivative of the first equation, and $(k_i - 1)!$ times the last equation is the $(k_i - 1)$-th derivative of the first equation.

Gathering together all of these equations for each eigenvalue λ_i, $i = 1, 2, \ldots, t$, one can get a system of s linear equations with s unknowns x_j's having an invertible coefficient matrix. (It is an exercise to show the invertibility of the coefficient matrix). Therefore, the unknowns $x_0, \ldots, x_{s-1}$ are uniquely determined and so one can get the power A^k.

Example 169 (*Computing A^k by using the minimal polynomial when A is not diagonalizable*) (Example 158 again) Compute the power A^k and A^{-1} by using the minimal polynomial for

$$A = \begin{bmatrix} 1 & 0 & 0 & -1 \\ 1 & 2 & 0 & 1 \\ 1 & 2 & 2 & 0 \\ 1 & 0 & 0 & 3 \end{bmatrix}.$$

Solution: In Example 167, the minimal polynomial of A is determined to be

$$m(\lambda) = (\lambda - 2)^3 = \lambda^3 - 6\lambda^2 + 12\lambda - 8.$$

Hence, one can write

$$A^k = x_0 I + x_1 A + x_2 A^2$$

with unknown coefficients x_i's and

$$A^2 = \begin{bmatrix} 0 & 0 & 0 & -4 \\ 4 & 4 & 0 & 4 \\ 5 & 8 & 4 & 1 \\ 4 & 0 & 0 & 8 \end{bmatrix}.$$

Now, for the eigenvalue $\lambda = 2$, one has

$$\begin{array}{lrcl} & 2^k & = & x_0 + 2x_1 + 2^2 x_2, \\ \text{take } \partial_\lambda, & k2^{k-1} & = & x_1 + 2 \cdot 2x_2, \\ \text{take } \partial_\lambda^2, & k(k-1)2^{k-2} & = & 2x_2. \end{array}$$

The solution is

$$x_2 = k(k-1)2^{k-3}, \quad x_1 = k(2-k)2^{k-1}, \quad x_0 = 2^k\left(\frac{1}{2}k^2 - \frac{3}{2}k + 1\right).$$

Thus, one has (the same power as in Example 158)

$$\begin{aligned} A^k &= x_0 I + x_1 A + x_2 A^2 \\ &= \begin{bmatrix} 2^k - k2^{k-1} & 0 & 0 & -k2^{k-1} \\ k2^{k-1} & 2^k & 0 & k2^{k-1} \\ 2^{k-3}(3k+k^2) & k2^k & 2^k & k(k-1)2^{k-3} \\ k2^{k-1} & 0 & 0 & k2^{k-1}+2^k \end{bmatrix}. \end{aligned}$$

By Corollary 8.8.1(2), A is invertible. To find A^{-1}, we use $m(A) = A^3 - 6A^2 + 12A - 8I = \mathbf{0}$. This gives

$$A^{-1} = \frac{1}{8}\left(A^2 - 6A + 12I\right) = \frac{1}{8}\begin{bmatrix} 6 & 0 & 0 & 2 \\ -2 & 4 & 0 & -2 \\ -1 & -4 & 4 & 1 \\ -2 & 0 & 0 & 2 \end{bmatrix},$$

which is A^k when $k = -1$ in our formula for A^k. □

Problem 211 (Problem 207 again) For the matrix $A = \begin{bmatrix} 1 & 0 & 0 \\ 0 & 2 & 0 \\ 5 & 0 & 2 \end{bmatrix}$, compute (1) the inverse A^{-1}, and (2) the matrix polynomial $A^5 + 3A^4 + A^3 - A^2 + 4A + 6I$ by using the minimal polynomial.

Problem 212 (Problem 203 again) Compute A^k by using the minimal polynomial for

$$(1)\ A = \begin{bmatrix} 3 & 1 & 0 & 0 \\ 0 & 3 & 1 & 0 \\ 0 & 0 & 3 & 0 \\ 0 & 0 & 0 & 1 \end{bmatrix}, \qquad (2)\ A = \begin{bmatrix} 4 & -5 & 3 & -2 \\ 2 & -1 & 1 & -2 \\ 2 & -1 & 1 & -2 \\ 2 & -5 & 3 & 0 \end{bmatrix}.$$

8.8.3 Computing e^A by using the minimal polynomial

The exponential matrix e^A can also be computed by using the minimal polynomial.

This method is quite similar to computing the power A^k discussed in Subsection 8.8.2. Because of the similarity to computing A^k, we state only the difference in computing the exponential matrix e^A in this section.

We also discuss how to compute e^A in two cases depending on the diagonalizability of A.

(1) First, let A be diagonalizable. Then its minimal polynomial is equal to $m(\lambda) = \prod_{i=1}^{t}(\lambda - \lambda_i)$, where $\lambda_1, \lambda_2, \ldots, \lambda_t$ are the distinct eigenvalues of A. Now, as before, one can write

$$e^A = x_0 I + x_1 A + \cdots + x_{t-1}A^{t-1}$$

with unknowns x_i's. This is actually possible even if e^A is a series, not a polynomial of the matrix A. By multiplying an eigenvector $\mathbf{x}$ of A belonging to each eigenvalue λ_i on the right of this equation, one can get

$$\begin{bmatrix} 1 & \lambda_1 & \cdots & \lambda_1^{t-1} \\ 1 & \lambda_2 & \cdots & \lambda_2^{t-1} \\ \vdots & \vdots & & \vdots \\ 1 & \lambda_t & \cdots & \lambda_t^{t-1} \end{bmatrix} \begin{bmatrix} x_0 \\ x_1 \\ \vdots \\ x_{t-1} \end{bmatrix} = \begin{bmatrix} e^{\lambda_1} \\ e^{\lambda_2} \\ \vdots \\ e^{\lambda_t} \end{bmatrix}.$$

Its unique solution x_i determines the matrix e^A.

(2) Next, let A be any $n \times n$ matrix, diagonalizable or not. Then, the minimal polynomial of A is $m(\lambda) = \prod_{i=1}^{t}(\lambda - \lambda_i)^{k_i}$, where $\lambda_1, \lambda_2, \ldots, \lambda_t$ are the distinct eigenvalues of A and k_i is the smallest positive integer ℓ such that rank $(A-\lambda_i I)^{\ell} + m_i = n$. Then $s = \sum_{i=1}^{t} k_i$ is the degree of $m(\lambda)$. Let

$$e^A = x_0 I + x_1 A + \cdots + x_{s-1}A^{s-1}$$

with unknown x_i's as before. For each eigenvalue λ_i and a chain of generalized eigenvectors $\{\mathbf{x}_1, \mathbf{x}_2, \ldots, \mathbf{x}_{k_i}\}$ belonging to λ_i, a parallel procedure for computing $A^k\mathbf{x}_j$ gives the following k_i equations:

$$\begin{array}{rcccccc} e^{\lambda_i} & = & x_0 & + & \lambda_i x_1 & + \cdots + & \lambda_i^{s-1}x_{s-1}, \\ \frac{1}{1!}e^{\lambda_i} & = & \binom{1}{1}x_1 & + & \binom{2}{1}\lambda_i x_2 & + \cdots + & \binom{s-1}{1}\lambda_i^{s-2}x_{s-1}, \\ \frac{1}{2!}e^{\lambda_i} & = & \binom{2}{2}x_2 & + & \binom{3}{2}\lambda_i x_3 & + \cdots + & \binom{s-1}{2}\lambda_i^{s-3}x_{s-1}, \\ & \vdots & & & & & \\ \frac{1}{(k_i-1)!}e^{\lambda_i} & = & \binom{k_i-1}{k_i-1}x_{k_i-1} & + & \binom{k_i}{k_i-1}\lambda_i x_{k_i} & + \cdots + & \binom{s-1}{k_i-1}\lambda_i^{s-k_i}x_{s-1}. \end{array}$$

Here, the last $k_i - 1$ equations are also equivalent to the successive derivatives of the first equation with respect to λ_i. For example, 2! times the third equation is just the second derivative of the first equation.

Gathering together all of these equations for each λ_i, one can get a system of s equations with s unknowns x_j's having an invertible coefficient matrix. Now, the unique solution for the x_i's determines the exponential matrix e^A.

Example 170 (*Computing e^A by using the minimal polynomial when A is diagonalizable*) Compute the exponential matrix e^A for

$$A = \begin{bmatrix} 5 & -4 & 6 \\ 12 & -9 & 12 \\ 6 & -4 & 5 \end{bmatrix}.$$

Solution: This matrix A is the coefficient matrix of the LDE given in Example 123 on page 293:

$$\begin{cases} y_1' &= 5y_1 - 4y_2 + 6y_3 \\ y_2' &= 12y_1 - 9y_2 + 12y_3 \\ y_3' &= 6y_1 - 4y_2 + 5y_3 \end{cases}$$

It can be shown that A is diagonalizable with the eigenvalues $\lambda_1 = \lambda_2 = -1$ and $\lambda_3 = 3$. Hence, the minimal polynomial of A is

$$m(\lambda) = (\lambda + 1)(\lambda - 3),$$

and one can write

$$e^A = x_0 I + x_1 A$$

with unknowns x_i's. Then

$$\begin{array}{rcrcl} \text{for } \lambda = -1, & e^{-1} & = & x_0 & - \quad x_1, \\ \text{and for } \lambda = 3, & e^3 & = & x_0 & + \quad 3x_1. \end{array}$$

By solving it, one can get $x_0 = \frac{1}{4}(3e^{-1} + e^3)$, $x_1 = \frac{1}{4}(-e^{-1} + e^3)$, which gives

$$e^A = x_0 I + x_1 A = \begin{bmatrix} \frac{1}{2}(-e^{-1} + 3e^3) & e^{-1} - e^3 & \frac{3}{2}(-e^{-1} + e^3) \\ -3e^{-1} + 3e^3 & 3e^{-1} - 2e^3 & -3e^{-1} + 3e^3 \\ \frac{3}{2}(-e^{-1} + e^3) & e^{-1} - e^3 & \frac{1}{2}(-e^{-1} + 3e^3) \end{bmatrix},$$

in expanded form. □

Example 171 (Example 158 again) (*Computing e^A by using the minimal polynomial when A is not diagonalizable*) Compute the exponential matrix e^A for

$$A = \begin{bmatrix} 1 & 0 & 0 & -1 \\ 1 & 2 & 0 & 1 \\ 1 & 2 & 2 & 0 \\ 1 & 0 & 0 & 3 \end{bmatrix}.$$

Solution: As in Example 167, the minimal polynomial of A is

$$m(\lambda) = (\lambda - 2)^3.$$

Hence, one can set as in Example 169,

$$e^A = x_0 I + x_1 A + x_2 A^2$$

with unknown coefficients x_i's and

$$A^2 = \begin{bmatrix} 0 & 0 & 0 & -4 \\ 4 & 4 & 0 & 4 \\ 5 & 8 & 4 & 1 \\ 4 & 0 & 0 & 8 \end{bmatrix}.$$

Now, for the eigenvalue $\lambda = 2$, one has

$$\begin{array}{rrcrrrrr} & e^2 & = & x_0 & + & 2x_1 & + & 2^2 x_2, \\ \text{take } \partial_\lambda, & e^2 & = & & & x_1 & + & 2 \cdot 2x_2, \\ \text{take } \partial_\lambda^2, & e^2 & = & & & & & 2x_2. \end{array}$$

The solution is $x_2 = \frac{1}{2}e^2,\ x_1 = -e^2$, and $x_0 = e^2$. Hence,

$$e^A = x_0 I + x_1 A + x_2 A^2 = e^2 \left(I - A + \frac{1}{2} A^2 \right) = \begin{bmatrix} 0 & 0 & 0 & -e^2 \\ e^2 & e^2 & 0 & e^2 \\ \frac{3}{2}e^2 & 2e^2 & e^2 & \frac{1}{2}e^2 \\ e^2 & 0 & 0 & 2e^2 \end{bmatrix},$$

as is shown in Example 158. □

Problem 213 (Problem 203 again) Compute e^A by using the minimal polynomial for

$$(1)\ A = \begin{bmatrix} 3 & 1 & 0 & 0 \\ 0 & 3 & 1 & 0 \\ 0 & 0 & 3 & 0 \\ 0 & 0 & 0 & 1 \end{bmatrix}, \qquad (2)\ A = \begin{bmatrix} 4 & -5 & 3 & -2 \\ 2 & -1 & 1 & -2 \\ 2 & -1 & 1 & -2 \\ 2 & -5 & 3 & 0 \end{bmatrix}.$$

8.8.4 LdE $\mathbf{x}_n = A\mathbf{x}_{n-1}$ for any square matrix A

An LdE is a matrix equation $\mathbf{x}_n = A\mathbf{x}_{n-1}$ with a $k \times k$ matrix A. If an initial vector $\mathbf{x}_0$ is given, then $\mathbf{x}_n = A^n\mathbf{x}_0$ is the solution.

If the matrix A is diagonalizable with k linearly independent eigenvectors $\mathbf{v}_1, \mathbf{v}_2, \ldots, \mathbf{v}_k$ belonging to the eigenvalues $\lambda_1, \lambda_2, \ldots, \lambda_k$, respectively, a general solution to an LdE $\mathbf{x}_n = A\mathbf{x}_{n-1}$ is

$$\mathbf{x}_n = c_1\lambda_1^n\mathbf{v}_1 + c_2\lambda_2^n\mathbf{v}_2 + \cdots + c_k\lambda_k^n\mathbf{v}_k$$

with constants $c_1, c_2, \ldots, c_k$. (See Theorem 6.13.)

In fact, an LdE $\mathbf{x}_n = A\mathbf{x}_{n-1}$ can be solved for any square matrix A (not necessarily diagonalizable) by using the power A^n, whose computation has been discussed in several ways.

Example 172 (*LdE $\mathbf{x}_n = A\mathbf{x}_{n-1}$ when A is not diagonalizable*) Solve the LdE $\mathbf{x}_n = A\mathbf{x}_{n-1}$, where

$$A = \begin{bmatrix} 1 & 0 & 0 & -1 \\ 1 & 2 & 0 & 1 \\ 1 & 2 & 2 & 0 \\ 1 & 0 & 0 & 3 \end{bmatrix}.$$

Solution: In Example 158, it is shown that the matrix A has an eigenvalue 2 of multiplicity 4 and A is not diagonalizable. However, the solution of $\mathbf{x}_n = A\mathbf{x}_{n-1}$ is $\mathbf{x}_n = A^n\mathbf{x}_0$, where

$$A^n = \begin{bmatrix} -n2^{n-1} - 2^n + 2^{n+1} & 0 & 0 & -n2^{n-1} \\ n2^{n-1} & 2^n & 0 & n2^{n-1} \\ 2^{n-3}(3n + n^2) & n2^n & 2^n & n(n-1)2^{n-3} \\ n2^{n-1} & 0 & 0 & n2^{n-1} + 2^n \end{bmatrix},$$

which is computed in Examples 158, 162, and 169 in three different ways. □

Problem 214 Compute A^n by using the minimal polynomial and use it to solve the LdE $\mathbf{x}_n = A\mathbf{x}_{n-1}$ for

$$(1)\ A = \begin{bmatrix} 3 & 1 & 0 & 0 \\ 0 & 3 & 0 & 0 \\ 0 & 0 & 3 & 0 \\ 0 & 0 & 0 & 3 \end{bmatrix}, \quad (2)\ A = \begin{bmatrix} 3 & 1 & 0 & 0 \\ 0 & 3 & 1 & 0 \\ 0 & 0 & 3 & 0 \\ 0 & 0 & 0 & 3 \end{bmatrix}, \quad (3)\ A = \begin{bmatrix} 3 & 1 & 0 & 0 \\ 0 & 3 & 1 & 0 \\ 0 & 0 & 3 & 1 \\ 0 & 0 & 0 & 3 \end{bmatrix}.$$

8.8.5 LDE $\mathbf{y}' = A\mathbf{y}$ for any square matrix A

We go back again to an LDE

$$\mathbf{y}' = A\mathbf{y} \quad \text{with initial condition} \quad \mathbf{y}(0) = \mathbf{y}_0.$$

The solution always exists and is known to be $\mathbf{y}(t) = e^{tA}\mathbf{y}_0$, as a linear combination of the columns of e^{tA}. (See Theorem 6.26.)

If A is diagonal or diagonalizable, a general solution to $\mathbf{y}' = A\mathbf{y}$ is

$$\mathbf{y}(t) = e^{tA}\mathbf{y}_0 = c_1 e^{\lambda_1 t}\mathbf{v}_1 + c_2 e^{\lambda_2 t}\mathbf{v}_2 + \cdots + c_n e^{\lambda_n t}\mathbf{v}_n,$$

where $\mathbf{v}_1, \mathbf{v}_2, \ldots, \mathbf{v}_n$ are n linearly independent eigenvectors belonging to the eigenvalues $\lambda_1, \lambda_2, \ldots, \lambda_n$ of A, respectively. (See Theorem 6.20.)

For any square matrix A (not necessarily diagonalizable), the matrix e^{tA} can be computed by using the Jordan canonical form of A, or by using the minimal polynomial of A.

Firstly, with the Jordan canonical form $Q^{-1}AQ = J$ of A, the solution $\mathbf{y}(t)$ is

$$\begin{aligned} e^{tA}\mathbf{y}_0 &= Qe^{tJ}Q^{-1}\mathbf{y}_0 \\ &= [\mathbf{u}_1\ \mathbf{u}_2\ \cdots\ \mathbf{u}_n] \begin{bmatrix} e^{tJ_1} & 0 & \cdots & 0 \\ 0 & e^{tJ_2} & \ddots & \vdots \\ \vdots & \ddots & \ddots & 0 \\ 0 & \cdots & 0 & e^{tJ_s} \end{bmatrix} \begin{bmatrix} c_1 \\ c_2 \\ \vdots \\ c_n \end{bmatrix}, \end{aligned}$$

where $Q^{-1}\mathbf{y}_0 = (c_1, c_2, \ldots, c_n)$ and the $\mathbf{u}_i$'s are generalized eigenvectors of A.

To compute the diagonal blocks e^{tJ_j}, let $J = \lambda I + N$ be a single Jordan block as one of $J_1, J_2, \ldots, J_s$. Then

$$\begin{aligned} e^{tJ} &= e^{t(\lambda I + N)} = e^{\lambda t I} e^{tN} \\ &= e^{\lambda t} \sum_{k=0}^{n-1} \frac{(tN)^k}{k!} \\ &= e^{\lambda t} \begin{bmatrix} 1 & \frac{t}{1!} & \frac{t^2}{2!} & \cdots & \frac{t^{n-1}}{(n-1)!} \\ 0 & 1 & \frac{t}{1!} & \cdots & \frac{t^{n-2}}{(n-2)!} \\ 0 & 0 & \ddots & \ddots & \vdots \\ \vdots & \cdots & & 1 & \frac{t}{1!} \\ 0 & \cdots & \cdots & 0 & 1 \end{bmatrix}. \end{aligned}$$

That is,

$$e^{tJ} = \begin{bmatrix} e^{\lambda t} & \frac{\partial_\lambda e^{\lambda t}}{1!} & \frac{\partial_\lambda^2 e^{\lambda t}}{2!} & \cdots & \frac{\partial_\lambda^{n-1} e^{\lambda t}}{(n-1)!} \\ 0 & e^{\lambda t} & \frac{\partial_\lambda e^{\lambda t}}{1!} & \vdots & \frac{\partial_\lambda^{n-2} e^{\lambda t}}{(n-2)!} \\ 0 & 0 & \ddots & \ddots & \vdots \\ \vdots & \cdots & \ddots & e^{\lambda t} & \frac{\partial_\lambda e^{\lambda t}}{1!} \\ 0 & \cdots & \cdots & 0 & e^{\lambda t} \end{bmatrix}. \tag{8.4}$$

As another and simpler way of computing e^{tA}, one can use the minimal polynomial of tA as discussed in Subsection 8.8.3. If $m(\lambda)$ is the minimal polynomial of A, then $m(t\lambda)$ is the minimal polynomial of the matrix tA.

The following two examples illustrate this case.

Example 173 (*Solve the LDE* $\mathbf{y}' = A\mathbf{y}$ *by using the minimal polynomial when* A *is diagonalizable*) (Example 123 again) Solve the LDE

$$\begin{cases} y_1' = 5y_1 - 4y_2 + 6y_3 \\ y_2' = 12y_1 - 9y_2 + 12y_3 \\ y_3' = 6y_1 - 4y_2 + 5y_3 \end{cases}$$

and also find its particular solution satisfying the initial conditions $y_1(0) = 0$, $y_2(0) = 1$, and $y_3(0) = 2$.

Solution: The matrix form of the LDE is $\mathbf{y}' = A\mathbf{y}$ with

$$A = \begin{bmatrix} 5 & -4 & 6 \\ 12 & -9 & 12 \\ 6 & -4 & 5 \end{bmatrix},$$

and its general solution is $\mathbf{y} = e^{tA}\mathbf{y}_0$. It is shown in Example 123 on page 293 that A is diagonalizable and the minimal polynomial of A is

$$m(\lambda) = (\lambda + 1)(\lambda - 3).$$

Hence, one can write

$$e^{tA} = x_0(t)I + x_1(t)A$$

with unknown functions $x_i(t)$'s. Then

$$\begin{array}{lcl} \text{for } \lambda = -1, & e^{-t} & = x_0(t) - x_1(t), \\ \text{and for } \lambda = 3, & e^{3t} & = x_0(t) + 3x_1(t). \end{array}$$

(Compare this with Example 170). By solving it, $x_0(t) = \frac{1}{4}(3e^{-t} + e^{3t})$, $x_1(t) = \frac{1}{4}(-e^{-t} + e^{3t})$, and then

$$e^{tA} = x_0(t)I + x_1(t)A = \begin{bmatrix} \frac{1}{2}(-e^{-t} + 3e^{3t}) & e^{-t} - e^{3t} & \frac{3}{2}(-e^{-t} + e^{3t}) \\ -3e^{-t} + 3e^{3t} & 3e^{-t} - 2e^{3t} & -3e^{-t} + 3e^{3t} \\ \frac{3}{2}(-e^{-t} + e^{3t}) & e^{-t} - e^{3t} & \frac{1}{2}(-e^{-t} + 3e^{3t}) \end{bmatrix}.$$

With the initial conditions $y_1(0) = 0$, $y_2(0) = 1$, and $y_3(0) = 2$, a particular solution is

$$\begin{aligned} \mathbf{y}(t) = e^{tA}\mathbf{y}_0 &= \begin{bmatrix} \frac{1}{2}(-e^{-t} + 3e^{3t}) & e^{-t} - e^{3t} & \frac{3}{2}(-e^{-t} + e^{3t}) \\ -3e^{-t} + 3e^{3t} & 3e^{-t} - 2e^{3t} & -3e^{-t} + 3e^{3t} \\ \frac{3}{2}(-e^{-t} + e^{3t}) & e^{-t} - e^{3t} & \frac{1}{2}(-e^{-t} + 3e^{3t}) \end{bmatrix} \begin{bmatrix} 0 \\ 1 \\ 2 \end{bmatrix} \\ &= \begin{bmatrix} -2e^{-t} + 2e^{3t} \\ -3e^{-t} + 4e^{3t} \\ 2e^{3t} \end{bmatrix}. \end{aligned}$$

This is a linear combination of the three columns of e^{tA}. One might compare this method with the previous method given in Example 123. □

Note: In Example 173, we compute e^{tA} by determining the unknown functions $x_i(t)$'s in $e^{tA} = x_0(t)I + x_1(t)A$. However, in Example 170, we have determined $e^A = x_0I + x_1A$ with $x_0 = (3e + e^{-3})/4$ and $x_1 = (e - e^{-3})/4$. Hence, it seems to be true that $e^{tA} = x_0I + x_1(tA)$ with the same $x_0 = (3e + e^{-3})/4$ and $x_1 = (e - e^{-3})/4$. But, this is *not* true, because if we put $e^{tA} = x_0I + x_1A$, then x_0 and x_1 should be functions of t, while they are constants in $e^A = x_0I + x_1A$.

Example 174 (*Solve the LDE* $\mathbf{y}' = A\mathbf{y}$ *by using the minimal polynomial when* A *is not diagonalizable*) Find a general solution to the LDE $\mathbf{y}'(t) = A\mathbf{y}(t)$, where

$$A = \begin{bmatrix} 1 & 0 & 0 & -1 \\ 1 & 2 & 0 & 1 \\ 1 & 2 & 2 & 0 \\ 1 & 0 & 0 & 3 \end{bmatrix}.$$

This matrix A is given in Examples 158 and 167.

Solution: In Example 167, the minimal polynomial of A is computed as

$$m(\lambda) = (\lambda - 2)^3.$$

Hence, as in Example 171, one can set

$$e^{tA} = x_0(t)I + x_1(t)A + x_2(t)A^2$$

with unknown functions $x_i(t)$'s. Now, for the eigenvalue $\lambda = 2$, one gets

$$\begin{array}{rrcrcrcr} & e^{2t} & = & x_0(t) & + & 2x_1(t) & + & 2^2x_2(t), \\ \text{by taking } \partial_\lambda, & te^{2t} & = & & & x_1(t) & + & 2\cdot 2x_2(t), \\ \text{and by taking } \partial_\lambda^2, & t^2e^{2t} & = & & & & & 2x_2(t). \end{array}$$

The solution is

$$x_2(t) = \frac{1}{2}t^2e^{2t}, \quad x_1(t) = te^{2t} - 2t^2e^{2t}, \quad x_0(t) = e^{2t} - 2te^{2t} + 2t^2e^{2t}.$$

Hence,

$$\begin{aligned} e^{tA} &= x_0(t)I + x_1(t)A + x_2(t)A^2 \\ &= \begin{bmatrix} e^{2t} - te^{2t} & 0 & 0 & -te^{2t} \\ te^{2t} & e^{2t} & 0 & te^{2t} \\ te^{2t} + \frac{1}{2}t^2e^{2t} & 2te^{2t} & e^{2t} & \frac{1}{2}t^2e^{2t} \\ te^{2t} & 0 & 0 & e^{2t} + te^{2t} \end{bmatrix}. \end{aligned}$$

Now, the four columns of e^{tA} are solutions of the given $\mathbf{y}'(t) = A\mathbf{y}(t)$, and a linear combination of these four columns is a general solution. See Theorem 6.26. □

Throughout the last four subsections, we have seen that the minimal polynomial is a quite convenient tool to compute the matrix power A^n, the exponential matrix e^A, and also to solve the LdE or LDE.

Problem 215 By using the minimal polynomial of A, solve the LDE $\mathbf{y}' = A\mathbf{y}$ for

$$(1)\ A = \begin{bmatrix} 3 & 1 & 0 & 0 \\ 0 & 3 & 0 & 0 \\ 0 & 0 & 3 & 0 \\ 0 & 0 & 0 & 3 \end{bmatrix}, \quad (2)\ A = \begin{bmatrix} 3 & 1 & 0 & 0 \\ 0 & 3 & 1 & 0 \\ 0 & 0 & 3 & 0 \\ 0 & 0 & 0 & 3 \end{bmatrix}, \quad (3)\ A = \begin{bmatrix} 3 & 1 & 0 & 0 \\ 0 & 3 & 1 & 0 \\ 0 & 0 & 3 & 1 \\ 0 & 0 & 0 & 3 \end{bmatrix}.$$

which are given in Problem 214.

Problem 216 By using the minimal polynomial of A, solve the LDE $\mathbf{y}' = A\mathbf{y}$ with the initial condition $\mathbf{y}(0) = \mathbf{y}_0$, where

$$A = \begin{bmatrix} 1 & 0 & 0 \\ 0 & 2 & 0 \\ 5 & 0 & 2 \end{bmatrix}, \quad \mathbf{y}_0 = \begin{bmatrix} 0 \\ 1 \\ 1 \end{bmatrix}.$$

8.9 Exercises

8.1. *A Jordan canonical matrix J is a block diagonal matrix.* Find the Jordan blocks and linearly independent eigenvectors of each of the following Jordan matrices.

$$(1)\ \begin{bmatrix} 1 & 1 & 0 & 0 & 0 \\ 0 & 1 & 1 & 0 & 0 \\ 0 & 0 & 1 & 0 & 0 \\ 0 & 0 & 0 & 3 & 1 \\ 0 & 0 & 0 & 0 & 3 \end{bmatrix}, \qquad (2)\ \begin{bmatrix} 2 & 0 & 0 & 0 & 0 \\ 0 & 2 & 0 & 0 & 0 \\ 0 & 0 & 2 & 0 & 0 \\ 0 & 0 & 0 & 5 & 1 \\ 0 & 0 & 0 & 0 & 5 \end{bmatrix}, \qquad (3)\ \begin{bmatrix} 2 & 1 & 0 & 0 & 0 \\ 0 & 2 & 0 & 0 & 0 \\ 0 & 0 & 3 & 0 & 0 \\ 0 & 0 & 0 & 3 & 0 \\ 0 & 0 & 0 & 0 & 5 \end{bmatrix}.$$

8.2. *For each eigenvalue λ of A, the Jordan blocks J_λ belonging to the eigenvalue λ are completely determined by the sequence of ranks,* $\text{rank}\,(A - \lambda I)^\ell$, $\ell = 1, 2, \ldots$. For each Jordan matrix J given in Exercise **8.1** and for each eigenvalue λ of J, find the sequence $\text{rank}(J - \lambda I)^\ell$ for $\ell = 1, 2, \ldots$.

Hint: See Example 150 on page 370. Also, see Example 152 on page 373.

8.3. *If a Jordan block is invertible, its inverse is also a Jordan block of the same size.*

For $J = \begin{bmatrix} \lambda & 1 & 0 & 0 \\ 0 & \lambda & 1 & 0 \\ 0 & 0 & \lambda & 1 \\ 0 & 0 & 0 & \lambda \end{bmatrix}$ $(\lambda \neq 0)$, show that $J^{-1} = \begin{bmatrix} \frac{1}{\lambda} & -\frac{1}{\lambda^2} & \frac{1}{\lambda^3} & -\frac{1}{\lambda^4} \\ 0 & \frac{1}{\lambda} & -\frac{1}{\lambda^2} & \frac{1}{\lambda^3} \\ 0 & 0 & \frac{1}{\lambda} & -\frac{1}{\lambda^2} \\ 0 & 0 & 0 & \frac{1}{\lambda} \end{bmatrix}$.

(1) Find the sequences of $\text{rank}\,(J - \lambda I)^\ell$ and of $\text{rank}\,(J^{-1} - \frac{1}{\lambda} I)^\ell$, $\ell = 1, 2, \ldots$.
(2) Find the Jordan canonical form of J^{-1}.

In general, *if a Jordan matrix J is invertible, then the Jordan canonical matrix of its inverse J^{-1} has diagonal blocks of the same sizes as J has.*

Hint: Use $J\mathbf{x} = \lambda\mathbf{x} \Leftrightarrow J^{-1}\mathbf{x} = \lambda^{-1}\mathbf{x}$ and $\text{rank}\,(J - \lambda I) = \text{rank}\,((J - \lambda I)J^{-1}) = \text{rank}\,(J^{-1} - \lambda^{-1} I)$.

8.4. Show that if a matrix A is invertible, then A^{-1} *has the same size block diagonals in its Jordan canonical form as A does.*

Hint: Use Exercise **8.3** for each Jordan block of the Jordan canonical form of A. Also, A and A^{-1} have exactly the same number of distinct eigenvalues of the same multiplicities.

8.5. Show that if matrix A is idempotent, (*i.e.*, $A^2 = A$), then its eigenvalue should be either 0 or 1.

8.6. Show that a Jordan block is idempotent if and only if it is 1×1 matrix $[0]$ or $[1]$.

Hint: Use Exercise **8.5** for a Jordan block $J_\lambda = \lambda I + N$.

8.7. Show that if A is an idempotent matrix, then $\text{tr}(A) = \text{rank}\,A$.

8.8. Show that a 2×2 matrix $A = \begin{bmatrix} a & b \\ c & d \end{bmatrix}$ is diagonalizable if and only if either $(a-d)^2 + 4bc \neq 0$ or $a = d, b = c = 0$.

Hint: See Exercise **6.2** on page 303 again. If A has two distinct eigenvalues, A is clearly diagonalizable. If A has a single eigenvalue λ, then A is diagonalizable if and only if $\operatorname{rank}(A - \lambda I) = \operatorname{rank}(J - \lambda I) = 0$, where J is the Jordan canonical form of A.

8.9. *Every square matrix has a Jordan decomposition.* Find the Jordan canonical form J of each of the following real matrices.

$$(1) \begin{bmatrix} 2 & 0 \\ 1 & 2 \end{bmatrix}, \qquad (2) \begin{bmatrix} k & 20 & 0 \\ 0 & k & 20 \\ 0 & 0 & k \end{bmatrix}, \qquad (3) \begin{bmatrix} -2 & 0 & -2 \\ -1 & 1 & -2 \\ 0 & 1 & -1 \end{bmatrix}.$$

8.10. *Every square matrix has a Jordan decomposition.* Determine a Jordan decomposition $A = QJQ^{-1}$ of each matrix A given in Exercise **8.9**.

8.11. *Every square matrix has a Jordan decomposition.* Determine a Jordan decomposition $A = QJQ^{-1}$ of each of the following complex matrices.

$$(1) \begin{bmatrix} i & 0 \\ 1 & i \end{bmatrix}, \qquad (2) \begin{bmatrix} i & 20 & 0 \\ 0 & i & 20 \\ 0 & 0 & i \end{bmatrix}, \qquad (3) \begin{bmatrix} i & -1 & 1 \\ 1 & i & 0 \\ 1 & 0 & i \end{bmatrix}.$$

8.12. Determine the Jordan canonical form of

$$A = \begin{bmatrix} 0 & 1 & 0 & 0 \\ 0 & 0 & 1 & 0 \\ 0 & 0 & 0 & 1 \\ -1 & 4 & -6 & 4 \end{bmatrix}.$$

Compare your answer with J in Example 149 on page 369.

8.13. Compute J^{-1} and J^n for each of Jordan matrices J listed in Exercise **8.1**.

Hint: Use Eq (8.1) on page 384 for each Jordan block.

8.14. Compute e^J and e^{tJ} for each of Jordan matrices J listed in Exercise **8.1**.

Hint: Use Eq (8.2) on page 385 for each Jordan block.

8.15. For a Jordan block $J = \begin{bmatrix} \lambda & 1 & & \mathbf{0} \\ & \ddots & \ddots & \\ & & \ddots & 1 \\ \mathbf{0} & & & \lambda \end{bmatrix}$ with a single eigenvalue λ of multiplicity n, Eq. (8.1) says

$$J^k = \begin{bmatrix} \lambda^k & \frac{\partial_\lambda \lambda^k}{1!} & \frac{\partial_\lambda^2 \lambda^k}{2!} & \cdots & \frac{\partial_\lambda^{n-1} \lambda^k}{(n-1)!} \\ 0 & \lambda^k & \frac{\partial_\lambda \lambda^k}{1!} & \vdots & \frac{\partial_\lambda^{n-2} \lambda^k}{(n-2)!} \\ 0 & 0 & \ddots & \ddots & \vdots \\ \vdots & \cdots & \ddots & \lambda^k & \frac{\partial_\lambda \lambda^k}{1!} \\ 0 & \cdots & \cdots & 0 & \lambda^k \end{bmatrix}.$$

(1) Show that for any polynomial $p(\lambda)$,

$$p(J) = \begin{bmatrix} p(\lambda) & \frac{\partial_\lambda p(\lambda)}{1!} & \frac{\partial_\lambda^2 p(\lambda)}{2!} & \cdots & \frac{\partial_\lambda^{n-1} p(\lambda)}{(n-1)!} \\ 0 & p(\lambda) & \frac{\partial_\lambda p(\lambda)}{1!} & \vdots & \frac{\partial_\lambda^{n-2} p(\lambda)}{(n-2)!} \\ 0 & 0 & \ddots & \ddots & \vdots \\ \vdots & \cdots & \ddots & p(\lambda) & \frac{\partial_\lambda p(\lambda)}{1!} \\ 0 & \cdots & \cdots & 0 & p(\lambda) \end{bmatrix}. \tag{8.5}$$

(2) For the characteristic polynomial $f(\lambda)$ of J, show that $f(J) = \mathbf{0}$ by using Eqs. (8.5).

8.16. Find the characteristic polynomial and the minimal polynomial for each of Jordan matrices J listed in Exercise **8.1**.

Hint: See Example 165.

8.17. For each of the following matrices, find its Jordan canonical form and the minimal polynomial.

$$(1) \begin{bmatrix} 1 & 2 \\ 0 & 1 \end{bmatrix}, \quad (2) \begin{bmatrix} 2 & -3 \\ 7 & -4 \end{bmatrix}, \quad (3) \begin{bmatrix} 1 & 4 & -3 \\ 0 & 3 & 1 \\ 0 & 2 & -1 \end{bmatrix}.$$

8.18. Find the minimal polynomial of each of the following matrices and use your answer to compute A^{-1} if there exists.

$$(1) \begin{bmatrix} 1 & 0 \\ 0 & 4 \end{bmatrix}, \quad (2) \begin{bmatrix} 1 & 2 \\ 0 & 1 \end{bmatrix}, \quad (3) \begin{bmatrix} 1 & 0 & 1 \\ 0 & 2 & 0 \\ 0 & 0 & 2 \end{bmatrix}.$$

Hint: All are upper triangular and hence the diagonal entries are eigenvalues.

8.19. Find the minimal polynomial of each A for the following complex matrices and use it to compute A^{10}.

$$(1) \begin{bmatrix} i & 0 & 0 \\ 0 & i & 0 \\ 0 & 0 & i \end{bmatrix}, \quad (2) \begin{bmatrix} i & 1 & 0 & 0 \\ 0 & i & 1 & 0 \\ 0 & 0 & i & 1 \\ 0 & 0 & 0 & i \end{bmatrix}, \quad (3) \begin{bmatrix} 1 & 0 & 0 & 0 \\ 1 & i & 0 & 0 \\ 0 & 1 & 1 & 0 \\ 0 & 0 & 1 & i \end{bmatrix}.$$

Hint: Use the Jordan canonical form.

8.20. Compute A^{-1}, A^n, and e^A for

$$(1)\ \begin{bmatrix} i & 0 \\ 1 & i \end{bmatrix}, \qquad (2)\ \begin{bmatrix} 4 & 1 & 2 \\ 0 & 4 & 2 \\ 0 & 0 & 4 \end{bmatrix}, \qquad (3)\ \begin{bmatrix} 1 & 0 & 1 \\ 0 & 2 & 1 \\ 0 & 0 & 1 \end{bmatrix},$$

8.21. Compute e^A and e^{tA} for each matrix A listed in Exercise **8.19**.

Hint: Use Eq. (8.2) for each Jordan block, or use the minimal polynomials.

8.22. Show that every Jordan block J is similar to its transpose, in fact $J^T = P^{-1}JP$ by the permutation matrix $P = [\mathbf{e}_t\ \mathbf{e}_{t-1}\ \cdots\ \mathbf{e}_1]$, where t is the order of J. See Exercise **4.25** on page 187. Deduce that *every square matrix A is similar to its transpose A^T.*

Hint: Use permutation matrices $P_1, \ldots, P_s$ for the Jordan blocks $J_1, \ldots, J_s$ of the Jordan canonical form J of A which induce $J_i^T = P_i^{-1}J_iP_i$ for each i. Let $Q^{-1}AQ = J$ and let

$$P = \begin{bmatrix} P_1 & & & \mathbf{0} \\ & P_2 & & \\ & & \ddots & \\ \mathbf{0} & & & P_s \end{bmatrix} \quad \text{and} \quad J = \begin{bmatrix} J_1 & & & \mathbf{0} \\ & J_2 & & \\ & & \ddots & \\ \mathbf{0} & & & J_s \end{bmatrix}.$$

Then, $J^T = P^{-1}JP$ for the Jordan canonical form J of A. Now, apply this similarity to $Q^{-1}AQ = J$.

8.23. A square matrix A is a **circulant matrix** if the i-th row of A is obtained from the first row of A by a cyclic shift of $i-1$ steps. As shown in Example 34 on page 84,

$$A = \begin{bmatrix} a_1 & a_2 & a_3 & \cdots & a_n \\ a_n & a_1 & a_2 & \cdots & a_{n-1} \\ a_{n-1} & a_n & a_1 & \cdots & a_{n-2} \\ \vdots & \vdots & \vdots & \ddots & \vdots \\ a_2 & a_3 & a_4 & \cdots & a_1 \end{bmatrix}.$$

(1) Show that every circulant matrix is normal. (2) Find all eigenvalues of the $n \times n$ circulant matrix

$$W = \begin{bmatrix} 0 & 1 & 0 & \cdots & 0 \\ 0 & 0 & 1 & \cdots & 0 \\ \vdots & \vdots & \vdots & \ddots & \vdots \\ 0 & 0 & 0 & \cdots & 1 \\ 1 & 0 & 0 & \cdots & 0 \end{bmatrix}.$$

(3) Find all eigenvalues of the circulant matrix A by showing that

$$A = \sum_{i=1}^{n} a_i W^{i-1}.$$

(4) Compute $\det A$. (Hint: It is the product of all its eigenvalues.)
(5) Use your answer to find the eigenvalues of

$$B = \begin{bmatrix} 0 & 1 & 1 & \cdots & 1 \\ 1 & 0 & 1 & \cdots & 1 \\ \vdots & \vdots & \ddots & \ddots & \vdots \\ 1 & 1 & \cdots & 0 & 1 \\ 1 & 1 & \cdots & 1 & 0 \end{bmatrix}_{n\times n}.$$

8.24. Compute $\det T_n$ for a *tridiagonal matrix*

$$T_n = \begin{bmatrix} b & b & 0 & 0 & \cdots & 0 & 0 & 0 & 0 \\ b & b & b & 0 & \cdots & 0 & 0 & 0 & 0 \\ 0 & b & b & b & \cdots & 0 & 0 & 0 & 0 \\ \vdots & \vdots & \vdots & \vdots & \ddots & \vdots & \vdots & \vdots & \vdots \\ 0 & 0 & 0 & 0 & \cdots & b & b & b & 0 \\ 0 & 0 & 0 & 0 & \cdots & 0 & b & b & b \\ 0 & 0 & 0 & 0 & \cdots & 0 & 0 & b & b \end{bmatrix}, \quad b > 0.$$

Hint: Set $D_0 = 1$ and $D_n = \det T_n$ for $n \geq 1$. By taking the cofactor expansion of $\det T_n$ along the last row, one has an LRR $D_n = bD_{n-1} - b^2 D_{n-2}$ for $n \geq 3$. Solve this LRR.

8.25. Find a general solution to each of the following LRRs.

(1) $x_n = -x_{n-1} + x_{n-2} + x_{n-3}, \quad n \geq 3,$
(2) $x_n = 4x_{n-1} - 6x_{n-2} + 4x_{n-3} - x_{n-4}, \quad n \geq 4,$
(3) $x_n = (1+i)x_{n-1} - ix_{n-2}, \quad n \geq 2.$

For the companion matrix A in each of (1), (2), and (3), find a general solution to an LdE $\mathbf{x}_n = A\mathbf{x}_{n-1}$.

8.26. Solve the LdE $\begin{cases} x_{n+1} = x_n + y_n \\ y_{n+1} = y_n + z_n \\ z_{n+1} = z_n, \end{cases}$ where $n = 0, 1, 2, \ldots$.

8.27. Solve the LDE $\mathbf{y}' = A\mathbf{y}$, where $A = \begin{bmatrix} -1 & 0 & 2 \\ 2 & 1 & -2 \\ -2 & 0 & 3 \end{bmatrix}$ and $\mathbf{y}(0) = \begin{bmatrix} -2 \\ 0 \\ -1 \end{bmatrix}$.

8.28. Find an SVD of a Jordan block $J = \begin{bmatrix} 0 & 1 & 0 & 0 \\ 0 & 0 & 1 & 0 \\ 0 & 0 & 0 & 1 \\ 0 & 0 & 0 & 0 \end{bmatrix}$.

8.29. Show that $\exp \begin{bmatrix} 0 & \pi \\ -\pi & 0 \end{bmatrix} = \begin{bmatrix} -1 & 0 \\ 0 & -1 \end{bmatrix} = \exp \begin{bmatrix} 0 & -\pi \\ \pi & 0 \end{bmatrix}$. (This shows that the function exp is not injective.)

8.30. Try to find 2×2 matrices X as many as you can which satisfy $X^2 = \begin{bmatrix} 1 & 0 \\ 0 & 1 \end{bmatrix}$.

8.31. Determine whether the following statements are true or false, in general, and justify your answers.

(1) Every (real or complex) square matrix is similar to at least one Jordan matrix even it is not diagonalizable.
(2) Two square matrices have the same Jordan canonical form if and only if they are similar.
(3) If a matrix A has exactly k linearly independent eigenvectors, then the Jordan canonical form of A has exactly k Jordan blocks.
(4) If a matrix A has exactly k distinct eigenvalues, then the Jordan canonical form of A has exactly k Jordan blocks.
(5) For any two square matrices A and B of the same order, if $\operatorname{rank} A^\ell \neq \operatorname{rank} B^\ell$ for some ℓ, then the matrices A, B cannot be similar.
(6) If two square matrices A and B have the same characteristic polynomial $\det(\lambda I - A) = \det(\lambda I - B)$ and for each eigenvalue λ the dimensions of their eigenspaces $\mathcal{N}(\lambda I - A)$ and $\mathcal{N}(\lambda I - B)$ are the same, then A and B are similar.
(7) If a 4×4 matrix A has eigenvalues 1 and 2, each of multiplicity 2, such that $\dim E(1) = 2$ and $\dim E(2) = 1$, then the Jordan canonical form of A has three Jordan blocks.
(8) If there is an eigenvalue λ of A with multiplicity m_λ and $\dim E(\lambda_i) \neq m_\lambda$, then A is not diagonalizable.
(9) In any Jordan decomposition of a matrix, the columns of a change of basis matrix Q are always linearly independent.
(10) For every Jordan block J belonging to an eigenvalue λ, $\operatorname{rank}(J - \lambda I)^\ell$ for $\ell = 1, 2, \ldots$ is a decreasing sequence.
(11) Every Jordan block J is a normal matrix.
(12) Every Jordan block J cannot be diagonalized if its size is greater than 2.
(13) For each Jordan block J belonging to an eigenvalue λ, J^k is a triangular matrix with diagonal entries λ^k.
(14) For each Jordan block J belonging to an eigenvalue λ, e^J is a triangular matrix with diagonal entries e^λ.
(15) For every Jordan block J belonging to an eigenvalue λ, $\det e^J = e^\lambda$.
(16) If $f(x)$ is a polynomial and A is a square matrix such that $f(A) = \mathbf{0}$, then $f(x)$ is a multiple of the characteristic polynomial of A.
(17) The minimal polynomial of a Jordan canonical matrix J is the product of the minimal polynomials of its Jordan blocks J_i.
(18) If the degree of the minimal polynomial of A is equal to the number of distinct eigenvalues of A, then A is diagonalizable.
(19) Any two $n \times n$ matrices A and B such that $AB = BA$ are simultaneously diagonalizable.
(20) Every LdE has always at least one solution, but some LDE cannot have any

solution.

Hint: Exactly eight of them are false.

Chapter 9

Quadratic equations and Quadratic Forms with Matrices

For a linear equation

$$a_1x_1 + a_2x_2 + \cdots + a_nx_n = b,$$

the left-hand side $a_1x_1 + a_2x_2 + \cdots + a_nx_n = \mathbf{a}^T\mathbf{x}$ is a homogeneous polynomial [1] of degree one in n real variables $x_1, x_2, \ldots, x_n$, called a *linear form* or 1*-form*.

A *quadratic form* (or 2*-form*) is a homogeneous polynomial of degree two in a number of variables. For example, $q(x) = ax^2$ is a quadratic form in one variable x, $q(x, y) = ax^2 + bxy + cy^2$ is a quadratic form in two variables x, y, and $q(x, y, z) = ax^2 + by^2 + cz^2 + dxy + eyz + fzx$ is a quadratic form in three variables x, y, z, and so on.

In this chapter, we study a (real or complex) quadratic form, in which matrices also play an important role. As a matter of fact, a real or complex quadratic form can be associated with a real symmetric or a complex Hermitian matrix, and vice-versa.

A more general type of a quadratic form is a *bilinear form* which will be introduced in Section 9.9, as an abstraction of an inner product.

Quadratic forms arise in a variety of applications, including geometry, number theory, vibrations of mechanical systems, statistics, electrical engineering, etc.

---

[1]A (multivariate) polynomial $f(x_1, x_2, \ldots, x_n)$ is said to be *homogeneous* of degree k if all of its nonzero terms have the same degree k, or equivalently $f(ax_1, ax_2, \ldots, ax_n) = a^k f(x_1, x_2, \ldots, x_n)$ for a fixed integer k. Simply, it is called a k*-form*.

9.1 Can you solve (multivariate) quadratic equations?

The solution set (or the roots) of a quadratic equation $ax^2 + bx + c = 0$ can be found by writing the polynomial $ax^2 + bx + c$ as

$$ax^2 + bx + c = a\left(\left(x - \frac{b}{2a}\right)^2 - \frac{b^2 - 4ac}{4a}\right),$$

that is in the form $a\left(x-p\right)^2 - s$.

A quadratic equation in two variables x and y is

$$ax^2 + 2bxy + cy^2 + dx + ey + f = 0,$$

where a constant f means a constant function.

Can you find its solution set? It doesn't seem easy. By assuming one of the constants a, b, c, d, e, f is zero, can you find it? If $b = 0$, the quadratic equation can be written in the form $a\left(x-p\right)^2 + c\left(y-q\right)^2 = t$. What is its solution set? What is the geometric type of the solution set?

As special cases, the solution set of $\left(x-p\right)^2 + \left(y-q\right)^2 = 1$ is a unit circle centered at the point (p, q), and the solution set of $\left(x-p\right)^2 - \left(y-q\right)^2 = 1$ is a hyperbola.

In the given quadratic equation above, we often say that

a constant term f is a constant form or a 0-form;
a linear term $dx + ey$ is a linear form or a 1-form; and
a quadratic term $ax^2 + 2bxy + cy^2$ is a quadratic form or a 2-form.

Also, one may consider a quadratic equation in three variables. As special cases, what is the solution set of $\left(x-p\right)^2 + \left(y-q\right)^2 + \left(z-r\right)^2 = 1$? Or, what is the solution set of $\left(x-p\right)^2 + \left(y-q\right)^2 - \left(z-r\right)^2 = 1$?

Certainly, some quadratic equations may not have any solutions. However, in most cases, the solution sets of quadratic equations in two variables are curves, and the solution sets of quadratic equations in three variables are surfaces. Such a curve and a surface are called a *quadratic curve*[2] and a *quadratic surface*, respectively.

[2]A quadratic curve can also be obtained as a *conic section* or simply *conic* which is the intersection of the surface of a cone with a plane.

The solution set of a quadratic equations in more than three variables is called a *quadratic hypersurface.*

In coming sections, a quadratic equation or quadratic form will be written in matrix notation and it will be used to identify the solution set or its geometric type.

9.2 Quadratic forms; $q(\mathbf{x}) = \mathbf{x}^H A\mathbf{x}$ with symmetric or Hermitian A

The quadratic form in two real variables x and y can be written in matrix notation as

$$ax^2 + 2bxy + cy^2 = \begin{bmatrix} x & y \end{bmatrix} \begin{bmatrix} a & b \\ b & c \end{bmatrix} \begin{bmatrix} x \\ y \end{bmatrix} = \mathbf{x}^T A\mathbf{x},$$

where

$$\mathbf{x} = \begin{bmatrix} x \\ y \end{bmatrix} \quad \text{and} \quad A = \begin{bmatrix} a & b \\ b & c \end{bmatrix}.$$

In matrix notation of a quadratic form, the matrix A is always taken to be a (real) symmetric matrix.

Definition 9.1 **(1)** A (**real**) **linear form** (or 1-**form**) on the Euclidean n-space $\mathbb{R}^n$ is a homogeneous polynomial of degree one in n (real) variables $x_1, x_2, \ldots, x_n$. In matrix notation, it is

$$\mathbf{b}^T\mathbf{x} = \sum_{i=1}^{n} b_i x_i,$$

where $\mathbf{b}^T = [b_1\ b_2\ \cdots\ b_n]$ and $\mathbf{x} = [x_1\ x_2\ \cdots\ x_n]^T$ are in $\mathbb{R}^n$.

(2) A (**real**) **quadratic form** (or 2-**form**) on the Euclidean n-space $\mathbb{R}^n$ is a homogeneous polynomial of degree two in n (real) variables $x_1, x_2, \ldots, x_n$. In matrix notation, it is

$$q(\mathbf{x}) = \mathbf{x}^T A\mathbf{x} = [x_1\ x_2\ \cdots\ x_n][a_{ij}] \begin{bmatrix} x_1 \\ x_2 \\ \vdots \\ x_n \end{bmatrix} = \sum_{i=1}^{n}\sum_{j=1}^{n} a_{ij}x_i x_j, \qquad (9.1)$$

where $\mathbf{x} = [x_1\ x_2\ \cdots\ x_n]^T \in \mathbb{R}^n$ and $A = [a_{ij}]$ is a real $n \times n$ symmetric matrix.

It is also possible to define a linear form and a quadratic form on the Euclidean complex n-space $\mathbb{C}^n$ instead of the Euclidean real n-space $\mathbb{R}^n$.

Definition 9.2 **(1)** A **(complex) linear form** (or **complex** 1**-form**) on the complex n-space $\mathbb{C}^n$ is a homogeneous polynomial of degree one in n complex variables $x_1, x_2, \ldots, x_n$. In matrix notation, it is

$$\mathbf{b}^H\mathbf{x} = \sum_{i=1}^{n} \overline{b_i}x_i,$$

where $\mathbf{b}^H = \overline{\mathbf{b}}^T = [\overline{b_1}\ \overline{b_2}\ \cdots\ \overline{b_n}]$ and $\mathbf{x} = [x_1\ x_2\ \cdots\ x_n]^T$ are in $\mathbb{C}^n$.

(2) A **complex quadratic form** (or **complex** 2**-form**) on the complex n-space $\mathbb{C}^n$ is a homogeneous polynomial of degree two in n complex variables $x_1, x_2, \ldots, x_n$. In matrix notation, it is

$$q(\mathbf{x}) = \mathbf{x}^H A\mathbf{x} = [\overline{x_1}\ \overline{x_2}\ \cdots\ \overline{x_n}][a_{ij}]\begin{bmatrix} x_1 \\ x_2 \\ \vdots \\ x_n \end{bmatrix} = \sum_{i=1}^{n}\sum_{j=1}^{n} a_{ij}\overline{x_i}x_j, \tag{9.2}$$

where $\mathbf{x}^H = [\overline{x_1}\ \overline{x_2}\ \cdots\ \overline{x_n}] \in \mathbb{C}^n$ and $A = [a_{ij}]$ is an $n \times n$ Hermitian matrix.

For example, $[\bar{z}\ \bar{w}]\begin{bmatrix} 1 & i \\ -i & 1 \end{bmatrix}\begin{bmatrix} z \\ w \end{bmatrix} = z\bar{z} - iz\bar{w} + i\bar{z}w + w\bar{w}$ is a complex quadratic form in two complex variables z and w.

The real quadratic form on $\mathbb{R}^n$ and the complex quadratic form on $\mathbb{C}^n$ can both be denoted as $q(\mathbf{x}) = \langle \mathbf{x}, A\mathbf{x}\rangle$ by using the dot product on the real n-space $\mathbb{R}^n$ or on the complex n-space $\mathbb{C}^n$.

The matrix A in the definition of a (real or complex) quadratic form $q(\mathbf{x}) = \langle \mathbf{x}, A\mathbf{x}\rangle$ is called the **associated matrix** for the quadratic form q or the **matrix representation** of the quadratic form q.

Remark: (1) All of 0-forms, 1-forms, and 2-forms can be considered as scalar-valued functions, but only 1-forms are linear functions.

(2) A 1-form is just the dot product on the real or complex n-space with a fixed vector $\mathbf{b}$; *i.e.*, $\mathbf{x} \mapsto \mathbf{b}^H\mathbf{x}$.

(3) The matrix A in the definition of a real quadratic form can be any square matrix. In fact, a square matrix A can be uniquely expressed as the sum of a symmetric part B and a skew-symmetric part C, say

$$A = B + C, \quad \text{where } B = \frac{1}{2}(A + A^T) \text{ and } C = \frac{1}{2}(A - A^T).$$

For the skew-symmetric matrix C, we have

$$\mathbf{x}^T C\mathbf{x} = \left(\mathbf{x}^T C\mathbf{x}\right)^T = \mathbf{x}^T C^T\mathbf{x} = -\mathbf{x}^T C\mathbf{x}.$$

Hence, as a real number, $\mathbf{x}^T C\mathbf{x} = 0$. Therefore,

$$q(\mathbf{x}) = \mathbf{x}^T A\mathbf{x} = \mathbf{x}^T (B + C)\mathbf{x} = \mathbf{x}^T B\mathbf{x}.$$

This means that, without any loss of generality, one may assume that the matrix A in the definition of a real quadratic form is a symmetric matrix.

(4) Similarly, for a complex quadratic form, let A be any $n \times n$ complex matrix. Then, for any $\mathbf{x} \in \mathbb{C}^n$, the quadratic form $\mathbf{x}^H A\mathbf{x}$ is a complex number, and the matrix A can be written as a sum $A = B + iC$, where B and C are Hermitian matrices. (See Theorem 7.15.) Hence,

$$\mathbf{x}^H A\mathbf{x} = \mathbf{x}^H (B + iC)\mathbf{x} = \mathbf{x}^H B\mathbf{x} + i\mathbf{x}^H C\mathbf{x},$$

in which $\mathbf{x}^H B\mathbf{x}$ and $\mathbf{x}^H C\mathbf{x}$ are both real numbers. Hence, it is enough to be concerned with the quadratic forms $\mathbf{x}^H B\mathbf{x}$ and $\mathbf{x}^H C\mathbf{x}$ separately.

Therefore, for a complex quadratic form $\mathbf{x}^H A\mathbf{x}$ on $\mathbb{C}^n$, one can *always* assume A to be Hermitian from the definition so that $\mathbf{x}^H A\mathbf{x}$ *is always real for any* $\mathbf{x} \in \mathbb{C}^n$.

A **(real) quadratic equation** on $\mathbb{R}^n$ is an equation in n variables $x_1, x_2, \ldots,$ x_n of the type

$$f(\mathbf{x}) = \sum_{i=1}^{n}\sum_{j=1}^{n} a_{ij}x_i x_j + \sum_{i=1}^{n} b_i x_i + c = 0,$$

where a_{ij}, b_j, and c are all real constants. In matrix notation,

$$f(\mathbf{x}) = \mathbf{x}^T A\mathbf{x} + \mathbf{b}^T\mathbf{x} + c = 0,$$

where $A = [a_{ij}]$, $\mathbf{x} = [x_1\ x_2\ \cdots\ x_n]^T$ and $\mathbf{b} = [b_1\ b_2\ \cdots\ b_n]^T$ in $\mathbb{R}^n$.

A linear or quadratic equation $f(\mathbf{x})$ is said to be **consistent** if it has a solution, *i.e.,* there is a vector $\mathbf{x} \in \mathbb{R}^n$ such that $f(\mathbf{x}) = 0$. Otherwise, it is said to be **inconsistent**. For instance, the equation $2x^2 + 3y^2 = -1$ in $\mathbb{R}^2$ is inconsistent.

In the following, we consider only consistent equations. The solution set of a linear equation $ax+by+cz=d$, $(a,b,c)\neq(0,0,0)$, in three unknowns represents a plane in the 3-space $\mathbb{R}^3$. See page 5.

In general, the solution set of a consistent linear equation $\mathbf{b}^T\mathbf{x}+c=0$ in n unknowns represents an $(n-1)$-dimensional hyperplane in $\mathbb{R}^n$. (This is an $(n-1)$-dimensional subspace of $\mathbb{R}^n$ if $c=0$.)

The solution set of a consistent quadratic equation $f(\mathbf{x})=\mathbf{x}^TA\mathbf{x}+\mathbf{b}^T\mathbf{x}+c=0$ is a kind of *level surface* in $\mathbb{R}^n$.[3] This is a curved surface that can be parameterized in $n-1$ variables in $\mathbb{R}^n$.

Problem 217 What is the general equation for real quadratic equation in 3 variables x,y,z? What are the 0-form, 1-form, and 2-form in your equation? Express the 1-form and 2-form in matrix notation.

Problem 218 Which one of the following equations is a quadratic equation? For each quadratic equation, express it in matrix notation.

(1) $x^2+6x-y=1$,
(2) $x^2-y^2=1$,
(3) $x^2+y^2+xy=1$,
(4) $x^2+y^2+z^2+xyz=1$.

9.3 Classification of quadratic curves and quadratic surfaces

The solution set of a quadratic equation in two variables is a quadratic curve, which is commonly known as a conic section.

Example 175 (*Three different types of quadratic curves* $\mathbf{x}^TA\mathbf{x}+\mathbf{b}^T\mathbf{x}+c=0$ *in the plane* $\mathbb{R}^2$)

(1) (circle or ellipse) $\frac{x^2}{a^2}+\frac{y^2}{b^2}=1$ with $A=\begin{bmatrix}\frac{1}{a^2} & 0\\ 0 & \frac{1}{b^2}\end{bmatrix}$.

(2) (hyperbola) $\frac{x^2}{a^2}-\frac{y^2}{b^2}=1$ or $\frac{y^2}{a^2}-\frac{x^2}{b^2}=1$ with

$$A=\begin{bmatrix}\frac{1}{a^2} & 0\\ 0 & -\frac{1}{b^2}\end{bmatrix} \quad\text{or}\quad A=\begin{bmatrix}-\frac{1}{b^2} & 0\\ 0 & \frac{1}{a^2}\end{bmatrix}.$$

[3]For a consistent real-valued function $f(x_1,x_2,\ldots,x_n)$, the solution set of $f(x_1,x_2,\ldots,x_n)=c$ is called a *level set* of f. When $n=2$, it is a *level curve*, when $n=3$, it is a *level surface*, and when $n\geq 4$, it is a *level hypersurface*.

(3) (parabola) $x^2 = ay$ or $y^2 = bx$ with $A = \begin{bmatrix} 1 & 0 \\ 0 & 0 \end{bmatrix}$ or $A = \begin{bmatrix} 0 & 0 \\ 0 & 1 \end{bmatrix}$.

All of these cases are illustrated in Figure 9.1 as conic sections. For an ellipse or a hyperbola, the matrix A is invertible, but for a parabola, A is not invertible. □

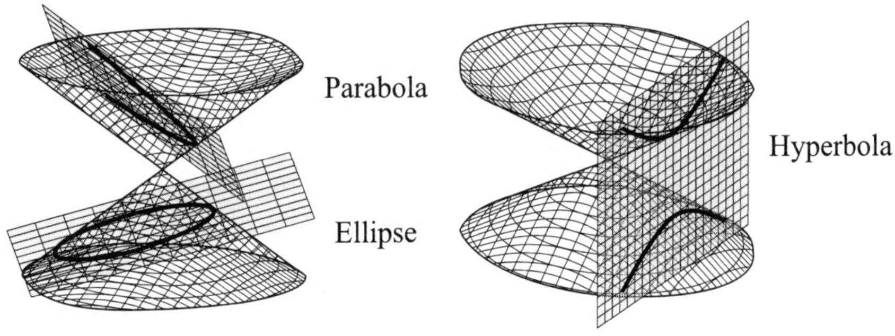

Figure 9.1: Conic sections

A *conic section* is an intersection of a plane, called the cutting plane, with the surface of a double right circular cone $x^2 + y^2 - z^2 = 0$ as illustrated in Figure 9.1. Planes that pass through the vertex of the cone will intersect the cone in a point, a line or a pair of intersecting lines.

These are called *degenerate conics* and sometimes they are not considered as conics at all. Unless otherwise stated, we assume that a conic section refers to a *nondegenerate conic*.

Example 176 (*Four different types of quadratic surfaces* $\mathbf{x}^T A\mathbf{x} + \mathbf{b}^T\mathbf{x} + c = 0$ *in the* 3*-space* $\mathbb{R}^3$)

(1) (sphere or ellipsoid) $\frac{x^2}{a^2} + \frac{y^2}{b^2} + \frac{z^2}{c^2} = 1$ with $A = \begin{bmatrix} \frac{1}{a^2} & 0 & 0 \\ 0 & \frac{1}{b^2} & 0 \\ 0 & 0 & \frac{1}{c^2} \end{bmatrix}$.

(2) (hyperboloid of one or two sheets) $\frac{x^2}{a^2} + \frac{y^2}{b^2} - \frac{z^2}{c^2} = 1$ (of one sheet) or $-\frac{x^2}{a^2} - \frac{y^2}{b^2} + \frac{z^2}{c^2} = 1$ (of two sheets) with

$$A = \begin{bmatrix} \frac{1}{a^2} & 0 & 0 \\ 0 & \frac{1}{b^2} & 0 \\ 0 & 0 & -\frac{1}{c^2} \end{bmatrix} \quad \text{or} \quad A = \begin{bmatrix} -\frac{1}{a^2} & 0 & 0 \\ 0 & -\frac{1}{b^2} & 0 \\ 0 & 0 & \frac{1}{c^2} \end{bmatrix}.$$

A hyperboloid of one sheet is also called a *hyperbolic hyperboloid*, while a hyperboloid of two sheets is sometimes called an *elliptic hyperboloid.*

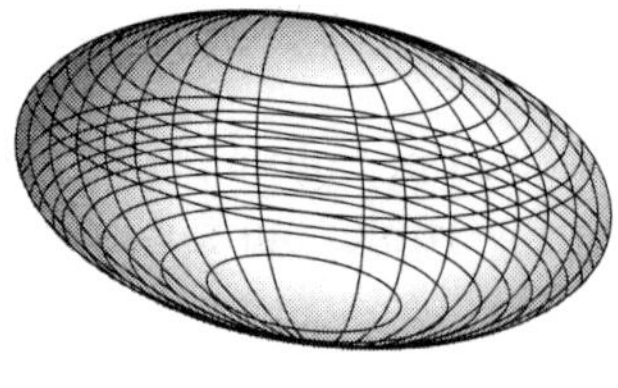

Figure 9.2: Ellipsoid $\frac{x^2}{a^2}+\frac{y^2}{b^2}+\frac{z^2}{c^2}=1$

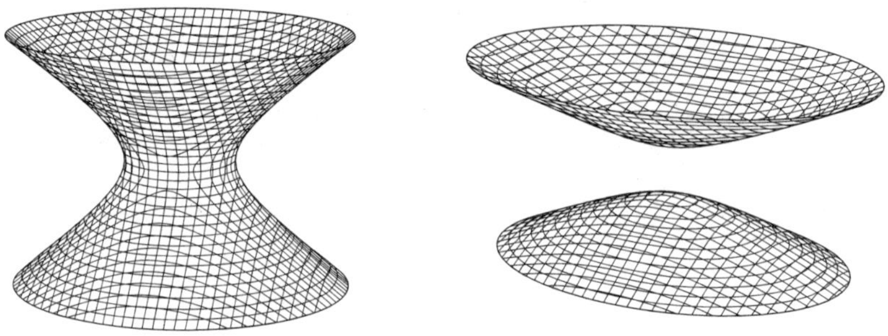

Figure 9.3: Hyperboloid of one sheet (hyperbolic hyperboloid) $\frac{x^2}{a^2}+\frac{y^2}{b^2}-\frac{z^2}{c^2}=1$, and of two sheets (elliptic hyperboloid) $-\frac{x^2}{a^2}-\frac{y^2}{b^2}+\frac{z^2}{c^2}=1$

(3) (cone) $\frac{x^2}{a^2}+\frac{y^2}{b^2}-\frac{z^2}{c^2}=0$ with $A=\begin{bmatrix}\frac{1}{a^2} & 0 & 0\\ 0 & \frac{1}{b^2} & 0\\ 0 & 0 & -\frac{1}{c^2}\end{bmatrix}$.

(4) (paraboloid; elliptic or hyperbolic) $\frac{x^2}{a^2}+\frac{y^2}{b^2}-\frac{z}{c}=0$ (elliptic) or $\frac{x^2}{a^2}-\frac{y^2}{b^2}-\frac{z}{c}=0$, $c>0$ (hyperbolic) with

$$A=\begin{bmatrix}\frac{1}{a^2} & 0 & 0\\ 0 & \frac{1}{b^2} & 0\\ 0 & 0 & 0\end{bmatrix} \quad \text{or} \quad A=\begin{bmatrix}\frac{1}{a^2} & 0 & 0\\ 0 & -\frac{1}{b^2} & 0\\ 0 & 0 & 0\end{bmatrix}.$$

All of these cases are illustrated in Figures 9.2– 9.5. For the first three cases, the matrices A are invertible, but for the last paraboloid case, A is not invertible. □

One can see that a hyperboloid of one sheet $\frac{x^2}{a^2}+\frac{y^2}{b^2}-\frac{z^2}{c^2}=1$ and a hyperboloid of two sheets $-\frac{x^2}{a^2}-\frac{y^2}{b^2}+\frac{z^2}{c^2}=1$ asymptotically approach the same conical surface $\frac{x^2}{a^2}+\frac{y^2}{b^2}-\frac{z^2}{c^2}=0$ as x or y approaches infinity.

A *(complex) quadratic equation* on $\mathbb{C}^n$

$$f(\mathbf{x})=\mathbf{x}^H A\mathbf{x}+\mathbf{b}^H\mathbf{x}+c=0$$

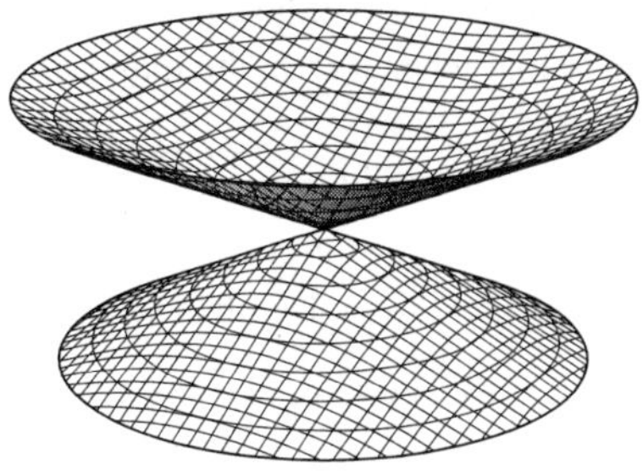

Figure 9.4: Cone $\frac{x^2}{a^2} + \frac{y^2}{b^2} - \frac{z^2}{c^2} = 0$

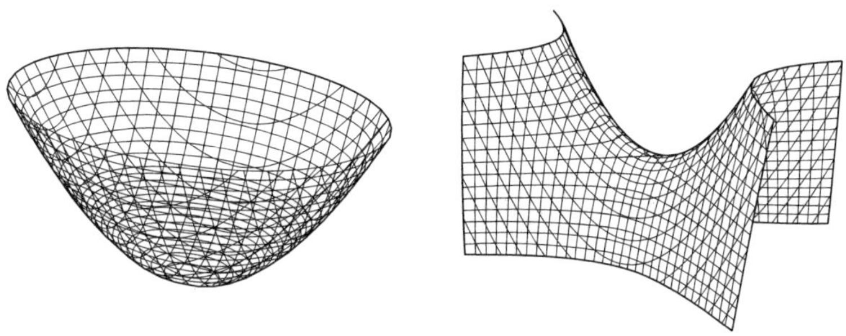

Figure 9.5: Elliptic paraboloid $\frac{x^2}{a^2} + \frac{y^2}{b^2} = \frac{z}{c}$ and hyperbolic paraboloid $\frac{x^2}{a^2} - \frac{y^2}{b^2} = \frac{z}{c}, c > 0$

becomes much more complicated. For example, over the complex numbers, ellipses and hyperbolas are not distinct; the ellipse $x^2 + y^2 = 1$ becomes a hyperbola under the substitution $y = iw$, geometrically taking a complex rotation yields $x^2 - w^2 = 1$, a hyperbola. We will not deal with geometric types of the complex case in detail.

Problem 219 Sketch the solution set (*i.e.*, level surface) of each of the following equations in the 3-space $\mathbb{R}^3$.

(1) $\frac{x^2}{a^2} + \frac{y^2}{b^2} + z = 0$, (2) $\frac{x^2}{a^2} + \frac{y^2}{b^2} + z = 1$,
(3) $\frac{x^2}{a^2} - \frac{y^2}{b^2} + z = 0$, (4) $\frac{x^2}{a^2} - \frac{y^2}{b^2} + z = 1$.

Hint: (1) and (2) are elliptic paraboloids, while (3) and (4) are hyperbolic paraboloids. (1) and (3) are illustrated as the upside down of Figure 9.5. (2) is a translation of (1) along the z-axis.

Problem 220 For an ellipse $\frac{x^2}{a^2} + \frac{y^2}{b^2} = 1$ with $A = \begin{bmatrix} \frac{1}{a^2} & 0 \\ 0 & \frac{1}{b^2} \end{bmatrix}$, assume that the two eigenvalues are $0 < \frac{1}{a^2} \le \frac{1}{b^2}$ in order, or $a \ge b > 0$. Set $c^2 = a^2 - b^2$. Show that this ellipse is the set of all points (x, y) the sum of whose distances from two *foci points*

$(\pm c, 0)$ equals $2a$. The *major axis* [4] of an ellipse is its longest diameter and the *minor axis* is its shortest diameter. Show that when $a \geq b$, the major axis lies along the x-axis and the minor axis lies along the y-axis.

Hint: Simplify $\sqrt{(x+c)^2+y^2} = 2a - \sqrt{(x-c)^2+y^2}$ with $c^2 = a^2 - b^2$.

Problem 221 For a hyperbola $\frac{x^2}{a^2} - \frac{y^2}{b^2} = 1$ with $A = \begin{bmatrix} \frac{1}{a^2} & 0 \\ 0 & -\frac{1}{b^2} \end{bmatrix}$, assume that $a, b > 0$ and set $c^2 = a^2 + b^2$. Show that this hyperbola is the set of all points (x, y) the difference of whose distances from two *foci points* $(\pm c, 0)$ equals $\pm 2a$. The *transverse axis* of the hyperbola is the line segment passing through the center of the hyperbola and connecting two points $(\pm a, 0)$. Its *conjugate axis* is the line segment connecting two points $(0, \pm b)$ which is perpendicular to the transverse axis.

Hint: Simplify $\sqrt{(x+c)^2+y^2} = 2a + \sqrt{(x-c)^2+y^2}$ with $c^2 = a^2 + b^2$.

9.4 Diagonalization of quadratic forms

How to determine the geometric type of a quadratic surface or hypersurface? How to sketch their level surfaces?

In this section, we first consider a special case of the type $\mathbf{x}^T A\mathbf{x} = c$ without the linear form in the quadratic equation $f(\mathbf{x}) = \mathbf{x}^T A\mathbf{x} + \mathbf{b}^T\mathbf{x} + c = 0$.

A quadratic form $\mathbf{x}^T A\mathbf{x}$ on $\mathbb{R}^n$ can be written as the sum of two parts:

$$q(\mathbf{x}) = \mathbf{x}^T A\mathbf{x} = \sum_{i=1}^{n} a_{ii}x_i^2 + 2\sum_{i<j} a_{ij}x_ix_j,$$

in which the first part $\sum_{i=1}^{n} a_{ii}x_i^2$ is called the **(perfect) square terms** and the second part $\sum_{i\neq j} a_{ij}x_ix_j$ is called the **cross-product terms.**

Actually, what makes it hard to sketch the level surface of a quadratic equation is the cross-product terms. However, the quadratic form $q(\mathbf{x}) = \mathbf{x}^T A\mathbf{x}$ can be transformed into a new quadratic form without the cross-product terms by a suitable change of variables (or by a change of coordinates). It can be done because the matrix A is symmetric and so orthogonally diagonalizable, *i.e.,* there exists an

[4]More precisely, the *major axis* of the ellipse is the chord that passes through its foci and has its endpoints on the ellipse. The *minor axis* of the ellipse is the chord that contains the center of the ellipse, has its endpoints on the ellipse and is perpendicular to the major axis.

orthogonal matrix P such that

$$P^T AP = P^{-1}AP = D = \begin{bmatrix} \lambda_1 & & & \mathbf{0} \\ & \lambda_2 & & \\ & & \ddots & \\ \mathbf{0} & & & \lambda_n \end{bmatrix}.$$

Here, the diagonal entries λ_i's are the eigenvalues of A and the column vectors of P are their associated (orthonormal) eigenvectors. Now, by setting $\mathbf{x} = P\mathbf{y}$, (that is, by a change of variables), we have

$$\mathbf{x}^T A\mathbf{x} = \mathbf{y}^T(P^T AP)\mathbf{y} = \mathbf{y}^T D\mathbf{y} = \lambda_1 y_1^2 + \lambda_2 y_2^2 + \cdots + \lambda_n y_n^2,$$

which is a quadratic form without the cross-product terms.

This is also true for a complex quadratic form $q(\mathbf{x}) = \mathbf{x}^H A\mathbf{x}$ with a Hermitian matrix A. Since every Hermitian matrix is unitarily diagonalizable, there exists a unitary matrix U such that $U^H AU = D$ is a diagonal matrix. Hence, by a change of variables $\mathbf{x} = U\mathbf{y}$, the quadratic form $q(\mathbf{x}) = \mathbf{y}^H D\mathbf{y}$ has only (perfect) square terms.

In either the real or complex case, we have the following theorem.

Theorem 9.3 (Principal Axes theorem) *Let A be a symmetric (or a Hermitian) matrix with the eigenvalues $\lambda_1, \lambda_2, \ldots, \lambda_n$ (possibly repeated). Then we have the following.*

(1) *For a real quadratic form $\mathbf{x}^T A\mathbf{x}$ in $\mathbf{x} = [x_1\ x_2\ \cdots\ x_n]^T \in \mathbb{R}^n$ with a symmetric matrix A, there is a change of coordinates of $\mathbf{x}$ into $\mathbf{y} = P^T\mathbf{x} = [y_1\ y_2\ \cdots\ y_n]^T$ such that*

$$\mathbf{x}^T A\mathbf{x} = \mathbf{y}^T D\mathbf{y} = \lambda_1 y_1^2 + \lambda_2 y_2^2 + \cdots + \lambda_n y_n^2,$$

where P is an orthogonal matrix which diagonalizes A as $P^T AP = D$.

(2) *For a complex quadratic form $\mathbf{x}^H A\mathbf{x}$ in $\mathbf{x} = [x_1\ x_2\ \cdots\ x_n]^T \in \mathbb{C}^n$ with a Hermitian matrix A, there is a change of coordinates of $\mathbf{x}$ into $\mathbf{y} = U^H\mathbf{x} = [y_1\ y_2\ \cdots\ y_n]^T$ such that*

$$\mathbf{x}^H A\mathbf{x} = \mathbf{y}^H D\mathbf{y} = \lambda_1|y_1|^2 + \lambda_2|y_2|^2 + \cdots + \lambda_n|y_n|^2,$$

where U is a unitary matrix which diagonalized A as $U^H AU = D$.

The (real or complex) quadratic form $\mathbf{y}^T D\mathbf{y} = \lambda_1 y_1^2 + \lambda_2 y_2^2 + \cdots + \lambda_n y_n^2$ or $\mathbf{y}^H D\mathbf{y} = \lambda_1|y_1|^2 + \lambda_2|y_2|^2 + \cdots + \lambda_n|y_n|^2$ with a diagonal matrix D is called the **normal form** of a (real or complex) quadratic form $\mathbf{x}^T A\mathbf{x}$ or $\mathbf{x}^H A\mathbf{x}$, respectively.

The columns $\mathbf{u}_i$ of the matrix P (U, respectively) form an orthonormal basis for $\mathbb{R}^n$ (for $\mathbb{C}^n$, respectively). The (new coordinate) axis spanned by the vector $\mathbf{u}_i$ is called the i**-th principal axis** of the quadratic form. The principal axes give a rectangular coordinate system for $\mathbb{R}^n$ or for $\mathbb{C}^n$, and the vector $\mathbf{y}$ is just the coordinate vector of $\mathbf{x}$ with respect to the principal axes.

The geometric type of a quadratic surface or a quadratic hypersurface on $\mathbb{R}^n$, which is an ellipsoid, a hyperboloid or the others, depends on the signs of the eigenvalues of A. Of course, a paraboloid is also a quadratic surface, but it appears only when a linear form is present in the quadratic equation.

Example 177 (*Determine the geometric type via the diagonalization of a quadratic form*) Determine the geometric type of a quadratic surface $2x^2+2xy+2y^2-9=0$ on the plane $\mathbb{R}^2$.

Solution: In matrix notation, this is

$$\begin{bmatrix} x & y \end{bmatrix} \begin{bmatrix} 2 & 1 \\ 1 & 2 \end{bmatrix} \begin{bmatrix} x \\ y \end{bmatrix} = 9.$$

The matrix $A = \begin{bmatrix} 2 & 1 \\ 1 & 2 \end{bmatrix}$ has eigenvalues

$$\lambda_1 = 1 \quad \text{and} \quad \lambda_2 = 3$$

with associated unit eigenvectors

$$\mathbf{v}_1 = \left(\frac{1}{\sqrt{2}}, -\frac{1}{\sqrt{2}}\right) \quad \text{and} \quad \mathbf{v}_2 = \left(\frac{1}{\sqrt{2}}, \frac{1}{\sqrt{2}}\right),$$

respectively, which form an orthonormal basis, say β. If α denotes the standard basis, then the change of basis matrix

$$P = [id]_\beta^\alpha = [\mathbf{v}_1 \ \mathbf{v}_2] = \frac{1}{\sqrt{2}} \begin{bmatrix} 1 & 1 \\ -1 & 1 \end{bmatrix} = \begin{bmatrix} \cos 45^\circ & \sin 45^\circ \\ -\sin 45^\circ & \cos 45^\circ \end{bmatrix},$$

which is a rotation through 45° in the clockwise direction and $P^T = P^{-1}$. This gives a change of coordinates, $\mathbf{x} = P\mathbf{y}$, *i.e.*,

$$\begin{bmatrix} x \\ y \end{bmatrix} = \frac{1}{\sqrt{2}} \begin{bmatrix} 1 & 1 \\ -1 & 1 \end{bmatrix} \begin{bmatrix} x' \\ y' \end{bmatrix} = \begin{bmatrix} \frac{1}{\sqrt{2}}x' + \frac{1}{\sqrt{2}}y' \\ -\frac{1}{\sqrt{2}}x' + \frac{1}{\sqrt{2}}y' \end{bmatrix}.$$

This implies that

$$\begin{aligned} 2x^2 + 2xy + 2y^2 &= \mathbf{x}^T A\mathbf{x} = \mathbf{y}^T P^T AP\mathbf{y} \\ &= \mathbf{y}^T \begin{bmatrix} 1 & 0 \\ 0 & 3 \end{bmatrix} \mathbf{y} = (x')^2 + 3(y')^2 = 9, \end{aligned}$$

or

$$\frac{(x')^2}{9} + \frac{(y')^2}{3} = 1.$$

This is an ellipse with the principal axes $\mathbf{v}_1 = P^T\mathbf{e}_1$ and $\mathbf{v}_2 = P^T\mathbf{e}_2$. □

To determine the geometric type of quadratic (hyper-)surface, we introduce the following definition for a symmetric matrix A or a quadratic form $\mathbf{x}^T A\mathbf{x}$.

Definition 9.4 Let $A = [a_{ij}] \in M_{n\times n}(\mathbb{R})$ be a symmetric matrix. Then, the matrix A, or a quadratic form $q(\mathbf{x}) = \mathbf{x}^T A\mathbf{x}$, is said to be

(1) positive-definite if $\mathbf{x}^T A\mathbf{x} = \sum_{i,j} a_{ij}x_i x_j > 0$ for all nonzero $\mathbf{x}$ in $\mathbb{R}^n$,

(2) positive-semidefinite if $\mathbf{x}^T A\mathbf{x} = \sum_{i,j} a_{ij}x_i x_j \geq 0$ for all $\mathbf{x}$,

(3) negative-definite if $\mathbf{x}^T A\mathbf{x} = \sum_{i,j} a_{ij}x_i x_j < 0$ for all nonzero $\mathbf{x}$,

(4) negative-semidefinite if $\mathbf{x}^T A\mathbf{x} = \sum_{i,j} a_{ij}x_i x_j \leq 0$ for all $\mathbf{x}$,

(5) indefinite if $\mathbf{x}^T A\mathbf{x}$ takes both positive and negative values.

Similarly, one can define the same terminology for a Hermitian matrix or for a complex quadratic form $\mathbf{x}^H A\mathbf{x}$ on $\mathbb{C}^n$.

Example 178 (*Definite type of symmetric matrix*) The real symmetric matrix

$$A = \begin{bmatrix} 2 & -1 & 0 \\ -1 & 2 & -1 \\ 0 & -1 & 2 \end{bmatrix}$$

is positive-definite because the quadratic form satisfies

$$\begin{aligned} \mathbf{x}^T A\mathbf{x} &= [x_1\ x_2\ x_3] \begin{bmatrix} 2 & -1 & 0 \\ -1 & 2 & -1 \\ 0 & -1 & 2 \end{bmatrix} \begin{bmatrix} x_1 \\ x_2 \\ x_3 \end{bmatrix} \\ &= [x_1\ x_2\ x_3] \begin{bmatrix} 2x_1 - x_2 \\ -x_1 + 2x_2 - x_3 \\ -x_2 + 2x_3 \end{bmatrix} \\ &= x_1(2x_1 - x_2) + x_2(-x_1 + 2x_2 - x_3) + x_3(-x_2 + 2x_3) \\ &= 2x_1^2 - 2x_1x_2 + 2x_2^2 - 2x_2x_3 + 2x_3^2 \\ &= x_1^2 + (x_1 - x_2)^2 + (x_2 - x_3)^2 + x_3^2 > 0 \end{aligned}$$

unless $x_1 = x_2 = x_3 = 0$.

As an equivalent condition, A has three eigenvalues $\lambda_1 = 2 + \sqrt{2}$, $\lambda_2 = 2$, $\lambda_3 = 2 - \sqrt{2}$, and all of them are positive. See the next Theorem 9.5. □

The following characterizations follow from the Principal Axes theorem.

Theorem 9.5 *A real symmetric or a Hermitian matrix A is*

(1) *positive-definite if and only if the eigenvalues of A are all positive,*
(2) *positive-semidefinite if and only if the eigenvalues of A are all nonnegative,*
(3) *negative-definite if and only if the eigenvalues of A are all negative,*
(4) *negative-semidefinite if and only if the eigenvalues of A are all nonpositive,*
(5) *indefinite if and only if A has both positive and negative eigenvalues.*

If A is positive-definite, $\det A > 0$ (as the product of all eigenvalues). If the eigenvalues of A are all negative, then $-A$ should be positive-definite and consequently A is negative-definite. If A has eigenvalues that differ in sign, then A is indefinite. Indeed, if λ_1 is a positive eigenvalue of A and $\mathbf{x}_1$ is an eigenvector belonging to λ_1, then

$$q(\mathbf{x}_1) = \langle \mathbf{x}_1, A\mathbf{x}_1 \rangle = \lambda_1 \langle \mathbf{x}_1, \mathbf{x}_1 \rangle = \lambda_1 \|\mathbf{x}_1\|^2 > 0,$$

and if λ_2 is a negative eigenvalue with associated eigenvector $\mathbf{x}_2$, then

$$q(\mathbf{x}_2) = \langle \mathbf{x}_2, A\mathbf{x}_2 \rangle = \lambda_2 \langle \mathbf{x}_2, \mathbf{x}_2 \rangle = \lambda_2 \|\mathbf{x}_2\|^2 < 0.$$

If A is definite, then $\mathbf{0}$ is the only critical point of a quadratic form $q(\mathbf{x}) = \langle \mathbf{x}, A\mathbf{x} \rangle$, and $q(\mathbf{0}) = 0$ is the *global minimum* if A is positive-definite and the *global maximum* if A is negative-definite.

If A is indefinite, then $\mathbf{0}$ is a *saddle point*, and the geometric types of $\mathbf{x}^T A\mathbf{x} = c$ are divided into several cases depending on the number of positive eigenvalues and the number of negative eigenvalues.

Problem 222 What is the matrix A for each of the following quadratic forms? Diagonalize the matrix A and determine its definite type.

(1) $x^2 - y^2 + 4z^2 + 6xy - 8xz + 2yz$,
(2) $3x^2 + 2y^2 + 4z^2 - xy + xz$.

Hint: For the definite types, use Theorem 9.5.

9.5 The inertia classifies level surfaces

In Section 9.4, we see that the geometric type of a quadratic (hyper-)surface $\mathbf{x}^T A\mathbf{x} = c$ (without the linear form) depends on the signs of the eigenvalues of A. In fact, it is completely determined by the numbers of positive, negative, and zero eigenvalues of A.

Definition 9.6 The **inertia** of a symmetric (or a Hermitian) matrix A is a triple of integers denoted by $\text{In}(A) = (p, q, k)$, where p, q, and k are the numbers of positive, negative, and zero eigenvalues of A, respectively.

The inertia In(A) determines the geometric type of a quadratic (hyper-)surface $\mathbf{x}^T A\mathbf{x} = c$ on $\mathbb{R}^n$. Clearly, if $\text{In}(A) = (p, q, k)$, then $\text{In}(-A) = (q, p, k)$, and if $p = 0$ and $c > 0$, then the equation $\mathbf{x}^T A\mathbf{x} = c$ is inconsistent. Hence, it suffices to consider the cases of $c \geq 0$ and $p > 0$.

Excluding the inconsistent cases, we have the following classification of the solution sets for $n = 2$ and 3.

For $n = 2$, there are only three possible cases for $\mathbf{x}^T A\mathbf{x} = c$:

In(A)	The solution of $\mathbf{x}^T A\mathbf{x} = c$	
(p, q, k)	$c > 0$	$c = 0$
(2,0,0)	ellipse	a point
(1,1,0)	hyperbola	two lines crossing at $\mathbf{0}$
(1,0,1)	two parallel lines	a line

For $n = 3$, there are six possibilities:

In(A)	The solution of $\mathbf{x}^T A\mathbf{x} = c$	
(p, q, k)	$c > 0$	$c = 0$
(3,0,0)	ellipsoid	a point
(2,1,0)	one-sheeted hyperboloid	elliptic cone
(2,0,1)	elliptic cylinder	a line
(1,2,0)	two-sheeted hyperboloid	elliptic cone
(1,1,1)	hyperbolic cylinder	two planes crossing in a line
(1,0,2)	two parallel planes	a plane

In general, for an $n \times n$ symmetric matrix A, $\text{In}(A) = (p, q, k)$ has $n(n+1)/2$ different possibilities with $p > 0$, which induce all possible different geometric types of quadratic equations $\mathbf{x}^T A\mathbf{x} = c$ in n variables, (see Problem 244 on page 462).

For example, if $\text{In}(A) = (n, 0, 0)$ and $c > 0$, *i.e.*, the eigenvalues of A are all positive, then the quadratic equation describes an ellipsoid in $\mathbb{R}^n$.

Example 179 (*The inertia classifies the geometric type of a quadratic surface*) Determine the geometric type of a quadratic surface for $2xy + 2xz = 1$ on $\mathbb{R}^3$.

Solution: The matrix for the given quadratic form is

$$A = \begin{bmatrix} 0 & 1 & 1 \\ 1 & 0 & 0 \\ 1 & 0 & 0 \end{bmatrix},$$

which is not invertible. The eigenvalues of A are

$$\lambda_1 = \sqrt{2}, \quad \lambda_2 = -\sqrt{2}, \quad \lambda_3 = 0$$

with associated orthonormal eigenvectors

$$\mathbf{v}_1 = \left(\frac{1}{\sqrt{2}}, \frac{1}{2}, \frac{1}{2}\right), \quad \mathbf{v}_2 = \left(-\frac{1}{\sqrt{2}}, \frac{1}{2}, \frac{1}{2}\right), \quad \mathbf{v}_3 = \left(0, -\frac{1}{\sqrt{2}}, \frac{1}{\sqrt{2}}\right),$$

respectively. Hence, an orthogonal matrix P that diagonalizes A is

$$P = \frac{1}{2}\begin{bmatrix} \sqrt{2} & -\sqrt{2} & 0 \\ 1 & 1 & -\sqrt{2} \\ 1 & 1 & \sqrt{2} \end{bmatrix}.$$

The change of coordinates $\mathbf{x} = P\mathbf{y}$, that is,

$$x = \frac{1}{\sqrt{2}}(x' - y'), \quad y = \frac{1}{2}(x' + y' - \sqrt{2}z'), \quad z = \frac{1}{2}(x' + y' + \sqrt{2}z'),$$

transforms the equation to

$$\sqrt{2}(x')^2 - \sqrt{2}(y')^2 = 1.$$

This is a **hyperbolic cylinder**, see Figure 9.6. In this case, $\text{In}(A) = (1, 1, 1)$ and A is not invertible. □

There are other kinds of cylinders in the 3-space $\mathbb{R}^3$. One can see that $\frac{x^2}{a^2}+\frac{y^2}{b^2}=1$ with $A=\begin{bmatrix}\frac{1}{a^2} & 0 & 0\\ 0 & \frac{1}{b^2} & 0\\ 0 & 0 & 0\end{bmatrix}$ and $x^2=ay$ with $A=\begin{bmatrix}1 & 0 & 0\\ 0 & 0 & 0\\ 0 & 0 & 0\end{bmatrix}$ in the 3-space $\mathbb{R}^3$ are an **elliptic cylinder** and a **parabolic cylinder**, respectively. In both cases, z can be any value and both matrices A are not invertible.

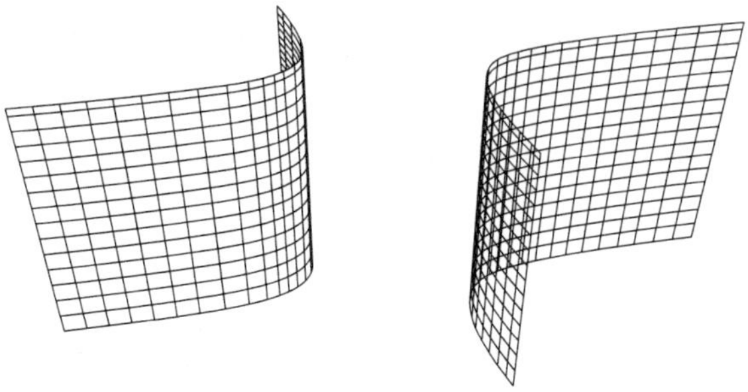

Figure 9.6: Hyperbolic cylinder $2xy+2xz=1$

Now, consider a general form of a quadratic equation on the n-space $\mathbb{R}^n$

$$\mathbf{x}^T A\mathbf{x}+\mathbf{b}^T\mathbf{x}=c.$$

(1) If the equation does not have a linear form, *i.e.,* $\mathbf{b}=\mathbf{0}$, then, as shown already, a *parabolic* surface cannot appear as the solution set of the quadratic equation.

(2) Suppose that the equation has a nonzero linear form, *i.e.,* $\mathbf{b}\neq\mathbf{0}$. If the (symmetric) matrix A is invertible, then by taking a change of variables (it is a translation) to be

$$\mathbf{y}=\mathbf{x}+\tfrac{1}{2}A^{-1}\mathbf{b} \quad \text{or} \quad \mathbf{x}=\mathbf{y}-\tfrac{1}{2}A^{-1}\mathbf{b},$$

the given quadratic equation is transformed into a new one $\mathbf{y}^T A\mathbf{y}=d$ without a linear form, where $d=c+\frac{1}{4}\mathbf{b}^T A^{-1}\mathbf{b}$.

However, if A is not invertible, the solution set of the quadratic equation depends not only on the inertia of A, but also on the type of linear form, and a parabolic surface appears as the solution set of a quadratic equation with a nonzero linear form.

For example, the quadratic equation $x^2-z=c$ in three variables x,y,z has a quadratic form for which A is not invertible with $\text{In}(A)=(1,0,2)$ and the nonzero linear form z cannot be eliminated by any change of variables. The solution set of this equation is a parabolic cylinder.

Example 180 (*Eliminating a linear form by a change of variables*) Determine the geometric type of the level surface of a quadratic equation

$$2x^2 - 4xy + 3y^2 + 2x - 2y = 0.$$

Solution: The matrix for the quadratic form $2x^2 - 4xy + 3y^2$ is

$$A = \begin{bmatrix} 2 & -2 \\ -2 & 3 \end{bmatrix}.$$

Its inverse is

$$A^{-1} = \frac{1}{2}\begin{bmatrix} 3 & 2 \\ 2 & 2 \end{bmatrix} \quad \text{and} \quad \mathbf{b} = \begin{bmatrix} 2 \\ -2 \end{bmatrix} \quad \text{for the linear form.}$$

With the change of variables $\mathbf{y} = \mathbf{x} + \frac{1}{2}A^{-1}\mathbf{b}$ (a translation by $\frac{1}{2}A^{-1}\mathbf{b}$), that is

$$x' = x + \frac{1}{2}, \qquad y' = y,$$

the equation is transformed to a new equation

$$2(x')^2 - 4x'y' + 3(y')^2 = \frac{1}{2}.$$

Clearly, the matrix representation of the new quadratic form is also A, and its eigenvalues are

$$(5 + \sqrt{17})/2 \quad \text{and} \quad (5 - \sqrt{17})/2.$$

Therefore, $\text{In}(A) = (2, 0, 0)$ and the solution of the equation is an ellipse. □

A (real or complex) quadratic form $q(\mathbf{x}) = \langle \mathbf{x}, A\mathbf{x} \rangle$ is said to be **nondegenerate** if the matrix A is invertible, and **degenerate** otherwise.

Equivalently, a quadratic form $q(\mathbf{x}) = \langle \mathbf{x}, A\mathbf{x} \rangle$ is degenerate if and only if $k \neq 0$ in $\text{In}(A) = (p, q, k)$. Degenerate forms include (elliptic, parabolic, or hyperbolic) cylinders, planes, lines, points, or even no points at all.

Problem 223 [Problem 222 continued] Sketch the level surface of each of the following quadratic equations.

(1) $x^2 - y^2 + 4z^2 + 6xy - 8xz + 2yz = 0,$

(2) $3x^2 + 2y^2 + 4w^2 - xy + xz = t, \quad (t \geq 0)$.

Problem 224 Sketch the level surface of each of the following quadratic equations.

(1) $x^2 + y^2 + z^2 + 2xy = 1$,
(2) $20x^2 + y^2 + z^2 + 2xy = 1$,
(3) $x^2 + y^2 + z^2 + 20xy = 1$.

With the matrix notation $\mathbf{x}^T A\mathbf{x} = 1$ in each case, explain how the level surfaces $\mathbf{x}^T A\mathbf{x} = t$ for various t vary with $t > 0$.

9.6 Characterizations of definite forms

For the geometric type of a quadratic (hyper-)surface $\mathbf{x}^T A\mathbf{x} = c$, it is frequently needed to determine whether a symmetric (or Hermitian) matrix A is positive-definite, negative-definite, or indefinite.

In most cases, the definition does not help much for such criteria.

But, the Principal Axes theorem (or Theorem 9.5) gives us a practical characterization of positive-definite matrices: *A is positive-definite if and only if all eigenvalues of A are positive.*

There exists another criteria in terms of the determinant of the matrix. To show this, we again look at the quadratic form in two real variables, $q(x, y) = ax^2 + 2bxy + cy^2$, which can be written in the perfect square form as

$$q(\mathbf{x}) = ax^2 + 2bxy + cy^2 = a\left(x + \frac{b}{a}y\right)^2 + \left(c - \frac{b^2}{a}\right)y^2.$$

Hence, q is positive-definite, *i.e.*, $q(\mathbf{x}) = \mathbf{x}^T A\mathbf{x} > 0$ for any nonzero vector $\mathbf{x} = (x, y) \in \mathbb{R}^2$, if and only if $a > 0$ and $ac > b^2$. In other words, the determinants of

$$[a] \quad \text{and} \quad \begin{bmatrix} a & b \\ b & c \end{bmatrix}$$

are both positive.

A generalization of these conditions will involve all n submatrices of A, called the **principal submatrices** of A, which are defined as the upper left square submatrices

$$A_1 = \begin{bmatrix} a_{11} \end{bmatrix}, \; A_2 = \begin{bmatrix} a_{11} & a_{12} \\ a_{21} & a_{22} \end{bmatrix}, \; A_3 = \begin{bmatrix} a_{11} & a_{12} & a_{13} \\ a_{21} & a_{22} & a_{23} \\ a_{31} & a_{32} & a_{33} \end{bmatrix}, \; \ldots, \; A_n = A.$$

With these submatrices A_k, one can have the following characterization of positive-definite matrices.

Theorem 9.7 *The following are equivalent for a real symmetric (or a Hermitian) matrix A.*

(1) *A is positive-definite, i.e., $\langle \mathbf{x}, A\mathbf{x}\rangle > 0$ for all nonzero vectors $\mathbf{x}$,*
(2) *all of the eigenvalues of A are positive,*
(3) *all of the principal submatrices A_k have positive determinants,*
(4) *A can be reduced to an upper triangular matrix by forward elimination performed by only the type III elementary row operations of "adding a constant multiple of a row to another", (without row interchanges), and all of the pivots are positive,*
(5) *there exists a unique lower triangular matrix L with positive diagonal entries such that $A = LL^T$ $(= LL^H$ if A is Hermitian$)$ (called the* **Cholesky decomposition** *or* **Cholesky factorization***),*
(6) *there exists a nonsingular matrix W such that $A = W^TW$, $(= W^HW$ if A is Hermitian), or equivalently $A = WW^T$.*

The Cholesky decomposition $A = LL^T$ $(= LL^H$ if A is Hermitian) of a positive-definite symmetric (or a Hermitian) matrix A is a special kind of LU decomposition of a symmetric (or a Hermitian) matrix A.

Proof: (1) ⇔ (2) has already been shown in Theorem 9.5.

(1) ⇒ (3) First, we prove the real case. Let A_k be a principal submatrix of A and let $\mathbf{x}_k$ be any nonzero vector in $\mathbb{R}^k$. Then $\mathbf{x} = \begin{bmatrix}\mathbf{x}_k\\ \mathbf{0}\end{bmatrix}$ is nonzero in $\mathbb{R}^n$ and

$$\mathbf{x}^TA\mathbf{x} = \begin{bmatrix}\mathbf{x}_k^T & \mathbf{0}\end{bmatrix}\begin{bmatrix}A_k & *\\ * & *\end{bmatrix}\begin{bmatrix}\mathbf{x}_k\\ \mathbf{0}\end{bmatrix} = \mathbf{x}_k^TA_k\mathbf{x}_k.$$

Since $\mathbf{x}^TA\mathbf{x}$ is assumed to be positive by (1), $\mathbf{x}_k^TA_k\mathbf{x}_k > 0$ for any nonzero vector $\mathbf{x}_k$; that is, A_k is positive-definite. Since (1) ⇔ (2) is true for any symmetric matrix, all of the eigenvalues of A_k are positive. Hence, $\det A_k$ is positive as the product of all eigenvalues.

The complex case with a Hermitian matrix A can be proved by the same argument except for using ^{H} instead of ^{T}.

(3) ⇒ (4) Let $A = [a_{ij}]$. Then the first principal submatrix $A_1 = [a_{11}]$ has positive determinant, *i.e.*, $a_{11} > 0$. Thus, a_{11} can be used as a pivot for forward elimination performed by only the type III elementary row operations to make all

the other first column entries below a_{11} zero. Let $a_{22}^{(1)}$ denote the $(2,2)$-entry of the resulting matrix, so that the principal submatrix A_2 has been transformed into the matrix

$$\begin{bmatrix} a_{11} & a_{12} \\ 0 & a_{22}^{(1)} \end{bmatrix}.$$

Since the type III elementary row operation does not change the determinant, we have

$$\det A_2 = a_{11}a_{22}^{(1)} \quad \text{and} \quad a_{22}^{(1)} = \frac{\det A_2}{a_{11}} = \frac{\det A_2}{\det A_1} > 0.$$

Since $a_{22}^{(1)} \neq 0$, it can be used as a pivot in the second forward elimination, to reduce the principal submatrix A_3 into

$$\begin{bmatrix} a_{11} & a_{12} & a_{13} \\ 0 & a_{22}^{(1)} & a_{23}^{(1)} \\ 0 & 0 & a_{33}^{(2)} \end{bmatrix}.$$

Also, one can show that

$$\det A_3 = a_{11}a_{22}^{(1)}a_{33}^{(2)} \quad \text{and} \quad a_{33}^{(2)} = \frac{\det A_3}{a_{11}a_{22}^{(1)}} = \frac{\det A_3}{\det A_2} > 0.$$

A similar process can be repeated to get an upper triangular matrix as a row-echelon form of A with the k-th pivot $a_{kk}^{(k-1)}$. This is exactly the ratio of $\det A_k$ to $\det A_{k-1}$. Hence, all of the pivots are positive.

$(4) \Rightarrow (5)$ First, consider the real case. By the hypothesis (4) A has an LU or an LDU factorization. By the uniqueness of the LDU factorization of a nonsingular matrix, the symmetric matrix A can be uniquely factored as a product LDL^T with[5]

$$D = \begin{bmatrix} d_1 & & & \mathbf{0} \\ & d_2 & & \\ & & \ddots & \\ \mathbf{0} & & & d_n \end{bmatrix}, \quad d_i > 0.$$

Define

$$\sqrt{D} = \begin{bmatrix} \sqrt{d_1} & & & \mathbf{0} \\ & \sqrt{d_2} & & \\ & & \ddots & \\ \mathbf{0} & & & \sqrt{d_n} \end{bmatrix}.$$

[5]This is called an **LDL^{T} decomposition** or an **LDL^{T} factorization** of a symmetric matrix A. Refer to Exercise **1.27**.

Then, clearly $\det(\sqrt{D}) > 0$, $D = \sqrt{D}\sqrt{D}$ and $(\sqrt{D})^T = \sqrt{D}$. Hence,

$$A = LDL^T = (L\sqrt{D})(\sqrt{D}L^T) = (L\sqrt{D})(L\sqrt{D})^T,$$

as desired.

For the complex case, use ^{H} instead of ^{T}.

$(5) \Rightarrow (6)$ is trivial.

$(6) \Rightarrow (1)$ Let $A = W^T W$ for a nonsingular matrix W. Then, for $\mathbf{x} \neq \mathbf{0}$,

$$\mathbf{x}^T A\mathbf{x} = \mathbf{x}^T (W^T W)\mathbf{x} = (W\mathbf{x})^T (W\mathbf{x}) = \|W\mathbf{x}\|^2 > 0,$$

because $W\mathbf{x} \neq \mathbf{0}$.

Similar for a complex case. □

Problem 225 Determine which of the following matrices A and B are positive-definite. For the positive-definite ones, find a nonsingular matrix W such that it is $W^T W$.

$$A = \begin{bmatrix} 2 & 1 & 1 \\ 1 & 2 & 1 \\ 1 & 1 & 2 \end{bmatrix}, \qquad B = \begin{bmatrix} 2 & -1 & 0 \\ -1 & 2 & 1 \\ 0 & 1 & 2 \end{bmatrix}.$$

Since a symmetric matrix A is negative-definite if and only if $-A$ is positive-definite, the next theorem follows from Theorem 9.7.

Theorem 9.8 *The following statements are equivalent for a real symmetric (or a complex Hermitian) matrix A.*

(1) *A is negative-definite, i.e.,* $\langle \mathbf{x}, A\mathbf{x} \rangle < 0$ *for all nonzero vectors* $\mathbf{x}$,

(2) *all of the eigenvalues of A are negative,*

(3) *the determinants of the principal submatrices* A_k*'s alternate in sign: i.e.,* $\det A_1 < 0$, $\det A_2 > 0$, $\det A_3 < 0$, *and so on,*

(4) *A can be reduced to an upper triangular matrix by forward elimination performed by only the type III elementary row operations of "adding a constant multiple of a row to another", (without row interchanges), and all of the pivots are negative,*

(5) *there exists a unique lower triangular matrix L with positive diagonal entries such that* $A = -LL^T$ $(= -LL^H$ *if A is Hermitian) (called the* **Cholesky decomposition** *or* **Cholesky factorization**),

(6) *there exists a nonsingular matrix W such that* $A = -W^T W$, $(= -W^H W$ *if A is Hermitian).*

Problem 226 Show that the determinant of a negative-definite $n \times n$ symmetric matrix is positive if n is even, and negative if n is odd.

Hint: The determinant is the product of the eigenvalues.

One can establish the following analogous theorem for semidefinite matrices.

Theorem 9.9 *The following statements are equivalent for a real symmetric (or a complex Hermitian) matrix A.*

(1) *A is positive-semidefinite, i.e., $\langle \mathbf{x}, A\mathbf{x} \rangle \geq 0$ for all nonzero vectors $\mathbf{x}$,*
(2) *all of the eigenvalues of A are nonnegative,*
(3) *A can be reduced to an upper triangular matrix by forward elimination performed by only the type III elementary row operations of "adding a constant multiple of a row to another", (without row interchanges), and all of the pivots are nonnegative,*
(4) *there exists a matrix W, possibly singular, such that $A = W^T W$, $(= W^H W$ if A is Hermitian$)$.*

Problem 227 Which of the following matrices are positive-definite, negative-definite, or indefinite?

$$(1) \begin{bmatrix} 1 & 0 \\ 0 & 1 \end{bmatrix}, \quad (2) \begin{bmatrix} 0 & 1 \\ 1 & 0 \end{bmatrix}, \quad (3) \begin{bmatrix} 1 & 3 & 1 \\ 3 & 1 & 1 \\ 1 & 1 & 0 \end{bmatrix}, \quad (4) \begin{bmatrix} -2 & -1 & 1 \\ -1 & -2 & -1 \\ 1 & -1 & -3 \end{bmatrix}.$$

Problem 228 Is it true or false: A symmetric matrix A is positive-semidefinite if and only if for every principal submatrix A_k, $\det A_k \geq 0$.

Hint: False. Why?

Problem 229 State the corresponding conditions to the ones in Theorem 9.9 for the negative-semidefinite forms.

9.7 Change of bases for quadratic forms and the congruence relation

Let $q(\mathbf{x}) = \langle \mathbf{x}, A\mathbf{x} \rangle$ be any (real or complex) quadratic form with $\mathbf{x} = [\mathbf{x}]_\alpha$ with respect to the standard basis $\alpha = \{\mathbf{e}_1, \mathbf{e}_2, \ldots, \mathbf{e}_n\}$ for $\mathbb{R}^n$ or for $\mathbb{C}^n$. Let $\beta = \{\mathbf{e}'_1, \mathbf{e}'_2, \ldots, \mathbf{e}'_n\}$ be another basis, not necessarily orthonormal.

Now, a vector $\mathbf{x}$ has two coordinate representations $[\mathbf{x}]_\alpha$ and $[\mathbf{x}]_\beta$ through the equations

$$x_1\mathbf{e}_1 + x_2\mathbf{e}_2 + \cdots + x_n\mathbf{e}_n \;=\; \mathbf{x} \;=\; y_1\mathbf{e}'_1 + y_2\mathbf{e}'_2 + \cdots + y_n\mathbf{e}'_n.$$

They are related as $[\mathbf{x}]_\alpha = [id]_\beta^\alpha[\mathbf{x}]_\beta$, where $P = [id]_\beta^\alpha$ is the change of basis matrix from β to α.

If we set $\mathbf{x} = [\mathbf{x}]_\alpha$ and $\mathbf{y} = [\mathbf{x}]_\beta$, then the quadratic form q can be written as

$$q(\mathbf{x}) = \langle \mathbf{x}, A\mathbf{x}\rangle = \langle P\mathbf{y}, AP\mathbf{y}\rangle = \langle \mathbf{y}, P^H AP\mathbf{y}\rangle = \langle \mathbf{y}, B\mathbf{y}\rangle,$$

where

$$B = P^H AP,$$

and $\langle \mathbf{y}, B\mathbf{y}\rangle$ is the expression of $q(\mathbf{x}) = \langle \mathbf{x}, A\mathbf{x}\rangle$ in the new basis β. Here, the matrix $P = [id]_\beta^\alpha$ is invertible, but not necessarily orthogonal (not necessarily $P^T = P^{-1}$) or unitary (not necessarily $P^H = P^{-1}$), and the basis β may not give a rectangular coordinate system for $\mathbb{R}^n$ (or for $\mathbb{C}^n$).

Definition 9.10 **(1)** Two real $n \times n$ matrices A and B are said to be **congruent** if there exists an invertible real matrix P such that $P^T AP = B$.

(2) Two complex $n \times n$ matrices A and B are said to be **Hermitian congruent** if there exists an invertible complex matrix P such that $P^H AP = B$.

The congruence relation is clearly an equivalence relation on the matrix space $M_{n\times n}(\mathbb{R})$.

It is shown that

any two matrix representations of a quadratic form on $\mathbb{R}^n$ with respect to different bases are congruent, and the converse is also true.
That is, any two congruent symmetric matrices can be matrix representations of the same quadratic form with respect to different bases.

A similar statement also holds for a complex quadratic form in terms of a Hermitian congruence on $M_{n\times n}(\mathbb{C})$.

Remark: **(1)** *Two orthogonally similar real matrices are congruent, but the converse is not true in general.* Let real matrices A and B be orthogonally similar, say $P^{-1}AP = B$ with an orthogonal matrix P. Since $P^T = P^{-1}$, one can get $P^T AP = B$, and hence A, B are congruent.

But, the converse is not true in general. As an example, for any identity matrix I_n and any positive k, $kI_n = (\sqrt{k}I_n)^T I_n(\sqrt{k}I_n)$, and hence the matrices I_n and kI_n are congruent but not similar if $k \neq 1$.

(2) Sylvester's law of inertia (Theorem 9.18 in Subsection 9.11.2) says that even though a real symmetric matrix A may be congruent to various diagonal matrices, the numbers of positive, negative, and zero diagonal entries are invariant under the congruence relation. That is, *any two symmetric matrices which are congruent have the same inertia.* Moreover, its converse is also true, and hence Sylvester's law of inertia can also be stated as follows.

Two symmetric matrices of the same size have the same numbers of positive, negative, and zero eigenvalues if and only if they are congruent.
As a consequence, any two matrix representations $[q]_\alpha$ and $[q]_\beta$ of a quadratic form q have the same inertia, that is $\mathrm{In}([q]_\alpha) = \mathrm{In}([q]_\beta)$.

Similar for Hermitian matrices: *any two Hermitian matrices which are Hermitian congruent have the same inertia.* (See Corollary 9.18.1.)

Certainly, the inertia of a symmetric matrix (of a Hermitian matrix in the complex case) can be found by computing the eigenvalues, but in most cases, it is not easy to find the roots of the characteristic equation $\det(\lambda I - A) = 0$ even for 3×3 or 4×4 matrices A.

However, (with a help of Sylvester's law of inertia) there is a practical way of computing the inertia using the congruence relation (the Hermitian congruence in the complex case), which repeatedly uses only a kind of elementary row and column operation of type $B \to EBE^T$ with an elementary matrix E of type III.

The detail will be shown in the next section.

Problem 230 Prove that *a square matrix is symmetric if and only if it is congruent to a diagonal matrix.*

Problem 231 Determine whether each of the following statements is true or false:

(1) If A and B are similar and A is symmetric, then B is also symmetric.
(2) If A and B are congruent and A is symmetric, then B is also symmetric.
(3) Any two similar matrices are congruent.
(4) Any two congruent matrices are similar.

9.8 Computing the inertia using the congruence relation

Every real symmetric matrix A is orthogonally diagonalizable, and hence A is also congruent to a diagonal matrix, say $P^TAP = D$ for an invertible matrix P. Since P is invertible, so is P^T. Hence, P^T can be expressed as a product of elementary matrices, say $P^T = E_k \cdots E_2E_1$. Then we have

$$D = P^TAP = E_k \cdots E_2E_1AE_1^TE_2^T \cdots E_k^T.$$

Recall that for any elementary matrix E, the product EA is the matrix that is obtained from A when the same elementary row operation to get E is applied to A. Clearly, if E is an elementary matrix, so is E^T.

Moreover, if an elementary matrix E is obtained by applying an elementary operation to the i-th row, then the product EAE^T is the matrix obtained from A when the same elementary operation is applied to both the i-th row (by E) and the i-th column (by E^T).

Since A is symmetric, the operation EAE^T will have the same effect on the diagonally opposite entries of A simultaneously, so that the resulting matrix EAE^T is also symmetric.

For instance, if

$$A = \begin{bmatrix} 1 & 2 & 2 \\ 2 & 0 & 4 \\ 2 & 4 & 6 \end{bmatrix} \quad \text{and} \quad E = \begin{bmatrix} 1 & 0 & 0 \\ -2 & 1 & 0 \\ 0 & 0 & 1 \end{bmatrix},$$

where E is the elementary matrix which adds -2 times the first row to the second row, then EA is the matrix obtained from A by replacing the second row $[2 \ \ 0 \ \ 4]$ by $[0 \ -4 \ 0]$. Now, the matrix EAE^T is obtained from EA by replacing the second column by $[0 \ -4 \ 0]^T$ to get the symmetry of the matrix EAE^T. In fact,

$$EAE^T = \begin{bmatrix} 1 & 0 & 0 \\ -2 & 1 & 0 \\ 0 & 0 & 1 \end{bmatrix} \begin{bmatrix} 1 & 2 & 2 \\ 2 & 0 & 4 \\ 2 & 4 & 6 \end{bmatrix} \begin{bmatrix} 1 & -2 & 0 \\ 0 & 1 & 0 \\ 0 & 0 & 1 \end{bmatrix} = \begin{bmatrix} 1 & 0 & 2 \\ 0 & -4 & 0 \\ 2 & 0 & 6 \end{bmatrix}.$$

In a process of diagonalizing

$$D = P^TAP = E_k \cdots E_2E_1AE_1^TE_2^T \cdots E_k^T$$

by using the congruence relation, it is not necessary to use elementary row/column operations of Type I, 'multiply a row/column by a nonzero constant', because it is not needed to derive the leading 1 in any row, and needed only a diagonal matrix D which is congruent to the matrix A in order to get the signs of the diagonal entries of D. Moreover, the operations of Type II, 'interchange two rows', is not required either.

As an example, for the matrix

$$A = \begin{bmatrix} 0 & 1 \\ 1 & 0 \end{bmatrix},$$

one might try to apply an elementary matrix $E = \begin{bmatrix} 0 & 1 \\ 1 & 0 \end{bmatrix}$ for interchanging two rows, but $EAE^T = EAE = A$, nothing changed. However, by applying an elementary matrix E of Type III which adds 1 times the second row to the first row, one can get

$$EAE^T = \begin{bmatrix} 1 & 1 \\ 0 & 1 \end{bmatrix} \begin{bmatrix} 0 & 1 \\ 1 & 0 \end{bmatrix} \begin{bmatrix} 1 & 0 \\ 1 & 1 \end{bmatrix} = \begin{bmatrix} 2 & 1 \\ 1 & 0 \end{bmatrix}.$$

Now, by applying once more an elementary row/column operation of type III for a forward elimination, one can reach a diagonal matrix.

Therefore, by applying only elementary row/column operations of type III, one can follow through the following process:

$$\begin{aligned} [A \mid I] &\to [E_1AE_1^T \mid E_1I] \to [E_2E_1AE_1^TE_2^T \mid E_2E_1I] \to \cdots \\ &\to [E_k \cdots E_1AE_1^T \cdots E_k^T \mid E_k \cdots E_1I] = [D \mid P^T]. \end{aligned}$$

In the complex case, since every Hermitian matrix A is unitarily diagonalizable, A is Hermitian congruent to a diagonal matrix, say $P^HAP = D$. An argument similar to the real case can be applied in parallel with ^{H} instead of ^{T}.

If EA is obtained from A by applying an elementary operation to the i-th row, then the product EAE^H is the matrix obtained from EA by replacing the i-th column with the conjugates of i-th row of EA so that the resulting matrix EAE^H is also Hermitian.

The following two examples show how to compute the inertia of a real symmetric or a Hermitian matrix A by using the congruence relation (instead of computing the eigenvalues).

Example 181 (*Computing In(A) by using the congruence relation*) Compute the inertia of the symmetric matrix

$$A = \begin{bmatrix} 1 & 0 & 2 \\ 0 & 3 & 1 \\ 2 & 1 & 6 \end{bmatrix}.$$

Is it positive-definite, negative-definite, or indefinite?

Solution: The preceding method produces

$$\begin{aligned}
[A \mid I] &= \left[\begin{array}{ccc|ccc} 1 & 0 & 2 & 1 & 0 & 0 \\ 0 & 3 & 1 & 0 & 1 & 0 \\ 2 & 1 & 6 & 0 & 0 & 1 \end{array}\right] \\
\longrightarrow [E_1AE_1^T \mid E_1I] &= \left[\begin{array}{ccc|ccc} 1 & 0 & 0 & 1 & 0 & 0 \\ 0 & 3 & 1 & 0 & 1 & 0 \\ 0 & 1 & 2 & -2 & 0 & 1 \end{array}\right] \\
\longrightarrow [E_2E_1AE_1^TE_2^T \mid E_2E_1I] &= \left[\begin{array}{ccc|ccc} 1 & 0 & 0 & 1 & 0 & 0 \\ 0 & 3 & 0 & 0 & 1 & 0 \\ 0 & 0 & 5/3 & -2 & -1/3 & 1 \end{array}\right] \\
&= [D \mid P^T],
\end{aligned}$$

where

$$E_1 = \begin{bmatrix} 1 & 0 & 0 \\ 0 & 1 & 0 \\ -2 & 0 & 1 \end{bmatrix}, \quad E_2 = \begin{bmatrix} 1 & 0 & 0 \\ 0 & 1 & 0 \\ 0 & -1/3 & 1 \end{bmatrix}.$$

Since the diagonal entries of D are $1, 3$, and $5/3$, it follows $\text{In}(A) = (3, 0, 0)$ and A is positive-definite. One can check that $P^TAP = D$ by a direct computation.

Even one gets a diagonal matrix D which is congruent to the matrix A, as a repeated reminder, this does *not* mean that D is similar to the matrix A. The diagonal entries $1, 3, 5/3$ of D are *not* the eigenvalues of A. The readers are suggested to check it with the characteristic polynomial $\lambda^3 - 10\lambda^2 + 22\lambda - 5$ of A.

Moreover, one might try to solve the characteristic equation $\lambda^3 - 10\lambda^2 + 22\lambda - 5 = 0$ to find the eigenvalues of A; try it yourself to see the difficulty. This shows that even if it is difficult to find the eigenvalues from the characteristic polynomial, one can compute $\text{In}(A)$ by using the congruence relation. □

Example 182 (*Computing In(A) by using the Hermitian congruence relation*) Compute the inertia of the Hermitian matrix

$$A = \begin{bmatrix} 1 & i & 2i \\ -i & 0 & 3i \\ -2i & -3i & 6 \end{bmatrix}.$$

Is it positive-definite, negative-definite, or indefinite?

Solution: In the same way,

$$\begin{aligned}
[A \mid I] &= \left[\begin{array}{ccc|ccc} 1 & i & 2i & 1 & 0 & 0 \\ -i & 0 & 3i & 0 & 1 & 0 \\ -2i & -3i & 6 & 0 & 0 & 1 \end{array}\right] \\
\longrightarrow [E_1AE_1^H \mid E_1I] &= \left[\begin{array}{ccc|ccc} 1 & 0 & 2i & 1 & 0 & 0 \\ 0 & -1 & -2+3i & i & 1 & 0 \\ -2i & -2-3i & 6 & 0 & 0 & 1 \end{array}\right] \\
\longrightarrow [E_2E_1AE_1^HE_2^H \mid E_2E_1I] &= \left[\begin{array}{ccc|ccc} 1 & 0 & 0 & 1 & 0 & 0 \\ 0 & -1 & -2+3i & i & 1 & 0 \\ 0 & -2-3i & 2 & 2i & 0 & 1 \end{array}\right] \\
\longrightarrow [E_3E_2E_1AE_1^HE_2^HE_3^H \mid E_3E_2E_1I] &= \left[\begin{array}{ccc|ccc} 1 & 0 & 0 & 1 & 0 & 0 \\ 0 & -1 & 0 & i & 1 & 0 \\ 0 & 0 & 13 & 3 & -2-3i & 1 \end{array}\right] \\
&= [D \mid P^H],
\end{aligned}$$

where

$$E_1 = \begin{bmatrix} 1 & 0 & 0 \\ i & 1 & 0 \\ 0 & 0 & 1 \end{bmatrix}, \quad E_2 = \begin{bmatrix} 1 & 0 & 0 \\ 0 & 1 & 0 \\ 2i & 0 & 1 \end{bmatrix}, \quad E_3 = \begin{bmatrix} 1 & 0 & 0 \\ 0 & 1 & 0 \\ 0 & -2-3i & 1 \end{bmatrix}.$$

It follows that $\text{In}(A) = (2, 1, 0)$ and A is indefinite. □

Problem 232 Find an invertible (real) matrix P such that P^TAP is diagonal for each of the following symmetric matrices, and compute $\text{In}(A)$.

$$(1)\ A = \begin{bmatrix} 0 & 1 \\ 1 & 0 \end{bmatrix}, \quad (2)\ A = \begin{bmatrix} 0 & 0 & 1 \\ 0 & 1 & 0 \\ 1 & 0 & 0 \end{bmatrix}, \quad (3)\ A = \begin{bmatrix} 0 & 1 & -1 \\ 1 & 1 & 0 \\ -1 & 0 & 2 \end{bmatrix}.$$

Problem 233 For each of the following Hermitian matrices A, find an invertible (complex) matrix P such that P^HAP is diagonal, and compute $\text{In}(A)$.

$$(1)\ A = \begin{bmatrix} 0 & 1 & i \\ 1 & 1 & 0 \\ -i & 0 & 2 \end{bmatrix}, \quad (2)\ A = \begin{bmatrix} 1 & -3i & 1 \\ 3i & 4 & 2i \\ 1 & -2i & 5 \end{bmatrix}, \quad (3)\ A = \begin{bmatrix} 0 & 0 & i \\ 0 & 0 & 0 \\ -i & 0 & 0 \end{bmatrix}.$$

9.9 Bilinear and sesquilinear forms; $b(\mathbf{x}, \mathbf{y}) = \mathbf{x}^H A\mathbf{y}$

Every inner product on a real vector space is bilinear and every Hermitian inner product on a complex vector space is sesquilinear.

In this section we examine such two forms, *bilinear form* and *sesquilinear form*, and find their matrix representations. Later in Subsection 9.11.3, it will be shown when and how these two forms can be associated to real and complex quadratic forms, respectively.

Definition 9.11 A **bilinear form** on a real vector space V is a real-valued function $b : V \times V \to \mathbb{R}$ that is bilinear:

(1) $b(k\mathbf{x} + \ell\mathbf{x}', \mathbf{y}) = k\, b(\mathbf{x}, \mathbf{y}) + \ell\, b(\mathbf{x}', \mathbf{y})$ (linear in 1st variable)
(2) $b(\mathbf{x}, k\mathbf{y} + \ell\mathbf{y}') = k\, b(\mathbf{x}, \mathbf{y}) + \ell\, b(\mathbf{x}, \mathbf{y}')$ (linear in 2nd variable)

for any $\mathbf{x}, \mathbf{x}', \mathbf{y}, \mathbf{y}'$ in V and any real scalars k, ℓ.

A bilinear form b on a real vector space V is said to be **symmetric** if $b(\mathbf{x}, \mathbf{y}) = b(\mathbf{y}, \mathbf{x})$ for any $\mathbf{x}, \mathbf{y} \in V$, and is **skew-symmetric** if $b(\mathbf{x}, \mathbf{y}) = -b(\mathbf{y}, \mathbf{x})$ for any $\mathbf{x}, \mathbf{y} \in V$.

Example 183 (*A real inner product is a positive-definite symmetric bilinear form*)

(1) The dot product on $\mathbb{R}^n$ is a positive-definite symmetric bilinear form.

(2) For any (symmetric or not) square matrix A, $b(\mathbf{x}, \mathbf{y}) = \mathbf{x}^T A\mathbf{y}$ is a bilinear form. Hence, if A is positive-definite and real symmetric, then $b(\mathbf{x}, \mathbf{y}) = \mathbf{x}^T A\mathbf{y}$ is an inner product on $\mathbb{R}^n$.

By noting that

$$\mathbf{e}_i^T A\mathbf{e}_j \quad \text{is the } (i, j)\text{-entry of a matrix } A,$$

one can see that the bilinear form $b : \mathbb{R}^n \times \mathbb{R}^n \to \mathbb{R}$ defined by $b(\mathbf{x}, \mathbf{y}) = \mathbf{x}^T A\mathbf{y}$ is symmetric (skew-symmetric, respectively) if and only if the matrix A is symmetric (skew-symmetric, respectively).

Problem 234 A bilinear form b on $\mathbb{R}^n$ is **alternating** if $b(\mathbf{x}, \mathbf{x}) = 0$ for all $\mathbf{x} \in \mathbb{R}^n$. Show that a bilinear form b on $\mathbb{R}^n$ is skew-symmetric if and only if it is alternating.
Hint: $b(\mathbf{x}+\mathbf{y}, \mathbf{x}+\mathbf{y}) = b(\mathbf{x}, \mathbf{x}) + b(\mathbf{x}, \mathbf{y}) + b(\mathbf{y}, \mathbf{x}) + b(\mathbf{y}, \mathbf{y})$.

Definition 9.12 A **sesquilinear form** on a complex vector space V is a complex-valued function $b : V \times V \to \mathbb{C}$ that is *sesquilinear*; [6] by definition,

(1) $b(k\mathbf{x} + \ell\mathbf{x}', \mathbf{y}) = \overline{k}\, b(\mathbf{x}, \mathbf{y}) + \overline{\ell}\, b(\mathbf{x}', \mathbf{y})$ (semilinear in 1st variable)

(2) $b(\mathbf{x}, k\mathbf{y} + \ell\mathbf{y}') = k\, b(\mathbf{x}, \mathbf{y}) + \ell\, b(\mathbf{x}, \mathbf{y}')$ (linear in 2nd variable)

for any $\mathbf{x}, \mathbf{x}', \mathbf{y}, \mathbf{y}'$ in V and any complex scalars k, ℓ.

A sesquilinear form b on a complex vector space V is said to be **Hermitian** if $b(\mathbf{x}, \mathbf{y}) = \overline{b(\mathbf{y}, \mathbf{x})}$ for any $\mathbf{x}, \mathbf{y} \in V$, and is **skew-Hermitian** if $b(\mathbf{x}, \mathbf{y}) = -\overline{b(\mathbf{y}, \mathbf{x})}$ for any $\mathbf{x}, \mathbf{y} \in V$.

Clearly, a sesquilinear form b is Hermitian form if and only if ib is a skew-Hermitian form.

Example 184 (*A complex inner product is a positive-definite Hermitian sesquilinear form*)

For any $n \times n$ complex matrix A, the function $b : \mathbb{C}^n \times \mathbb{C}^n \to \mathbb{C}$ defined by $b(\mathbf{x}, \mathbf{y}) = \mathbf{x}^H A\mathbf{y}$ for $\mathbf{x}, \mathbf{y} \in \mathbb{C}^n$ is a sesquilinear form.

It is a Hermitian sesquilinear form (skew-Hermitian sesquilinear form, respectively) if and only if A is a Hermitian matrix (skew-Hermitian matrix, respectively).

In fact, $\mathbf{y}^H A\mathbf{x} = \overline{\mathbf{x}^H A\mathbf{y}}$ for all $\mathbf{x}, \mathbf{y} \in \mathbb{C}^n$ if and only if A is Hermitian. As a special case, if one takes $A = I_n$, the identity matrix, then

(1) the dot product on $\mathbb{C}^n$ is a positive-definite Hermitian sesquilinear form.

(2) For any positive-definite Hermitian matrix A, $b(\mathbf{x}, \mathbf{y}) = \mathbf{x}^H A\mathbf{y}$ is a (Hermitian) inner product on $\mathbb{C}^n$. This is actually a positive-definite Hermitian sesquilinear form. □

Like the matrix representation $\langle \mathbf{x}, \mathbf{y} \rangle = \mathbf{x}^T A\mathbf{y}$ of an inner product in Section 5.3, every bilinear form can also be expressed as a matrix product.

Let $b : V \times V \to \mathbb{R}$ be a bilinear form on a real vector space V, and let $\alpha = \{\mathbf{v}_1, \mathbf{v}_2, \ldots, \mathbf{v}_n\}$ be any ordered basis for V. Then, the bilinear form b is completely determined by the values $b(\mathbf{v}_i, \mathbf{v}_j)$ of the vectors $\mathbf{v}_i, \mathbf{v}_j$ in the basis α because of the bilinearity.

[6] A prefix sesqui- is used in Latin combinations for $1\frac{1}{2}$. Here, sesquilinear means that it is $\frac{1}{2}$ linear for the first variable and is linear for the second variable.

In fact, if

$$\begin{aligned}\mathbf{x} &= x_1\mathbf{v}_1 + x_2\mathbf{v}_2 + \cdots + x_n\mathbf{v}_n &= \sum_{i=1}^{n} x_i\mathbf{v}_i,\\ \mathbf{y} &= y_1\mathbf{v}_1 + y_2\mathbf{v}_2 + \cdots + y_n\mathbf{v}_n &= \sum_{j=1}^{n} y_j\mathbf{v}_j\end{aligned}$$

are vectors in V, then

$$b(\mathbf{x},\mathbf{y}) = \sum_{i=1}^{n}\sum_{j=1}^{n} x_i y_j b(\mathbf{v}_i,\mathbf{v}_j) = [\mathbf{x}]_\alpha^T A[\mathbf{y}]_\alpha, \tag{9.3}$$

where $A = [a_{ij}]$ and $a_{ij} = b(\mathbf{v}_i,\mathbf{v}_j)$. (Compare this with Eq. (5.1).) The coefficient matrix A is called the **matrix representation** of b with respect to the basis α and is denoted by $[b]_\alpha$.

Let β be another basis for the vector space V and let $P = [id]_\beta^\alpha$ be the change of basis matrix from β to α. Then,

$$P[\mathbf{x}]_\beta = [id]_\beta^\alpha[\mathbf{x}]_\beta = [\mathbf{x}]_\alpha$$

for any $\mathbf{x}$ in V, and hence

$$b(\mathbf{x},\mathbf{y}) = [\mathbf{x}]_\alpha^T A[\mathbf{y}]_\alpha = [\mathbf{x}]_\beta^T(P^TAP)[\mathbf{y}]_\beta$$

for any $\mathbf{x}$ and $\mathbf{y}$ in V.

Consequently, we have,

$$[b]_\beta = P^TAP = ([id]_\beta^\alpha)^T[b]_\alpha[id]_\beta^\alpha. \tag{9.4}$$

Thus, in parallel with the case of quadratic form q,

any two matrix representations of a bilinear form b with respect to different bases are congruent, and conversely any two congruent matrices can be matrix representations of the same bilinear form with respect to different bases.

A similar process works in the complex case with ^{H} instead of ^{T} to get a matrix representation $[b]_\alpha$ of the sesquilinear form b with respect to a basis α.

As in the real case, one can say that *any two matrix representations of a sesquilinear form b with respect to different bases are Hermitian congruent, and conversely any two Hermitian congruent matrices can be matrix representations of the same sesquilinear form.*

Clearly, any real matrix which is congruent to a symmetric matrix is also symmetric and any complex matrix which is Hermitian congruent to a Hermitian matrix is also Hermitian.

It gives the following theorem.

Theorem 9.13 **(1)** *A bilinear form b is symmetric (or skew-symmetric, respectively) if and only if one (and hence, all) of its matrix representations is a symmetric matrix (or a skew-symmetric matrix, respectively).*

(2) *A sesquilinear form b is Hermitian (or skew-Hermitian, respectively) if and only if one (and hence, all) of its matrix representations is a Hermitian matrix (or a skew-Hermitian matrix, respectively).*

Problem 235 Prove directly that a sesquilinear form b is skew-Hermitian if and only if one (and hence, all) of its matrix representations is a skew-Hermitian matrix.

Hint: b is a skew-Hermitian form if and only if ib is a Hermitian form.

Problem 236 For a given bilinear form $b : V \times V \to \mathbb{R}$ on a real vector space V, we define a new bilinear form $b^T : V \times V \to \mathbb{R}$ by $b^T(\mathbf{x}, \mathbf{y}) = b(\mathbf{y}, \mathbf{x})$ for any $\mathbf{x}, \mathbf{y}$ in V. Show that

(1) $b + b^T$ is a symmetric bilinear form and $b - b^T$ is a skew-symmetric bilinear form,

(2) b can be expressed as the sum of the symmetric part $\frac{1}{2}(b + b^T)$ and the skew-symmetric part $\frac{1}{2}(b - b^T)$, so that $b = \frac{1}{2}(b + b^T) + \frac{1}{2}(b - b^T)$. Moreover, this expression is unique.

Similarly, a sesquilinear form b on a complex vector space V can be expressed as the sum of the Hermitian part $\frac{1}{2}(b + b^H)$ and the skew-Hermitian part $\frac{1}{2}(b - b^H)$, where $b^H(\mathbf{x}, \mathbf{y}) = \overline{b(\mathbf{y}, \mathbf{x})}$ by definition.

9.10 Diagonalization of symmetric bilinear or Hermitian sesquilinear forms

Every quadratic form can be represented as $\langle \mathbf{x}, A\mathbf{x} \rangle$, where A is a symmetric matrix, which is always diagonalizable. However, this is not true in general for a bilinear form $b(\mathbf{x}, \mathbf{y}) = \langle \mathbf{x}, A\mathbf{y} \rangle$, where A can be any square matrix.

If a real bilinear form b is symmetric, then its representation matrix A is symmetric and hence A is diagonalizable.

In this section we shall show that the converse is also true: *If a representation matrix of a real bilinear form b is diagonal, then b is symmetric.*

As a continuation of Theorem 9.13, one can have more characterizations of symmetric bilinear forms (or Hermitian sesquilinear forms).

Theorem 9.14 *The following are equivalent for a bilinear form b on a real vector space V.*

(1) *b is symmetric, i.e., $b(\mathbf{x}, \mathbf{y}) = b(\mathbf{y}, \mathbf{x})$ for any $\mathbf{x}, \mathbf{y} \in V$,*
(2) *one of the matrix representations of b is a symmetric matrix,*
(3) *all of the matrix representations of b are symmetric matrices,*
(4) *b is* **diagonalizable**, *i.e., by definition, one of the matrix representations of b is a diagonal matrix,*
(5) *all of the matrix representations of b are congruent to a diagonal matrix.*

Similar characterizations for a Hermitian sesquilinear form on a complex vector space can be stated by using Hermitian instead of symmetric.

Proof: $(1) \Leftrightarrow (2)$ and $(2) \Leftrightarrow (3)$ have already been shown in Theorem 9.13.

$(3) \Rightarrow (4)$ is clear because every symmetric matrix is orthogonally diagonalizable and hence congruent to a diagonal matrix.

$(4) \Rightarrow (3)$. Let $[b]_\alpha$ be a diagonal matrix. Then any other matrix representation of b is congruent to $[b]_\alpha$ and hence should be symmetric.

$(4) \Leftrightarrow (5)$ is also clear because any two matrix representations of b are congruent.

The complex case with a Hermitian matrix A can be proved by the same argument except for using the Hermitian congruence instead of the congruence. For example, a sesquilinear form on a complex vector space is Hermitian if and only if it is diagonalizable and all diagonal entries are real. □

Problem 237 Prove: A sesquilinear form b on a complex vector space V is Hermitian if and only if it is diagonalizable and all diagonal entries are real.

Problem 238 Determine whether each of the following statements is true or false:

(1) A bilinear form $b(\mathbf{x}, \mathbf{y}) = \mathbf{x}^T A\mathbf{y}$ is diagonalizable if and only if the matrix A is diagonalizable.
(2) A bilinear form $b(\mathbf{x}, \mathbf{y}) = \mathbf{x}^T A\mathbf{y}$ is diagonalizable if and only if the matrix A is symmetric.

Hint for (1): The matrix $\begin{bmatrix} 0 & 1 \\ 0 & 1 \end{bmatrix}$ is not symmetric and has two distinct eigenvalues.

To find a diagonal representation a symmetric bilinear form, certainly one may try to compute the eigenvalues and eigenvectors of a symmetric matrix but it may not be easy sometimes. Instead of such computing, one can also use the congruence relation of matrices.

Example 185 (*Diagonalizing a symmetric bilinear form*) Let $b : \mathbb{R}^3 \times \mathbb{R}^3 \to \mathbb{R}$ be the bilinear form defined by

$$b(\mathbf{x}, \mathbf{y}) = 3x_1y_3 - 2x_2y_2 + 2x_2y_3 + 3x_3y_1 + 2x_3y_2 - x_3y_3.$$

Clearly $b(\mathbf{x}, \mathbf{y}) = b(\mathbf{y}, \mathbf{x})$ and the matrix representation of b with respect to the standard basis $\alpha = \{\mathbf{e}_1, \mathbf{e}_2, \mathbf{e}_3\}$ is

$$[b]_\alpha = \begin{bmatrix} 0 & 0 & 3 \\ 0 & -2 & 2 \\ 3 & 2 & -1 \end{bmatrix},$$

which is symmetric. Hence, by Theorem 9.14, b is diagonalizable.

Certainly, the symmetric matrix $[b]_\alpha$ can be diagonalized through the similarity, that is, by finding the eigenvalues and the eigenvectors. Try to find the eigenvalues from its characteristic equation $\lambda^3 + 3\lambda^2 - 11\lambda - 18 = 0$. Can you do it?

However, one can easily find a diagonal matrix which is congruent to $[b]_\alpha$. In fact,

$$\begin{aligned}
[[b]_\alpha \mid I] &= \left[\begin{array}{ccc|ccc} 0 & 0 & 3 & 1 & 0 & 0 \\ 0 & -2 & 2 & 0 & 1 & 0 \\ 3 & 2 & -1 & 0 & 0 & 1 \end{array}\right] \\
\longrightarrow [E_1[b]_\alpha E_1^T \mid E_1 I] &= \left[\begin{array}{ccc|ccc} 0 & 0 & 3 & 1 & 0 & 0 \\ 0 & -2 & 0 & 0 & 1 & 0 \\ 3 & 0 & 1 & 0 & 1 & 1 \end{array}\right] \\
\longrightarrow [E_2E_1[b]_\alpha E_1^T E_2^T \mid E_2E_1 I] &= \left[\begin{array}{ccc|ccc} -9 & 0 & 0 & 1 & -3 & -3 \\ 0 & -2 & 0 & 0 & 1 & 0 \\ 0 & 0 & 1 & 0 & 1 & 1 \end{array}\right] \\
&= [D \mid P^T],
\end{aligned}$$

where

$$E_1 = \begin{bmatrix} 1 & 0 & 0 \\ 0 & 1 & 0 \\ 0 & 1 & 1 \end{bmatrix}, \qquad E_2 = \begin{bmatrix} 1 & 0 & -3 \\ 0 & 1 & 0 \\ 0 & 0 & 1 \end{bmatrix}.$$

By a direct computation, one can show that $P^T[b]_\alpha P = D$. Let $\beta = \{\mathbf{c}_1, \mathbf{c}_2, \mathbf{c}_3\}$ be a new basis consisting of the column vectors of the matrix P, then $P = [id]_\beta^\alpha$ and

$$b(\mathbf{x}, \mathbf{y}) = [\mathbf{x}]_\alpha^T [b]_\alpha [\mathbf{y}]_\alpha = ([id]_\beta^\alpha [\mathbf{x}]_\beta)^T [b]_\alpha ([id]_\beta^\alpha [\mathbf{y}]_\beta) = [\mathbf{x}]_\beta^T D [\mathbf{y}]_\beta.$$

Hence, $[b]_\beta = D$.

If one takes $[\mathbf{x}]_\beta = (x_1', x_2', x_3')$ and $[\mathbf{y}]_\beta = (y_1', y_2', y_3')$ as new variables, then the bilinear form b becomes

$$b(\mathbf{x}, \mathbf{y}) = -9x_1'y_1' - 2x_2'y_2' + x_3'y_3'.$$

So far we have found a matrix representation of the bilinear form b to be a diagonal matrix D by using the congruence relation. As mentioned before, the diagonal entries $-9, -2, 1$ of D need *not* to be the eigenvalues of $[b]_\alpha$, as one can check it using the characteristic polynomial $\lambda^3 + 3\lambda^2 - 11\lambda - 18$ of $[b]_\alpha$.

This is another example of computing the inertia $\text{In}(\cdot) = (p, q, k)$ of a symmetric matrix by using the congruence relation even if it is difficult to find the eigenvalues. $\square$

Even though a skew-symmetric bilinear form is not diagonalizable in general, the next theorem shows the structure of a skew-symmetric bilinear form through a matrix representation.

It is a fact that a bilinear form b is skew-symmetric if and only if $b(\mathbf{x}, \mathbf{x}) = 0$ for all $\mathbf{x}$ in V, (see Problem 234). Hence, for any basis β for V, the diagonal entries of the matrix representation $[b]_\beta$ of a skew-symmetric bilinear form b should be all zero.

Theorem 9.15 *Let $b : V \times V \to \mathbb{R}$ be a skew-symmetric bilinear form. Then there exists a basis α for V with respect to which the matrix representation $[b]_\alpha$ is of the type*

$$\begin{bmatrix} \begin{bmatrix} 0 & 1 \\ -1 & 0 \end{bmatrix} & & & & & \mathbf{0} \\ & \ddots & & & & \\ & & \begin{bmatrix} 0 & 1 \\ -1 & 0 \end{bmatrix} & & & \\ & & & 0 & & \\ & & & & \ddots & \\ & \mathbf{0} & & & & 0 \end{bmatrix}.$$

This is a block diagonal matrix in which the diagonal blocks are the (most simple nontrivial) skew-symmetric matrix $\begin{bmatrix} 0 & 1 \\ -1 & 0 \end{bmatrix}$ or the zero matrix $\begin{bmatrix} 0 \end{bmatrix}$.

A similar statement is also true for a skew-Hermitian sesquilinear form in which the diagonal blocks are also the (most simple nontrivial) skew-Hermitian matrix $\begin{bmatrix} 0 & 1 \\ -1 & 0 \end{bmatrix}$ *or the zero matrix* $\begin{bmatrix} 0 \end{bmatrix}$.

Proof: If $b = 0$, then $[b]_\alpha$ is the zero matrix. Also if $\dim V = 1$, then $b(\mathbf{x}, \mathbf{x}) = 0$ for any basis vector $\mathbf{x}$ in V, so $b = 0$.

Now we assume that $b \neq 0$ and prove the theorem by induction on $\dim V$. Since $b \neq 0$, there exist nonzero vectors $\mathbf{x}$ and $\mathbf{y}$ in V such that $b(\mathbf{x}, \mathbf{y}) \neq 0$. By the bilinearity of b, one can assume that

$$b(\mathbf{x}, \mathbf{y}) = 1.$$

Such vectors $\mathbf{x}$ and $\mathbf{y}$ should be linearly independent, because if $\mathbf{y} = k\mathbf{x}$, then $b(\mathbf{x}, \mathbf{y}) = kb(\mathbf{x}, \mathbf{x}) = 0$.

Let U be the subspace of V spanned by $\mathbf{x}$ and $\mathbf{y}$, and let

$$W = \{\mathbf{v} \in V : b(\mathbf{v}, \mathbf{u}) = 0 \text{ for any } \mathbf{u} \in U\}.$$

Then, one can easily show that W is also a subspace of V and $U \cap W = \{\mathbf{0}\}$. Moreover, $U + W = V$. The space W is called the *orthogonal complement* of the subspace U (in the space (V, b)) and written commonly by $U^\perp$. [7] See Problem 242.

In fact, for a given vector $\mathbf{v} \in V$, let $\mathbf{u} = b(\mathbf{v}, \mathbf{y})\mathbf{x} - b(\mathbf{v}, \mathbf{x})\mathbf{y}$. Since $b(\mathbf{x}, \mathbf{y}) = 1$ and $b(\mathbf{y}, \mathbf{x}) = -1$ are assumed, it follows that $\mathbf{u} \in U$ and $\mathbf{v} - \mathbf{u} \in W$. Thus $V = U \oplus W$, where $\dim W = n - 2$.

Clearly, the matrix representation of the restriction of b to U with respect to the basis $\{\mathbf{x}, \mathbf{y}\}$ is $\begin{bmatrix} 0 & 1 \\ -1 & 0 \end{bmatrix}$, and the restriction of b to W is also skew-symmetric. The induction hypothesis can be applied to W, which completes the proof.

We leave a proof of the complex case as an exercise. □

As a result, *any nontrivial skew-symmetric bilinear form cannot be diagonalizable, and any nontrivial skew-Hermitian sesquilinear form cannot be diagonalizable, either.*

[7] This is a generalization of the orthogonal complement of a subspace in an inner product space.

Example 186 (*Block diagonalization of a skew-symmetric bilinear form*) Let $b : \mathbb{R}^3 \times \mathbb{R}^3 \to \mathbb{R}$ be the bilinear form defined by

$$b(\mathbf{x}, \mathbf{y}) = x_1y_2 - x_2y_1 + 3x_3y_1 - 3x_1y_3 + x_2y_3 - x_3y_2.$$

Clearly, $b(\mathbf{x}, \mathbf{y}) = -b(\mathbf{y}, \mathbf{x})$, and the matrix representation of b with respect to the standard basis $\alpha = \{\mathbf{e}_1, \mathbf{e}_2, \mathbf{e}_3\}$ is

$$[b]_\alpha = \begin{bmatrix} 0 & 1 & -3 \\ -1 & 0 & 1 \\ 3 & -1 & 0 \end{bmatrix},$$

which is skew-symmetric. By a simple computation, $b(\mathbf{e}_1, \mathbf{e}_2) = 1 = -b(\mathbf{e}_2, \mathbf{e}_1)$, as a nontrivial case of b.

Let U be the subspace of $\mathbb{R}^3$ spanned by $\mathbf{e}_1$ and $\mathbf{e}_2$, *i.e.*, the xy-plane, and let $W = \{\mathbf{v} \in \mathbb{R}^3 : b(\mathbf{v}, \mathbf{u}) = 0 \text{ for any } \mathbf{u} \in U\}$. Then one can see that W is a subspace and is spanned by $\mathbf{z} = (1, 3, 1)$. That is, $W = \{\lambda\mathbf{z} : \lambda \in \mathbb{R}\}$.

Clearly, $\beta = \{\mathbf{e}_1, \mathbf{e}_2, \ \mathbf{z}\}$ is a basis for $\mathbb{R}^3$ and $b(\mathbf{z}, \mathbf{z}) = 0$ so that

$$[b]_\beta = \begin{bmatrix} 0 & 1 & 0 \\ -1 & 0 & 0 \\ 0 & 0 & 0 \end{bmatrix},$$

as a block diagonal matrix. □

Problem 239 Find an invertible matrix P such that P^TAP is diagonal for each of the following symmetric matrices A.

$$(1)\ A = \begin{bmatrix} 0 & 1 & 1 \\ 1 & 0 & 0 \\ 1 & 0 & 1 \end{bmatrix}, \qquad (2)\ A = \begin{bmatrix} 0 & 3 & 1 \\ 3 & 0 & 2 \\ 1 & 2 & 0 \end{bmatrix}, \qquad (3)\ A = \begin{bmatrix} 2 & 2 & 2 \\ 2 & 3 & 0 \\ 2 & 0 & 4 \end{bmatrix}.$$

Hint: One can find a diagonal matrix which is congruent to a symmetric matrix A by using the congruence relation even if it is difficult to find the eigenvalues.

Problem 240 Find an invertible matrix P such that P^TAP is block diagonal for each of the following skew-symmetric matrices A.

$$(1)\ A = \begin{bmatrix} 0 & 1 & 1 \\ -1 & 0 & 0 \\ -1 & 0 & 0 \end{bmatrix}, \quad (2)\ A = \begin{bmatrix} 0 & -3 & 1 \\ 3 & 0 & 2 \\ -1 & -2 & 0 \end{bmatrix}, \quad (3)\ A = \begin{bmatrix} 0 & 0 & 1 \\ 0 & 0 & 0 \\ -1 & 0 & 0 \end{bmatrix}.$$

Problem 241 Find an invertible matrix P such that $P^H AP$ is diagonal for each of the following Hermitian matrices A.

$$(1)\ A = \begin{bmatrix} 1 & i & 0 \\ -i & 1 & 0 \\ 0 & 0 & 1 \end{bmatrix}, \qquad (2)\ A = \begin{bmatrix} 0 & 1 & i \\ 1 & 0 & 0 \\ -i & 0 & 0 \end{bmatrix}.$$

9.11 Applications

9.11.1 Nondegenerate bilinear forms

Following the definition of nondegenerate for a quadratic form $q(\mathbf{x})$, we define a nondegeneracy for a bilinear form as follows.

Definition 9.16 A bilinear form $b : V \times V \to \mathbb{R}$ on a vector space V is said to be **nondegenerate** if it satisfies

$$\begin{array}{llll} b(\mathbf{v},\mathbf{w}) = 0 & \text{for all } \mathbf{w} \in V & \text{implies } \mathbf{v} = \mathbf{0}, & \text{and} \\ b(\mathbf{v},\mathbf{w}) = 0 & \text{for all } \mathbf{v} \in V & \text{implies } \mathbf{w} = \mathbf{0}. & \end{array}$$

A bilinear form that is not nondegenerate is said to be **degenerate**.

Clearly, the bilinearity of b implies that $b(\mathbf{0},\mathbf{w}) = b(\mathbf{v},\mathbf{0}) = 0$ for all $\mathbf{v},\mathbf{w} \in V$. The nondegeneracy condition asserts that the equation $b(\mathbf{v},\mathbf{w}) = 0$ for all $\mathbf{w}$ holds only in the case of $\mathbf{v} = \mathbf{0}$.

For example, every inner product on a vector space V is a nondegenerate bilinear form on V.

Theorem 9.17 *For a bilinear form $b : V \times V \to \mathbb{R}$, the following are equivalent.*

(1) *b is nondegenerate,*
(2) *the matrix representation $[b]_\alpha$ of b with respect to any basis α for V is invertible, so its determinant is not zero.*

Proof: (1)⇒(2). Suppose $[b]_\alpha[\mathbf{w}]_\alpha = \mathbf{0}$. Then, $b(\mathbf{v},\mathbf{w}) = [\mathbf{v}]_\alpha^T[b]_\alpha[\mathbf{w}]_\alpha = 0$ for all $\mathbf{v} \in V$. Since b is assumed to be nondegenerate, it follows $\mathbf{w} = \mathbf{0}$ and hence $[\mathbf{w}]_\alpha = \mathbf{0}$.

It means that the system $[b]_\alpha[\mathbf{w}]_\alpha = \mathbf{0}$ has only the trivial solution and the matrix $[b]_\alpha$ is invertible.

(2)⇒(1). Let $b(\mathbf{v},\mathbf{w}) = 0$ for all $\mathbf{w} \in V$. Then $[\mathbf{v}]_\alpha^T[b]_\alpha[\mathbf{w}]_\alpha = b(\mathbf{v},\mathbf{w}) = 0$ for all $\mathbf{w} \in V$. Then $[\mathbf{v}]_\alpha^T[b]_\alpha = \mathbf{0}$. Since the matrix $[b]_\alpha$ is assumed to be invertible, $[\mathbf{v}]_\alpha^T = \mathbf{0}$ and then $\mathbf{v} = \mathbf{0}$.

Similarly, one can show that $b(\mathbf{v},\mathbf{w}) = 0$ for all $\mathbf{v} \in V$ implies $\mathbf{w} = \mathbf{0}$. □

A complex version for a *nondegenerate* sesquilinear form b can be defined in a similar way and a similar characterization can be given as the real case in Theorem 9.17. For example, *a sesquilinear form b is nondegenerate if and only if the matrix representation $[b]_\alpha$ of b with respect to any basis α is invertible.*

Problem 242 Let V be a real vector space on which a symmetric (or skew-symmetric) bilinear form b is defined. Show that

(1) For any subspace W of V, $W^\perp = \{\mathbf{v} \in V : b(\mathbf{v},\mathbf{w}) = 0 \text{ for all } \mathbf{w} \in W\}$ is a subspace, called the *orthogonal complement* of W (in the space (V, b)).

(2) $V^\perp = \{\mathbf{v} \in V : b(\mathbf{v},\mathbf{w}) = 0 \text{ for all } \mathbf{w} \in V\}$ is $\{\mathbf{0}\}$ if and only if b is nondegenerate. The orthogonal complement $V^\perp$ is also called the *kernel*, or the *radical*, of the symmetric (or skew-symmetric) bilinear form b, and is also written as $\operatorname{Ker} b$ or $\operatorname{Rad} b$ sometimes.

9.11.2 Sylvester's law of inertia

In this section, we prove Sylvester's law of inertia. A reader may skip reading the proof if not interested in details.

Theorem 9.18 (Sylvester's law of inertia) [8]

(1) *Two real symmetric matrices of the same size have the same inertia, i.e., the same numbers of positive, negative, and zero eigenvalues if and only if they are congruent.*

(2) *Two Hermitian matrices of the same size have the same inertia, i.e., the same numbers of positive, negative, and zero eigenvalues if and only if they are Hermitian congruent.*

[8] This law is named after James Joseph Sylvester FRS (September 3, 1814– March 15, 1897) who published its proof in 1852.

Proof: We prove only (1) for a symmetric matrix because the complex case can be proved in a similar method.

($\Rightarrow$) Suppose that two symmetric matrices A and B have the same inertia $\text{In}(\cdot) = (p, q, k)$, and let D and E be diagonal matrices congruent to A and B, respectively. Without any loss of generality, one may choose D and E so that the diagonal entries are in the order of positive, negative, and zero. (See Problem 243.) Here, p and q become the numbers of positive and negative diagonal entries respectively, of both D and E.

Let d_i denote the i-th diagonal entry of D, and let Q be a diagonal matrix whose i-th diagonal entry q_i is given by

$$q_i = \begin{cases} 1/\sqrt{d_i} & \text{if } 1 \le i \le p, \\ 1/\sqrt{-d_i} & \text{if } p < i \le p+q, \\ 1 & \text{if } p+q < i \le n. \end{cases}$$

Then,

$$Q^T D Q = \begin{bmatrix} I_p & 0 & \mathbf{0} \\ 0 & -I_q & \mathbf{0} \\ \mathbf{0} & \mathbf{0} & \mathbf{0} \end{bmatrix} = J_{pq},$$

and so A is congruent to J_{pq}. In a similar way, B is also congruent to J_{pq}. Hence, A is congruent to B.

($\Leftarrow$) Let two symmetric matrices A and B be congruent. Then they can be matrix representations of the same quadratic form $q(\mathbf{x})$ with respect to different bases, see page 444. Let $\text{In}(A) = (p, q, k)$ and $\text{In}(B) = (p', q', k')$. And let $\alpha = \{\mathbf{x}_1, \ldots, \mathbf{x}_p, \mathbf{x}_{p+1}, \ldots, \mathbf{x}_n\}$ be an ordered basis for $\mathbb{R}^n$ for which $[q]_\alpha$ is diagonal and

$$\begin{aligned} q(\mathbf{x}_i) &> 0 \quad \text{for } i = 1, 2, \ldots, p, \text{ and} \\ q(\mathbf{x}_i) &\le 0 \quad \text{for } i = p+1, p+2, \ldots, n, \end{aligned}$$

and let $\beta = \{\mathbf{y}_1, \ldots, \mathbf{y}_{p'}, \mathbf{y}_{p'+1}, \ldots, \mathbf{y}_n\}$ be another ordered basis for $\mathbb{R}^n$ for which $[q]_\beta$ is diagonal and

$$\begin{aligned} q(\mathbf{y}_i) &> 0 \quad \text{for } i = 1, 2, \ldots, p', \text{ and} \\ q(\mathbf{y}_i) &\le 0 \quad \text{for } i = p'+1, p'+2, \ldots, n. \end{aligned}$$

To show $p = p'$, let U and W be subspaces of $\mathbb{R}^n$ spanned by $\{\mathbf{x}_1, \ldots, \mathbf{x}_p\}$ and $\{\mathbf{y}_{p'+1}, \ldots, \mathbf{y}_n\}$, respectively. Then, $q(\mathbf{u}) > 0$ for any nonzero vector $\mathbf{u} \in U$ and $q(\mathbf{w}) \le 0$ for any nonzero vector $\mathbf{w} \in W$. Thus, $U \cap W = \{\mathbf{0}\}$, and

$$\dim(U + W) = \dim U + \dim W - \dim(U \cap W) = p + (n - p') \le n,$$

or $p \le p'$. By taking the other basis vectors $\{\mathbf{y}_1, \dots, \mathbf{y}_{p'}\}$ and $\{\mathbf{x}_{p+1}, \dots, \mathbf{x}_n\}$, one can show $p' \le p$ to conclude that $p = p'$. Therefore, any two diagonal matrix representations of q have the same number of positive diagonal entries.

By considering the quadratic form $-q$ instead of q, one can show that any two diagonal matrix representations of q have the same number of negative diagonal entries.

As a consequence, one gets $\mathrm{In}(A) = (p, q, k) = (p', q', k')$ $=\mathrm{In}(B)$. □

Corollary 9.18.1 **(1)** *Let q be any real or complex quadratic form. Then, any two diagonal representations $[q]_\alpha$ and $[q]_\beta$ of q have the same inertia, that is,* $\mathrm{In}([q]_\alpha) = \mathrm{In}([q]_\beta)$.

(2) *Let b be any symmetric bilinear or a Hermitian sesquilinear form. Then, any two diagonal representations $[b]_\alpha$ and $[b]_\beta$ of b ave the same inertia, that is,* $\mathrm{In}([b]_\alpha) = \mathrm{In}([b]_\beta)$.

Problem 243 Show that if the diagonal entries of a diagonal matrix are permuted, then the resulting diagonal matrix is congruent to the original one.

Hint: To show that $D_1 = \begin{bmatrix} 1 & 0 \\ 0 & 2 \end{bmatrix}$ and $D_2 = \begin{bmatrix} 2 & 0 \\ 0 & 1 \end{bmatrix}$ are congruent, let $E = \begin{bmatrix} 0 & 1 \\ 1 & 0 \end{bmatrix}$ be the 'row-interchanging' elementary matrix. Then $D_2 = ED_1E^T$.

Problem 244 Show that the total number of distinct inertia (p, q, k) with $p > 0$ of all $n \times n$ real symmetric matrices is equal to $n(n+1)/2$.

Problem 245 Find the inertia for each of the following symmetric matrices.

$$(1) \begin{bmatrix} -1 & 1 & 2 \\ 1 & -20 & 3 \\ 2 & 3 & 40 \end{bmatrix}, \qquad (2) \begin{bmatrix} 1 & 2 & 3 \\ 2 & 4 & 1 \\ 3 & 1 & 6 \end{bmatrix}, \qquad (3) \begin{bmatrix} 2 & 0 & 1 \\ 0 & 1 & 5 \\ 1 & 5 & 1 \end{bmatrix}.$$

Which are congruent to each other?

Hint: Don't need to find the eigenvalues; use the congruence relation to find $\mathrm{In}(\cdot)$.

9.11.3 Real quadratic forms versus symmetric bilinear forms

We now show how real quadratic forms and symmetric bilinear forms can be associated with each other.

Theorem 9.19 *If b is a symmetric bilinear form on $\mathbb{R}^n$, then the function $q(\mathbf{x}) = b(\mathbf{x}, \mathbf{x})$ for $\mathbf{x} \in \mathbb{R}^n$ is a quadratic form.*

Conversely, the symmetric bilinear form b can be recovered from the quadratic $q(\mathbf{x})$. In other words, for every real quadratic form q, there is a unique symmetric bilinear form b such that $q(\mathbf{x}) = b(\mathbf{x}, \mathbf{x})$ for all $\mathbf{x}$ in $\mathbb{R}^n$.

Proof: If a symmetric bilinear form $b(\mathbf{x}, \mathbf{y}) = \mathbf{x}^T A\mathbf{y}$ is given, then $q(\mathbf{x}) = b(\mathbf{x}, \mathbf{x}) = \mathbf{x}^T A\mathbf{x}$ is clearly a quadratic form.

To show the converse, first note that every symmetric bilinear form b satisfies the identity

$$b(\mathbf{x} + \mathbf{y}, \mathbf{x} + \mathbf{y}) = b(\mathbf{x}, \mathbf{x}) + 2b(\mathbf{x}, \mathbf{y}) + b(\mathbf{y}, \mathbf{y}).$$

Hence, for any given quadratic form $q(\mathbf{x})$, one can define a function b by

$$b(\mathbf{x}, \mathbf{y}) = \frac{1}{2}[q(\mathbf{x} + \mathbf{y}) - q(\mathbf{x}) - q(\mathbf{y})], \tag{9.5}$$

which is called the *polar form* of b. Then b is clearly a symmetric bilinear form and $b(\mathbf{x}, \mathbf{x}) = q(\mathbf{x})$. The uniqueness is clear from this identity. □

Problem 246 Show that for any given quadratic form $q(\mathbf{x})$,

$$b(\mathbf{x}, \mathbf{y}) = \frac{1}{2}[q(\mathbf{x} + \mathbf{y}) - q(\mathbf{x}) - q(\mathbf{y})]$$

is a symmetric bilinear form.

Hint: For a quadratic form $q(\mathbf{x}) = \mathbf{x}^T A\mathbf{x}$, $q(\mathbf{x}_1 + \mathbf{x}_2) = q(\mathbf{x}_1) + q(\mathbf{x}_2) + 2\mathbf{x}_1^T A\mathbf{x}_2$.

Example 187 (*The quadratic form corresponding to the symmetric bilinear form*)

(1) If $b(\mathbf{x}, \mathbf{y}) = \mathbf{x}^T\mathbf{y} = x_1y_1 + x_2y_2$ is a dot product, then the corresponding quadratic form is $q(\mathbf{x}) = \|\mathbf{x}\|^2$. The polar form of b says

$$b(\mathbf{x}, \mathbf{y}) = x_1y_1 + x_2y_2 = \frac{1}{2}(\|\mathbf{x} + \mathbf{y}\|^2 - \|\mathbf{x}\|^2 - \|\mathbf{y}\|^2).$$

(2) If $b(\mathbf{x}, \mathbf{y}) = x_1y_1 - x_2y_2$, then b is a symmetric bilinear form and the corresponding quadratic form is $q(\mathbf{x}) = x_1^2 - x_2^2$.

(3) For a symmetric matrix $\begin{bmatrix} 2 & 1 \\ 1 & 2 \end{bmatrix}$, $b(\mathbf{x}, \mathbf{y}) = \mathbf{x}^T A\mathbf{y} = 2x_1y_1 + x_1y_2 + x_2y_1 + 2x_2y_2$ is a symmetric bilinear form, and the corresponding quadratic form is $q(\mathbf{x}) = 2x_1^2 + 2x_1x_2 + 2x_2^2$.

In each of the three cases, the symmetric bilinear form b can be recovered from the quadratic form $q(\mathbf{x})$. □

The following theorem shows how complex quadratic forms and Hermitian sesquilinear forms are associated with each other.

Theorem 9.20 *If b is a Hermitian sesquilinear form on $\mathbb{C}^n$, then the function $q(\mathbf{x}) = b(\mathbf{x}, \mathbf{x})$ for $\mathbf{x} \in \mathbb{C}^n$ is a complex quadratic form.*

Conversely, the Hermitian sesquilinear form b can be recovered from the complex quadratic $q(\mathbf{x})$. In other words, for every complex quadratic form q, there is a unique Hermitian sesquilinear form b such that $q(\mathbf{x}) = b(\mathbf{x}, \mathbf{x})$ for all $\mathbf{x}$ in $\mathbb{C}^n$.

Proof: If $b(\mathbf{x}, \mathbf{y}) = \mathbf{x}^H A\mathbf{y}$ is a Hermitian sesquilinear form with $A = [b]_\alpha$, then $q(\mathbf{x}) = b(\mathbf{x}, \mathbf{x}) = \mathbf{x}^H A\mathbf{x}$ is clearly a complex quadratic form.

Conversely, if b is a Hermitian sesquilinear form, then it follows [9]

$$\begin{aligned} b(\mathbf{x}+\mathbf{y}, \mathbf{x}+\mathbf{y}) &= b(\mathbf{x}, \mathbf{x}) + b(\mathbf{y}, \mathbf{y}) + b(\mathbf{x}, \mathbf{y}) + \overline{b(\mathbf{x}, \mathbf{y})}, \\ b(\mathbf{x}-\mathbf{y}, \mathbf{x}-\mathbf{y}) &= b(\mathbf{x}, \mathbf{x}) + b(\mathbf{y}, \mathbf{y}) - b(\mathbf{x}, \mathbf{y}) - \overline{b(\mathbf{x}, \mathbf{y})}, \end{aligned}$$

and

$$\begin{aligned} b(i\mathbf{x}+\mathbf{y}, i\mathbf{x}+\mathbf{y}) &= b(\mathbf{x}, \mathbf{x}) + b(\mathbf{y}, \mathbf{y}) - ib(\mathbf{x}, \mathbf{y}) + i\overline{b(\mathbf{x}, \mathbf{y})}, \\ b(i\mathbf{x}-\mathbf{y}, i\mathbf{x}-\mathbf{y}) &= b(\mathbf{x}, \mathbf{x}) + b(\mathbf{y}, \mathbf{y}) + ib(\mathbf{x}, \mathbf{y}) - i\overline{b(\mathbf{x}, \mathbf{y})}. \end{aligned}$$

Hence, for any given complex quadratic form $q(\mathbf{x}) = \mathbf{x}^H A\mathbf{x}$ with a Hermitian matrix A, a Hermitian sesquilinear form b can be defined by

$$b(\mathbf{x}, \mathbf{y}) = \frac{1}{4}[q(\mathbf{x}+\mathbf{y}) - q(\mathbf{x}-\mathbf{y})] + \frac{i}{4}[q(i\mathbf{x}+\mathbf{y}) - q(i\mathbf{x}-\mathbf{y})], \qquad (9.6)$$

which is called the *polar form* of a Hermitian sesquilinear form b. This form b is clearly Hermitian and $b(\mathbf{x}, \mathbf{x}) = q(\mathbf{x})$, which implies the uniqueness of b. □

By Theorems 9.19 and 9.20, the real quadratic forms are in a one-to-one correspondence with the symmetric bilinear forms, and the complex quadratic forms are in a one-to-one correspondence with the Hermitian sesquilinear forms.

In fact, with a fixed basis for $\mathbb{R}^n$ or $\mathbb{C}^n$, a quadratic form

$$q(\mathbf{x}) = \mathbf{x}^T A\mathbf{x} = \sum_{i=1}^{n}\sum_{j=1}^{n} a_{ij}x_i x_j$$

[9]For a complex number $z = x + iy$, its real part x denoted by $\mathrm{Re}(z)$ and its imaginary part y is denoted by $\mathrm{Im}(z)$. For a complex number $b(\mathbf{x}, \mathbf{y})$, we have $b(\mathbf{x}, \mathbf{y}) + \overline{b(\mathbf{x}, \mathbf{y})} = 2\,\mathrm{Re}(b(\mathbf{x}, \mathbf{y}))$ and $b(\mathbf{x}, \mathbf{y}) - \overline{b(\mathbf{x}, \mathbf{y})} = 2i\,\mathrm{Im}(b(\mathbf{x}, \mathbf{y}))$.

with a symmetric matrix $A = [a_{ij}]$ is paired with exactly one symmetric bilinear form

$$b(\mathbf{x}, \mathbf{y}) = \mathbf{x}^T A\mathbf{y} = \sum_{i=1}^{n}\sum_{j=1}^{n} a_{ij}x_i y_j.$$

A similar pairing is also given between the complex quadratic forms and the Hermitian sesquilinear forms. In this case, $q(\mathbf{x}) = \mathbf{x}^H A\mathbf{x} = \sum_{i=1}^{n}\sum_{j=1}^{n} a_{ij}\overline{x_i}x_j$ is paired with $b(\mathbf{x}, \mathbf{y}) = \mathbf{x}^H A\mathbf{y} = \sum_{i=1}^{n}\sum_{j=1}^{n} a_{ij}\overline{x_i}y_j$ with a Hermitian matrix A.

Problem 247 Show that

(1) real *nondegenerate* quadratic forms are in a one-to-one correspondence with *nondegenerate* symmetric bilinear forms,
(2) complex *nondegenerate* quadratic forms are in a one-to-one correspondence with *nondegenerate* Hermitian sesquilinear forms,
(3) real *positive-definite* quadratic forms are in a one-to-one correspondence with *positive-definite* symmetric bilinear forms,
(4) complex *positive-definite* quadratic forms are in a one-to-one correspondence with *positive-definite* Hermitian sesquilinear forms.

Hint for (1) and (2): Nondegenerate $\Leftrightarrow$ $\det A \neq 0 \Leftrightarrow k = 0$ in $\text{In}(A) = (p, q, k)$.
Hint for (3) and (4): Positive-definite $\Leftrightarrow q = k = 0$ in $\text{In}(A) = (p, q, k)$.

9.11.4 MinMax of quadratic forms and estimate eigenvalues

How can one find the eigenvalues of a given matrix?

If the matrix is of low order like $n = 2$ or 3, one can find them by solving the characteristic equation.

How about matrices of higher order? Most cases are almost impossible by hand. Moreover, it is shown in Lemma 6.9 (1) that any monic polynomial can be the characteristic polynomial of a certain companion matrix. Therefore, any algorithm for finding eigenvalues of a matrix could also be used to find the roots of polynomials. In fact, this is still not an easy algorithm problem.

In this last section we consider an optimization problem (minimization or maximization) for a quadratic form in n variables and show how one can use this problem to estimate the (unknown) eigenvalues.

As a special case, we are interested in the following optimization problem with a constraint.

Find the maximum and minimum values of a (real or complex) quadratic form $q(\mathbf{x}) = \langle \mathbf{x}, A\mathbf{x} \rangle$ subject to the constraint $\|\mathbf{x}\| = 1$.

Advanced calculus tells us that such constrained maximum or minimum of $q(\mathbf{x})$ always exists.

Example 188 (*The extreme values of a quadratic form $\mathbf{x}^T A\mathbf{x}$ subject to $\|\mathbf{x}\| = 1$ are the extreme eigenvalues of A*) Find the maximum and minimum values of the quadratic form

$$q(\mathbf{x}) = 3x^2 + y^2$$

subject to the constraint $x^2 + y^2 = 1$, and determine the vector $\mathbf{x} = (x, y)$ at which the maximum and minimum occur.

Solution: The quadratic form can be written as

$$q(\mathbf{x}) = 3x^2 + y^2 = \mathbf{x}^T A\mathbf{x} = \begin{bmatrix} x & y \end{bmatrix} \begin{bmatrix} 3 & 0 \\ 0 & 1 \end{bmatrix} \begin{bmatrix} x \\ y \end{bmatrix}.$$

The eigenvalues of A are $\lambda_1 = 3$ and $\lambda_2 = 1$, which are the maximum and minimum eigenvalues, respectively. Their associated unit eigenvectors are $\pm\mathbf{e}_1$ and $\pm\mathbf{e}_2$, respectively.

Since $\text{In}(A) = (2, 0, 0)$, the level curve $q(\mathbf{x}) = t, t > 0$ is an ellipse centered at the origin $(0, 0)$, with major axis along the y-axis and minor axis along the x-axis. When t increases from a very small $t > 0$, these level curves, *i.e.*, the ellipses, become bigger and bigger from a very small ellipse, and eventually surround the unit circle $\|\mathbf{x}\| = 1$.

It means that, $q(\mathbf{x}) = t$ is the minimum value with $\|\mathbf{x}\| = 1$ when the level curve $q(\mathbf{x}) = t$ inscribes the unit circle $\|\mathbf{x}\| = 1$ and is tangent to it at $\mathbf{x} = (0, \pm 1) = \pm\mathbf{e}_2$. At this time, $\mathbf{x}^T A\mathbf{x} = 1 = \lambda_2$, the minimum eigenvalue of A.

By the similar reason, $q(\mathbf{x}) = t$ is the maximum value with $\|\mathbf{x}\| = 1$ when the level curve $q(\mathbf{x}) = t$ circumscribes the unit circle $\|\mathbf{x}\| = 1$ and is tangent to it at $\mathbf{x} = (\pm 1, 0) = \pm\mathbf{e}_1$. At this time, $\mathbf{x}^T A\mathbf{x} = 3 = \lambda_1$, the maximum eigenvalue of A.

Consequently, the extreme values of the quadratic form $q(\mathbf{x}) = \mathbf{x}^T A\mathbf{x}$ subject to $\|\mathbf{x}\| = 1$ are the extreme eigenvalues of A. □

This can be generalized to the following theorem.

Theorem 9.21 *Let A be a symmetric or Hermitian matrix, and let the eigenvalues of A be $\lambda_{\max} = \lambda_1 \geq \lambda_2 \geq \cdots \geq \lambda_n = \lambda_{\min}$ in descending order. Then,*

(1) $\lambda_{\min} \|\mathbf{x}\|^2 \leq \langle \mathbf{x}, A\mathbf{x} \rangle \leq \lambda_{\max} \|\mathbf{x}\|^2$ *for all* $\mathbf{x}$.

(2) $\langle \mathbf{x}, A\mathbf{x} \rangle = \lambda \|\mathbf{x}\|^2$ *if* $\mathbf{x}$ *is an eigenvector of A belonging to an eigenvalue* λ.

(3) $\lambda_{\max} = \max_{\mathbf{x} \neq \mathbf{0}} \dfrac{\langle \mathbf{x}, A\mathbf{x} \rangle}{\langle \mathbf{x}, \mathbf{x} \rangle} = \max_{\|\mathbf{x}\|=1} \langle \mathbf{x}, A\mathbf{x} \rangle$, *and for a unit vector* $\mathbf{x}$, $\lambda_{\max} = \langle \mathbf{x}, A\mathbf{x} \rangle$ *if and only if* $\mathbf{x}$ *is an eigenvector belonging to* $\lambda_{\max}$.

(4) $\lambda_{\min} = \min_{\mathbf{x} \neq \mathbf{0}} \dfrac{\langle \mathbf{x}, A\mathbf{x} \rangle}{\langle \mathbf{x}, \mathbf{x} \rangle} = \min_{\|\mathbf{x}\|=1} \langle \mathbf{x}, A\mathbf{x} \rangle$, *and for a unit vector* $\mathbf{x}$, $\lambda_{\min} = \langle \mathbf{x}, A\mathbf{x} \rangle$ *if and only if* $\mathbf{x}$ *is an eigenvector belonging to* $\lambda_{\min}$.

In particular, the maximum and minimum values of a (real or complex) quadratic form $q(\mathbf{x}) = \langle \mathbf{x}, A\mathbf{x} \rangle$ subject to the constraint $\|\mathbf{x}\| = 1$ is the maximum and minimum eigenvalues of A, respectively.

Proof: (1) We prove this only for a Hermitian matrix A and the other case of a real symmetric matrix is left as an exercise.

If A is Hermitian, there is a unitary matrix U such that $U^H AU = D$ is a diagonal matrix with $\lambda_1, \lambda_2, \ldots, \lambda_n$ as its diagonal entries. Moreover, with a change of coordinates $\mathbf{y} = U^H\mathbf{x} = [y_1\ y_2\ \cdots\ y_n]^T$ we have

$$\mathbf{x}^H A\mathbf{x} = \mathbf{y}^H D\mathbf{y} = \lambda_1|y_1|^2 + \lambda_2|y_2|^2 + \cdots + \lambda_n|y_n|^2,$$

by the Principal Axes theorem, and $\|\mathbf{x}\| = \|\mathbf{y}\|$ because U is unitary. This implies that

$$\begin{aligned}
\langle \mathbf{x}, A\mathbf{x} \rangle = \mathbf{x}^H A\mathbf{x} &= \lambda_1|y_1|^2 + \lambda_2|y_2|^2 + \cdots + \lambda_n|y_n|^2 \\
&\leq \lambda_1|y_1|^2 + \lambda_1|y_2|^2 + \cdots + \lambda_1|y_n|^2 \\
&= \lambda_1(|y_1|^2 + |y_2|^2 + \cdots + |y_n|^2) \\
&= \lambda_1\|\mathbf{y}\|^2 = \lambda_{\max} \|\mathbf{x}\|^2
\end{aligned}$$

since $\lambda_1 = \lambda_{\max}$ is the maximum eigenvalue.

Similarly, one can show $\lambda_{\min} \leq \langle \mathbf{x}, A\mathbf{x} \rangle$ for all $\mathbf{x}$. It proves (1).

(2) If $\mathbf{x}$ is an eigenvector of A belonging to λ, then

$$\langle \mathbf{x}, A\mathbf{x} \rangle = \langle \mathbf{x}, \lambda\mathbf{x} \rangle = \lambda\langle \mathbf{x}, \mathbf{x} \rangle = \lambda\|\mathbf{x}\|^2.$$

In particular, if $\mathbf{x}$ is an eigenvector of A belonging to $\lambda_{\max}$ ($\lambda_{\min}$, respectively) and $\|\mathbf{x}\| = 1$, then $\langle \mathbf{x}, A\mathbf{x} \rangle = \lambda_{\max}$ ($\lambda_{\min}$, respectively).

(3) We only prove the necessity part of the second assertion, because all the other parts are clear from (1) and (2). To show this, suppose that $\lambda_{\max} = \langle \mathbf{x}, A\mathbf{x} \rangle$ for a Hermitian matrix A and a unit vector $\mathbf{x}$. Let $\lambda_1, \lambda_2, \ldots, \lambda_n$ be the eigenvalues of A with associated eigenvectors $\mathbf{v}_1, \mathbf{v}_2, \ldots, \mathbf{v}_n$, respectively. One can assume that $\lambda_{\max} = \lambda_1$ and the eigenvectors $\mathbf{v}_1, \mathbf{v}_2, \ldots, \mathbf{v}_n$ are orthonormal since A is Hermitian. Let $\mathbf{x} = \sum_j a_j \mathbf{v}_j$. Then, we have

$$\begin{aligned}\lambda_1 = \langle \mathbf{x}, A\mathbf{x} \rangle &= \left\langle \sum_{j=1}^n a_j \mathbf{v}_j, A\left(\sum_{j=1}^n a_j \mathbf{v}_j\right)\right\rangle = \left\langle \sum_{j=1}^n a_j \mathbf{v}_j, \sum_{j=1}^n a_j \lambda_j \mathbf{v}_j \right\rangle \\ &= \sum_{j=1}^n |a_j|^2 \lambda_j \le \sum_{j=1}^n |a_j|^2 \lambda_1 = \lambda_1\end{aligned}$$

since $\lambda_{\max} = \lambda_1$. Hence, $a_j = 0$ whenever $\lambda_j \neq \lambda_1$, which implies that $\mathbf{x}$ is an eigenvector belonging to λ_1.

(4) can be proved in a similar way to (3). □

Definition 9.22 The **Rayleigh quotient** [10] of a symmetric or Hermitian matrix A is the real-valued function R_A defined for $\mathbf{x} \neq \mathbf{0}$ by

$$R_A(\mathbf{x}) = \frac{\langle \mathbf{x}, A\mathbf{x} \rangle}{\langle \mathbf{x}, \mathbf{x} \rangle} = \frac{\langle \mathbf{x}, A\mathbf{x} \rangle}{\|\mathbf{x}\|^2} \quad \text{for } \mathbf{x} \neq \mathbf{0}.$$

By definition, *the Rayleigh quotient $R_A(\mathbf{x})$ is the quadratic form $\langle \mathbf{x}, A\mathbf{x} \rangle$ on the unit sphere.* Hence, it follows from Theorem 9.21 that, subject to the constraint $\|\mathbf{x}\| = 1$, the quadratic form $\langle \mathbf{x}, A\mathbf{x} \rangle$ has the maximum value $\lambda_{\max}$ and the minimum value $\lambda_{\min}$. This means that

the minimum and the maximum eigenvalues of the matrix A are characterized as the solutions of a minimum and maximum problem of the Rayleigh quotient or the quadratic form $\langle \mathbf{x}, A\mathbf{x} \rangle$ on the unit sphere.

The Rayleigh quotient is often used in the min-max problem to estimate the true eigenvalues of a matrix or their approximations.

[10] John William Strutt, 3rd Baron Rayleigh (November 12, 1842– June 30, 1919) was a physicist who, with William Ramsay, discovered argon, an achievement for which he earned the Nobel Prize for Physics in 1904. He also discovered the phenomenon now called Rayleigh scattering, which can be used to explain why the sky is blue.

Example 189 (*The extreme values of a quadratic form* $\mathbf{x}^T A\mathbf{x}$ *subject to* $\|\mathbf{x}\| = 1$ *are the extreme eigenvalues of* A) Find the maximum and minimum values of the quadratic form

$$q(\mathbf{x}) = x^2 + y^2 + 4xy$$

subject to the constraint $\|\mathbf{x}\| = 1$, and determine the vector $\mathbf{x} = (x, y)$ at which the maximum and minimum occur.

Solution: The quadratic form can be written as

$$q(\mathbf{x}) = x^2 + y^2 + 4xy = \mathbf{x}^T A\mathbf{x} = \begin{bmatrix} x & y \end{bmatrix} \begin{bmatrix} 1 & 2 \\ 2 & 1 \end{bmatrix} \begin{bmatrix} x \\ y \end{bmatrix}.$$

The eigenvalues of A are

$$\lambda_{\max} = 3 \quad \text{and} \quad \lambda_{\min} = -1,$$

which are the maximum and minimum eigenvalues, respectively. Their associated unit eigenvectors are

$$\pm\left(\frac{1}{\sqrt{2}}, \frac{1}{\sqrt{2}}\right), \qquad \pm\left(\frac{1}{\sqrt{2}}, -\frac{1}{\sqrt{2}}\right),$$

respectively.

Since $\text{In}(A) = (1, 1, 0)$, the level curve $q(\mathbf{x}) = t$ is a hyperbola centered at the origin $(0, 0)$, with transverse axis along the line $y = x$ if $t > 0$, and with transverse axis along the line $y = -x$ if $t < 0$. See Figure 9.7.

When t increases from a very small $t > 0$, these level curves, *i.e.,* the hyperbolas are getting further away from the origin along the diagonal line $y = x$. In this case, the maximum value of the quadratic form $q(\mathbf{x}) = \mathbf{x}^T A\mathbf{x}$ subject to $\|\mathbf{x}\| = 1$ is $\lambda_{\max} = 3$, which occurs at the unit eigenvector $\mathbf{x} = \pm(1/\sqrt{2},\ 1/\sqrt{2})$.

Similarly, the minimum value of $q(\mathbf{x}) = \mathbf{x}^T A\mathbf{x}$ subject to $\|\mathbf{x}\| = 1$ is $\lambda_{\min} = -1$, which occurs at $\mathbf{x} = \pm(1/\sqrt{2},\ -1/\sqrt{2})$.

Clearly, both extreme values occur when a hyperbola and the unit circle intersect tangentially as in Figure 9.7. □

Problem 248 (Example 189 continued) For the quadratic form $q(\mathbf{x}) = x^2 + y^2 + 4xy = (2x + y)^2 - (\sqrt{3}x)^2$, which is a difference of two perfect squares,

(1) what is the solution set of $q(\mathbf{x}) = 0$?

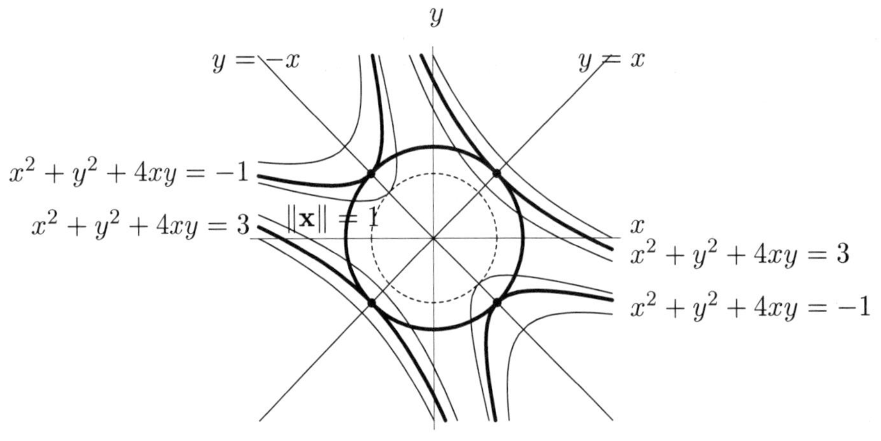

Figure 9.7: Extreme values of the constraint quadratic form

(2) When $t > 0$, explain how the level curves $q(\mathbf{x}) = x^2 + y^2 + 4xy = t$ for $t = 1, 10, 100, \ldots,$ vary with t.
(3) When $t < 0$, explain how the level curves $q(\mathbf{x}) = x^2 + y^2 + 4xy = t$ for $t = -1, -10, -100, \ldots,$ vary with t.

Conclude that $q(\mathbf{x}) = x^2 + y^2 + 4xy = t$ is a hyperbolic paraboloid in the (x, y, t)-space $\mathbb{R}^3$ with a saddle point at the origin, despite the fact that the coefficients of $q(\mathbf{x})$ are all positive.

As the final statement in this book, we introduce an algorithm, called *Rayleigh quotient iteration*, which is an iterative method by using the Rayleigh quotient to estimate accurate approximations of the maximum eigenvalue $\lambda_{\max}$.

Example 190 (*Estimate the maximum eigenvalue $\lambda_{\max}$ by Rayleigh quotient iteration*) Consider again the symmetric matrix A given in Example 117 (2) on page 280:

$$A = \begin{bmatrix} 100 & 0 & 1 & 0 \\ 0 & 200 & 2 & 1 \\ 1 & 2 & 300 & 0 \\ 0 & 1 & 0 & 400 \end{bmatrix}.$$

The true eigenvalues of A are approximated by

$$\lambda_1 \approx 99.995, \quad \lambda_2 \approx 199.955, \quad \lambda_3 \approx 300.045, \quad \lambda_4 \approx 400.005.$$

Even the true eigenvalues of A are not known, one can guess that $\lambda_{\max}$ is close to the $(4, 4)$-entry of A by the Gershgorin theorem. The exact value of $\lambda_{\max}$ can be estimated by an iteration of Rayleigh quotient as follows.

Since the matrix A is quite close in entrywise comparison to the diagonal matrix with diagonal entries $100, 200, 300, 400$ which are eigenvalues with associated eigenvectors $\mathbf{e}_1, \mathbf{e}_2, \mathbf{e}_3, \mathbf{e}_4$, respectively, one may choose the unit vector $\mathbf{x}_{(0)} = \mathbf{e}_4$, as an approximate eigenvector of A belonging to $\lambda_{\max}$. Then the Rayleigh quotient

$$R_A(\mathbf{x}_{(0)}) = \frac{\langle \mathbf{x}_{(0)}, A\mathbf{x}_{(0)} \rangle}{\langle \mathbf{x}_{(0)}, \mathbf{x}_{(0)} \rangle} = 400,$$

(say $\lambda_{(0)}$). Then, the pair $(400, \mathbf{e}_4)$ of the Rayleigh quotient $R_A(\mathbf{x}_{(0)}) = 400$ and the vector $\mathbf{x}_{(0)} = \mathbf{e}_4$ gives an approximation of the pair $(\lambda_{\max}, \mathbf{x})$ of the eigenvalue $\lambda_{\max}$ and an associated unit eigenvector $\mathbf{x}$.

To estimate more accurate approximations of the maximum eigenvalue $\lambda_{\max}$, we use *Rayleigh quotient iteration* as follows.

With the pair $(\lambda_{(0)}, \mathbf{x}_{(0)}) = (400, \mathbf{e}_4)$, define

$$\mathbf{x}_{(1)} = \frac{(A - \lambda_{(0)} I)^{-1} \mathbf{x}_{(0)}}{\|(A - \lambda_{(0)} I)^{-1} \mathbf{x}_{(0)}\|}, \qquad \lambda_{(1)} = R_A(\mathbf{x}_{(1)}) = \frac{\langle \mathbf{x}_{(1)}, A\mathbf{x}_{(1)} \rangle}{\langle \mathbf{x}_{(1)}, \mathbf{x}_{(1)} \rangle}.$$

(Here, if $A - \lambda_{(0)} I$ is not invertible, $\lambda_{(0)}$ is an eigenvalue.) By a numerical computing, one can get the pair $(\lambda_{(1)}, \mathbf{x}_{(1)})$

$$\mathbf{x}_{(1)} = \frac{1}{5998675}(2, 29999, 600, 5998600), \qquad \lambda_{(1)} = 400.005.$$

This $\lambda_{(1)}$ is already the same as the mentioned approximation of $\lambda_{\max}$. In this case, the vector $\mathbf{x}_{(1)}$ is also a good approximation of a true eigenvector belonging to $\lambda_{\max}$. □

As shown in Example 190, in the choice of the initial pair $(\lambda_{(0)}, \mathbf{x}_{(0)})$ for Rayleigh quotient iteration, the initial vector $\mathbf{x}_{(0)}$ should be chosen to be sufficiently close to an associated eigenvector $\mathbf{x}$ belonging to the true maximum eigenvalue $\lambda_{\max}$.

We give another example with a 2×2 symmetric matrix.

Example 191 (*Estimate the maximum eigenvalue $\lambda_{\max}$ by Rayleigh quotient iteration*) Consider a symmetric matrix

$$A = \begin{bmatrix} 2 & 3 \\ 3 & 1 \end{bmatrix}.$$

The eigenvalues of A are

$$\lambda_1 = \tfrac{1}{2}(3+\sqrt{37}) \approx 4.54138 \quad \text{and} \quad \lambda_2 = \tfrac{1}{2}(3-\sqrt{37}) \approx -1.54138$$

with associated eigenvectors

$$\mathbf{x}_1 = \left(\frac{1}{6}(1+\sqrt{37}), 1\right) \quad \text{and} \quad \mathbf{x}_2 = \left(\frac{1}{6}(1-\sqrt{37}), 1\right),$$

respectively.

Rayleigh quotient iteration starts from a choice of an initial pair $(\lambda_{(0)}, \mathbf{x}_{(0)})$ as a guessed eigenvalue and a guessed associated eigenvector. Choose it as

$$(\lambda_{(0)}, \mathbf{x}_{(0)}) = (10, \mathbf{e}_1)).$$

As the first iteration, define

$$\mathbf{x}_{(1)} = \frac{(A-\lambda_{(0)}I)^{-1}\mathbf{x}_{(0)}}{\|(A-\lambda_{(0)}I)^{-1}\mathbf{x}_{(0)}\|}, \qquad \lambda_{(1)} = R_A(\mathbf{x}_{(1)}) = \frac{\langle \mathbf{x}_{(1)}, A\mathbf{x}_{(1)}\rangle}{\langle \mathbf{x}_{(1)}, \mathbf{x}_{(1)}\rangle}.$$

By a numerical computing, one can get the pair $(\lambda_{(1)}, \mathbf{x}_{(1)})$ as

$$\mathbf{x}_{(1)} = \frac{(-1/7, -1/21)}{\|(-1/7, -1/21)\|} = (-3\sqrt{10}/10, -\sqrt{10}/10) \quad \lambda_{(1)} = 3.7.$$

As the second iteration, define

$$\mathbf{x}_{(2)} = \frac{(A-\lambda_{(1)}I)^{-1}\mathbf{x}_{(1)}}{\|(A-\lambda_{(1)}I)^{-1}\mathbf{x}_{(1)}\|}, \qquad \lambda_{(2)} = R_A(\mathbf{x}_{(2)}) = \frac{\langle \mathbf{x}_{(2)}, A\mathbf{x}_{(2)}\rangle}{\langle \mathbf{x}_{(2)}, \mathbf{x}_{(2)}\rangle}.$$

A numerical computing gives

$$\mathbf{x}_{(2)} = \frac{(-0.795948, -0.767265)}{1.10554} = (-0.719963, -0.694018) \quad \lambda_{(2)} = 4.47967.$$

To estimate more accurate approximations of $\lambda_{\max}$, iterate once more:

$$\mathbf{x}_{(3)} = \frac{(A-\lambda_{(2)}I)^{-1}\mathbf{x}_{(2)}}{\|(A-\lambda_{(2)}I)^{-1}\mathbf{x}_{(2)}\|}, \qquad \lambda_{(3)} = R_A(\mathbf{x}_{(3)}) = \frac{\langle \mathbf{x}_{(3)}, A\mathbf{x}_{(3)}\rangle}{\langle \mathbf{x}_{(3)}, \mathbf{x}_{(3)}\rangle}.$$

Also, a computing gives

$$\mathbf{x}_{(3)} = \frac{(-12.3458, -10.4445)}{16.1712} = (-0.763444, -0.64587) \quad \lambda_{(3)} = 4.54137.$$

This $\lambda_{(3)}$ is a quite good approximation of $\lambda_{\max} = \lambda_1 = \frac{1}{2}(3+\sqrt{37})$.

As another possible choice of an initial pair, let

$$(\lambda_{(0)}, \mathbf{x}_{(0)}) = (40, \mathbf{e}_1 + \mathbf{e}_2).$$

As before, define

$$\mathbf{x}_{(1)} = \frac{(A-\lambda_{(0)}I)^{-1}\mathbf{x}_{(0)}}{\|(A-\lambda_{(0)}I)^{-1}\mathbf{x}_{(0)}\|}, \qquad \lambda_{(1)} = R_A(\mathbf{x}_{(1)}) = \frac{\langle \mathbf{x}_{(1)}, A\mathbf{x}_{(1)}\rangle}{\langle \mathbf{x}_{(1)}, \mathbf{x}_{(1)}\rangle}.$$

A numerical computing gives

$$\mathbf{x}_{(1)} = \frac{(-0.0285132, -0.0278344)}{0.0398467} = (-0.715572, -0.698537),$$

$$\lambda_{(1)} = 4.51116.$$

As the second iteration, define

$$\mathbf{x}_{(2)} = \frac{(A-\lambda_{(1)}I)^{-1}\mathbf{x}_{(1)}}{\|(A-\lambda_{(1)}I)^{-1}\mathbf{x}_{(1)}\|}, \qquad \lambda_{(2)} = R_A(\mathbf{x}_{(2)}) = \frac{\langle \mathbf{x}_{(2)}, A\mathbf{x}_{(2)}\rangle}{\langle \mathbf{x}_{(2)}, \mathbf{x}_{(2)}\rangle}.$$

A computing gives

$$\mathbf{x}_{(2)} = \frac{(-25.1925, -21.326)}{33.007} = (-0.763247, -0.646105) \quad \lambda_{(2)} = 4.54137.$$

This $\lambda_{(2)}$ is the same approximation of $\lambda_{\max} = \frac{1}{2}(3+\sqrt{37})$ as a given before. Here, the vector $\mathbf{x}_{(2)}$ is also a good approximation of a true eigenvector belonging to $\lambda_{\max}$. □

Problem 249 Find the maximum and the minimum values of the Rayleigh quotient $R_A(\mathbf{x})$ for

$$(1)\ A = \begin{bmatrix} 2 & 1 & 0 \\ 1 & 2 & 0 \\ 0 & 0 & 6 \end{bmatrix}, \qquad (2)\ A = \begin{bmatrix} 2 & 0 & -i \\ 0 & 0 & i \\ i & -i & 1 \end{bmatrix}.$$

Hint: (1) A is symmetric. (2) A is Hermitian.

Problem 250 Find the maximum and minimum values of the quadratic form

$$2x^2 + 2y^2 - 3xy$$

subject to the constraint $x^2 + y^2 = 1$, and determine values of x and y at which the maximum and minimum occur.

It is shown that the Rayleigh quotient characterizes the maximum and the minimum eigenvalues and their associated eigenvectors of a real symmetric or a Hermitian matrix A in terms of a constrained optimization problem.

All other eigenvalues and their associated eigenvectors can be characterized in a similar way.

For example, the second largest eigenvalue can be characterized as the maximum value of the quadratic form $\langle \mathbf{x}, A\mathbf{x} \rangle$ subject to the constraint $\mathbf{x}^H \mathbf{v}_1 = 0$, where $\mathbf{v}_1$ is a good estimate of an eigenvector belonging to the maximum eigenvalue $\lambda_{\max}$, *i.e.*, $R_A(\mathbf{v}_1) \approx \lambda_{\max}$.

For a further discussion on estimating the eigenvalues, it is suggested to refer to some advanced linear algebra books. For example, see Applied Linear Algebra by Noble and Daniel [Noble & Daniel] or Numerical Linear Algebra and Applications by Datta [Datta].

9.12 Exercises

9.1. *High-school geometry?* Can you sketch the solution set of each of the following equations? Which are curves?

(1) $x^2 + 6x - y = 1$,
(2) $x^2 - y^2 = 1$,
(3) $x^2 + y^2 + xy = 1$,
(4) $x^2 + y^2 + z^2 + xyz = 1$.

Which one is easy? Which one is difficult? For which equation is the left-hand side homogeneous?

Hint: For (3), what happens if $y = \pm x$? For (4); If one of x, y, z is zero, what?

9.2. *The matrix is also useful in studying a quadratic form or quadratic equation.* Express each of the following quadratic forms in matrix notation with a *symmetric* matrix.

(1) $x^2 + 4xy + 3y^2$,
(2) $x^2 - y^2 + z^2 + 4xz + 4yz$,
(3) $2x^2 - 2xy + 6yx + 2yz + 4zy - 5z^2$,
(4) $[x\ y] \begin{bmatrix} 2 & 0 \\ 4 & 1 \end{bmatrix} \begin{bmatrix} x \\ y \end{bmatrix}$.

Hint: In (3), the given form equals $2x^2 + 4xy + 6yz - 5z^2$.

9.3. *In a quadratic equation $\mathbf{x}^T A\mathbf{x} + \mathbf{b}^T\mathbf{x} = c$, if A is invertible, then by a translation, a linear term $\mathbf{b}^T\mathbf{x}$ can be eliminated.* By a translation, eliminate the linear term of each of the following quadratic equations.

(1) $3x^2 + 2xy + y^2 - 2x + y = -1$,
(2) $4x^2 - 2xy + y^2 - 4x - 2y = 1$,
(3) $4x^2 + y^2 - 2z^2 - 2xy - 4x - 2y + z = 4$.

Hint: Take a translation $\mathbf{x} = \mathbf{y} - \frac{1}{2}A^{-1}\mathbf{b}$ as stated on page 437.

9.4. *What makes it hard to sketch the level curve (or the level surface) of a quadratic equation $q(\mathbf{x}) = c$ is the cross-product terms.* Diagonalize each of the following quadratic forms to eliminate the cross-product terms and determine their geometric types.

(1) $3x^2 + 2xy + y^2 = 1$,
(2) $4x^2 - 2xy + y^2 = 1$,
(3) $4x^2 + y^2 - 2z^2 - 2xy = 1$.

Note: Here, each quadratic form is a quadratic term (2-form) of the quadratic equation given Exercise **9.3**.

9.5. *The cross-product terms in a quadratic equation $q(\mathbf{x}) = c$ can be eliminated by a diagonalization of $q(\mathbf{x})$.* Diagonalize the following quadratic forms to eliminate the cross-product terms.

(1) $x^2 - y^2$,
(2) $x^2 + 2xy + y^2$,
(3) $x^2 + y^2 + z^2 - 2xy - 2yz$,
(4) $[x\ y]\begin{bmatrix} 2 & 0 \\ 4 & 1 \end{bmatrix}\begin{bmatrix} x \\ y \end{bmatrix}$.

How the level curves (or surfaces) $q(\mathbf{x}) = t$ for $t = 1, 10, 100$ vary with $t > 0$ in each case.

9.6. Let q be a quadratic form on $\mathbb{R}^3$ represented by $A = \begin{bmatrix} 7 & 4 & -5 \\ 4 & -2 & 4 \\ -5 & 4 & 7 \end{bmatrix}$ with respect to the standard basis $\alpha = \{(1,0,0), (0,1,0), (0,0,1)\}$.

(1) Diagonalize q orthogonally, *i.e.*, find an orthogonal matrix P so that $P^T AP$ is a diagonal matrix.
(2) Construct a basis γ for $\mathbb{R}^3$ in which each basis vector spans a principal axis of the quadratic form q.

9.7. *The geometric type of the level surface of a quadratic equation $\mathbf{x}^T A\mathbf{x} = c, (c \geq 0)$ is completely determined by the number of positive, negative, and zero eigenvalues of A.* Determine the geometric types of the following quadratic curves on the plane.

(1) $xy = 2$,

(2) $2x^2 - 2xy - 2y^2 = 1$,
(3) $3x^2 + 2xy + y^2 - 2x + y = 4$,
(4) $4x^2 - 2xy + y^2 - 4x - 2y = 1$.

Hint: Compute $\text{In}(A)$ by using the congruence relation. Don't need to diagonalize A as in Exercises **9.4**, **9.5**, **9.6**. Is it easier?

9.8. *The geometric type of the level surface of a quadratic equation* $\mathbf{x}^T A\mathbf{x} = c$, $(c \geq 0)$ *is completely determined by the inertia In*(A) Determine the geometric types of the following quadratic surfaces on the space $\mathbb{R}^3$.

(1) $2xy + 2yz + 2xz = 1$,
(2) $3x^2 + y^2 + 2z^2 + 2xy + 4yz = 10$,
(3) $4x^2 + y^2 - 2z^2 - 2xy - 4x - 2y + z = 1$.

Hint: Compute $\text{In}(A)$. Don't try to diagonalize A.

9.9. For each of the following quadratic forms, determine its definite form; positive-definite, negative-definite, or indefinite. (Use the congruence relation.)

(1) $q(\mathbf{x}) = 2x^2 + 6xy + 2y^2$,
(2) $q(\mathbf{x}) = x^2 + y^2 + z^2 + 2(xy + xz + yz)$,
(3) $q(\mathbf{x}) = x^2 + y^2 + z^2 - 4(xy + xz + yz)$.

9.10. Use the Gershgorin Theorem to determine whether or not each of the following symmetric matrices is positive-definite:

$$(1)\ \begin{bmatrix} 2 & 1 & 0 \\ 1 & 3 & -1 \\ 0 & -1 & 2 \end{bmatrix}, \qquad (2)\ \begin{bmatrix} 3 & 1 & 1 \\ 1 & 3 & 1 \\ 1 & 1 & 3 \end{bmatrix}, \qquad (3)\ \begin{bmatrix} 3 & 2 & 0 \\ 2 & 3 & 0 \\ 0 & 0 & -400 \end{bmatrix}.$$

9.11. Show that if A and B are both positive-definite, so are A^2, A^{-1}, and $A + B$.

9.12. Let A be a positive-definite real symmetric matrix. Show that $P^T AP$ is also positive-definite for any nonsingular matrix P. Also, if A is a positive-definite Hermitian matrix, then $P^H AP$ is also positive-definite for any nonsingular complex matrix P.

9.13. Which of the following functions b on $\mathbb{R}^2$ are bilinear forms?

(1) $b(\mathbf{x}, \mathbf{y}) = 1$,
(2) $b(\mathbf{x}, \mathbf{y}) = (x_1 - y_1)^2 + x_2 y_2$,
(3) $b(\mathbf{x}, \mathbf{y}) = (x_1 + y_1)^2 - (x_2 - y_2)^2$,
(4) $b(\mathbf{x}, \mathbf{y}) = x_1 y_2 - x_2 y_1$.

9.14. Let b be a bilinear form on $\mathbb{R}^2$ defined by

$$b(\mathbf{x}, \mathbf{y}) = 2x_1 y_1 - 3x_1 y_2 + x_2 y_2.$$

(1) Find the matrix A of b with respect to the basis $\alpha = \{(1, 0), (1, 1)\}$.
(2) Find the matrix B of b with respect to the basis $\beta = \{(2, 1), (1, -1)\}$.

(3) Find the change of basis matrix P from the basis β to the basis α and verify that $B = P^T AP$.

Hint: Use Eq. (9.4) on page 452.

9.15. For a bilinear form on $\mathbb{R}^2$ defined by $b(\mathbf{x}, \mathbf{y}) = x_1y_1 + x_2y_2$, find the matrix representation of b with respect to each of the following bases:

$$\alpha = \{(1,0),(0,1)\}, \quad \beta = \{(1,-1),(1,1)\}, \quad \gamma = \{(1,2),(3,4)\}.$$

9.16. Which one of the following bilinear forms on the space $\mathbb{R}^3$ are symmetric? Which one are skew-symmetric? For each symmetric one, find its matrix representation of the diagonal form, and for each skew-symmetric one, find its matrix representation of the block diagonal form as in Theorem 9.15.

(1) $b(\mathbf{x}, \mathbf{y}) = x_1y_3 + x_3y_1$,
(2) $b(\mathbf{x}, \mathbf{y}) = x_1y_1 + 2x_1y_3 + 2x_3y_1 - x_2y_2$,
(3) $b(\mathbf{x}, \mathbf{y}) = x_1y_2 + 2x_1y_3 - x_2y_3 - x_2y_1 - 2x_3y_1 + x_3y_2$,
(4) $b(\mathbf{x}, \mathbf{y}) = \sum_{i,j=1}^{3}(i-j)x_iy_j$.

9.17. Find the inertia for each of the following symmetric matrices.

$$(1)\ \begin{bmatrix} 0 & 1 & 2 \\ 1 & -1 & 3 \\ 2 & 3 & 0 \end{bmatrix}, \qquad (2)\ \begin{bmatrix} 2 & 3 & 0 \\ 3 & -1 & -2 \\ 0 & -2 & 0 \end{bmatrix}, \qquad (3)\ \begin{bmatrix} 1 & -1 & 1 \\ -1 & 2 & 1 \\ 1 & 1 & 0 \end{bmatrix}.$$

Hint: Compute $\text{In}(A)$ by using the congruence relation.

9.18. Find the inertia for each of the following Hermitian matrices.

$$(1)\ \begin{bmatrix} 0 & 1 & i \\ 1 & 1 & 0 \\ -i & 0 & 2 \end{bmatrix}, \qquad (2)\ \begin{bmatrix} 1 & -3i & 1 \\ 3i & 4 & 2i \\ 1 & -2i & 5 \end{bmatrix}, \qquad (3)\ \begin{bmatrix} 0 & 0 & i \\ 0 & 0 & 0 \\ -i & 0 & 0 \end{bmatrix},$$

which are given in Problem 233. Which are Hermitian congruent to each other?

Hint: Compute $\text{In}(A)$ by using the congruence relation.

9.19. Let $V = M_{2\times 2}(\mathbb{R})$ be the vector space of 2×2 matrices, and let $b : V \times V \to \mathbb{R}$ be defined by $b(A, B) = \text{tr}(A) \cdot \text{tr}(B)$. Is this a bilinear form? If yes, find the matrix representation of b with respect to the basis

$$\alpha = \left\{ E_1 = \begin{bmatrix} 1 & 0 \\ 0 & 0 \end{bmatrix}, E_2 = \begin{bmatrix} 0 & 1 \\ 0 & 0 \end{bmatrix}, E_3 = \begin{bmatrix} 0 & 0 \\ 1 & 0 \end{bmatrix}, E_4 = \begin{bmatrix} 0 & 0 \\ 0 & 1 \end{bmatrix} \right\}.$$

Find its inertia.

9.20. Find the eigenvalues of each of the following symmetric matrices and the maximum value of its associated quadratic form on the unit sphere $||\mathbf{x}|| = 1$.

$$(1)\ \begin{bmatrix} -1 & 1 & 0 \\ 1 & -2 & 1 \\ 0 & 1 & -1 \end{bmatrix}, \qquad (2)\ \begin{bmatrix} -2 & 1 & 0 \\ 1 & -2 & 1 \\ 0 & 1 & -2 \end{bmatrix}, \qquad (3)\ \begin{bmatrix} 3 & -2 & 0 \\ -2 & 3 & 0 \\ 0 & 0 & 5 \end{bmatrix}.$$

9.21. Find the maximum and minimum of each of the following quadratic forms subject to the constraint $x_1^2 + x_2^2 + x_3^2 = 1$ and determine the values of x_1, x_2, and x_3 at which the maximum and minimum occur:

(1) $x_1^2 + x_2^2 + 2x_3^2 - 2x_1x_2 + 4x_1x_3 + 4x_2x_3$,
(2) $2x_1^2 + x_2^2 + x_3^2 + 2x_1x_3 + 2x_1x_2$.

Hint: The extreme values of a constrained quadratic form $\mathbf{x}^T A\mathbf{x}$ subject to $||\mathbf{x}|| = 1$ are the extreme eigenvalues of A.

9.22. Find the maximum and minimum of the quadratic forms listed in Exercise **9.5** subject to the constraint $||\mathbf{x}|| = 1$ and determine the vectors $\mathbf{x}$ at which the maximum and minimum occur.

Hint: The extreme values of a constrained quadratic form $\mathbf{x}^T A\mathbf{x}$ subject to $||\mathbf{x}|| = 1$ are the extreme eigenvalues of A, to which occur at the eigenvectors belong.

9.23. In Theorem 5.28, we have seen that a symmetric matrix A is idempotent if and only if it is the projection matrix of an orthogonal projection. Prove that for any $n \times n$ symmetric or Hermitian matrix A, the following statements are equivalent.

(1) A is idempotent,
(2) A is the projection matrix of an orthogonal projection,
(3) Among the eigenvalues of A, r of them equal 1, and the remaining $n - r$ equal zero, where r is the rank of A.

9.24. (1) Try to find the eigenvalues of each of the following symmetric matrices.
(2) Using the Rayleigh quotient R_A, try to find any lower bound for the maximum eigenvalue and any upper bound for the minimum eigenvalue of each of the following symmetric matrices. What are your best lower/upper bounds?

$$(1)\ A = \begin{bmatrix} 1 & 1 & 0 \\ 1 & 2 & 1 \\ 0 & 1 & 2 \end{bmatrix}, \qquad (2)\ A = \begin{bmatrix} 1 & 0 & 1 \\ 0 & 2 & -1 \\ 1 & -1 & -3 \end{bmatrix}.$$

Hint: Compute $\text{In}(A)$. Use Rayleigh quotient iteration to estimate $\lambda_{\max}$, or the Gershgorin theorem.
(1) The characteristic polynomial of A is $-\lambda^3 + 5\lambda^2 - 6\lambda + 1$. The eigenvalues of A are approximated by (use a calculator for computing λ_i)

$$\lambda_1 \approx 3.24698; \quad \lambda_2 \approx 1.55496; \quad \lambda_3 \approx 0.198062.$$

(2) The eigenvalues of A are approximated by (use a calculator)

$$\lambda_1 \approx -3.41147; \quad \lambda_2 \approx 2.22668; \quad \lambda_3 \approx 1.18479.$$

9.25. Determine whether the following statements are true or false, in general, and justify your answers.

(1) $x^3 + y^3 + z^3 + xy + yz + zx + xyz$ is a homogeneous polynomial in three variables.
(2) A quadratic form $q(\mathbf{x}) = \mathbf{x}^T A\mathbf{x}$ on $\mathbb{R}^n$ is a linear transformation from $\mathbb{R}^n$ to $\mathbb{R}^1$.
(3) For every real square matrix A, $q(\mathbf{x}) = \mathbf{x}^T A\mathbf{x}$ is a quadratic form.
(4) For every real square matrix A, $b(\mathbf{x}, \mathbf{y}) = \mathbf{x}^T A\mathbf{y}$ is a bilinear form.
(5) For any quadratic form q on $\mathbb{R}^n$, there exists a basis α for $\mathbb{R}^n$ with respect to which the matrix representation of q is diagonal.
(6) For any bilinear form b on $\mathbb{R}^n$, there exists a basis α for $\mathbb{R}^n$ with respect to which the matrix representation of b is diagonal.
(7) If A is positive-definite symmetric matrix, then every square submatrix of A has a positive determinant.
(8) A symmetric matrix A is negative-definite if and only if all the principal submatrices A_k's have negative determinants.
(9) The sum of a positive-definite quadratic form and a negative-definite quadratic form is indefinite.
(10) Any two matrix representations of a quadratic form have the same inertia.
(11) The geometric type of the level (hyper)surface of a quadratic equation $\mathbf{x}^T A\mathbf{x} = c, (c \geq 0)$ on $\mathbb{R}^n$ is completely determined by the inertia $\text{In}(A) = (p, q, k)$.
(12) For any real symmetric or skew-symmetric matrix A, the matrix representations of the bilinear form $b(\mathbf{x}, \mathbf{y}) = \mathbf{x}^T A\mathbf{y}$ on the space $\mathbb{R}^n$ with respect to two different bases have the same inertia, that is, $\text{In}([b]_\alpha) = \text{In}([b]_\beta)$ for any two different bases α and β for $\mathbb{R}^n$.
(13) For any real square matrix A and for any $k > 0$, the matrices A and kA may not be similar but are always congruent to each other.
(14) For any nontrivial bilinear form $b \neq \mathbf{0}$ on a vector space V, if $b(\mathbf{v}, \mathbf{v}) = 0$, then $\mathbf{v} = \mathbf{0}$.
(15) Any real symmetric matrix is congruent to at least one diagonal matrix.
(16) Any two congruent matrices have the same eigenvalues.
(17) The sum of two bilinear forms on V is also a bilinear form.
(18) For every bilinear form b on $\mathbb{R}^n$, the function $q(\mathbf{x}) = b(\mathbf{x}, \mathbf{x})$ is a quadratic form on $\mathbb{R}^n$.
(19) For every bilinear form $b : V \times V \to \mathbb{R}$, if $b(\mathbf{v}, \mathbf{w}) = 0$ for all $\mathbf{w} \in V$, then $\mathbf{v} = \mathbf{0}$.
(20) For a symmetric matrix A, the maximum value of the Rayleigh quotient $R_A(\mathbf{x}) = \frac{\langle \mathbf{x}, A\mathbf{x}\rangle}{\langle \mathbf{x}, \mathbf{x}\rangle}$ is the maximum eigenvalue of A.

Selected Answers and Hints

Chapter 1

Problems

1 Figure 9.8 illustrates three possible cases.

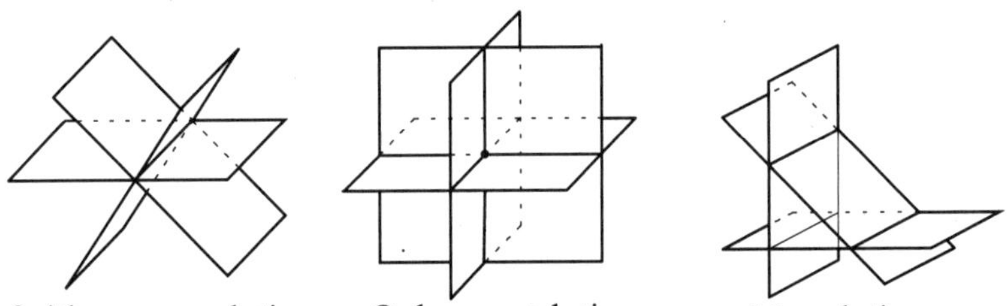

Infinitly many solutions Only one solution No solutions

Figure 9.8: Three planes in $\mathbb{R}^3$

2 (1) $(x_1, x_2, x_3) = (5, -2, 2)$. (2) Inconsistent. (3) $(x_1, x_2, x_3) = (1, 3, 2)$.

4 (1) $(x_1, x_2, x_3, x_4) = (10 - 4t, 20 - 5t,\ 30 - 6t, t)$ for any $t \in \mathbb{R}$.
(2) $(x_1, x_2, x_3, x_4) = (10 - 2t, 30 - t,\ t, 50)$ for any $t \in \mathbb{R}$.

5 (2) $(x, y, z) = (t, -t, t)$, t is any real. Try to write the solutions in matrix notation.

6 (2) $(w, x, y, z) = (2 + s + t, -2s - t, s, t)$, s, t are any real. In matrix notation,

$$\begin{bmatrix} w \\ x \\ y \\ z \end{bmatrix} = \begin{bmatrix} 2 \\ 0 \\ 0 \\ 0 \end{bmatrix} + s \begin{bmatrix} 1 \\ -2 \\ 1 \\ 0 \end{bmatrix} + t \begin{bmatrix} 1 \\ -1 \\ 0 \\ 1 \end{bmatrix},$$

where s, t are any real numbers.

7 (3) For any (i, j), $[A + (-A)]_{ij} = [A]_{ij} + [-A]_{ij} = [A]_{ij} + (-1)[A]_{ij} = 0$.

8 Let $B = \begin{bmatrix} x & y \\ u & v \end{bmatrix}$. Then, the given identity is

$$\begin{bmatrix} 1 & 2 \\ 3 & 4 \end{bmatrix} + \begin{bmatrix} x & u \\ y & v \end{bmatrix} = \begin{bmatrix} 1 & 3 \\ 2 & 4 \end{bmatrix} - \begin{bmatrix} x & u \\ y & v \end{bmatrix}.$$

By comparing the both sides entrywise, one can get B.

9 Check the matrices $A = \begin{bmatrix} 2 & 4 \\ 3 & 6 \end{bmatrix}$, $B = \begin{bmatrix} 2 & 1 \\ 3 & 4 \end{bmatrix}$, $C = \begin{bmatrix} 8 & 7 \\ 0 & 1 \end{bmatrix}$.

10 (1), (2) are clear. For the uniqueness in (3), use the fact that $B_1 - B_2 = C_2 - C_1$ are both symmetric and skew-symmetric, and hence it should be a zero matrix.

11 (3) $I = I^T = (AA^{-1})^T = (A^{-1})^T A^T$ means by definition $(A^T)^{-1} = (A^{-1})^T$.

12 Use Problem 11(3).

14 (1) Compute (i, j)-entries of AB for $i < j$.
(2) A^{-1} can be expressed as a product of lower triangular elementary matrices. Use (1). Or use Eq. (1.3) for computing A^{-1}.

15 Compare this with the elementary row operations. Use $(EA)^T = A^T E^T$.

16 The product of their inverses in reverse order.

17 Examine the case that some d_i is zero.

18 Follow all elementary row operations appeared in the solution of Example 14.

19 Not true in general, but true if $A\mathbf{x} = \mathbf{b}$ is consistent.

22 $AB = \left[\begin{array}{cc|c} 3 & 5 & 12 \\ 0 & 2 & 7 \\ \hline 1 & 8 & 7 \end{array}\right]$.

25 With $L = \begin{bmatrix} 1 & 0 & 0 \\ 1 & 1 & 0 \\ 1 & 2 & 1 \end{bmatrix}$, $U = \begin{bmatrix} 1 & 1 & 1 \\ 0 & 1 & 0 \\ 0 & 0 & -1 \end{bmatrix}$, solve $L\mathbf{y} = \mathbf{b}$ and $U\mathbf{x} = \mathbf{y}$.

Exercises

1.1 (2) $x = 17/2, y = 3, z = -4$. (3) $x_1 = 0, x_2 = 1, x_3 = -1, x_4 = 2$.

1.3 (1) and (2).

1.4 Row-echelon forms are (1), (2), (4), (6). Reduced row-echelon forms are (1), (2), (6). The cases (1), (2), (3), (4), (5) have infinitely many solutions, while (6) has no solutions.

1.5 (1) Reduced one is: $\begin{bmatrix} 1 & 0 & 0 & 11 \\ 0 & 1 & 0 & 4 \\ 0 & 0 & 1 & -17 \end{bmatrix}$. (2) Unique solution: All x_i's are 1.

1.8 $ABC = \begin{bmatrix} 3 & -3 \\ 1 & -1 \end{bmatrix}$, $BCA = \begin{bmatrix} 0 & 2 & -2 \\ 0 & 1 & -1 \\ 0 & -1 & 1 \end{bmatrix}$.

1.9 $(A - B)C^T = \begin{bmatrix} 0 & 1 & 0 \\ 0 & 0 & 2 \\ 0 & 0 & 0 \end{bmatrix}$. $B^T B = \begin{bmatrix} 5 & -2 & 0 \\ -2 & 1 & 0 \\ 0 & 0 & 1 \end{bmatrix}$.

1.10 (2) $\begin{bmatrix} 3 & -22 & 101 \\ 0 & 25 & -60 \\ 0 & 0 & 85 \end{bmatrix}$.

1.11 First try it with the diagonal matrix B with diagonal entries x, y, z to show that A should be diagonal. Next try it with the matrix $B = [b_{ij}]$, where $b_{ij} = 1$ if $j = 1$ and $b_{ij} = 0$ otherwise.

1.12 ($\Leftarrow$) is clear. ($\Rightarrow$): Compare the (i, j)-entries; $[DA]_{ij} = [AD]_{ij}$, giving $\sum_k d_{ik}a_{kj} = \sum_k a_{ik}d_{kj}$. Since $D = [d_{ij}]$ is diagonal, it follows $d_{ii}a_{ij} = a_{ij}d_{jj}$. But, $d_{ii} \neq d_{jj}$ if $i \neq j$. Hence, $a_{ij} = 0$ if $i \neq j$.

1.13 $A^k = \begin{bmatrix} 1 & 2k & 3k(k-1) \\ 0 & 1 & 3k \\ 0 & 0 & 1 \end{bmatrix}$.

1.14 Use: If $A = [a_{ij}], B = [b_{ij}]$ are symmetric, then $[AB]_{ij} = \sum_k a_{ik}b_{kj} = \sum_k b_{jk}a_{ki} = [BA]_{ji}$.

1.16 $AA^T = \begin{bmatrix} 74 & 51 & -31 \\ 51 & 146 & 52 \\ -31 & 52 & 95 \end{bmatrix}$ and $A^TA = \begin{bmatrix} 1 & 3 & 9 & 2 \\ 3 & 90 & 36 & 78 \\ 9 & 36 & 107 & -9 \\ 2 & 78 & -9 & 117 \end{bmatrix}$.

1.18 (1) $A^{-1}AB = B$. (2) $C = A^{-1}AC = A - A^T$.

1.19 $a = 0,\ c^{-1} = b \neq 0$.

1.20 $A^{-1} = A = A^T$. $B^{-1} = B = B^T$.

1.21 $B^{-1} = \dfrac{1}{4}\begin{bmatrix} -2 & 1 & 2 \\ 0 & 2 & 0 \\ 6 & -3 & -2 \end{bmatrix}$, $C^{-1} = \begin{bmatrix} 1 & -1 & 0 & 0 \\ 0 & 1/2 & -1/2 & 0 \\ 0 & 0 & 1/3 & -1/3 \\ 0 & 0 & 0 & 1/4 \end{bmatrix}$.

1.22 $A^{-1} = \dfrac{1}{6}\begin{bmatrix} 18 & -9 & 2 \\ -15 & 12 & -3 \\ 3 & -3 & 1 \end{bmatrix}$. $C^{-1} = \begin{bmatrix} 1/k & -1/k^2 & 1/k^3 & -1/k^4 \\ 0 & 1/k & -1/k^2 & 1/k^3 \\ 0 & 0 & 1/k & -1/k^2 \\ 0 & 0 & 0 & 1/k \end{bmatrix}$.

1.23 AB is 2×2 matrix, and the first row is a scalar multiple of the second row.

1.24 (1) $A^{-1} = \begin{bmatrix} 1/3 & 1/6 & 1/6 \\ -4/3 & -5/3 & 4/3 \\ -1/3 & -2/3 & 1/3 \end{bmatrix}$. (2) $A^{-1} = \dfrac{1}{2}\begin{bmatrix} 1 & 1 & 0 \\ 3 & -1 & -2 \\ 4 & -2 & -2 \end{bmatrix}$.

1.25 (1) $LU = \begin{bmatrix} 1 & 0 & 0 \\ 3 & 1 & 0 \\ 2 & 0 & 1 \end{bmatrix}\begin{bmatrix} 1 & 2 & 3 \\ 0 & -3 & -5 \\ 0 & 0 & -6 \end{bmatrix}$,

(2) $LU = \begin{bmatrix} 1 & 0 & 0 & 0 \\ 3 & 1 & 0 & 0 \\ 1 & 0 & 1 & 0 \\ 2 & 0 & 3 & 1 \end{bmatrix}\begin{bmatrix} 1 & 2 & 3 & 0 \\ 0 & -3 & -7 & 4 \\ 0 & 0 & -2 & 2 \\ 0 & 0 & 0 & -4 \end{bmatrix}$.

1.26 Use the LU decompositions in Exercise **1.25** to find $(LU)^{-1} = U^{-1}L^{-1}$.

1.27 (1) $LDL^T = \begin{bmatrix} 1 & 0 \\ b/a & 1 \end{bmatrix}\begin{bmatrix} a & 0 \\ 0 & d - b^2/a \end{bmatrix}\begin{bmatrix} 1 & b/a \\ 0 & 1 \end{bmatrix}$.

(2) $LDL^T = \begin{bmatrix} 1 & 0 & 0 \\ 2 & 1 & 0 \\ 3 & 1 & 1 \end{bmatrix} \begin{bmatrix} 1 & 0 & 0 \\ 0 & -3 & 0 \\ 0 & 0 & -6 \end{bmatrix} \begin{bmatrix} 1 & 2 & 3 \\ 0 & 1 & 1 \\ 0 & 0 & 1 \end{bmatrix}$.

1.28 $\mathbf{y} = [2\ -1\ 3]^T$, $\mathbf{x} = [4\ 2\ 3]^T$.

1.29 (2) $A = LU = \begin{bmatrix} 1 & 0 & 0 \\ 1 & 1 & 0 \\ 1 & 1 & 1 \end{bmatrix} \begin{bmatrix} 1 & 1 & 1 \\ 0 & 3 & 4 \\ 0 & 0 & 2 \end{bmatrix}$.

(4) $A^{-1} = U^{-1}L^{-1} = \frac{1}{6} \begin{bmatrix} 8 & -3 & 1 \\ -2 & 6 & -4 \\ 0 & -3 & 3 \end{bmatrix}$.

1.31 (1) $(A^k)^{-1} = (A^{-1})^k$. (2) $A^{n-1} = \mathbf{0}$ if $A \in M_{n\times n}$ is triangular with zero diagonal entries. (3) comes from (2). (4) $(I-A)(I+A+\cdots+A^{k-1}) = I - A^k$.

1.32 (1) $A = \begin{bmatrix} 1 & 1 \\ 0 & 0 \end{bmatrix}$. (3) $A = A^{-1}A^2 = A^{-1}A = I$.

1.33 $1, 3, 5, 7, 10, 11, 13, 14, 15, 18, 19$ are true.

(3) See Theorem 1.22. (4) One-sided identity? (5) $P^{-1} = P^T$.

(6) Consider $A = \begin{bmatrix} 1 & 0 \\ 0 & 0 \end{bmatrix}$ or $A = \begin{bmatrix} 0 & 1 \\ 0 & 1 \end{bmatrix}$.

(7) $A^2 = I \Rightarrow A = A^{-1}$. (8) If $A = -B$? (10) $AB = (AB)^T = B^TA^T = BA$.

(12) $\begin{bmatrix} 0 & 0 & 1 \\ 0 & 0 & 0 \\ -1 & 0 & 0 \end{bmatrix} \begin{bmatrix} 0 & 0 & 0 \\ 0 & 0 & 1 \\ 0 & -1 & 0 \end{bmatrix} = \begin{bmatrix} 0 & -1 & 0 \\ 0 & 0 & 0 \\ 0 & 0 & 0 \end{bmatrix}$.

(13) $(A^{-1})^T = (A^T)^{-1} = A^{-1}$. (14) If A^{-1} exists, $A^{-1} = A^{-1}(AB^T) = B^T$.

(15) If AB has the (right) inverse C, then $A^{-1} = BC$.

(16) Consider $E_1 = \begin{bmatrix} 0 & 1 \\ 1 & 0 \end{bmatrix}$ and $E_2 = \begin{bmatrix} 1 & 1 \\ 1 & 0 \end{bmatrix}$.

(17) Consider the identity $A^{-1} = E_k \cdots E_2E_1 = E_k \cdots E_2E_1I_n$.

(20) Consider a permutation matrix $\begin{bmatrix} 0 & 1 \\ 1 & 0 \end{bmatrix}$.

Chapter 2

Problems

29 (2) Note: 2nd column − 1st column = 3rd column − 2nd column.

31 (1) $\det A = 2x - 16$. (2) B is skew-symmetric and $\det B = \det B^T = \det(-B) = (-1)^3 \det B$, implying $\det B = 0$. (3) $\det C = 0$.

33 (1) -12. (3) $x^3 - 11x^2 + 52x - 84$.

37 $\det(\cdot) = 34$.

38 $\det B = 180$. $\det C = 12$.

39 Use $A \cdot \text{adj}\, A = (\det A)I$ or Eq. (2.7).

40 (1) $x = 3, y = -1, z = 2$. (2) $x = 40,\ y = 10,\ z = -20$.
41 det can be altered by the sign.
42 $p(x) = -5 + 3x - 4x^2 + 2x^3$.
43 $\det(A^T A) = 0$ but $\det(AA^T) = 251$.

Exercises

2.2 $\det B = -54x - 70$.
2.4 $x = -1$ or 4.
2.6 $\det A = 56, \quad \det B = -327$.
2.7 (1) $\det A = (-1)^{n-1}(n-1)$. (2) 0 if $n \geq 3$.
2.8 (2) $(a_{11} - x)((a_{22} - x)(x - a_{33}) + a_{23}a_{32}) - a_{12}(a_{21}(x - a_{33}) + a_{23}a_{31})$, a polynomial in x of degree 3, (3) x^4.
2.10 $\det A = a_{11}a_{22}a_{33} + a_{12}a_{23}a_{31} + a_{13}a_{21}a_{32} - a_{11}a_{23}a_{32} - a_{12}a_{21}a_{33} - a_{13}a_{22}a_{31}$.

$\det A_{\pi/2} = a_{31}a_{22}a_{13} + a_{21}a_{12}a_{33} + a_{11}a_{32}a_{23} - a_{31}a_{12}a_{23} - a_{21}a_{32}a_{13} - a_{11}a_{22}a_{33} = -\det A$.

$\det A_{\pi} = a_{33}a_{22}a_{11} + a_{32}a_{21}a_{13} + a_{31}a_{23}a_{12} - a_{33}a_{21}a_{12} - a_{32}a_{23}a_{11} - a_{31}a_{22}a_{13} = \det A$.

Anti-diagonal reflection: $\det \begin{bmatrix} a_{33} & a_{23} & a_{13} \\ a_{32} & a_{22} & a_{12} \\ a_{31} & a_{21} & a_{11} \end{bmatrix} = a_{33}a_{22}a_{11} + a_{23}a_{12}a_{31} + a_{13}a_{32}a_{21} - a_{33}a_{12}a_{21} - a_{23}a_{32}a_{11} - a_{13}a_{22}a_{31} = \det A$.

2.11 Use the LU decompositions in Exercise **1.25** and $\det(LU) = \det L \det U = \det U$.
2.13 $\det(ABC) = 0$ implies $\det A = 0$, $\det B = 0$, or $\det C = 0$. But, $\det C = 0$ has no solutions.
2.14 Use $\det A = \det A^T = \det(-A) = (-1)^n \det A$. The matrix in Exercise **2.8**(3) gives an example of 4×4 skew-symmetric matrices with $\det A \neq 0$.
2.15 (3) $A_{11} = -2, \quad A_{12} = 7, \quad A_{13} = -8, \quad A_{33} = 3$.
2.17 The solution of the system $I_n\mathbf{x} = \mathbf{x}$ is $x_i = \frac{\det C_i}{\det I_n} = \det A$. Or, use the cofactor expansion of $\det A$ along the i-th row.
2.18 $k = 0$ or ± 1.
2.19 $\mathbf{x} = (-5, 1, 2, 3)^T$.
2.20 (2) $\text{adj } A = \begin{bmatrix} 2 & -7 & -6 \\ 1 & -7 & -3 \\ -4 & 7 & 5 \end{bmatrix}$, $\det A = -7$, $\det(\text{adj } A) = 49$, $A^{-1} = -\frac{1}{7}\text{adj } A$.

(3) $\text{adj } A = \begin{bmatrix} 1 & 1 & -1 \\ -10 & 4 & 2 \\ 7 & -3 & -1 \end{bmatrix}$, $\det A = 2$, $\det(\text{adj } A) = 4$, $A^{-1} = \frac{1}{2}\text{adj } A$.

2.21 Suppose that A is symmetric. Then the minors satisfy $M_{ij}^T = M_{ji}$ and so $\det M_{ij} = \det M_{ji}$. Therefore, the cofactors satisfy $A_{ij} = A_{ji}$, *i.e.*, adj A is symmetric. If A is invertible, it can be easily shown by taking the transpose of both sides of $A \cdot \text{adj } A = (\det A)I$.

2.23 If A and B are invertible matrices, then $(AB)^{-1} = B^{-1}A^{-1}$. Since for any invertible matrix A we have adj $A = (\det A)A^{-1}$, (1) and (2) are obvious. To show (3), let $AB = BA$ and A be invertible. Then $A^{-1}B = A^{-1}(BA)A^{-1} = A^{-1}(AB)A^{-1} = BA^{-1}$, which gives (3).

2.26 $\det B = (w-x)(w-y)(w-z)(z-x)(z-y)(y-x)(x+y+z+w)$, as the coefficient of $-a^3$ in the determinant of the Vandermonde matrix

$$\begin{bmatrix} 1 & 1 & 1 & 1 & 1 \\ x & y & z & w & a \\ x^2 & y^2 & z^2 & w^2 & a^2 \\ x^3 & y^3 & z^3 & w^3 & a^3 \\ x^4 & y^4 & z^4 & w^4 & a^4 \end{bmatrix}.$$

2.27 $\det A = -x^{12} + 3x^8 - 3x^4 + 1 = (1-x^4)^3$.

2.28 $\det A = b^2c^2 - 3bcx^2 + x^4$. $\det B = wxyz + wx + wz + xy + 1$.

2.29 $\det A = b^2c^2 - 3bcx^2 + x^4$. $\det B = wxyz + wx + wz + xy + 1$. These are the same as the determinants in Exercise **2.28**.

2.30 (1) 1, (2) 24.

2.31 The area is equal to $\sqrt{\det(AA^T)} = \sqrt{74}$.

2.33 $3, 5, 9, 11, 12, 15, 16$ are true.

(2) If $A = B = I_2$, both are the identity matrix ?

(3) $\det(AB) = \det A \det B$. (4) See Problem 43. (6) If $A = \begin{bmatrix} 1 & 1 \\ 0 & 0 \end{bmatrix}$?

(8) If $c = 0$? (9) $(cI_n - A)^T = cI_n - A^T$. (10) Consider $E = \begin{bmatrix} 3 & 0 \\ 0 & 1 \end{bmatrix}$.

(11) and (12) Compare their determinants. (13) $\det A = 1$ or 0.

(15) $\det(\mathbf{u}\mathbf{v}^T) = \det(\mathbf{u}[v_1 \cdots v_n]) = \det[v_1\mathbf{u} \cdots v_n\mathbf{u}] = 0$. See Exercise **1.23**.

(16) Use $A^{-1} = \frac{1}{\det A}\,\mathrm{adj}A$. (17) Consider $\begin{bmatrix} 1 & 0 & 1 \\ 1 & 1 & 0 \\ 0 & 1 & 1 \end{bmatrix}$. (18) If $\det A = 0$?

(19) $\det A^T = \det A$ for any square matrix A. (20) $\det(-A) = -\det A$?

Chapter 3

Problems

47 Hint: Which pair can be the zero vector? Also, check the commutativity of the sum.

49 (2), (3).

50 (1), (3) are subspaces.

51 (3) For any $\mathbf{u} \in U$ and any $\mathbf{w} \in W$, both $\mathbf{u}$ and $\mathbf{w}$ are in Z. Hence, $\mathbf{u} + \mathbf{w} \in Z$, implying that $U + W$ as a subspace of Z.

53 No.

56 ($\Rightarrow$) is clear. For ($\Leftarrow$), try to prove by yourself. Or see the proof of Theorem 3.15.

57 Every vector $\mathbf{v}$ in W is of the form $a_1\mathbf{x}_1 + a_2\mathbf{x}_2 + \cdots + a_m\mathbf{x}_m$, which also belongs to U.

61 $(1, 4, 4, -3)$ for $\mathbf{u}$ and $(1, 5, -1, -5)$ for $\mathbf{v}$.

65 Let A be the 3×5 matrix whose row vectors are $\mathbf{x}_1, \mathbf{x}_2, \mathbf{x}_3$. Use its reduced row-echelon form $U = \begin{bmatrix} 1 & 0 & 0 & 2 & 10/3 \\ 0 & 1 & 0 & 2 & -28/3 \\ 0 & 0 & 1 & 0 & 5/3 \end{bmatrix}$.

66 (1) $\{(-5, 3, 1)\}$. (2) dim 3.

67 Use the reduced row-echelon form $U = \begin{bmatrix} 1 & 0 & 3 & 2 & 1 \\ 0 & 1 & -3/2 & -1 & 0 \\ 0 & 0 & 0 & 0 & 1 \end{bmatrix}$.

69 (1) rank = 3, nullity = 1. (2) rank = 3, nullity = 2.

70 By (2) of Theorem 3.25 and Corollary 3.23.1, a matrix A of rank r must have an invertible submatrix C of rank r. By (1) of the same theorem, the rank of C should be the largest.

71 (1) rank $A = 3$. An invertible submatrix is $\begin{bmatrix} 1 & 4 & 9 \\ 1 & 5 & 0 \\ 1 & 7 & 7 \end{bmatrix}$.

73 (1) rank $A = 3$, a wide matrix, so A has only a right inverse. (2) rank $A = 2$, a long matrix, so A has only a left inverse. (3) A is a square matrix of full rank.

74 Challenge yourself.

Exercises

3.1 (1) (Commutativity in sum): $\mathbf{x}\dot{+}\mathbf{y} = \mathbf{y}\dot{+}\mathbf{x}$? Is $0 \cdot (1, 1) = (0, 1)$ acceptable?

3.2 (8) $1 \cdot \mathbf{x} = \mathbf{x}$?

3.3 (2), (3) are vector spaces. Both are subspaces of the vector space of all continuous functions defined on the real line $\mathbb{R}$. For (1), if $f(0) = 1$, then $2f(0) = 2$.

3.4 (1) and (4) are subspaces.

3.5 Note that $\mathrm{tr}(AB - BA) = 0$.

3.6 $p(x) = -p_1(x) + 3p_2(x) - 2p_3(x)$.

3.7 Linearly dependent. Their sum is the zero vector.

3.8 Linearly independent.

3.9 (1) $AB\mathbf{x} = \mathbf{0} \Rightarrow B\mathbf{x} = \mathbf{0} \Rightarrow B\mathbf{x} = \mathbf{0}$.
(2) Use (1) with $(AB)^T = B^T A^T$.
(3) If the column vectors of B are linearly dependent, $B\mathbf{x} = \mathbf{0}$ has a nontrivial solution $\mathbf{x}$. For such $\mathbf{x} \neq \mathbf{0}$, $AB\mathbf{x} = \mathbf{0}$. So the column vectors of AB are linearly dependent.
(4) Use (3) with $(AB)^T = B^T A^T$.

3.10 5!=120. Since $\dim M_{4\times4}(\mathbb{R}) = 4 \times 4$ (see Exercise 52(3)), they are linearly dependent. All matrices of the form $\sum_i c_i P_i$ as linear combinations of the 4×4 permutation matrices have the same row sums and also the same column sums. So, they cannot span the space $M_{4\times4}(\mathbb{R})$.

3.11 Two vectors $(2, 1, 0)$, $(3, 0, 1)$ are in W and they form a basis.

3.12 If $\mathbf{b}_1, \ldots, \mathbf{b}_n$ denote the column vectors of B, then $AB = [A\mathbf{b}_1 \ \cdots \ A\mathbf{b}_n]$.

3.13 If $W_i \cap W_j \neq \{\mathbf{0}\}$, then $\mathbf{v}_i, \mathbf{v}_j$ cannot be linearly independent.

3.14 Let $\dim(V \cap W) = m$ and take a basis $\omega = \{\mathbf{u}_1, \mathbf{u}_2, \ldots, \mathbf{u}_m\}$ for $V \cap W$. Extend ω to a basis for V and for W, respectively; say $\alpha = \{\mathbf{u}_1, \ldots, \mathbf{u}_m, \mathbf{v}_{m+1}, \ldots, \mathbf{v}_k\}$ for V and $\beta = \{\mathbf{u}_1, \ldots, \mathbf{u}_m, \mathbf{w}_{m+1}, \ldots, \mathbf{w}_\ell\}$ for W. By showing that the set $\{\mathbf{v}_{m+1}, \ldots, \mathbf{v}_k, \mathbf{w}_{m+1}, \ldots, \mathbf{w}_\ell\}$ of extended vectors is linearly independent, one can show that $\alpha \cup \beta = \{\mathbf{u}_1, \ldots, \mathbf{u}_m, \mathbf{v}_{m+1}, \ldots, \mathbf{v}_k, \mathbf{w}_{m+1}, \ldots, \mathbf{w}_\ell\}$ is a basis for $V + W$.

Consequently, $k + \ell = \dim(V + W) + \dim(V \cap W)$.

3.15 A basis for V is $\{(1,0,0,0), (0,-1,1,0), (0,-1,0,1)\}$, so $\dim V = 3$. $\{(-1,1,0,0), (0,0,2,1)\}$ is a basis for W and $\{(3,-3,2,1)\}$ is a basis for $V \cap W$. Thus, $\dim(V + W) = 4$ by Exercise **3.14**, which means $V + W = \mathbb{R}^4$ and any basis for $\mathbb{R}^4$ works for $V + W$.

3.16 (1) dim$= n - 1$, (2) dim$= n(n + 1)/2$, (3) dim$= n(n - 1)/2$.

3.18 $\mathcal{R}(B) = \langle (1,2,0,0), (0,0,1,0), (0,0,0,1) \rangle$, $\mathcal{N}(B) = \langle (-2,1,0,0) \rangle$,
$\mathcal{C}(B) = \langle (1,2,1), (1,3,1)(0,0,1) \rangle$.
$\mathcal{R}(C) = \langle (0,-2,2,0), (1,3,6,8), (1,6,4,4) \rangle$,
$\mathcal{C}(C) = \langle (0,1,1), (-2,3,6), (2,6,4) \rangle$,
$\mathcal{N}(C) = \langle (-44,4,4,1) \rangle$.

3.19 rank $= 2$ when $x = 12$, rank $= 3$ when $x \neq 12$.

3.20 rank $A = 2$ and any 2×2 submatrix. rank $B = 3$ and the left 3×3 submatrix. rank $C = 3$ and the left 3×3 submatrix.

3.21 (1) rank $A = 5$; rank $A^2 = 4$; rank $A^3 = 3$; rank $A^4 = 2$; rank $A^5 = 1$; rank $A^n = 0$ for $n \geq 6$.

3.22 Let A be $m \times n$ and B be $n \times \ell$ matrices. If the rows of B are linearly independent, $\dim \mathcal{C}(B) = \text{rank } B = n$, implying that any vector $\mathbf{y} \in \mathbb{R}^n$ is a linear combination of columns of B. Hence, $\mathcal{C}(AB) = \{AB\mathbf{x} : \mathbf{x} \in \mathbb{R}^\ell\} = \{A\mathbf{y} : \mathbf{y} \in \mathbb{R}^n\} = \mathcal{C}(A)$. See the proof of Theorem 3.24(3).

3.23 See Exercises **1.23** and **2.33**. Each column vector of $\mathbf{u}\mathbf{v}^T$ is of the form $v_i\mathbf{u}$, that is, $\mathbf{u}$ spans the column space. Or, use Exercise **3.22**. Conversely, if A is of rank one, then the column space is spanned by a one column of A, say the first column $\mathbf{u}$ of A. Then, all of the remaining columns except $\mathbf{u}$ are of the form $v_i\mathbf{u}$, $i = 2, \ldots, n$. Take $\mathbf{v} = [1 \ v_2 \ \cdots \ v_n]^T$. Then one can easily see that $A = \mathbf{u}\mathbf{v}^T$.

3.25 (1) $\begin{bmatrix} 1 & 0 & 0 \\ 0 & 1/2 & 0 \\ 0 & 0 & 1/3 \\ 0 & 0 & 0 \end{bmatrix}$ is a right inverse. (2) $\begin{bmatrix} 1 & 0 & 0 & 0 \\ 0 & 1/2 & 0 & 0 \\ 0 & 0 & 1/3 & 0 \end{bmatrix}$ is a left inverse.

(3) $\dfrac{1}{4}\begin{bmatrix} 2 & 2 & -2 \\ 1 & -1 & 1 \\ -2 & 2 & 2 \end{bmatrix}$ is the two-sided inverse.

3.26 (1) $\det(A^TA) = 0$, $\det(AA^T) = 144$. (2) $\det(AA^T) = 0$. (3) $\det(A^TA) = \det(AA^T) = 16$.

3.27 $1, 5, 8, 11, 12, 15, 16, 17, 20$ are true.

(3) $\mathbf{0}^T = \mathbf{0}^{-1}$?. The identity I_3 satisfies $A^T = A^{-1}$ but $3I_3$ does not.

(7) Even for $U = W$, $\alpha \cap \beta$ can be an empty set.

(8) How can you find a basis for $\mathcal{C}(A)$. See Example 57.

(9) Consider $A = \begin{bmatrix} 0 & 0 \\ 1 & 0 \end{bmatrix}$. (10) Consider $A = \begin{bmatrix} 1 & 0 \\ 0 & 0 \end{bmatrix}$ and $U = \begin{bmatrix} 0 & 0 \\ 1 & 0 \end{bmatrix}$.

(11) See Theorem 3.27. (12) See Theorem 3.28. (13) If $\mathbf{x} = -\mathbf{y}$,

(14) In $\mathbb{R}^2$, Consider $U = \mathbb{R}^2 \times \{\mathbf{0}\}$ and $V = \{\mathbf{0}\} \times \mathbb{R}^2$. (15) See Theorem 3.29.

(17) $\dim \mathcal{C}(A^T) = \dim \mathcal{R}(A) = \dim \mathcal{C}(A)$. Recall the Rank-Nullity Theorem 3.23.

(18) Recall the Rank-Nullity Theorem 3.23 again. (20) See Theorem 3.26.

Chapter 4

Problems

75 In the plane $\mathbb{R}^2$, $\begin{bmatrix} 0 & 1 \\ 1 & 0 \end{bmatrix} \begin{bmatrix} x \\ y \end{bmatrix} = \begin{bmatrix} y \\ x \end{bmatrix}$.

77 W is the kernel of tr, so is a subspace by Theorem 4.4. Let E_{ij} be the $n \times n$ matrix whose (i, j)-th entry is 1 and all the others are zero. Let F_k be the matrix whose (k, k)-th entry is 1, (n, n)-th entry is -1 and all the others are zero. Then the set $\{E_{ij}, F_k : 1 \le i \ne j \le n,\ k = 1, \ldots, n-1\}$ is a basis for W. Thus $\dim W = n^2 - 1$.

78 $\operatorname{tr}(AB) = \sum_{i=1}^m \sum_{k=1}^n a_{ik} b_{ki} = \sum_{k=1}^n \sum_{i=1}^m b_{ki} a_{ik} = \operatorname{tr}(BA)$.

79 If yes, $(1, 0) = T(2, 4, 6) = 2T(1, 2, 3) = (0, 2)$. Contradiction.

80 Let $a_1\mathbf{v}_1 + a_2\mathbf{v}_2 + \cdots + a_k\mathbf{v}_k = \mathbf{0}$. Then $\mathbf{0} = T(a_1\mathbf{v}_1 + a_2\mathbf{v}_2 + \cdots + a_k\mathbf{v}_k) = a_1\mathbf{w}_1 + a_2\mathbf{w}_2 + \cdots + a_k\mathbf{w}_k$ implies $a_i = 0$ for $i = 1, 2, \ldots, k$.

81 (1) If $T(\mathbf{x}) = T(\mathbf{y})$, then $S \circ T(\mathbf{x}) = S \circ T(\mathbf{y})$ implies $\mathbf{x} = \mathbf{y}$. (4) If A and B are invertible, so is AB.

82 (1) $T(\mathbf{x}) = T(\mathbf{y})$ if and only if $T(\mathbf{x} - \mathbf{y}) = \mathbf{0}$, *i.e.*, $\mathbf{x} - \mathbf{y} \in \operatorname{Ker}(T)$.

(2) Let $\{\mathbf{v}_1, \ldots, \mathbf{v}_n\}$ be a basis for V. If T is injective, then the set $\{T(\mathbf{v}_1), \ldots, T(\mathbf{v}_n)\}$ is linearly independent as the proof of Theorem 4.9 shows. Corollary 3.19.1 shows that it is another basis for V. Thus, for any $\mathbf{y} \in V$, we can write $\mathbf{y}$ as $\mathbf{y} = \sum_{i=1}^n a_i T(\mathbf{v}_i) = T(\sum_{i=1}^n a_i \mathbf{v}_i)$. So T is surjective.

Conversely, if T is surjective, then for each $i = 1, 2, \ldots, n$ there exists $\mathbf{x}_i \in V$ such that $T(\mathbf{x}_i) = \mathbf{v}_i$. Then the set $\{\mathbf{x}_1, \mathbf{x}_2, \ldots, \mathbf{x}_n\}$ is linearly independent in V, (see Problem 80). Thus it is a basis by using Corollary 3.19.1 again. If $T(\mathbf{x}) = \mathbf{0}$ for $\mathbf{x} = \sum_{i=1}^n a_i \mathbf{x}_i \in V$, then $\mathbf{0} = T(\mathbf{x}) = \sum_{i=1}^n a_i T(\mathbf{x}_i) = \sum_{i=1}^n a_i \mathbf{v}_i$ implies $a_i = 0$ for all $i = 1, 2, \ldots, n$, that is $\mathbf{x} = \mathbf{0}$. Thus Ker $(T) = \{\mathbf{0}\}$.

83 (1) (5, 2, 3). (2) (2, 3, 0).

84 $[T]_\alpha^\beta = \begin{bmatrix} 2 & 2 & 0 & 2 \\ 3 & 0 & -3 & 3 \\ 4 & 5 & 6 & 7 \end{bmatrix}$.

85 (1) $[T]_\alpha = \begin{bmatrix} 2 & 3 & 4 \\ 5 & 6 & 7 \\ -8 & -9 & 0 \end{bmatrix}$, $[T]_\beta = \begin{bmatrix} 0 & -9 & -8 \\ 7 & 6 & 5 \\ 4 & 3 & 2 \end{bmatrix}$, $[T]_\alpha^\beta = \begin{bmatrix} 4 & 3 & 2 \\ 7 & 6 & 5 \\ 0 & -9 & -8 \end{bmatrix}$.

86 Use the rotation $R_{\frac{\pi}{3}}$ and reflection $T = \begin{bmatrix} 1 & 0 \\ 0 & -1 \end{bmatrix}$ across the x-axis. Then, the matrix is $R_{\frac{\pi}{3}} \circ T \circ R_{-\frac{\pi}{3}}$.

87 Clearly, $[id]_\beta^\alpha = [[\mathbf{x}_1]_\alpha \ \cdots \ [\mathbf{x}_n]_\alpha] = [\mathbf{x}_1 \ \mathbf{x}_2 \ \cdots \ \mathbf{x}_n]$ and $[id]_\alpha^\beta = ([id]_\beta^\alpha)^{-1}$. See Eq (4.1)

89 $[id]_\gamma^\beta = \dfrac{1}{2}\begin{bmatrix} 0 & -1 & 5 \\ 4 & 3 & -1 \\ 2 & 1 & 1 \end{bmatrix}$, $[id]_\beta^\gamma = \begin{bmatrix} -2 & -3 & 7 \\ 3 & 5 & -10 \\ 1 & 1 & -2 \end{bmatrix}$.

90 $[T]_\alpha = \begin{bmatrix} 1 & 2 & 1 \\ 0 & -1 & 0 \\ 1 & 0 & 4 \end{bmatrix}$, $[T]_\beta = \begin{bmatrix} 1 & 4 & 5 \\ -1 & -2 & -6 \\ 1 & 1 & 5 \end{bmatrix}$.

91 $(Q^{-1}AQ)^n = Q^{-1}A^nQ$ for all $n \geq 2$.

92 $[S+T]_\alpha = \begin{bmatrix} 3 & 0 & 0 \\ 2 & 2 & 3 \\ 2 & 3 & 3 \end{bmatrix}$, $\quad [T \circ S]_\alpha = \begin{bmatrix} 3 & 2 & 0 \\ 3 & 3 & 3 \\ 6 & 5 & 3 \end{bmatrix}$.

93 By BA, get a tilting along the x-axis, while by B^TA, get a tilting along the y-axis.

94 $R_{(\mathbf{u}, \frac{\pi}{4})}$ is

$$\begin{bmatrix} \cos\frac{\pi}{4} & -\sin\frac{\pi}{4} & 0 \\ \sin\frac{\pi}{4} & \cos\frac{\pi}{4} & 0 \\ 0 & 0 & 1 \end{bmatrix} \begin{bmatrix} \cos\frac{\pi}{4} & 0 & -\sin\frac{\pi}{4} \\ 0 & 1 & 0 \\ \sin\frac{\pi}{4} & 0 & \cos\frac{\pi}{4} \end{bmatrix} \begin{bmatrix} 1 & 0 & 0 \\ 0 & \cos\frac{\pi}{4} & -\sin\frac{\pi}{4} \\ 0 & \sin\frac{\pi}{4} & \cos\frac{\pi}{4} \end{bmatrix}$$

$$\times \begin{bmatrix} \cos(-\frac{\pi}{4}) & 0 & -\sin(-\frac{\pi}{4}) \\ 0 & 1 & 0 \\ \sin(-\frac{\pi}{4}) & 0 & \cos(-\frac{\pi}{4}) \end{bmatrix} \begin{bmatrix} \cos(-\frac{\pi}{4}) & -\sin(-\frac{\pi}{4}) & 0 \\ \sin(-\frac{\pi}{4}) & \cos(-\frac{\pi}{4}) & 0 \\ 0 & 0 & 1 \end{bmatrix}$$

$$= \frac{1}{2^4}\begin{bmatrix} 1 & -1 & 0 \\ 1 & 1 & 0 \\ 0 & 0 & 1 \end{bmatrix} \begin{bmatrix} 1 & 0 & -1 \\ 0 & 1 & 0 \\ 1 & 0 & 1 \end{bmatrix} \begin{bmatrix} 1 & 0 & 0 \\ 0 & 0 & -1 \\ 0 & 1 & 0 \end{bmatrix} \begin{bmatrix} 1 & 0 & 1 \\ 0 & 1 & 0 \\ -1 & 0 & 1 \end{bmatrix} \begin{bmatrix} 1 & 1 & 0 \\ -1 & 1 & 0 \\ 0 & 0 & 1 \end{bmatrix}.$$

Exercises

4.1 Injective: (1),(3),(4). Surjective: (1),(3).

4.2 (2) ($\Leftarrow$) is easy. ($\Rightarrow$): For each $y \in Y$, there exists an element $x \in X$ such that $f(x) = y$. Define $g(y) = x$ by choosing an x. Then $f \circ g = id_Y$.

4.3 For (1) and (4), what are $T(0,0)$? (2), (5), and (6) are linear transformations.

4.4 $T(a,b,c) = cx^3 + bx^2 + (a+c)x + (a+c)$; not invertible.

4.5 $T(ax^2+bx+c) = aT(x^2) + bT(x) + cT(1) = ax^3 + bx^2 + (a+c)x$.

4.6 (3) $T \circ S(x,y,z) = T(2x,y) = (2x, 3y+2x, 4x-4y, y)$.

4.7 (1) $\text{Ker}(T) = \{(2x, x, -2x) \in \mathbb{R}^3 : x \in \mathbb{R}\}$. (3) Not onto.

4.8 (1) U and W are clearly subspaces. (2) Assume $U = \{\mathbf{0}\}$. Then for all $\mathbf{v} \neq \mathbf{0}$, $T(\mathbf{v}) \neq \mathbf{v}$ and $T(T(\mathbf{v}) - \mathbf{v}) = \mathbf{v} - T(\mathbf{v})$. Hence, $T(\mathbf{v}) - \mathbf{v}$ is a nonzero vector in W. (3) is clear. (4) Use the decomposition of $\mathbf{v} = \frac{\mathbf{v}+T(\mathbf{v})}{2} + \frac{\mathbf{v}-T(\mathbf{v})}{2}$.

4.9 Show $\text{Ker}(T) = \{\mathbf{0}\}$ for each case. (3) $T^{-1}(r, s, t) = (\frac{1}{2}r,\ 2r - s, 7r - 3s - t)$.

4.10 Enough to show $(S \circ T)^{-1} = T^{-1} \circ S^{-1}$.

4.11 T cannot be one-to-one and S cannot be onto.

4.12 $T + T^2 = T \circ (id + T) = id$, so $id + T$ is the inverse of T.

4.13 (3) (5, 2, 0).

4.14 Use $[\mathbf{x}]_\beta = [id]_\alpha^\beta [\mathbf{x}]_\alpha$ for the standard ordered basis α for $\mathbb{R}^3$.

4.15 (2) $\begin{bmatrix} 0 & 2 & 0 \\ 3 & -1 & 0 \\ 1 & 0 & 2 \end{bmatrix}$, (3) $\begin{bmatrix} 3 & -4 & 1 \\ 1 & 5 & 1 \\ 0 & 1 & 1 \end{bmatrix}$.

4.16 $\begin{bmatrix} 5 & 4 & -6 \\ -4 & -3 & -2 \\ 0 & 0 & 1 \end{bmatrix}$.

4.17 (1) $[T]_\alpha = \begin{bmatrix} 0 & -1 \\ 1 & 0 \end{bmatrix}$. (2) $[T]_\beta = [id]_\alpha^\beta [T]_\alpha [id]_\beta^\alpha = \begin{bmatrix} -1/3 & 2/3 \\ -5/3 & 1/3 \end{bmatrix}$.

(3) If $c = 0$, it is trivial. Suppose $c \neq 0$ and let $(T - c\, id)(x, y) = (0, 0)$. Then one can show $y = 0, x = 0$; $\text{Ker}(T - c\, id) = \{\mathbf{0}\}$. So, $T - c\, id$ is injective and hence is an isomorphism.

4.18 (1) $T(1, 0, 0) = (4, 0)$, $T(1, 1, 0) = (3, 1)$, $T(1, 1, 1) = (4, 3)$.
(2) $T(x, y, z) = (4x - y + z, y + 2z)$.

4.19 (2) $[id]_\alpha^\beta = \begin{bmatrix} 3 & 1 \\ 1 & 2 \end{bmatrix}$.

4.20 (1) $Q = \begin{bmatrix} 1 & 1 & 1 \\ 1 & 1 & 0 \\ 1 & 0 & 0 \end{bmatrix}$. (2) $P = Q^{-1}$. (3) Use $[T]_\alpha = \begin{bmatrix} 1 & 2 & 0 \\ 1 & 1 & 1 \\ 3 & 0 & 1 \end{bmatrix}$.

4.21 For the identity matrix I_n, the parallelepiped $\mathcal{P}(I_n)$ is the n-cube, and its volume is 1. See page 91. It is known already that

$$\text{vol}(\mathcal{P}(B)) = |\det B|\ \text{vol}(\mathcal{P}(I_n)) \ \text{ and } \ \text{vol}(\mathcal{P}(C)) = |\det C|\ \text{vol}(\mathcal{P}(I_n)).$$

Now, use $\det C = \det A \det B$.

4.23 (1) Yes. (2) Compute the trace of $AB - BA$.

4.26 Write $B = Q^{-1}AQ$ for some invertible matrix Q.
(1) $\det B = \det(Q^{-1}AQ) = \det Q^{-1} \det A \det Q = \det A$.
(2) $\text{tr}(B) = \text{tr}(Q^{-1}AQ) = \text{tr}(QQ^{-1}A) = \text{tr}(A)$ (see Problem 78).
(3) Use Exercise **3.22**.

4.27 Suppose that $Q^{-1}AQ = \begin{bmatrix} d_1 & 0 \\ 0 & d_2 \end{bmatrix}$ is diagonal, call it D. Then A and D have the same determinant and the same trace. Hence, $d_1 = d_2 = 1$ and $A = QDQ^{-1}$ is the identity matrix. Contradiction.

4.28 2, 7, 9, 13, 15, 18, 19 are false.
(2) Consider $T : \mathbb{R}^2 \to \mathbb{R}^3$ defined by $T(x, y) = (x, 0, 0)$.
(3) $\dim \text{Ker}(T) + \dim \text{Im}(T) = \dim \mathbb{R}^3 = 3$. (4) $\dim \text{Ker}(T) = \dim \mathcal{N}([T]_\alpha^\beta)$.
(5) $\dim \text{Im}(T) = \dim \mathcal{C}([T]_\alpha) = \dim \mathcal{R}([T]_\alpha)$. (6) $\text{Ker}(T) \subset \text{Ker}(S \circ T)$.

(7) $p(x) = 2$ is not linear. (8) By definition of a linear transformation.

(9) and (11) See Remark (1) in Section 4.5.

(10) $T : \mathbb{R}^n \to \mathbb{R}^n$ is injective if and only if T is an isomorphism. See Remark in Subsection 4.8.2 and Theorem 4.20.

(12) By Definition 4.11 and Theorem 4.20. (13) What are their determinants?

(15) $\det(A + B) \neq \det A + \det B$ in general.

(18) $T(\mathbf{0}) = (\mathbf{0})$? (19) If $A = B = I_n$? (20) See page 181.

Chapter 5

Problems

98 Let $\langle \mathbf{x}, \mathbf{y}\rangle = \mathbf{x}^T A\mathbf{y}$ be an inner product. Then, for $\mathbf{x} = (1, 0)$, $\langle \mathbf{x}, \mathbf{x}\rangle = ax_1y_1 + c(x_1y_2+x_2y_1)+bx_2y_2 > 0$ implies $a > 0$. Similarly, for any $\mathbf{x} = (x, 1)$, $\langle \mathbf{x}, \mathbf{x}\rangle > 0$ implies $ab - c^2 > 0$.

99 If $d_1 < 0$, then for $\mathbf{x} = (1, 0, 0)$, $\langle \mathbf{x}, \mathbf{x}\rangle = d_1 < 0$; cannot be positive-definite.

100 $\langle \mathbf{x}, \mathbf{y}\rangle^2 = \langle \mathbf{x}, \mathbf{x}\rangle\langle \mathbf{y}, \mathbf{y}\rangle$ if and only if $\langle t\mathbf{x} + \mathbf{y}, t\mathbf{x} + \mathbf{y}\rangle = \langle \mathbf{x}, \mathbf{x}\rangle t^2 + 2\langle \mathbf{x}, \mathbf{y}\rangle t + \langle \mathbf{y}, \mathbf{y}\rangle = 0$ has a repeated real root t_0.

105 (1) Use the Cauchy-Schwarz inequality, Theorem 5.2. (2) Use the triangle inequality in Problem 101.

106 $a_{13} = a_{31} = \langle f_1, f_3\rangle = \int_0^1 f_1(x)f_3(x)dx = \int_0^1 1 \cdot x^2 dx = \frac{1}{3}$.

107 (1): Orthogonal, (2) and (3): Neither, (4) and (5): Orthonormal.

108 $\mathbf{u}_1 = (0, 1/\sqrt{2}, 1/\sqrt{2}, 0)$, $\mathbf{u}_2 = (-\frac{\sqrt{2}}{\sqrt{3}}, \frac{1}{\sqrt{6}}, -\frac{1}{\sqrt{6}}, 0)$, $\mathbf{u}_3 = (\frac{1}{2}, \frac{1}{2}, -\frac{1}{2}, -\frac{1}{2})$, $\mathbf{u}_4 = (-\frac{1}{2\sqrt{3}}, -\frac{1}{2\sqrt{3}}, \frac{1}{2\sqrt{3}}, -\frac{\sqrt{3}}{2})$. And, in α, $(0, 1, 1, 0) = \sqrt{2}\mathbf{u}_1$, $(-1, 1, 0, 0) = \frac{1}{\sqrt{2}}\mathbf{u}_1 + \frac{\sqrt{3}}{\sqrt{2}}\mathbf{u}_2$, $(2, 4, 0, -2) = 2\sqrt{2}\mathbf{u}_1 + 4\mathbf{u}_3$, $(-6, 0, 0, -6) = 2\sqrt{6}\mathbf{u}_2 + 4\sqrt{3}\mathbf{u}_4$.

109 Clearly, $\mathbf{x}_1 = (1, 0, 1), \mathbf{x}_2 = (0, 1, 2)$ are in W. The Gram-Schmidt orthogonalization gives $\mathbf{u}_1 = \frac{\mathbf{x}_1}{\|\mathbf{x}_1\|} = \frac{1}{\sqrt{2}}(1, 0, 1)$, $\mathbf{u}_2 = \frac{1}{\sqrt{3}}(-1, 1, 1)$, which form a basis for W.

110 For a unit vector $\mathbf{u} = (1/\sqrt{2}, 1/\sqrt{2})$ in W, Proj$(\mathbf{x}) = (\mathbf{u} \cdot \mathbf{x})\mathbf{u} = (1/\sqrt{2})(x + y)\mathbf{u}$ $= ((x + y)/2, (x + y)/2)$.

111 $\mathbf{c}_2 = \langle \mathbf{u}_1, \mathbf{c}_2\rangle\mathbf{u}_1 + \langle \mathbf{u}_2, \mathbf{c}_2\rangle\mathbf{u}_2 + \langle \mathbf{u}_3, \mathbf{c}_2\rangle\mathbf{u}_3 = 2\mathbf{u}_1 + \sqrt{2}\mathbf{u}_2 + 0\mathbf{u}_3$.

113 (4) $\mathrm{Im}(id_V - T) \subseteq \mathrm{Ker}(T)$ because $T(id_V - T)(\mathbf{x}) = T(\mathbf{x}) - T^2(\mathbf{x}) = \mathbf{0}$. $\mathrm{Im}(id_V - T) \supseteq \mathrm{Ker}(T)$ because if $T(\mathbf{x}) = \mathbf{0}$, $\mathbf{x} = \mathbf{x} - \mathbf{0} = \mathbf{x} - T(\mathbf{x}) = (id_V - T)\mathbf{x}$. The others are clear.

114 (1) $\langle \mathbf{x}, \mathbf{x}\rangle = 0$ only for $\mathbf{x} = \mathbf{0}$. For (2), use Definition 5.13(2).

116 Let W be the plane. Then W is a subspace. Clearly, $\mathbf{x}_1 = (1, 1, 0), \mathbf{x}_2 = (1, 0, 1)$ are in W. The Gram-Schmidt orthogonalization gives $\mathbf{u}_1 = \frac{\mathbf{x}_1}{\|\mathbf{x}_1\|} = \frac{1}{\sqrt{2}}(1, 1, 0)$, $\mathbf{u}_2 = \frac{1}{\sqrt{6}}(1, -1, 2)$, which form an orthonormal basis for W. Find $\mathrm{Proj}_W(\mathbf{x})$.

119 $-1/6 + x$.

121 (1) $\mathbf{b} \in \mathcal{C}(A)$ and $\mathbf{y} \in \mathcal{N}(A^T)$.

122 The null space of $\begin{bmatrix} 1 & 2 & 1 & 2 \\ 0 & -1 & -1 & 1 \end{bmatrix}$ is $\mathbf{x} = t[1\ -1\ 1\ 0]^T + s[-4\ 1\ 0\ 1]^T$ for $t, s \in \mathbb{R}$.

124 Hint: Use Example 96. The reflection across the x-axis (and also across the diagonal line) transforms the square $-1 \le x, y \le 1$ onto itself. There are four rotations and four reflections.

125 (3) is orthogonal. (2) and (4) are not orthogonal.

127 $\mathbf{x}_0 = (1, -1, 0) + t(2, 1, -1)$ for any number t.

128 $(A^TA)^{-1} = \frac{1}{9}\begin{bmatrix} 29 & -7 \\ -7 & 2 \end{bmatrix}$; $A^T\mathbf{b} = \begin{bmatrix} 7 \\ 23 \end{bmatrix}$; $\mathbf{x} = (A^TA)^{-1}A^T\mathbf{b} = \begin{bmatrix} 14/3 \\ -1/3 \end{bmatrix}$.

129 $(A^TA)^{-1} = \begin{bmatrix} 21/5 & -17/5 \\ -17/5 & 14/5 \end{bmatrix}$; $A^T\mathbf{b} = \begin{bmatrix} 15 \\ 15 \end{bmatrix}$; $\mathbf{x} = (A^TA)^{-1}A^T\mathbf{b} = \begin{bmatrix} 12 \\ -9 \end{bmatrix}$.

134 $P = \dfrac{1}{3}\begin{bmatrix} 2 & 1 & -1 \\ 1 & 2 & 1 \\ -1 & 1 & 2 \end{bmatrix}$.

Exercises

5.1 Inner products are (2), (3), (5), (6).

5.2 For the positive-definiteness, note that $\langle A, A\rangle = \text{tr}(A^TA) = \sum_{i,j} a_{ij}^2 = 0$ if and only if $a_{ij} = 0$ for all i, j.

5.4 $\cos\theta = \frac{\mathbf{x}\cdot\mathbf{y}}{\|\mathbf{x}\|\|\mathbf{y}\|} = \frac{15}{\sqrt{5}\sqrt{55}} = \frac{3}{\sqrt{11}}$. So, $\theta \approx 0.4405$.

5.5 (1) $k = 0$. (2) $k = -3/2$.

5.8 $a_{ij} = \mathbf{v}_i^T\mathbf{v}_j$. So, $a_{11} = n$, $a_{12} = a_{22} = n - 1$, etc.

5.9 $\mathbf{u}_1 = \frac{1}{\sqrt{2}}(1, 0, 1)$, $\mathbf{u}_2 = \frac{1}{\sqrt{2}}(1, 0, -1)$, $\mathbf{u}_3 = (0, 1, 0)$.

5.10 Find a unit vector $\mathbf{u} = (x, y, z)$ such that $\mathbf{v}_1 \cdot \mathbf{u} = x + y + 2z = 0, \mathbf{v}_2 \cdot \mathbf{u} = 0x + y + 3z = 0$. Then $\mathbf{u} = (1/\sqrt{11}, -3/\sqrt{11}, 1/\sqrt{11})$.

5.11 (1) $\mathbf{u} = \mathbf{v}_2/\|\mathbf{v}_2\| = (1/\sqrt{6}, 1/\sqrt{6}, 2/\sqrt{6})$; $\text{Proj}_{\mathbf{u}}(\mathbf{v}_1) = \langle\mathbf{u}, \mathbf{v}_1\rangle\mathbf{u} = (3/2, 3/2, 3)$.
(2) $\mathbf{u} = \mathbf{v}_2/\|\mathbf{v}_2\| = (2/\sqrt{6}, 1/\sqrt{6}, -1/\sqrt{6})$;
$$\text{Proj}_{\mathbf{u}}(\mathbf{v}_1) = \langle\mathbf{u}, \mathbf{v}_1\rangle\mathbf{u} = (1, 1/2, -1/2).$$
(3) $\mathbf{u} = \mathbf{v}_2/\|\mathbf{v}_2\| = (0, \frac{1}{\sqrt{2}}, \frac{1}{\sqrt{2}}, 0)$; $\text{Proj}_{\mathbf{u}}(\mathbf{v}_1) = \langle\mathbf{u}, \mathbf{v}_1\rangle\mathbf{u} = (0, \frac{1}{2}, \frac{1}{2}, 0)$.

5.13 A basic vector is (1) $\mathbf{x} = (1, 0, 0)$, (2) $\mathbf{x} = (1, -1, 0)$, (3) $\mathbf{x} = (-3, 2, -1, 2)$, respectively.

5.14 $\langle\mathbf{x} + \mathbf{y}, \mathbf{x} - \mathbf{y}\rangle = \langle\mathbf{x}, \mathbf{x}\rangle - \langle\mathbf{x}, \mathbf{y}\rangle + \langle\mathbf{y}, \mathbf{x}\rangle - \langle\mathbf{y}, \mathbf{y}\rangle = 0 \Leftrightarrow \|\mathbf{x}\| = \|\mathbf{y}\|$.

5.15 (1) comes from the definition, and use (1) to prove (2).

5.16 (1) A basis is $\{(0, 1, 0, 1), (0, 0, 1, 1), (1, 1, 0, 0)\}$. (2) A basis for $U^\perp$ is $\{(0, 0, 0, 1), (-1, 1, 1, 0)\}$. (3) A basis for $W^\perp$ is $\{(0, -1, -1, 1), (1, 0, 0, 0)\}$. So, a basis for $U^\perp + W^\perp$ is $\{(0, 0, 0, 1), (-1, 1, 1, 0), (0, -1, -1, 1)\}$. (4) A basis for $U \cap W = (U^\perp + W^\perp)^\perp$ is $\{(0, -1, 1, 0)\}$.

5.17 Take two points $Q_1 = (4, -2, 1)$ and $Q_2 = (3, 1, 3)$ on the line L by taking $t = 0, 1$ respectively. Then the vector $\overrightarrow{Q_1Q_2} = (-1, 3, 2)$ is parallel to the line L.

(1) The vector $\mathbf{w} = \overrightarrow{Q_1Q_2}/||\overrightarrow{Q_1Q_2}|| = (-1/\sqrt{14}, 3/\sqrt{14}, 2/\sqrt{14})$ forms an orthonormal basis for W, and the projection matrix of Proj_W is

$$\mathbf{w}\mathbf{w}^T = \begin{bmatrix} 1/14 & -3/14 & -1/7 \\ -3/14 & 9/14 & 3/7 \\ -1/7 & 3/7 & 2/7 \end{bmatrix}.$$

(2) By the Pythagorean theorem,

$$\begin{aligned} d^2 &= ||\overrightarrow{Q_1P}||^2 - ||\mathrm{Proj}_W(\overrightarrow{Q_1P})||^2 = ||\overrightarrow{Q_1P}||^2 - \langle \overrightarrow{Q_1P}, \mathbf{w}\rangle \\ &= [(x_0-4)^2 + (y_0+2)^2 + (z_0-1)^2] \\ &\quad - \left[\frac{-1}{\sqrt{14}}(x_0-4) + \frac{3}{\sqrt{14}}(y_0+2) + \frac{2}{\sqrt{14}}(z_0-1)\right]. \end{aligned}$$

(3) Clearly $U = W^\perp$; If we let $A = \mathbf{w}^T = \begin{bmatrix} -1/\sqrt{14} & 3/\sqrt{14} & 2/\sqrt{14} \end{bmatrix}$, then $W = \mathcal{R}(A)$ the row space and $U = \mathcal{N}(A)$ the null space. By computing an orthonormal basis $\{\mathbf{u}_1 = (\frac{3}{\sqrt{10}}, \frac{1}{\sqrt{10}}, 0), \mathbf{u}_2 = (\frac{1}{\sqrt{35}}, \frac{-3}{\sqrt{35}}, \frac{\sqrt{35}}{7})\}$ for $U = \mathcal{N}(A)$, one can get the projection matrix of Proj_U as

$$\mathbf{u}_1\mathbf{u}_1^T + \mathbf{u}_2\mathbf{u}_2^T = \begin{bmatrix} 13/14 & 3/14 & 1/7 \\ 3/14 & 5/14 & -3/7 \\ 1/7 & -3/7 & 5/7 \end{bmatrix}.$$

5.18 Take noncollinear three points $Q_1 = (1,0,0), Q_2 = (0,2,0)$ and $Q_3 = (0,0,2)$ on the plane Y. Then two vectors $\overrightarrow{Q_1Q_2} = (-1,2,0)$ and $\overrightarrow{Q_1Q_3} = (-1,0,2)$ form a basis for the subspace W.

(1) Orthonormalize two vectors $\overrightarrow{Q_1Q_2}$ and $\overrightarrow{Q_1Q_3}$ to get $\mathbf{w}_1 = \overrightarrow{Q_1Q_2}/||\overrightarrow{Q_1Q_2}|| = \frac{1}{\sqrt{5}}(-1,2,0)$. and

$$\mathbf{w}_2 = \frac{\overrightarrow{Q_1Q_3} - \mathrm{Proj}_{\langle \mathbf{w}_1\rangle}\overrightarrow{Q_1Q_3}}{||\overrightarrow{Q_1Q_3} - \mathrm{Proj}_{\langle \mathbf{w}_1\rangle}\overrightarrow{Q_1Q_3}||} = \frac{(-1,0,2) - \frac{1}{\sqrt{5}}\mathbf{w}_1}{||(-1,0,2) - \frac{1}{\sqrt{5}}\mathbf{w}_1||} = \frac{1}{\sqrt{30}}(-2,-1,5).$$

Then, the projection matrix of Proj_W is

$$\mathbf{w}_1\mathbf{w}_1^T + \mathbf{w}_2\mathbf{w}_2^T = \begin{bmatrix} 1/3 & -1/3 & -1/3 \\ -1/3 & 5/6 & -1/6 \\ -1/3 & -1/6 & 5/6 \end{bmatrix}.$$

(2) $d = ||\overrightarrow{Q_1P} - \mathrm{Proj}_W\overrightarrow{Q_1P}|| = ||(0,1,1) - (-2/3, 2/3, 2/3)|| = \sqrt{6}/3$.

(3) Clearly $U = W^\perp$; of dimension one. In several ways, one can choose a unit vector $\mathbf{u}$ in U. For example, the vector $\mathbf{u} = (2,1,1)/\sqrt{6}$ is orthogonal to the two basis vectors $\overrightarrow{Q_1Q_2} = (-1,2,0)$ and $\overrightarrow{Q_1Q_3} = (-1,0,2)$ for W. Hence, the projection matrix of Proj_U is

$$\mathbf{u}\mathbf{u}^T = \frac{1}{6}\begin{bmatrix} 4 & 2 & 2 \\ 2 & 1 & 1 \\ 2 & 1 & 1 \end{bmatrix}.$$

5.19 (1) Let $\mathbf{c}_1 = (1,0,0,0)$, $\mathbf{c}_2 = (0,1,2,2)$, and $\mathbf{c}_3 = (0,0,1,2)$ and let A be 4×3 matrix with columns $\mathbf{c}_1$, $\mathbf{c}_2$, and $\mathbf{c}_3$. Then the volume of the parallelepiped determined by the side vectors $\mathbf{c}_1$, $\mathbf{c}_2$, and $\mathbf{c}_3$ is $\sqrt{\det(A^TA)}$ and hence the volume of the tetrahedron is $\frac{1}{6}\sqrt{\det(A^TA)} = 1/2$.

(2) Let $\mathbf{c}_1 = (1,0,0,0,0)$, $\mathbf{c}_2 = (0,1,2,2,2))$, and $\mathbf{c}_3 = (0,0,1,2,2)$ and let A be 5×3 matrix with columns $\mathbf{c}_1$, $\mathbf{c}_2$, and $\mathbf{c}_3$. Then the volume of the tetrahedron is $\frac{1}{6}\sqrt{\det(A^TA)} = \sqrt{17}/6$.

5.20 Orthogonal: (4). Not orthogonal: (1), (2), (3).

5.21 $A^TA = I$ and $\det A^T = \det A$ imply $\det A = \pm 1$.
The matrix $A = \begin{bmatrix} 0 & 1 \\ 1 & 0 \end{bmatrix}$ is orthogonal with $\det A = -1$.

5.22 (1) Its reduced row-echelon form is $U = \begin{bmatrix} 1 & 0 & 0 \\ 0 & 1 & 0 \\ 0 & 0 & 1 \end{bmatrix}$. Its row vectors clearly form an orthonormal basis for $\mathcal{R}(A) = \mathcal{R}(U)$. The null space is trivial.

(2) Its reduced row-echelon form is $U = \begin{bmatrix} 1 & 0 & 1/2 \\ 0 & 1 & 1/2 \\ 0 & 0 & 0 \end{bmatrix}$. Its first two row vectors form a basis for $\mathcal{R}(A) = \mathcal{R}(U)$. From this, one can find an orthonormal basis $\{\frac{1}{2\sqrt{5}}(2,0,1), \frac{1}{\sqrt{30}}(-1,5,2)\}$. By solving $U\mathbf{x} = \mathbf{0}$, one can get orthonormal basis $\{\frac{1}{\sqrt{6}}(1,1,-2)$ for the null space $\mathcal{N}(A)$.

(3) Its reduced row-echelon form is $U = I_4$; the row vectors clearly form an orthonormal basis for $\mathcal{R}(A) = \mathcal{R}(U)$. The null space is trivial.

5.23 (1) For $A = \begin{bmatrix} 1 & 1 \\ 1 & 0 \\ 1 & 2 \end{bmatrix}$, the two columns are linearly independent, and its column space is W. The projection matrix of the orthogonal projection Proj_W is

$$A(A^TA)^{-1}A^T = A\begin{bmatrix} 5/6 & -1/2 \\ -1/2 & 1/2 \end{bmatrix} A^T = \frac{1}{6}\begin{bmatrix} 2 & 2 & 2 \\ 2 & 5 & -1 \\ 2 & -1 & 5 \end{bmatrix}.$$

(2) For $A = \begin{bmatrix} 1 & 1 \\ 1 & 1 \\ 2 & -1 \end{bmatrix}$, the two columns are linearly independent, and its column space is W. The projection matrix of the orthogonal projection Proj_W is

$$A(A^TA)^{-1}A^T = A\begin{bmatrix} 6 & 0 \\ 0 & 3 \end{bmatrix}^{-1} A^T = \begin{bmatrix} 1/2 & 1/2 & 0 \\ 1/2 & 1/2 & 0 \\ 0 & 0 & 1 \end{bmatrix}.$$

5.24 (1) The row-echelon form is $\begin{bmatrix} 1 & 0 \\ 0 & 1 \end{bmatrix}$. The row space is $\mathbb{R}^2$ and the null space is trivial.

(2) For the row space, an orthonormal basis is $\{\frac{1}{\sqrt{21}}(2,4,1), \frac{1}{\sqrt{6}}(1,-1,\sqrt{4})\}$; For the null space, an orthonormal basis is $\{\frac{1}{\sqrt{14}}(-3,1,\sqrt{4})\}$.

(3) The row space is $\mathbb{R}^3$ and the null space is trivial.

5.25 (1) Let $h(x) = \frac{1}{2}(f(x) + f(-x))$ and $g(x) = \frac{1}{2}(f(x) - f(-x))$. Then $f = h + g$.
(2) For $f \in U$ and $g \in V$, $\langle f, g \rangle = \int_{-1}^{1} f(x)g(x)dx = -\int_{1}^{-1} f(-t)g(-t)dt$
$= -\int_{-1}^{1} f(t)g(t)dt = -\langle f, g \rangle$, by change of variable $x = -t$.

5.26 Find a least squares solution of $\begin{bmatrix} 1 & 0 \\ 1 & 1 \\ 1 & 2 \\ 1 & 3 \end{bmatrix} \begin{bmatrix} a \\ b \end{bmatrix} = \begin{bmatrix} 1 \\ 3 \\ 4 \\ 4 \end{bmatrix}$ for (a, b) in $y = a + bx$. Then $y = x + \frac{3}{2}$.

5.27 $A = \begin{bmatrix} 1 & -1 & 1 & -1 \\ 1 & 0 & 0 & 0 \\ 1 & 1 & 1 & 1 \\ 1 & 2 & 4 & 8 \\ 1 & 3 & 9 & 27 \end{bmatrix}$. $y = 2x^2 + 3.8x - 9.8$.

5.28 Solve the normal equation $A^T A\mathbf{x} = A^T\mathbf{b}$.
(1) $x = 55/14, y = 9/7$. (2) $x = 1993/182, y = 61/13, z = -68/13$.

5.29 $A = QR = \begin{bmatrix} 1 & 0 & 0 \\ 0 & 1 & 0 \\ 0 & 0 & 1 \end{bmatrix} \begin{bmatrix} 1 & 2 & 3 \\ 0 & 2 & 3 \\ 0 & 0 & 3 \end{bmatrix}$.

$B = QR = \begin{bmatrix} \sqrt{2}/2 & 0 \\ \sqrt{2}/2 & 0 \\ 0 & 1 \end{bmatrix} \begin{bmatrix} \sqrt{2} & 2\sqrt{2} \\ 0 & 2 \end{bmatrix}$.

5.30 $1, 3, 6, 7, 8, 10, 11, 14, 17, 19$ are true.
(1) Possible by a natural isomorphism. See Theorem 5.10.
(2) Consider $\begin{bmatrix} 1 & 0 \\ 0 & -1 \end{bmatrix}$. (4) Consider $(1, 0)$ and $(-1, 0)$.
(5) Consider two subspaces U and W of $\mathbb{R}^3$ spanned by $\mathbf{e}_1$ and $\mathbf{e}_2$, respectively.
(6) $||\mathbf{x} - \mathbf{y}|| + ||\mathbf{y}|| \geq ||\mathbf{x}||$ by the triangle inequality. (8) See Theorem 5.10.
(10) A is a projection if and only if $A^2 = A$.
(11) By Corollary 5.12.1 and Definition 5.15.
(12) $\mathcal{R}(A)$ and $\mathcal{C}(A)$ are not subspaces of the same vector space.
(13) $\begin{bmatrix} 1 & 0 \\ 0 & 0 \end{bmatrix}$ is an orthogonal projection of $\mathbb{R}^2$ onto the x-axis. This is not an orthogonal matrix.
(15) A dilation does not preserve the lengths. (16) A dilation is an isomorphism.
(17) The columns (or the rows) of any permutation matrix P are $\{\mathbf{e}_1, \mathbf{e}_2, \ldots, \mathbf{e}_n\}$ in some order. Also, see Lemma 1.25.
(18) For any unit vector $\mathbf{u}$ in $\mathbb{R}^2$ or in $\mathbb{R}^3$, $\mathbf{u}\mathbf{u}^T$ is an orthogonal projection, but not an orthogonal matrix. (19) $A^T\mathbf{b} \in \mathcal{C}(A^T A)$ always.
(20) The solution set is $\mathbf{x}_0 + \mathcal{N}(A)$ by Corollary 5.19.1.

Chapter 6

Problems

136 $\det\begin{bmatrix}\lambda-a_{11} & -a_{12} & -a_{13}\\ -a_{21} & \lambda-a_{22} & -a_{23}\\ -a_{31} & -a_{32} & \lambda-a_{33}\end{bmatrix} = (\lambda-a_{11})(\lambda-a_{22})(\lambda-a_{33}) - a_{12}a_{23}a_{31} - a_{13}a_{21}a_{32} + (\lambda-a_{11})a_{23}a_{32} + a_{12}a_{21}(\lambda-a_{33}) + a_{13}(\lambda-a_{22})a_{31} = \lambda^3 - (a_{11}+a_{22}+a_{33})\lambda^2 + \cdots$ by the Sarrus' method.

139 Use $\det(\lambda I - A) = \det(\lambda I - A)^T$. Check with $A = \begin{bmatrix}1 & 1\\ 0 & 1\end{bmatrix}$.

140 Hint: $A\mathbf{x} = \lambda\mathbf{x}$ if and only if $\mathbf{x} = \lambda A^{-1}\mathbf{x}$.

141 Consider the matrices $\begin{bmatrix}1 & 1\\ 0 & 1\end{bmatrix}$ and $\begin{bmatrix}1 & 0\\ 0 & 1\end{bmatrix}$.

142 (1) If there were a $Q = [\mathbf{x}_1\ \mathbf{x}_2\ \mathbf{x}_3]$ which diagonalizes A, then the diagonal matrix should be λI and $AQ = \lambda QI$. Expand this equation and compare the corresponding columns on each side to find a contradiction on the invertibility of Q.

143 $Q = \begin{bmatrix}2 & 3\\ 1 & 2\end{bmatrix}$, $D = \begin{bmatrix}2 & 0\\ 0 & 3\end{bmatrix}$. Then $A = QDQ^{-1} = \begin{bmatrix}-1 & 6\\ -2 & 6\end{bmatrix}$.

144 By an EVD of A,

$$\begin{aligned} A^{100} &= Q\begin{bmatrix}5^{100} & 0\\ 0 & (-2)^{100}\end{bmatrix}Q^{-1} = \begin{bmatrix}1 & -4\\ 1 & 3\end{bmatrix}\begin{bmatrix}5^{100} & 0\\ 0 & (-2)^{100}\end{bmatrix}\begin{bmatrix}3/7 & 4/7\\ -1/7 & 1/7\end{bmatrix}\\ &= \frac{1}{7}\begin{bmatrix}3\cdot 5^{100}+4\cdot 2^{100} & 4\cdot 5^{100}-4\cdot 2^{100}\\ 3\cdot 5^{100}-3\cdot 2^{100} & 4\cdot 5^{100}+3\cdot 2^{100}\end{bmatrix}. \end{aligned}$$

145 (1) The eigenvalues of A are $1, 1, -3$, and their associated eigenvectors are $(1,1,0)$, $(-1,0,1)$ and $(1,3,1)$, respectively.
(2) If $f(x) = x^{10} + x^7 + 5x$, then $f(1)$, $f(1)$ and $f(-3)$ are the eigenvalues of $A^{10} + A^7 + 5A$.

147 Fibonacci sequence: $a_{n+1} = a_n + a_{n-1}$ with $a_1 = 2$ and $a_2 = 3$.

148 $\begin{bmatrix}x_{n+2}\\ x_{n+1}\\ x_n\end{bmatrix} = \begin{bmatrix}2 & 1 & -2\\ 1 & 0 & 0\\ 0 & 1 & 0\end{bmatrix}\begin{bmatrix}x_{n+1}\\ x_n\\ x_{n-1}\end{bmatrix}$. The eigenvalues are $1, 2, -1$, and $x_n = 2\cdot 1^n - 2^n/3 - (-1)^n 2/3$.

149 $\begin{bmatrix}x_{n+2}\\ x_{n+1}\\ x_n\end{bmatrix} = \begin{bmatrix}6 & -11 & 6\\ 1 & 0 & 0\\ 0 & 1 & 0\end{bmatrix}\begin{bmatrix}x_{n+1}\\ x_n\\ x_{n-1}\end{bmatrix}$. The eigenvalues are $1, 2, 3$, and $x_n = -3 + 5\times 2^n - 2\times 3^n$.

150 By Lemma 6.9, $\det(\lambda I - A) = \lambda^4 - \lambda^3 - 7\lambda^2 + \lambda + 6 = (\lambda-1)(\lambda+1)(\lambda+2)(\lambda-3)$ with four distinct eigenvalues, so that the geometric sequences $\{1^n\}, \{(-1)^n\}$, $\{(-2)^n\}$, $\{3^n\}$ are linearly independent and a general solution is $x_n = c_1 1^n + c_2(-1)^n + c_3(-2)^n + c_4 3^n$. By computing c_i, $x_n = \frac{5}{12}1^n - \frac{1}{8}(-1)^n - \frac{4}{15}(-2)^n - \frac{1}{40}3^n$ is a particular solution.

152 Hint: The characteristic polynomial of the associated companion matrix is $f(\lambda) = (\lambda+1)(\lambda-2)^2$. So $(-1)^n, 2^n, n2^n$ form a fundamental set of solutions.

153 (1)

$$\mathbf{x}_n = c_1(1.2)^n \begin{bmatrix}1\\0\end{bmatrix} + c_2(0.8)^n \begin{bmatrix}0\\1\end{bmatrix} = \begin{bmatrix}c_1(1.2)^n\\c_2(0.8)^n\end{bmatrix}.$$

This system is unstable if $c_1 \neq 0$. See Figure 9.9

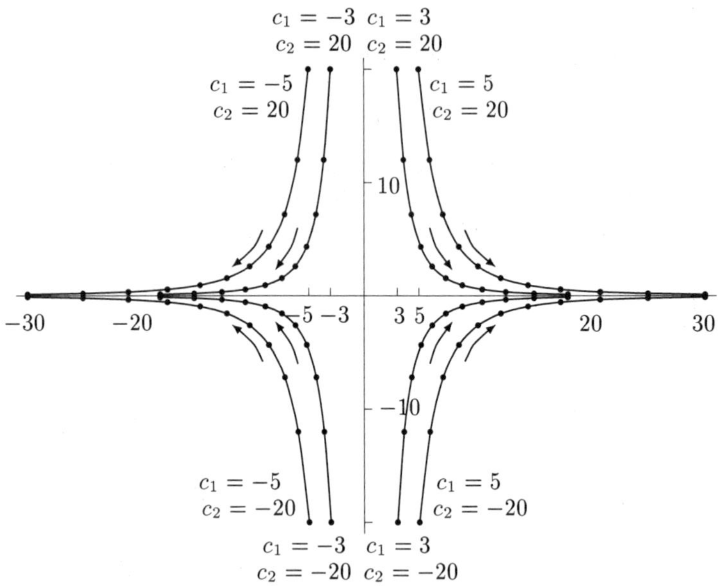

Figure 9.9: An unstable dynamical system

154 Hint: The eigenvalues of A are $1, 0.6$, and 0.5.

155 $A^{10} = \frac{1}{1024}\begin{bmatrix}342 & 0 & 341 & 0\\0 & 683 & 0 & 682\\682 & 0 & 683 & 0\\0 & 341 & 0 & 342\end{bmatrix}$.

156 The eigenvalues of $A = \begin{bmatrix}0 & 1\\1 & 0\end{bmatrix}$ are $\lambda_1 = 1$ and $\lambda_2 = -1$ with associated eigenvectors $\mathbf{v}_1 = (1,1)$ and $\mathbf{v}_2 = (1,-1)$, respectively. A general solution is

$$\mathbf{y}(t) = c_1 e^t \begin{bmatrix}1\\1\end{bmatrix} + c_2 e^{-t}\begin{bmatrix}1\\-1\end{bmatrix}.$$

The initial conditions $y_1(0) = 1, y_2(0) = 0$ determine $c_1 = c_2 = \frac{1}{2}$.

158 Write $A = I + N$ with $N = \begin{bmatrix}0 & 1\\0 & 0\end{bmatrix}$. Then $A^n = (I+N)^n = \begin{bmatrix}1 & n\\0 & 1\end{bmatrix}$.

(2) $e^A = e(I+N) = \begin{bmatrix} e & e \\ 0 & e \end{bmatrix}$. (3) A has two eigenvalues $\lambda_1 = \lambda_2 = 1$. Suppose that A is diagonalizable, then $A = QDQ^{-1}$ with $D = I$. Contradiction.

159 For (1), use $(A+B)^k = \sum_{i=0}^{k} \binom{k}{i} A^i B^{k-i}$ if $AB = BA$. For (2) and (3), use the definition of e^A. Use (1) for (4).

160 $\begin{bmatrix} y_1 \\ y_2 \end{bmatrix} = c_1 e^{2t} \begin{bmatrix} 1 \\ 1 \end{bmatrix} + c_2 e^{-3t} \begin{bmatrix} -\frac{1}{4} \\ 1 \end{bmatrix}$; $y_1 = c_1 e^{2t} - \frac{1}{4} c_2 e^{-3t}$, $y_2 = c_1 e^{2t} + c_2 e^{-3t}$.

161 (1) $\lambda_i = 1, -1$ with $\mathbf{v}_1 = (1,0), \mathbf{v}_2 = (1,1)$. $\exp tA = \begin{bmatrix} e^t & e^{-t} - e^t \\ 0 & e^{-t} \end{bmatrix}$.

(2) $\lambda_i = 1, 0, -1$ with $\mathbf{v}_1 = (1,0,0), \mathbf{v}_2 = (-1,1,0), \mathbf{v}_3 = (0,-1,1)$, $\exp tA = \begin{bmatrix} e^t & e^t - 1 & e^t - 1 \\ 0 & 1 & 1 - e^{-t} \\ 0 & 0 & e^{-t} \end{bmatrix}$.

Exercises

6.1 (2) This is a triangular matrix. The eigenvalues are $1, 2$, and 3. Their associated eigenvectors are $\mathbf{x}_1 = (1,0,0)$, $\mathbf{x}_2 = (2,1,0)$, $\mathbf{x}_3 = (9,6,2)$, respectively.

(3) $f(\lambda) = (\lambda+2)(\lambda^2 - 8\lambda + 15)$, $\lambda_1 = -2$, $\lambda_2 = 3$, $\lambda_3 = 5$. Their associated eigenvectors are $\mathbf{x}_1 = (-35, 12, 19)$, $\mathbf{x}_2 = (0, 3, 1)$, $\mathbf{x}_3 = (0, 1, 1)$, respectively.

(4) Eigenvalues are $0, 0, 2$, and -2, with associated eigenvectors $\mathbf{x}_1 = (-1,0,1,0)$, $\mathbf{x}_2 = (0,-1,0,1)$, $\mathbf{x}_3 = (1,1,1,1)$, $\mathbf{x}_4 = (-1,1,-1,1)$, respectively.

(5) $\lambda = 0$ of multiplicity 3, $\lambda = 4$ of multiplicity 1. Eigenvectors are $\mathbf{e}_i - \mathbf{e}_{i+1}$ for $1 \le i \le 3$ and $\sum_{i=1}^{4} \mathbf{e}_i = (1,1,1,1)$.

(6) This has two block matrices, each of which is skew-symmetric. No real eigenvalues; the complex eigenvalues are $\lambda = \pm 5i, \pm 2i$.

6.3 $A = QDQ^{-1} = \dfrac{1}{2}\begin{bmatrix} 1 & -2 & -1 \\ 1 & 4 & -1 \\ 1 & 2 & 7 \end{bmatrix}$.

6.4 $\{\mathbf{v}\}$ is a basis for $\mathcal{N}(A)$ and $\{\mathbf{u}, \mathbf{w}\}$ is a basis for $\mathcal{C}(A) = \mathcal{R}(A)$.

6.6 The order in the product doesn't matter, and any eigenvector of A is killed by $B = (A-I)(A-2I)(A-3I)$. Since the eigenvalues are all different, the eigenvectors belonging to $1, 2, 3$ form a basis. Thus $B = \mathbf{0}$.

6.7 If A is invertible, then $AB = A(BA)A^{-1}$. For two singular matrices $A = \begin{bmatrix} 1 & 0 \\ 0 & 0 \end{bmatrix}$ and $B = \begin{bmatrix} 0 & 1 \\ 0 & 0 \end{bmatrix}$, AB and BA are not similar, but they have the same eigenvalues.

6.8 Hint: The discriminant of the characteristic polynomial of a skew-symmetric matrix is nonpositive always.

6.9 $\mathbb{R}^n = W \oplus W^\perp$ and $P(\mathbf{w}) = \mathbf{w}$ for $\mathbf{w} \in W$ and $P(\mathbf{v}) = \mathbf{0}$ for $\mathbf{v} \in W^\perp$. Thus, the eigenspace belonging to $\lambda = 1$ is W, and for $\lambda = 0$ it is $W^\perp$.

6.10 For any $\mathbf{w} \in \mathbb{R}^n$, $A\mathbf{w} = \mathbf{u}(\mathbf{v}^T\mathbf{w}) = (\mathbf{v} \cdot \mathbf{w})\mathbf{u}$. Thus $A\mathbf{u} = (\mathbf{v} \cdot \mathbf{u})\mathbf{u}$, so $\mathbf{u}$ is an eigenvector belonging to the eigenvalue $\lambda = \mathbf{v} \cdot \mathbf{u}$. The other eigenvectors are those in $\mathbf{v}^\perp$ with eigenvalue zero. Thus, A has either two eigenspaces $E(\lambda)$ that are one-dimensional spanned by $\mathbf{u}$ and $E(0) = \mathbf{v}^\perp$ if $\mathbf{v} \cdot \mathbf{u} \neq 0$, or just one eigenspace $E(0) = \mathbb{R}^n$ if $\mathbf{v} \cdot \mathbf{u} = 0$.

6.13 Use Gram-Schmidt orthogonalization with (1) eigenvalues $\lambda_i = 0, 1, 2$ with associated eigenvectors $\mathbf{v}_1 = \frac{1}{\sqrt{2}}(0, -1, 1), \mathbf{v}_2 = (1, 0, 0), \mathbf{v}_3 = \frac{1}{\sqrt{2}}(0, 1, 1)$, respectively.

(2) eigenvalues $\lambda_i = -1, 2, 5$ with associated eigenvectors $\mathbf{v}_1 = \frac{1}{3}(2, -2, 1), \mathbf{v}_2 = \frac{1}{3}(-2, -1, 2), \mathbf{v}_3 = \frac{1}{3}(1, 2, 2)$, respectively.

(3) eigenvalues $\lambda_i = -7, -7, 2$ with associated eigenvectors $\mathbf{v}_1 = (-1, 0, 1), \mathbf{v}_2 = (-1, 2, 0), \mathbf{v}_3 = (2, 1, 2)$, respectively.

6.14 (1) $\begin{bmatrix} -4 \times 2^{100} - 5 \times 7^{100} & 2 \times 2^{100} - 2 \times 7^{100} & 4 \times 2^{100} - 4 \times 7^{100} \\ 2 \times 2^{100} - 2 \times 7^{100} & 2^{100} & 2 \times 2^{100} - 2 \times 7^{100} \\ 4 \times 2^{100} - 4 \times 7^{100} & 2 \times 2^{100} - 2 \times 7^{100} & -4 \times 2^{100} - 5 \times 7^{100} \end{bmatrix}$.

(2) $\begin{bmatrix} 4 + 4 \times 2^{100} + 5^{100} & -4 + 2 \times 2^{100} + 2 \times 5^{100} & 2 - 4 \times 2^{100} + 2 \times 5^{100} \\ -4 + 2 \times 2^{100} + 2 \times 5^{100} & 4 + 2^{100} + 4 \times 5^{100} & -2 - 2 \times 2^{100} + 4 \times 5^{100} \\ 2 - 4 \times 2^{100} + 2 \times 5^{100} & -2 - 2 \times 2^{100} + 4 \times 5^{100} & 1 + 4 \times 2^{100} + 4 \times 5^{100} \end{bmatrix}$.

(3) $\begin{bmatrix} 1 & 0 & 0 \\ 0 & 2^{99} & 2^{99} \\ 0 & 2^{99} & 2^{99} \end{bmatrix}$.

6.15 (1) $f(\lambda) = \lambda^3 - 3\lambda - 2$; the eigenvalues are $2, -1, -1$.
(2) $f(\lambda) = \lambda^3 - 7\lambda - 6$; the eigenvalues are $3, -1, -2$.

6.16 Eigenvalues are $1, 1, 2$ and eigenvectors are $(1, 0, 0)$, $(0, 1, 2)$ and $(1, 2, 3)$, respectively. $A^{10}\mathbf{x} = (1025,\ 2050,\ 3076)$.

6.17 Use Theorem 6.13 with (1) $\lambda_1 = 1 + \sqrt{2}, \lambda_2 = 1 - \sqrt{2}$; eigenvectors $\mathbf{v}_1 = (1 + \sqrt{2}, 1), \mathbf{v}_2 = (1 - \sqrt{2}, 1)$, respectively. (2) $\lambda_1 = 2, \lambda_2 = -1, \lambda_3 = 1$; eigenvectors $\mathbf{v}_1 = (4, 2, 1), \mathbf{v}_2 = (1, -1, 1), \mathbf{v}_3 = (1, 1, 1)$, respectively. (3) $\lambda_1 = 3, \lambda_2 = 2, \lambda_3 = -1, \lambda_4 = 1$; eigenvectors $\mathbf{v}_1 = (27, 9, 3, 1), \mathbf{v}_2 = (8, 4, 2, 1), \mathbf{v}_3 = (-1, 1, -1, 1), \mathbf{v}_4 = (1, 1, 1, 1)$, respectively.

6.18 $\lambda_1 = 2, \lambda_2 = 4$ with associated eigenvectors $\mathbf{v}_1 = \begin{bmatrix} -1 \\ 1 \end{bmatrix}, \mathbf{v}_2 = \begin{bmatrix} 1 \\ 1 \end{bmatrix}$, respectively Hence, $\mathbf{x}_n = c_1 2^n \mathbf{v}_1 + c_2 4^n \mathbf{v}_2$ by Eq. (6.2).

6.19 $\lambda_1 = 2, \lambda_2 = -2, \lambda_3 = -4$ with associated eigenvectors $\mathbf{v}_1 = \begin{bmatrix} 2 \\ 1 \\ -1 \end{bmatrix}, \mathbf{v}_2 = \begin{bmatrix} 2 \\ 1 \\ -2 \end{bmatrix}, \mathbf{v}_3 = \begin{bmatrix} 4 \\ 1 \\ -2 \end{bmatrix}$, respectively. Hence, $\mathbf{x}_n = c_1 2^n \mathbf{v}_1 + c_2 (-2)^n \mathbf{v}_2 + c_3 (-4)^n \mathbf{v}_3$ by Eq. (6.2). Determine c_i.

6.20 The characteristic equation is $\lambda^2 - x\lambda - 0.18 = 0$. Since $\lambda = 1$ is a solution, $x = 0.82$. The eigenvalues are now $1, -0.18$ and the eigenvectors are $(-0.3, -1)$ and $(1, -0.6)$, respectively.

6.21 $e^A = \begin{bmatrix} e & -e+e^2 \\ 0 & e^2 \end{bmatrix}$.

6.22 (1) $\mathbf{y}(t) = e^{4t}\begin{bmatrix} 1 \\ 1 \end{bmatrix} + e^{2t}\begin{bmatrix} 1 \\ -1 \end{bmatrix}$.

(2) Hint: Two eigenvalues $2+\sqrt{2}$ and $2-\sqrt{2}$ with associated eigenvectors $(1+\sqrt{2}, 1)$ and $(1-\sqrt{2}, 1)$, respectively. $\mathbf{y}(t) = e^{\lambda_i t}\mathbf{v}_i$ for LDE: $y_1(t) = \frac{2+\sqrt{2}}{4}e^{(2-\sqrt{2})t} + \frac{2-\sqrt{2}}{4}e^{(2+\sqrt{2})t}$, $y_2(t) = \frac{\sqrt{2}}{4}e^{(2-\sqrt{2})t} - \frac{\sqrt{2}}{4}e^{(2+\sqrt{2})t}$.

6.23 (1) $\lambda_1 = 2, \lambda_2 = -2, \lambda_3 = -4$ with associated eigenvectors $\mathbf{v}_1 = \begin{bmatrix} 2 \\ 1 \\ -1 \end{bmatrix}$, $\mathbf{v}_2 = \begin{bmatrix} 2 \\ 1 \\ -2 \end{bmatrix}$, $\mathbf{v}_3 = \begin{bmatrix} 4 \\ 1 \\ -2 \end{bmatrix}$, respectively. Hence, $\mathbf{y}(t) = c_1e^{2t}\mathbf{v}_1 + c_2e^{-2t}\mathbf{v}_2 + c_3e^{-4t}\mathbf{v}_3$ by Eq. (6.6). The initial condition $\mathbf{y}(0)$ gives $c_1 = 2, c_2 = -1, c_3 = 0$.

(2) Hint: Three eigenvalues $1, 2, 3$ with associated eigenvectors $(1,0,0)$, $(1,1,0)$ and $(1,2,2)$, respectively. $y_1(t) = y_2(t) = e^{2t}$, $y_3(t) = 0$.

6.24 $e^{(A^T)} = (e^A)^T$ by definition (thus, if A is symmetric, so is e^A), and use Theorem 6.24(2).

6.25 (1) $\begin{bmatrix} e & e-1 \\ 0 & 1 \end{bmatrix}$. (2) $\begin{bmatrix} \frac{1}{2}(e^2+e^4) & \frac{1}{2}(-e^2+e^4) \\ \frac{1}{2}(-e^2+e^4) & \frac{1}{2}(e^2+e^4) \end{bmatrix}$. (3) $\begin{bmatrix} e^3 & e^3t \\ 0 & e^3 \end{bmatrix}$

(4) $A = 2I + N$ with $N = \begin{bmatrix} 0 & 3 & 0 \\ 0 & 0 & 3 \\ 0 & 0 & 0 \end{bmatrix}$. Then $N^3 = \mathbf{0}$. $e^A = e^2\begin{bmatrix} 1 & 3 & \frac{3^2}{2!} \\ 0 & 1 & 3 \\ 0 & 0 & 1 \end{bmatrix}$.

(5) $e^A = e^\lambda \begin{bmatrix} 1 & 1 & \frac{1}{2!} & \frac{1}{3!} \\ 0 & 1 & 1 & \frac{1}{2!} \\ 0 & 0 & 1 & 1 \\ 0 & 0 & 0 & 1 \end{bmatrix}$.

6.26 $\mathbf{y}(t) = e^{tA}\mathbf{y}_0 = e^{2t}\begin{bmatrix} 1 & 3t & \frac{3^2}{2!}t^2 \\ 0 & 1 & 3t \\ 0 & 0 & 1 \end{bmatrix}\mathbf{y}_0$.

Similarly, one can show that $[D]_\alpha^2$ has an eigenvalue 0 of multiplicity 4 with two associated (linearly independent) eigenvectors $(1,0,0,0)$ and $(0,1,0,0)$ in $\mathbb{R}^4$, and so the transformation D^2 has an eigenvalue 0 of multiplicity 4 with an associated (linearly independent) eigenvector $f(x) = 1$ and $g(x) = x$ in $P_3(\mathbb{R})$.

6.29 $5, 7, 13, 14, 15, 16, 18, 19, 20$ are true.

(1) Consider $\begin{bmatrix} 0 & 1 \\ 0 & 0 \end{bmatrix}$, See Example 109. (2) Consider $\begin{bmatrix} 0 & 1 \\ -1 & 0 \end{bmatrix}$.

(3) Consider $A = \begin{bmatrix} 1 & 1 \\ 0 & 1 \end{bmatrix}$ and $B = \begin{bmatrix} 1 & 0 \\ 1 & 1 \end{bmatrix}$. (4) See Example 107.

(5) See Theorem 6.7. (6) Consider I_2.

(8) See Example 111. (9) $\begin{bmatrix} 0 & 0 \\ 0 & 1 \end{bmatrix}$ and $\begin{bmatrix} 0 & 1 \\ 0 & 0 \end{bmatrix}$ have the different eigenvalues.

(10) If $A = \begin{bmatrix} 1 & 1 \\ 0 & 0 \end{bmatrix}, B = \begin{bmatrix} 0 & 1 \\ 0 & 1 \end{bmatrix}$? (11) $\begin{bmatrix} \cos\theta & -\sin\theta \\ \sin\theta & \cos\theta \end{bmatrix}$? (12) $\begin{bmatrix} 1 & 0 \\ 0 & 0 \end{bmatrix}$?
(16) See Lemma 6.9 (1). (17) Use Theorem 6.24 with $A = I_n$.
(18) Use the definition of e^A. (19) See Lemma 6.9 (2). (20) Is it clear?

Chapter 7

Problems

162 (1) Eigenvalues are $i, -i$ with associated eigenvectors $(2i, 1)$ and $(0, 1)$, respectively.
(2) Eigenvalues are $2, -1+i, 1+i$ with associated eigenvectors $(0, 1, 0), (-1, 0, 1)$ and $(1, 0, 1)$, respectively.

164 Suppose that $\mathbf{x}$ and $\mathbf{y}$ are linearly independent, and consider the linear dependence $a(\mathbf{x}+\mathbf{y}) + b(\mathbf{x}-\mathbf{y}) = \mathbf{0}$ of $\mathbf{x}+\mathbf{y}$ and $\mathbf{x}-\mathbf{y}$. Then $\mathbf{0} = (a+b)\mathbf{x} + (a-b)\mathbf{y}$. Since $\mathbf{x}$ and $\mathbf{y}$ are linearly independent, we have $a+b = 0$ and $a-b = 0$ which are possible only for $a = 0 = b$. Thus $\mathbf{x}+\mathbf{y}$ and $\mathbf{x}-\mathbf{y}$ are linearly independent. Conversely, if $\mathbf{x}+\mathbf{y}$ and $\mathbf{x}-\mathbf{y}$ are linearly independent, then the linear dependence $a\mathbf{x} + b\mathbf{y} = \mathbf{0}$ of $\mathbf{x}$ and $\mathbf{y}$ gives $\frac{1}{2}(a+b)(\mathbf{x}+\mathbf{y}) + \frac{1}{2}(a-b)(\mathbf{x}-\mathbf{y}) = \mathbf{0}$. Thus we get $a = 0 = b$. Thus $\mathbf{x}$ and $\mathbf{y}$ are linearly independent.

165 (1) $\mathbf{u}\cdot\mathbf{v} = \overline{\mathbf{u}}^T\mathbf{v} = \sum_i \overline{u_i}v_i = \overline{\sum_i \overline{v_i}u_i} = \overline{\mathbf{v}\cdot\mathbf{u}}$.
(3) $(k\mathbf{u})\cdot\mathbf{v} = \sum_i \overline{ku_i}v_i = \overline{k}\sum_i \overline{u_i}v_i = \overline{k}(\mathbf{u}\cdot\mathbf{v})$.
(4) $\mathbf{u}\cdot\mathbf{u} = \sum_i |u_i|^2 \geq 0$, and $\mathbf{u}\cdot\mathbf{u} = 0$ if and only if $u_i = 0$ for all i.

169 $(A^H)(A^{-1})^H = (A^{-1}A)^H = I$.

170 $\pm AB = (AB)^H = B^HA^H = (\pm B)(\pm A) = BA$, where $+$ if they are Hermitian, $-$ if they are skew-Hermitian.

171 To prove (3) directly, show that $\overline{\lambda}(\mathbf{x}\cdot\mathbf{y}) = \overline{\mu}(\mathbf{x}\cdot\mathbf{y})$ by using the fact that $A^H\mathbf{y} = (-A)\mathbf{y} = -\mu\mathbf{y}$ when $A\mathbf{y} = \mu\mathbf{y}$.

172 (1) is true: The determinant is just the product of the eigenvalues and a Hermitian matrix has only real eigenvalues, but (2) is false.

173 ($\Leftarrow$): Use $A^H = B^H + (iC)^H = B^T - iC^T = -B - iC = -A$.

176 (1) If $A^{-1} = A^H$ and $B^{-1} = B^H$, then $(AB)^HAB = I$.
(2) If U is unitary, then $U\cdot\text{adj}\,U = (\det U)I = I$. So, adj $U = U^{-1} = U^H$, which is also unitary.

178 (1) $\lambda = 1 \pm i$ with eigenvectors $\frac{1}{\sqrt{2}}(1, 1), \frac{1}{\sqrt{2}}(-1, 1)$, respectively.
(2) Two eigenvalues -1 and $2i$ (of multiplicity 2) with two (linearly independent) eigenvectors $(0, 1, 0), (0, 1, -2+i)$, respectively. This is a non-diagonalizable case. To find columns of U, fix one eigenvector $(0, 1, 0)$ and use Gram-Schmidt orthogonalization to extend it to an orthonormal basis. See the proof of Schur's triangulation Theorem 7.13. Then, we have two more columns: $(0, 0, \frac{-2+i}{\sqrt{5}}), (1, 0, 0)$. In this case,

$$U = \begin{bmatrix} 0 & 0 & 1 \\ 1 & 0 & 0 \\ 0 & \frac{-2+i}{\sqrt{5}} & 0 \end{bmatrix}, \quad \text{and} \quad U^{-1}AU = \begin{bmatrix} -1 & \frac{-1-2i}{\sqrt{5}} & i \\ 0 & 2i & \frac{2+i}{\sqrt{5}} \\ 0 & 0 & 2i \end{bmatrix}$$

184 (3) AB and BA are of different sizes. E.g., $A = \begin{bmatrix} 1 & 0 & 0 \\ 0 & 1 & 0 \end{bmatrix}$ and $B = \begin{bmatrix} 1 & 0 \\ 0 & 1 \\ 0 & 0 \end{bmatrix}$.

(4) Clearly, both matrices are Hermitian. If $(A^H A)\mathbf{x} = \lambda\mathbf{x}$ with a nonzero eigenvalue λ, then $AA^H(A\mathbf{x}) = \lambda(A\mathbf{x})$ with nonzero $A\mathbf{x}$ (since $\lambda \neq 0$). Moreover, $\mathbf{x}^H(A^H A)\mathbf{x} = (A\mathbf{x})^H(A\mathbf{x}) \geq 0$ for any nonzero $\mathbf{x}$ and hence λ is nonnegative.

179 (1) $\lambda_1 = 2, \lambda_2 = 1, \lambda_3 = 1$ with $\mathbf{x}_1 = (-1, 0, 1), \mathbf{x}_2 = (0, 1, 0), \mathbf{x}_3 = (1, 0, 0)$, respectively.

(2) $\lambda_1 = x, \lambda_2 = x + 1, \lambda_3 = x + 2$ with $\mathbf{x}_1 = (1, 0, 0), \mathbf{x}_2 = (0, 1, 0), \mathbf{x}_3 = (1, 0, 2)$, respectively.

180 (1) Hermitian; (2) symmetric, so Hermitian; (3) skew-Hermitian; (4) neither; (5) unitary; (6) Hermitian. All are normal except (4).

182 A is normal. From a direct computation, one can find the eigenvalues, $1 - i, 1 - i$ and $1 + 2i$, and the associated eigenvectors: $(-1, 0, 1)$, $(-1, 1, 0)$ and $(1, 1, 1)$, respectively, which are not orthogonal. But, by an orthonormalization, one can get a unitary matrix U so that $U^H AU$ is diagonal.

183 $A^H A = (H_1 - iH_2)(H_1 + iH_2) = (H_1 + iH_2)(H_1 - iH_2) = AA^H$ if and only if $H_1H_2 - H_2H_1 = \mathbf{0}$.

185 (1) $A = \frac{-1}{2}\begin{bmatrix} 1 & -1 \\ -1 & 1 \end{bmatrix} + \frac{5}{2}\begin{bmatrix} 1 & 1 \\ 1 & 1 \end{bmatrix}$, (2) $B = \frac{-1}{2}\begin{bmatrix} 1 & -i \\ i & 1 \end{bmatrix} + \frac{5}{2}\begin{bmatrix} 1 & i \\ -i & 1 \end{bmatrix}$.

Hint for (3): $\lambda = 2$ of multiplicity 2, with two associated orthogonal eigenvectors $\mathbf{v}_1 = (1, 0, 0, 2), \mathbf{v}_2 = (0, 1, 1, 0)$; and $\lambda = -2$ of multiplicity 2, with two associated orthogonal eigenvectors $\mathbf{v}_3 = (-1, 0, 0, 2), \mathbf{v}_4 = (0, -1, 1, 0)$.

186 $Q = \frac{1}{\sqrt{6}}\begin{bmatrix} \sqrt{3} & -\sqrt{2} & -1 \\ 0 & \sqrt{2} & -2 \\ \sqrt{3} & \sqrt{2} & 1 \end{bmatrix} = [\mathbf{u}_1\ \mathbf{u}_2\ \mathbf{u}_3]$.

$$A = \lambda_1\mathbf{u}_1\mathbf{u}_1^H + \lambda_2\mathbf{u}_2\mathbf{u}_2^H + \lambda_3\mathbf{u}_3\mathbf{u}_3^H = (-1)\left(\mathbf{u}_1\mathbf{u}_1^H + \mathbf{u}_2\mathbf{u}_2^H\right) + 5\mathbf{u}_3\mathbf{u}_3^H.$$

187 (1) A has a zero singular value $\Leftrightarrow$ $A^H A$ is singular $\Leftrightarrow$ A is singular.

188

$$A = U\Sigma V^T = \begin{bmatrix} 1/\sqrt{2} & -1/\sqrt{2} \\ 1/\sqrt{2} & 1/\sqrt{2} \end{bmatrix}\begin{bmatrix} \sqrt{3} & 0 & 0 \\ 0 & 1 & 0 \end{bmatrix}\begin{bmatrix} 1/\sqrt{6} & 1/\sqrt{2} & -1/\sqrt{3} \\ \sqrt{2}/\sqrt{3} & 0 & 1/\sqrt{3} \\ -1/\sqrt{6} & 1/\sqrt{2} & 1/\sqrt{3} \end{bmatrix}^T$$

191 Use an SVD $A = U\Sigma V^T$, where

$$V = \frac{1}{2}\begin{bmatrix} -1 & 0 & \sqrt{2} & -1 \\ 1 & \sqrt{2} & 0 & -1 \\ 1 & 0 & \sqrt{2} & 1 \\ -1 & \sqrt{2} & 0 & 1 \end{bmatrix}, \quad \Sigma = \begin{bmatrix} 2 & 0 & 0 & 0 \\ 0 & \sqrt{2} & 0 & 0 \\ 0 & 0 & \sqrt{2} & 0 \end{bmatrix}, \quad U = \begin{bmatrix} 0 & 0 & 1 \\ 0 & 1 & 0 \\ 1 & 0 & 0 \end{bmatrix}.$$

Exercises

7.1 (2) $(x, y) = \frac{1}{2}(4 + i, i)$.

7.2 Complex inner products are (3) and (4).

7.3 (1) $\sqrt{2}$, (2) $\sqrt{6}$, (3) 4.

7.5 $\mathbf{u}_1 = \frac{1}{\sqrt{5}}(i, 2i)$, $\mathbf{u}_2 = \frac{1}{5}(2 + 4i, -1 - 2i)$.

7.6 (1) Use $U = \begin{bmatrix} 1 & 0 & (-1-i)/4 \\ 0 & 1 & (3-3i)/4 \end{bmatrix}$. rank $A = 2$.

(2) Use $U = \begin{bmatrix} 1 & 0 & 0 \\ 0 & 1 & 0 \\ 0 & 0 & 1 \end{bmatrix}$. rank $A = 3$.

7.7 $A = QDQ^{-1}$, where

$$Q = \begin{bmatrix} 0 & (-\sqrt{3}+i)/2 & (\sqrt{3}+i)/2 \\ 0 & 1 & 1 \\ 1 & 0 & 0 \end{bmatrix},$$

$$D = \begin{bmatrix} 1 & 0 & 0 \\ 0 & (-\sqrt{3}+i)/2 & 0 \\ 0 & 0 & (\sqrt{3}+i)/2 \end{bmatrix},$$

$$Q^{-1} = \begin{bmatrix} 0 & 0 & 1 \\ -1/\sqrt{3} & (3+i\sqrt{3})/6 & 0 \\ 1/\sqrt{3} & (3-i\sqrt{3})/6 & 0 \end{bmatrix}.$$

7.8 $A + cI$ is invertible if and only if $\det(A + cI) \neq 0$. However, for every matrix A, $\det(A + cI) = 0$ has at least one real or complex solution c. For the real matrix $A = \begin{bmatrix} \cos\theta & -\sin\theta \\ \sin\theta & \cos\theta \end{bmatrix}$ $(\theta \neq n\pi)$, $A + rI$ is invertible for every real number r since A has no real eigenvalues. Also, the same is true for any 2×2 nonzero skew-symmetric matrix $A = \begin{bmatrix} 0 & -x \\ x & 0 \end{bmatrix}$ $(x \neq 0)$.

7.9 rank $A = 1$. An eigenvector is $\mathbf{x}$ belonging to $\lambda = ||\mathbf{x}||^2$. For $\lambda = 0$, there are $n - 1$ linearly independent eigenvectors all of which belong to the orthogonal complement of $\mathbf{x}$.

7.10 (1) Since the eigenvalues of a skew-Hermitian matrix are always purely imaginary, ± 1 cannot be an eigenvalue.

(2) For any skew-Hermitian matrix A, $(e^A)^H = e^{A^H} = e^{-A} = (e^A)^{-1}$.

7.11 (1) $\frac{1}{\sqrt{3}}\begin{bmatrix} 1 & 1-i \\ 1-i & -1 \end{bmatrix}$, (2) $\frac{1}{2}\begin{bmatrix} 1 & i & 1-i \\ \sqrt{2}i & \sqrt{2} & 0 \\ 1 & i & -1+i \end{bmatrix}$.

7.13 For A: $f(\lambda) = (\lambda + 2)(\lambda^2 - 8\lambda + 15)$, $\lambda_1 = -2$, $\lambda_2 = 3$, $\lambda_3 = 5$.

For B: $f(\lambda) = \lambda^3 - 19\lambda - 30 = (\lambda+2)(\lambda-5)(\lambda+3)$, $\lambda_1 = -2$, $\lambda_2 = -3$, $\lambda_3 = 5$.

For C: $f(\lambda) = \lambda^3 - 2\lambda^2 - 5\lambda + 30$, $\lambda_1 = -3$, $\lambda_2 = \frac{1}{2}(5 + i\sqrt{15})$, $\lambda_3 = \frac{1}{2}(5 - i\sqrt{15})$.

7.14 (1) $\lambda_1 = 1$, $\mathbf{v}_1 = (i,\ 1)$, $\lambda_2 = -1$, $\mathbf{v}_2 = (-i,\ 1)$. Normalize $\mathbf{v}_i$ for U.

(2) $\lambda_1 = \frac{1}{2}(3+\sqrt{5})$, $\lambda_2 = \frac{1}{2}(3-\sqrt{5})$ with $\mathbf{v}_1 = (\frac{1}{2}i(-1+\sqrt{5}), 1)$, $\mathbf{v}_2 = (-\frac{1}{2}i(1+\sqrt{5}), 1)$. Normalize $\mathbf{v}_i$ for U.

(3) $\lambda_1 = -\sqrt{14}$, $\lambda_2 = \sqrt{14}$ with $\mathbf{v}_1 = (-\frac{1}{13}(2+3i))(-1+\sqrt{14}), 1)$, $\mathbf{v}_2 = (\frac{1}{13}(2+3i)(1+\sqrt{14}), 1)$. Normalize $\mathbf{v}_i$ for U.

7.15 (1) $\lambda_1 = i\sqrt{14}, \lambda_2 = -i\sqrt{14}, \lambda_3 = 0$. $\mathbf{v}_1 = (-\frac{3}{13} - \frac{2i\sqrt{14}}{13}, \frac{2}{13} - \frac{3i\sqrt{14}}{13}, 1)$, $\mathbf{v}_2 = (-\frac{3}{13} + \frac{2i\sqrt{14}}{13}, \frac{2}{13} + \frac{3i\sqrt{14}}{13}, 1)$, $\mathbf{v}_3 = (3, -2, 1)$. Normalize $\mathbf{v}_i$ for U.
(2) $\lambda_1 = 2i, \lambda_2 = -i, \lambda_3 = -i$. $\mathbf{v}_1 = (1,1,1), \mathbf{v}_2 = (-1,0,1)$, $\mathbf{v}_3 = (-1,1,0)$, respectively. Normalize $\mathbf{v}_i$ for U.
(3) $\lambda_1 = i\sqrt{2}, \lambda_2 = -i\sqrt{2}, \lambda_3 = 0$. $\mathbf{v}_1 = (\sqrt{2}, i, 1), \mathbf{v}_2 = (-\sqrt{2}, i, 1)$, $\mathbf{v}_3 = (0, -i, 1)$, respectively. Normalize $\mathbf{v}_i$ for U.

7.16 (1) $\lambda_1 = 1, \lambda_2 = \frac{1}{2}i(i+\sqrt{3}), \lambda_3 = -\frac{1}{2}i(-i+\sqrt{3})$. $\mathbf{v}_1 = (1,1,1), \mathbf{v}_2 = (\frac{1}{2}i(i+\sqrt{3}), -\frac{1}{2}i(-i+\sqrt{3}), 1), \mathbf{v}_3 = (-\frac{1}{2}i(-i+\sqrt{3}), \frac{1}{2}i(i+\sqrt{3}), 1)$ respectively. Normalize $\mathbf{v}_i$ for U.
(2) $\lambda_1 = -1, \lambda_2 = \frac{1}{2}(1+i\sqrt{3}), \lambda_3 = \frac{1}{2}(1-i\sqrt{3})$. $\mathbf{v}_1 = (-1,-i,1), \mathbf{v}_2 = (\frac{1}{2}(1+i\sqrt{3}), \frac{1}{2}(i+\sqrt{3}), 1), \mathbf{v}_3 = (\frac{1}{2}(1-i\sqrt{3}), \frac{1}{2}(i-\sqrt{3}), 1)$ respectively. Normalize $\mathbf{v}_i$ for U.
(3) $\lambda_1 = i, \lambda_2 = e^{-i\theta}, \lambda_3 = e^{i\theta}$. $\mathbf{v}_1 = (0,1,0), \mathbf{v}_2 = (-i,0,1)$, $\mathbf{v}_3 = (i,0,1)$, respectively. Normalize $\mathbf{v}_i$ for U.

7.17 (1) Unitary; diagonal entries are $\{1, i\}$. (2) Orthogonal (unitary); $\{\lambda\} = \{e^{i\theta} = \cos\theta + i\sin\theta, e^{-i\theta} = \cos\theta - i\sin\theta\}$, where $\theta = \cos^{-1}(0.6)$. (3) Hermitian; $\{\lambda\} = \{1, 1+\sqrt{2}, 1-\sqrt{2}\}$.

7.18 The eigenvalues are $\lambda_1 = 2+i$ and $\lambda_2 = 2-i$. Their associated eigenvectors are $\mathbf{v}_1 = (1,1)$ and $\mathbf{v}_2 = (-1,1)$, respectively. It gives $U = \frac{1}{\sqrt{2}}\begin{bmatrix} 1 & -1 \\ 1 & 1 \end{bmatrix}$, $D = U^HAU = \begin{bmatrix} 2+i & 0 \\ 0 & 2-i \end{bmatrix}$.

7.19 The normality of A is clear. The eigenvalues of A are $\lambda_1 = 1-(1+i)/\sqrt{2}$ and $\lambda_2 = 1+(1+i)/\sqrt{2}$ with associated eigenvectors $(1, (-1+i)/\sqrt{2})$ and $(1, (1-i)/\sqrt{2})$, respectively. Since these eigenvectors are orthogonal, by normalization, we have orthonormal eigenvectors $(1/\sqrt{2}, (-1+i)/2)$ and $(1/\sqrt{2}, (1-i)/2)$. If $A = UDU^H$, $AA^H = UDD^HU^H$, where DD^H is diagonal with (positive) real diagonal entries.

7.21 ($\Rightarrow$) is well known. For ($\Leftarrow$), if $\det A = \det B$ and $\mathrm{tr}(A) = \mathrm{tr}(B)$, then A and B have the same characteristic polynomial, see Problem 138. Hence, the two symmetric matrices have the same (real) eigenvalues, and they are similar to the same diagonal matrix.

7.22 If λ is an eigenvalue of A, then λ^n is an eigenvalue of A^n. Thus, if $A^n = \mathbf{0}$, then $\lambda^n = 0$ or $\lambda = 0$. Conversely, by Schur's triangulation Theorem, A is similar to an upper triangular matrix whose diagonal entries are eigenvalues and they are all zero. Now it is easy to conclude that A is nilpotent.

7.23 The eigenvalues are $1, 1, 4$, and the associated orthonormal eigenvectors are $(1/\sqrt{2}, -1/\sqrt{2}, 0)$, $(-1/\sqrt{6}, -1/\sqrt{6}, \sqrt{2}/\sqrt{3})$ and $(1/\sqrt{3}, 1/\sqrt{3}, 1/\sqrt{3})$, respectively.

Therefore,

$$A = \frac{1}{3}\begin{bmatrix} 2 & -1 & -1 \\ -1 & 2 & -1 \\ -1 & -1 & 2 \end{bmatrix} + \frac{4}{3}\begin{bmatrix} 1 & 1 & 1 \\ 1 & 1 & 1 \\ 1 & 1 & 1 \end{bmatrix},$$

$$A^n = \frac{1^n}{3}\begin{bmatrix} 2 & -1 & -1 \\ -1 & 2 & -1 \\ -1 & -1 & 2 \end{bmatrix} + \frac{4^n}{3}\begin{bmatrix} 1 & 1 & 1 \\ 1 & 1 & 1 \\ 1 & 1 & 1 \end{bmatrix}.$$

7.24 Use Theorem 6.13 with (1) Eigenvalues: $\lambda_1 = (1+\sqrt{7}i)/2, \lambda_2 = (1-\sqrt{7}i)/2$.
(2) Eigenvalues: $\lambda_1 = 3, \lambda_2 = -1+i, \lambda_3 = -1-i$.
(3) Eigenvalues: $\lambda_1 = 1, \lambda_2 = 1+i, \lambda_3 = 1-i$.

7.25 (1) Eigenvalues are $\lambda_i = 2 \pm i$ with associated eigenvectors $\mathbf{v}_i = (-1 \pm i, 2)$.

$$\begin{cases} y_1(x) &= e^{2x}(\cos x - \sin x) \\ y_2(x) &= 2e^{2x}\sin x. \end{cases}$$

(2) Three distinct eigenvalues $2, (1+i\sqrt{3})/2, (1-i\sqrt{3})/2$ with associated eigenvectors $\mathbf{v}_1 = (1,1,1), \mathbf{v}_2 = ((-1+i\sqrt{3})/2, (-1-i\sqrt{3})/2, 1), \mathbf{v}_3 = ((-1-i\sqrt{3})/2, (-1+i\sqrt{3})/2, 1)$, respectively.

7.26 (1)

$$A = LDU = \begin{bmatrix} 1 & 0 \\ i & 1 \end{bmatrix}\begin{bmatrix} 1 & 0 \\ 0 & 1 \end{bmatrix}\begin{bmatrix} 1 & 0 & i \\ 0 & 0 & 2 \end{bmatrix}$$

$$A = U\Sigma V^H = \begin{bmatrix} 0 & 1 \\ 1 & 0 \end{bmatrix}\begin{bmatrix} \sqrt{2} & 0 & 0 \\ 0 & \sqrt{2} & 0 \end{bmatrix}\begin{bmatrix} -i/\sqrt{2} & 1/\sqrt{2} & 0 \\ 0 & 0 & 1 \\ 1/\sqrt{2} & -i/\sqrt{2} & 0 \end{bmatrix}^H$$

(2)

$$A = LDU = \begin{bmatrix} 1 & 0 & 0 \\ 0 & 1 & 0 \\ -1 & 1 & 1 \end{bmatrix}\begin{bmatrix} 1 & 0 & 0 \\ 0 & 1 & 0 \\ 0 & 0 & 2 \end{bmatrix}\begin{bmatrix} 1 & 0 & 1 & 0 \\ 0 & 1 & 0 & 1 \\ 0 & 0 & 1 & -1 \end{bmatrix}$$

$$A = U\Sigma V^T = \begin{bmatrix} 0 & 0 & 1 \\ 0 & 1 & 0 \\ 1 & 0 & 0 \end{bmatrix}\begin{bmatrix} 2 & 0 & 0 & 0 \\ 0 & \sqrt{2} & 0 & 0 \\ 0 & 0 & \sqrt{2} & 0 \end{bmatrix}\begin{bmatrix} -1/2 & 0 & 1/\sqrt{2} & -1/2 \\ 1/2 & 1/\sqrt{2} & 0 & -1/2 \\ 1/2 & 0 & 1/\sqrt{2} & 1/2 \\ -1/2 & 1/\sqrt{2} & 0 & 1/2 \end{bmatrix}^T$$

7.27 $V\Sigma^{\text{right}}U^H$ is a right inverse of A with U, V in the SVD given in Exercise **7.26**.

7.29 ($\Rightarrow$): $A = QD_1Q^{-1}$ and $B = QD_2Q^{-1}$ imply $AB = BA$ since $D_1D_2 = D_2D_1$. ($\Leftarrow$): Suppose $AB = BA$. For an eigenvalue λ of A, let $E(\lambda) = \mathcal{N}(\lambda I - A)$ be the eigenspace of A belonging to λ, which might not be one-dimensional, say of dimension k. For any $\mathbf{x} \in E(\lambda)$, $AB\mathbf{x} = BA\mathbf{x} = \lambda B\mathbf{x}$ implies $B\mathbf{x} \in E(\lambda)$. This shows that $E(\lambda)$ is an invariant subspace under the matrix B (also under the matrix A). Since B is diagonalizable, B has k linearly independent eigenvectors in the subspace $E(\lambda)$. Therefore, these k vectors become the common eigenvectors of A and B. Since the eigenvalue λ is chosen arbitrary, A and B have the same (full) set of eigenvectors and hence are simultaneously diagonalizable.

7.30 $1, 2, 3, 5, 7, 8, 10, 11, 12, 13, 15, 16, 19$ are true.

(3) See Theorem 7.14. (4) Consider $\begin{bmatrix} \cos\theta & -\sin\theta \\ \sin\theta & \cos\theta \end{bmatrix}$ with $\theta \neq k\pi$.

(5) See Corollary 6.4.3. (6) Consider $\begin{bmatrix} 1 & 1 \\ 0 & 2 \end{bmatrix}$. (7) Such a matrix A is unitary.

(8) and (9) A permutation matrix is an orthogonal matrix, but need not be symmetric.

(11) If A is Hermitian, by Schur's triangulation Theorem, A is orthogonally similar to an upper triangular matrix T. If A is nilpotent, the eigenvalues of A, as the diagonal entries of T, are all zero. By showing $T^H = T$, one can conclude that such A should be the zero matrix.

(12) See Schur's triangulation Theorem.

(13) For any Hermitian A, $-i$ cannot be an eigenvalue of A. So, $\det(A + iI) \neq 0$.

(14) Consider $A = \begin{bmatrix} 1 & -1 \\ 2 & -1 \end{bmatrix}$. (15) Modify (13).

(16) If $U^H AU = D$ with real diagonal entries, then $A^H = A$.

(18) $\begin{bmatrix} 1 & 0 \\ 0 & 1 \end{bmatrix}$ and $\begin{bmatrix} 0 & 1 \\ 1 & 0 \end{bmatrix}$. (19) See Corollary 7.23.1 on page 339. (20) If $A = [i]$?

Chapter 8

Problems

194 $1, 2, 3, 5, 7, 11, 15$. Such numbers are equal to the number of ways of writing the number n as a sum of positive integers, called a *partition of a positive integer n*.

196 Six different possibilities.

197 (1) $\begin{bmatrix} i & 1 \\ 0 & i \end{bmatrix}$, (2) $\begin{bmatrix} 2 & 1 & 0 \\ 0 & 2 & 0 \\ 0 & 0 & 3 \end{bmatrix}$, (3) $\begin{bmatrix} 2 & 0 & 0 & 0 \\ 0 & 4 & 1 & 0 \\ 0 & 0 & 4 & 1 \\ 0 & 0 & 0 & 4 \end{bmatrix}$.

199 Similar to the proof of Theorem 8.5.

198 (1) $Q = \begin{bmatrix} 0 & 1 \\ 1 & 0 \end{bmatrix}$, (2) $Q = \begin{bmatrix} 1 & 0 & 1 \\ 0 & 1 & 1 \\ 0 & 0 & 1 \end{bmatrix}$, (3) $Q = \begin{bmatrix} 1 & 1 & 0 & 0 \\ 2 & 0 & 1 & -1 \\ -2 & 0 & 0 & 1/2 \\ 2 & 0 & 0 & 0 \end{bmatrix}$.

201 Immediate by Eqs. (8.1) and (8.2).

203 (1) Use Eqs. (8.1) and (8.2) with $A = J$.

(2)

$$A = QJQ^{-1} = \begin{bmatrix} 1 & 0 & 1 & 1 \\ 1 & 1 & 0 & 1 \\ 1 & 2 & 0 & 1 \\ 1 & 0 & 1 & 0 \end{bmatrix} \begin{bmatrix} 0 & 1 & 0 & 0 \\ 0 & 0 & 0 & 0 \\ 0 & 0 & 2 & 0 \\ 0 & 0 & 0 & 2 \end{bmatrix} \begin{bmatrix} -1 & 2 & -1 & 1 \\ 0 & -1 & 1 & 0 \\ 1 & -2 & 1 & 0 \\ 1 & 0 & 0 & -1 \end{bmatrix},$$

and

$$J^k = \begin{bmatrix} 0 & 0 & 0 & 0 \\ 0 & 0 & 0 & 0 \\ 0 & 0 & 2^k & 0 \\ 0 & 0 & 0 & 2^k \end{bmatrix}; \quad e^J = \begin{bmatrix} 1 & 1 & 0 & 0 \\ 0 & 1 & 0 & 0 \\ 0 & 0 & e^2 & 0 \\ 0 & 0 & 0 & e^2 \end{bmatrix}.$$

205 $f(A) \neq \det(AI - A)$.

206 Let $\lambda_1, \ldots, \lambda_n$ be the eigenvalues of A. Then $f(\lambda) = \det(\lambda I - A) = (\lambda - \lambda_1)\cdots(\lambda - \lambda_n)$. Thus, $f(B) = (B - \lambda_1 I_m)\cdots(B - \lambda_n I_m)$ is nonsingular if and only if $B - \lambda_i I_m$, $i = 1, \ldots, n$, are all nonsingular. That is, none of the λ_i's is an eigenvalue of B.

207 (1) $A^{-1} = \frac{1}{2}\begin{bmatrix} 2 & 0 & 0 \\ 0 & 1 & 0 \\ -5 & 0 & 1 \end{bmatrix}$.

(2) The characteristic polynomial of A is $f(\lambda) = (\lambda - 1)(\lambda - 2)^2$, and the remainder is $104A^2 - 228A + 138I = \begin{bmatrix} 14 & 0 & 0 \\ 0 & 98 & 0 \\ 420 & 0 & 98 \end{bmatrix}$.

209 ($\Leftarrow$) is clear. For ($\Rightarrow$), use Theorem 8.8.

210 For $J^{(2)}$, $m(x) = (x - \lambda)^3$. For $J^{(3)}$, $m(x) = (x - \lambda)^4$.

211 $m(\lambda) = (\lambda - 1)(\lambda - 2)$.

212 (1) $m(\lambda) = (\lambda - 1)(\lambda - 3)^3$.

(2) $m(\lambda) = \lambda^2(\lambda - 2)$. $A^k = \begin{bmatrix} 2^{k+1} & -2^{k+1} & 2^k & -2^k \\ 2^k & 0 & 0 & -2^k \\ 2^k & 0 & 0 & -2^k \\ 2^k & -2^{k+1} & 2^k & 0 \end{bmatrix}$ $k \geq 2$.

213 (2) $e^A = \begin{bmatrix} -1 + 2e^2 & 1 - 2e^2 & e^2 & 1 - e^2 \\ -1 + e^2 & 0 & 1 & 1 - e^2 \\ -1 + e^2 & -1 & 2 & 1 - e^2 \\ -1 + e^2 & 1 - 2e^2 & e^2 & 1 \end{bmatrix}$.

214 $m(\lambda) = (\lambda - 3)^2, (\lambda - 3)^3, (\lambda - 3)^4$, respectively.

215 Easily, one can get the minimal polynomial of A as in Problem 214.

216 The minimal polynomial of A is $m(\lambda) = (\lambda - 1)(\lambda - 2)$. The solution is

$$\mathbf{y}(t) = \begin{bmatrix} y_1(t) \\ y_2(t) \\ y_3(t) \end{bmatrix} = e^{2t}\begin{bmatrix} 0 \\ 1 \\ 1 \end{bmatrix}.$$

Exercises

8.1 (1) Can take $\{\mathbf{e}_1, \mathbf{e}_4\}$, (3) Can take $\{\mathbf{e}_1, \mathbf{e}_3, \mathbf{e}_4, \mathbf{e}_5\}$.

8.3 For J^{-1}, use Eq. (1.3) in Page 36 or see Eq. (8.1) in Page 384. Since $J\mathbf{x} = \lambda\mathbf{x} \Leftrightarrow J^{-1}\mathbf{x} = \lambda^{-1}\mathbf{x}$, rank $(J - \lambda I)$ = rank $(J^{-1} - \lambda^{-1}I)$. Hence, the Jordan canonical form of J^{-1} is $\begin{bmatrix} \frac{1}{\lambda} & 1 & 0 & 0 \\ 0 & \frac{1}{\lambda} & 1 & 0 \\ 0 & 0 & \frac{1}{\lambda} & 1 \\ 0 & 0 & 0 & \frac{1}{\lambda} \end{bmatrix}$.

8.5 $\lambda\mathbf{v} = A\mathbf{v} = A^2\mathbf{v} = \lambda^2\mathbf{v}$ implies $\lambda(\lambda - 1) = 0$.

8.7 Use a Jordan decomposition of A; $A = QJQ^{-1}$ with Exercise **8.6**. Clearly, $\text{tr}(A) = \text{tr}(J)$ and rank A = rank J.

8.9 (1) $\lambda = 2$ of multiplicity 2. rank $(A - \lambda I) = 1$.
(2) $\lambda = k$ of multiplicity 3. rank $(A - kI) = 2$.
(3) For $\lambda = -1$ of multiplicity 2. rank $(A - \lambda I) = 2$.

$$(1)\ J = \begin{bmatrix} 2 & 1 \\ 0 & 2 \end{bmatrix}, \quad (2)\ J = \begin{bmatrix} k & 1 & 0 \\ 0 & k & 1 \\ 0 & 0 & k \end{bmatrix}, \quad (3)\ J = \begin{bmatrix} -1 & 1 & 0 \\ 0 & -1 & 0 \\ 0 & 0 & 0 \end{bmatrix}.$$

8.10 (1) $A = QJQ^{-1} = \begin{bmatrix} 0 & 1 \\ 1 & 0 \end{bmatrix} \begin{bmatrix} 2 & 1 \\ 0 & 2 \end{bmatrix} \begin{bmatrix} 0 & 1 \\ 1 & 0 \end{bmatrix}$.

$$(2)\ A = QJQ^{-1} = \begin{bmatrix} 1 & 0 & 0 \\ 0 & 1/20 & 0 \\ 0 & 0 & 1/400 \end{bmatrix} \begin{bmatrix} k & 1 & 0 \\ 0 & k & 1 \\ 0 & 0 & k \end{bmatrix} \begin{bmatrix} 1 & 0 & 0 \\ 0 & 20 & 0 \\ 0 & 0 & 400 \end{bmatrix}.$$

$$(3)\ A = QJQ^{-1} = \begin{bmatrix} -2 & 2 & -1 \\ 0 & 1 & 1 \\ 1 & 0 & 1 \end{bmatrix} \begin{bmatrix} -1 & 1 & 0 \\ 0 & -1 & 0 \\ 0 & 0 & 0 \end{bmatrix} \begin{bmatrix} 1 & -2 & 3 \\ 1 & -1 & 2 \\ -1 & 2 & -2 \end{bmatrix}.$$

8.11 (1) $A = QJQ^{-1} = \begin{bmatrix} 0 & 1 \\ 1 & 0 \end{bmatrix} \begin{bmatrix} i & 1 \\ 0 & i \end{bmatrix} \begin{bmatrix} 0 & 1 \\ 1 & 0 \end{bmatrix}$.

$$(2)\ A = QJQ^{-1} = \begin{bmatrix} 1 & 0 & 0 \\ 0 & 1/20 & 0 \\ 0 & 0 & 1/400 \end{bmatrix} \begin{bmatrix} i & 1 & 0 \\ 0 & i & 1 \\ 0 & 0 & i \end{bmatrix} \begin{bmatrix} 1 & 0 & 0 \\ 0 & 20 & 0 \\ 0 & 0 & 400 \end{bmatrix}.$$

$$(3)\ A = QJQ^{-1} = \begin{bmatrix} 0 & 1 & 0 \\ 1 & 0 & -1 \\ 1 & 0 & 0 \end{bmatrix} \begin{bmatrix} i & 1 & 0 \\ 0 & i & 1 \\ 0 & 0 & i \end{bmatrix} \begin{bmatrix} 0 & 0 & 1 \\ 1 & 0 & 0 \\ 0 & -1 & 1 \end{bmatrix}.$$

8.12 $J = \begin{bmatrix} 1 & 1 & 0 & 0 \\ 0 & 1 & 1 & 0 \\ 0 & 0 & 1 & 1 \\ 0 & 0 & 0 & 1 \end{bmatrix}$.

8.13

$$(1)\ J^{-1} = \begin{bmatrix} 1 & -1 & 1 & 0 & 0 \\ 0 & 1 & -1 & 0 & 0 \\ 0 & 0 & 1 & 0 & 0 \\ 0 & 0 & 0 & 1/3 & -1/9 \\ 0 & 0 & 0 & 0 & 1/3 \end{bmatrix} \quad J^n = \begin{bmatrix} 1^n & \binom{n}{1} & \binom{n}{2} & 0 & 0 \\ 0 & 1^n & \binom{n}{1} & 0 & 0 \\ 0 & 0 & 1^n & 0 & 0 \\ 0 & 0 & 0 & 3^n & n3^{n-1} \\ 0 & 0 & 0 & 0 & 3^n \end{bmatrix}.$$

$$(2)\ J^{-1} = \begin{bmatrix} 1/2 & 0 & 0 & 0 & 0 \\ 0 & 1/2 & 0 & 0 & 0 \\ 0 & 0 & 1/2 & 0 & 0 \\ 0 & 0 & 0 & 1/5 & -1/25 \\ 0 & 0 & 0 & 0 & 1/5 \end{bmatrix} \quad J^n = \begin{bmatrix} 2^n & 0 & 0 & 0 & 0 \\ 0 & 2^n & 0 & 0 & 0 \\ 0 & 0 & 2^n & 0 & 0 \\ 0 & 0 & 0 & 5^n & n \cdot 5^{n-1} \\ 0 & 0 & 0 & 0 & 5^n \end{bmatrix}.$$

$$(3)\ J^{-1} = \begin{bmatrix} 1/2 & -1/4 & 0 & 0 & 0 \\ 0 & 1/2 & 0 & 0 & 0 \\ 0 & 0 & 1/3 & 0 & 0 \\ 0 & 0 & 0 & 1/3 & 0 \\ 0 & 0 & 0 & 0 & 1/5 \end{bmatrix} \quad J^n = \begin{bmatrix} 2^n & n \cdot 2^{n-1} & 0 & 0 & 0 \\ 0 & 2^n & 0 & 0 & 0 \\ 0 & 0 & 3^n & 0 & 0 \\ 0 & 0 & 0 & 3^n & 0 \\ 0 & 0 & 0 & 0 & 5^n \end{bmatrix}.$$

8.14

$$(1)\ e^{tJ} = \begin{bmatrix} e^t & e^t & 0 & 0 & 0 \\ 0 & e^t & e^t & 0 & 0 \\ 0 & 0 & e^t & 0 & 0 \\ 0 & 0 & 0 & e^{3t} & e^t \\ 0 & 0 & 0 & 0 & e^{3t} \end{bmatrix} \quad (2)\ e^{tJ} = \begin{bmatrix} e^{2t} & 0 & 0 & 0 & 0 \\ 0 & e^{2t} & 0 & 0 & 0 \\ 0 & 0 & e^{2t} & 0 & 0 \\ 0 & 0 & 0 & e^{5t} & e^t \\ 0 & 0 & 0 & 0 & e^{5t} \end{bmatrix}$$

$$(3)\ e^{J} = \begin{bmatrix} e^2 & e & 0 & 0 & 0 \\ 0 & e^2 & 0 & 0 & 0 \\ 0 & 0 & e^3 & 0 & 0 \\ 0 & 0 & 0 & e^3 & 0 \\ 0 & 0 & 0 & 0 & e^5 \end{bmatrix}, \quad e^{tJ} = \begin{bmatrix} e^{2t} & e^t & 0 & 0 & 0 \\ 0 & e^{2t} & 0 & 0 & 0 \\ 0 & 0 & e^{3t} & 0 & 0 \\ 0 & 0 & 0 & e^{3t} & 0 \\ 0 & 0 & 0 & 0 & e^{5t} \end{bmatrix}$$

8.16 (1) $m(\lambda) = (\lambda - 1)^3(\lambda - 3)^2$, (2) $(\lambda - 2)(\lambda - 5)^2$, (3) $(\lambda - 2)^2(\lambda - 3)(\lambda - 5)$.

8.17

$$(2)\ J = \begin{bmatrix} -1 - 2i\sqrt{3} & 0 \\ 0 & -1 + 2i\sqrt{3} \end{bmatrix} \quad (3)\ J = \begin{bmatrix} 1 & 0 & 0 \\ 0 & 1 - \sqrt{6} & 0 \\ 0 & 0 & 1 + \sqrt{6} \end{bmatrix}$$

(1) $m(\lambda) = (\lambda - 1)^2$, (2) $\lambda^2 + 2\lambda + 13$, (3) $(\lambda - 1)(\lambda^2 - 2\lambda - 5)$.

8.18 Hint: (3) is diagonalizable.

8.19 (2) The minimal polynomial is $(\lambda - i)^4$ and $A^{10} = \begin{bmatrix} i^{10} & 10i^9 & \binom{10}{2}i^8 & \binom{10}{3}i^7 \\ 0 & i^{10} & 10i^9 & \binom{10}{2}i^8 \\ 0 & 0 & i^{10} & 10i^9 \\ 0 & 0 & 0 & i^{10} \end{bmatrix}$

(3) Use the Jordan decomposition,

$$A^{10} = QJ^{10}Q^{-1} = \begin{bmatrix} 0 & 0 & 0 & -2i \\ 0 & i-1 & 0 & 1-i \\ 0 & 1 & 1-i & 1 \\ 1 & 0 & 1 & 0 \end{bmatrix} \begin{bmatrix} i & 1 & 0 & 0 \\ 0 & i & 0 & 0 \\ 0 & 0 & 1 & 1 \\ 0 & 0 & 0 & 1 \end{bmatrix}^{10}$$

$$\times \begin{bmatrix} -1/2+i/2 & -i/2 & -1/2+i/2 & 1 \\ i/2 & -1/2-i/2 & 0 & 0 \\ 1/2+i/2 & i/2 & 1/2+i/2 & 0 \\ i/2 & 0 & 0 & 0 \end{bmatrix}.$$

Or, its minimal polynomial is $(\lambda - i)^2(\lambda - 1)^2$, and hence one has a system of four equations

$$\begin{aligned} i^{10} &= c_1 i^3 + c_2 i^2 + c_3 i + c_4, \\ 10 \cdot i^9 &= 3c_1 i^2 + 2c_2 i + c_3, \\ 1^{10} &= c_1 1^3 + c_2 1^2 + c_3 1 + c_4, \\ 10 \cdot 1^9 &= 3c_1 1^2 + 2c_2 1 + c_3. \end{aligned}$$

By solving these equations, one can get c_i for $A^{10} = c_1A^3 + c_2A^2 + c_3A + c_4I$.

8.20 (1) $A^{-1} = \begin{bmatrix} -i & 0 \\ 1 & -i \end{bmatrix}$, $A^n = \begin{bmatrix} i^n & 0 \\ \binom{n}{1} i^{n-1} & i^n \end{bmatrix}$, $e^A = e^i \begin{bmatrix} 1 & 0 \\ 1 & 1 \end{bmatrix}$.

(2) $A^{-1} = \frac{1}{32}\begin{bmatrix} 8 & -2 & -3 \\ 0 & 8 & -4 \\ 0 & 0 & 8 \end{bmatrix}$, $A^n = 4^{n-2}\begin{bmatrix} 16 & 4n & n^2+7n \\ 0 & 16 & 8n \\ 0 & 0 & 16 \end{bmatrix}$,

$$e^A = \begin{bmatrix} e^4 & e^4 & 3e^4 \\ 0 & e^4 & 2e^4 \\ 0 & 0 & e^4 \end{bmatrix}.$$

(3) $A^{-1} = \begin{bmatrix} 1 & 0 & -1 \\ 0 & \frac{1}{2} & -\frac{1}{2} \\ 0 & 0 & 1 \end{bmatrix}$, $A^n = \begin{bmatrix} 1 & 0 & n \\ 0 & 2^n & 2^n - 1 \\ 0 & 0 & 1 \end{bmatrix}$, $e^A = \begin{bmatrix} e & 0 & e \\ 0 & e^2 & -e+e^2 \\ 0 & 0 & e \end{bmatrix}$.

8.21

(1) $e^A = e^i \begin{bmatrix} 1 & 0 & 0 \\ 0 & 1 & 0 \\ 0 & 0 & 1 \end{bmatrix}$. (2) $e^{tA} = e^{it}\begin{bmatrix} 1 & t & \frac{t^2}{2} & \frac{t^3}{3!} \\ 0 & 1 & t & \frac{t^2}{2} \\ 0 & 0 & 1 & t \\ 0 & 0 & 0 & 1 \end{bmatrix}$.

(3) Use $e^{tA} = Qe^{tJ}Q^{-1}$.

8.23 (2) The characteristic polynomial of W is $f(\lambda) = \lambda^n - 1$. So, its eigenvalues are $1, \omega, \omega^2, \ldots, \omega^{n-1}$, where $\omega = e^{2\pi i/n}$.

(3) Eigenvalues of A are $\lambda_k = \sum_{i=1}^n a_i\omega^{(i-1)k} = a_1 + a_2\omega^k + a_3\omega^{2k} + \cdots + a_n\omega^{(n-1)k}$, $k = 0, 1, \ldots, n-1$.

(4) $\det A = \prod_{k=0}^{n-1}(a_1 + a_2\omega^k + a_3\omega^{2k} + \cdots + a_n\omega^{(n-1)k})$.

(5) The characteristic polynomial of B is $f(\lambda) = (\lambda - n + 1)(\lambda + 1)^{n-1}$. Since $\omega^{kn} = 1$, one can show $\omega^k + \omega^{2k} + \cdots + \omega^{(n-1)k} = -1$ for $k = 1, \ldots, n-1$.

8.24 For the LRR $D_n = bD_{n-1} - b^2D_{n-2}$ for$n \geq 3$, its companion matrix has an eigenvalue $\lambda = \frac{1}{2}(b \pm i\sqrt{3}b) = b(\frac{1}{2} \pm i\frac{\sqrt{3}}{2})$. So $\lambda_1 = b(\cos\frac{\pi}{3} + i\sin\frac{\pi}{3}) = be^{\pi i/3}$ and $\lambda_2 = b(\cos\frac{\pi}{3} - i\sin\frac{\pi}{3}) = be^{-\pi i/3}$. A general solution to the LRR is $D_n =$

$c_1\lambda_1^n + c_2\lambda_2^n$. By a direct computation, $D_1 = b$, $D_2 = 0$, $D_3 = -b^3, D_4 = -b^4, D_5 = 0, \ldots$. By determining the constants c_1, c_2, one can get

$$D_n = b^n \left[\cos\frac{n\pi}{3} + \frac{1}{\sqrt{3}}\sin\frac{n\pi}{3}\right].$$

8.25 Use Eqs. (6.2), (6.3), (6.4). Also, see Subsection 8.8.4.

(1) Eigenvalues $\lambda_1 = -1, \lambda_2 = 1, \mathbf{v}_1 = (1,-1,1), \mathbf{v}_2 = (1,1,1)$, Associated generalized eigenvectors: $\lambda = -1, \mathbf{v} = (-2,1,0)$. Hence, $x_n = c_1(-1)^n + c_2 n(-1)^n + c_3 1^n$. For $\mathbf{x}_n$, compute A^n using the minimal polynomial $m(\lambda) = (\lambda+1)^2(\lambda-1)$. Also, one can use a Jordan decomposition of A or the Cayley-Hamilton theorem.

(2) Eigenvalues $\lambda_1 = 1$ of multiplicity 4; Associated eigenvector: $\mathbf{v}_1 = (1,1,1,1)$, Associated generalized eigenvectors: $\mathbf{v}_2 = (3,2,1,0), \mathbf{v}_3 = (3,1,0,0), \mathbf{v}_4 = (1,0,0,0)$. Hence, $x_n = c_1 1^n + c_2 n 1^n + c_3 n^2 1^n + c_4 n^3 1^n$ by Eq. (6.4). For $\mathbf{x}_n$, compute A^n using the minimal polynomial $m(\lambda) = (\lambda-1)^4$.

(3) The characteristic polynomial of A is $f(\lambda) = \lambda^2 - (1+i)\lambda + i = (\lambda-1)(\lambda-i)$. So, $\lambda_1 = 1, \lambda_2 = i$ with associated eigenvectors $\mathbf{v}_1 = \begin{bmatrix}1\\1\end{bmatrix}, \mathbf{v}_2 = \begin{bmatrix}i\\1\end{bmatrix}$. Hence, $x_n = c_1 + c_2 i^n$ and $\mathbf{x}_n = c_1\mathbf{v}_1 + c_2 i^n \mathbf{v}_2$.

8.26 The coefficient matrix is $A = \begin{bmatrix}1&1&0\\0&1&1\\0&0&1\end{bmatrix}$, which is a Jordan matrix. Compute $A^n = (I+N)^n$.

8.27 $A = QJQ^{-1} = \begin{bmatrix}0&2&0\\1&-2&0\\0&2&1\end{bmatrix}\begin{bmatrix}1&0&0\\0&1&1\\0&0&1\end{bmatrix}\begin{bmatrix}1&1&0\\1/2&0&0\\-1&0&1\end{bmatrix}$. One can use this decomposition of A or the minimal polynomial $m(\lambda) = (\lambda-1)^2$ to get $y_1(t) = 2(t-1)e^t$, $y_2(t) = -2te^t$, $y_3(t) = (2t-1)e^t$.

8.28

$$J = U\Sigma V^T = \begin{bmatrix}0&0&1&0\\0&1&0&0\\1&0&0&0\\0&0&0&1\end{bmatrix}\begin{bmatrix}1&0&0&0\\0&1&0&0\\0&0&1&0\\0&0&0&0\end{bmatrix}\begin{bmatrix}0&0&0&1\\0&0&1&0\\0&1&0&0\\1&0&0&0\end{bmatrix}^T,$$

where U and V are both permutation matrices, and Σ is a diagonal matrix.

8.30 Clearly,

$$\begin{bmatrix}1&0\\0&1\end{bmatrix}, \begin{bmatrix}1&0\\0&-1\end{bmatrix}, \begin{bmatrix}-1&0\\0&1\end{bmatrix}, \begin{bmatrix}-1&0\\0&-1\end{bmatrix}$$

are roots of $X^2 = I_2$. These are the only roots having integer entries. However, there are many more other roots besides these four roots. In fact, there are infinitely many symmetric rational roots, given by

$$\frac{1}{t}\begin{bmatrix}s&r\\r&-s\end{bmatrix}, \frac{1}{t}\begin{bmatrix}s&-r\\-r&-s\end{bmatrix}, \frac{1}{t}\begin{bmatrix}-s&r\\r&s\end{bmatrix}, \frac{1}{t}\begin{bmatrix}-s&-r\\-r&s\end{bmatrix},$$

where (r, s, t) is any *Pythagorean triple*, that is, any triple of positive integers such that $r^2 + s^2 = t^2$, like $(3, 4, 5), (5, 12, 13), (8, 15, 17), (7, 24, 25), (20, 21, 29)$, etc. Interesting! In addition, there are infinitely many non-symmetric ones.

8.31 $1, 2, 3, 5, 7, 8, 9, 10, 12, 13, 14, 18$ are true.

(1)-(4) See Theorem 8.1. (4) and (6): Check it with Example 147.

(7) and (8) See Examples 147 and 151. (10) $J - \lambda I$ is nilpotent.

(11) $AA^H \neq A^H A$ if its size is greater than 2. See Lemma 7.21. All normal matrices are always diagonalizable.

(12) It has only one linearly independent eigenvector.

(13) and (14) See Eqs. (8.1) and (8.2). (15) $\det e^{I_n} = e^n$.

(16) What is the minimal polynomial of I_n? (17) See Example 165(2).

(18) See Corollary 8.8.1.

(19) Check with $A = \begin{bmatrix} 1 & 0 \\ 0 & 1 \end{bmatrix}$, $B = \begin{bmatrix} 1 & 1 \\ 0 & 1 \end{bmatrix}$. However, it is true if A and B are both diagonalizable. See Exercise **7.29**.

(20) Every LDE also has at least one solution. See Subsection 8.8.5 or Theorem 6.26.

Chapter 9

Problems

222 (1) is indefinite, (2) is positive-definite.

223 (1) infinite elliptic cone, (2) a point if $t = 0$, an ellipsoid if $t > 0$.

224 (1) $x^2 + y^2 + z^2 + 2xy = (x + y)^2 + z^2$. The eigenvalues of A are $0, 1, 2$. It is an elliptic cylinder.
(2) Three positive eigenvalues and it is an ellipse.
(3) The eigenvalues of A are $-9, 1, 11$. It is a one-sheeted hyperboloid.

Level surface with various $t > 0$: (1) If $t > 0$ increases, the radius of the elliptic cylinder increases, too. (2) If $t > 0$ increases, the ellipse becomes bigger and bigger. (3) When $t > 0$ increases, the hyperboloid becomes fat and fat.

225 A is positive-definite with eigenvalues $1, 1, 4$. B is positive-definite with eigenvalues $2, 2 + \sqrt{2}$, and $2 - \sqrt{2}$. $W = (L\sqrt{D})^T = \begin{bmatrix} \sqrt{2} & -1/\sqrt{2} & 0 \\ 0 & \sqrt{3/2} & \sqrt{2/3} \\ 0 & 0 & 2/\sqrt{3} \end{bmatrix}$.

227 (2) is indefinite. (3) is indefinite. (4) is negative-definite.

228 False with a counter-example $A = \begin{bmatrix} 0 & 0 \\ 0 & -1 \end{bmatrix}$; $\det A_k = 0$ for all k, but A is not positive-semidefinite.

229 For example, A is negative-semidefinite if and only if the eigenvalues of A are all nonpositive.

231 (2) is true. (3) is false by the following counter-example; $A = \begin{bmatrix} 1 & 1 \\ 0 & 2 \end{bmatrix}$ is similar (not orthogonally) to $D = \begin{bmatrix} 1 & 0 \\ 0 & 2 \end{bmatrix}$, because A is diagonalized to D. However, they are not congruent because A is not symmetric.

(4) False with a counter-example; $A = \begin{bmatrix} 1 & 0 \\ 0 & 1 \end{bmatrix}$ is congruent to any of $\begin{bmatrix} \lambda & 0 \\ 0 & \lambda \end{bmatrix}$, $\lambda > 0$. But, they are not similar if $\lambda \neq 1$.

232 (1) $[D \mid P^T] = \left[\begin{array}{cc|cc} 1 & 0 & 1 & 1 \\ 0 & -1 & -1 & 0 \end{array}\right]$. $\text{In}(A) = (1, 1, 0)$.

(2) $[D \mid P^T] = \left[\begin{array}{ccc|ccc} 1 & 0 & 0 & 1 & 0 & 1 \\ 0 & 1 & 0 & 0 & 1 & 0 \\ 0 & 0 & -1 & -1 & 0 & 0 \end{array}\right]$. $\text{In}(A) = (2, 1, 0)$.

(3) $[D \mid P^T] = \left[\begin{array}{ccc|ccc} 1 & 0 & 0 & 1 & 1 & 0 \\ 0 & -3 & 0 & -2 & -1 & 0 \\ 0 & 0 & 7/3 & -1/3 & 1/3 & 1 \end{array}\right]$. $\text{In}(A) = (2, 1, 0)$.

233 (1) $[D \mid P^H] = \left[\begin{array}{ccc|ccc} 1 & 0 & 0 & 1 & 1 & 0 \\ 0 & -3 & 0 & -2 & -1 & 0 \\ 0 & 0 & -\frac{5}{3} & -i/3 & i/3 & 1 \end{array}\right]$. $\text{In}(A) = (1, 2, 0)$.

(3) $[D \mid P^H] = \left[\begin{array}{ccc|ccc} 1 & 0 & 0 & 1 & 0 & i \\ 0 & 0 & 0 & 0 & 1 & 0 \\ 0 & 0 & -1 & i & 0 & 0 \end{array}\right]$. $\text{In}(A) = (1, 1, 1)$.

237 ($\Rightarrow$) is clear: If $b(\mathbf{x}, \mathbf{y}) = \mathbf{x}^H A\mathbf{y}$ is Hermitian, $\mathbf{y}^H A\mathbf{x} = \overline{\mathbf{x}^H A\mathbf{y}}$ for all $\mathbf{x}, \mathbf{y} \in \mathbb{C}^n$, then $A = [b]_\alpha$ is Hermitian, and so diagonalizable and all diagonal entries are real. ($\Leftarrow$): Let there exist a basis $\alpha = \{\mathbf{v}_1, \ldots, \mathbf{v}_n\}$ such that $a_{ij} = b(\mathbf{v}_i, \mathbf{v}_j)$ gives a diagonal matrix with real diagonal entries. Then b is Hermitian.

238 (1) False. ($\Rightarrow$): If b is diagonalizable, A is congruent to a diagonal matrix. Hence, A is symmetric and diagonalizable.

($\not\Leftarrow$): Choose a diagonalizable but not symmetric matrix A, like $\begin{bmatrix} 0 & 1 \\ 0 & 1 \end{bmatrix}$. Then A cannot be congruent to a diagonal matrix, and then b cannot be diagonalizable.

(2) True. ($\Leftarrow$) is clear. ($\Rightarrow$): If b is diagonalizable, A is congruent to a diagonal matrix. So A is symmetric.

239 Several different types of solutions are possible. However, (1) $\text{In}(A) = (2, 1, 0)$, (2) $\text{In}(A) = (1, 2, 0)$, (3) $\text{In}(A) = (2, 1, 0)$.

240 (2) For a basis $\beta = \{\mathbf{e}_1, \mathbf{e}_3, \mathbf{x}\}$ with $\mathbf{x} = (-2, 1, 3)$ for $\mathbb{R}^3$, $[b]_\beta = \begin{bmatrix} 0 & 1 & 0 \\ -1 & 0 & 0 \\ 0 & 0 & 0 \end{bmatrix}$

(3) For a basis $\beta = \{\mathbf{e}_1, \mathbf{e}_3, \mathbf{e}_2\}$ for $\mathbb{R}^3$, $[b]_\beta = \begin{bmatrix} 0 & 1 & 0 \\ -1 & 0 & 0 \\ 0 & 0 & 0 \end{bmatrix}$.

242 (1) from the bilinearity of b. (2) from the symmetry of b.

244 Count the number of distinct inertia (p, q, k). For a fixed n, the number of inertia with $p = \ell$ is $n - \ell + 1$ for $\ell = 1, 2, \ldots, n$.

245 (1) $\text{In}(\cdot) = (1, 2, 0)$. (2) $\text{In}(\cdot) = (2, 1, 0)$. (3) $\text{In}(\cdot) = (2, 1, 0)$. Hence, (2) and (3) are congruent by Theorem 9.18.

249 The maximum value of $R(\mathbf{x})$ is the maximum eigenvalue of A, and similarly for the minimum value.

(2) Eigenvalues are $1 + \sqrt{3}, 1$, and $1 - \sqrt{3}$.

248 (1) $q(\mathbf{x}) = 0$ is two crossing lines.
(2) $q(\mathbf{x}) = 1$ is a hyperbola with transverse axis along the line $y = x$ and passing through $(\pm 1, 0)$ and $(0, \pm 1)$. When t increases to ∞, the level curves, *i.e.*, the hyperbolas are getting further away from the origin.
(3) $q(\mathbf{x}) = -1$ is also a hyperbola, but with transverse axis along the line $y = -x$ and lying in the region $xy < 0$. When t approaches to $-\infty$, the level curves, *i.e.*, the hyperbolas are getting further away from the origin.

250 $\max = 7/2$ at $\pm(1/\sqrt{2}, -1/\sqrt{2})$, $\min = 1/2$ at $\pm(1/\sqrt{2}, 1/\sqrt{2})$.

Exercises

9.1 (1) parabola, (2) hyperbola, (3) ellipse. (2), (3) are homogeneous. (4) is difficult.

9.2 (1) $\begin{bmatrix} 1 & 2 \\ 2 & 3 \end{bmatrix}$, (2) $\begin{bmatrix} 1 & 0 & 2 \\ 0 & -1 & 2 \\ 2 & 2 & 1 \end{bmatrix}$, (3) $\begin{bmatrix} 2 & 2 & 0 \\ 2 & 0 & 3 \\ 0 & 3 & -5 \end{bmatrix}$, (4) $\begin{bmatrix} 2 & 2 \\ 2 & 1 \end{bmatrix}$.

9.3 (1) $3x'^2 + 2x'y' + y'^2 = 3/8$, (2) $4x'^2 - 2x'y' + y'^2 = 5$, (3) $4x'^2 - 2x'y' + y'^2 - 2z'^2 = -1/8$

9.4 (1) Ellipse, (2) Ellipse, (3) One-sheeted hyperboloid.

9.5 (1) $\begin{bmatrix} 1 & 0 \\ 0 & -1 \end{bmatrix}$, giving $x^2 - y^2$. (2) $\begin{bmatrix} 0 & 0 \\ 0 & 2 \end{bmatrix}$, giving $2(y')^2$.

(3) $\begin{bmatrix} 1 & 0 & 0 \\ 0 & 1-\sqrt{2} & 0 \\ 0 & 0 & 1+\sqrt{2} \end{bmatrix}$, giving $(x')^2 + (1 - \sqrt{2})(y')^2 + (1 + \sqrt{2})(z')^2$.

(4) $\begin{bmatrix} \frac{1}{2}(3+\sqrt{17}) & 0 \\ 0 & \frac{1}{2}(3-\sqrt{17}) \end{bmatrix}$, giving $(\frac{1}{2}(3+\sqrt{17}))(x')^2 + (\frac{1}{2}(3-\sqrt{17}))(y')^2$.

When $t = 1$, (1) Hyperbola, (2) Two parallel lines, (3) One-sheeted hyperboloid, (4) Hyperbola.

9.6 (1) Eigenvectors of $A = [q]_\alpha$ are $\mathbf{x}_1 = (-1, 0, 1), \mathbf{x}_2 = (1, 1, 1), \mathbf{x}_3 = (1, -2, 1)$, with eigenvalues $\lambda_1 = 12, \lambda_2 = 6, \lambda_3 = -6$. $P = [\mathbf{x}_1\ \mathbf{x}_2\ \mathbf{x}_3]$.
(2) $\gamma = \{\mathbf{x}_1, \mathbf{x}_2, \mathbf{x}_3\}$.

9.7 (1) $\text{In}(\cdot) = (1, 1, 0)$, hyperbola. (2) $\text{In}(\cdot) = (1, 1, 0)$, hyperbola.
(3) $\text{In}(\cdot) = (2, 0, 0)$, ellipse. (4) $\text{In}(\cdot) = (2, 0, 0)$, ellipse.

9.8 (1) $\text{In}(\cdot) = (1, 2, 0)$, two-sheeted hyperboloid. (2) $\text{In}(\cdot) = (2, 1, 0)$, one-sheeted hyperboloid. (3) $\text{In}(\cdot) = (2, 1, 0)$, one-sheeted hyperboloid.

9.9 (1) Indefinite with $\text{In}(\cdot) = (1,1,0)$; (2) Positive-semidefinite with $\text{In}(\cdot) = (1,0,2)$; (3) Indefinite with $\text{In}(\cdot) = (1,2,0)$.

9.10 The first two are positive-definite.

9.11 If λ is an eigenvalue of A, then λ^2 and $\frac{1}{\lambda}$ are eigenvalues of A^2 and A^{-1}, respectively. Use $\mathbf{x}^T(A+B)\mathbf{x} = \mathbf{x}^T A\mathbf{x} + \mathbf{x}^T B\mathbf{x}$.

9.12 A and P^TAP are congruent and hence they have the same inertia. Or, one can use $\mathbf{x}^T(P^TAP)\mathbf{x} = (P\mathbf{x})^TA(P\mathbf{x})$ for any $\mathbf{x}$.

9.13 Only (4).

9.14 (1) $A = \begin{bmatrix} 2 & -1 \\ 2 & 0 \end{bmatrix}$, (2) $B = \begin{bmatrix} 3 & 9 \\ 0 & 6 \end{bmatrix}$, (3) $[b]_\beta = [id]_\alpha^\beta [b]_\alpha [id]_\beta^\alpha$; $P = \begin{bmatrix} 1 & 2 \\ 1 & -1 \end{bmatrix}$.

9.15 $[b]_\alpha = \begin{bmatrix} 1 & 0 \\ 0 & 1 \end{bmatrix}$, $[b]_\beta = \begin{bmatrix} 2 & 0 \\ 0 & 2 \end{bmatrix}$, $[b]_\gamma = \begin{bmatrix} 5 & 11 \\ 11 & 25 \end{bmatrix}$.

9.16 (1) Symmetric, (2) Neither, (3) Skew-symmetric, (4) Skew-symmetric.

9.17 Compute $\text{In}(A)$ by using the congruence relation. (1) $\text{In}(A) = (1,2,0)$. (3) $\text{In}(A) = (2,1,0)$.

9.18 (1) $[D \mid P^H] = \left[\begin{array}{ccc|ccc} 1 & 0 & 0 & 1 & 1 & 0 \\ 0 & -3 & 0 & -2 & -1 & 0 \\ 0 & 0 & 7/3 & -i/3 & i/3 & 1 \end{array}\right]$. (2) $\text{In}(\cdot) = (2,1,0)$. (3) $\text{In}(\cdot) = (1,1,1)$.

9.19 In $[b]_\alpha = [b_{ij}]$, $b_{11} = b_{14} = b_{41} = b_{44} = 1$, all the others are zero. b is symmetric and the eigenvalues are $\lambda_1 = 2, \lambda_2 = \lambda_3 = \lambda_4 = 0$.

9.20 (1) $\lambda_1 = 0$, $\lambda_2 = -1$, and $\lambda_3 = -3$. $\lambda_{\max} = 0$.
(2) $\lambda_1 = \sqrt{2} - 2$, $\lambda_2 = -2$, and $\lambda_3 = -2 - \sqrt{2}$.
(3) $\lambda_1 = 5$, $\lambda_2 = 5$, and $\lambda_3 = 1$. $\lambda_{\max} = 5$.

9.21 (1) $\max = 4$ at $\pm(1,\ 1,\ 2)/\sqrt{6}$, $\min = -2$ at $\pm(-1,\ -1,\ 1)/\sqrt{3}$;
(2) $\max = 3$ at $\pm(2, 1,\ 1)/\sqrt{6}$, $\min = 0$ at $\pm(1,\ -1,\ -1)/\sqrt{3}$.

9.22 (1) The maximum is 1 at $(\pm1, 0)$; the minimum is -1 at $(0, \pm1)$.
(2) The maximum is 2 at $\pm(1,1)$; the minimum is 0 at $\pm(-1,1)$.
(4) The maximum is $(3+\sqrt{17})/2$ at $\pm((1+\sqrt{17})/4, 1)$; the minimum $(3-\sqrt{17})/2$ at $\pm((1-\sqrt{17})/4, 1)$.

9.23 (1)⇔(2): See the proof of Theorem 5.28. (1)⇔(3): Use Sylvester's law of inertia.

9.25 $3, 4, 5, 10, 11, 13, 15, 17, 18, 20$ are true.
(1) It has terms of degrees 2 and 3. (2) $q(k\mathbf{x}) = kq(\mathbf{x})$?
(3) See Remark (2) on page 424. (4) See Example 183. (5) See Theorem 9.3.
(6) If b is not symmetric, what? (7) Consider $A = \begin{bmatrix} 1 & 0 & 1 \\ 0 & 1 & 0 \\ 1 & 0 & 2 \end{bmatrix}$.
(8) See Problem 228 or check with $-I_n$. (9) Consider $I + (-\frac{1}{2})I$ and $I + (-2)I$.
(10) A matrix representation of a quadratic form is always symmetric. Any two congruent matrices have the same inertia.

(12) For every square matrix A, $b(\mathbf{x}, \mathbf{y}) = \mathbf{x}^T A\mathbf{y}$ is a bilinear form. However, $\text{In}[b]$ is defined only when A is symmetric. For example, a skew-symmetric matrix does not have any real positive or negative eigenvalue.

(14) If $b(\mathbf{v}, \mathbf{v}) = 0$ for all $\mathbf{v}$, then b is alternating.

(15) Every symmetric matrix is orthogonally diagonalizable.

(16) The identity I is congruent to kI for all positive $k \in \mathbb{R}$.

(18) For any bilinear form b on $\mathbb{R}^n$, $b(\mathbf{x}, \mathbf{y}) = \mathbf{x}^T A\mathbf{y}$ for some matrix A. Let $A = B + C$ with the symmetric part B and the skew-symmetric part C. Then $q(\mathbf{x}) = b(\mathbf{x}, \mathbf{x}) = \mathbf{x}^T A\mathbf{x} = \mathbf{x}^T B\mathbf{x}$ is a quadratic form on $\mathbb{R}^n$.

(19) Yes if it is nondegenerate. (20) See Theorem 9.21.

Bibliography

[Artin] M. Artin, *Algebra*, Prentice-Hall, Inc., Englewood Cliffs, N.J., 1991.

[Braun] M. Braun, *Differential Equations and Their Applications,* 4th Edition, Springer-Verlag, New York, 1993.

[Datta] B.N. Datta, *Numerical Linear Algebra and Applications*, 2nd Edition, SIAM, Philadelphia, PA, 2010.

[Gantmacher] F.R. Gantmacher, *The Theory of Matrices, I, II,* Chelsea, New York, 1959.

[Horn & Johnson] R.A. Horn and C.R. Johnson, *Matrix Analysis*, 2nd Edition, Cambridge University Press, Cambridge, 2013.

[Kwak & Hong] J.H. Kwak and S. Hong, *Linear Algebra,* 2nd Edition, Birkhäuser, Boston, MA, 2004.

[Noble & Daniel] B. Noble and J.W. Daniel, *Applied Linear Algebra*, 3rd Edition, Prentice Hall/Englewood Cliffs, New Jersey, 1988.

[Roman] S. Roman, *Advanced linear algebra*, 3rd Edition, Springer, New York, NY, 2008

[Strang] G. Strang, *Linear Algebra and Its Applications,* 3rd Edition, Harcourt Brace Jovanovich, San Diego, CA, 1998.

[Prasolov] V.V. Prasolov, *Problems and Theorems in Linear Algebra,* Translated from the Russian manuscript by D.A. Leĭtes, American Mathematical Society, Providence, RI, 1994.

[Wikipedia] https : //en.wikipedia.org/wiki/Linear_algebra

Index

저자와의
합의하에
인지를
생략합니다.

FUNDAMENTALS OF
LINEAR ALGEBRA

핵심 선형 대수학 제3판

지은이 곽진호
펴낸이 조경희
펴낸곳 경문사
펴낸날 2019년 3월 5일 1판 1쇄
2020년 3월 2일 개정판 1쇄
2023년 3월 2일 3판 1쇄
등 록 1979년 11월 9일 제1979-000023호
주 소 04057, 서울특별시 마포구 와우산로 174
전 화 (02)332-2004 팩스 (02)336-5193
이메일 kyungmoon@kyungmoon.com

값 37,000원

ISBN 979-11-6073-624-3

★ 경문사의 다양한 도서와 콘텐츠를 만나보세요!

홈페이지 www.kyungmoon.com
포스트 post.naver.com/kyungmoonbooks
북이오 buk.io/@pa9309
페이스북 facebook.com/kyungmoonsa
블로그 blog.naver.com/kyungmoonbooks
유튜브 https://www.youtube.com/channel/UClDC8x4xvA8eZlrVaD7QGoQ

경문사 출간 도서 중 수정판에 대한 **정오표**는 **홈페이지 자료실**에 있습니다.